Satellite Remote Sensing of Terrestrial Hydrology

Christopher Ndehedehe

Satellite Remote Sensing of Terrestrial Hydrology

Springer

Christopher Ndehedehe
Griffith University
Brisbane, QLD, Australia

ISBN 978-3-030-99576-8 ISBN 978-3-030-99577-5 (eBook)
https://doi.org/10.1007/978-3-030-99577-5

This Springer imprint is published by the registered company Springer Nature Switzerland AG
The registered company address is: Gewerbestrasse 11, 6330 Cham, Switzerland

Foreword

Remote sensing using satellites with optical detection capability has been possible for some decades now, since the first Landsat missions of the 1970s. However, the uptake and use of satellite remote sensing for environmental applications has not been commensurate with emerging technological improvements in sensor spectral, spatial, radiometric, and geometric capability. Additionally, some of the potential of satellite remote sensing for environmental monitoring is only now starting to be realized with the development of coordinated strategies to support data acquisition, processing (e.g., using 'artificial intelligence') and visualization products. This book on Satellite Remote Sensing of Terrestrial Hydrology by Dr. Christopher Ndehedehe is timely and relevant because it provides a state-of-the-art synthesis of the field of remote sensing hydrology. It should be essential reading for many interested students, educationalists, and practitioners; from undergraduate students who seek fundamental knowledge about how remote sensing can improve their understanding of hydrological systems, to advanced practitioners who wish to update their technological, analytical, and/or hydrological skills. In between is a large and diverse group who will benefit from the many components of remote sensing hydrology described in this book, with application for hydrologists, graduate students studying in the area, and environmental managers who are interested in utilizing the potential of remote sensing to assist with improved decision-making for water management.

Changes in global surface water storage in the 'Era of the Anthropocene' is an area that is reiterated throughout this book, including impacts on vegetation cover, climate change and relationships to floods and droughts, and the impact of large surface water reservoirs. A series of case studies from across the globe reinforces the seriousness of the rate of change in our hydrological systems. These case studies have been selected as global hotspots to demonstrate the speed of hydrological change, issues of water scarcity and drought, and the geopolitical context that influences the quality of decision-making for water management.

Satellite Remote Sensing of Terrestrial Hydrology presents exciting new opportunities from satellite technological advances and analytical platforms. For example, the book provides background information and case studies related to the Gravity Recovery and Climate Experiment satellite observations. The resulting geospatial data can provide regional-scale assessments of groundwater storage that can be linked to surface water availability and vegetation cover. The Google Earth Engine and other cloud-based platforms discussed in the book are making satellite data increasingly accessible and are being used to perform critical pre-processing steps to ease the barrier to entry in the exploration of satellite data. The book also includes a forward-looking perspective in the discussion of new satellite hydrology missions like the Surface Water Ocean Topography mission scheduled for launch in 2022. The combination of theoretical understanding, synthesis of technological advances and case studies makes this book especially useful.

Dr. Christopher Ndehedehe is an academic in the Australian Rivers Institute at Griffith University who, despite being in the early stages of his career, has established himself as a global leader in hydrological remote sensing and environmental geoscience. He is pioneering the development of new geospatial tools to support assessments of tropical floodplain hydrology and the impacts of climate change on surface water, groundwater, and ecological resources. Dr. Ndehedehe is a prolific writer who has bridged multiple scientific disciplines in compiling chapters in Satellite Remote Sensing of Terrestrial Hydrology. The book provides all the tools required to successfully apply remote sensing in terrestrial hydrological assessments and should be essential reading for anyone with an interest in this subject.

March 2021

David Hamilton
Deputy Director – Australian Rivers Institute
Griffith University
Brisbane, QLD, Australia

Preface

The use of multi-mission optical satellite systems in environmental monitoring, including applications in hydrology and water resources has been the focuse in the last few decades. These satellite technologies provided a continuous and systematic stream of geospatial information across different time scales that underpinned an improved understanding of environmental processes and systems. Without any doubt, the application of satellite technologies in global environmental monitoring has contributed to new directions in policy, creating innovative opportunities to address several socio-economic and environmental issues confronting society. However, the potential of recent technological advances in satellite remote sensing, be it optical or geodetic systems, is still not well articulated and not fully understood. Dedicated satellite hydrology missions for direct monitoring of terrestrial water storage and fluxes have only recently emerged and others are scheduled to be launched in the coming months. But the opportunities that exist through the use of existing innovative satellite methods and frameworks to improve understanding, management, and monitoring of natural systems and key Earth resources are yet to be fully explored. The concept of satellite hydrology only became more pronounced after the launch of the Gravity Recovery and Climate Experiment (GRACE) in 2002. Satellite hydrology missions like GRACE and its successor (GRACE-Follow On) and radar altimetry missions as well as the anticipated Surface Water Ocean Topography mission, among other existing missions lead to a new satellite hydrology concept called remote sensing hydrology. This concept as further advocated in this book encapsulates the cross-disciplinary aspects of environmental geoscience and remote sensing of the environment, as well as integrating several geospatial tools and methods, including relevant cognate disciplines to provide quantitative evaluation of terrestrial hydrology. The growing need to understand hydrological processes and impacts of climate change as well as human actions (e.g., dams, water diversion projects, land-use change, intensive groundwater extraction, deforestation, surface water resources schemes, etc.) on global freshwater systems necessitated the discussion on GRACE observations in this book. Being an important tool in the box to navigate the challenges posed by lack of or limited observational data for freshwater monitoring (including groundwater, lakes, and reservoirs), its processing chains and hydrological applications are also detailed in this book.

As discussed in this book, the rise in improved sensor capabilities in satellite missions resulted in big remote sensing data. On the one hand, innovative pattern recognition and machine learning algorithms are now increasingly being used to improve key information extraction from the plenitude of big geospatial data. On the other hand, cloud computing platforms, have facilitated the use of data from these satellite systems, contributing to the growing applications of remote sensing observations in environmental monitoring in recent times. These cloud-based platforms are now easy pathways to remote sensing data retrieval and processing as well as geospatial analyses. In this day and age, large earth observation datasets are not only available and freely accessible from global repositories but can also be quickly processed to generate timely information that underpins important organizational decisions. These opportunities have facilitated informed decisions by government institutions, scientists, and the public on the environment. This book discusses, explores, and demonstrates key applications of various earth observations and earth resources satellites ranging from optical remote sensing systems to satellite geodetic systems (e.g., gravity and radar missions) in advancing our fundamental understanding of environmental systems (e.g., vegetation, freshwater ecosystems, droughts, groundwater, lakes, agriculture, etc.). The complexities of hydrological systems and global environmental change in this era of the Anthropocene due to human activities and the impact emanating from variability in natural systems are discussed using a suite of quantitative and empirical region-specific case studies. The book has 20 chapters covering several multidisciplinary aspects of remote sensing hydrology and environmental monitoring. The focus on satellite-based environmental monitoring in this book ensures that opportunities through investments in innovative satellite programmes to optimize key environmental decisions and policies are maximized.

Brisbane, QLD, Australia
March 2021

Christopher Ndehedehe

Acknowledgements

I am very grateful to the Almighty God who through His Son, Jesus Christ, my Redeemer and personal Saviour, gave me abundant life and preserved me to see this day. I'm highly appreciative of Griffith University's Areas of Strategic Investment fund, which provided a career pathway for me at the Australian Rivers Institute (ARI) and School of Environment and Science at Griffith University, Queensland. My sincere gratitude also goes to Prof. David Hamilton, Deputy Director, ARI, for his great support as well as his strong leadership and mentorship. Prof. Stuart Bunn, Director of ARI, and the entire ARI family are acknowledged for the conducive and enabling research environment and support that led to my notable accomplishments since joining the institute in June 2018. I also thank Prof. Mark Kennard for his guidance and mentorship as well as the numerous support and help he has offered on several occasions. My work on the National Environmental Science Programme (NESP) projects undertaken in Northern Australia (Mitchel, Gilbert, and Flinders catchments) led to great collaborations with great team leaders and experts in freshwater ecology like Ben Stewart-Koster and Prof Stuart Bunn. The NESP project and international collaborations with Prof. Vagner G. Ferreira of Hohai University, Dr. Onuwa Okwuashi, and Dr. Nathan Agutu, among others, and other freshwater-based consultancies funded by the Asian Development Bank and Australian Water Partnership provided a platform and opportunity that underpinned my pioneering effort in developing the science of remote sensing hydrology. Moreover, working with A/Prof Brian McIntosh, the Education Director for the Master of Catchment Science programme (MCS), International Water Center, Griffith University, on capacity building projects was also very valuable. My postgraduate students (Masters and Ph.D.s) here at Griffith University and the pioneer MCS master's students I taught were also the key motivation for me on the need for an academic textbook and literature that simplifies the concept and provides easy-to-use resources in the field of environmental remote sensing and satellite hydrology. Special thanks to all my collaborators around the world in, the USA, Canada, Australia, Asia, Africa, Europe, and South America, you are all indeed an amazing team of diligent researchers. Collaborating with these excellent brands and teams of academics, which are too numerous to mention here is one of the best things that happened to me in the last 5 years. Their novel and ground-breaking contributions in the fields of water resources, hydro-geodesy, environmental geoscience, climate change, machine learning applications, and remote sensing hydrology also strengthened the concept advocated in this book.

My deepest appreciation goes to my beloved family. To my wife and special love, Mmayen Ndehedehe, her understanding, patience, and support are invaluable. To my three adorable, amazing, and beautiful kids; Koabasi, Kendarabasi, and Kereseabsi, I say thanks for your understanding and patience and God bless you. I'm also grateful to Ms. Enobong Akpan for reading the draft of this book (Chap. 1) just to ensure a non-remote sensing expert can also understand and get the message in the book. Mr. Ikechukwu Kalu is also acknowledged for his similar role among other things, having assisted with reading the draft of this book to ensure undergraduates and postgraduate students will find it an interesting read. I thank my late dad, Mr. Edet John Ndehedehe, who went to be with the Lord in 2018 shortly after I joined ARI. The foundation he laid and his numerous encouragement and prayers contributed immensely to the career path I forged and my success over the years. You are always remembered and may your soul continue to rest in peace. I also appreciate my mum for her consistent and patient love, and to all my siblings for their love and prayers.

Contents

Part XI Impacts of Global Climate Teleconnections

Part I
Global Terrestrial Hydrology

Chapter 1
Remote Sensing Hydrology

Abstract The emergence of new optical sensors provides unique capability in comprehensive monitoring of the environment. As opposed to well-known optical satellite Earth observation systems such as the Landsat and Sentinel missions, there are now satellite systems dedicated to hydrological applications. The combination of geospatial science with these systems provides an innovative satellite hydrology framework that underpins our capability to directly monitor and assess changes in terrestrial stored water and how they are impacted by human actions and climate change. In this chapter, the concept of *remote sensing hydrology* has an emerging discipline is discussed. A distinction between *remote sensing hydrology* and *application of remote sensing in hydrology* is articulated. The latter has been the focus in the last few decades, leveraging several methods to aid the processing of Earth resources satellite for hydrological applications. However, new investments in satellite geodetic programmes (e.g., Gravity Recovery and Climate Experiment mission) and the scheduled launch of the Surface Water and Ocean Topography mission in November 2022 gave birth to the former. These satellite geodetic systems are discipline-specific and dedicated hydrology missions, providing level 1 information on the quantitative aspects of freshwater resources. Together with novel computing technologies for big Earth observation data, they provide improved monitoring capabilities of the earth system that underpins our understanding of environmental processes and changes in hydrological systems.

1.1 Introduction

Several improvements in remote sensing optical systems have occurred in recent years. The use of these technologies in environmental monitoring and impact analysis from extreme events such as droughts, floods, and wildfires has been widely documented. New satellite technologies with capabilities to monitor changes in terrestrial water storage have emerged, complementing the well-known optical systems in baseline studies and freshwater ecosystem assessment. With the large amounts of geospatial data generated from satellite Earth observation systems, cloud computing platforms for big Earth observation data management and analysis such as Google

C. Ndehedehe, *Satellite Remote Sensing of Terrestrial Hydrology*,
https://doi.org/10.1007/978-3-030-99577-5_1

Earth Engine, Sentinel Hub, Open Data Cube, and pipsCloud, among several others emerged (e.g., Gomes et al. 2020; Wang et al. 2018; Gorelick et al. 2017) to facilitate timely access and ease the application of these observations for decision support system. These novel computing technologies for big Earth observation data, though with varied levels of sophistication, have since enhanced access to archived petabytes of publicly available Earth observation data, as well as simplifying processing chains of satellite remote sensing observations. The fundamental objective of Earth observation systems in providing a continuous and systematic stream of geospatial information across different time scales to improve understanding of environmental processes and systems has been fostered by the emergence of cloud-based computing platforms. Additionally, these platforms have emerged as excellent alternatives to conventional infrastructures for spatial data sharing and processing. The dissemination of a new generation of digital infrastructure for planetary-scale computing is now today's reality, creating opportunities to address the diverse socio-economic and environmental issues confronting society. The efficacy, sophistication, and innovation behind these technologies and systems are apparently evidenced in the rise of satellite remote sensing applications in global environmental monitoring based on data retrieved and processed with these platforms. This engagement and enthusiasm reaffirm these cutting-edge platforms have been well received by the larger community of experts, students, researchers, and relevant stakeholders who rely on Earth observation for key decisions and regulations.

The application of these satellite technologies as well as cloud-based platforms in monitoring as a key foundation for planning and supervision, as well as for the systematic development of targeted water service delivery and policies, underpin their value as valuable tools in freshwater accounting and management. State-of-the-art satellite hydrology missions such as the Gravity Recovery and Climate Experiment (GRACE, Landerer et al. 2020; Tapley et al. 2004) are such technologies that have enabled the monitoring of freshwater changes and water accounting, including assessment of important contributions to improve our understanding of global water cycle and its responses to changes in climate and anthropogenic actions (Ndehedehe and Ferreira 2020; Tapley et al. 2019; Thomas and Famiglietti 2019; Rodell et al. 2018; Famiglietti and Rodell 2013). Several of these human influence or activities, including the construction of dams and river impoundments, water transfer projects, land-use change, and extensive groundwater extraction for irrigation and industrial applications have been quantified by combining multi-satellite hydrology missions with novel computational models (e.g., Ndehedehe et al. 2021; Getirana et al. 2020; Long et al. 2020; Zarfl et al. 2015). In the light of this recent advances, it is therefore important to facilitate the dialogue between knowledge providers and stakeholders through the use of this technology and geospatial tools. Existing opportunities in the use of satellite Earth observations continue to grow, ensuring that the advancement in space science and human capacity to fully utilize these data and information in directing policy and environmental analysis, and implementation of action plans is actualized.

Therefore, the use of sophisticated space-based observations depicting the biophysical and human dimensions of freshwater across the globe can expand our knowledge base on the causes and cascading impacts of climate change and the human actions that drive environmental and hydrological processes. Satellite data has been used to fill some of the existing data gaps, especially in emerging and some advanced economies where such observations are increasingly lacking. In addition to this, several data obtained from the model and by combining ground observations with simulations (reanalysis products) have shown great potential in environmental monitoring. The expectation of science at least by stakeholders and the end-users of scientific products is that it must take the lead in providing new information and evidence that advance human societies and improves relevant conditions. Earth resources satellites and the innovative capabilities behind them and satellite geodetic missions like GRACE and radar remote sensing techniques make this achievable. With this in mind, this book aims to provide an understanding of this emerging and novel field of study, called *remote sensing hydrology*. This concept and its practical applications in terrestrial hydrology as well as remote sensing of the environment and impact analysis are also detailed in this book.

The overarching goal of this book is to lay bare the concepts of *remote sensing hydrology* as it relates to the applications of optical (e.g., Sentinel, Landsat, SPOT, MODIS, etc.) and satellite geodetic (e.g., satellite gravity observations, radar altimetry, etc.) systems in quantitative and multidisciplinary aspects of land water storage (e.g., rivers, lakes, soil moisture, groundwater, floodplain wetlands, etc.) and assessing the impacts of climate change on freshwater ecosystems. The use of observations from these sensors, either independently or in combination with other data (e.g., reanalysis and model outputs) in remote sensing hydrology and impact assessment from extreme climate events such as droughts, floods, and wildfires is highlighted and demonstrated. With water being a key challenge in the twenty-first century, improving our contemporary understanding of global terrestrial hydrology, through the use of these observations and space-borne measurements is crucial. Another important aspect of this book is that it details the broad range of geospatial and relevant statistical methods and tools that are valuable in the analyses of hydrological time series and multi-mission satellite observations. With the growing applications of GRACE observations by Earth scientists, the book details practical utility of GRACE observations, including processing chains and demonstrating the contribution of climate change to the increased acceleration of global water cycle. In this regard, issues associated with freshwater changes and availability in the Anthropocene are highlighted and discussed using a plethora of region-specific case studies across the globe. Together, this book provides a range of opportunities for practitioners and relevant agencies to benefit from satellite-based information.

1.2 What is Remote Sensing Hydrology?

The emergence of new remote sensing platforms and sensors (Fig. 1.1) has revolutionized the applications of remote sensing in general. In the context of quantitative hydrology, new satellite systems like the European Space Agency state-of-art Sen-

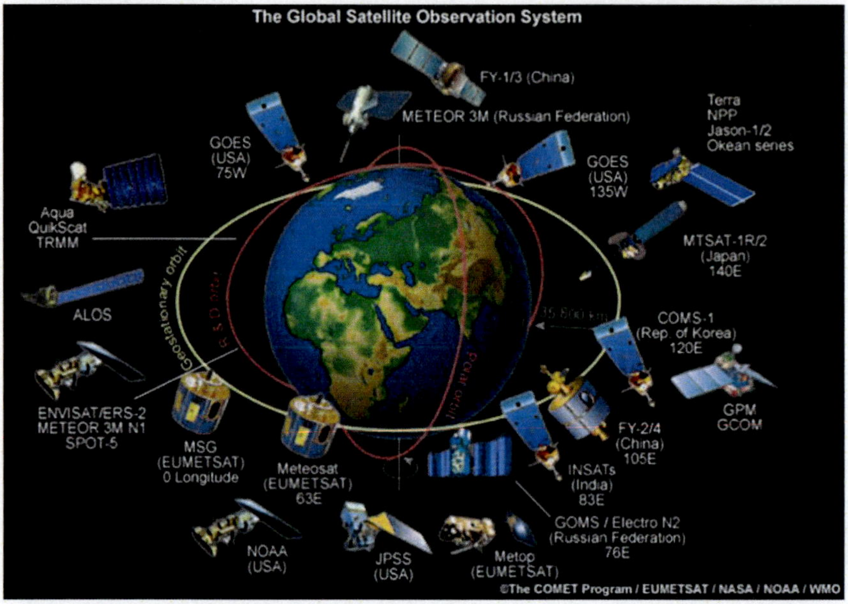

Fig. 1.1 Global satellite observation systems

tinel products, CryoSat-2, Ice, Cloud, and land Elevation Satellite (ICESat)-2 and Soil Moisture and Ocean Salinity, among other satellite missions provide unique and improved capability in comprehensive monitoring of the environment ranging from surface water in lakes, rivers, and floodplains to land surface conditions (evapotranspiration, temperature, etc.), land cover states (e.g., vegetation, urban settlements), soil moisture conditions, and extreme climate conditions such as floods, droughts, and wildfires. As opposed to previous optical satellite Earth observation systems such as the Landsat missions, there are now remote sensing systems dedicated to hydrological and meteorological applications. Satellites such as the Tropical Rainfall Measuring Mission (TRMM) provided extensive spatial coverage of rainfall across the globe on a relatively higher temporal (hourly, daily, monthly) and spatial resolution (0.25°). TRMM's operation commenced in 1997, providing critical inputs to tropical cyclone forecasting, numerical weather prediction, and precipitation climatologies, among several other contributions and societal applications. NASA's CloudSat mission also provides information on the Earth's atmosphere, oceans, land surface, polar ice regions, and solid Earth. This satellite sends out pulses of microwave energy through the clouds, and some of the energy in the pulses is reflected back to the spacecraft (https://cloudsat.atmos.colostate.edu/overview). The phase difference or time lag between when the pulse is sent and when the reflected energy is received back at the spacecraft is the basis for quantifying how much water or ice is contained in the cloud. Observations from CloudSat have been used to improve understanding of the Earth's water budget, climatology, land surface processes, tropical cyclones, among other atmospheric circulation patterns, which drive the global water cycle. In addition to these measurements, other advanced satellite systems such as GRACE enabled the assessment of changes in global terrestrial water storage (sum of soil moisture, groundwater, canopy water storage, ice/snow, and surface water storage in rivers, reservoirs, lakes and wetlands/floodplains, etc).

1.2.1 Satellite Hydrology

With the opportunity of combining satellite hydrology missions such as altimetry and GRACE missions with other Earth resource satellites (e.g., Landsat), direct monitoring of surface water hydrology (e.g., changes of lake water levels and volumes) and fluxes are gradually emerging. The combination of the second generation of Ice, Cloud, and Land Elevation Satellite (ICESat-2) and Lidar (Light detection and ranging) datasets with Landsat imagery in long-term monitoring of annual changes in lake water levels and volumes (Fig. 1.2) is a good example of satellite hydrology applications. One of the recent satellite hydrology missions is the ICESat, which measures ice sheet mass balance, cloud and aerosol heights, as well as land topography and vegetation characteristics (https://icesat.gsfc.nasa.gov/). The first generation of ICESat was launched in January 2003, and it collected multi-year data on the elevation of the Earth's ice sheets, clouds, vegetation, and the thickness of sea ice off and on until October 2009. Its precise observation of the surface elevation of the Arctic sea ice enabled the measurement of ice thickness and volume. Figure 1.3a–d illustrate the sea ice thickness in the Northern Hemisphere, excluding the Baltic Sea and the Pacific for the months of February 2021 (3 and 25) and March 2021 (23 and 25). The insert on the upper right corner of Fig. 1.3a–d show the Arctic Sea Ice volumes from January to December. The maps in Fig. 1.3 show fluctuations in the thickness of ice. Sea ice is frozen seawater that floats on the ocean surface. It forms in both the Arctic and the Antarctic during winter periods. They retreat during summer but do not completely disappear. This floating ice has a profound influence on the polar environment, influencing ocean circulation, weather, and regional climate (https://earthobservatory.nasa.gov/features/SeaIce). As illustrated in Fig. 1.3, their observation and monitoring are possible and can be achieved using the ICESat-2 mission.

However, future space agency missions such as Surface Water and Ocean Topography (SWOT) mission, which is currently under development (www.jpl.nasa.gov/missions/surface-water-and-ocean-topography-swot) will directly provide this capability, serving as another dedicated satellite hydrology mission and tool in box for *remote sensing hydrology*, in addition to GRACE. The SWOT mission (see more details in Sect. 1.3.2) is expected to provide detailed measurements of surface water storage variations (i.e., wetlands, lakes, or reservoirs), complementing the GRACE mission. As indicated in the SWOT mission document (https://swot.jpl.nasa.gov/files/swot/SWOT_MSD_1202012.pdf), the scientific rationale for the development of SWOT is twofold. Firstly, to make high-resolution, wide-swath altimetric

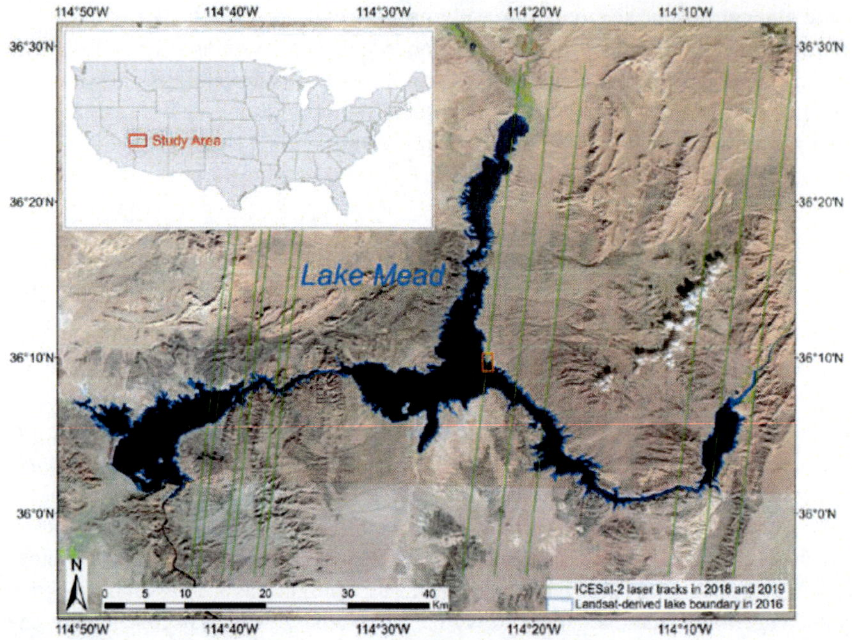

Fig. 1.2 Satellite monitoring of annual changes of lake water levels and volumes using Ice, Cloud, and land Elevation Satellite (ICESat)-2 and Landsat. Source Xu et al. (2020)

measurements of the ocean surface topography will advance the understanding of the oceanic mesoscale and sub-mesoscale processes. Secondly, the quest for a mission that will measure the elevation of water on land that will boost fundamental advances in the knowledge of the spatio-temporal distribution of storage and discharge of terrestrial water. Nevertheless, some other space agency missions relevant to hydrology and water resources have also been flagged for deployment in the coming years. They include, the Copernicus Sentinel-6, and Water Cycle Observation Mission, which is aimed at measuring snow water equivalent, soil moisture, precipitation, atmospheric water vapour, and other state variables. ICESat-2 is another sophisticated hydrology mission primarily designed to map ice sheet and in addition monitor surface water elevations (e.g., Ndehedehe 2019; McCabe et al. 2017). ICESat-2 mentioned earlier is second-generation of the laser altimeter ICESat mission and is optimized to use a micro-pulse multi-beam approach. It is designed to determine ice sheet mass balance as well as cloud property information, provide topography and vegetation data around the globe, as well as the polar-specific coverage over the Greenland and Antarctic ice sheets (https://icesat.gsfc.nasa.gov/). All these missions and others provide substantial context to what is now known as *remote sensing hydrology*.

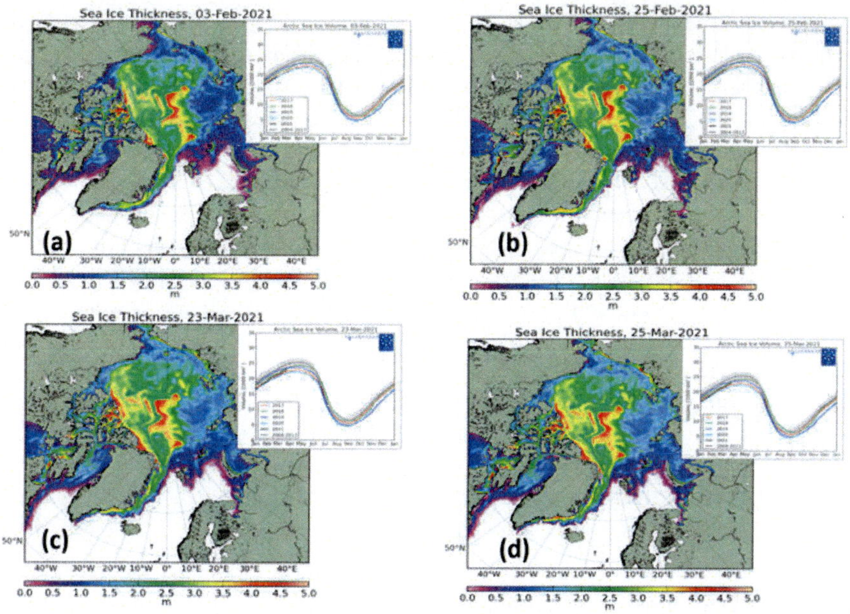

Fig. 1.3 Sea Ice thickness and volume observed by the Ice, Cloud, and Land Elevation Satellite (ICESat). Source, Polar Portal (http://polarportal.dk/en/home/)

1.2.2 *The Concept of Remote Sensing Hydrology*

Remote sensing hydrology is therefore the science that relies on space technology to monitor the occurrence, distribution, changes, and the characteristics of terrestrial stored water in all its forms (snow, ice, lakes, rivers, groundwater, canopy water, etc.) and their relationship with the environment and the Earth's climate system. This concept encapsulates the multidisciplinary aspects of environmental geoscience, and remote sensing of the environment, integrating several geospatial tools and methods, including allied sciences (e.g., statistics) to provide quantitative evaluation of terrestrial hydrology. Within this context, statistical and geospatial methods, using remote sensing and reanalysis data (a synthesis of satellite and ground observations), or even the use of platforms for big data analysis (e.g., Google Earth Engine) to support decision-making and management of water resources are integral aspects of remote sensing hydrology. Apparently, the application of satellite products in water resources and hydrology, including soil moisture and groundwater has been reported (e.g., Rango 1994). Over the years, these products have arguably been useful in understanding land surface hydrology and key state variables like soil moisture, especially in regions where in situ networks are sparse or lacking. However, the combination of geoinformation systems with new remotely sensed observations provides an innovative remote sensing hydrology framework that underpins improved water-related decisions. For example, with different digital image processing techniques, including image enhancement and directional Sobel filters, accurate fracture maps, and derived products (fracture density, coincidence map, and cross-points) have been generated using Landsat 8 (visible and infrared bands) in combination with advanced digital elevation models (Oussou et al. 2020). These fractured maps of aquifers are critical to improving drinking water access and can directly feed into water resources planning on a broad or local scale. While remote sensing has facilitated the successful location of important groundwater resources (Waters et al. 1990), the advent of space science in the last two decades contributed immensely to the progress in Earth system science, making it possible for frequent observations of large-scale, key water budget quantities such as rainfall, evapotranspiration, and terrestrial stored water from space. However, as discussed further in Sect. 1.3, some new remote sensing technologies are discipline-specific, providing measurements on surface water elevation and changes in freshwater bodies.

1.3 Remote Sensing Hydrology Systems

The revolution in space technology ushered in a new era of sophisticated satellite measurements of these quantities on several temporal (weekly, bimonthly, monthly, etc.) and spatial scales (high, medium, and low resolution). Dedicated remote sensing systems or platforms that will or can measure changes in surface water hydrology or provide the capability to directly monitor terrestrial stored water fall into the categories of what is here described as *remote sensing hydrology systems*. Apparently, the Landsat systems, SPOT, and some European Space Agency-based Sentinel products (e.g., Sentinel-2) are Earth resources satellites that can be used to monitor changes in surface water, among several other quantities such as vegetation and land surface conditions. But they are not dedicated systems that directly measure changes in surface water. Monitoring changes in surface water from these optical satellite systems require the use of indicators and metrics, especially to help with quantifying surface water extents, floods, water quality, etc. As global freshwater is undergoing substantial changes because of the impacts of climate change, measurements from remote sensing hydrology systems provide an increased opportunity to monitor such impacts. Satellite observations of freshwater (groundwater, surface water in lake, rivers, and floodplains) from new advanced satellite geodetic missions like the Gravity Recovery and Climate Experiment (GRACE) is therefore important to keep track of its variability. This mission has been providing an unparalleled perspective on global freshwater changes through quantitative estimates of variations in land water storage since its launch in 2002. The GRACE mission and its Follow-On are discussed in detail in Chap. 4. In this chapter, examples of remote sensing hydrology missions, like those in satellite radar family, which are dedicated satellite hydrology missions are highlighted.

With the anticipated launch of a key satellite hydrology program, the Surface Water Ocean Topography (SWOT) mission, the concept of *remote sensing hydrology* will be even more pronounced. The mission objectives of the SWOT program detailed in Sect. 1.3.2 clearly identify it as a perfect example of remote sensing hydrology system in modern human history. The idea advocated in this book is that as with satellite radar altimetry and geodetic missions, the SWOT program fits into this family of dedicated satellite hydrology missions. That brings us to the hypothesis that *remote sensing hydrology* is not the same thing as *application of remote sensing in hydrology*. The latter has been the focus in the last few decades, emphasizing, exploring, and developing new methods to aid the processing of Earth resources satellites systems for hydrological applications, while recent investments in new satellite geodetic platforms gave birth to the former. For example, scientists rely on the combination of visible and short-wave infra-red regions of the electromagnetic spectrum to detect the presence of water on landscapes using optical systems like Landsat, Sentinel-2, MODIS, and even the French SPOT missions. Observations from these same missions can also be processed and used to monitor several other variables, resources, and environmental changes including, water quality, urban sprawl, crop health, droughts, and wildfires, among others. In each of these applications, indicators and algorithms (Okwuashi et al. 2020; Feyisa et al. 2014; Xu 2006) have been developed to enhance the processing and extraction of these quantities from satellite images. For instance, Pekel et al. (2016) developed an expert system to detect and map the presence of water across the globe using about 35 years of archived Landsat data. But *remote sensing hydrology* platforms are different. They are dedicated and discipline-specific, providing direct and level 1 information on the quantitative aspects of freshwater resources. GRACE satellites, for example, is a satellite geodetic-based mission that detects changes in continental freshwater based on observed changes in surface mass variations caused by changes in gravity fields of the Earth. All GRACE products (i.e., levels 1–4), directly provide information on changes in land water storage while satellite altimetry missions (Sect. 1.3.1) generate data on water level fluctuations. The SWOT mission will complement existing satellite hydrology missions like GRACE by facilitating large-scale monitoring of freshwater bodies to aid the understanding of global climate change and anthropogenic contributions.

1.3.1 Satellite Radar Missions

Satellite radar altimetry measurements are now available to track changes in surface water levels. By measuring the satellite-to-surface round-trip time of a radar pulse, satellite altimetry determine the distance from the satellite to a target surface (Fig. 1.4). This implies that the altimeter emits a radar wave and analyses the return signal, which bounces off the surface. The surface height is the difference between the satellite's position in orbit with respect to an assumed surface, which best approximates the Earth's surface. This surface is called the line of best fit or the ellipsoid. The satellite to surface range in Fig. 1.4 is calculated by measuring the time taken for the signal to make the round trip. The illustration in Fig. 1.4 also shows that ocean surface topography can be derived from the relationship between sea surface height and the geoid leads. Although other quantities are measured in the process (e.g.,

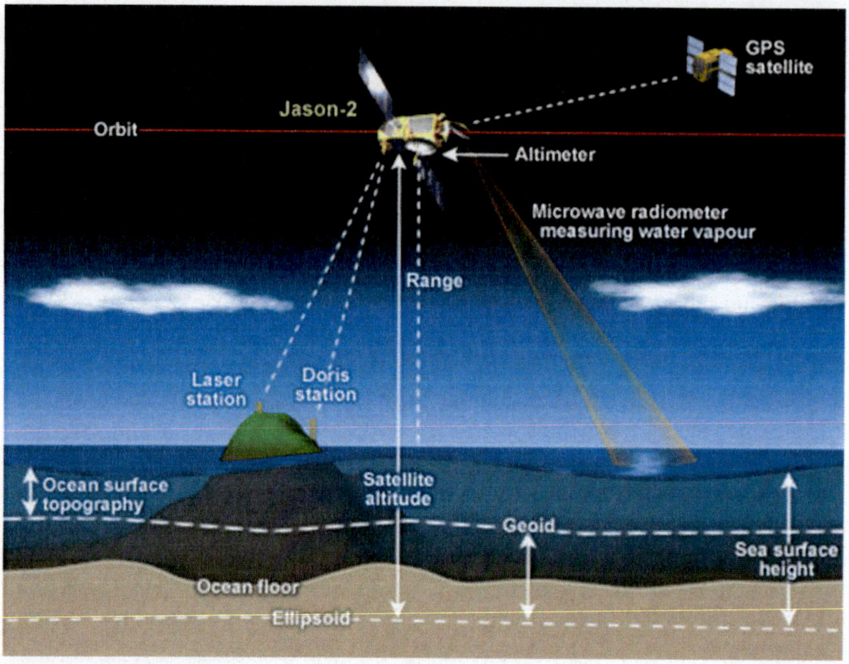

Fig. 1.4 Satellite Altimetry mission. Source, Wikimedia Commons

wind speed, backscatter coefficient, and surface roughness), they also provide information on the Earth 's gravity field, the shape and structure of the ocean bottom, and geostrophic ocean currents (Cheney 2001). Lake height variations computed from TOPEX/POSEIDON (T/P), Jason-1 and Jason-2/OSTM altimetry are now provided by the United States Department of Agriculture (USDA) as auxiliary information to analyse water reservoirs across the globe (Fig. 1.5). These data have been widely used in monitoring variations in lake levels such as the Lake Volta (Fig. 1.6). They can be downloaded from the United States Department of Agriculture portal (www.pecad.fas.usda.gov/cropexplorer/globalreservoir) and the hydroweb Laboratoire dEtude en Geophysique et Oceanographie Spatiale (LEGOS) database. Several globally important large lakes have been monitored using altimetry. Figure 1.5 shows the average lake height fluctuations where blue dots show increasing heights (water gain) and red dots show decreasing heights (water loss). The fluctuation in lake heights in lakes with yellow dots are relatively stable and not strong. The deep blue dots have the strongest lake height fluctuations largely due to human water management practices such as dams and they spread across North America, Africa, and Central Asia. As further illustrated in Fig. 1.6, Lake Volta is one of the lakes with deep blue dots and strongest variation in lake heights. Despite being smaller in surface area compared to Lake Victoria in East Africa where lake height variations range from approximately 2–3 m, Lake Volta shows about 10 m in lake height variations between 2010 and

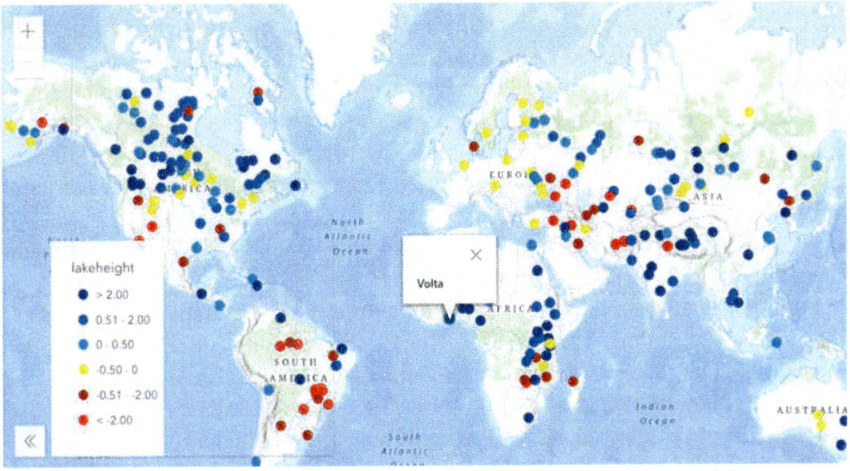

Fig. 1.5 Spatial distribution of Global Reservoirs and Lakes Monitor (G-REALM) satellite altimetry products. Dots show average variation in lake height. Source, USDA/NASA G-REALM program

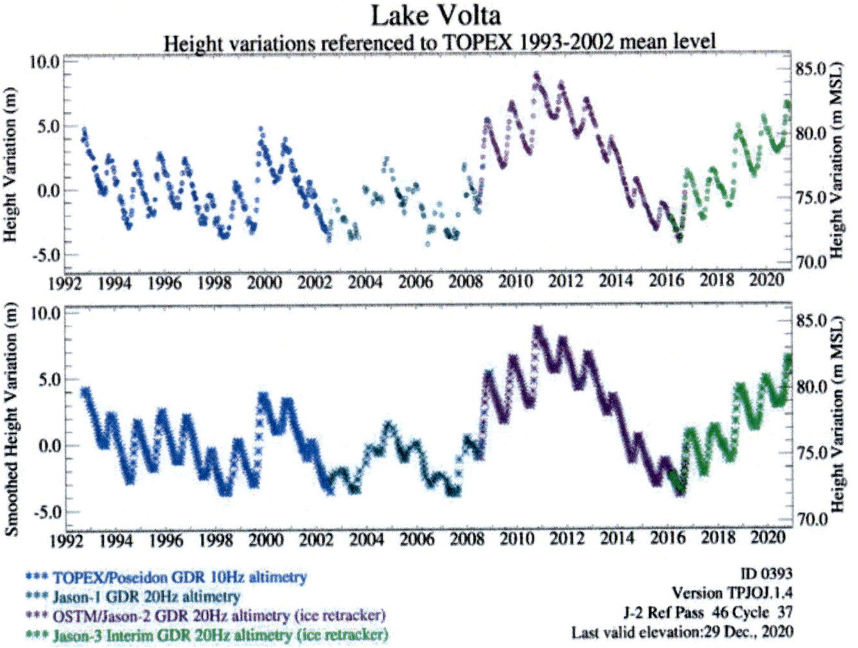

Fig. 1.6 Lake Volta Height Variations from TOPEX/POSEIDON and Jason series Altimetry. Source, USDA/NASA G-REALM program

2016 (Fig. 1.6). These operational satellite altimetry products (TOPEX/POSEIDON, Jason-1, Jason-2/OSTM, and Jason-3) are regularly updated and data is even available in near-real-time and can be retrieved with ease. Moreover, in combination with GRACE observations, the utility of these satellite altimetry measurements has been extended to include the localization and separation of surface water storage changes to understand variability in other hydrological stores like groundwater and basin terrestrial water storage (Agutu et al. 2019; Ndehedehe et al. 2017; Moore and Williams 2014). This has enabled the monitoring of anthropogenic and climate change influence on surface water hydrology.

Through near-real-time data assimilation, altimetry observations can be used to forecast large-scale climatic processes like the El-Ninõ Southern Oscillation, monitor coastal circulation, and predict hurricane intensity (Cheney 2001). As with GRACE mission, satellite altimetry has been employed to support the evaluation of floods in data-deficient regions and integrated with other satellite data to improve flood models (Tarpanelli and Benveniste 2019). New operational sensors with improved technology such as CryoSat-2 provide precise measurement of ice thickness on both land and sea to improve the understanding of the impacts of global climate on ice. Sentinel-3 satellite radar mission, though primarily an ocean mission, however, it also provides atmospheric and land applications. The Sentinel-3 mission is equipped with a SAR altimeter and its capability to monitor water surface elevation and increase data availability over inland water bodies and reservoirs have been illustrated (Zhang et al. 2020). Its mission objectives show it is substantially a *remote sensing hydrology system*. These objectives have been summarized by European Space Agency and include[1];

- Measuring sea surface topography, sea surface height, and significant wave height;
- Measuring ocean and land surface temperature;
- Measuring ocean and land surface colour;
- Monitoring sea and land ice topography;
- Seawater quality and pollution monitoring;
- Inland water monitoring, including rivers and lakes;
- Aid ocean forecasts with acquired data.

With capabilities in weather forecasting, accurate flood forecasting is also feasible with Sentinel-3 radar mission. Furthermore, new Copernicus Sentinel-6 is the latest satellite radar altimetry mission in the new generation of radar altimetry missions. As anticipated, the mission will extend the legacy of sea surface height measurements into the year 2030. Other pioneering radar missions from the European Space Agency family, the Sentinel-1 mission provides an operational interferometry capability and generates C-band synthetic aperture radar imagery regardless of weather conditions. Interferometric synthetic aperture radar imagery from this mission (Sentinel-1) and other satellite radar observations (e.g., RADARSAT-1 and 2, Phased Array L-band Synthetic Aperture Radar-PALSAR) have been combined with those of satellite gravity systems such as GRACE to monitor changes in groundwater and detect areas

[1] https://earth.esa.int/web/guest/missions/esa-eo-missions/sentinel-3.

where excessive groundwater use and extraction has resulted in land subsidence and other damages to public infrastructure such as roads, pipelines, buildings, etc. (Smith and Majumdar 2020; Ojha et al. 2018; Castellazzi et al. 2018, 2016).

1.3.2 The Surface Water Ocean Topography Mission

Mapping the global distribution of surface water fluxes in rivers, lakes, reservoirs, and wetlands and quantifying its changes in time and space is a key challenge. While they represent key components of the global water cycle and affect human societies, monitoring infrastructures to improve understanding of their variability on different spatial and temporal scales is limited. As discussed in preceding sections, there are numerous developments of multi-mission sensors that provide useful complementary observations of Earth's resources. However, dedicated satellite missions for surface water storage change and fluxes are lacking (Biancamaria et al. 2015). To address this gap in global surface water measurements, and better understand global oceans and its terrestrial surface waters, the Surface Water Ocean Topography (SWOT) mission has been proposed and is scheduled for launch in February 2022. The SWOT mission is a collaboration between the US National Aeronautics November and Space Administration collaboration and the Centre National d'E' tudes Spatiales, and includes contributions from the Canadian Space Agency and the United Kingdom Space Agency. Key science objectives of the SWOT mission as highlighted by the Canadian Space Agency are:

1. Evaluate discharge variations in rivers.
2. Monitor coastal and high-latitude tides and internal tides.
3. Monitor changes in water stored in lakes, reservoirs, and wetlands.
4. Obtain precise measurements of ocean elevation features (i.e., high latitude tides).
5. Measure small- and medium-scale features of the ocean's surface such as waves.

In anticipation of the SWOT system, its land hydrology capabilities, benefits, and measurement potentials have already been detailed in several literatures (e.g., Biancamaria et al. 2015). Expected outcomes include improved estimation of surface water availability, enhanced global water balance studies, and better representation of processes such as flood dynamics and anthropogenic contributions to the global water cycle. With these, the SWOT mission will be the latest of a family of sophisticated space programs that will provide unique capabilities that underpins effort in understanding climate change. With about 90% coverage of the globe, and collecting data over the world's freshwater stores (lakes, rivers, reservoirs) at least twice every 21 days, weather and climate predictions will be improved and freshwater management will be optimized.[2] This means that several important lakes and continental freshwater bodies where gauged observations are lacking or not publicly available will benefit from SWOT operations. Currently, the water budget approach to quantifying groundwater from GRACE satellites relies on outputs from hydrological models and altimetry measurements. While uncertainties in estimated groundwater could be

[2] https://www.jpl.nasa.gov/missions/surface-water-and-ocean-topography-swot.

large for regions where surface water is largely substantial, opportunities exist for the SWOT data to be integrated with GRACE for improved groundwater assessment. The remaining chapters outline important geospatial methods, machine learning, and multi-satellite missions to further highlight the concept of remote sensing hydrology and its applications and benefits in environmental geoscience in more detail.

References

Agutu N, Awange J, Ndehedehe C, Kirimi F, Kuhn M (2019) GRACE-derived groundwater changes over Greater Horn of Africa: temporal variability and the potential for irrigated agriculture. Sci Total Environ 693:133467. https://doi.org/10.1016/j.scitotenv.2019.07.273

Biancamaria S, Lettenmaier D, Pavelsky T (2015) The SWOT mission and its capabilities for land hydrology. Surv Geophys 37:307–337. https://doi.org/10.1007/s10712-015-9346-y

Castellazzi P, Longuevergne L, Martel R, Rivera A, Brouard C, Chaussard E (2018) Quantitative mapping of groundwater depletion at the water management scale using a combined GRACE/InSAR approach. Remote Sens Environ 205:408–418. https://doi.org/10.1016/j.rse.2017.11.025

Castellazzi P, Martel R, Galloway DL, Longuevergne L, Rivera A (2016) Assessing groundwater depletion and dynamics using GRACE and InSAR: potential and limitations. Groundwater 54(6):768–780. https://doi.org/10.1111/gwat.12453

Cheney R (2001) Satellite altimetry. Academic Press, pp 2504–2510. https://doi.org/10.1006/rwos.2001.0340

Famiglietti JS, Rodell M (2013) Water in the balance. Science 340(6138):1300–1301. https://doi.org/10.1126/science.1236460

Feyisa GL, Meilby H, Fensholt R, Proud SR (2014) Automated water extraction index: a new technique for surface water mapping using landsat imagery. Remote Sens Environ 140:23–35. https://doi.org/10.1016/j.rse.2013.08.029

Getirana A, Jung HC, Hoek JVD, Ndehedehe CE (2020) Hydropower dam operation strongly controls Lake Victoria's freshwater storage variability. Sci Total Environ 726:138343. https://doi.org/10.1016/j.scitotenv.2020.138343

Gomes VCF, Queiroz GR, Ferreira KR (2020) An overview of platforms for big earth observation data management and analysis. Remote Sens 12(8). https://doi.org/10.3390/rs12081253

Gorelick N, Hancher M, Dixon M, Ilyushchenko S, Thau D, Moore R (2017) Google Earth engine: planetary-scale geospatial analysis for everyone. Remote Sens Environ 202:18–27. https://doi.org/10.1016/j.rse.2017.06.031

Landerer FW, Flechtner FM, Save H, Webb FH et al (2020) Extending the global mass change data record: GRACE follow-on instrument and science data performance. Geophys Res Lett 47(12):e2020GL088306. https://doi.org/10.1029/2020GL088306

Long D, Yang W, Scanlon B et al (2020) South-to-north water diversion stabilizing Beijing's groundwater levels. Nat Commun 11(3665). https://doi.org/10.1038/s41467-020-17428-6.

McCabe MF, Rodell M, Alsdorf DE, Miralles DG, Uijlenhoet R, Wagner W, Lucieer A, Houborg R, Verhoest NEC, Franz TE, Shi J, Gao H, Wood EF (2017) The future of Earth observation in hydrology. Hydrol Earth Syst Sci 21(7):3879–3914. https://doi.org/10.5194/hess-21-3879-2017

Moore P, Williams SDP (2014) Integration of altimetry lake levels and GRACE gravimetry over Africa: inferences for terrestrial water storage change 2003–2011. Water Resour Res 50:9696–9720. https://doi.org/10.1002/2014WR015506

Ndehedehe CE (2019) The water resources of tropical West Africa: problems, progress and prospect. Acta Geophys 67(2):621–649. https://doi.org/10.1007/s11600-019-00260-y

Ndehedehe CE, Awange J, Kuhn M, Agutu N, Fukuda Y (2017) Analysis of hydrological variability over the Volta river basin using in-situ data and satellite observations. J Hydrol Reg Stud 12:88–110. https://doi.org/10.1016/j.ejrh.2017.04.005

Ndehedehe CE, Ferreira VG (2020) Assessing land water storage dynamics over Southern America. J Hydrol 580:124339. https://doi.org/10.1016/j.jhydrol.2019.124339

Ndehedehe CE, Ferreira VG, Agutu NO, Onojeghuo AO, Okwuashi O, Kassahun HT, Dewan A (2021) What if the rains do not come? J Hydrol 595:126040. https://doi.org/10.1016/j.jhydrol.2021.126040

Ojha C, Shirzaei M, Werth S, Argus DF, Farr TG (2018) Sustained groundwater loss in California's central valley exacerbated by intense drought periods. Water Resour Res 54(7):4449–4460. https://doi.org/10.1029/2017WR022250

Okwuashi O, Ndehedehe CE, Attai H (2020) Tide modeling using partial least squares regression. Ocean Dyn 70:1089–1101. https://doi.org/10.1007/s10236-020-01385-1.

Oussou FE, Yalo N, Ndehedehe CE, Oloukoi J, Alassane A, Boukari M, Gbewezoun VHG (2020) Fracture aquifers identification in the Zou basin (West Africa) using remote sensing and GIS. Geocarto Int 1–24. https://doi.org/10.1080/10106049.2020.1852613

Pekel J-F, Cottam A, Gorelick N, Belward AS (2016) High-resolution mapping of global surface water and its long-term changes. Nature 540:418–422. https://doi.org/10.1038/nature20584

Rango A (1994) Application of remote sensing methods to hydrology and water resources. Hydrol Sci J 39(4):309–320. https://doi.org/10.1080/02626669409492752

Rodell M, Famiglietti JS, Wiese DN, Reager JT, Beaudoing HK, Landerer FW, Lo M-H (2018) Emerging trends in global freshwater availability. Nature 557:651–659. https://doi.org/10.1038/s41586-018-0123-1

Smith RG, Majumdar S (2020) Groundwater storage loss associated with land subsidence in western United States mapped using machine learning. Water Resour Res 56(7):e2019WR026621. https://doi.org/10.1029/2019WR026621

Tapley B, Bettadpur S, Watkins M, Reigber C (2004) The gravity recovery and climate experiment: mission overview and early results. Geophys Res Lett 31:1–4. https://doi.org/10.1029/2004GL019920

Tapley B, Watkins M, Flechtner F et al (2019) Contributions of GRACE to understanding climate change. Nat Clim Change 9:358–369. https://doi.org/10.1038/s41558-019-0456-2

Tarpanelli A, Benveniste J (2019) On the potential of altimetry and optical sensors for monitoring and forecasting river discharge and extreme flood events. Elsevier Book, pp 267–287. https://doi.org/10.1016/B978-0-12-814899-0.00011-0

Thomas B, Famiglietti J (2019) Identifying climate-induced groundwater depletion in GRACE. Sci Rep 9:4124. https://doi.org/10.1038/s41598-019-40155-y

Wang L, Ma Y, Yan J, Chang V, Zomaya AY (2018) pipsCloud: high performance cloud computing for remote sensing big data management and processing. Future Gener Comput Syst 78:353–368. https://doi.org/10.1016/j.future.2016.06.009

Waters P, Greenbaum D, Smart PL, Osmaston H (1990) Applications of remote sensing to groundwater hydrology. Remote Sens Rev 4(2):223–264. https://doi.org/10.1080/02757259009532107

Xu H (2006) Modification of normalised difference water index (NDWI) to enhance open water features in remotely sensed imagery. Int J Remote Sens 27(14):3025–3033. https://doi.org/10.1080/01431160600589179

Xu N, Ma Y, Zhang W, Wang XH, Yang F, Su D (2020) Monitoring annual changes of lake water levels and volumes over 1984-2018 using landsat imagery and ICESat-2 data. Remote Sens 12(23). https://doi.org/10.3390/rs12234004.

Zarfl C, Lumsdon AE, Berlekamp J, Tydecks L, Tockner K (2015) A global boom in hydropower dam construction. Aquat Sci 77:161–170. https://doi.org/10.1007/s00027-014-0377-0

Zhang X, Jiang L, Kittel CMM, Yao Z, Nielsen K, Liu Z, Wang R, Liu J, Andersen OB, Bauer-Gottwein P (2020) On the performance of Sentinel-3 Altimetry over new reservoirs: Approaches to determine onboard a priori elevation. Geophys Res Lett 47(17):e2020GL088770. https://doi.org/10.1029/2020GL088770

Chapter 2
Global Freshwater Systems

Abstract Water availability is an important factor that drives global patterns of agricultural production and trade and the peaceful coexistence among riparian communities with shared water resources (Odorico et al. 2020; Ndehedehe 2019). Water crises and scarcity could trigger increases in prices of groceries due to price volatility (Ndehedehe et al. 2021). This is because water is considerably vital to production and manufacturing process. For instance, about 8000 L of water is required to produce a pair of leather shoes while the production of a kilogramme of cheese requires about 5000 L of water (Prosser 2011). The transactional aspects of water resources come with significant input to national income, contributing to gross domestic product of such countries.

2.1 Global Water Cycle

Water is a fundamental natural resource that is essential for the survival of humanity and ecosystems. The world's water exists naturally in various forms (solid, liquid, and gaseous) due to variations in climatic condition and is present in oceans, seas, lakes, rivers, ice caps, in the atmosphere, and as groundwater. The primary components of the global water cycle, a mechanism by which water moves over, above, and below the Earth's surface, are precipitation, infiltration, runoff, groundwater, discharge, and evapotranspiration. This bulk movement of water by these components redistributes water around the globe, making it present almost everywhere. By approximate estimate, water makes up 71% of the Earth's surface, while the remaining 29% are land territories, comprising continents and islands. Although a reliable assessment of global water storage is difficult because of its dynamic nature, an appraisal and assessment of global water resources by Shiklomanov (2000) indicates that of the world's total water supply of about 1,386 million cubic kilometres, about 97.5% is saline while only 2.5% is freshwater. About 68.7% of total freshwater is locked up in ice and glaciers while 29.9 and 0.26% represent groundwater resources and surface waters, respectively. These surface waters are concentrated in lakes, reservoirs, and

C. Ndehedehe, *Satellite Remote Sensing of Terrestrial Hydrology*,
https://doi.org/10.1007/978-3-030-99577-5_2

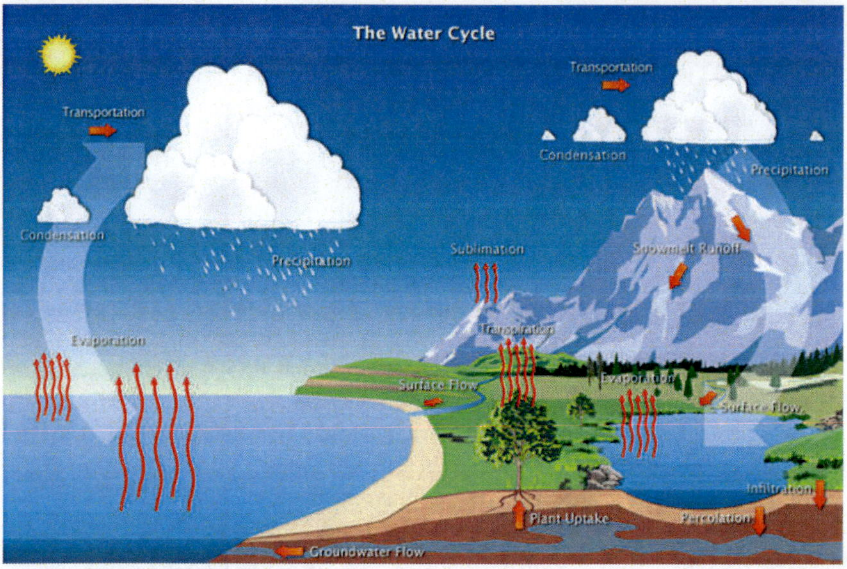

Fig. 2.1 Illustration of the global water cycle. *Source NOAA National Weather Service*

river systems. Indeed, as popularly known, 'Earth is the water planet'. The illustration in Fig. 2.1 confirms that freshwater is not static-it changes from one form to another. For example, it changes from liquid (e.g., water in oceans, seas, lakes, rivers, streams) to gas (water in the form of molecules). This process is called evaporation. While 90% of water moisture in the atmosphere emanates from water bodies, about 10% of water in the atmosphere comes from plants through transpiration. Small proportion of water in solid state (ice or snow) can also change to gas through a process called sublimation. All these processes (evaporation, transpiration, and sublimation) are natural channels that transports or converts liquid water from the land surface back to atmosphere. Precipitation is the main input to the water cycle, transporting water from the atmosphere to the Earth's surface. This then transitions into freshwater in lakes, rivers, water holes, floodplains, soil moisture, and groundwater, which occur through infiltration and recharge. Some proportion of this freshwater returns to the atmosphere through evaporation, transpiration, and sublimation (Fig. 2.1). However, humans also intercept some of the available water for several uses.

Available freshwater in the form of lakes, reservoirs, groundwater, and river systems is extremely limited and yet critical for terrestrial life and ecosystem functioning. The dynamics of freshwater withdrawal in the world (e.g., Alcamo et al. 2003; Shiklomanov 2000; Gleick 2000) underscores its importance as a life-sustaining resource. This is because freshwater is a valuable resource to billions of people in the world, and plays key roles in all ecological and societal activities such as agriculture, human health, socio-economic development, and security. Globally, irrigated agriculture is the biggest consumer of global water resources (Fig. 2.2). While this is true, the proportion of water used for livestock is substantially different across several regions of the world with African sub-regions indicating the highest consumption of water for Livestock (Fig. 2.2). With projected availabilities of freshwater being less than 1000 m^3/capita/yr in the world's densely populated arid areas, water scarcity will be a major restriction to agricultural development in the coming decades (Rijsberman 2006). Because of increasing global population, rapid urbanization, and industrialization, the demand for freshwater is inevitably expected to rise. For example, a large number of river basins that makes up 12% of the total global area of basins with limited annual available water are already under pressure because of increased water withdrawals (Alcamo et al. 1997). Water availability is more likely to restrict economic development in these basins, and an estimated 1.7 billion people live in regions where groundwater resources and groundwater dependent ecosystems are under threat (Gleeson et al. 2012). Globally, more than 1.5 billion people rely on groundwater as a major source of drinking water and are the only source of water in most arid and desert regions. Given the importance of groundwater, in some arid regions, for example, in Israel, groundwater management is integrated as part of the operation of the national system. Groundwater extraction has tripled and depletion rates have doubled between 1960 and 2000 in Asia and north America (Gleick 2000).

There are about 40% increase in the number of people living under absolute water scarcity, up from 15% of the global population with a severe decrease in water resources (Schewe et al. 2013). As at the year 2000, total water withdrawals amounted to 8.4% of global water resources, and this amount is expected to increase up to 12.2% (i.e., between 4600-5800 km^3/yr) by 2025 (Shiklomanov 2000). Vörösmarty et al. (2000) projected estimate of global water use by 2025 is conservative and puts the value at 4700 km^3/yr. The rise in urban population and changing climate will lead to more withdrawals of groundwater as surface water could be limited due to extreme drought. Given the increasing reliance on groundwater for urban water supply and agriculture during times of droughts and water shortage, global water use will increase in the coming decades. In the next section some issues associated with freshwater changes and availability in the Anthropocene are highlighted. Some of the challenges in assessing terrestrial water systems are also mentioned, and the opportunities that exist through the use of new satellite remote sensing methods and frameworks are also discussed.

2.1.1 Water Scarcity and Crises

Climate and human actions drive environmental change (e.g., Ndehedehe et al. 2021). Many regions across the globe are now facing threats to food and water security because of increased acceleration of the water cycle. In regions that will be impacted by extreme drought, there is the likelihood such regions could run out of existing freshwater culminating in water scarcity problems. Globally, water scarcity is becoming a major challenge and concern around the world. Water scarcity occurs when water demand exceeds the available water supply. One indicator of water

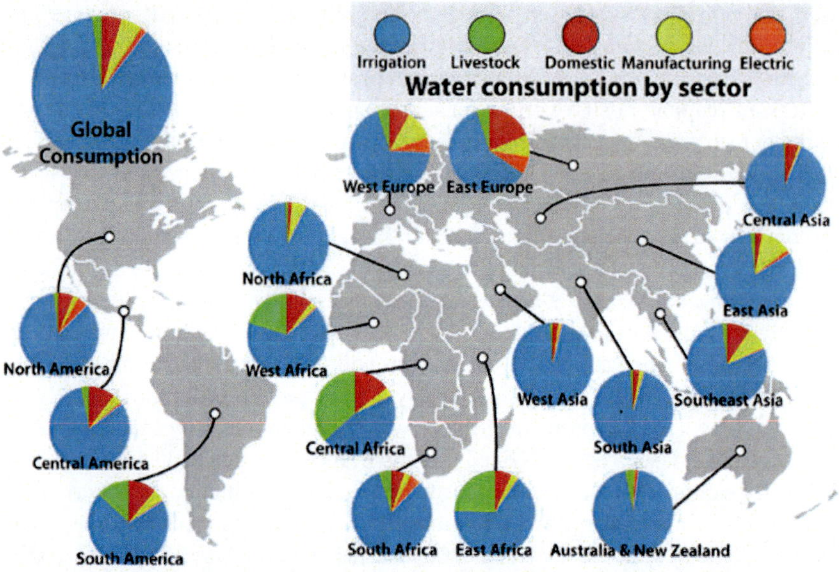

Fig. 2.2 Global water consumption by sector. Irrigation is the leading water consumption sector for all sub-regions of the world. *Source* Brauman et al. (2016)

scarcity is extensive declines in groundwater levels. For example, India's widespread groundwater depletion is alarming due to extensive annual withdrawal for irrigating agricultural lands. Groundwater is an important source of freshwater, especially in arid and semi-arid regions of the world where surface water is considerably limited (e.g., Australia and Middle East). Regions that are heavily dependent on groundwater such as India, the world's largest user of groundwater (Chindarkar and Grafton 2019) are at risk of water scarcity or could be at risk of acute water shortage. The International Water Management Institute estimates that nearly 20% of the world's population live in areas of physical water scarcity, where water use for various purposes exceed 75% of river flows (Gleick 2000). This underscores the risk of water scarcity in these regions. Water scarcity projections based on other metrics estimates that about 57% and 69% of the world's population will be living in countries facing high water stress by 2025 and 2075, respectively (Alcamo et al. 1997). These values may be somewhat too high as argued by Gleick (2000), they, however, suggest the urgent conversations needed to raise awareness on the susceptibility of global water resources. The numerical projections and perspectives on water scarcity and anticipated future use underscore the threats to food production and livelihood, and for the non-industrialized regions of the world (e.g., South Africa), stormwater harvesting has been suggested as possible alternative water resource that could supplement traditional urban water supplies, cushion the impacts of climate change, and improve water security (Fisher-Jeffes et al. 2017). Water scarcity by itself represents one of the characteristics (e.g., degree of scarcity) that could make water a potential source of strategic rivalry (Gleick 1993). This is true for Texas where there are some indications that the water requirements for hydraulic fracturing are already creating conflicts with other uses and could restrict future natural gas production in some areas (see, Gleick 2000).

The World Economic Forum defines water crises as *a significant decline in the available quality and quantity of freshwater resulting in harmful effects on human health and/or economic activity*. With a focus on the quantitative aspect of freshwater, water crises are arguably the result of water scarcity. The latter is expressed as the ratio between water demand and water availability. Following this ratio, a threshold of 0.4 is usually adopted as severe water scarcity. On the one hand, a failing climate as was observed in the Lake Chad Basin will limit freshwater availability. On the other hand, access to water resources could become difficult due to rising population growth and pressure from increased water needs. Apparently, about 90% decline of the Lake's surface area declined due to prolonged below-normal rainfall (Fig. 2.3). Inflows from major river systems (Chari-Logone River) that provide about 95% freshwater of the Lake were restricted due to this change in rainfall. Due to climate change and human influence, the present Lake Chad as shown in Fig. 2.3 is segmented to two pools with the northern pool completely dried up during drought periods. This led to water crises and armed conflicts caused by competition for the limited and shared water resources in the region. However, water withdrawals from the Lake for irrigation purposes during this water deficit period continued unabated. With increasing population accompanied by a rise in water demand, the northern portion of the Lake dried up with little water left in the southern pool. This condition created water crises that have escalated to agitations, tensions, and even armed conflicts among local communities from the riparian countries sharing the Lake's water resources. Several other freshwater bodies such as lakes and rivers are at the risk of desiccation while others occasionally run dry. The Aral Sea has experienced a gradual decline in the last decade and may be completely desiccated with time. The Colorado River in the US, Indu River in Pakistan, and the Hai and Huai Rivers in China are some examples of highly vulnerable freshwater bodies that run dry and diminish due to the combined effects of climate and human actions such as water diversion. With increasing human population and water demand, all of these scenarios could create potential water crises and scarcity, especially during prolonged drought periods.

2.1.2 Water and Agricultural Production

There is a significant relationship between hydrology and national income. Global food production benefits from 65% of annual rainfall while irrigation contributes 35% agricultural development. In rainfall-fed agricultural systems, food production is largely reliant on rainfall. Arguably, the impacts of below-normal rainfall or drought on crop yield will lead to major losses in national income. In Australia, farms and freshwater ecosystems are increasingly being affected by frequent drought events in the last 20 years. Changes in climatic conditions have reduced average profits

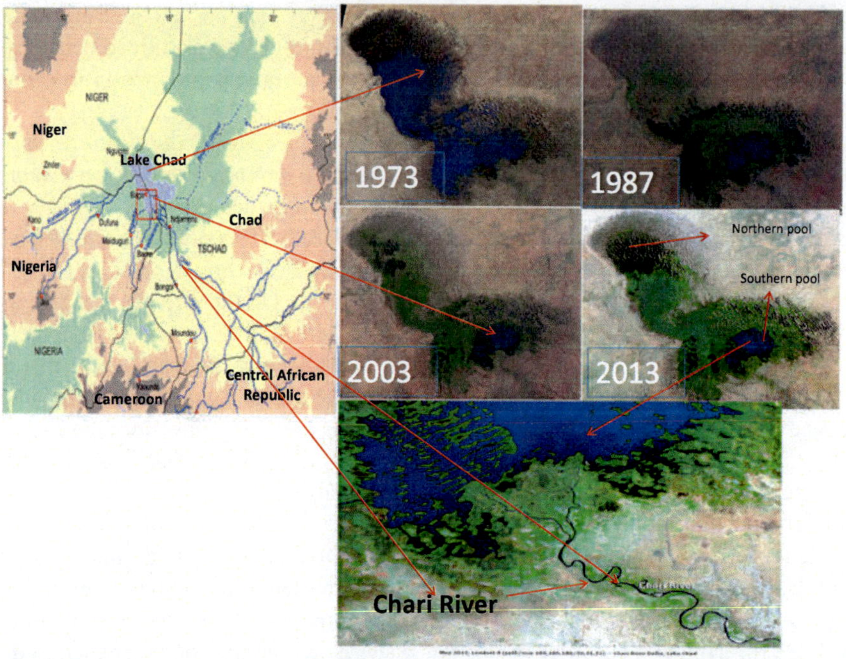

Fig. 2.3 Morphological changes of the Lake Chad's freshwater during the last four decades. *Source* Ndehedehe et al. (2016b)

by 35% for a typical cropping farm in Australia, resulting in national average loss in revenue (1.1 billion dollars per year based on 2015 – 2018 production levels) in the cropping industry (Hughes et al. 2019). Furthermore, in East Africa, famine and food insecurity (e.g., Agutu et al. 2017) have been the results of stakeholders inability to manage drought-risk and conflicts among communities over grazing land and available freshwater (Ndehedehe et al. 2020). Similarly, due to prolonged drought-inflicted water deficits, farmers and pastoralists in the Lake Chad Basin have been engaged in intense competition over the limited water resources in the region. Water availability is an important factor that drives global patterns of agricultural production and trade and the peaceful coexistence among riparian communities with shared water resources (Odorico et al. 2020; Okewu et al. 2019; Ndehedehe 2019). Moreover, agricultural production in water scarce region relies on about 90% of surface and groundwater withdrawals.

Water crises and scarcity could trigger increases in the prices of groceries due to price volatility. This is because water is considerably vital to the production and manufacturing process. For instance, about 8000 L of water is required to produce a pair of leather shoes while the production of a kilogramme of cheese requires about 5000 L of water (Prosser 2011). The transactional aspects of water resources come with significant input to national income, contributing to the gross domestic product of such countries. Australia exports much of its agricultural products and the West African sub-region produces an estimated 70% of the world's cocoa (Schroth et al. 2016). Countries with limited portions of arable land tend to import more water because of their considerable food imports but rather export large quantities of manufactured goods, which require less water (Prosser 2011). This value of water helps creates a global supply chain that contributes to national income. In other words, the impacts of droughts on Australia's water resources could affect countries that benefit from its exported food through a disruption of this chain or price volatility. Similarly, in cases of subsistence agriculture in emerging economies where high proportion of small-scale farmers are heavily reliant on rainfed agriculture, prolonged rainfall deficit could trigger systemic acute food insecurity. Water availability and its vulnerability to climate variability is therefore a key risk to global economy. Sustainable pathways to predict impacts on food production are thus important to guide future investments in drought resilience and prudent management of water resources.

2.1.3 Water Infrastructure and Diplomacy

Climate-risk continue to amplify in regions with critical water challenges. Arid regions, e.g., Australia, Israel, the African Sahel, and the entire Middle East face these challenges due to aridity, which limits the availability of surface water. Groundwater abstraction in these countries results in drastic declines in aquifer levels. With rapidly expanding agricultural activities in most water deficit regions, their strategic water-management policies have favoured the development of several water infrastructures. These were targeted efforts aimed at addressing the susceptibility of these regions to dangerous climatic thresholds and aridity. The investments in hydraulic and water infrastructures serve as risk-based decisions, mitigation strategies, and science-based adaptive approaches to boost water availability. These investments are some traditional approaches to solving the problem of water insecurity and they include dams, reservoirs, pipes, canals, drains, and cement-lined streams for the movement of water. Notably, the economic basis to invest in water infrastructures by developed nations has improved their capacity to cushion the impacts of extreme drought events and threats to freshwater resources (Vorosmarty et al. 2010). However, regions with no infrastructure and economic instruments to manage and develop mechanisms to share risks associated with the cascading impacts of extreme droughts on water resources continue to be vulnerable to the impacts of climate variability (e.g., Vorosmarty et al. 2010).

In geopolitical zones with shared trans-boundary water resources, there have been recent reports and recommendations on the use of diplomacy and institutional frameworks to foster cooperation and management of water-related conflicts (e.g.,Akhmadiyeva and Abdullaev 2019; Atef et al. 2019). Water diplomacy is now perceived as a relevant tool for the cooperative management of shared water resources, multi-stakeholder dialogues, policy and political engagement to support productive discussions related to the legal nature of water sharing, and increase the effectiveness of trans-boundary water cooperation, among others (see, e.g., Honkonen and Lipponen 2018; Grech-Madin et al. 2018; Schmeier and Shubber 2018). As a template that can be extended to support regional cooperation on trans-boundary water sharing in other African regions, Al-Saidi and Hefny (2018) highlights the institutional arrangements and the merits of regional cooperation in addressing the underlying issues of water diplomacy in the Eastern Nile Basin. Water diplomacy and negotiation have been advocated as vehicle to foster cooperation over trans-boundary freshwater resources. Inter-basin water transfers, the movement of water from one watershed to another, are now emerging as part of a risk-base solution to cushion the effects of climate change and water crises.

2.2 Challenges in Environmental Monitoring

2.2.1 Changes in Terrestrial Hydrology

The underlying issues in global water systems are multifaceted. First, the impact of climate change and/or indices of climate variability (e.g., El-Niño Southern Oscillation-ENSO) affects the spatial and temporal distribution of rainfall. These changes in precipitation contribute significantly to rapid changes in terrestrial water cycle (i.e., hydrological cycle) in many parts of the world (Phillips et al. 2012; Hurkmans et al. 2009; Vörösmarty et al. 2000). Second, poor understanding of large-scale dynamics in terrestrial water systems due to insufficient observational networks for its monitoring (e.g., Hall et al. 2014; Alsdorf et al. 2007). Third, excessive human water abstraction of groundwater for irrigation purposes. This human footprint results in aquifer depletion and loss of groundwater and changes in land use and land surface conditions may complicate natural hydrologic processes such as groundwater recharge, infiltration, discharge and runoff, and water availability.

Reoccurring drought episodes and the incessant impacts of climate variability on socio-economic systems are major triggers to the plenitude of climate research, which focused have mainly on analysing multiple climate variables such as precipitation, runoff, soil moisture, temperature, and zonal winds, among others. Using historical gauge data, outputs from global and regional climate models (i.e., GCMs and RCMs), and a host of optical remote sensing data, hydro-climatic research at all levels (region-specific and large scales) has heavily explored and diagnosed the mechanisms of meteorological patterns and how it drives other fluxes (runoff and evapotranspiration) and hydrological state variables (soil moisture). However, there are still knowledge gaps, particularly in areas regarding the temporal and spatial dynamics of freshwater and the impacts of both climate change and water resources development (e.g., irrigation schemes and dam constructions for hydro-power) on hydrological systems. This is true for several regions that lack comprehensive observational networks. Major water development schemes such as the Lake Volta in West

Africa impact on natural hydrology making it difficult to understand the impacts of climate on hydrological conditions. Another contemporary problem is poor understanding of hydrological variability and what drives changes in surface hydrology. The human use of freshwater lakes around the world for hydro-power and other purposes makes it even more difficult to assess key drivers of variability. This poses significant challenge to risk management and the prediction of extreme weather events (see, Hall et al. 2014). In addition to anthropogenic impacts on hydrological changes, observed devastating extremes in global climate drive terrestrial hydrology in ways that raises some concerns not just for water availability but also for food security, policy and risk-management strategies, socio-economic challenges (e.g., migration), and human livelihood. Given that changes in future hydrological conditions and water-management practices are complex and will influence agricultural systems in terms of adaptation measures and mitigation strategies (e.g., Paolino et al. 2012; Roudier et al. 2011; Falloon and Betts 2010), understanding hydrological variability becomes essential.

2.2.2 *Monitoring Architecture and Observational Networks*

Changes in terrestrial water storage, be they groundwater, soil moisture, or surface waters (i.e., lakes, wetlands, and rivers) remain one of the most critical components of the hydrological cycle. However, estimating these changes in stored water remains a major challenge due to sparsity of in-situ monitoring stations and lack of large-scale hydrological data, and unreliable field measurements among others. In some instances, degraded hydrological infrastructures contribute to creating gaps in routine measurements of relevant hydrological variables (e.g., river discharge, groundwater). The poor or lack of investments in observational data in several regions has necessitated the need to investigate how decision-makers can leverage on geospatial information from space-borne measurements to improve knowledge on surface water hydrology, and support intervention measures and risk assessments from extreme droughts and their impacts. Knowledge gaps in large-scale terrestrial water storage changes and the impacts of climate on the hydrology exist primarily because of limited or lack of observational networks to provide the current state of hydrologic information. Gauge stations for rainfall and river discharge measurements are in decline globally (e.g., Alsdorf et al. 2007; Alsdorf and Lettenmaier 2003; Lettenmaier 2005) and it is even worse in non-industrialized nations where gauge networks are not just extremely sparse, discontinuous and lacking, but their density falls far below the World Meteorological Organization guidelines (e.g., Vörösmarty et al. 2001). Some investments in gauge measurements for meteorological and hydrological applications exist in non-industrialized nations but some regions have little or no history of hydrological measurements due to incessant political instabilities, lack of civilization, and the costs and logistics implications for the installation of gauges to characterize flow dynamics. The lack of capacity to face the future challenges of climate change impact on water resources, owing to sparse observational record in some gauged river basins, is intensified by political concerns, legal and institutional constraints that have hindered the acquisition of existing data for scientific purposes. Research institutions in these regions find it difficult to engage in regional water management and trans-boundary water sharing conversations because hydro-climatic data from national archives and government institutions are frequently withheld mostly for political and security reasons. This could only be more damaging for countries where water sharing across political boundaries is vastly enormous. As widely reported (see, e.g., Ndehedehe et al. 2016a; Todd et al. 2011; Farnsworth et al. 2011; Conway et al. 2009), the paucity of high quality base line data remains a major constraint to understanding the climatic influence on changes in the hydrological cycle of several regions. The consistent monitoring of the environment to quantify these impacts is required for risk-management solutions to support adaptation plans and increase resilience to climate change. Ultimately, it is also anticipated that such monitoring will lead to a better understanding of climate change processes and the cascading effects of extreme climate events such as droughts and floods on agricultural and socio-economic systems.

2.2.3 *Empirical Methods for Assessment of Hydrological Variability*

The lack of a suitable framework to improve the characterization of space-time evolutions of hydrological quantities and extreme weather events is another problem that limits climate forecasting systems, understanding of changes in land water storage, and the holistic assessment of key hydrological metrics (e.g., magnitude, frequency, duration, predictability, etc.). The impacts of climate variability and other processes of oceanic inter-annual variability (e.g., ENSO, PDO, etc.) could result in contrasting outcomes on hydrological regimes. One of such example is the debate on rainfall recovery over West Africa. Nicholson (2005) and Lebel and Ali (2009) have contrary opinions on rainfall recovery possibly due to different reference periods in the data used. But essentially, they both show that while the recovery was marked in some areas, rainfall deficit continued unabated in other Sahelian areas. The failure of rainfall regime in one region does not translate to failure in other sub-regions (e.g., Ndehedehe et al. 2016c), probably owing to different drivers of rainfall, local and land surface conditions. Assuming drought and wet conditions vary in space and time, then the existing methods (e.g., van Huijgevoort 2014; Panthou et al. 2014) used to analyse extreme weather events and hydrological conditions are insufficient, warranting further studies.

As elaborated further by Ndehedehe et al. (2016c), regional average time series of drought indices such as effective drought index, standardized precipitation index, groundwater drought index, and Palmer Drought Severity Index, among several other indices (see, e.g., Li and Rodell 2015; Vicente-Serrano 2006; Heim 2002; McKee et al. 1993) hides the underlying spatial variability of the index and may lead to generalizations of wet and dry regimes. As this may not be very helpful in predicting ecological responses, adequate evaluation of drought impacts, crafting of drought policies, and in the fostering of campaign on regional adaptation strategies, the spatio-temporal evolutions of drought/wet conditions, an approach where the various drought indices are localized over the region would adequately improve our knowledge of hydrological variability. As will be demonstrated in subsequent chapters, a range of methods and framework exist, including the suitability of advanced multivariate statistics and integration of remote sensing methods with model and gauged data to characterize droughts and assess key hydrological metrics.

References

Agutu N, Awange J, Zerihun A, Ndehedehe C, Kuhn M, Fukuda Y (2017) Assessing multi-satellite remote sensing, reanalysis, and land surface models' products in characterizing agricultural drought in East Africa. Remote Sens Environ 194:287–302. https://doi.org/10.1016/j.rse.2017.03.041

Akhmadiyeva Z, Abdullaev I (2019) Water management paradigm shifts in the Caspian Sea region: review and outlook. J Hydrol 568:997–1006. https://doi.org/10.1016/j.jhydrol.2018.11.009

Al-Saidi M, Hefny A (2018) Institutional arrangements for beneficial regional cooperation on water, energy and food priority issues in the Eastern Nile Basin. J Hydrol 562:821–831. https://doi.org/10.1016/j.jhydrol.2018.05.009

Alcamo J, Döll P, Henrichs T, Kaspar F, Lehner B, Rösch T, Siebert S (2003) Global estimates of water withdrawals and availability under current and future 'business-as-usual' conditions. Hydrol Sci J 48(3):339–348. https://doi.org/10.1623/hysj.48.3.339.45278

Alcamo J, Döll P, Kaspar F, Siebert S (1997) Global change and global scenarios of water use and availability: an application of WaterGAP1.0. Center for environmental systems research (CESR), University of Kassel, Germany. https://pdfs.semanticscholar.org/a788/80a9f9952f0e04bc88c7c5699ea9c9e0fa30.pdf. Accessed 18th Aug 2016

Alsdorf DE, Lettenmaier DP (2003) Tracking fresh water from space. Science 301(5639):1491–1494. https://doi.org/10.1126/science.1089802

Alsdorf DE, Rodríguez E, Lettenmaier DP (2007) Measuring surface water from space. Rev Geophys 45(2):RG2002. https://doi.org/10.1029/2006RG000197

Atef SS, Sadeqinazhad F, Farjaad F, Amatya DM (2019) Water conflict management and cooperation between Afghanistan and Pakistan. J Hydrol 570:875–892. https://doi.org/10.1016/j.jhydrol.2018.12.075

Brauman KA, Richter BD, Postel S, Malsy M, Flörke M (2016) Water depletion: an improved metric for incorporating seasonal and dry-year water scarcity into water risk assessments. Elementa 4(83). https://doi.org/10.12952/journal.elementa.000083

Chindarkar N, Grafton RQ (2019) India's depleting groundwater: when science meets policy. Asia Pacif Polic Stud 6(1):108–124. https://doi.org/10.1002/app5.269

Conway D, Persechino A, Ardoin-Bardin S, Hamandawana H, Dieulin C, Mahé G (2009) Rainfall and water resources variability in Sub-Saharan Africa during the twentieth century. J Hydrometeorol 10(1):41–59. https://doi.org/10.1175/2008JHM1004.1

Falloon P, Betts R (2010) Climate impacts on European agriculture and water management in the context of adaptation and mitigation-the importance of an integrated approach. Sci Total Environ 408(23):5667–5687. https://doi.org/10.1016/j.scitotenv.2009.05.002

Farnsworth A, White E, Williams CJ, Black E, Kniveton DR (2011) Understanding the large scale driving mechanisms of rainfall variability over central Africa, in African climate and climate

change: physical, social and political perspectives. African climate and climate change: physical, social and political perspectives, pp 101–122. https://doi.org/10.1007/978-90-481-3842-5_5
Fisher-Jeffes L, Carden K, Armitage NP, Winter K (2017) Stormwater harvesting: improving water security in South Africa's urban areas. South African J Sci 113(1/2):4. https://doi.org/10.17159/sajs.2017/20160153
Gleeson T, Wada Y, Bierkens MFP, van Beek LPH (2012) Water balance of global aquifers revealed by groundwater footprint. Nature 488:197–200. https://doi.org/10.1038/nature11295
Gleick PH (1993) Water and conflict: fresh water resources and international security. Int Sec 18(1):79–112. https://doi.org/10.2307/2539033
Gleick PH (2000) The world's water: biennial report on freshwater resources 2000–2001. Island Press, Washington, DC
Grech-Madin C, Döring S, Kim K, Swain A (2018) Negotiating water across levels: a peace and conflict "toolbox" for water diplomacy. J Hydrol 559:100–109. https://doi.org/10.1016/j.jhydrol.2018.02.008
Hall JW, Grey D, Garrick D, Fung F, Brown C, Dadson SJ, Sadoff CW (2014) Coping with the curse of freshwater variability. Science 346(6208):429–430. https://doi.org/10.1126/science.1257890
Heim RR (2002) A review of twentieth-century drought indices used in the United States. Bul Am Meteorol Soc 83(8):1149–1165
Honkonen T, Lipponen A (2018) Finland's cooperation in managing transboundary waters and the UNECE Principles for Effective Joint Bodies: value for water diplomacy? J Hydrol 567:320–331. https://doi.org/10.1016/j.jhydrol.2018.09.062
Hughes N, Galeano D, Hattfield-Dodds (2019) The effects of drought and climate variability on Australian farms. Australian Bureau of agricultural and resource economics and sciences, Canberra. http://doi.org/10.25814/5de84714f6e08. Accessed 16th Oct 2020
Hurkmans R, Troch PA, Uijlenhoet R, Torfs P, Durcik M (2009) Effects of climate variability on water storage in the Colorado River Basin. J Hydrometeorol 10:1257–1270. https://doi.org/10.1175/2009JHM1133.1
Lebel T, Ali A (2009) Recent trends in the Central and Western Sahel rainfall regime (1990–2007). J Hydrol 375(1–2):52–64. https://doi.org/10.1016/j.jhydrol.2008.11.030
Lettenmaier DP (2005) Observations of the global water cycle—global monitoring networks. In: Anderson MG (ed) Encyclopedia of hydrological sciences, vol 5. Wiley, pp 2719–2732
Li B, Rodell M (2015) Evaluation of a model-based groundwater drought indicator in the conterminous US. J Hydrol 526:78–88. https://doi.org/10.1016/j.jhydrol.2014.09.027
McKee TB, Doeskin NJ, Kieist J (1993) The relationship of drought frequency and duration to time scales. In: Conference on applied climatology, American Meteorological Society,Boston, Massachusetts, pp 179–184. www.ccc.atmos.colostate.edu/relationshipofdroughtfrequency.pdf. Accessed 27 June 2014
Ndehedehe C, Awange J, Agutu N, Kuhn M, Heck B (2016) Understanding changes in terrestrial water storage over West Africa between 2002 and 2014. Adv Water Res 88:211–230. https://doi.org/10.1016/j.advwatres.2015.12.009
Ndehedehe CE (2019) The water resources of tropical West Africa: propblems, progress and prospect. Acta Geophysica 67(2):621–649. https://doi.org/10.1007/s11600-019-00260-y
Ndehedehe CE, Agutu N, Ferreira VG, Getirana A (2020) Evolutionary drought patterns over the Sahel and their teleconnections with low frequency climate oscillations. Atmos Res 233:104700. https://doi.org/10.1016/j.atmosres.2019.104700
Ndehedehe CE, Agutu NO, Okwuashi OH, Ferreira VG (2016) Spatio-temporal variability of droughts and terrestrial water storage over Lake Chad Basin using independent component analysis. J Hydrol 540:106–128. https://doi.org/10.1016/j.jhydrol.2016.05.068
Ndehedehe CE, Awange JL, Corner R, Kuhn M, Okwuashi O (2016) On the potentials of multiple climate variables in assessing the spatio-temporal characteristics of hydrological droughts over the Volta Basin. Sci Total Environ 557–558:819–837. https://doi.org/10.1016/j.scitotenv.2016.03.004

Ndehedehe CE, Ferreira VG, Agutu NO, Onojeghuo AO, Okwuashi O, Kassahun HT, Dewan A (2021) What if the rains do not come? J Hydrol 595:126040. https://doi.org/10.1016/j.jhydrol.2021.126040

Nicholson S (2005) On the question of the "recovery" of the rains in the West African Sahel. J Arid Environ 63(3):615–641. https://doi.org/10.1016/j.jaridenv.2005.03.004

Odorico P, Chiarelli DD, Rosa Lorenzo et al (2020) The global value of water in agriculture. In: Proceedings of the national academy of sciences, vol 117(36), pp 21985–21993. https://doi.org/10.1073/pnas.2005835117

Okewu E, Misra S, Sanz LF, Ayeni F, Mbarika V, Damaševičius R (2019) Deep Neural Networks for curbing climate change-induced farmers-herdsmen clashes in a sustainable social inclusion initiative. Probl Sustain Dev 14(2):143–155

Panthou G, Vischel T, Lebel T (2014) Recent trends in the regime of extreme rainfall in the central sahel. Int J Climatol 34(15):3998–4006. https://doi.org/10.1002/joc.3984

Paolino DA, III Kinter JL, Kirtman BP, Min D, Straus DM (2012) The impact of land surface and atmospheric initialization on seasonal forecasts with CCSM. J Clim 25(3):1007–1021. https://doi.org/10.1175/2011JCLI3934.1

Phillips T, Nerem RS, Fox-Kemper B, Famiglietti JS, Rajagopalan B (2012) The influence of ENSO on global terrestrial water storage using GRACE. Geophys Res Lett 39(L16705):2012. https://doi.org/10.1029/2012GL052495

Prosser IP (2011) Water: science and solutions for Australia. CSIRO. https://www.csiro.au/en/research/natural-environment/water/Water-Book

Rijsberman FR (2006) Water scarcity: fact or fiction? Agric Water Manag 80(1–3):5–22. https://doi.org/10.1016/j.agwat.2005.07.001

Roudier P, Sultan B, Quirion P, Berg A (2011) The impact of future climate change on West African crop yields: what does the recent literature say? Glob Environ Change 21:1073–1083. https://doi.org/10.1016/j.gloenvcha.2011.04.007

Schewe J, Heinke J, Gerten D, Haddeland I, N. W. A. et al (2013) Multimodel assessment of water scarcity under climate change. PNAS 111(9):3245–3250. https://doi.org/10.1073/pnas.1222460110

Schmeier S, Shubber Z (2018) Anchoring water diplomacy—the legal nature of international river basin organizations. J Hydrol 567:114–120. https://doi.org/10.1016/j.jhydrol.2018.09.054

Schroth G, Läderach P, Martinez-Valle AI, Bunn C, Jassogn L (2016) Vulnerability to climate change of cocoa in West Africa: patterns, opportunities and limits to adaptation. Sci Total Environ 556:231–241. https://doi.org/10.1016/j.scitotenv.2016.03.024

Shiklomanov IA (2000) Appraisal and assessment of world water resources. Water Int 25(1):11–32. https://doi.org/10.1080/02508060008686794

Todd M, Andersson L, Ambrosino C, Hughes D, Kniveton DR, Mileham L, Murray-Hudson M, Raghavan S, Taylor R, Wolski P (2011) Climate change impacts on hydrology in Africa: case studies of river basin water resources, in African climate and climate change: physical, social and political perspectives. African climate and climate change: physical, social and political perspectives, pp 123–153. https://doi.org/10.1007/978-90-481-3842-5_5

van Huijgevoort M (2014) Hydrological drought: characterisation and representation in large-scale models. PhD thesis,Wageningen University, Wageningen, NL. http://library.wur.nl/WebQuery/wda/2057611. Accessed 1st Feb 2017

Vicente-Serrano SM (2006) Spatial and temporal analysis of droughts in the Iberian Peninsula (1910–2000). Hydrol Sci J 51(1):83–97. https://doi.org/10.1623/hysj.51.1.83

Vörösmarty C, Askew A, Grabs W, Barry RG, Birkett C, Döll P, Goodison B, Hall A, Jenne R, Kitaev L, Landwehr J, Keeler M, Leavesley G, Schaake J, Strzepek K, Sundarvel SS, Takeuchi K, Webster F (2001) Global water data: a newly endangered species. EOS Trans Am Geophys Union 82(5):54–58. https://doi.org/10.1029/01EO00031

Vörösmarty CJ, Green P, Salisbury J, Lammers RB (2000) Global water resources: vulnerability from climate change and population growth. Science 289(5477):284–288. https://doi.org/10.1126/science.289.5477.284

Vorosmarty CJ, McIntyre PB, Gessner MO, Dudgeon D, Prusevich A, Green P, Glidden S, Bunn SE, Sullivan CA, Liermann CR, Davies PM (2010) Global threats to human water security and river biodiversity. Nature 467:555–555–561. https://doi.org/10.1038/nature09440

Part II
Earth Observations

Chapter 3
Optical Remote Sensing Systems

Abstract This chapter focuses on multi-mission optical satellite remote sensing and the characteristics of data they collect. The concept of satellite optical remote sensing systems and their benefits in environmental monitoring are discussed. Given the innovation and advancement in optical systems and platforms, these systems have great potential in the multidisciplinary aspects of remote sensing of the environment. This includes, inventorying of wetlands, mapping of floodplain productivity, urbanization, crop mapping, disaster management (e.g., earthquake, desertification), quantifying changes in extreme events (wildfire, floods, droughts) and their impacts on the environment and national income. Additionally, the legacy of the Landsat systems as important open access Earth resources satellites, which have contributed to understanding key drivers and processes surrounding human civilization is discussed.

3.1 Introductory Remarks

Remote sensing is defined as the measurement of object properties on the Earth's surface using data acquired from aircraft, satellites, unmanned aerial vehicles, or any other platform that does not require physical contact. In other words, it is an attempt to measure something or an object at a distance without coming into direct or physical contact with the said object. Since we are not in direct contact with the object of interest, we must rely on propagated signals of some sort, for example, optical, acoustical, or microwave (Schowengerdt 2007). As an old concept introduced by Evelyn Pruitt of the US Office of Naval Research in the 1950s (Dong and Chen 2018), the term *'remote sensing'*, apparently, has several definitions, which have been detailed in several literatures (e.g., Lavender and Andrew 2016; Cracknell and Hayes 2007). The fundamental idea and concept of remote sensing is that it involves an interaction between incident radiation (or electromagnetic radiation) and the targets of interest (Earth resources). Collecting information from targets of interest can only be possible by sensing and recording reflected or emitted energy from these targets. These information are then processed, analysed, and applied in relevant areas. But the

C. Ndehedehe, *Satellite Remote Sensing of Terrestrial Hydrology*,
https://doi.org/10.1007/978-3-030-99577-5_3

sensing of emitted energy and the use of non-imaging sensors is also another aspect of remote sensing that will be discussed further in this chapter. This chapter describes the basic characteristics of multi-mission optical satellite remote sensing and includes their benefits. These optical systems have great potential in multidisciplinary aspects of remote sensing hydrology such as inventorying of wetlands, mapping of aquatic biomass distribution in intermittent water holes, and assessment of freshwater aquatic ecosystem services.

Remote sensing is based on detecting the energy reflected from the surface of the Earth as electromagnetic energy or radiation. Generally, remote sensing systems provide a repetitive and consistent view of the Earth that is invaluable to monitoring short-term and long-term changes and anthropogenic influence. Improved technological capabilities of remote sensing systems have resulted in tremendous progress in the studies of environmental systems. These capabilities are usually reflected in advancements made in remote sensing sensors and platforms. Remote sensing systems provide a range of benefits. This includes a repetitive and consistent view of the Earth that is invaluable to monitoring short-term and long-term changes and anthropogenic influence. Other key benefits of remote sensing include:

- Multi-scale observation;
- Repetitive coverage;
- Global coverage;
- Direct observation;
- Non-destructive observation;
- Change detection;
- Elevation retrieval.

Although the operational application of remote sensing methods to hydrology and water resources is an old concept (Rango 1994), it is now gradually increasing because of the sophistication and advances in satellite technology. With the advent of advanced sensors currently onboard new satellite systems, there have been improved approaches for estimating hydro-meteorological states and biophysical variables such as surface water extent, land surface temperature, soil moisture, land use, vegetation cover, evapotranspiration, etc. With these improvements in sensor capabilities, environmental monitoring and remote sensing of the environment can be undertaken even at relatively large spatial scales and at high temporal frequencies. For example, large-scale monitoring of vegetation activity can be based on Normalized Difference Vegetation Index (NDVI) or Enhanced Vegetation Index time series datasets acquired by sensors onboard satellites with short revisit periods, in particular, the National Oceanic and Atmospheric Administration (NOAA) Advanced Very High-Resolution Radiometer (AVHRR) (Beck et al. 2006). A couple of these sensors (e.g., Moderate Resolution Imaging Spectroradiometer (MODIS)) have a spatial resolution ranging from 500m to 1 km by 1 km. And their large swath width can provide global daily estimates of these vegetation indices and other biophysical variables. Even at catchment or smaller scales, vegetation conditions can be monitored using high and moderate-spatial resolution satellite imageries such as Sentinel-1 and Landsat-8. By simply applying a mathematical formula to Landsat-8 or Sentinel-1 using the Red and NIR bands of these sensors (i.e., $NDVI = b_{NIR} - b_R/b_{NIR} + b_R$), pixels containing vegetation can be easily highlighted and extracted. Moreover, surface water extent or floodplain inundation can be mapped using a different set of band information from these satellites. The use of satellite remote sensing indicators or metrics such as the Modified Normalized Difference Water Index (MNDWI, Xu 2006) makes this possible. This remote sensing indicator leverage the Mid-Infrared (MIR) and the Green (G) bands in the visible and short-wave infrared regions of the electromagnetic spectrum, respectively, to detect open water features on landscapes (i.e., MNDWI=$b_G - b_{MIR}/b_G + b_{MIR}$). While monitoring vegetation dynamics using vegetation indices (e.g., NDVI) can improve knowledge of the impacts of climate variability and anthropogenic pressures on the environment, the large-scale assessment of freshwater habitats and floodplain productivity has been studied using remote sensing water-related indicators (Ndehedehe et al. 2020, 2021).

3.2 Optical Satellite Remote Sensing Systems

3.2.1 Sensors

Remote sensing of the surface of the Earth is carried out using several instruments. These instruments are known as sensors, and they use a variety of different wavelengths of electromagnetic (EM) radiation, which can either be in the visible, near-infrared (or reflected-infrared), thermal infrared, microwave, or radio wave part of the electromagnetic spectrum (Cracknell and Hayes 2007). EM radiation is the energy that moves through space and matter in the form of both waves and a stream of particles called photons, with the range of EM radiation making up the EM spectrum (Lavender and Andrew 2016). Remote sensing systems are designed to record EM radiation emitted from an object remotely. Several segments of the EM spectrum can be used as a basis for remote sensing measurements and each portion offers the potential to interpret something differently. The specific portions of EM spectrum where the various sensors operate are controlled by two key factors. They are: (i) the availability of radiant energy to record and (ii) the nature of surface objects to be recorded. The nature of these objects (such as their physical, chemical, and biological properties) to be recorded can be influenced by the manner through which radiant energy is absorbed, received, and emitted (Klein 2012). Additional details on EM radiation and the various portions frequently used in remote sensing are available in the literature (Lavender and Andrew 2016).

3.2.1.1 Remote Sensing Imaging Instruments

The device that detects EM energy, quantifies, and records is called a sensor. In other words, sensors are detectors and devices that detect electromagnetic radiation (e.g., camera, scanner, etc.). They measure the amount of electromagnetic energy that's

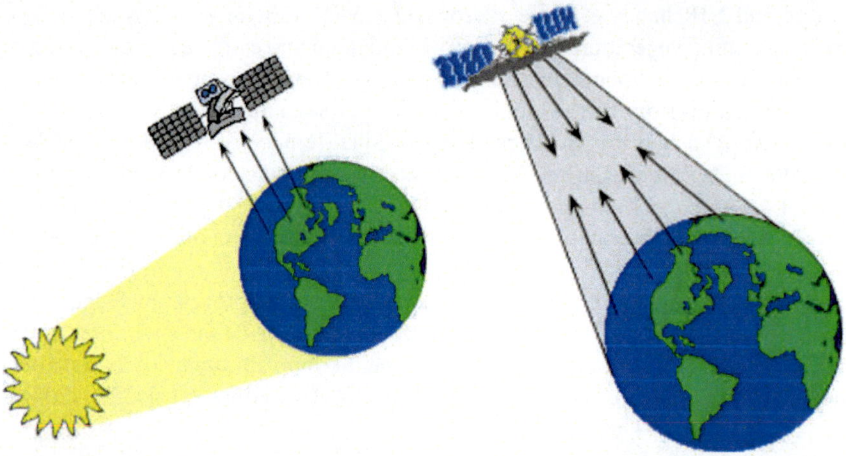

Fig. 3.1 Active and passive remote sensing systems. *Source*, Canada Center for Remote Sensing

scattered or reflected from a given surface. These sensors have different capabilities and are designed to either transmit recorded electromagnetic signals or detect the energy emitted from targets or the Earth surface. There are two main kinds of remote sensing imaging instruments (sensors), the passive and active (Fig. 3.1). A passive sensor is one that simply responds to the radiation that is incident on the instrument. Passive remote sensor uses energy from the Sun, which is emitted towards the surface of the Earth (i.e., solar irradiance) and is subsequently absorbed or scattered. These kinds of sensors are non-operational at night time and do have problems collecting data when weather conditions are poor. Another limitation associated with these kinds of sensing instruments is that when the Sun does not shine brightly, in some areas, such regions come under cloud cover (e.g., Tempfli et al. 2009). In an active instrument (Active sensor), the radiation is generated by the instrument, transmitted downward to the surface of the Earth, and reflected back to the sensor (Fig. 3.2). Unlike the passive sensing systems, active sensors are operational on a 24-hour basis. Meaning they have the capability to detect and collect information during day and at night times and are also less dependent on weather conditions (Fig. 3.2). This is because an active sensor generates its own source of energy. A good example is the radar remote sensing system.

The different surfaces that EM interacts with determine the strength, direction, spectral shape, and property received by that sensor. The remotely sensed signals detected by the sensors on a satellite or aircraft are usually influenced by the intervening atmosphere. In the process of capturing EM emanating from various surfaces or substances, some of the energy will either be absorbed or scattered by the atmosphere as the signal moves to and from the sensor. The magnitude of influence depends on the altitude of the platform carrying the sensor and on the atmospheric conditions prevailing at the time among other factors (Tempfli et al. 2009). For these reasons,

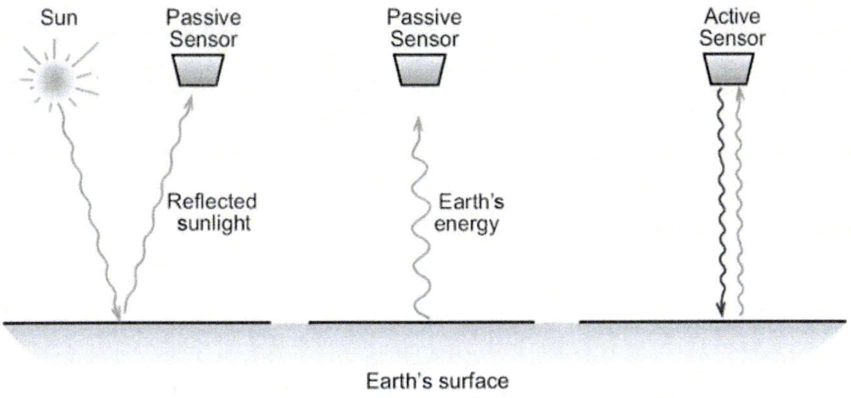

Fig. 3.2 Illustration of a traditional remote sensing systems. *Source* Tempfli et al. (2009)

some corrections (e.g., geometric, radiometric, etc.) are applied to remotely-sensed images to accommodate the variability.

3.2.1.2 Satellite Remote Sensing Systems and Data

This chapter focuses on optical satellite systems and the characteristics of data they collect. Based on imaging capabilities and sensor design characteristics, the data collected by remote sensing systems (active and passive) are different. Examples of passive systems, include sensors (or optical cameras) onboard satellite platforms, which are Landsat/TM, SPOT/HRV, TERRA/ASTER, MODIS, Quick Bird, IKONOS, and MERIS, and they present a few spectral bands and with broad bandwidths (Camps-Valls et al. 2012). To undertake measurements of Earth resources, these systems rely on the use of different scanners, which can be multispectral (e.g., Landsats 4 and 5), whisk broom (Landsat-7), and push broom (SPOT) system). In the early stage of spaceborne Earth observation, traditional film-based imaging devices, return beam vidicon television cameras, and optical scanners were the main devices used for Earth observation (e.g., Landsats 1-3). Subsequent Landsat missions used Multispectral scanners. Scanners are sensing systems used to scan the Earth's surface in a contiguous series of narrow strips by using a rotating/oscillating mirror. A good example of this is the whisk broom scanners. However, in other scanners such as those used in the SPOTS satellite mission, the sensor field of view is detected at once by a linear array of detectors. Passive remote sensing methods (Fig. 3.1) can be classified as

- Multispectral remote sensing: visible and reflected infrared (near-infrared and mid-infrared) images are collected by recording the reflection of solar radiation from the Earth using airborne and spaceborne sensors (Tempfli et al. 2009). Thermal infrared images are collected by recording emitted thermal radiation from the

Earth. Data from multispectral systems are widely used and also the most popular satellite observations.

- Hyperspectral remote sensing: uses imaging spectrometers that measure near-laboratory-quality spectra in narrow spectral bands. In other words, they collect data (reflected infrared radiation) from hundreds if not thousands of continuous wavelengths or narrow spectral bands. Hyperspectral images are highly innovative remote sensing data and unlike the conventional panchromatic and multispectral imageries, they do enable a better distinct discrimination of Earth resources or targets (Okwuashi and Ndehedehe 2020).

As earlier mentioned in this chapter, passive electro-optical sensors, be it airborne (when aerial cameras are used) or spaceborne (satellite systems involving the application of scanners), are limited by cloud cover and Sun illumination conditions. This is because of their dependence on an external source of energy (Sun). Radar imaging systems are excellent substitutes because they operate independently of lighting conditions and largely independently of weather (e.g., Dong and Chen 2018). The active remote sensing systems (Fig. 3.1) on the other hand can be grouped into

- Radar remote sensing system: uses generates its own electromagnetic energy in the microwave region to "illuminate" the terrain and detects the reflected energy or signal from the Earth surface (Fig. 3.2). Synthetic Aperture Radar observations emanate from radar remote sensing.
- LiDAR: uses a technology that measures distances (or ranges) based on the time between transmitting and receiving laser signals. Examples include airborne Lidar sensors and Terrestrial Laser Scanners, which are ground-based Lidar systems.

Through transmitting and receiving in multiple polarizations, the radar system can detect unique and distinct features of both natural (e.g., water bodies) and man-made targets (e.g., buildings). One good example of radar remote sensing is the Polarimetric Synthetic Aperture Radar (PolSAR). This is an optimized version of the well-known Synthetic Aperture Radar (SAR) system. PolSAR can collect information regardless of weather conditions by transmitting and receiving electromagnetic waves in diverse polarimetric states. This makes them very useful in several fields, including land cover and land use, oceanography, geology, forestry, and mapping oil spill, flooding, among others. The drawbacks associated with its usage and new machine learning approaches to overcome key challenges (e.g., degraded by speckles) to optimized its application have been fully detailed in the literature (Okwuashi and Ndehedehe 2021).

3.2.2 Platforms

Sensors used in Earth observation can operate in space or at an altitude that is just a few centimetres above ground. These sensors are usually mounted on a moving vehicle called the platform. Platforms carry the sensor, such as aircrafts, space shuttles, polar orbiting, and Geo-stationary satellites. Hence, platforms can be space-based (satellites) or airborne (aircraft). There are also ground-based platforms, for example, mounting spectrometer on a pole to measure the changing reflectance of a specific crop during the day or season, or the use of photo-theodolite and total stations in feature mapping. There are also remotely operated aircrafts of ultra-lightweight and load called Unmanned Aerial Vehicles (UAV) that fly at much lower altitudes as opposed to airplanes (20 km). They are called remotely piloted aircraft systems or popularly known as drones. UAV platforms are now very popular remote sensing instruments with several key benefits. These advantages have been summarized by McCabe et al. (2017) and they include the capacity to

- collect spatial resolution imagery ranging between 1 and 20 cm pixel size,
- acquire cost-effective, high temporal resolution data on demand at any time,
- operate as both active and passive sensing system,
- be employed as in situ data to calibrate and validate satellite products,
- complement, extend, or potentially replace field surveys, especially in regions with difficult terrain, and
- provide a scaling tool between field and satellite data.

The rapid developments and broad applications of UAVs in precision environmental monitoring, including wildfire, freshwater ecosystems, floods, precision, and multi-disciplinary aspects of environmental geoscience are indications they are emerging remote sensing platforms for the future. The applications of these remotely piloted aircraft systems are discussed in detail in Chap. 8.

3.2.3 Characteristics of Remotely Sensed Systems

Environmental remote sensing activities are now being out carried with satellite-based sensors because of their capacity for continuous data acquisition, high temporal, spatial and spectral resolution, broad aerial coverage, and easy data processing with computer (Klein 2012). There are different satellite systems, including Earth observation (e.g., environmental, meteorological, and hydrological), and communication satellites, among others. A number of different multispectral scanners are carried on these satellites. The three key characteristics that should be taken into consideration when using satellite images are spatial, spectral, and temporal resolutions. With reference to typical Earth observation satellites such Landsat program, each of these quantities is briefly considered in this section. Spectral resolution is determined by the design of the sensor system while spatial and temporal resolutions are determined both by the construction of the sensor system and by the conditions under which it is flown. To some extent, there is a trade-off between spatial and temporal resolution; good spatial resolution tends to be associated with low temporal resolution.

1. Spatial resolution: This is the size of the smallest object that can be seen on the Earth surface. It is also called ground sampling distance. For raster images, spatial resolution is similar to pixel size and is a function of sensor altitude, detector size, focal size, and system configuration. The implication of this in an Earth observation satellite system is that when viewing an image with a spatial resolution of 30 m as is the case with Landsat multispectral scanner, it will be impossible to see any object smaller than 30 m. To identify objects with less than 30 m in size, the sensor must have a spatial resolution of less than 30 m. This leads to the concept of high and low spatial resolution in satellite images. The data from commercial satellites (e.g., IKONOS and Quickbird), which have a spatial resolution of 1 m or even less are usually referred to as high spatial resolution imagery. The spatial resolution of Landsat MSS and SPOT (medium resolution) is not as good as IKONOS and Quickbird which have a high spatial resolution.
2. Spectral resolution: This is the number of available wavebands present in the EM spectrum that an optical sensor is taking measurements over. This means the number of bands or channels provided by the sensor as well as their spectral bandwidths or channels. Spectral resolution can be panchromatic (use of a single broad spectral waveband); multispectral (use of multispectral channels or several wavebands at the same time); and hyperspectral (the use of several hundreds of wavebands and allows much more spectral details to be seen in the image). The sensing system or Multispectral Instrument with more bands or channels has a higher spectral resolution than those with few bands. In this regard, Sentinel-2 satellite system, which has about 12 bands is of a higher spectral resolution compared to IKONOS with only 5 multispectral bands. The latter has a higher spatial resolution (3.2 m for the multi-spectral) with applications extending to urban and rural mapping of natural resources and disasters, agriculture and forestry analysis, mining, and change detection. With approximately 3 days of revisit cycle, it can provide important information for impact assessment such as quantifying areas affected by floods. Sentinel-2, on the other hand, has a moderate spatial resolution (10 m and 20m for most channels) but with about 13 spectral channels, much more geospatial information can be collected from targets, this can be, for example, collecting information on different vegetation states or water quality types.
3. Temporal resolution: This is also the revisit cycle or frequency of coverage of each satellite system. It is the overall time taken for a satellite to return to collect data from the same location on Earth. In the development of the technology, the constraints imposed by the sensor design, the onboard electronics, and the data link to the ground limit the total amount of data that can be obtained. temporal resolution is dependent on the design of the satellite system (e.g., the satellite's orbit and swath width), and is expressed in terms of time in hours, days, or months. Thus, the smaller the spatial resolution, the more data there are to be handled for any given area on the ground, and the less frequently data will be available for a given area (Cracknell and Hayes 2007).

3.2.4 The Concept of Remote Sensing

With remote sensing, one can determine the characteristics of Earth resources (e.g., lakes, rivers, vegetation, buildings, etc.) in a considerably large geographic space simply by relying on reflected or emitted electromagnetic radiation from the object. To this end, an energy source, a sensor, a platform, and a medium for propagation of emitted and reflected signals will be required to interact with the target object. As illustrated in Fig. 3.3, an energy source (A) illuminates and provides electromagnetic energy to the target of interest on the Earth surface (C). The emitted energy interacts with the atmosphere (B) it passes through and the target before getting back to the sensor (D). The sensor collects and records the electromagnetic radiation emanating from the target. As part of the key component of a remote sensing system, Fig. 3.3 also shows that the energy recorded by the sensor has to be transmitted to a ground station (E) where the electronic or digital data (reflected signals) are processed into an image. After the data processing, experts interpret, extract and analyse (F) the data for new information that will be useful in other applications (G) or assist in solving a particular problem (Fig. 3.3). More details on optical remote sensing systems and their characteristics are available in relevant texts (Cracknell and Hayes 2007). The concept behind remote sensing measurements is briefly discussed in the sections that follow.

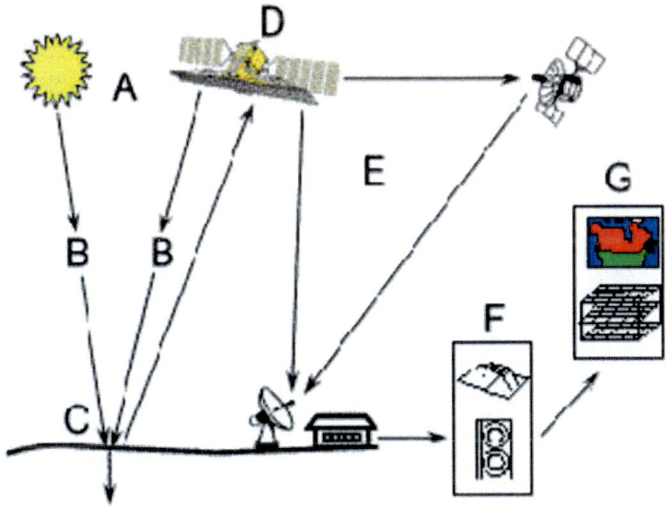

Fig. 3.3 Concept of remote sensing. *Source*, Canada Center for Remote Sensing

3.2.4.1 Electromagnetic Radiation

Energy is the ability to do work and it exists in various forms, e.g., electrical, chemical, and mechanical energy. Energy can be transferred from one body or one place to another. Such transfers are effected by conduction, convection, and radiation. In remote sensing of the Earth, we are primarily concerned with radiation. Radiation is the transfer of energy that requires no physical medium. EM energy can be transmitted through a medium or a vacuum (space). The Sun is a major source of Electromagnetic energy but it is not the only one. All bodies with temperatures above absolute zero emit energy in radiant form. The characteristics of the emitted radiation can be assessed in terms of their wavelengths and intensities.

Electromagnetic radiation is therefore the energy that moves through space and matter in the form of both waves and a stream of particles called photons, with the range of EM radiation making up the EM spectrum. All electromagnetic radiations have fundamental properties and are predictable. For example, electromagnetic radiations emanating from water bodies are apparently different from vegetation surface. This is because water reflects strongly in visible regions of the electromagnetic spectrum (Fig. 3.4) and poorly in the short-wave infrared regions of the electromagnetic spectrum, unlike vegetation. The other concept used in radiation studies is Black Body (BB), a model absorber and radiator of EM radiation. A BB body is conceived to be a substance that absorbs all radiation incident upon it and emits the maximum amount of radiation at all temperatures. Such substance does not exist in real world but the concept is invaluable for the formulation of laws by comparison with which the behaviour of actual radiators may be assessed. For example, the practical significance of Stefan–Boltzmann's law is that hot radiators emit more energy per unit area compared to cooler ones.

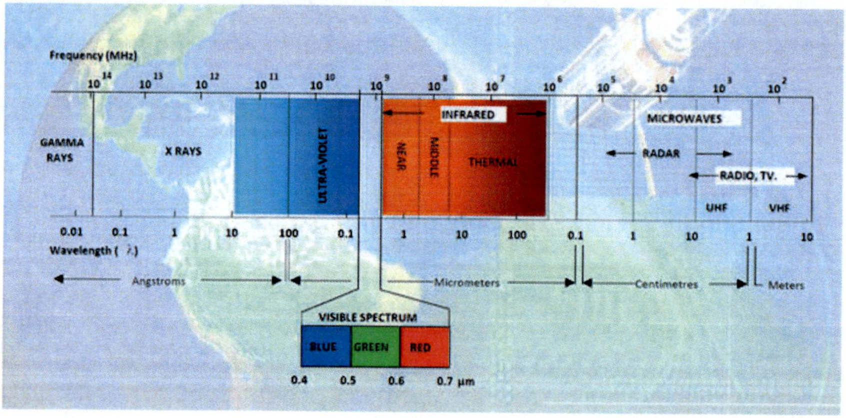

Fig. 3.4 The electromagnetic spectrum

3.2.4.2 Electromagnetic Spectrum

The electromagnetic spectrum (EMS) concept is nearly similar to that of DNA (deoxyribonucleic acid). As an illustration, a plot of the intensities of the emitted radiation against appropriate wavelength in the EMS gives a characteristic curve or spectral signature for each remote sensing target (object). This signature is very unique for every object (e.g., water bodies, plants, buildings, roads, farmlands, etc.) and it is the basis for mapping and quantifying changes in the environment.

Remote sensing of the Earth surface is carried out using several instruments. These instruments, in turn, use a variety of different wavelengths of electromagnetic (EM) radiation, which can either be in the visible, near-infrared (or reflected-infrared), thermal infrared, microwave, or radio wave part of the electromagnetic spectrum (Fig. 3.4). The total range of these wavelengths of electromagnetic radiation is called the electromagnetic spectrum. The different portions of the electromagnetic spectrum as illustrated in Fig. 3.4 include X-rays, gamma rays, UV radiation, visible radiation (light), infrared radiation, microwaves, and radio waves. Specific portions of the EMS where the various sensors operate are controlled by two key factors. The availability of radiant energy to record, and the nature of surface objects to be recorded. The nature of these objects (such as their physical, chemical, and biological properties) to be recorded can influence the manner by which radiant energy is absorbed, received, and emitted. Different portions of EMS are very unique because they have different relevance to the acquisition of Earth observation. This can be in terms of the type of information collected or the quantity of geospatial data acquired.

3.2.5 Remote Sensing Image Processing

Satellite data used in remote sensing exists in the form of images and contains a lot of spatial information. Image processing involves the manipulation of images and is used to extract information and emphasize or de-emphasize certain aspects of the information contained in the image and perform statistical or other analyses to extract non-image information. A digital image consists of an array of numbers. Although arrays may be square, they are quite commonly rectangular. Some common examples of image processing operations are image enhancement, density slicing histogram stretching, distogram equalization, and filtering. The manipulation of images using these methods requires image processing applications and software such as ENVI and ERDAS Imagine. An image processing system basically comprises computer hardware and software to perform the necessary image processing and display operations. Image processing can also involve the application of mathematical and statistical techniques to the multispectral bands of satellite images to extract information not visible to the human eye. Such operations may include the application of mathematical methods (e.g., maximum likelihood, support vector machine) to identify different land cover (e.g., water bodies, built-up areas, agricultural, and drylands, etc.) classes (multispectral image classification) in an image or the use of statistical methods (e.g., principal component analysis) to extract the maximum information content of an image. A good example is the derivation of Normalized Difference Vegetation Index (NDVI) from satellite images. Vegetation has a much higher reflectance in the near-infrared (NIR) regions of the EM relative to the red wavelengths as opposed to soil and man-made materials. This difference can be leveraged to map or quantify the distribution of healthy and non-healthy vegetation on a large scale both in time and space. As will be discussed later in this book, this capability has been explored in agriculture to monitor crop health and productivity.

3.2.6 Applications of Remote Sensing

The emergence of remote sensing platforms and sensors between the 1960s and 1970s revolutionized the applications of Earth resources satellites in environmental monitoring. Over the years, remote sensing has served as the basis for informed decisions and management of Earth resources across several regions of the world. The capability of satellite remote sensing to collect a broad range of information on various geographic entities, including the physical, chemical, biological, and human elements within Earth's environmental system has made that possible (Klein 2012). Investments in new satellite systems such as the Sentinel, Cryosat, and Landsat-8 missions as well as other Earth resources satellites provide unique and improved capabilities in comprehensive monitoring of the environment. Some of these capabilities include assessing land surface conditions (e.g., temperature), land cover states (e.g., urban sprawl, vegetation), water quality, floodplains, soil moisture as well as extreme climate conditions such as floods, droughts, and wildfires. Apparently, the use of Earth resources satellites in remote sensing of the environment has been widely reported, and together with new satellite technologies provide the opportunity to make informed decisions at regional and global scales. Even though the use of these systems to support environmental assessment and decision-making in various capacities seem to be increasing, some important applications of satellite remote sensing include

- Coastal studies; erosion, accretion, bathymetry.
- Oceanography; sea topography and land surface temperature mapping.
- Land use; urban land use, agricultural land use, vegetation health, crop health, and wetland monitoring.
- Disaster monitoring; floods, droughts, cyclones, forest fires, landslides, earthquakes, Tsunamis, etc.
- Water resources; continental freshwater changes, water quality.
- Meteorology; Weather systems tracking, precipitation monitoring, cloud monitoring to help predict the weather.
- Climatology; Desertification, surface albedo, monitoring dust storms.

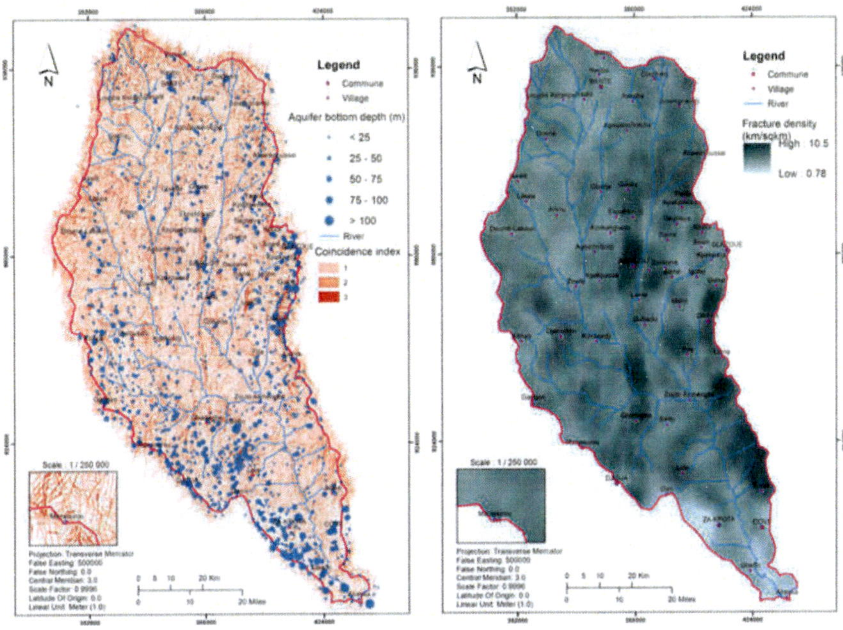

Fig. 3.5 Coincidence (left) and fracture density (right) maps over the the Zou basin in Benin republic, West Africa. *Source*, Oussou et al. (2020)

Additionally, novel hydro-geological applications of satellite remote sensing further highlight the importance of remote sensing as an emerging geospatial tool not only in the aforementioned areas but also in geophysical exploration studies. For instance, Landsat-7 ETM has been successfully used to isolate hydrothermally altered rocks by applying the Crosta techniques (Aydal et al. 2007). Moreover, a combination of Landsat-8 with satellite-derived elevation models has been used to identify fractured aquifers (Oussou et al. 2020). Figure 3.5 shows coincidence (left) and fracture density (right) maps over the Zou basin in Benin republic, West Africa. On average, the highest density of 6.7 km/sqkm for the PALSAR-derived fracture map is observed and occurs in the quartzite formations of the region, while the lowest average is 5.5 km/sqkm recorded in the charnockite (Fig. 3.5). As discussed further by Oussou et al. (2020), the coincidence map is a key tool that supports better and more precise fracture identification. A considerable large proportion (more than 80%) of the Republic of Benin where this study was undertaken are characterized by fractured aquifers thus warranting an improved understanding of their geometry for more reliable exploitation. Although fractures beneath the weathered layer are difficult to detect (Oussou et al. 2020), satellite remote sensing methods provide opportunities to extract substantial information regarding their hidden geometries.

3.3 Earth Resources Satellites: The Landsat Legacy

The Landsat system is fundamentally optical in its sensor design and is the biggest family among several Earth resources satellites since the advent of space mission. The continued observation of Earth's resources and acquiring continuous multispectral data every 16 days has been made possible since the launch of the first Landsat on 23 July 1972. This was followed by Landsat-2 (1975) and Landsat-3 (1978), which all used the same sensors or television cameras (return beam vidicon). Ever since, the Landsat program has progressed with several more improvements in sensor capabilities in successive missions—Landsat-4 (1982), Landsat-5 (1984), Landsat-7 (1999), Landsat-8 OLI (2013), and now Landsat-9 (September 2021). Figure 3.6 provides an overview of the Landsat program, indicating the evolution of different Landsat missions, including launch dates and the planned Landsat-9, which is scheduled for launch in September 2021. Although early Landsat missions have been terminated (Fig. 3.6), other missions are still operational, given their unique sensor and orbital properties. Together, they have provided a huge repository of Earth observations (1972-present) essential for understanding evolutionary changes in land cover as well as other important information.

The rise in human population from around 2.6 billion in the early 1950s to more than 7.7 billion in the present age came with substantial changes within the planetary space. Changes in land cover linked to a plethora of human actions resulting from increased urbanization as well as agricultural, and industrial development have been key drivers in the evolutionary patterns of several Earth resources as well as natural and man-made targets (water, vegetation, ice, human settlements, forest ecosystems, etc.). Apart from human alterations of the landscape, climate change through increased frequency of extreme events (prolonged extreme droughts, wildfires) also contributed to observed changes in terrestrial landscapes. The application

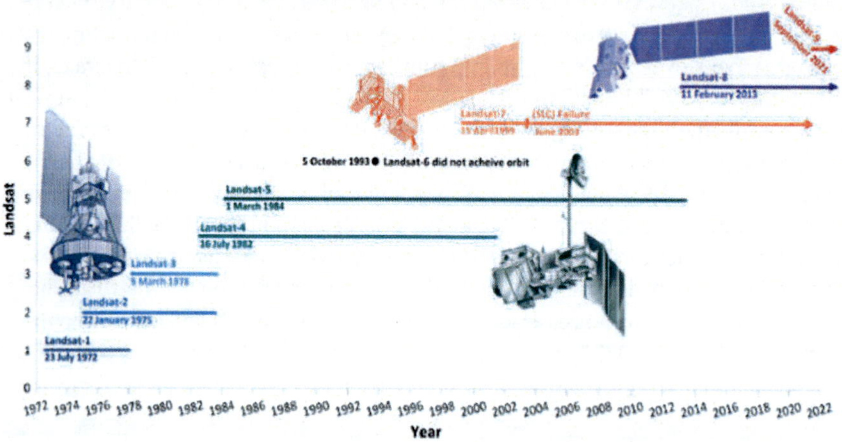

Fig. 3.6 Evolution of the Landsat program. *Source*, Hemati et al. (2021)

of these Landsat observations in global environmental monitoring therefore emerged as innovative approaches underpinning planning and management. They have contributed to new directions in policy, creating novel opportunities to address several socio-economic and environmental issues confronting society. The trends in the use of Landsat-derived information in the last 50 years (Hemati et al. 2021) to improve understanding of global environmental changes at multiple temporal and spatial scales relevant to management decisions, which underpin new policy attest to this wonderful legacy of the Landsat program.

3.3.1 Access, Availability, and Retrieval of Landsat Data

Due to NASA's open data policy, the Landsat data is freely available and hosted in the United States Geological Survey (USGS) data repository, which can be accessed or downloaded using this link (https://earthexplorer.usgs.gov/). Access requires free registration and the interface for data collection is user-friendly. Figure 3.7 shows part of the interface of this USGS explorer and access to archived level 2 Landsat data and other collections of satellite Earth observations like Sentinel, MODIS, etc., are facilitated through this USGS data portal. Potential study sites or study areas can be uploaded in a shapefile format or points or polygons representing study sites can be defined or manually drawn over the map area to extract relevant images (Fig. 3.7). Although there were some limitations surrounding the use of level 1 raw Landsat data in the past by non-remote sensing experts due to inexperience in pre-processing operations (e.g., atmospheric, geometric, and radiometric calibrations), level 2 data and other preprocessed products archived by USGS are now available based on request. Moreover, cloud computing platforms like Google Earth Engine (Chap. 5) and the availability of automated cloud detection algorithms have further

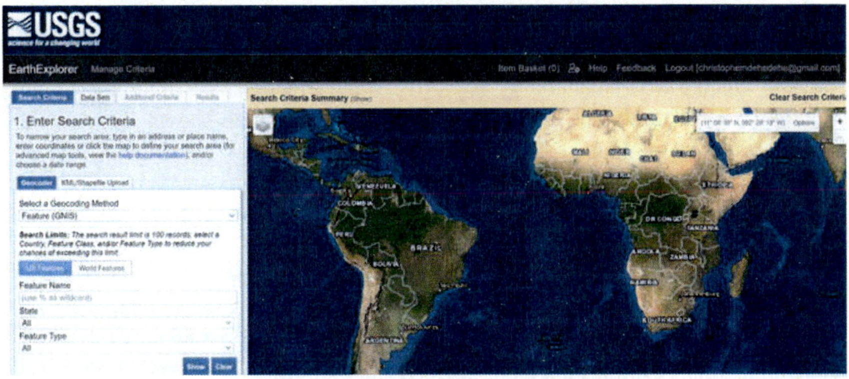

Fig. 3.7 The United States Geological Survey explorer for access to archived Landsat data. *Source*, USGS

simplified both access and processing of Landsat data for several applications. In the USGS Landsat repository, there are about 9.2 million Level 1 Landsat images as of December 2020 with about 1200 new satellite observations from Landsat-7 and -8 missions being added to the USGS repository daily (Hemati et al. 2021).

The availability of these data has made global-scale analysis feasible, thus facilitating an understanding of global change. For instance, Hansen et al. (2013) undertook global-scale forest cover change analyses (2000–2012), relying on 654,178 Landsat-7 scenes. The study by Pekel et al. (2016) on the hand focused on detecting the presence of surface water across the globe. The Global Surface Water Dataset generated under this project provided long-term statistics on the extent and transition of water based on historical Landsat (1984–2015), which used 4,185,439 scenes from Landsats 5, 7, and 8 acquired between March 1984 and December 2019 (Pekel et al. 2016). On a continental scale, the availability of Landsat has also aided the feasibility of continent-wide studies and data curation for other purposes. The Australian Geoscience Data Cube is one of such and the Water Observations from space provide historical Landsat-derived surface water over Australia from 1987 to date demonstrating the feasibility of continent-wide, multi-temporal analysis of surface water distribution (Lewis 2017; Mueller et al. 2016).

Recently, these algorithms have been improved in the Digital Earth Australia Waterbodies owned by Geoscience Australia and over 30 years of Landsat satellite-derived water bodies over the Australian landscape have been identified (Krause et al. 2021). Further discussion on the Digital Earth Australia Waterbodies has been provided in Chap. 6. Ultimately, the continuity, open data policy, and easy access are some of the most valuable legacies of the Landsat program.

3.3.2 Applications in Environmental Monitoring

Without doubt, Landsat data is the most widely used satellite data globally. The use of Landsat data has tremendously increased over the years and applications range from environmental monitoring and change detection, be it urban sprawl or land cover change to understanding key drivers and processes surrounding human civilization. For example, the study by Hemati et al. (2021) has shown a substantial rise in the average number of Landsat images used in studies, increasing from 10 images prior to 2008 to about 100,000 in 2020. This usage is strongly evidenced in the astronomical increase in the number of published records of Landsat-related studies. Notably, these increased publications and usage of Landsat data were also facilitated largely by the emergence of open-source new computational platforms (Gorelick et al. 2017), which aided access petabytes of publicly available remote sensing images and the high-speed processing infrastructure and suites of machine learning algorithms (Tamiminia et al. 2020; Kumar and Mutanga 2018).

From a cohort of 349 published articles between 2010 and October 2019 across 146 different journals, Landsat data remain the most widely used with Landsat-8 accounting for approximately 65% of such publications (Tamiminia et al. 2020). An upward trend in the use of Landsat images since 1994 for change detection studies has also recently been reported following a review of 490 studies (Hemati et al. 2021). New Landsat sensors come with new capabilities and applications. The use of Landsat-8 might slow down when Landsat-9 is launched and made available to the public. This expected rise in Landsat-9 will largely be due to new capabilities in sensor. Today, the application of Landsat data is skyrocketing. By discipline and change detection, applications include, wetland, water resources, urbanization, crop mapping, disaster management (e.g., earthquake, desertification), land cover land use, climate change (e.g., droughts, wildfire, floods), soil, geomorphology, among several others.

References

Aydal D, Ardal E, Dumanlilar Ã (2007) Application of the Crosta technique for alteration mapping of granitoidic rocks using ETM+ data: case study from eastern Tauride belt (SE Turkey). Int J Remote Sens 28(17):3895–3913. https://doi.org/10.1080/01431160601105926

Beck PS, Atzberger C, Hgda KA, Johansen B, Skidmore AK (2006) Improved monitoring of vegetation dynamics at very high latitudes: a new method using MODIS NDVI. Remote Sens Environ 100(3):321–334. https://doi.org/10.1016/j.rse.2005.10.021

Camps-Valls G, Tuia D, Gmez-Chova L, Jimnez S, Malo J (2012) Remote sensing image processing. Morgan and Claypool Publishers

Cracknell AP, Hayes L (2007) Introduction to remote sensing, 2nd edn. Taylor and Francis, CRC Press

Dong P, Chen Q (2018) Lidar remote sensing and applications. Taylor & Francis, CRC Press, US

Gorelick N, Hancher M, Dixon M, Ilyushchenko S, Thau D, Moore R (2017) Google Earth engine: planetary-scale geospatial analysis for everyone. Remote Sens Environ 202:18–27. https://doi.org/10.1016/j.rse.2017.06.031

Hansen MC, Potapov PV, Moore R, Hancher M et al (2013) High-resolution global maps of 21st-century forest cover change. Science 342(6160):850–853. https://doi.org/10.1126/science.1244693

Hemati M, Hasanlou M, Mahdianpari M, Mohammadimanesh F (2021) A systematic review of landsat data for change detection applications: 50 years of monitoring the earth. Remote Sens 13(15). https://doi.org/10.3390/rs13152869

Klein JK (2012) Environmental sensing; analytical techniques for earth observation. Springer, New York

Krause CE, Newey V, Alger MJ, Lymburner L (2021) Mapping and monitoring the multi-decadal dynamics of Australia's open waterbodies using landsat. Remote Sens 13(8). https://doi.org/10.3390/rs13081437

Kumar L, Mutanga O (2018) Google Earth Engine applications since inception: usage, trends, and potential. Remote Sens 10(10). https://doi.org/10.3390/rs10101509

Lavender S, Andrew L (2016) Practical handbook of remote sensing. Taylor and Francis, CRC Press, New York

Lewis K (2017) Understanding climate as a driver of food insecurity in Ethiopia. Clim Change 144(2):317–328. https://doi.org/10.1007/s10584-017-2036-7

McCabe MF, Rodell M, Alsdorf DE, Miralles DG, Uijlenhoet R, Wagner W, Lucieer A, Houborg R, Verhoest NEC, Franz TE, Shi J, Gao H, Wood EF (2017) The future of Earth observation in hydrology. Hydrol Earth Syst Sci 21(7):3879–3914. https://doi.org/10.5194/hess-21-3879-2017

Mueller N, Lewis A, Roberts D, Ring S, et al (2016) Water observations from space: mapping surface water from 25years of landsat imagery across Australia. Remote Sens Environ 174:341–352. https://doi.org/10.1016/j.rse.2015.11.003

Ndehedehe CE, Burford MA, Stewart-Koster B, Bunn SE (2020) Satellite-derived changes in floodplain productivity and freshwater habitats in northern Australia (1991–2019). Ecol Indic 114:106320. https://doi.org/10.1016/j.ecolind.2020.106320

Ndehedehe CE, Onojeghuo AO, Stewart-Koster B, Bunn SE, Ferreira VG (2021) Upstream flows drive the productivity of floodplain ecosystems in tropical queensland. Ecol Indic. https://doi.org/10.1016/j.ecolind.2021.107546

Okwuashi O, Ndehedehe CE (2020) Deep support vector machine for hyperspectral image classification. Pattern Recognit 103:107298. https://doi.org/10.1016/j.patcog.2020.107298

Okwuashi O, Ndehedehe CE (2021) Integrating machine learning with Markov chain and cellular automata models for modelling urban land use change. Remote Sens Appl Soc Environ 21:100461. https://doi.org/10.1016/j.rsase.2020.100461

Oussou FE, Yalo N, Ndehedehe CE, Oloukoi J, Alassane A, Boukari M, Gbewezoun VHG (2020) Fracture aquifers identification in the Zou basin (West Africa) using remote sensing and GIS. Geocarto Inte 1–24. https://doi.org/10.1080/10106049.2020.1852613

Pekel J-F, Cottam A, Gorelick N, Belward AS (2016) High-resolution mapping of global surface water and its long-term changes. Nature 540:418–422. https://doi.org/10.1038/nature20584

Rango A (1994) Application of remote sensing methods to hydrology and water resources. Hydrol Sci J 39(4):309–320. https://doi.org/10.1080/02626669409492752

Schowengerdt R (2007) Remote sensing: remote sensing: models and methods for image processing, 3rd edn. Academic Press

Tamiminia H, Salehi B, Mahdianpari M, Quackenbush L, Adeli S, Brisco B (2020) Google Earth Engine for geo-big data applications: a meta-analysis and systematic review. ISPRS J Photogramm Remote Sens 164:152–170. https://doi.org/10.1016/j.isprsjprs.2020.04.001

Tempfli K, Kerle N, Huurneman GC, Janssen LL (2009) Principles of remote sensing. In The international institute for geo-information science and Earth observation, Netherlands. Accessed from https://webapps.itc.utwente.nl/librarywww/papers_2009/general/principlesremotesensing.pdf on 3rd March 2021

Xu H (2006) Modification of normalised difference water index (NDWI) to enhance open water features in remotely sensed imagery. Int J Remote Sens 27(14):3025–3033. https://doi.org/10.1080/01431160600589179

Chapter 4
Satellite Geodetic Missions

Abstract The launch of the Gravity Recovery and Climate Experiment (GRACE) satellite in 2002 resulted in new capabilities to track changes in continental water storage and the acceleration of the global water cycle. It contributed to addressing the challenges posed by lack of or limited observational data for freshwater (including groundwater, soil moisture, lakes, and reservoirs) monitoring, and is now considered an important "tool in the box" in quantitative large-scale hydrology. GRACE satellite observations are now helping with freshwater accounting and improving our understanding of climate change and anthropogenic influence on changes in terrestrial water storage on different spatial scales (local, regional, and global). The use of GRACE to strengthen capacity in water resources assessment and to identify risk associated with climate change and human water consumption means it will continue to be an important tool for the future. The broad applications of GRACE-satellite in water resources, Earth and planetary science, and climate science attest to this. This chapter discusses GRACE as one of the novel satellite hydrology missions upon which the concept and definition of remote sensing hydrology is based. The principles behind the GRACE mission, its processing chains, and hydrological applications are also detailed.

4.1 Introductory Remarks

Satellite observations of different components of the hydrological cycle and their dynamics are crucial to assessing the impacts of climate variability and predicting extreme events such as floods and droughts. Traditional gauge-based measurements of water resources are usually restricted due to difficult terrain and the sheer size of several regions as well as the unavailability of monitoring wells in many parts of the world. This has necessitated the use of spaceborne geodetic sensors in monitoring changes in terrestrial hydrology. Satellite radar, altimetry, and gravimetry are good examples of active geodetic missions that measure changes in key components of the hydrological cycle. This include water-level variations in lakes, rivers, and terrestrial stored water, including groundwater variations. As noted by Ndehedehe (2019), apart from the growing applications of the Gravity Recovery and Climate Experi-

C. Ndehedehe, *Satellite Remote Sensing of Terrestrial Hydrology*,
https://doi.org/10.1007/978-3-030-99577-5_4

ment (GRACE) satellite observations in hydrology and Earth science, other sustained investments in satellite geodetic programmes have provided large coverage of dynamics in terrestrial water bodies. For example, measurements obtained from interferometric synthetic aperture radar (InSAR), radar altimetry (TOPEX/Poseidon), ERS satellites, Envisat and Jason1 missions, Ice, Cloud and land elevation satellite (ICESat), Global positioning System, Japan's Advance Land Observing Satellite, among others have proved to be useful in understanding the impacts of climate variability on the temporal and spatial dynamics of surface water resources (e.g., Ndehedehe and Ferreira 2020; Tourian et al. 2017; Alsdorf et al. 2007; Frappart et al. 2006; Lettenmaier 2005; Alsdorf et al. 2003, 2001). All of these satellite programmes and global land surface schemes have ignited a plethora of scientific findings that are not only informative but also useful for public policy and management decisions related to water resources (e.g., Ndehedehe 2017).

After the GRACE mission was decommissioned in December 2017 after more than 15 years of gravity observations across the globe, it was replaced by the launch of a next-generation time-variable gravity satellite pair, the GRACE Follow-On (GRACE-FO) mission. In non-industrialized regions where hydrological infrastructures and observational networks are in decline, lacking, non-operational or sparse (e.g., Alsdorf et al. 2003; Alsdorf and Lettenmaier 2003), the GRACE-FO mission would be the only operational, state-of-the-art, and most sophisticated gravity mission that would benefit hydrological and climate research tremendously. The finer spatial resolution of GRACE-FO mission ($< 50,000$ km^2) would allow not just the quantification of the terrestrial water budget components but also the closure of water balance with limited uncertainties. This improved spatial resolution would be the aftermath of an optimized system design and the inclusion of an experimental laser ranging system to complement the K-band microwave ranging system in the previous GRACE mission. Already, the GRACE-FO mission continues with the extremely successful work of its predecessor, providing measurements of changes in gravity fields across the globe. With the growing applications of GRACE observations in hydrology and Earth science, this chapter provides more details on GRACE, including concepts, processing chains, and hydrological applications.

4.2 Gravity Recovery and Climate Experiment

The Gravity Recovery and Climate Experiment (GRACE, Tapley et al. 2004) satellite mission, which has been in operation since March 2002, provides an integrated sum of changes in catchment stores (e.g., groundwater, soil moisture, etc.), surface waters, Ice sheets, and canopy water based on monthly observations of the Earth's time-variable gravity fields. The GRACE mission comprises two identical satellites in near-circular orbits with an altitude of $\sim$ 500 km and an inclination angle of 89.5° that are separated by $\sim$ 220 km. The two identical twin satellites popularly known as *Tom* and *Jerry* are linked with inter-satellite ranging system and equipped with global positioning system receivers and attitude sensors and high-precision accelerometers

Fig. 4.1 Gravity Recovery and Climate Experiment satellite mission. *Source* NASA

(Tapley et al. 2004). The ranging system is super sensitive and can detect separation changes about one-tenth the width of a human hair over a distance of 220 kilometers (NASA 2002). The GRACE satellites are designed to map the global gravity field every 30 days. Unlike optical remote sensing platforms, which use visible, near-infrared and, short-wave infrared sensors to collect details of the Earth surface, GRACE satellites (Fig. 4.1), which are equipped with a K-band microwave ranging system and high-precision accelerometers, use the satellite-to-satellite tracking system. Changes in the Earth's gravity field are caused mostly by dynamic processes that include mass redistribution of water (i.e., continental water storage, ocean, and atmosphere), gravitational tide in the solid Earth, post-glacial rebound, and variations in Antarctic and Greenland ice volumes (e.g., Swenson and Wahr 2002; Wahr et al. 1998). Accurate observations of these gravity fields are obtained from precise measurements of the distance between the two GRACE satellites as they orbit the Earth (see, e.g., Chen et al. 2005; Swenson and Wahr 2002; Wahr et al. 1998).

4.2.1 Standard GRACE Observations

The standard GRACE data (i.e., GRACE Level 02) is available online (http://icgem.gfz-potsdam.de/ICGEM/shms/monthly/) as monthly geopotential spherical harmonics Stokes coefficients from three well-known processing centres (Center for Space

Research-CSR, GeoForschungsZentrum-GFZ, and the Jet Propulsion Laboratory-JPL). These centres use different algorithms and background models to compute gravity fields from these dimensionless Stokes coefficient ($C_{l,m}$, $S_{l,m}$) to a maximum degree (l) (lies between $60 \, and \, 120$, corresponding to a spatial resolution of 150–300 km (Wouters et al. 2014)) and order (m). The Earth's gravity field is commonly represented in terms of the shape of the geoid because of its close approximation to the mean sea level. As detailed in Wahr et al. (1998), the geoid shape N can be expanded as a sum of spherical harmonics

$$N(\phi, \lambda) = R \sum_{l=0}^{\infty} \sum_{m=0}^{l} P_{lm}(\cos\phi)(C_{l,m}\cos(m\lambda) + S_{l,m}\sin(m\lambda)), \tag{4.1}$$

where R is the radius of the Earth, ϕ and λ are co-latitudes and east longitudes, and P_{lm} are the normalized associated Legendre functions. Assuming there is a time-dependent change in the geoid ΔN, then this change in N can be expressed in terms of changes, $\Delta C_{l,m}$ and $\Delta S_{l,m}$, in the spherical harmonic geoid coefficients as

$$N(\phi, \lambda) = R \sum_{l=0}^{\infty} \sum_{m=0}^{l} P_{lm}(\cos\phi)(\Delta C_{l,m}\cos(m\lambda) + \Delta S_{l,m}\sin(m\lambda)). \tag{4.2}$$

The change in surface mass density $\Delta\sigma$, in time t, from changes in the Stokes coefficients is given as

$$\Delta\sigma(\phi, \lambda, t) = \frac{R\rho_{ave}}{3} \sum_{l=0}^{\infty} \sum_{m=0}^{l} P_{lm}(\cos\phi)\frac{2l+1}{1+k_l}(\Delta C_{l,m}\cos(m\lambda) + \Delta S_{l,m}\sin(m\lambda)), \tag{4.3}$$

where R is the mean radius of the Earth (i.e., 6378.137 km), ρ_{ave} is the average density of the Earth (5515 kg/m^3) and k_l is the load Love numbers of degree l, provided in Wahr et al. (1998) and represents the effects of the Earth's response to surface loads (Swenson and Wahr 2002). The change in surface density $\Delta\sigma$, is expressed in equivalent water heights (EWT) or TWS by dividing with the average density of water as, $TWS(\phi, \lambda) = \frac{\Delta\sigma(\phi,\lambda)}{\varrho_w}$, where ϱ_w is 1000 kg/m^3. Since changes in Earth's gravity field are mostly caused by redistribution of water concentrated within a thin layer of thickness at the Earth's surface (Wahr et al. 1998), EWT would represent changes in stored water within this column.

4.2.2 *Correcting for Errors in GRACE Observations*

The spherical harmonic coefficients do suffer from signal attenuation and satellite measurement errors (e.g., system noise error in the inter-satellite range rate, accelerometer error, orbit error, etc.) causing noise in the higher degree coefficients (e.g., Landerer and Swenson 2012; Swenson and Wahr 2002). As detailed in Ndehedehe (2017), this systematic noise in the GRACE geopotential solutions typically manifest as pseudo-North–South striping patterns and have been attributed to correlations in the gravity field coefficients (e.g., Wouters and Schrama 2007; Swenson and Wahr 2006). This has resulted in improved reprocessed GRACE solutions that explored several approaches to reduce the effect of noise in GRACE data (e.g., Wouters et al. 2014). This includes (i) spatial averaging (see, Swenson and Wahr 2002) where an averaging kernel function is employed to convert global spherical harmonics into vertically integrated water storage over a region of interest and (ii) smoothing the GRACE coefficients through filtering either in the spectral or spatial domain (e.g., Belda et al. 2015; Swenson and Wahr 2006). GRACE filters use isotropic averaging functions, e.g., the popular Gaussian kernel (Swenson and Wahr 2002; Jekeli 1981) or non-isotropic averaging functions, which use symmetric or non-symmetric kernels (e.g., Chen et al. 2007; Kusche 2007; Swenson and Wahr 2006). The non-isotropic filters and others have been listed and categorized (see, Duan et al. 2009; Werth et al. 2009) and are based on different decorrelation schemes for a trade-off between the elimination of correlated errors and the minimization of signal distortion. The case studies in this book used the decorrelation filter of Kusche (2007) called DDK2. Decorrelation filter is a more efficient non-symmetric based, non-isotropic filter, which uses a set of regularizations and error covariance matrix of the spherical harmonic coefficients. Although several other studies have also explored Chen et al. (2007) decorrelation filter, practically, non-isotropic filters such as the DDK2 (Kusche 2007) accommodate better the GRACE error structure when compared to the conventional isotropic Gaussian filter (e.g., Werth et al. 2009), and is viewed as an improvement to the Gaussian filter (Duan et al. 2009; Kusche 2007). In fact, Klees et al. (2007) argued that the bias in continental water storage caused by non-isotropic filters are likely to be smaller compared to isotropic filters. It is worth noting, that there are no standard filtering procedures as most GRACE users would largely prefer the most convenient and easy-to-implement approaches and the fact that all filtering methods significantly attenuate the original signals. Otherwise, the choice of filtering would depend largely on the objectives of the studies (Duan et al. 2009). However, it is generally agreed that the filtering process can be significantly improved by using decorrelation filters, which unlike the Gaussian filter, fits better the noise in the data and extracts more information from the GRACE Level 02 products (e.g., Belda et al. 2015; Duan et al. 2009; Kusche 2007; Swenson and Wahr 2006). When the smoothing kernel (W_l) is applied on the standard GRACE solutions, the surface mass grids in Eq. 4.3 becomes

$$\Delta\sigma(\phi, \lambda, t) = \frac{R\rho_{ave}}{3} \sum_{l=0}^{\infty} \sum_{m=0}^{l} W_l P_{lm}(\cos\phi) \frac{2l+1}{1+k_l} (\Delta C^*_{l,m} \cos(m\lambda) + \Delta S^*_{l,m} \sin(m\lambda)), \tag{4.4}$$

where $\Delta C^*_{l,m}$ and $\Delta S^*_{l,m}$ are the Stokes coefficients after the decorrelation filter has been used. Regional averages of TWS can also be obtained by using the area weighted average (e.g., Tourian et al. 2015; Sneeuw et al. 2014):

$$\Delta TWS(\chi; t) = \sum_{l=1}^{n} \Delta W(\phi_i, \lambda_i, t) \frac{A_i}{A_\chi}, \tag{4.5}$$

where χ is the basin index, n is the number of pixels in the basin, A_i is the area of the grid cell i in χ, and A_χ is the total area of χ. Apart from truncating GRACE Stokes coefficient at low degrees, the filtering of monthly GRACE solutions affects the geophysical signals, especially at high latitudes, resulting in a significant loss of the original signal (e.g., Landerer and Swenson 2012; Duan et al. 2009). This causes bias and leakage errors in the observed GRACE-derived TWS (e.g., Baur et al. 2009; Klees et al. 2007; Swenson and Wahr 2006, 2007). The use of a scale factor (e.g., Long et al. 2015; Landerer and Swenson 2012; Fenoglio-Marc et al. 2012; Velicogna and Wahr 2006) and forward modelling (e.g., Long et al. 2016; Baur et al. 2009; Chen et al. 2014) are conventional methods that have been employed to restore the loss of signals caused by the effects of filtering and leakage. The scaling factor approach is widely used in global studies due to its simplicity. Other methods, which are based on iterative processes or regularization procedures are somewhat complicated and computationally expensive (e.g., Long et al. 2016; Chen et al. 2014). Scale factor approach can be used to correct for the bias in GRACE estimates due to filtering and truncation (Landerer and Swenson 2012), which usually results in a considerable leakage and attenuation of the original signal amplitude (e.g., Chen et al. 2016). This approach derives a gain factor k by minimizing the misfit between the unfiltered (ΔS_u), and filtered (ΔS_f) global hydrological model time series through a simple least square regression as

$$\gamma = \sum_{t=1}^{n} (\Delta S_u - k \Delta S_f)^2, \tag{4.6}$$

where γ is the leakage, t ranges from the first month to n (151 months). This scaling factor is the multiplicative factor (e.g., Long et al. 2015; Landerer and Swenson 2012) that reduces the difference between ΔS_u and ΔS_f modelled derived monthly changes in TWS. After scaling (Sect. 4.2.2.1), the measurement error is obtained as the product of the root mean square (RMS) of the leakage and the ratio of the RMSs of GRACE data to the model

$$\xi = RMS_{(\Delta S_u - k\Delta S_f)} \times \frac{RMS_{GRACE}}{RMS_{model}}. \tag{4.7}$$

4.2.2.1 Scaling of Terrestrial Water Storage Grids

When the gain factor, k in Eq. 4.6 is applied to the gridded fields of GRACE-derive terrestrial water storage observations, a substantial portion of the signal attenuation is restored. The regional averages of these re-scaled GRACE-solutions compares well with regional averages obtained (i.e., using Eq. 4.5 and Swenson and Wahr (2002) basin scale approach) by applying a single gain factor to unscaled data. Scale factors can be derived from land surface model data like the Global Land data Assimilation System (GLDAS) data (Rodell et al. 2004b). GLDAS drives four land surface models (Mosaic, Community Land Model, Variable Infiltration Capacity, and Noah) to produce different fields of the land surface (e.g., Hassan and Jin 2014). This data is freely available and covers the years 2001–2020 and can be obtained from the open access file available at http://grace.jpl.nasa.gov/data/get-data/land-water-content/. As illustration, using GLDAS-NOAH model data over West and Central Africa, the filtered and unfiltered model data were compared based on Pearson's correlation coefficient to evaluate the impact of the filter (Fig. 4.2a–d). Gridded scale factors below latitude 15° in the region are mostly greater than 1 and higher than 2 in West Sahel, and some locations in the Congo basin because of the strong signal damping (Fig. 4.2a). Strong correlations approaching one indicate the signal shape is not affected by the filter while lower correlations suggest the shape of the signal has been modified by the filter (Fig. 4.2b). As reported by Landerer and Swenson (2012), this happens in transition zones between plain and high elevation areas where spatial decorrelation length is usually shorter than what GRACE can resolve. Beyond reducing the variance of the leakage error by more than 80%, re-scaling GRACE data in order to remedy signal loss caused by the DDK filter is critical for regional water balance and hydrological applications. As Landerer and Swenson (2012) emphasized, if the effect of the filter is not accounted for in the transformed GRACE observations, the signal attenuation will become an error in the residual in the regional water balance or will serve as a constraint in water budget closure.

In particular, the effect of the DDK2 filter on computed GRACE-derived TWS before and after restoring the geophysical signal loss can be measured by comparing it with model-derived TWS. Although differences between smoothed and unsmoothed GRACE products (CSR, GFZ, and JPL) are not more than $10 - 15$ mm in some regions as noted by Grippa et al. (2011) who used a 300 km Gaussian filter in the post processing of GRACE solutions, Ferreira et al. (2012) observed consistency between the time series of GRACE-derived TWS changes using GFZ and CSR coefficients in the region. The time series in Fig. 4.2d captures the water storage change in the region, consistent with rainfall and water level fluctuations as earlier reported by Ndehedehe et al. (2016). CSR or JPL spherical harmonic solutions, which have been widely used in other regional and global studies are thus suitable GRACE level 2 data for hydrological applications. In the process of accounting for signal attenuation in the filtered GRACE solution, the regularization filter of Kusche (2007) is also applied to the model derived TWS content to derive a synthetic signal that reflects the inherent impact of the filter. The derived scale factor is independent of the GRACE-derived TWS as it is based on simulated TWS changes and does not match the GRACE-derived TWS to those of GLDAS, rather it only restores the relative amplitude of the 'original' data (e.g., Andam-Akorful et al. 2015; Ferreira et al. 2013; Landerer and Swenson 2012). Consequently, GRACE solutions resulting from this process can be averaged over arbitrary regions and compared to other gridded climate data (e.g., precipitation) without having to apply the regularization filter to

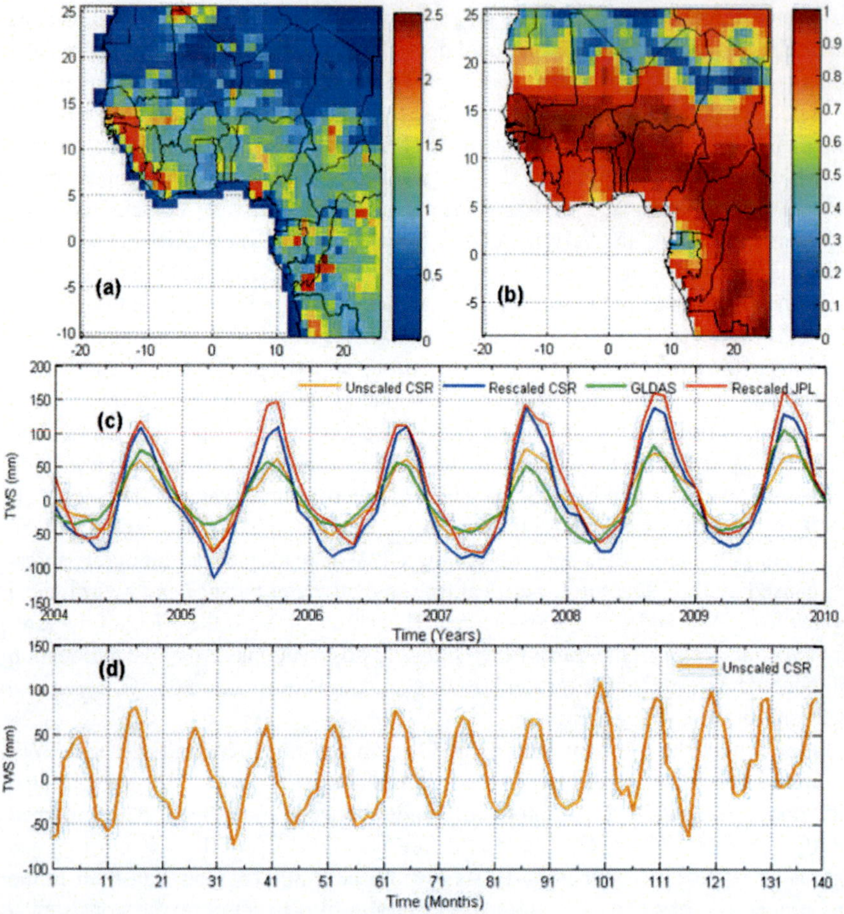

Fig. 4.2 **a** Estimated gridded scale factors (gain factors) based on the least square approach of Landerer and Swenson (2012), **b** correlation between filtered and unfiltered monthly TWS variations derived from GLDAS-NOAH model data. The monthly regional averaged time series of TWS from CSR, JPL, and GLDAS were also compared. **c** The effect of the DDK2 filter on computed GRACE-derived TWS before and after restoring the geophysical signal loss is compared with the GLDAS-TWS. This comparison is done for the common time period were there are no data gaps in GRACE-derived TWS (i.e., 2004/01–2010/12) and **d** the original time series of GRACE-derived TWS (i.e., without restoring the signal loss caused by the DDK2 filter) before interpolation

that data in the spherical harmonic domain (Landerer and Swenson 2012). In addition, a recent study that employed three-cornered hat method to examine uncertainties in GRACE coefficients from the three centres showed that CSR coefficients provided the lowest uncertainties for almost all global river basins with even JPL performing a bit better than GFZ (Ferreira et al. 2016). The rescaled monthly TWS grids usually have a few random gaps of up to 12 months in between and can be filled through interpolation. Even though this is a common reconstruction method for hydrological time series (e.g., Santos et al. 2010), other gap-filling methods have been used to predict missing GRACE observations. This is particularly important for multivariate statistical methods, which require continuous spatio-temporal data. This has been discussed further in Chap. 13 of this book.

4.2.2.2 Replacing Missing Gravity Coefficients

Other pre-processing details of GRACE level 2 product ensures the use of degree one gravity coefficients (i.e., C_{10}, C_{11}, and S_{11}) determined from ocean and atmospheric models (Swenson et al. 2008) and the replacement of degree two coefficients (i.e., C_{20}) with estimates from satellite laser ranging (SLR) solutions (Cheng et al. 2013). These are conventional procedures aimed at improving GRACE estimates of mass variability because

- GRACE does not provide changes in degree one gravity coefficients, which represent changes in the Earth's centre of mass relative to the geocentre motion (e.g., Chen et al. 2005),
- degree one Stokes coefficient are critical to the accurate and complete recovery of mass variability along with high-latitude areas and large-scale inter-basin mass exchange (Swenson et al. 2008),
- GRACE degree two Stoke coefficients are affected by large tide-like aliases caused by variations in geophysical models of ocean bottom pressure, tides, and atmospheric surface pressure (e.g., Seo et al. 2008a, b).

For these reasons, several studies (Ferreira et al. 2020; Ndehedehe et al. 2016) use the degree one coefficients provided by Swenson et al. (2008) while also replacing the degree two zonal gravitation change (C_{20}) with the improved and accurate estimates of SLR solutions obtained from NASA website (ftp://podaac.jpl.nasa.gov/allData/grace/docs/TN-07_C20_SLR.txt).

4.2.2.3 GRACE Monthly Mass Grids Level 3 Solutions

Gridded scaled observations following the processing discussed in Sect. 4.2.1 provide non-tradition users who lack expertise in processing and filtering the standard GRACE spherical harmonic geopotential coefficients to undertake hydrological analysis with GRACE-derived terrestrial water storage. The pioneering work of Landerer and Swenson (2012), who introduced a processing scheme for global-gridded terrestrial water storage estimates from temporal gravity field variations observed by the GRACE satellite made this possible. In addition, they also provide gridded fields of leakage and GRACE measurement errors to enable users estimate uncertainties in TWS. Their processed GRACE data, now popularly known as Monthly Mass Grids—Land (or GRACETellus grid) are available for download from the

GRACE data portal (https://grace.jpl.nasa.gov/data/get-data/monthly-mass-grids-land/). These monthly grids represent the surface mass deviation for that month relative to the baseline average, which is during 2004–2009. For comparisons against other hydrological data or models, it is ideal to use anomalies relative to the same time average for comparison. For instance, if the new baseline is 2004–2006, average the monthly grids over January 2004–December 2006, and subtract this average grid from all other monthly grids. Information about data processing chains, as well as grid scaling, units, format, error & uncertainty estimates, and suitability for Cryospheric studies have been detailed in this portal. As preprocessing is no longer required with the use of this GRACE monthly mass grids, hydrological applications of GRACE in the last five years skyrocketed. With the monthly mass grids now popularly used, non-geodesists and environmental experts have leveraged the freely available gridded GRACE solutions for climate change studies, including droughts, floods, and monitoring of groundwater changes, among several others. With the decommissioning of the GRACE mission, the GRACE Follow-on mission continues to provide these observations into the future (Landerer et al. 2020), including GRACE mass concentration solutions, which is discussed in Sect. 4.2.3.

4.2.3 GRACE Mass Concentration Solutions

Gravity Recovery and Climate Experiment terrestrial water storage-based mascon products are also available to study changes in land water storage across the globe (Fig. 4.3). To facilitate regional applications of GRACE mascon products, the Center for Space Research (CSR) GRACE mass concentration (mascons) Release 06 solutions are now provided as 0.25 arc-degree resolution. The data represent the equal-area geodesic grid of size $1.0° \times 1.0°$ at the equator, which is the current native resolution of CSR RL06 mascon solutions. The RL06 grid files, which are now released as 1/4° (or 0.25 arc-degree) ensure that the coastlines as defined in the new RL06 mascon grid could be represented properly (Save et al. 2016). The data can be accessed at the CSR data portal (http://www2.csr.utexas.edu/grace/RL06_mascons.html). GRACE mascons are available and provided by CSR, GFSC, and JPL processing centres. The CSR solution describes the global mass changes expressed in terrestrial water storage solved for 40,962 cells with each having an approximately 12,400 km^2 and an average distance of about 120 km between the cells (Save et al. 2016). The GFSC mascon solution is solved for 1°-by-1° equal-area grid blocks in which there are 41,168 mascon blocks covering the entire globe with a mean area of 12,389 km (Luthcke et al. 2013). The JPL mascon solution solves for monthly gravity field variations in terms of 4,551 equal-area 3-° spherical cap mascons. Both JPL and GFSC are resampled into a resolution of 0.5°-by-0.5° (Watkins et al. 2015).

GRACE mascon solutions are generally similar to the level 3 solutions (i.e., those processed from level 2 GRACE spherical harmonic discussed in Sect. 4.2.1) but their processing is very different. Mascon product solves for monthly gravity field variations in terms of 120 km wide mascon block (Save et al. 2016; Wiese et al. 2016;

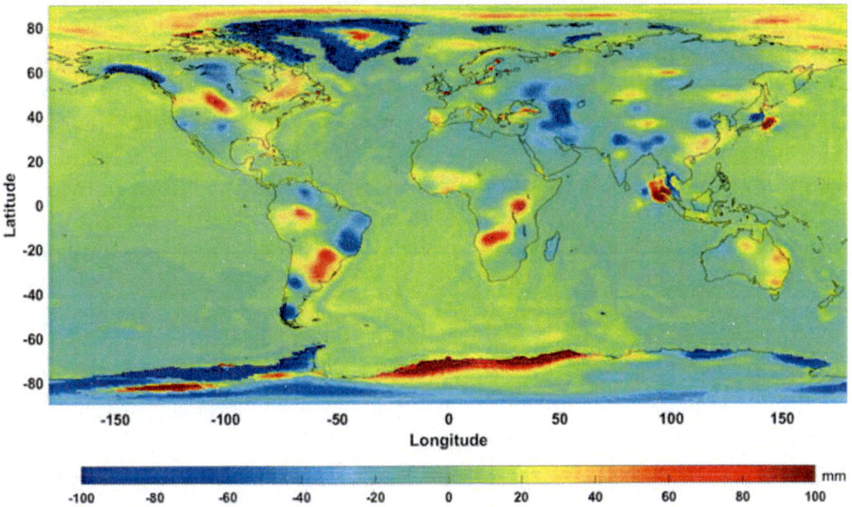

Fig. 4.3 Mean global terrestrial water storage from Gravity Recovery and Climate Experiment-based mascon solutions during 2002–2017 period

Watkins et al. 2015). That is, equivalent water heights resulting from mass changes are predicted in concentrated blocks on the surface of the Earth (Save et al. 2016; Luthcke et al. 2013). These global mascon solutions provide accurate surface-based grids of terrestrial water storage (Fig. 4.3), which require no further processing (e.g., empirical scaling factors) as it is devoid of stripe errors and capture all hydrological signals observed by GRACE within the measurement noise level (Save et al. 2016). As opposed to the level 2 products, this is a key benefit derived from the use of mascon solutions as they are ready-made solutions for applications in several other disciplines, including hydrology, climatology, oceanography, land surface hydrology, and cryosphere. Apart from eliminating post-processing protocols, they also have better resolution, making them suitable for hydrological applications in relatively smaller spatial areas though still require caution and understanding of the native resolution and hydrological signals that can be resolved (Watkins et al. 2015).

4.2.4 *Satellite Hydrological Applications*

Due to its ability to detect changes in deep subsurface and sensitivity to water stress, GRACE has the unique infrastructure to support freshwater monitoring and impacts of changing climate on water availability. Since its inception, GRACE has given an unparalleled perspective to global terrestrial hydrology by providing quantitative estimates of monthly changes in TWS (soil moisture; groundwater; surface water lakes, rivers, wetlands; snow; and canopy) over large spatial scales. Because of its spatial resolution (200,000 km^3), the dynamics in multi-layered land water storage (i.e., all aspects of TWS) can be measured at global or regional scales with an accuracy of 1.5 cm equivalent water height (Famiglietti and Rodell 2013). GRACE provides an observation-based assessment of global terrestrial hydrology, thus improving our ability to predict emerging threats to water and food security. Accurate accounting of changes in freshwater availability makes this possible including, human and ecosystem health, energy generation, and social unrest (Rodell et al. 2018). Apart from its broad applications in droughts, floods, terrestrial water budget, and ecosystem assessments (e.g., Zhang et al. 2016; Thomas et al. 2014; Reager et al. 2014; Yang et al. 2014; Long et al. 2013; Houborg et al. 2012; Yirdaw et al. 2008), it is now one of the most vital tools in hydrological research, specifically in monitoring sub-surface water storage, aquifer system processes, and evaluating groundwater resources (e.g., Chen et al. 2016; Castellazzi et al. 2016; Famiglietti et al. 2015; Alley and Konikow 2015; Swenson and Wahr 2007). The applications of GRACE data in hydrology research globally are growing and well documented (e.g., Humphrey et al. 2016; Wouters et al. 2014) and are only summarized here. Additionally, GRACE has been used to study and estimate hydrological sinks and other hydrological units, e.g., runoff (Andam-Akorful et al. 2015; Senay et al. 2014; Sneeuw et al. 2014; Rodell et al. 2004a). Furthermore, the illustration in Fig. 4.3 shows global hotspots of hydrological changes. These changes are those emanating from either strong variations in surface water, groundwater, or melting of glaciers and ice caps. GRACE-satellite hydrological applications in droughts, floods, terrestrial water budget, and ecosystem assessments, among several others have been detailed in chapters 7–9, 12, 14–17 and 20 of this book, and their practical applications and implementation will be explored further.

4.2.5 Crustal Deformations and Subsidence

In regions of the world with complex physical dynamic processes such as surface deformations and crustal deformations, GRACE is increasingly becoming valuable, especially when combined with vertical crustal displacements obtained from Global Positioning System (GPS) and model data (Fig. 4.4). By combining GPS and model-derived vertical crustal displacements with level 2 GRACE spherical harmonic solutions, the study by Ferreira et al. (2019) demonstrated this potential. They showed that the highest observed amplitudes of vertical crustal displacements of the annual signals are concentrated mainly over the Amazon Rainforest and Brazilian Highlands. Figure 4.4a illustrates this further indicating strong agreement between GRACE and GPS series. While output from models and reanalysis data also show substantial association, the correlation between GPS with GRACE time series revealed that approximately 45% of the GPS stations had correlation coefficients higher than 0.60 Fig. 4.4a–c. As the hydrological cycle is changing with impacts on water availability and the Earth's surface due to its elastic response to the surface mass loading/unloading (Ferreira et al. 2019), using GRACE in such geodetic exercise to assess

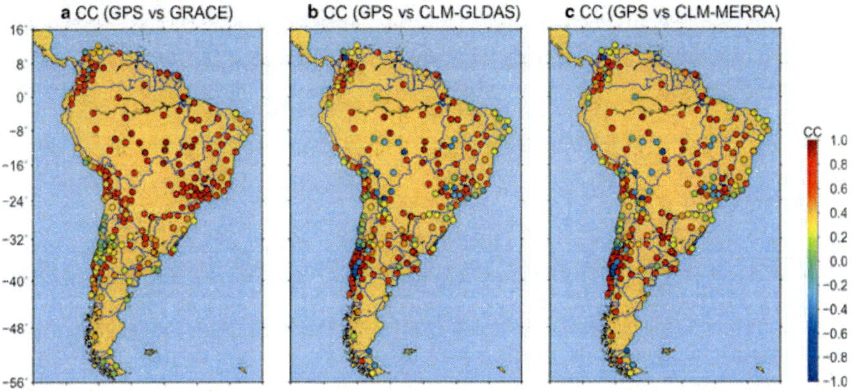

Fig. 4.4 Correlation coefficients (CCs) for **a** Global Positioning Systems-GPS versus GRACE, **b** GPS versus land surface models (CLM-GLDAS), and **c** GPS versus CLM-MERRA. *Source* Ferreira et al. (2019)

vertical crustal displacements is thus crucial to improve understanding of the impacts of extreme events like drought on water availability (Ferreira et al. 2018).

InSAR data derived from radar observation can deliver detailed measurements of large-scale ground deformations associated with earthquakes, prolonged groundwater withdrawal due to drought, aquifer systems, resource extraction, and landslides (Smith and Majumdar 2020). InSAR-derived vertical ground displacement measurements are assumed to be completely associated with aquifer compaction in areas with extensive groundwater extraction. As a result, excessive groundwater extraction from compressible aquifers leads to land subsidence that will significantly cause damage to buildings and infrastructure. But GRACE is also becoming useful in the studies of land subsidence and deformations. For instance, GRACE has been combined with InSAR (Interferometry of Synthetic Aperture Radar) techniques to assess the sensitivities of aquifer systems to climate processes (e.g., inter-annual variations in rainfall) and water extraction. On the one hand, this can lead to groundwater drought (Chap. 11). These factors, on the other hand, can also result in geodetic disturbance that leads to surface deformation and subsidence. InSAR measures ground displacement due to aquifer response to fluid-pressure changes and is based on the difference between the phase from two processed SAR (Synthetic Aperture Radar) complex images acquired from two different positions. Its combined use with GRACE in deformation studies and mapping of groundwater-deficit regions have increased in recent years, complementing the lack of observational groundwater data (Ojha et al. 2020; Castellazzi et al. 2018). A GRACE–InSAR combination is thus useful in generating deformation maps and time series of surface displacements for hydro-geologic river basins, water planning areas, and hydrologic hotspots where groundwater extraction is intensive. To validate InSAR deformation, the relationship between the rates of vertical land motion measured by InSAR and Global Positioning System (GPS) can be examined using statistical methods. However, GRACE data has been found to be valuable in the validation of vertical deformations (Ojha et al. 2020; Ferreira et al. 2018).

References

Alley WM, Konikow LF (2015) Bringing GRACE down to earth. Groundwater 53(6):826–829. https://doi.org/10.1111/gwat.12379

Alsdorf D, Birkett C, Dunne T, Melack J, Hess L (2001) Water level changes in a large amazon lake measured with spaceborne radar interferometry and altimetry. Geophys Res Lett 28(14):2671–2674. https://doi.org/10.1029/2001GL012962

Alsdorf D, Lettenmaier D, Vörösmarty C (2003) The need for global, satellite-based observations of terrestrial surface waters. Eos, Trans Am Geophys Union 84(29):269–276. https://doi.org/10.1029/2003EO290001

Alsdorf DE, Lettenmaier DP (2003) Tracking fresh water from space. Science 301(5639):1491–1494. https://doi.org/10.1126/science.1089802

Alsdorf DE, Rodríguez E, Lettenmaier DP (2007) Measuring surface water from space. Rev Geophys 45(2):RG2002. https://doi.org/10.1029/2006RG000197

Andam-Akorful SA, Ferreira VG, Awange JL, Forootan E, He XF (2015) Multi-model and multi-sensor estimations of evapotranspiration over the Volta Basin, West Africa. Int J Climatol 35(10):3132–3145. https://doi.org/10.1002/joc.4198

Baur O, Kuhn M, Featherstone WE (2009) GRACE-derived ice-mass variations over Greenland by accounting for leakage effects. J Geophys Res: Solid Earth 114(B6):B06407. https://doi.org/10.1029/2008JB006239

Belda S, García-García D, Ferrándiz JM (2015) On the decorrelation filtering of RL05 GRACE data for global applications. Geophys J Int 200:173–184. https://doi.org/10.1093/gji/ggu386

Castellazzi P, Longuevergne L, Martel R, Rivera A, Brouard C, Chaussard E (2018) Quantitative mapping of groundwater depletion at the water management scale using a combined GRACE/InSAR approach. Remote Sens Environ 205:408–418. https://doi.org/10.1016/j.rse.2017.11.025

Castellazzi P, Martel R, Galloway DL, Longuevergne L, Rivera A (2016) Assessing groundwater depletion and dynamics using GRACE and InSAR: potential and limitations. Groundwater 54(6):768–780. https://doi.org/10.1111/gwat.12453

Chen J, Famigliett JS, Scanlon BR, Rodell M (2016) Groundwater storage changes: present status from GRACE observations. Surv Geophys 37(2):397–417. https://doi.org/10.1007/s10712-015-9332-4

Chen J, Li J, Zhang Z, Ni S (2014) Long-term groundwater variations in Northwest India from satellite gravity measurements. Glob Plan Change 116:130–138. https://doi.org/10.1016/j.gloplacha.2014.02.007

Chen JL, Rodell M, Wilson CR, Famiglietti JS (2005) Low degree spherical harmonic influences on Gravity Recovery and Climate Experiment (GRACE) water storage estimates. Geophys Res Lett 32(14):L14405. https://doi.org/10.1029/2005GL022964

Chen JL, Wilson CR, Tapley BD, Grand S (2007) GRACE detects coseismic and postseismic deformation from the Sumatra-Andaman earthquake. Geophys Res Lett 34(13):L13302. https://doi.org/10.1029/2007GL030356

Cheng MK, Tapley BD, Ries JC (2013) Deceleration in the Earth's oblateness. J Geophys Res 118:1–8. https://doi.org/10.1002/jgrb.50058

Duan XJ, Guo JY, Shum CK, van der Wal W (2009) On the postprocessing removal of correlated errors in GRACE temporal gravity field solutions. J Geod 83(11):1095. https://doi.org/10.1007/s00190-009-0327-0

Famiglietti JS, Cazenave A, Eicker A, Reager JT, Rodell M, Velicogna I (2015) Satellites provide the big picture. Science 349(6249):684–685. https://doi.org/10.1126/science.aac9238

Famiglietti JS, Rodell M (2013) Water in the balance. Science 340(6138):1300–1301. https://doi.org/10.1126/science.1236460

Fenoglio-Marc L, Rietbroek R, Grayek S, Becker M, Kusche J, Stanev E (2012) Water mass variation in the Mediterranean and Black Seas. J Geodyn 59–60:168–182

Ferreira V, Montecino H, Ndehedehe C, Heck B, Gong Z, Westerhaus M, de Freitas S (2018) Space-based observations of crustal deflections for drought characterization in brazil. Sci Total Environ 644:256–273. https://doi.org/10.1016/j.scitotenv.2018.06.277

Ferreira V, Yong B, Tourian M, Ndehedehe C, Shen Z, Seitz K, Dannouf R (2020) Characterization of the hydro-geological regime of Yangtze river basin using remotely-sensed and modeled products. Sci Total Environ 718:137354. https://doi.org/10.1016/j.scitotenv.2020.137354

Ferreira VG, Gong Z, Andam-Akorful SA (2012) Monitoring mass changes in the Volta River Basin Using GRACE satellite gravity and TRMM precipitation. Boletim De Ciencias Geodesicas 18(4):549–563. WOS:000313106300003

Ferreira VG, Gong Z, He X, Zhang Y, Andam Akorful SA (2013) Estimating total discharge in the Yangtze River basin using satellite-based observations. Remote Sens 5(7):3415–3430. https://doi.org/10.3390/rs5073415

Ferreira VG, Montecino HD, Ndehedehe CE, del Rio RA, Cuevas A, de Freitas SRC (2019) Determining seasonal displacements of Earth's crust in South America using observations from space-borne geodetic sensors and surface-loading models. Earth Planets Space 71(1):84. https://doi.org/10.1186/s40623-019-1062-2

Ferreira VG, Montecino HDC, Yakubu CI, Heck B (2016) Uncertainties of the Gravity Recovery and Climate Experiment time-variable gravity-field solutions based on three-cornered hat method. J Appl Remote Sens 10(1):015015. https://doi.org/10.1117/1.JRS.10.015015

Frappart F, Calmant S, Cauhopé M, Seyler F, Cazenave A (2006) Preliminary results of ENVISAT RA-2-derived water levels validation over the Amazon basin. Remote Sens Environ 100(2):252–264. https://doi.org/10.1016/j.rse.2005.10.027

Grippa M, Kergoat L, Frappart F, Araud Q, Boone A et al (2011) Land water storage variability over West Africa estimated by Gravity Recovery and Climate Experiment (GRACE) and land surface models. Water Res Res 47(5):W05549. https://doi.org/10.1029/2009wr008856

Hassan AA, Jin S (2014) Water cycle and climate signals in Africa observed by Satellite Gravimetry. IOP Conf Ser: Earth Environ Sci 17:1–6. https://doi.org/10.1088/1755-1315/17/1/012149

Houborg R, Rodell M, Li B, Reichle R, Zaitchik BF (2012) Drought indicators based on model-assimilated Gravity Recovery and Climate Experiment (GRACE) terrestrial water storage observations. Water Res Res 48(7):W07525. https://doi.org/10.1029/2011WR011291

Humphrey V, Gudmundsson L, Seneviratne SI (2016) Assessing global water storage variability from GRACE: trends, seasonal cycle, subseasonal anomalies and extremes. Surv Geophys 37(2):357–395. https://doi.org/10.1007/s10712-016-9367-1

Jekeli C (1981) Alternative methods to smooth the Earth's gravity field. Department of civil and environmental engineering and geodetic science, Ohio State Univiversity, Columbus, Report 327. https://earthsciences.osu.edu/sites/earthsciences.osu.edu/files/report-327.pdf. Accessed 14th Feb 2017

Klees R, Zapreeva EA, Winsemius HC, Savenije HHG (2007) The bias in GRACE estimates of continental water storage variations. Hydrol Earth Syst Sci 11(4):1227–1241. https://doi.org/10.5194/hess-11-1227-2007

Kusche J (2007) Approximate decorrelation and non-isotropic smoothing of time-variable GRACE-type gravity field models. J Geod 81(11):733–749. https://doi.org/10.1007/s00190-007-0143-3

Landerer FW, Flechtner FM, Save H, Webb FH et al (2020) Extending the global mass change data record: GRACE Follow-On instrument and science data performance. Geophys Res Lett 47(12):e2020GL088306. https://doi.org/10.1029/2020GL088306

Landerer FW, Swenson SC (2012) Accuracy of scaled GRACE terrestrial water storage estimates. Water Res Res 48(4):W04531. https://doi.org/10.1029/2011WR011453

Lettenmaier DP (2005) Observations of the global water cycle—global monitoring networks. In: Anderson MG (ed) Encyclopedia of hydrological sciences, vol 5. Wiley, pp 2719–2732

Long D, Chen X, Scanlon BR, Wada Y, Hong Y, Singh VP, Chen Y, Wang C, Han Z, Yang W (2016) Have GRACE satellites overestimated groundwater depletion in the Northwest India aquifer? Sci Rep 6:24398. https://doi.org/10.1038/srep24398

Long D, Longuevergne L, Scanlon BR (2015) Global analysis of approaches for deriving total water storage changes from GRACE satellites. Water Res Res 51(4):2574–2594. https://doi.org/10.1002/2014WR016853

Long D, Scanlon BR, Longuevergne L, Sun AY, Fernando DN, Save H (2013) GRACE satellite monitoring of large depletion in water storage in response to the 2011 drought in Texas. Geophys Res Lett 40(13):3395–3401. https://doi.org/10.1002/Grl.50655

Luthcke SB, Sabaka T, Loomis B, Arendt A, McCarthy J, Camp J (2013) Antarctica, Greenland and Gulf of Alaska land-ice evolution from an iterated GRACE global mascon solution. J Glaciol 59(216):613–631. https://doi.org/10.3189/2013JoG12J147

NASA (2002). Grace launch. National aeronautics and space administration. http://www.jpl.nasa.gov/news/press-kits/gracelaunch.pdf. Accessed 9th Feb 2017

Ndehedehe C, Awange J, Agutu N, Kuhn M, Heck B (2016) Understanding changes in terrestrial water storage over West Africa between 2002 and 2014. Adv Water Res 88:211–230. https://doi.org/10.1016/j.advwatres.2015.12.009

Ndehedehe CE (2017) Remote sensing of West Africa's water resources using multi-satellites and models. PhD thesis, Curtin University, Bentley, Perth, Western Australia. http://hdl.handle.net/20.500.11937/59637. Accessed 12th Jan 2018

Ndehedehe CE (2019) The water resources of tropical West Africa: propblems, progress and prospect. Acta Geophysica 67(2):621–649. https://doi.org/10.1007/s11600-019-00260-y

Ndehedehe CE, Ferreira VG (2020) Identifying the footprints of global climate modes in time-variable gravity hydrological signals. Clim Change 159:481–502. https://doi.org/10.1007/s10584-019-02588-2

Ojha C, Werth S, Shirzaei M (2020) Recovery of aquifer-systems in Southwest US following 2012–2015 drought: Evidence from InSAR, GRACE and groundwater level data. J Hydrol 587. https://doi.org/10.1016/j.jhydrol.2020.124943

Reager JT, Thomas BF, Famiglietti JS (2014) River basin flood potential inferred using GRACE gravity observations at several months lead time. Nat Geosc 7(8):588–592. https://doi.org/10.1038/ngeo2203

Rodell M, Famiglietti JS, Chen J, Seneviratne SI, Viterbo P, Holl S, Wilson CR (2004) Basin scale estimates of evapotranspiration using GRACE and other observations. Geophys Res Lett 31(20):L20504. https://doi.org/10.1029/2004GL020873

Rodell M, Famiglietti JS, Wiese DN, Reager JT, Beaudoing HK, Landerer FW, Lo M-H (2018) Emerging trends in global freshwater availability. Nature 557:651–659. https://doi.org/10.1038/s41586-018-0123-1

Rodell M, Houser PR, Jambor U, Gottschalck J, Mitchell K, Mitchell, K et al (2004) The global land data assimilation system. Bull Am Meteorol Soc 85(3):381–394. https://doi.org/10.1175/BAMS-85-3-381.R

Santos JF, Pulido-Calvo I, Portela MM (2010) Spatial and temporal variability of droughts in Portugal. Water Res Res 46(3):W03503. https://doi.org/10.1029/2009WR008071

Save H, Bettadpur S, Tapley BD (2016) High-resolution CSR GRACE RL05 mascons. J Geophys Res: Solid Earth 121(10):7547–7569. https://doi.org/10.1002/2016JB013007

Senay GB, Velpuri NM, Bohms S, Demissie Y, Gebremichael M (2014) Understanding the hydrologic sources and sinks in the Nile Basin using multisource climate and remote sensing data sets. Water Res Res 50(11):8625–8650. https://doi.org/10.1002/2013WR015231

Seo K-W, Wilson CR, Chen J, Waliser DE (2008) GRACE's spatial aliasing error. Geophys J Int 172(1):41–48. https://doi.org/10.1111/j.1365-246X.2007.03611.x

Seo KW, Wilson CR, Han SC, Waliser DE (2008) Gravity recovery and climate experiment (GRACE) alias error from ocean tides. J Geophysical Res: Solid Earth 113(B3):B03405. https://doi.org/10.1029/2006JB004747

Smith RG, Majumdar S (2020) Groundwater storage loss associated with land subsidence in western United States mapped using machine learning. Water Res Res 56(7):e2019WR026621. https://doi.org/10.1029/2019WR026621

Sneeuw N, Lorenz C, Devaraju B, Tourian M, Riegger J, Kunstmann H, Bárdossy A (2014) Estimating runoff using hydro-geodetic approaches. Surv Geophys 35(6):1333–1359. https://doi.org/10.1007/s10712-014-9300-4

Swenson S, Chambers D, Wahr J (2008) Estimating geocenter variations from a combination of GRACE and ocean model output. Geophys Res Lett 31:1–4. https://doi.org/10.1029/2004GL019920

Swenson S, Wahr J (2002) Methods for inferring regional surface-mass anomalies from Gravity Recovery and Climate Experiment (GRACE) measurements of time-variable gravity. J Geophys Res-Solid Earth 107(B9). https://doi.org/10.1029/2001jb000576

Swenson S, Wahr J (2006) Post-processing removal of correlated errors in GRACE data. Geophys Res Lett 33(8):L08402. https://doi.org/10.1029/2005GL025285

Swenson S, Wahr J (2007) Multi-sensor analysis of water storage variations of the Caspian Sea. Geophys Res Lett 34(16). https://doi.org/10.1029/2007gl030733

Tapley B, Bettadpur S, Watkins M, Reigber C (2004) The gravity recovery and climate experiment: mission overview and early results. Geophys Res Lett 31:1–4. https://doi.org/10.1029/2004GL019920

Thomas AC, Reager JT, Famiglietti JS, Rodell M (2014) A GRACE-based water storage deficit approach for hydrological drought characterization. Geophys Res Lett 41(5):1537–1545. https://doi.org/10.1002/2014GL059323

Tourian M, Elmi O, Chen Q, Devaraju B, Roohi S, Sneeuw N (2015) A spaceborne multisensor approach to monitor the desiccation of Lake Urmia in Iran. Remote Sens Environ 156:349–360. https://doi.org/10.1016/j.rse.2014.10.006

Tourian M, Schwatke C, Sneeuw N (2017) River discharge estimation at daily resolution from satellite altimetry over an entire river basin. J Hydrol 546:230–247. https://doi.org/10.1016/j.jhydrol.2017.01.009

Velicogna I, Wahr J (2006) Acceleration of Greenland ICE mass loss in spring 2004. Nature 443(7109):329–331. https://doi.org/10.1038/nature05168

Wahr J, Molenaar M, Bryan F (1998) Time variability of the Earth's gravity field: hydrological and oceanic effects and their possible detection using GRACE. J Geophys Res-Solid Earth 103(B12):30205–30229. https://doi.org/10.1029/98jb02844

Watkins MM, Wiese DN, Yuan D, Boening C, Landerer FW (2015) Improved methods for observing earth's time variable mass distribution with GRACE using spherical cap Mascons. J Geophys Res: Solid Earth 120(4):2648–2671. https://doi.org/10.1002/2014JB011547

Werth S, Göntner A, Schmidt R, Kusche J (2009) Evaluation of GRACE filter tools from a hydrological perspective. Geophys J Int 179(3):1499–1515. https://doi.org/10.1111/j.1365-246X.2009.04355.x

Wiese DN, Landerer FW, Watkins MM (2016) Quantifying and reducing leakage errors in the JPL RL05M GRACE mascon solution. Water Res Res 52(9):7490–7502. https://doi.org/10.1002/2016WR019344

Wouters B, Bonin JA, Chambers DP, Riva REM, Sasgen I, Wahr J (2014) GRACE, time-varying gravity, Earth system dynamics and climate change. Rep Prog Phys 77(11):116801. https://doi.org/10.1088/0034-4885/77/11/116801

Wouters B, Schrama EJO (2007) Improved accuracy of GRACE gravity solutions through empirical orthogonal function filtering of spherical harmonics. Geophys Res Lett 34(23):L23711. https://doi.org/10.1029/2007GL032098

Yang Y, Long D, Guan H, Scanlon BR, Simmons CT, Jiang L, Xu X (2014) GRACE satellite observed hydrological controls on interannual and seasonal variability in surface greenness over mainland Australia. J Geophys Res: Biogeosci 119(12):2245–2260. https://doi.org/10.1002/2014JG002670

Yirdaw SZ, Snelgrove K, Agboma C (2008) GRACE satellite observations of terrestrial moisture changes for drought characterization in the Canadian Prairie. J Hydrol 356:84–92

Zhang D, Zhang Q, Werner AD, Liu X (2016) GRACE-based hydrological drought evaluation of the Yangtze River Basin, China. J Hydrometeorol 17(3):811–828. https://doi.org/10.1175/JHM-D-15-0084.1

Part III
Cloud Computing for Big Data

Chapter 5
Cloud-Based Geospatial Analysis

Abstract The developments and improvements in satellite remote sensing sensors have contributed to the emergence of big geospatial data, which are now freely available for several applications in remote sensing of the environment. The rich repository of geospatial data and their easy access through effective data management and storage, revolutionized the use of big earth observation data. The future of remote sensing data analysis is expected to be dominated by cloud computing frameworks like the Google Earth Engine (GEE), which provides an interface for application programming and aids easy access to several geospatial data and remotely sensed observations across the globe. But will cloud computing completely replace traditional desktop remote sensing software applications in the processing of satellite earth observation? Do we expect more computing platforms for big remote sensing data processing? With strategic partnerships between large research funding organizations and the private sector, more of these technological innovations may emerge in the future to support the handling of big data. In this chapter, cloud computing platforms for big data are discussed but with a focus on the GEE platform and its contributions to aiding environmental monitoring and assessment. Some examples of its usage are demonstrated and briefly discussed.

5.1 Introductory Remarks

With the emergence of big data in remote sensing, cloud computing is increasingly becoming popular given its crucial role in understanding processes, creating synergy among cognate disciplines, and improving decision-making through rapid environmental monitoring. Big data processing of satellite Earth observations and novel machine learning techniques are at the heart of this monitoring. The research community, stakeholders, and users of satellite Earth observations are currently experiencing an outburst of upsurge in improved sensor capabilities in satellite missions, aerial, ground, and mobile applications. Such capabilities are not unconnected with big geospatial data, which are computationally expensive and intensive to process, manage, and perform spatial analysis because of the high computing power required. However, advancements in the use of these technologies and big remote sensing data

C. Ndehedehe, *Satellite Remote Sensing of Terrestrial Hydrology*,
https://doi.org/10.1007/978-3-030-99577-5_5

have been optimized by innovative pattern recognition and machine learning algorithms. Robust satellite image processing capacities relying on these algorithms (e.g., Okwuashi et al. 2020) and integrated digital technology such as cloud computing platforms have contributed to the growing applications of remote sensing observations in environmental monitoring in recent times (e.g., Tamiminia et al. 2020; Kumar and Mutanga 2018). As earlier postulated by Craglia et al. (2012) in their discussion of scientific and societal drivers for the development of digital Earth, they envisaged that these technologies would produce globally seamless and constantly updated global human settlement layer.[1] Indeed, today's reality is that of a global village where large Earth observation datasets are not only available and freely accessible from global repositories but can be quickly processed to generate timely information that underpins important organizational decisions. These opportunities have enabled government institutions, scientists, and the public to monitor and analyse environmental changes through rapid and repeatable large-scale analyses of big geospatial data in time and space (Lewis 2017).

As with Spatial Data Infrastructure, which enabled users to exchange and share geospatial data, including geographic information, through technologies and institutional frameworks (Gomes et al. 2020), cloud-based platform is a prototype of this kind of Infrastructure but with the capability for big data processing and management. Essentially, cloud-based computational platforms facilitate not only access to big data but integration of multi-source planetary scale data leveraging innovative, state of the art, and sophisticated technologies. The developments and improvements in satellite remote sensing sensors have contributed to the emergence of big data. These Earth observation systems have resulted in big geospatial data that are now freely available for several applications in the remote sensing of environment. In other words, with the advent of big data came the development of cloud computing platforms to aid their processing, analyses, and data management. The rich repository of geospatial data and easy access available in cloud computing through effective data management and storage revolutionized the use of big Earth observation data. Among these cloud computing platforms, Google Earth Engine (GEE, Gorelick et al. 2017) is very prominent, robust, and the most widely used in the processing of large remote sensing data over different spatial scales. However, other platforms such as Sentinel Hub, Open Data Cube (ODC), System for Earth Observation Data Access, Processing and Analysis for Land Monitoring (SEPAL), openEO, JEODPP, and pipsCloud have also been used. Given the pros and cons of these platforms for big data processing (e.g., Gomes et al. 2020), the future of remote sensing data analysis could be dominated by these cloud computing frameworks. Importantly, the increasing application of GEE in geospatial data and environmental monitoring by research and educational institutions in recent years raise important questions. Will cloud computing replace traditional remote sensing software in the processing of satellite Earth observation? Do we expect more computing platforms for big remote sensing data process? With strategic partnerships between large research funding organizations and the private

[1] Multi-scale volumetric information layer describing artificial spaces where people live, work, and enjoy life.

sector, more key technological innovations along these lines are expected. In what follows, cloud computing platforms for big data are elaborately discussed. The GEE platform is detailed and some examples of its usage are demonstrated and briefly discussed.

5.2 Cloud Computing Platforms for Big Data

5.2.1 Google Earth Engine Infrastructure

GEE (Google Earth Engine) is an open-source computational platform that provides an interface for application programming and aids easy access to several geospatial data and remotely sensed observations across the globe (Gorelick et al. 2017). As a cloud-based platform for planetary-scale geospatial analysis, GEE provides highly sophisticated computational capabilities that facilitate global environmental monitoring, be it climate change impacts on water resources or the environment in general. The GEE is a cloud-based computational platform that enables data harmonization, integration, and development of large-scale cloud computing resources for the advancement of Earth science and environmental monitoring applications on a broad range of scales and relevant disciplines. Several characteristics of the GEE makes it very unique. As an open-source platform, it currently hosts petabyte scales of more than 40 years of remotely sensed Earth observations (Kumar and Mutanga 2018). These observations are those from optical and geodetic platforms and include, Landsat, Sentinel products, MODIS, Gravity Recovery and Climate experiment, National Oceanographic and Atmospheric Administration Advanced Very High Resolution Radiometer, and Advanced Land Observing Satellite, and preprocessed geospatial data, among others. The GEE repository also consists of multi-resolution satellite climate observations (tropical rainfall measuring mission) and several model-generated outputs for hydrological applications, among which include the Global Land and Data Assimilation System. With Google's computational infrastructure, access to petabytes of publicly available Earth observation data, high-speed parallel processing and machine learning algorithms, and a plethora of libraries of Application Programming Interfaces (APIs) (Yang et al. 2020; Gorelick et al. 2017) is now possible even with a desktop computer.

The GEE combines both JavaScript API and a Python API for data management and analysis. For example, the web Integrated Development Environment (https://code.earthengine.google.com), which provides users with easy access to available data, routines, and real-time visualization of big data is based on a JavaScript. The Python API can also be used but is available through a JavaScript-like module structured. The capacities of GEE for big geospatial data management and analysis with respect to other cloud computing platforms have been evaluated. Several parameters, including data abstraction, data access interoperability, infrastructure replicability, storage and processing scalability, physical infrastructure abstraction, and extensibility, among others were used, and GEE appears to be relatively better across these metrics (Gomes et al. 2020). With this impressively great and remarkable advancement, GEE has indeed created the opportunity for the geoscience and remote sensing community not only to engage with Earth observation data but in sharing of computational resources and routines for problem solving. Apart from robust data sharing, GEE provides simplified user interface that boost interaction among users. For example, data visualization and basic analyses that allow users to implement simple routines are underpinned by the easy interface of GEE's explorer. The GEE application provides access to a client-side user interface API for developers to build and publish their own applications, and with over 800 functions for handling big data sets that both experts and non-experts can leverage and develop custom applications (e.g., Gomes et al. 2020; Tamiminia et al. 2020; Kumar and Mutanga 2018).

5.2.1.1 Google Earth Engine Code Editor

Users with valid Google accounts can use Google Earth Engine to run algorithms on geospatial data stored on Google's infrastructure. this can be achieved using the Code editor. As shown in Fig. 5.1, the Code Editor is an interactive environment for developing Earth Engine applications and is available at code.earthengine.google.com. The centre panel provides a JavaScript code editor where you can write and develop your scripts, run it, save, and clear the map. The Get Link button generates a unique URL for the script in the address bar. The map in the bottom panel contains the

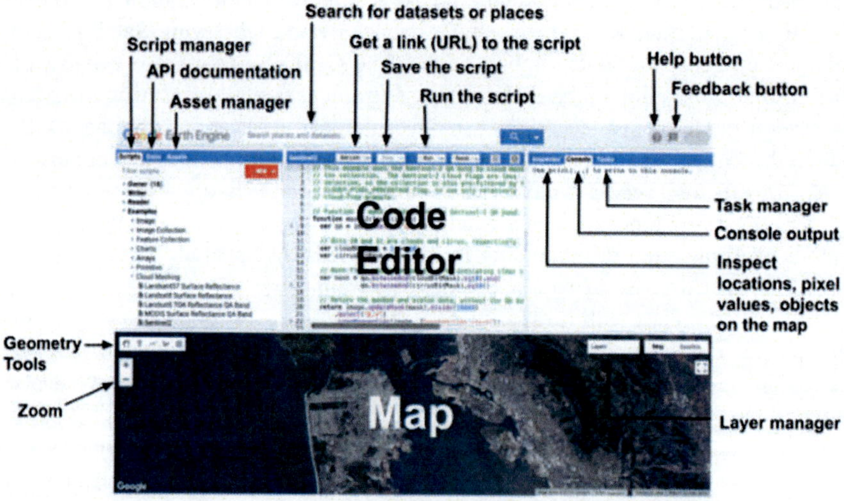

Fig. 5.1 The Google Earth Engine code editor. Source, https://developers.google.com/earth-engine/guides/playground

layers added by the script and can be switched to either a satellite or map mode. There is Search Box to enable access to all archived geospatial or remote sensing datasets. The left panel contains code examples, your saved scripts, a searchable API reference, and an asset manager for private data import and export. On the right panel there is an inspector for querying the map, an output console, and a manager for long-running tasks. The illustration in Fig. 5.1 shows key elements of the GEE code editor interface as summarized at https://developers.google.com/earth-engine/guides/playground.

1. JavaScript code editor.
2. Map display for visualizing geospatial datasets.
3. API reference documentation. This available on the Docs tab to the left.
4. Git-based Script Manager and is available on the Scripts tab on the left panel.
5. Console output is available on the Console tab on the right.
6. Task Manager (Tasks tab) to handle long-running queries (on the right panel).
7. Interactive map query (Inspector tab).
8. Search of the data archive or saved scripts.
9. Geometry drawing tools.

5.2.1.2 Google Earth Engine API

The process for signing up for Earth Engine up a GEE account is straightforward and simple and has been detailed online. (https://developers.google.com/earth-engine). With an email confirmation of your acceptance you will now have access to the Earth Engine Code Editor (Sect. 5.2.1.1), the IDE for writing Earth Engine JavaScript code in a web browser. The example in Fig. 5.2 shows a Java script developed with the aid

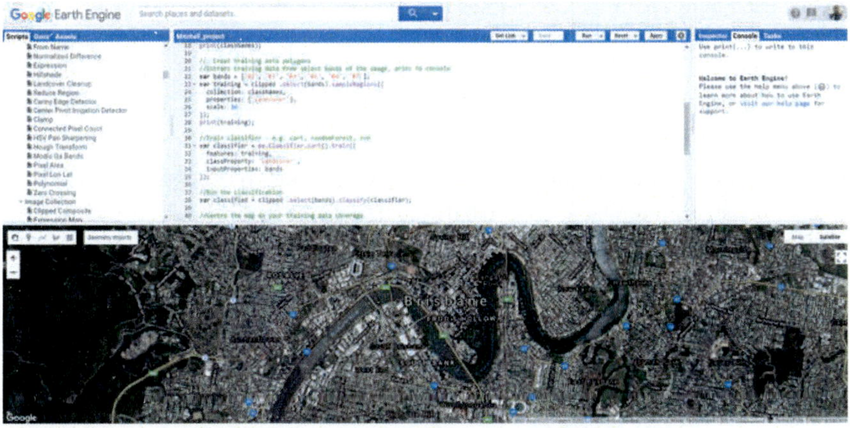

Fig. 5.2 The Google Earth Engine code editor interface

of JavaScript editor for land cover classification over Brisbane city. The JavaScript editor plays key role, which is detailed in the GEE online documentation and reproduced here (https://developers.google.com/earth-engine/guides/playground):

1. Helps with formatting and highlighting your code as you type.
2. identifies problems with codes, offer fixes and other hints for correct syntax.
3. Autocomplete pairs of quotes, brackets, and parentheses.
4. Provides hint on code completion for Earth Engine functions.

Typical exercises, guidelines, functionalities, computational resources, including APIs and custom JavaScripts to help new users have been documented in the online resources available at the Earth Engine Code Editor webpage (https://developers.google.com/earth-engine/guides/playground). As illustrated in Fig. 5.2, the classifier used in line 31 of the JavaScript is the Classification and Regression Trees technique that was taken from existing API documentation in the 'Doc'tab on the left panel of the Code Editor (Fig. 5.1). There are myriads of JavaScript API documentation for use in several analysis and applications. For example, in satellite image processing and analysis, the *ee.ImageCollection* script helps in stacking times series of satellite Earth observation data like Landsat, Sentinel, etc. Several processing operations can be applied to these images, e.g., they can be filtered by boundaries (*filterBounds()*), dates (*filterDate()*), sorted for clouds (.sort('CLOUD_COVER')), and added as a map using *Map.addLayer*. In the case of multi-spectral images like Landsat and Sentinel any three different band combinations (Any of visible-red, green, blue, or Near Infrared and Short wave infrared bands) can be used to display a colour composite in Earth Engine. For example, if you add Landsat 8 OLI image scene to the map, you can have different composites based on your band combinations. GEE provides an easy platform to compute different remote indicators like the Normalized Difference Vegetation Index (NDVI). In estimating NDVI based on Landsat 8 OLI in GEE over a specific region, e.g., Kabul city in Afghanistan for the 2013–2020 period, here are simple steps, which can be coded as detailed below. Note that each bullet point below defines each line of the code.

Step 1: We will retrieve all Landsat 8 OLI collections over Kabul city starting from 2013 to 2020. A polygon shapefile representing the boundary of Kabul city will be used to limit these collections over our region of interest (roi). The JavaScript begins with declaring a variable written as 'var'.

- var collection = ee.ImageCollection('LANDSAT/LC08/C01/T1_SR');
- .filterBounds(roi).filterDate('2013-01-01', '2020-12-31');
- print(collection);

Step 2: In this step, you will use an existing API to create a function that will mask clouds, cloud shadows, and snow. Thereafter, the 'map' function will be applied to each image in the collection in the Landsat time series. In this function, we will make a new single band image from the pixel qa band and then retain clear (0) and water (1) pixels

- var maskClouds = function(image)

- var pixelqa = image.select('pixelqa');
- return image.updateMask(pixelqa.eq(322));

Step 3: The output 'maskClouds' can now be applied to the image collection variable ("collection") as

- var Landsatmasked = collection.map(maskClouds);
- print(Landsatmasked)

- var maskClouds = function(image)
- var pixelqa = image.select('pixelqa');
- return image.updateMask(pixelqa.eq(322));

Step 4: Using Landsat 8, NDVI is calculated using information from the near infrared (NIR) and red (R) bands, which are designated as bands 5 and 4, respectively. Note that when using other satellite data such as Sentinel, SPOT and even data from past Landsat missions (e.g., Enhanced thematic mapper plus), these bands are designated differently. A function to add NDVI using NIR (B5) and the red band (B4)is then created as

- var getNDVI = function(img)
- return img.addBands(img.normalizedDifference(['B5','B4']).rename('NDVI'));;

and the NDVI function can be mapped over the entire collection as

- var Landsat8NDVI = Landsatmasked.map(getNDVI);

Step 5: Since Landsat has a 16-day temporal resolution, for each pixel, we will select the "best" set of bands from available images based on the maximum NDVI/greenness as

- var Kablucomposite = Landsat8NDVI.qualityMosaic('NDVI').clip(roi);

Step 6: The mean NDVI composite can be added to the map as illustrated in Fig. 5.3 using the 'Map.addLayer' to display and 'Map.setCenter' to centre the displayed NDVI over your region of interest.

- Map.setCenter(69, 34.5, 9);
- var ndviParams = min: -1, max: 1, palette: ['blue', 'white', 'green'];
- Map.addLayer(Kablucomposite, ndviParams, 'Mean NDVI');

where ndviParams is the predefined visualization parameters for the NDVI. The spatial distribution of NDVI values are based on the ndviParams, which shows low ('red'), moderate ('yellow'), and high ('green') NDVI values. An alternative approach is to select the individual bands and directly calculate the NDVI as shown below.

- var nir = image.select('B5');
- var red = image.select('B4');
- var ndvi = nir.subtract(red).divide(nir.add(red)).rename('NDVI');

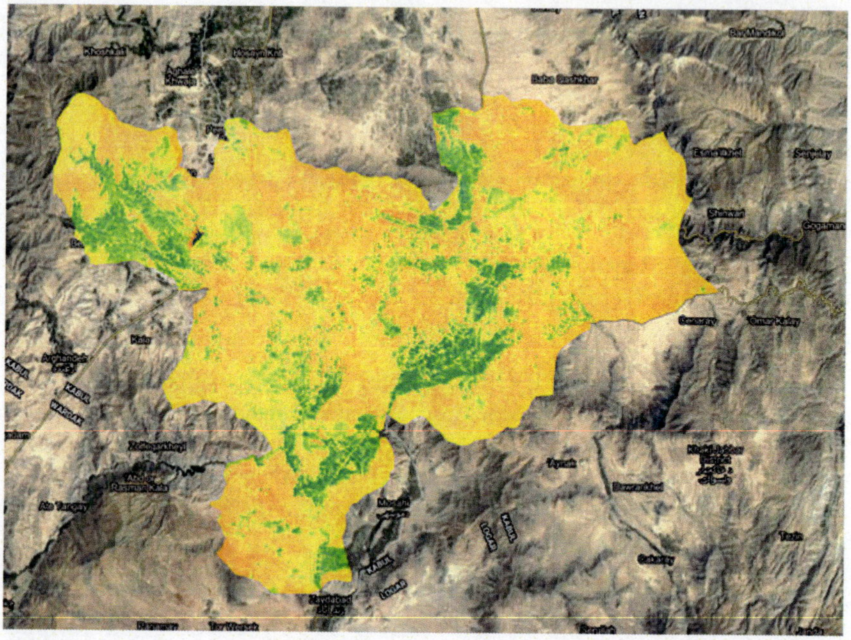

Fig. 5.3 Normalized Difference Vegetation Index over the Kabul region in Afghanistan using the Google Earth Engine. Greener areas have high NDVI values (ranging from 0.7 to 1.0) and are indicated as green on the map

where 'B5' and 'B4' are the near infrared (nir) and the red bands, respectively. The NDVI (i.e., ndvi) map can be centred over your region of interest using the JavaScript indicated above.

Moreover, aerial averaged time series of NDVI can also be plotted with the GEE platform. The GEE provides several time series analysis tools and this includes a function to plot a time series generated from satellite images. The capability to plot time series comes from the 'ui.Chart.feature' module, which contains a set of routines for rendering charts from *Feature* and *FeatureCollection* objects. Several choices can be made using any set of available routines to determine the arrangement of data in the chart or graph, i.e., what defines x- and y-axis values and what defines the series.

Step 7: To plot the corresponding time series of NDVI over Kabul (Fig. 5.3), the following lines can be added to the code as

- var NDVIgraph = ui.Chart.image.series(
- imageCollection: Landsat8NDVI.select('NDVI'),
- region:roi,reducer:ee.Reducer.first(),
- scale:30).setOptions(title:'NDVI time series (2013-2020)',);
- print(NDVIgraph);

Figure 5.4 shows the time series of changes in mean NDVI over Kabul city during the 2013–2020 period. Missing data points are those where the Landsat 8 collections

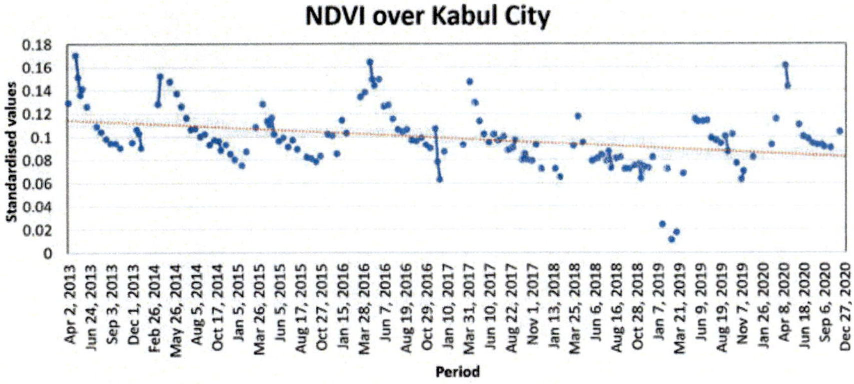

Fig. 5.4 Time series of Normalized Difference Vegetation Index (2013–2020) over the Kabul region in Afghanistan using Google Earth Engine

did not meet the criteria in Steps 2 and 3 above. In addition to having the capability for linear regression analysis based on the least squares estimate of a linear function of one variable with a constant term, one can also fit a trend line to time series based on least squares estimated slope. Trend analysis is important to show a variable of interest is increasing or decreasing and by how much. By fitting a trend line to the NDVI time series (Fig. 5.4), we can see that there is a decreasing trend in vegetation greenness over the period. This can be due to human activities such as land use change where agricultural and forested areas are converted to residential areas or used for other purposes. Another reason for this observed decline in NDVI could be due to decline in long term precipitation. Even though rainfall conditions in some areas have not sufficiently explained changes in surface vegetation greenness, other regions show that the latter is largely determined by both the combined influence of human activities and the impacts of climate change (Ndehedehe et al. 2019). The GEE platform thus provides a quick and easy to use technique to assess land surface conditions on a small and large scale.

5.2.1.3 Access to Other Geospatial Layers

Apart from allowing users to run geospatial analysis on Google's infrastructure, the Google Earth Engine computing platform also provides an interface to retrieve existing geospatial data layers and datasets. In the next example, a range of drainage dataset is retrieved from the HydroSHEDS repository. The HydroSHEDS data provide hydrographic information for regional and global-scale applications in a consistent format, offering a range of geo-referenced datasets (vector and raster) at various scales, including river networks, watershed boundaries, drainage directions, and flow accumulations. HydroSHEDS is based on elevation data obtained in 2000 by NASA's Shuttle Radar Topography Mission. An example of watershed data (elevation and

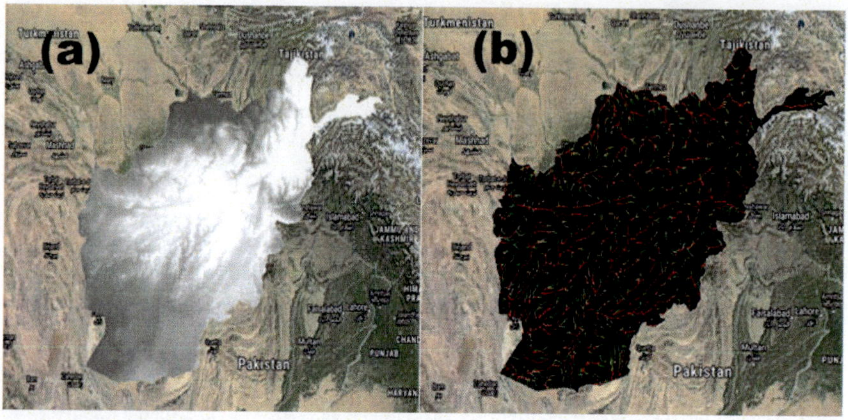

Fig. 5.5 Watershed data over Afghanistan retrieved from the Global hydroshed data using Google Earth Engine. **a** elevation dataset from NASA's Shuttle Radar Topography Mission and **b** flow accumulation dataset

flow accumulation) retrieved from the HydroSHEDS repository using GEE is shown in Fig. 5.5.

Step A: Elevation data retrieval. The itemized steps below shows how the elevation data in Fig. 5.5a was extracted from the HydroSHEDS data.

- var dataset = ee.Image('WWF/HydroSHEDS/30CONDEM');
- var elevation = dataset.select('b1').clip(Afghanistan);
- var elevationVis = min: -50.0, max: 3000.0, gamma: 2.0,;
- Map.addLayer(elevation, elevationVis, 'Elevation');
- Map.setCenter(69, 34.5, 9);

As highlighted earlier in Step 1 of Sect. 5.2.1.2, a polygon shapefile representing the boundary (extent) of Afghanistan will be used to clip our region of interest (roi) to limit our data retrieval to Afghanistan. In point item 2 of step B, 'b1' is the layer of interest representing elevation and 'Afghanistan' is the region of interest. The visualization parameter is defined as 'elevationVis' and 'Map.addLayer' function is used to display the elevation data indicated in Fig. 5.5a. In Step A above, the raster data or layer containing both the elevation and flow accumulation data is 'WWF/HydroSHEDS/30ACC'

Step B: This is the flow accumulation data retrieval.

- var Flowdata = ee.Image('WWF/HydroSHEDS/30ACC');
- var flowAccumulation = Flowdata.select('b1').clip(Afghanistan);
- var flowAccumulationVis = min: 0.0, max: 500.0,
- palette: ['000000', '023858', '006837', '1a9850', '66bd63', 'a6d96a', 'd9ef8b', 'ffffbf', 'fee08b', 'fdae61', 'f46d43', 'd73027'],;
- Map.setCenter(69, 34.5, 9);
- Map.addLayer(flowAccumulation, flowAccumulationVis, 'Flow Accumulation');

Noticed that visualization parameter for the flow accumulation data ('flowAccumulationVis') in step B is different from that of elevation. Visualization parameters for any layer of interest are usually defined based on the inherent data range. Figure 5.5b shows the flow accumulation over Afghanistan. The colour palette used in the visualization of the flow accumulation data is used to represent the range of flow accumulation values over Afghanistance and can be accessed from html colour codes available online (https://htmlcolorcodes.com/). The elevation dataset (Fig. 5.5a) extracted over Afghanistan is the result of an iterative conditioning and correction process such that Endorheic basins (inland sinks) are 'seeded' with a no-data cell at their lowest point in order to terminate the flow (Lehner et al. 2008). This elevation dataset is at 30 arc-second resolution. The flow accumulation dataset on the other hand defines the amount of upstream area (in number of cells) draining into each cell. The number of accumulated cells is a measure of the upstream catchment area with values ranging from 1 at topographic highs (river sources) to very large numbers at the mouths of rivers (Lehner et al. 2008). For instance, in Fig. 5.5b, flow accumulation values along main river channels or other forms of water bodies like lakes (red) have considerably higher values (more than 65,000) compared to the high elevation areas where values are 1 (blue) and less than 200 (green) in some areas. As with the elevation dataset, the flow accumulation data is at 30 arc-second resolution.

5.2.2 *The Australian Geoscience Data Cube*

The Australian Geoscience Data Cube (AGDC) is a 'Big Data' infrastructure designed to hold Australia's growing Earth observation data. It was an initiative following Geoscience Australia's collaboration with key government agencies, the National Computational Infrastructure (NCI), and the Australian Commonwealth Scientific Industrial Research Organization (Lewis 2017). The Data Cube, which has been revised and now referred to as Open Data Cube, is an open-source platform made available through the Apache License, Version 2.0. This has enabled easy access with support and cooperation by several international organizations, including The Committee on Earth Observation Satellites among other Australian federal agencies. The AGDC code is written in Python and freely available (https://github.com/data-cube/agdc-v2), having benefited from a plethora of open-source libraries for file attribute and database access, data creation and access, managing workflows, array handling, for geospatial data access, and analysis (Lewis 2017). Among several benefits provided such as pre-processing of satellite images to surface reflectance and storing higher level products within a high-performance data environment, it also provides support for multiple spatial resolutions and reprojection to allow data integration. Although the AGDC is scored high in data abstraction extensibility, unlike the GEE, it has some limitations, including medium capability in processing scalability (e.g., Gomes et al. 2020).

5.2.3 OpenEO

To address the challenges of several users of Earth observation data in data management and migration to cloud-based processing platforms, the openEO platform was created to simply the process. As an open application programming interface, it simplifies the process of connecting programmes like R, Python, and JavaScript to big Earth observation cloud back-ends (e.g., Gomes et al. 2020).

5.2.4 pipsCloud

Another new computing platform for big remote sensing data is pipsCloud (Wang et al. 2018). It is a proprietary solution that is owned by Chinese research institutions for the management and processing of large volumes of remotely sensed data (e.g., Gomes et al. 2020). With the increasing availability of large spatial coverage of remote sensing observations on several multi-temporal and multi-spectral scales, pipsCloud was proposed to address computation cost associated with large processing of Earth observation for regional environmental and disaster monitoring. As with other aforementioned cloud-based platforms, it has a web interface where dynamic data processing, querying, and retrieval for environmental monitoring can all be performed. These capabilities are delivered through the use of a Hilbert-R+-based data indexing and its architecture is fully described in Wang et al. (2018). The pipsCloud platform is known to be efficient, offering highly optimized performance cloud environment for big Earth observation data where multi-source level one or raw remote sensing and geospatial data products can be managed and accessed using robust data management services (Wang et al. 2018). In the handling of data processing on the server-side, pipsCloud uses C++ code templates for developing applications but lacks the functionality to export data using Open Geospatial Consortium standards. (e.g., Gomes et al. 2020; Wang et al. 2018). In addition to the non-reproducibility of science, a key limitation of pipscloud is that it is a closed platform that limits users to only institutions that participated in the project, and the source code is not available for use by other institutions. As further detailed by Gomes et al. (2020), the performance or capabilities of pipsCloud based on several metrics is low compared to other cloud-based platforms for planetary scale geospatial data analysis.

5.2.5 Sentinel Hub

The Sentinel is another repository of satellite data (Sentinels, Landsat, and among others) providing easily data accessible for analysis within a cloud geospatial environment. The Sentinel hub (https://www.sentinel-hub.com/) is not completely open access with functionality that limits user access based on different payment plans (e.g., Gomes et al. 2020). For instance, subscribers have access to data through OGC protocols and APIs that enable data processing, mobile application data access, higher resource access limits, and technical support while the open access restrict users to viewing, selection, and data download (https://www.sentinel-hub.com/pricing/). The similarity between GEE and Sentinel Hub platforms is embedded in the transparency in processing protocols though the use of a non-pixel-by-pixel processing approach is limited. Although other commonalities exist with other platforms, e.g., high abstraction of the physical infrastructure used for data storage and processing, Sentinel hub has some challenges (its reproducibility), which have been detailed in Gomes et al. (2020).

The Sentinel satellite products are owned by European Space Agency and they have great potentials in advancing research in remote sensing of the environment. With several state-of-the-art satellite missions of Sentinel (e.g., Sentinels 1–5) available on higher spatial and temporal resolutions since its launch, The Copernicus Open Access Hub (previously known as Sentinels Scientific Data Hub) is now available to provide complete, free, and open access to these Sentinel products, even for traditional users who cant use cloud computing platforms (https://scihub.copernicus.eu/). The Copernicus Australasia Regional Data Hub was therefore created as part of an institutional arrangement and collaboration with European Copernicus Programme to make its data available for applications in the Australian context. Under this arrangement and with support from several other multi-national agencies, the Data Hub provides free, open, trusted, and reliable access to Sentinel satellite data for Australasia, South-East Asia, the South Pacific, the Indian Ocean, and the Australian Antarctic Territory (http://www.copernicus.gov.au/). To this end, the Sentinel Australasia Regional Access (SARA) was designed as the main access point to the Hub data. SARA's interface is user-friendly with a download capability and a Python Application Programme Interface (API) for advanced user interaction (http://www.copernicus.gov.au/data-access). However, SARA is considerably limited compared to analytical platforms such as GEE where one can develop custom applications and processing chains data processing and analysis.

5.2.6 System for Earth Observation Data Access, Processing, and Analysis for Land Monitoring

The System for Earth Observation Data Access, Processing, and Analysis for Land Monitoring (SEPAL) is a cloud-based computing platform for geographical data processing resulting from FAO's collaboration with Norway (https://github.com/openforis/sepal/). SEPAL is a cloud-base computing platform developed for the automatic monitoring of land cover and uses Google Earth Engine as well as Amazon Web Services Cloud (AWS), and other open access software (e.g, GDAL, RStudio, R, SNAP Toolkit). The platform enables rapid access to users to process large amount of data without high network bandwidth requirements and as with GEE, high-performance computing infrastructure is not a requirement. This can be attributed to its use of the Google Earth Engine infrastructure for data processing and retrieval, and support by the Amazon Web Services Cloud for data storage and computation (Gomes et al. 2020). The SEPAL platform is freely available and can be accessed using a dedicated web portal (https://sepal.io).

5.3 Applications in Remote Sensing of the Environment

Big technological platforms like Google's computational infrastructure (i.e., the GEE platform) are increasingly being used as a major data repository for all Earth observation data, including NASA and European Space Agency satellite products. This platform has been very handy and helpful to researchers and students who use different Earth observation products to share computational resources and develop Application Programming Interface for additional data processing and analysis. Plans by relevant agencies are underway to integrate other data hubs with GEE to further facilitate access and use of these resources. Consequently, the interest in the use of GEE in monitoring and understanding changes in the environment is growing and is obviously evident in the myriads of emerging remote sensing studies that have recently employed GEE computational platform (e.g., Tamiminia et al. 2020; Kumar and Mutanga 2018). These studies arguably, benefited from the large repository of important APIs to aid pre-processing of satellite images to surface reflectance, visualizing, and analyzing of big data.

5.3.1 Detection and Multi-temporal Analysis of Surface Water Change

An expert system compatible with the GEE platform was developed by Pekel et al. (2016) for the parallel processing of three million Landsat images to quantify long term changes in global surface water. Apparently, a stand alone computer with central processing unit lacks the processing capability to sustain this kind of data processing. As Pekel et al. (2016) noted, it '*would have taken 1,212 years, but using 10,000 computers in Earth Engine the processing was completed in around 45 days*'. Under the funding and auspices of the European Joint Research Centre, Global Surface Water Dataset was generated from this project and is available in several geospatial layers containing maps of the location and distribution of surface water. The data provides statistics on the extent and change of water surfaces across the globe based on a 35-year historical Landsat data processed within the GEE environment using an expert system. From the compendium of inundation statistics embedded in this data, one can processed water occurrence (inundation frequency), change intensity, or transitions of surface water for any location of choice using new sets of routines. Along this line of epic implementation of parallel processing of big data, an automatic rule-based superpixel approach has been developed to generate monthly temporal and 10 m spatial scale surface water dynamics in France based on GEE cloud computing (Yang et al. 2020). Clearly, GEE leverages on Google's computational infrastructure, thus making the parallel processing of big data less cumbersome and with reduced computational time. Several studies have used this platform in assessing surface water distribution in lakes and river basins. Two examples from the Lake Chad and Murray–Darling Basins following the work of Ndehedehe et al. (2020b) and Ndehedehe et al. (2021a) are given below.

1. In the Murray–Darling Basin, a multi-temporal change detection analysis to assess the response of surface water distribution to hydro-climatic fluctuations was implemented Ndehedehe et al. (2021a) using the GEE. This was achieved through the classification of Landsat image over the basin into areas where open water gain and open water loss were observed. Areas where open water persisted were also mapped to understand water transitions and how extreme events regulate the distribution of open water in time and space during the period (2008–2012). This period was chosen because the period that preceded the big wet period of 2011/2012 was an extreme drought period for most Australian territories. Using this period as a case study over the MDB, open water transitions and distribution were estimated. The change detection result showed more surface water in the water years of 2010/2011 (when the wet condition began) and 2011/2012 (when the wet condition peaked) (Fig. 5.6a). Much of this water was gained during the water years of 2009/2010 (37,014,525.00 Ha) and 2010/2011 (39,848,031.00 Ha), contributing to the sustained rise in surface water distribution during the period (Fig. 5.6a). The temporal variations of surface water during 2008–2012 period confirm the impacts of strong anomalies in rainfall and climate variability on surface water (Fig. 5.6b). Apparently, hydrological conditions in terms of surface water extents in the dry 2008 (41,411,979 Ha) and 2009 (34,100,181 Ha) are in sharp contrast to the wet 2011 (95,930,370 Ha) and 2012 (101,890,566 Ha) periods (Fig. 5.6b). The rise in surface water extent starting in 2010 with peaks in 2012 coincided with the rise in TWS, GWS, and soil moisture in Australia in the same period.
2. Multi-temporal analysis of surface water change has been undertaken in the vicinity of the Lake Chad in West Africa. Both Landsat 8 and Sentinel-2 (S2) multi-spectral data were retrieved and processed using the Google Earth Engine (GEE) API. The S2 imagery downloaded from GEE was Top of Atmosphere multi-spectral data acquired during the wet season months of May to September of 2016 to 2019 while 2015 images were those of Landsat 8 OLI (Ndehedehe et al. 2020b). The open water maps of the southern pool in Lake Chad Basin were generated for 2015–2019 using Sentinel-2 by applying the modified Normalized Difference Vegetation Index (Fig. 5.7). Accuracy assessment for these maps was

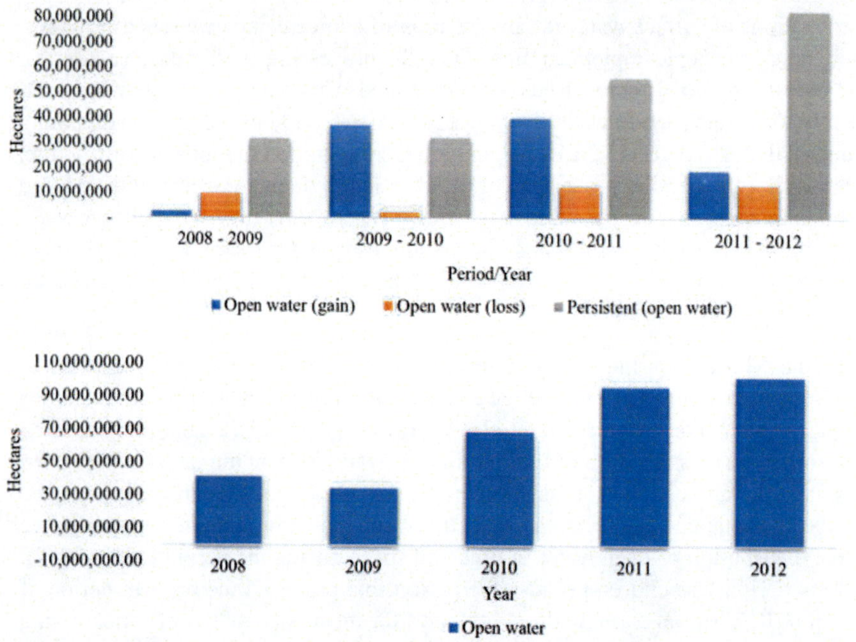

Fig. 5.6 Temporal changes in open water surface distribution between 2008 and 2012 in the Murray–Darling Basin showing, **a** change detection of open water features and **b** the temporal variations of surface water. Source Ndehedehe et al. (2021a)

based on kappa statistics. To this end, validation points were generated across the study area on the satellite image based on random stratified design (Fig. 5.8). The spatial extent of Sentinel-derived open water feature for the Lake Chad Basin (i.e., the Lake Chad and its environ) showed increase during the 2015–2017 period before a gradual decline in 2018 and 2019 (Fig. 5.7). Surface extent rose from 153,127.66 Ha in 2015 to 156,909 Ha in 2017. Thereafter, it fell to 153,523 Ha and 152,379 Ha in 2018 and 2019, respectively. The GEE platform provides as well as enhances the handling of big remote sensing data. For instance, the number of images overlapping the study area for each of these years was 284 in total. Processing these images with a desktop computer would have been computationally expensive and will require computers with advanced processing capabilities.

5.3.2 *Land Cover Mapping*

With increasing land cover changes in most regions where transitions to agricultural purposes are popular, space-based remote sensing is one of the most reliable approaches to obtain information detailing these changes. Be it monitoring of natural

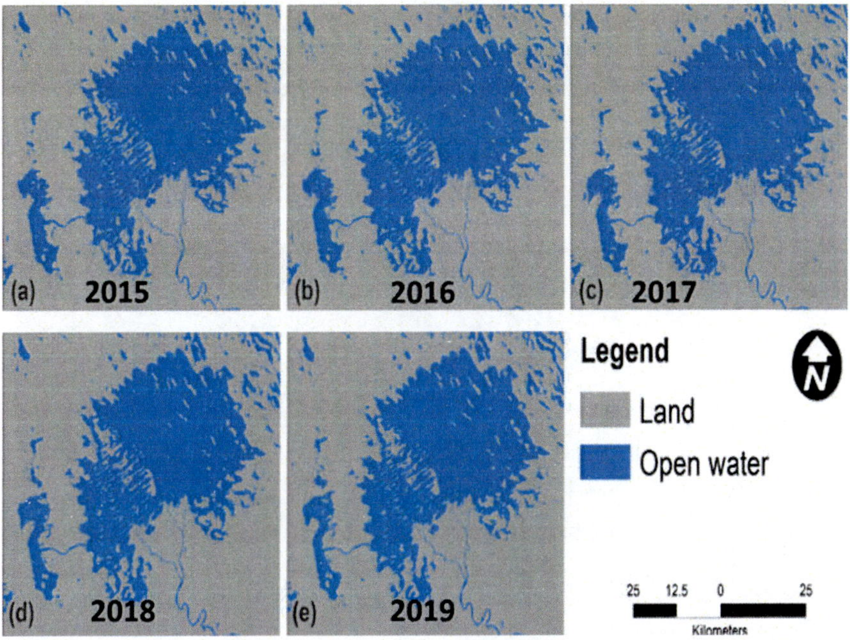

Fig. 5.7 Surface water distribution in the southern Lake Chad and environ in West Africa between 2015 and 2019. Source Ndehedehe et al. (2020b)

resources or land cover states, remote sensing observations offer valuable opportunities in rapid large-scale land resource mapping. This is largely because of their repetitive and large spatial coverage and availability. However, over vast and complex landscapes, detailed national scale land cover mapping is usually difficult due to the large amount of satellite data required. Adding to the cumbersome nature of this data, the limitations in desktop processing and framework to facilitate such data preprocessing are other key challenges that remote sensing experts are confronted with. Big geospatial data processing platform like the GEE offers excellent opportunity to process, store, and manage such data. The use of GEE in land cover applications is gradually emerging, especially in updating existing land cover map at a national scale. Using multi-temporal satellite data (Sentinel products), this potential of GEE in updating national land cover maps without the need for ground-truth data from the target year (Ghorbanian et al. 2020). Several other GEE applications in the mapping of land cover states include, forest degradation, river ice monitoring, wildfire severity mapping, wetland classification, cropland classification, burned area mapping, and urban land mapping (Amani et al. 2021; Ghorbanian et al. 2020).

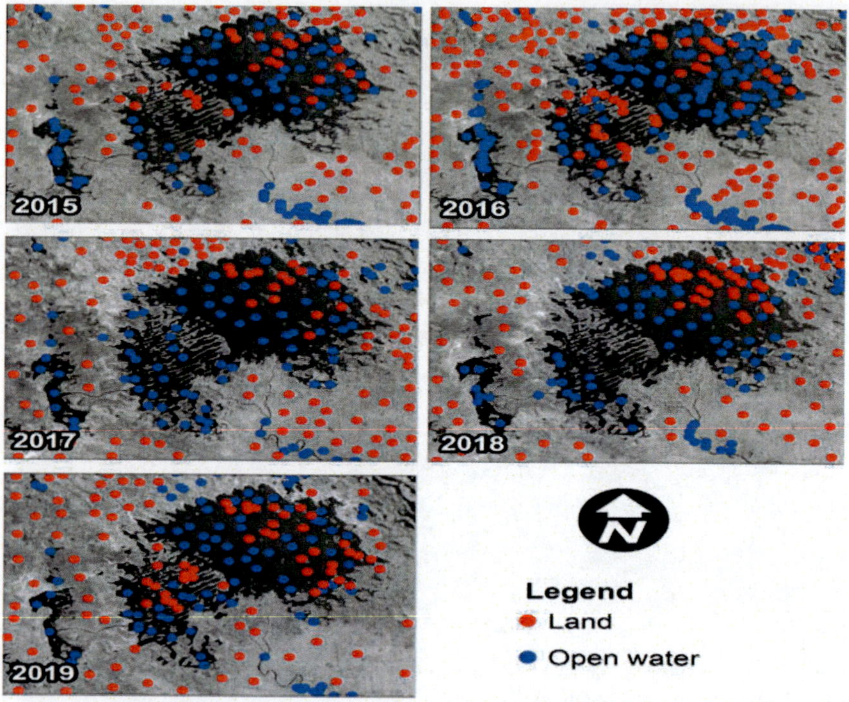

Fig. 5.8 Random stratified design of validation points used for accuracy assessment (background – Sentinel-2 NIR band – 2015–2019). Source Ndehedehe et al. (2020b)

5.3.3 Extreme Events: Wild Fires, Floods, and Droughts

Environmental change is arguably the aftermath of climate change and human actions on multiple spatial and temporal scales. As further discussed by Ndehedehe et al. (2021a), improving knowledge of this change, especially in the context of droughts, floods, and wildfires is required to predict environmental hazards, increase resilience to climate change, and ensure sustainable management of water resources. Climate is changing, and the rise in drought frequency and severity and their intensities would mean increased risk of food and water security, wildfires, and other environmental and life-threatening hazards (e.g., Ndehedehe et al. 2020c; Yuan et al. 2019; Spinoni et al. 2014). The occurrence of wildfire is increasing with more than a thousand events being reported annually across several regions of the world (Gibson et al. 2020; Seydi et al. 2021). Their far-reaching impacts on human lives, including the alteration of ecosystem structure, soil erosion, destructing wildlife habitats, flooding, and economic and ecological damages have been reported (Gibson et al. 2020). The mapping of disasters like wild fires is thus warranted and of substantial interest to environmentalist in supporting post-fire management and improved decision-making that prioritizes emergency for response. Their mapping can enhance the understanding of ecological implications of fire thus providing insights to key predictors of wildfire characteristics. Over Australia where wild fire events are prevalent, regions affected by wildfire have been quantified using Sentinel-2 imagery and Moderate Resolution Imaging Spectroradiometer (MODIS) products within the GEE cloud computing platform (Seydi et al. 2021). Utilizing automated framework (e.g., rule-based decision) in a GEE environment, global mapping and spatial distribution of wild fire has been undertaken using Landsat and Sentinel data (Barboza Castillo et al. 2020; Seydi et al. 2021). The development of novel remote sensing indicators or spectral indices such as the Normalized Burn Ratio, Normalized Burn Ratio2, and Mid-Infrared Burn Index, and the NDVI is making this possible. The use of GEE is now facilitating the routine monitoring of wildfires and burnt vegetation or forest cover as well as their reliable burnt area estimates. Moreover, there is also an increasing application of GEE in the mapping of large-scale flood events. An example is the floodplain inundation mapping of the Okavango Delta in Botswana. Floods and inundation extents for the peak flood season at a high spatial resolution (30 m) for the Okavango Delta were quantified using Landsat imagery in GEE (Inman and Lyons 2020). In the Lake Chad Basin, the impacts of droughts on surface extent have been quantified using the GEE to analyse multi-temporal Landsat images (Ndehedehe et al. 2020b). This change detection analysis of Lake Chad's surface extent showed that decreasing drought extents and intensities coincided with increasing surface water extent. The increasing applications of GEE platform in monitoring these extreme events are directly linked to key merits of the platform such as high-speed parallel processing, cloud-based computing scheme that does not require data download and its user-friendly interface.

5.3.4 *Assessment of Floodplain Productivity*

Tropical floodplains and wetlands are well known for their valuable ecosystem functions and services. Some of these services include improving water quality, shoreline protection, processing of carbon and other nutrients, food security, and the support of a broad range of aquatic biodiversity (e.g., Slagter et al. 2020; Ndehedehe et al. 2020a). Their mapping is thus essential to improving knowledge on the link between ecological impacts of human intervention and climate change influence on wetland hydrology. This knowledge can be directly incorporated into policy frameworks and used in conservation planning to optimize proactive water resources management. The use of GEE-based method in the mapping of Canadian Wetland proved not only to be computationally efficient but saved time and resources (Amani et al. 2021). Furthermore, recent reports have highlighted the use of GEE to improving understanding of the critical flow needs to sustain freshwater ecosystems by predicting impacts of future development on important ecosystem linkages between the river and its floodplain wetlands. One of the key focus of this study was to pioneer the use of GEE in assessing the productivity of floodplain ecosystems, and mapping hotspots of biomass accumulation in large floodplain river catchments in tropical

Queensland (Ndehedehe et al. 2021b). The GEE repository and infrastructure was used to show how remote sensing vegetation and inundation metrics can be computed and combined to create binary layers representing maps of hotspots of primary producers based on a sequence of logical operations. This application becomes a 'new tool in the box' for conservation planning and wetland management, which require spatial estimates of the aquatic habitats needed to maintain aquatic biodiversity (Ndehedehe et al. 2020d). This information is also vital to help inform decision-makers about water allocation that both enables agricultural development and protects environmental assets.

5.3.5 *Multidisciplinary Applications in Environmental Geoscience*

The preponderance of peer-reviewed journal articles shows a common pattern that suggests pioneering applications of the GEE platforms in accessing and processing Earth observation data were in the field of remote sensing (Fig. 5.9). These applications have increased exponentially in recent times, resulting in approximately 200 journal articles in remote sensing journals in 2020. GEE usage is still largely dominated by experts in remote sensing. Since its inception, Landsat data is the most widely used and the largest component of the GEE data portal (Kumar and Mutanga 2018). With a 100% rise in published remote sensing journal outputs in 2020 as against 2019 (Fig. 5.9), GEE is indeed the fastest growing cloud-based platform in human history in terms of usage. This is clearly indicated in the extensive application of GEE in a broad range of disciplines such as social sciences, environmental sciences engineering, biochemistry, agriculture, and among other fields (Fig. 5.10).

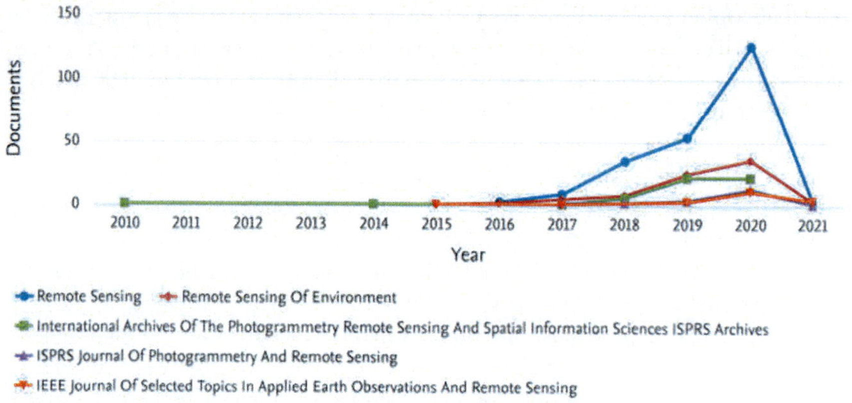

Fig. 5.9 Inter-annual variations of peer-reviewed publications in remote sensing journals that employed Google Earth Engine platform based on Scopus journal outputs of 2021

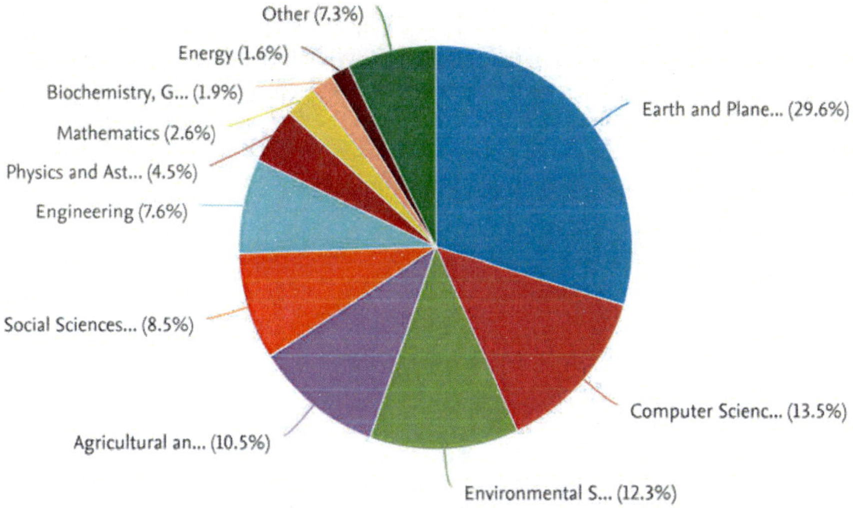

Fig. 5.10 Google Earth Engine-related publications in discipline-specific areas. Data is based on Scopus outputs of 2021

Apparently, GEE applications in these disciplines are very substantial with Earth and planetary sciences (29.6%), environmental sciences (12.3%), and agriculture (10.5%) leading. While the applications of GEE are considerably diverse in terms of subject areas (Fig. 5.10), GEE users from developed nations, be it user nationality or geographic location, were found to top the list (Kumar and Mutanga 2018). Notably, the multidisciplinary application of GEE shown in Fig. 5.10 is largely within environmental geoscience. The contributions of GEE in advancing global environmental monitoring through the use Earth observation are very substantial. Such progress involving the use of big geospatial data is evident in the application of various geospatial products in the last few years in several disciplines, including forest and vegetation, land cover, ecosystem and sustainability, eco-hydrology, agriculture, climate science, and urban studies among others. For example, 90% of studies used remote sensing datasets while others employed already processed products such as Normalized Difference Vegetation Index for vegetation, crop, land cover mapping, and drought monitoring (Tamiminia et al. 2020). There are no indications this trend will change soon because the relative ease in access to preprocess or ready-made data (e.g., vegetation indices, surface inundation, and model outputs), and undertaking large-scale geospatial analyses with limited computational resources tend to be attracting more experts and stakeholders into the GEE community. Another reason for this is the implementation of several machine learning algorithms and advanced statistical methods for geospatial data analysis and image pre-processing and classification within the GEE platform. For example, remote sensing image classification algorithms such as support vector machines, artificial neural network, Classification and Regression Trees, Decision Tree classifier, Naïve Bayes, K-nearest neighbour, and Random Forest are all available GEE scripts and routines to aid image analysis. Moreover, in addition to several other image processing functions such as band ratio techniques, feature extraction, image mosaicking, cloud, and error detection, routines for regression analysis are also embedded in the list of available GEE JavaScripts. These factors have contributed to optimizing image processing and analyses, thus making the GEE a sophisticated platform for big data analyses in environmental geoscience.

References

Amani M, Brisco B, Mahdavi S, Ghorbanian A, Moghimi A, DeLancey ER, Merchant M, Jahncke R, Fedorchuk L, Mui A, Fisette T, Kakooei M, Ahmadi SA, Leblon B, LaRocque A (2021) Evaluation of the Landsat-Based Canadian wetland inventory map using multiple sources: challenges of large-scale wetland classification using remote sensing. IEEE J Sel Top Appl Earth Obs Remote Sens 14:32–52. https://doi.org/10.1109/JSTARS.2020.3036802

Barboza Castillo E, Turpo Cayo EY, de Almeida CM, Salas López R, Rojas Briceño NB, Silva López JO, Barrena Gurbillón MÁ, Oliva M, Espinoza-Villar R (2020) Monitoring wildfires in the Northeastern Peruvian Amazon using Landsat-8 and Sentinel-2 imagery in the GEE platform. ISPRS Int J Geo-Inf 9(10). https://doi.org/10.3390/ijgi9100564

Craglia M, de Bie K, D J et al (2012) Digital earth 2020: towards the vision for the next decade. Int J Digit Earth 5(1):4–21. https://doi.org/10.1080/17538947.2011.638500

Ghorbanian A, Kakooei M, M A et al (2020) Improved land cover map of Iran using Sentinel imagery within Google Earth Engine and a novel automatic workflow for land cover classification using migrated training samples. ISPRS J Photogramm Remote Sens 167:276–288. https://doi.org/10.1016/j.isprsjprs.2020.07.013

Gibson R, Danaher T, Hehir W, Collins L (2020) A remote sensing approach to mapping fire severity in south-eastern Australia using sentinel 2 and random forest. Remote Sens Environ 240:111702. https://doi.org/10.1016/j.rse.2020.111702

Gomes VCF, Queiroz GR, Ferreira KR (2020) An overview of platforms for big earth observation data management and analysis. Remote Sens 12(8). https://doi.org/10.3390/rs12081253.

Gorelick N, Hancher M, Dixon M, Ilyushchenko S, Thau D, Moore R (2017) Google Earth Engine: Planetary-scale geospatial analysis for everyone. Remote Sens Environ 202:18–27. https://doi.org/10.1016/j.rse.2017.06.031

Inman VL, Lyons MB (2020) Automated inundation mapping over large areas using landsat data and Google Earth Engine. Remote Sens 12(8). https://doi.org/10.3390/rs12081348

Kumar L, Mutanga O (2018) Google Earth Engine applications since inception: usage, trends, and potential. Remote Sens 10(10). https://doi.org/10.3390/rs10101509.

Lehner B, Verdin K, Jarvis A (2008) New global hydrography derived from spaceborne elevation data. EoS 89(10):93–94

Lewis K (2017) Understanding climate as a driver of food insecurity in Ethiopia. Clim Change 144(2):317–328. https://doi.org/10.1007/s10584-017-2036-7

Ndehedehe CE, Burford MA, Stewart-Koster B, Bunn SE (2020) Satellite-derived changes in floodplain productivity and freshwater habitats in northern Australia (1991–2019). Ecol Indic 114:106320. https://doi.org/10.1016/j.ecolind.2020.106320

Ndehedehe CE, Ferreira VG, Agutu NO (2019) Hydrological controls on surface vegetation dynamics over West and Central Africa. Ecol Indic 103:494–508. https://doi.org/10.1016/j.ecolind.2019.04.032

Ndehedehe CE, Ferreira VG, Agutu NO, Onojeghuo AO, Okwuashi O, Kassahun HT, Dewan A (2021) What if the rains do not come? J Hydrol 595:126040. https://doi.org/10.1016/j.jhydrol.2021.126040

Ndehedehe CE, Ferreira VG, Onojeghuo AO, Agutu NO, Emengini E, Getirana A (2020) Influence of global climate on freshwater changes in Africa's largest endorheic basin using multi-scaled indicators. Sci Total Environ 737:139643. https://doi.org/10.1016/j.scitotenv.2020.139643

Ndehedehe CE, Haile GG, Agutu NO, Ferreira VG, Getirana A, Okwuashi O (2020) Hydrological hotspots of climatic influence in Brazil: a two-step regularization approach. Atmos Res 246:105116. https://doi.org/10.1016/j.atmosres.2020.105116

Ndehedehe CE, Onojeghuo AO, Stewart-Koster B, Bunn SE, Ferreira VG (2021) Upstream flows drive the productivity of floodplain ecosystems in tropical queensland. Ecol Indic. https://doi.org/10.1016/j.ecolind.2021.107546

Ndehedehe CE, Stewart-Koster B, Burford MA, Bunn SE (2020) Predicting hot spots of aquatic plant biomass in a large floodplain river catchment in the Australian wet-dry tropics. Ecol Indic 117:106616. https://doi.org/10.1016/j.ecolind.2020.106616

Okwuashi O, Ndehedehe CE, Attai H (2020) Tide modeling using partial least squares regression. Ocean Dyn 70:1089–1101. https://doi.org/10.1007/s10236-020-01385-1

Pekel J-F, Cottam A, Gorelick N, Belward AS (2016) High-resolution mapping of global surface water and its long-term changes. Nature 540:418–422. https://doi.org/10.1038/nature20584

Seydi ST, Akhoondzadeh M, Amani M, Mahdavi S (2021) Wildfire damage assessment over Australia using sentinel-2 imagery and MODIS land cover product within the Google Earth Engine cloud platform. Remote Sens 13(2). https://doi.org/10.3390/rs13020220.

Slagter B, Tsendbazar N-E, Vollrath A, Reiche J (2020) Mapping wetland characteristics using temporally dense Sentinel-1 and Sentinel-2 data: a case study in the St. Lucia wetlands, South Africa. Int J Appl Earth Obs Geoinf 86:102009. https://doi.org/10.1016/j.jag.2019.102009

Spinoni J, Naumann G, Carrao H, Barbosa P, Vogt J (2014) World drought frequency, duration, and severity for 1951–2010. Int J Climatol 34(8):2792–2804. https://doi.org/10.1002/joc.3875

Tamiminia H, Salehi B, Mahdianpari M, Quackenbush L, Adeli S, Brisco B (2020) Google Earth Engine for geo-big data applications: a meta-analysis and systematic review. ISPRS J Photogramm Remote Sens 164:152–170. https://doi.org/10.1016/j.isprsjprs.2020.04.001

Wang L, Ma Y, Yan J, Chang V, Zomaya AY (2018) pipsCloud: high performance cloud computing for remote sensing big data management and processing. Future Gener Comput Syst 78:353–368. https://doi.org/10.1016/j.future.2016.06.009

Yang X, Qin Q, Yésou H, Ledauphin T, Koehl M, Grussenmeyer P, Zhu Z (2020) Monthly estimation of the surface water extent in France at a 10-m resolution using Sentinel-2 data. Remote Sens Environ 244:111803. https://doi.org/10.1016/j.rse.2020.111803

Yuan X, Wang L, Wu P, et al (2019) Anthropogenic shift towards higher risk of flash drought over china. Nat Commun 10:4661. https://doi.org/10.1038/s41467-019-12692-7

Part IV
Remote Sensing Ecohydrology

Chapter 6
Remote Sensing of Freshwater Habitats

Abstract The monitoring of floodplain productivity and dynamics in freshwater habitats, be it, riverine (e.g., rivers streams), lacustrine (large water holes in low elevation areas), or palustrine (swamps, intermittent floodplain water channels) systems is required to predict the impacts of hydro-meteorological fluctuations and the consequences of changes in flows on floodplain wetland ecosystems. This is essential among other things, for proactive water resources planning and the development of climate change mitigation and resource management strategies. However, such monitoring and assessment are complicated by the inaccessibility of many large wetland systems during times of inundation, particularly in remote parts of the world. This can make in-situ sampling difficult at a time when high levels of aquatic primary producers are generating food and energy sources for higher order consumers. This chapter provides some highlights on the concept and satellite applications in ecohydrology, leveraging the outcome from the work undertaken by Ndehedehe et al. (2020a) over a large river floodplain in the Gulf region of northern Australia. This chapter also highlights critical aspects of mapping and monitoring changes in freshwater habitats and floodplains using satellite remote sensing. The overarching goal is to improve understanding of the link between ecological implications resulting from human intervention and climate change on wetland hydrology.

6.1 Introductory Remarks

Freshwater systems, including rivers, lakes, streams, and floodplain ecosystems, and vegetated wetlands are important environmental resources, which offer several ecosystem services and functions (e.g., Keddy et al. 2009). They are highly valued because of the important roles they play in tourism and recreation, improvement of water quality, provision of shelter for fish and native wildlife, drought relief for wildlife, flood water storage, as well as support for a range of terrestrial and aquatic biodiversity (e.g., Ward et al. 2014; Chen et al. 2014b; Tockner et al. 2010; Gidley 2009). Wetlands have been recognized as high priorities for conservation and management (e.g., Midwood and Chow-Fraser 2010) due to these multiple strings of ecological, cultural, and economic values and services that they provide. Studies

C. Ndehedehe, *Satellite Remote Sensing of Terrestrial Hydrology*,
https://doi.org/10.1007/978-3-030-99577-5_6

that look into the productivity of these systems are gradually emerging with focus on improving understanding of (i) their response to several stressors and impacts on ecosystem processes (e.g., Davranche et al. 2010; Tockner et al. 2010) and (ii) key biogeochemical processes on floodplains such as the interactions of green house gas with the atmosphere (e.g., Ward et al. 2014).

During periods of high amplitude floods and high flows, the plenitude of higher order organisms such as invertebrates, fish, and waterfowl; and aquatic plants like macrophytes and algae result in highly productive habitats and food web (e.g., Kingsford et al. 2014; Gidley 2009; Balcombe and Arthington 2009). However, droughts, human impacts (e.g., construction of dams and development of water infrastructures), and environmental factors like pollution threaten freshwater habitats, altering the composition of aquatic biodiversity in streams and riverine systems (e.g., Kingsford et al. 2014; Bunn and Arthington 2002). The observed global trends in degradation and destruction of floodplain wetland ecosystems, disruption of several physical processes and alteration of flow connectivity, that sustain primary and quaternary producers who depend on these freshwater habitats (e.g., Ward et al. 2013; Davranche et al. 2010; Bunn et al. 2006; Bunn and Arthington 2002) are some evidence of existential threats to these freshwater systems. This chapter provides some highlights on the concept and applications in remote sensing eco-hydrology leveraging the work by Ndehedehe et al. (2020a) over a large river floodplain in the gulf region of northern Australia. Important outcomes of this chapter include highlighting the use of satellite remote sensing monitoring framework to (i) assess changes in floodplain inundation in space and time, (ii) quantify the distribution of hot spots of primary producers in freshwater habitats, and (iii) assess key constraints on freshwater habitats and the growth of primary producers. The overarching goal is to improve understanding of the link between ecological implications resulting from human intervention and climate change influence on wetland hydrology.

6.2 Changes in Floodplain Wetlands and Freshwater Habitats

6.2.1 Freshwater Habitats: Mapping and Monitoring

Freshwater supports the sustainability of eco-hydrological systems and several environmental functions. With the provision of habitats to several flora and fauna, freshwater contributes to the preservation and nourishment of biodiversity in wetland ecosystems. It's a key component of the hydrologic cycle and provides numerous functions, including drinking water, irrigation agriculture, hydro-power, and transportation, among others. Although other forms of freshwater exist, surface water is crucial to several aspects of our daily life as well as being the most accessible. For instance, it impacts on the exchange of heat, gas, and water vapour between the land surface and atmosphere, driving the various mechanisms behind the movement, migration distribution of biodiversity in the human planet. Monitoring its variability is thus important to understand the impacts of global climate change and in guiding future expectations of droughts and unfavourable climatic conditions. However, its monitoring using conventional approaches is limited and satellite technologies are filling such gaps. With the advent of Earth resources satellites, its monitoring and mapping as evolved and more sophisticated approaches to quantifying freshwater, including surface water dynamics have emerged. This section will focus on the advances and progress made in surface water monitoring using remote sensing. Given its efficiency, repetitive coverage, high temporal and spatial resolution, and access to freely available satellite data, remote sensing data sets provide the opportunity to assess changes in surface water in time and space. For example, in large semi-arid domains like Australia, the potential and utility of satellite observations to monitor changes in surface water extent provides a capability that underpins water resources management for sustainable development and understanding climatic processes and several human actions that can limit its availability.

As one of the driest continents in the world, Australia's water requirement is one of the highest. While this results in substantial usage and reliance on groundwater resources, surface water availability is therefore a key aspect of its environmental and soci-economic prosperity. Several monitoring efforts have been in place with extensive surface water monitoring gauged networks distributed across the country. However, satellite technology has been deployed as part of key monitoring framework over Australia's vast mainland areas. The water observation from space product that leveraged 25 years of Landsat Imagery to map open water features across Australia is a good example (Mueller et al. 2016). With improved water detection algorithms now being routinely used in quantifying water occurrences on landscape, this effort has progressed further, resulting in the improved monitoring of spatial locations and extents of waterbodies across Australia (Krause et al. 2021). This advancement in continental or country-wide surface water monitoring is now a service application that can be accessed in the Digital Earth Australia's (DEA) data portal (http://maps.dea.ga.gov.au/). As shown in Fig. 6.1a, this service application is user-friendly, providing opportunities to download historical (since 1987) time series of water-related pixels in any location within the country with a simple click. Both water associated pixel counts and percentage of surface area observed as wet can be downloaded (Fig. 6.1b, c). The locations of waterbodies across Australia were mapped for every available Landsat satellite image, making it possible to provide a time series of water pixels for areas where water is present more than 10% of the time and is larger than 3125 m^2 (Krause et al. 2021). As indicated by Krause et al. (2021), about 295,906 polygons of waterbodies of varying sizes (ranging from 3125 m^2 to 4820 km^2) were generated over the Australian landscape. The illustration in Fig. 6.1 shows one location in the northern territory and the other water body in south Australia (Fig. 6.1b, c) where time series of changes in surface water extents of these lakes have been retrieved. Notably, the DEA waterbodies consists of natural and man-made waterbodies, including rivers, lakes, water supply dams, on farm storages, wetlands, oxbow lakes, and ponds. The Landsat programme is currently an ongoing mission and with Landsat 9 scheduled for launch in September 2021, this means that changes in surface

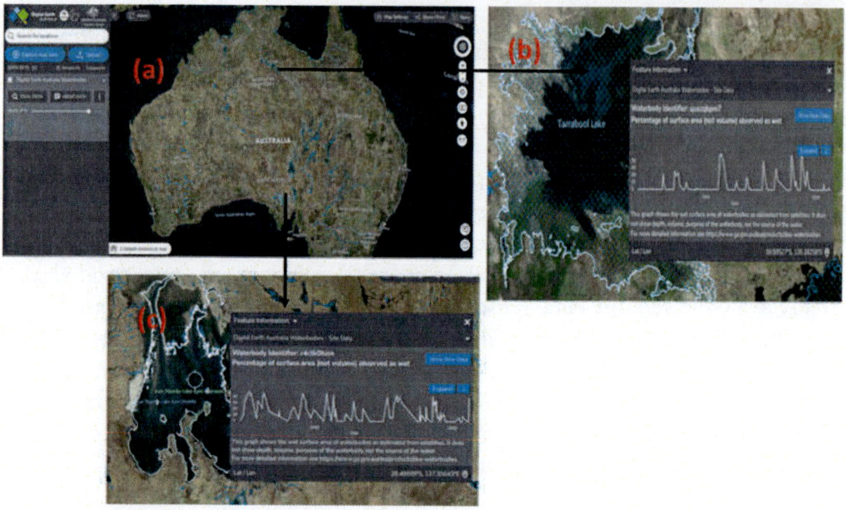

Fig. 6.1 Surface water monitoring over Australia from space. *Source* Digital Earth Australia

water can be continuously monitored even in near-real time, given its 16-day temporal resolution or frequency. As demonstrated in the Australian case, such changes can be placed into the context of historical surface water dynamics, facilitating change detection applications and providing insight into water resources availability. The DEA algorithm uses an objective, coded workflow, making it possible to integrate new observations and re-run it to provide updates on water presence (Krause et al. 2021).

On a global scale, Pekel et al. (2016) pioneered the use of expert system to quantify global surface water relying on about 35 years of Landsat data. Several surface water characteristics and map layers, including surface water occurrence, surface water seasonality, water recurrence as well as water transitions were quantified using satellite observations. Their work relied on or used acquired observations of historical Landsat 5, 7 and 8 satellites. With the support of a cloud computing platform, an expert system was developed to classify water-related pixels from three million Landsat satellite images Pekel et al. (2016). In Fig. 6.2a–d, the water transitions based on the global surface water data product (Pekel et al. (2016)) are illustrated for the Congo and Murray–Darling Basins. The maps on the right hand show polygons representing different transitions (permanent, new permanent, lost permanent, seasonal, seasonal to permanent, etc.) of water as indicated on the legend (Fig. 6.2b, d). The proportion of these transitions is also captured in the pie charts on the left (Fig. 6.2a, c).

The proportion of permanent water body in the last 3 decades over the Congo Basin (39.7%) as opposed to the Murray–Darling Basin (MDB) (5.6%) shows how typical climate features regulate surface water availability. Comparatively, the MDB is dry compared with the humid wet Congo Basin, which experiences up to eight

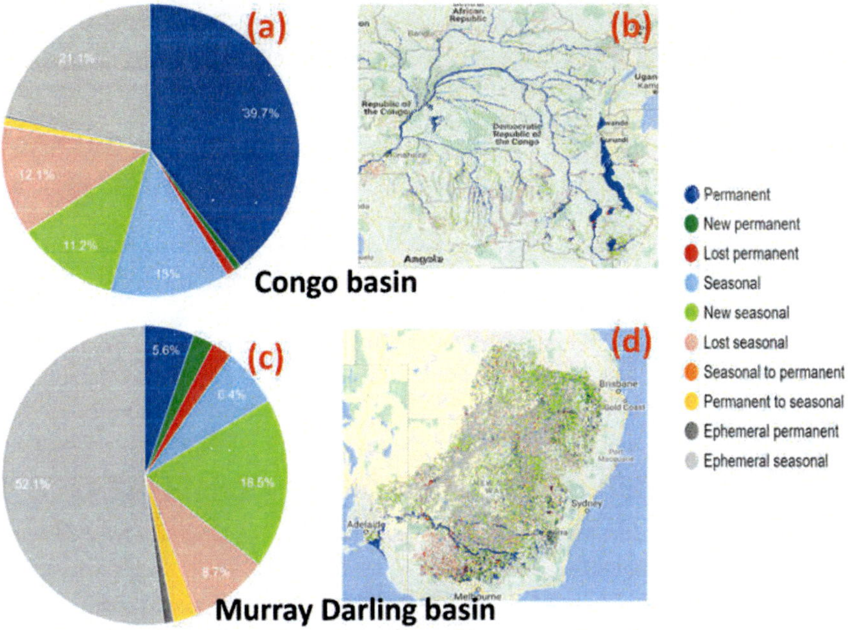

Fig. 6.2 Water transitions in the **a**, **b** Congo and **c**, **d** Murray–Darling Basins reprocessed from the global surface water data (March 1984 to October 2015

months of substantial rainfall every year (Ndehedehe et al. 2018). The spatial patterns of these water classes show wider spread of surface water distribution the Congo Basin than MDB (Fig. 6.2b, d). The MDB is largely characterized and dominated by ephemeral and seasonal changes in surface water. While this illustration summarizes the characteristics of surface water in extreme humid climate (Congo Basin) and semi-arid domains (Murray Daring Basin), it underscores the changing nature of water in time and through space. This can be the result of extreme climate events such as droughts and floods. Human actions such as water extraction and harvesting for irrigation agriculture can also impact on surface water availability and its distribution. On the one hand, the construction of new infrastructures to hold water for agricultural purposes could create the presence of water on the landscape in one location. These are cases with 'new permanent water' features and the proportion corresponding to this in the MDB is higher unlike in the Congo Basin (Fig. 6.2a, c). In the MDB, this 'new permanent water' features are more likely to be associated with the expansion of irrigated agriculture during the last three decades. However, in the Congo Basin where agriculture is largely rainfed, these 'new permanent water' features could be related to the development of dams for hydro-power purposes. On the other hand, water extraction from reservoirs and lakes during periods of rainfall deficit could lead to the disappearance of water in another location. Although not shown here, this is the case in Lake Chad where climate change and prolonged periods of rainfall deficits resulted in an empty lake or substantial loss of water availability (See Chapter 20). As further indicated in Fig. 6.2a, c, there have been a permanent water loss in both basins. Surface water can permanently disappear or be lost due to the composite influence of climate and human activities. This is currently the case in the Lake Chad and Lake Urmia Basins, which is further discussed in Chap. 12. Will these lakes ever returned to their status quo even if the climate becomes favourable? Furthermore, seasonal availability of water like the inundation of downstream floodplains and intermittent streams, channels, and water holes which persist can be permanently lost when there is prolonged drought. However, they can also transition to permanent water features when there is an increasing trend in rainfall or when flows are maintained. With rising population and unmitigated human actions contributing to climate change, surface water will continue to be considerably liable to change rapidly and unpredictably.

6.2.2 Remote Sensing Indicators for Detecting Surface Water

6.2.2.1 Landsat Missions

Earth resources satellites like the Landsat missions provide a wealth of information that supports an improved understanding of environmental processes and global climate change. Policy change and several other human advancements have benefitted from improved sensor technologies and sophistication in these missions. The first generation of space-borne Earth observation was given birth during the 1960–1972 and this includes CORONA, ARGON, etc. After this period came the Landsat systems (1972–1986), Earth resources satellites that were experimental and tentative in the nature of their application (second generation). While these satellite technologies were largely owned by the US, the third generation of satellite systems included the French-owned (SPOT system) and the European Space Agency ERS-1 systems (1986–1997). The fourth generation is largely characterized by high-resolution and hyperspectral imaging (1997–2010) such as the Ikonos, Quick Bird, and WorldView-3. This current age is characterized by new generation of satellite Earth resources for real-time environmental analysis of Earth. This is the age of Landsat 8, ICEsat, and the various multispectral Sentinel (1–6) missions of European Space Agency as well as radar imaging systems. These Sentinel missions have broad applications including, land monitoring, measuring sea-surface topography, sea and land surface temperature, ocean and land colour, air quality and weather monitoring, among others.

With the development and progress in the several Landsat missions (Fig. 6.3), a robust archive of historical images and satellite observations are now available to assist in understanding Earth and environmental processes and change that has occurred in space and time. In particular, this made the global observations of water from space possible as well as other regional or continent-wide applications of surface water detection (Krause et al. 2021; Pekel et al. 2016; Mueller et al. 2016).

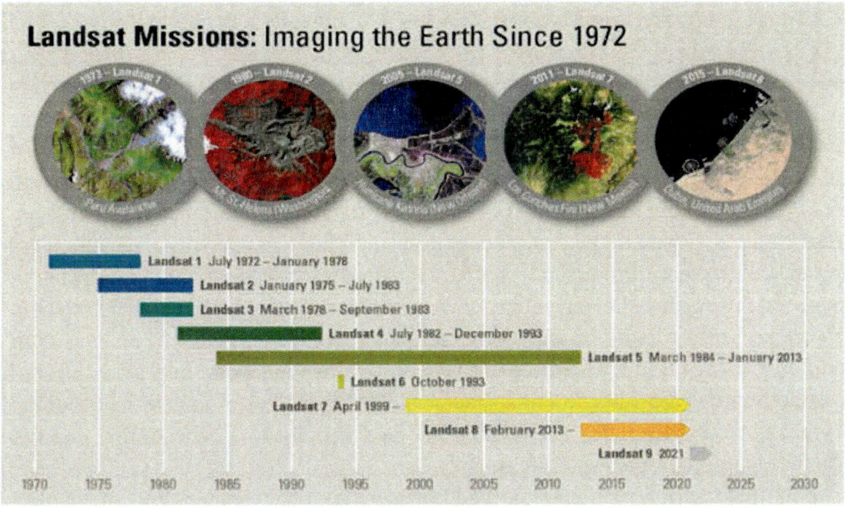

Landsat Mission Timeline.

Fig. 6.3 Landsat 9 mission. *Source* NASA/USGS

As indicated in Fig. 6.3, future satellite programmes like the Landsat 9 mission that seek to build on the current capabilities of existing Landsat missions (Landsats 1-8) will revolutionize satellite applications in water resources management, either in terms of improved radiometric and spatio-temporal accuracy as well as number of spectral channels. For instance, the Coastal/Aerosol Band (Band 1) in Landsat 9 will be optimized to aid the measurement of chlorophyll concentrations (ocean colour) in coastal regions. Phytoplankton, tiny plant-like organisms that live in surface waters are the major source of chlorophyll.

Additionally, this spectral channel will also be handy to help in the detection and mapping of aerosol. Moreover, as with Landsat 8, Landsat 9, will collect data for two distinct thermal infrared spectral regions. Both Landsat missions are however similar in their operational capabilities and will complement each other. They are more sensitive and reliable than previous Landsat missions due to their innovative design. They use a pushbroom scanner and this architecture gives them more dwell time over the ground, resulting in a better signal-to-noise ratio. What does this mean? More on-ground nuanced details can be seen and also detected by computers. This and several other sophistication that comes with the new Landsat programmes are available in NASA's home page. However, with this wealth of satellite-based information and sophistication comes the burden to improve feature extraction capabilities from satellite images to aid specific applications, be it in drought and flood monitoring or water resources management.

6.2.2.2 Remote Sensing Metrics

Several remote sensing metrics have been developed to aid the assessment and extraction of open water features from satellite imageries. The science of detecting surface water from optical remote sensing system or monitoring its fluctuations in terms of water level height or variation has been ongoing for a long time now (Alsdorf et al. 2007; Alsdorf and Lettenmaier 2003). However, given the importance of monitoring surface water dynamics as a key aspect of informing policies and water management, new indicators were introduced to help quantify surface water extent from satellite images. In this regard, the normalized difference water index was introduced to help map the presence of surface water on landscape (McFeeters 1996). This was later modified by Xu (2006) who used information from the Green and Mid-Infrared region to improve the detection of surface water. This index was called the Modified normalized difference index (MNDWI). The use of indicators is becoming popular in feature extraction from satellite images.

As with extraction of inundated and wet surfaces, the combination of low red reflectance with high (Near Infrared) NIR reflectance has been used to assess the water-driven changes in surface vegetation or healthy vegetation using the well known Normalized Difference Vegetation Index (NDVI). The use of indicators is also helpful in the mapping of wildfires and urban area estimation from remote sensing. The latter uses NIR and Mid Infrared while the former relies on NIR and Short wave Infrared (SWIR) to calculate the normalized burn ratio index. This index calculates burn severity and new methods based on Sentinel data like the Burned area for Sentinel 2 (Filipponi 2018) are emerging to improve the estimates of areas affected by fire. Radar satellites are also increasingly being used in the estimation of these variables. The Sentinel-1 Synthetic Aperture Radar data combined with a deep learning method has been developed to detect burnt areas, highlighting potential in near real-time wildfire monitoring (Ban et al. 2020). Remote sensing data has been very handy not only in the monitoring of surface water changes but in the mapping of lakes and reservoir as well as monitoring of glaciers and snows. For instance, MODIS data has been combined with Sentinel-1 radar observations to generate a monthly wet–dry snow map (Amani et al. 2020). Several approaches to the use of remote sensing data in water extraction are discussed below.

1. Normalized difference indicators: Moreover, the monitoring of surface water with remote sensing indicators is still evolving as they are particularly useful in flood and inundation mapping. Arguably, the MNDWI (Xu 2006) is usually a preferred metric to detect the presence of water from satellite images. It is very mainstream and has been well embraced by the remote sensing community and in several freshwater-based assessments, including flood and inundation mapping, aquatic, and other remote sensing hydrology applications that leverage satellite observations (e.g., Tulbure and Broich 2019; Masocha et al. 2018; Fisher et al. 2016; Davranche et al. 2010). As earlier mentioned it was an improvement of the earlier work of McFeeters (1996) and users find it not only simple but computational

Table 6.1 Remote sensing indices used in water resources and vegetation studies

Indicators	Algorithm	References
Modified Normalized Difference Water Index	MNDWI = $b_G - b_{MIR}/b_G + b_{MIR}$	Xu (2006)
Normalized Difference Water Index	$NDWI_{Gao} = b_{NIR} - b_{MIR}/b_{NIR} + b_{MIR})$	Gao (1996)
Optimized Soil Moisture Adjusted Index	OSAVI = $b_{NIR} - b_R/b_{NIR} + b_R + 0.16$	Rondeaux et al. (1996)
Automated Water Extraction Index-$_{sh}$	AWEI$_{sh}$ = $\rho b_1 + 2.5 * \rho b_2 - 1.5 * (\rho b_4 + \rho b_5) - 0.25 * \rho b_7$	Feyisa et al. (2014)
Automated Water Extraction Index-$_{nsh}$	AWEI$_{nsh}$ = $4 * (\rho b_2 - \rho b_5) - (0.25 * \rho b_4 + 2.75 * \rho b_7)$	Feyisa et al. (2014)

efficient in the identification and definition thresholds. In other words, after applying this metric, one can easily conclude that regardless of water properties and landscapes, values greater than one represent water-related pixels. In the identification of spectral channels required to compute water from satellite images, one must be careful to make the right choice. For example, in Landsat 5 TM, the green and MIR bands correspond to bands 2 and 5, respectively. But in the Landsat 8 OLI, these bands (green and MIR) are now bands 3 and 6, respectively (Table 6.1). It is unclear how these bands will be categorized or named in the anticipated Landsat 9 mission. However, in the multi-spectral Sentinel-2 mission, these bands are designated as bands 3 (green) and 11 (MIR or SWIR1). Prior to applying these bands in mapping inundation or quantifying open water presence, it is always important to look at the metadata of each satellite image to decipher what the individual band designations are. As we see in the successive Landsat programmes, the improvement in spectral channels from 7 bands in the ETM plus mission to 11 bands in Landsat 8 OLI resulted in changes in the nomenclature of previous band designations. The MNDWI appears ideal and appeals to several users, being spectral-based automated water extraction and simple, similar to other two-band remote sensing indices like the NDVI.

2. Multi-band indicators: New remote sensing indicators that are multi-band have also been developed to improve surface water detection on complex environmental landscapes (e.g., Khandelwal et al. 2017; Feng et al. 2016; Fisher et al. 2016; Feyisa et al. 2014; Davranche et al. 2013). The multi-band-derived Automated Water Extraction Indices ($AWEI_{nsh}$ and $AWEI_{sh}$) of Feyisa et al. (2014) are a very competitive indicator in this category and in relation to MNDWI. They have been summarized in Table 6.1. Feyisa et al. (2014) outlined in detail the environmental conditions, rules to guide decisions on the suitable application of these indices and which particular index should be used in a given location or situation. The threshold values of these indices are optimal seems to be rather

stable similar to the MNDW. Several studies have used this method and it was also employed over the Flinders catchment in northern Australia. As noted in the study area, the thresholds employed to detect the presence of water on landscape were values greater than -0.005 for both MNDWI and the AWEI following a visual comparison and data from site visits. The variability in threshold values has also been acknowledged and potential threshold values have been suggested depending on the metrics used (Worden and de Beurs 2020).

In principle, threshold values greater than zero are usually employed for detecting inundated surfaces from these indicators. In practice nonetheless, this could vary and sometime subjective depending on regions and perhaps the physical properties of water. In Australia, for example, water bodies are mostly brownish and in the wet–dry tropics characterized by large water holes and intermittent streams, water could be highly turbid in those areas. This is contrast to wet humid regions like the Amazon and Congo Basins where it is largely blue or green or a combination of both colours based on true colour composites of satellite image and high spatial resolution data like the Google Earth images. This is because water is a highly variable target and its spectral characteristics are predisposed to changes in space and time. The presence of chlorophyll concentration, suspended solids, dissolved organic matter, depths and bottom materials drive these changes and properties of water in different locations (Pekel et al. 2016). This makes the consistent discrimination of water from other surfaces across different regions a challenge. This is further compounded by the impacts of turbidity, water pollution, and other surface conditions (e.g., shallow waters), making visual analytic and expert knowledge of the target variable very crucial in the process of water detection from space. This is important to help the community of experts who use remote sensing data to decide on what the appropriate thresholds are given the characteristics of water in their region of interest. Table 6.1 summarizes some of these remote sensing metrics, which are based on the normalized surface reflectance of spectral bands (ρb_1(blue), ρb_2(G), ρb_4(NIR), ρb_5 (MIR or SWIR1), and ρb_7(SWIR2). The accuracy and performance of these indicators in the extraction of water-related pixels can be assessed using kappa statistics (Congalton and Green 2009). Existing data or high spatial resolution images like the Google Earth image can be used as referenced data to validate the predicted inundation data. For instance, digitized polygons of water features and non-water features from the Google Earth image can be used as reference data. This can then be compared with the open water or estimated water pixel using the indices, making it possible to generate a confusion matrix for accuracy assessment.

3. The thresholding of a single band (SWIR) is also another method to detect surface water from satellite images: This method was pioneered by Wolski et al. (2017) over the large wetlands of the Okavango Delta. The two bands and multi-band indicators discussed earlier seem to do well in open water mapping and not in complex landscapes like large wetlands in the Okavango Delta in southern Africa. The Delta covers approximately 22,000 km^2 alluvial fan, and water flows occur through a system of discontinuous channels and connected floodplains interspersed by islands (Wolski et al. 2017). The complexities of climate

and land surface conditions as well as its hydro-geologic uniqueness and the low propagation of floods through the wetlands of the Okavango Delta warranted new approaches to classifying water-related pixels in wetlands. Wolski et al. (2017) argued that inundation maps generated through the use of single SWIR band were superior to those predicted from two-band indices like the NDWI and MNDWI and also provided good accuracy and a suitable method for automated classification of water from satellite images. Although the study by Wolski et al. (2017) used Moderate Resolution Imaging Spectroradiometer (MODIS) and National Oceanic and Atmospheric Administration Advanced Very-High-Resolution Radiometer imagery, a follow-up study (Inman and Lyons 2020) in the same region based on historical Landsat data was undertaken using the same method. Inundation extent was underestimated with Landsat compared to the 500 m spatial resolution of MODIS used by Wolski et al. (2017), which resulted in higher number of water pixels and overestimation of total inundated areas (Inman and Lyons 2020). Even though the single band thresholding of the SWIR band appears to be simple, the accuracy of this approach could be relative to the spatial resolution of data used. A key advantage, however, is that this technique can be fully automated as well as its capability in flagging uncertain classifications using the value of internally determined spectral separation distance.

4. The progress in near real-time detection of open water features, be it in wetlands, rivers, lakes, or inland water bodies has been substantially enhanced by reliance on Synthetic Aperture Radar (SAR) satellite data. The use of European Space Agency's Sentinel-1 product in this regard as well as TerraSAR-X, Cosmo SkyMed, and Radarsat-2 has shown significant improvement in the mapping and classification accuracy of floods. The processing chains and conceptual framework for SAR applications in near real-time flood mapping have been detailed in Twele et al. (2016).

The performance of several well-known inundation metrics has been undertaken. The study by Fisher et al. (2016) focused on comparative analyses of these indices to provide an accurate method for automated water classification across different regions. Using surface reflectance Landsat TM/ETM+/OLI imagery across eastern Australia, seven water extraction metrics, including the two AWEI products, tasselled cap wetness, MNDWI, NDWI, and a water index created using canonical variates analysis were assessed. The composition of the validation pixels used contributed to the accuracy of each index and all indicators performed well across all water and non-water pixel types in the Landsat images. Nevertheless, issues regarding water detection in some areas is still a challenge. For example, Fisher et al. (2016) noted that clear-deep terrestrial and green-brown water pixels are two most difficult classes of water to detect. This leads to higher omission errors. Irrespective of water index used, the morphology of water bodies could also play a key role in the accuracy of water classification. Mixed pixels in small water bodies, or those with long perimeters tend to have greater omission errors. These instabilities and other complexities in water properties was the basis for a multi-band index like the AWEI products and others. Although Feyisa et al. (2014) recommended $AWEI_{nsh}$ for urban areas to address limitations caused by tall buildings, Fisher et al. (2016) still observed that the combination of dark surfaces, highly reflective surfaces, and shadows predominant in urban areas created problems across all metrics assessed. Expert systems and new algorithms facilitated by cloud-based computing platforms have been developed to improve the accuracy of open water mapping from satellite image. As seen in the case of Inman and Lyons (2020) and Wolski et al. (2017), it is important to also note that the spatial resolution of data used could also play a key role in the accuracy of predicted inundation.

6.2.3 Satellite Remote Sensing of Wetlands

6.2.4 Optical Remote Sensing Systems in Eco-Hydrology

As tropical rivers and wetlands are increasingly becoming threatened by climate change and various human actions, their monitoring is thus essential. But the inaccessibility of several large wetland systems in remote parts of the world during periods of extreme floods inhibits such monitoring and assessment. This can make ground sampling difficult, especially during periods when high levels of aquatic primary producers are generating food and energy sources for higher order consumers. However, the effects of changes in flow dynamics on floodplain wetland ecosystems and impacts of climatic fluctuations on productive ecosystems that shield communities from storms and serve as key reservoirs blue carbon can now be monitored from space. The pioneering work of Ward et al. (2013) in the northern catchment of Australia highlighted the potential of satellite remote sensing in studying changes in freshwater habitats, including palustrine (swamps, intermittent floodplain water channels, riverine (e.g., rivers streams), and lacustrine (large water holes in low elevation areas).

The progress made in satellite remote sensing systems is not only limited to making direct or indirect measurements of land surface conditions and natural (water bodies, mountains, and vegetation) and man-made (e.g., buildings) targets but also key water budget components, including precipitation, soil moisture, evapotranspiration, and land water storage. While there is a long history and legacy in the use of these satellite technologies to monitor vegetation conditions, plant productivity and health, new remote sensing indicators are being developed to support the assessment of water quality, floodplain productivity, and assessment of water resources development and climate change on the productivity of tropical floodplain rivers. The work of Ndehedehe et al. (2020a) has demonstrated this over the Gilbert catchment, showing how climate variability is a key constraint to the productivity and growth of primary producers. Satellite remote sensing is also becoming more valuable in ecological applications, including the mapping of spatial variation in algal primary productivity and turbidity, especially when combined with statistical methods. A case study in northern Australia using Landsat 8 data to predict turbidity and macrophytes cover to facilitate algal productivity predictions in floodplains highlights another opportunity to quantify changes in hotspots of production along floodplains (Molinari et al. 2021). Satellite techniques are thus important tools to identify these hotspots to aid their management and prioritization for conservation.

There are numerous benefits and advantages in using satellite technologies in environmental monitoring and assessment of natural systems, including floodplain. These technologies have advanced and more sophisticated methods like satellite gravity (e.g., Tapley et al. 2004) with profound capabilities in studying the global water cycle have emerged. Together, these optical remote sensing and time-variable satellite gravity methods have contributed substantially to the multidisciplinary aspects of remote sensing hydrology, enabling understanding of environmental changes and processes. Some applications of these satellite methods in environmental monitoring and water resources include:

- Inventorying of wetlands and mapping of aquatic plants (e.g., Ndehedehe et al. 2021b; Ward et al. 2016; Tockner et al. 2010).
- Assessment of floodplain inundation and connectivity (e.g., Karim et al. 2018; Ward et al. 2014; Sakamoto et al. 2007).
- Quantifying land water storage, including groundwater (e.g., Ndehedehe 2019; Kim et al. 2017; Ojha et al. 2018; Chen et al. 2016).
- Monitoring and extraction of surface water in extensive floodplains (e.g., Tulbure and Broich 2019; Normandin et al. 2018; Feyisa et al. 2014).
- Assessing hydrological controls on surface vegetation (e.g., Ndehedehe et al. 2019; Chen et al. 2014a).
- Analyses of extreme events such as droughts, floods, and wildfires (e.g., Seydi et al. 2021; Ferreira et al. 2018; Agutu et al. 2017; Ward et al. 2013).
- Assessing the response of hydrological stores (groundwater, soil moisture and surface water) to climate change (e.g., Ndehedehe et al. 2021a, 2020b).

There is an increasing threat to water resources due to multiple strings of climate and environmental stressors, including land use change and agricultural expansion (Tulbure and Broich 2019). The use of Earth resources satellites to assess the growth of aquatic primary producers during times of inundation is important to understanding the synergistic impacts of these stressors on floodplain rivers. For example, new indicators to detect the presence of surface water bodies have been introduced (Feyisa et al. 2014) to improve the mapping of open water features. But one of the most widely used indicators for this purpose and mapping of surface water distribution and flood events is the modified normalized difference water index (Xu 2006). This index combines the green and mid-infrared bands of multi-spectral Landsat images (or other satellite systems). Generally, the methodological developments around the classification of open water features from satellite data rely on two-band thresholding and multiple-band indices, resulting in optimized inundation mapping capability and (ii) in the tracking of flood hydrographs and sedimentation patterns. These remote sensing metrics and other widely used remote sensing indicators in water resources, flood mapping, and environmental monitoring are summarized in Ndehedehe et al. (2020c).

6.3 Assessing Floodplain Productivity in the Gilbert Catchment

In this section, a case study of the Gilbert catchment is presented (Fig. 6.4a–d). This catchment is located in the Southern Gulf region of Australia. It has an aerial extent of 46,200 km^2 and rainfall is strongly seasonal with 93% of rainfall happening during the summer (Waltham et al., 2013). Additionally, it is characterized by interconnected streams which normally result in high flows and severe flooding along with its floodplains. Rainfall is moderate (775 mm) and evaporation rates (1868 mm annual mean) exceed rainfall by a factor of 2.4 to 1, and 84% of rainfall gets evaporated while 13% are transitioned to streamflow (Ndehedehe et al. 2020a). Moreover, the region is characterized by isolated turbid waterholes, which thrives on seasonal rainfall and flow. During wet seasons, these water holes are continuously nourished by over bank flows from adjacent rivers, resulting in large distribution of aquatic plant biomass (Fig. 6.4c). This region is one of the areas identified for the expansion of agriculture and development of water resources. The key drivers of floodplain inundation patterns and distribution of aquatic plant biomass in the Gilbert catchment were studied using remote sensing methods.

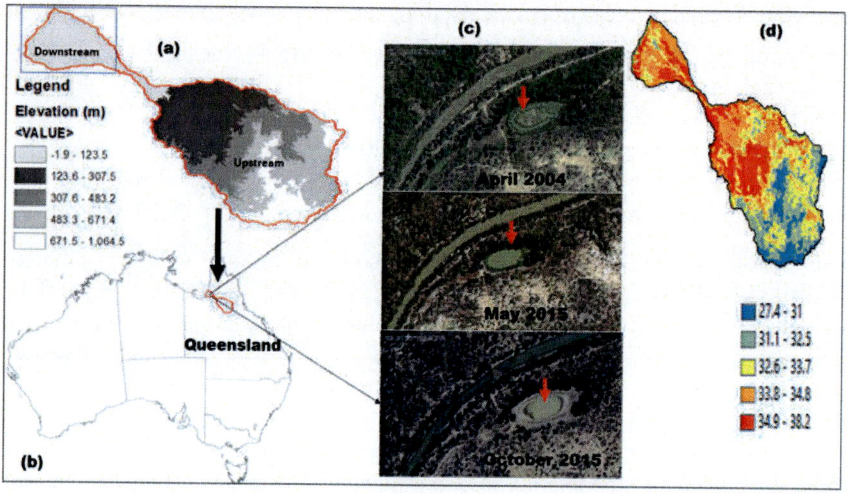

Fig. 6.4 The Gilbert catchment showing the **a** topographic distribution around the downstream and upstream Gilbert catchment based on corrected digital elevation model, **b** location of the Gilbert catchment in northern Queensland, Australia and, **c** illustration of the physical dynamics in aquatic primary producer (hot spots) centres under different seasonal conditions in a freshwater habitat (riverine, lacustrine, and palustrine systems) located south-east of the downstream catchment. The red arrow is a typical water hole covered with macrophytes (primary producers) as was the case on 2 April 2004. **d** Shows mean land surface temperature (DegC) estimated from MODIS (1 km spatial resolution and 8 day temporal resolution) surface reflectance (2000–2019) retrieved from Google Earth Engine platform. *Source* Ndehedehe et al. (2020a)

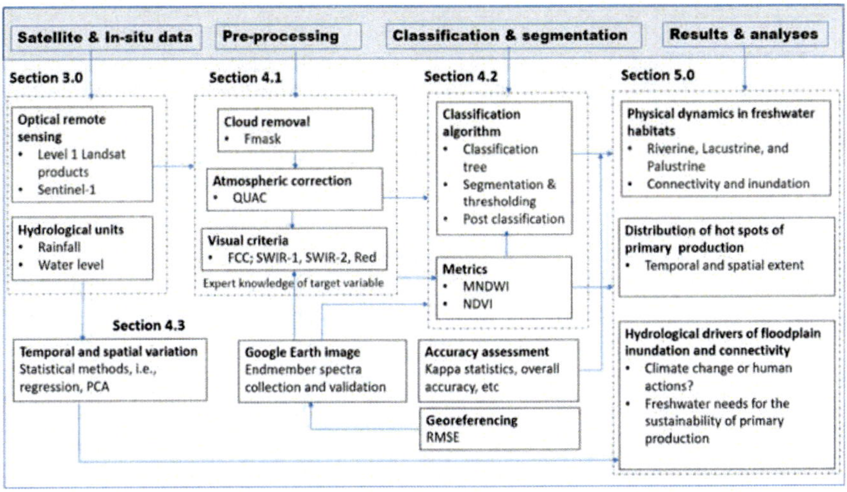

Fig. 6.5 Schematic representation of methodological framework to assess changes in floodplain productivity. *Source* Ndehedehe et al. (2020a)

6.3.1 Framework for Eco-Hydrological Assessment

A useful methodological framework to improve this understanding is summarized in Fig. 6.5. This framework was based on five steps following the work of Ndehedehe et al. (2020a); it combines multivariate analysis of hydrological indicators (e.g., rainfall) with satellite-derived biophysical indicators to assess the link between floodplain productivity and local hydrology.

(i) Data acquisition: Landsat 5 thematic mapper (TM) and Landsat 8 OLI imageries (1991–2019) were retrieved from the archives of the United States Geological Survey through an online data portal (https://earthexplorer.usgs.gov/). These imageries have temporal resolution of 16 days (i.e., overpass frequencies) and scene quality of nine. They were retrieved mostly for the wet seasons (January-March) to assess the changes in freshwater habitats and the impacts of climate variability on the inundation patterns of floodplain during the period. In this kind of assessment, relatively higher spatial resolution data like the European Space Agency Sentinel-2 products available at 10 and 20 m spatial resolution are preferred to Landsat data (30 m) and MODIS surface reflectance (500 m resolution) in quantifying total inundation and mapping the extent of hotspots of primary producers. This is because of the small surface extent of some existing freshwater habitats in the Gilbert catchment. The MODIS surface reflectance data will compromise the ability to accurately map these hotspots of aquatic plant biomass accumulation because of its coarse spatial resolution. As the Sentinel data is limited in terms of data length and or availability (mission started in 2015), the spatial scales in which these hotspots exist in the downstream areas of Gilbert can only be mapped with Landsat given its moderate spatial resolution and availability of historical data. Moreover in-situ hydrological units (rainfall and river discharge data) were accessed from the Queensland water monitoring portal.

(ii) Image pre-processing: The pre-processing can be divided into main steps. In the first step, images were flagged for clouds an then removed. The spectral bands of optical sensors are usually affected by clouds, cloud shadows, and snow. This impacts on the quality of images, necessitating the identification and removal of these clouds before use. In the second step, the cloud-free images were then calibrated from their raw digital numbers of Landsat imageries to standard surface reflectance. The QUick Atmospheric Correction Code (QUAC, Bernstein et al. 2005) module in ENVI 5.5 (Exelis Visual Information Solutions) was employed to calibrate each Landsat scene to standard surface reflectance. Several atmospheric correction algorithms that isolate clouds in images exist. These include (i) ACORN (Atmospheric CORrection Now), (ii) ATREM (The Atmospheric REMoval), (iii) FLAASH (Fast Line of sight Atmospheric Analysis of Spectral Hypercubes), (iv) QUAC (Quick Atmospheric Correction), (v) ATCOR (The Atmospheric Correction), and (iv) Rapid Atmospheric Correction. QUAC performs atmospheric correction on multi-band imageries in a more sophisticated manner such as using in-scene approach that relies on information about sensor band locations and their radiometric calibration. They also tend to show more advantages unlike other physics-based methods (see, Bernstein et al. 2005, 2012). QUAC is also computationally efficient and fast, and its approach to aerosol optical depth retrieval, which does not require the presence of dark pixels makes it a more desirable (Bernstein et al. 2005). The output from QUAC is atmospherically corrected multi-temporal imageries with similar radiometric scale, regardless of their image acquisition dates, which are usually accompanied by variations in solar irradiance and sensor geometry among other parameters (e.g., Chander et al. 2009).

(iii) Estimation of floodplain inundation: Floodplain inundation and water bodies were quantified by applying the widely used modified normalized difference water index (MNDWI, Xu 2006), which combines the green (G) and mid-infrared (MIR) bands as $MNDWI = b_G - b_{MIR}/b_G + b_{MIR}$. There have been several methodological developments around the automated extraction of surface water bodies (e.g., Fisher et al. 2016; Feyisa et al. 2014), which are usually compared against the MNDWI or other supervised classification methods. The MNDWI nonetheless, remains an ideal, mainstream automated water extraction metric because of its accuracy, computational efficiency, and simplicity, especially in threshold identification and definition (ranges from $0 - 1$ for water-related pixels).

(iv) Prediction of primary production centres: In the characterization of floating, emergent, and submerged vegetation, rule-based approach such as decision tree could give better results as opposed to conventional methods (Ozesmi and Bauer 2002). This is why some studies advocate the use of decision trees in quantifying aquatic vegetation (Midwood and Chow-Fraser 2010; Davranche et al. 2010;

Zhao et al. 2013). The distribution of these hotspots of primary producers, areas defined as inundated areas covered with different aquatic species such as macrophytes. These hotspots were identified by combining two remote sensing indices, the MDNWI and NDVI (Normalized Difference Vegetation Index) in a decision tree framework. The spectral properties of the red (R) and near infrared (NIR) bands of Landsat, which show relatively low reflectance and high reflectance, respectively, were used to estimate NDVI (i.e., $NDVI = b_{NIR} - b_R/b_{NIR} + b_R$) in each of the Landsat image.

(v) Model validation and accuracy assessment: To minimize the need for field campaigns and validation while making it an accurate general-purpose application suitable for other floodplain wetlands, our classification tree model is compared to a physically based spectral classification algorithm (Kruse et al. 1993) as summarized in Fig. 6.5. The output from the classification tree (CT) was compared with spectral angle mapper (SAM), a pixel-based classification algorithm that depends on the spectral similarity between two spectra (Kruse et al. 1993). So, to assess the efficacy of physically based spectral classification algorithm, several endmember spectra (i.e., spectral profiles) in the catchment were collected from hot spots of primary production and freshwater bodies using the Google Earth image as a reference. The classification of hot spots and inundation from the SAM (ten different endmember spectra were collected) and CT model based on the April 2005 Landsat data were subsequently compared with pure pixel validation data from Google Earth imageries. Notably, image-to-image co-registration of Google Earth imagery and Landsat image (23 April 2005) based on several control points was employed for the validation sites. The geo-referenced Google Earth data was then used as reference map for accuracy assessment based on kappa statistics (e.g., Congalton and Green 2009).

(vi) Multivariate analysis of hydrological indicators: The multivariate analyses of rainfall over the Gilbert catchment was achieved with the use of principal component analysis (PCA, Preisendorfer 1988). This method and other multivariate analysis tools are discussed in detail in Chap. 13. Monthly rainfall grids over the Gilbert catchment were statistically decomposed into temporal and spatial patterns using the PCA method. With this, the dominant rainfall patterns (annual changes) and short term (multi-annual variations) hydrological signals associated with floods and seasonal inundation on the Gilbert floodplain were identified. Cross-correlograms were employed to examine the relationship and similarity between hydrological indicators (rainfall, river discharge, and water level) and floodplain productivity (total inundation and hot spots of primary producers) while also estimating their corresponding lags (i.e., the number of time periods that separate the two variables. To further diagnose the relationship between monthly rainfall and total downstream inundation of freshwater habitats, averaged rainfall (2000–2018) over the Gilbert was integrated with observed water level in the downstream area based on a non-parametric, empirical joint probability framework to estimate a multivariate standardized indicator (e.g., Farahmand and AghaKouchak 2015).

6.3.2 *Spectral Separability and Optical Complexities of Aquatic Habitats*

The characteristics of surface water vary in time and space because they can be affected by several environmental conditions and factors, including algal blooms, tidal cycles, sediment inputs, and reflectance from depth and bottom (e.g., Fisher et al. 2016; Tulbure et al. 2016). In the wet–dry tropics of Australia, surface water is characterized by a range of turbidities and land cover states. The abundance of floodplain macrophytes and other forms of epiphytic algae within the vicinity of some freshwater ecosystems and other land surface characteristics contribute to this variability (e.g., Ward et al. 2016). In addition to the obvious presence of algae in larger waterholes in Australian inland floodplains, studies have shown that these waterholes are also highly turbid (e.g., Faggotter et al. 2013; Bunn et al. 2003). The range of turbidities in optically different freshwater bodies over the Gilbert catchment is evident in their endmember spectra profiles (Fig. 6.6). Open water features in locations with aquatic vegetation have different spectral properties and reflectance

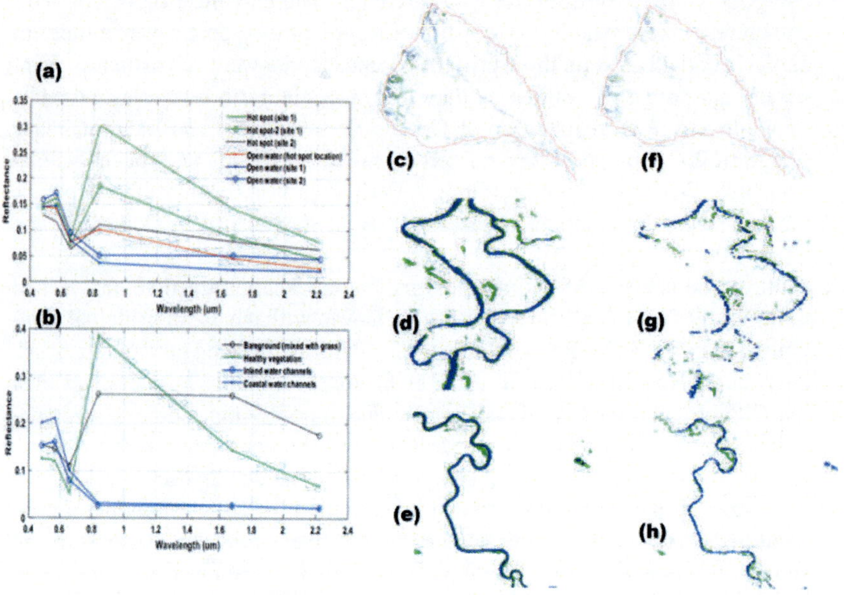

Fig. 6.6 Separability of freshwater habitats and hot spots of primary production. **a, b** Spectral profiles for land cover states associated with aquatic ecosystems and water bodies with a range of spectral properties in the Gilbert catchment, including the two reference locations. These endmember spectra (signature profiles) were collected from calibrated Landsat imagery for the reference locations and other randomly selected sites (include areas near the coast) and used in the Spectral Angle Mapper supervised algorithm. **c–e** and **f–h** are raster maps from classification tree and endmember collections of freshwater habitats (blue) and hot spots of primary production (green), respectively. *Source* Ndehedehe et al. (2020a)

compared with other open water channels. Unlike other regions like the humid Congo Basin or tropical South America where water is largely blue in colour, in Australia, water in several locations is brown in colour and turbid in the wet–dry tropics. The dynamics in water properties has been a challenge in identifying thresholds in the use of remotely sensed observations to map water bodies. This are largely caused by changes in atmospheric conditions, variations of the incident angle of the Sun, and changes in the chemical and biophysical properties of the water all of which influence reflected wavelengths seen in surface water bodies. By taking into account the spectral properties and environmental characteristics of water in specific locations, thresholds can vary from one region to another. For example, the performance of some remote sensing metrics to detect the presence water in a particular canal was restricted resulting in the modification of the water threshold value (Worden and de Beurs 2020).

The lack of a uniform threshold value for open water features across different water landscapes in different locations was also noticed in the Gilbert catchment. For instance, the spectral characteristics of inland and coastal waters within the same river channel were substantially different (Fig. 6.6). The considerable morphological differences in the riverine systems classified by using these endmember spectra collections in the spectral angle mapper algorithm (Fig. 6.6) are directly linked to this. Compared with this supervised algorithm, the classification tree model gave better results in terms of the extraction of these riverine systems and hot spots of primary production. Indeed, several landscape processes and conditions (erosion and pollution) could alter the spectral composition of fluxes within these freshwater habitats. Obstructions to the accuracy of mapping open water using automated water index detection methods could also be caused by urban areas and mountain shadows because of having positive index values, which may further introduce classification errors in the water features presented in the image scene (Worden and de Beurs 2020).

Overall, factors affecting the mapping of floodplain wetlands and freshwater habitats can be classified into environmental and anthropogenic. Environmental factors that impact the use of satellite remote sensing include

1. plant phenology,
2. shadows from tall objects, which can be trees or buildings,
3. changes in hydrological conditions such as flooding, drought, water content of the soil, and tide levels,
4. cloud cover, snow, and ice in higher latitudes and elevation,
5. growing seasons.

Human activities that have been identified as contributing to the challenges of water feature extraction from satellite image are land clearing, construction of roads, railroads, buildings; drainage, filling, etc. These factors have considerable potential to complicate the spectral separability of freshwater systems and wetlands. Such complication arises because the unique spectral property of water bodies are compromised by these factors, resulting in different levels of turbidities. This makes most supervised and unsupervised classification algorithms unsuitable for the mapping of aquatic habits and primary production centres (see, Gidley 2009). Physically based spectral techniques such as the spectral angle mapper have demonstrated this weakness. Rule-based approaches therefore are therefore emerging as reliable alternatives over traditional methods for the assessment of floodplain wetlands and mapping the spatial distribution of aquatic vegetation (e.g., Ward et al. 2014; Zhao et al. 2013; Davranche et al. 2010; Ozesmi and Bauer 2002).

6.3.3 Understanding Changes in Floodplain Productivity

6.3.3.1 Floodplain Inundation Patterns

Changes in the spatial patterns of floodplain inundation during wet season over the downstream Gilbert catchment showed marked variability (Fig. 6.7). The period between 1998 and 2010 is the wettest since 1991 while the years between 1993 and 1997 were found to be relatively dry (Fig. 6.7, cf. Fig. 6.9). The magnitude of the 1998 flood event was a considerable deviation from these previous dry periods or spells (excluding 1991) and led to about 1144 km^2 inundation extent. However, the hydrological records (discharge and water levels) of the downstream Gilbert catchment indicate that 2009 has been the wettest year in more than two decades. Although about 26% of the January 2009 Landsat imagery was affected by cloud (mostly around floodplains), inundation extent recovered within the main floodplain along the Gilbert River was about 708 km^2. The floodplain remained highly productive in March 2009, resulting in an inundation extent of approximately 940 km^2. Moreover, it seems the impacts of extreme wet season flood have considerable influence on the floodplain productivity of dry season within the same year. For example, comparing the floodplain inundation in May and July of 2009 and 2016, years with different levels of flood magnitude, 2009 was relatively more productive even in dry season unlike 2016 (Fig. 6.8). Similarly, April and July 2018 inundation was not very different given that the wet season of the same year was relatively wetter than 2016 (Fig. 6.8). Except for the 2016 – 2018 period (though rather wet, there was no cloud-free image for 2017 wet season), 2014 – 2019 is generally much similar to the dry cycles of the mid-1990s with 2015 being the driest (26.7 km^2) (Fig. 6.7, cf. Fig. 6.9).

Rainfall seasonality is an important indicator of the spatial distribution of floodplain inundation in the Gilbert downstream catchment. For example, floodplain inundation extent as at 31st of March 2009 was approximately 940 km^2. By 21st July of the same year when there is usually little or no rainfall, the estimated inundation extent was 108 km^2 (Fig. 6.8). Similarly, about 326 km^2 spatial inundation extent was estimated on 18 March 2016. This inundation extent receded to approximately 25 km^2 by 24 July 2016. Notably, the change in seasonal floodplain inundation has impact on the distribution of primary producers both in the wet and dry seasons (Figs. 6.8, 6.9 and 6.10). For example, the predicted extent of primary production

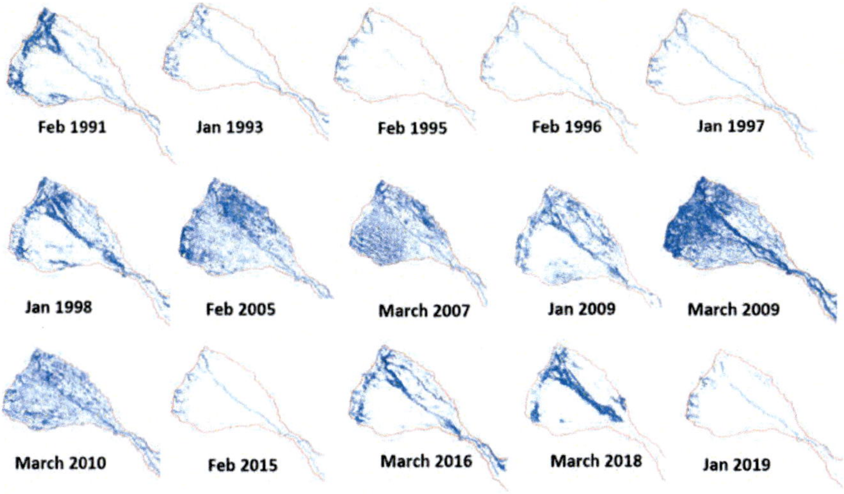

Fig. 6.7 Spatial patterns of Landsat-derived wet season (January, February, and March) floodplain inundation over the Gilbert catchment for selected years based on the MNDWI. Owing to the lack of cloud-free Landsat data for wet season in 2018, available cloud-free Sentinel 2 data covering most segments of downstream Gilbert catchment was used to estimate inundation for March 2018. *Source* Ndehedehe et al. (2020a)

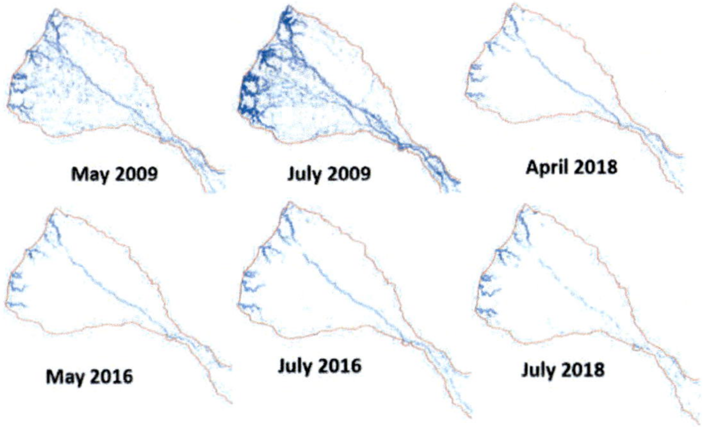

Fig. 6.8 Comparing floodplain inundation during dry season (e.g., April, May, and July) for years with high magnitude flood events and relatively dry periods. *Source* Ndehedehe et al. (2020a)

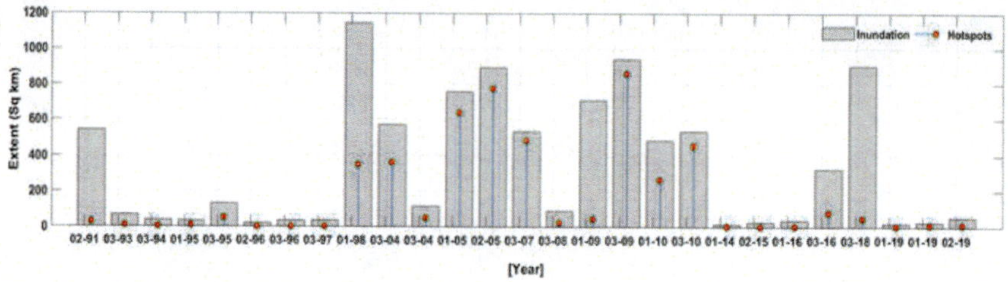

Fig. 6.9 Historical patterns of wet season floodplain inundation and predicted hot spots of primary production derived from a combination of NDVI and MNDWI. These vegetation and inundation metrics were employed in a decision tree framework to estimate hot spots of primary production and aquatic ecosystems. *Source* Ndehedehe et al. (2020a)

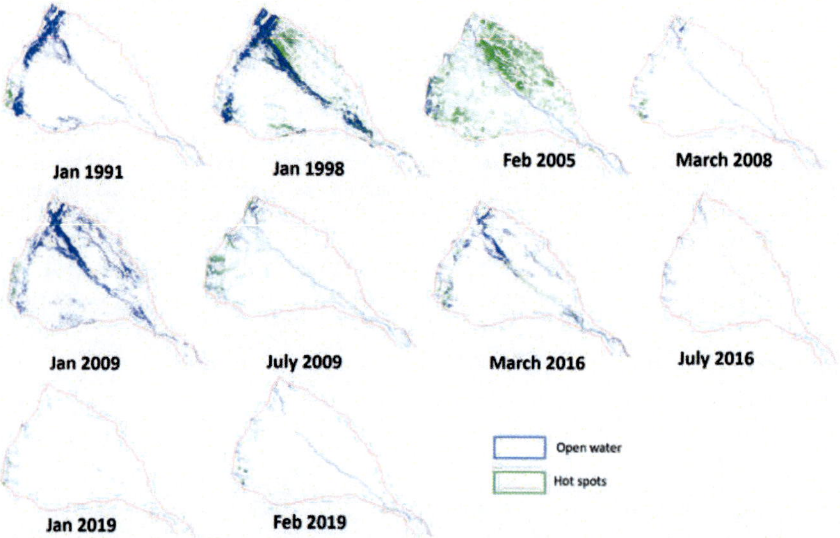

Fig. 6.10 Spatial patterns of hot spots of primary production for years with flood events and relatively dry years. *Source* Ndehedehe et al. (2020a)

reduced from 25 km^2 in early April 2018 to about 3 km^2 in mid-July of 2018. The spread of hot spots of primary producers were also revealed in their transitional patterns during wet seasons. In 2009 for instance, hot spot extent was about 44 km^2 as at January 26th (though this month's imagery was significantly affected by cloud) because of the heavy flood and standing water on the floodplain. By 31 March 2009 rainfall and discharge had reduced significantly, resulting in extensive evolutions of hot spots of primary production (859 km^2) (Fig. 6.9). These same patterns were observed in other imageries, including those of 2010 (13th January-265 km^2; 18th March-454 km^2), 2016 (14th January-4 km^2; 18th March-79 km^2), and 2019 when the estimated extents of hot spots rose from about 5 km^2 in January to about 11 km^2 in mid-March.

6.3.3.2 Spatial and Temporal Patterns of Aquatic Plant Biomass

The temporal patterns of the predicted hotspots of primary producers for the 1991–2019 period are embedded in Fig. 6.9. Except for overwhelming flood periods (e.g., February 1991, January 1998, January 2009, and March 2018), these patterns show that there is a reasonable agreement between floodplain inundation and the evolution of primary production ($r = 0.73$). While mutual information (MI) and maximum correlation between extents in hot spots of primary production and inundation are considerably higher ($r = 0.83$ and $MI = 0.87$, respectively) with a phase lag less than one month, the distribution of hot spots shows spatial heterogeneity over the Gilbert catchment. However, the drought in 2015 affected productivity on the floodplain and showed the lowest extent (about $2\,\text{km}^2$) since 1997, i.e., based on the available imageries (Fig. 6.9).

On a more localized scale, the influence of floodplain inundation and impacts on the spatial patterns of aquatic habitats, especially water holes and primary production centres are indicated (Fig. 6.11a–j). The biomass accumulation of aquatic canopy-covering plants in water holes (Fig. 6.11a) and the persistence of similar freshwater habitats depend on over bank spilling of floodplain rivers into these freshwater domains during extreme or severe wet periods (Fig. 6.11d, f, g, i). The connectivity of these rivers with slightly isolated water holes and other intermittent water bodies can last up to 4 months and more during such extreme wet periods (Fig. 6.11f, g).

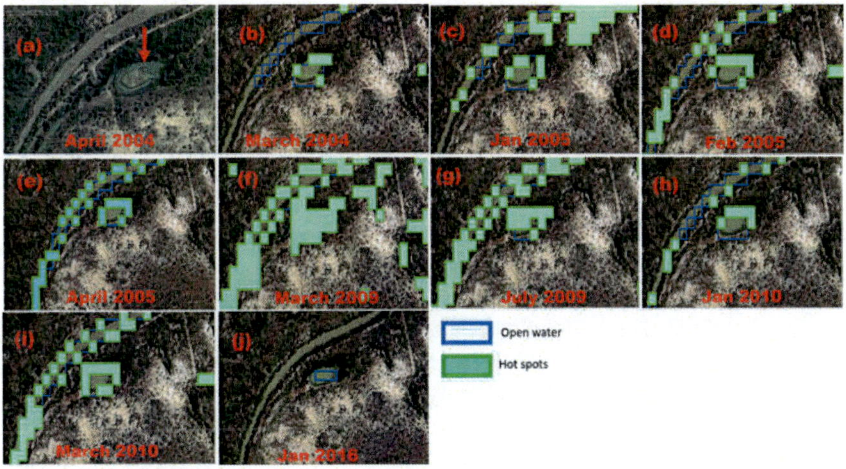

Fig. 6.11 Seasonal floodplain inundation and impacts on the physical dynamics of aquatic habitats and primary production centres. **a** Freshwater habitats showing intermittent floodplain water bodies in downstream Gilbert catchment (red arrow is a typical water hole covered with macrophytes-hot spot of primary production). **b–j** indicate the changing patterns of inundation and primary production in these freshwater habitats. The high-resolution satellite imagery in (**a**) is taken from Google Earth and was captured in April 2004 while others were taken in May 2015. *Source* Ndehedehe et al. (2020a)

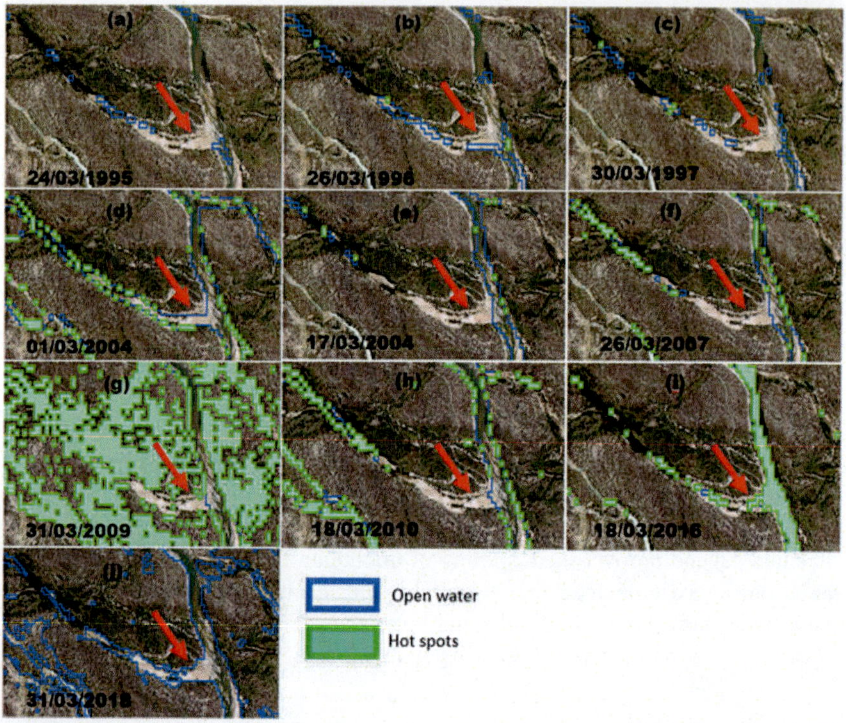

Fig. 6.12 Illustration of the spatial dynamics of floodplains in the riverine systems towards the end of wet seasons using available Landsat imageries. **a–j** show the inundation patterns (temporal and spatial) of the lower Gilbert River located downstream and their connectivity to the tributaries in March ending. *Source* Ndehedehe et al. (2020a)

At the start of the wet season, they exist independently and get connected as the wet season progresses or depending on the intensity and magnitude of flood (Fig. 6.11c, d, h, i), and later disconnect at the end of the wet season (Fig. 6.11e). Note that Fig. 6.11j does not imply the disappearance of the floodplain river but rather implies a contraction of the river beyond what the 30 m pixel size of Landsat can capture. This contraction could be caused by change in precipitation patterns, which in turn influence flow variability and subsequently floodplain inundation patterns. Further, the spatial patterns of floodplains in the riverine systems towards the end of wet season (March ending) also show interesting patterns (Fig. 6.12a–j). As the dry season approaches, some tributaries of the Gilbert River remain disconnected in relatively dry years with no significant floods (Fig. 6.12a–c) while the riverine systems in wet years remain connected with some wet years having excessive flood waters along the floodplain rivers as was the case in 2009 and 2018 (Fig. 6.12d, g, i, j).

6.3.4 *Influence of Climate Variability on the Productivity of Tropical Floodplains*

The knowledge of climate change as a concept is increasing but it likely impacts, for example, on freshwater ecosystems are still poorly understood. Some pioneering studies (e.g., Mulholland et al. 1997; Bazzaz 1990) have documented the influence of increasing greenhouse gas concentrations and global warming on the earth's energy budget and natural ecosystems. In the freshwater ecosystems of the south-eastern United States and the Gulf coast of Mexico, for example, climate change has been shown to have ecological consequences, including, rise in the rates of primary production, loss of organisms during strong flushing events in streams and wetlands, among several other effects (see, Mulholland et al. 1997). Unfortunately, the impacts of human-induced climate change are expected to exacerbate, resulting in the acceleration of the water cycle and other devastating effects on natural ecosystems, including wetlands, freshwater habitats, and species richness. Rivers, especially in wet–dry tropical savannas with similar topographic layout (Fig. 6.4a) and rainfall distribution could serve as key indicators of floodplain productivity. This argument is further underpinned by the fact that the biota in the wet–dry tropics are highly adapted to variable flow conditions (Kingsford et al. 2014).

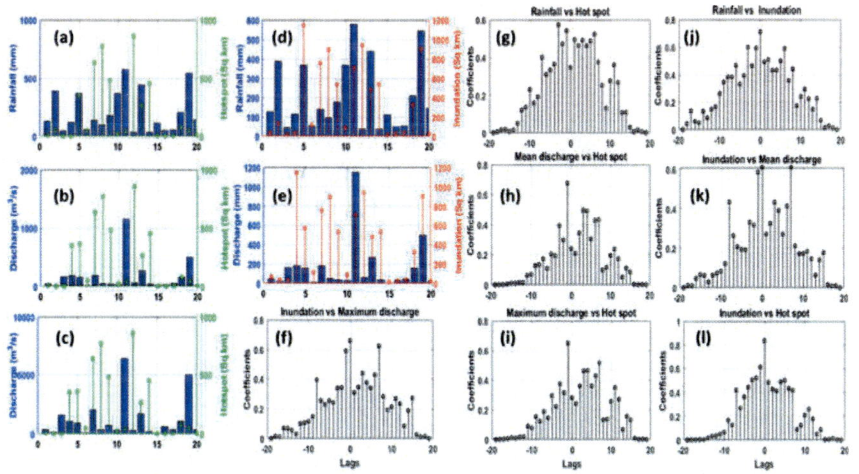

Fig. 6.13 Correlation of downstream wet season rainfall (monthly totals) and flow with floodplain productivity during the 1995–2019 period. The relationship of total monthly rainfall amounts with predicted extents of hot spots of primary producers and total inundation extent are indicated in (**a**–**e**). (**f**–**k**) Are their corresponding cross-correlograms during the same period with lags indicated in months. (f and i) Are the cross-correlograms of observed maximum discharge at downstream catchment with floodplain inundation and hot spot while (**l**) is the cross-correlogram between floodplain inundation and hot spot. Periods corresponding with missing total rainfall amounts (30/03/1997, 1/03/2004, 06/01/2019, and 23/02/2019) and discharge have been excluded in the analyses. *Source* Ndehedehe et al. (2020a)

Local rainfall also plays key roles in driving downstream hydrology. The gauged rainfall records of downstream Gilbert indicate that variations in floodplain inundation are also induced by the local rainfall amounts of previous months. We found this phenomenon to be true also for wet season discharge, for example, in March 2005, 2009, and 2010. While taking note of probable uncertainties in the estimated extents of hot spots due to impact of cloud, the flow-hot spot relationship ($r = 0.68$ at one month lag) of the Gilbert catchment (Fig. 6.13), is obviously moderately higher than the local rainfall-hot spot association ($r = 0.57$ at three months lag). In the event of major water resources development along the upstream floodplains or change in government water-related policies, we argue that such developments could amplify flow variability, resulting in significant impacts on freshwater habits and hot spots of primary producers.

6.3.5 Freshwater Needs for the Sustainability of Floodplain Productivity

The case-specific study undertaken in the Gilbert catchment shows river discharge and local rainfall are important drivers of floodplain inundation as well as contributing to the growth of aquatic plant biomass. The combined influence of local rainfall and flow (discharge) on changes in floodplain productivity indicates that these floodplains benefit from the frequent flooding of these freshwater floodplains, which occur during summer. For example, the summer floods in these catchments are usually large scale in some years with substantial impact on freshwater habitats in the downstream floodplains. Tropical rivers and freshwater habitats including water holes, and intermittent streams and channels can benefit from these moderately high magnitude flood events, when stored nutrients are mobilized during storms. When the water levels decline, these nutrients are then subsequently processed by primary producers like macrophytes, among others (e.g., Ndehedehe 2020). These nutrients later become an important source of support to aquatic ecosystem, stimulating primary production as well as providing support for higher order organisms, including fish, and invertebrates. However, large high-intensity floods can quickly move these nutrients from floodplain rivers to the surrounding oceans. This arguably will negatively affect the productivity of aquatic food webs. The pros and cons of floods on aquatic ecosystems have been detailed in Talbot et al. (2018). For instance, floods contribute to rivers, aiding their ability to create unique habitat and support biodiversity and biological productivity. Small-sized floods are crucial to improving access to recreation and food, in addition to supporting primary production and water regulation. Although the impacts of floods on the latter are different depending on floodplain conditions and catchment hydrology, large magnitude floods are known to contribute to groundwater recharge Ndehedehe et al. (2021b); Talbot et al. (2018). Floods are arguably devastating and key disturbance that impacts on humans as well as freshwater ecosystems and the different functions and services they provide. But their occurrence, be it moderate or extreme has important ecological value and function. While the inundation of floodplain is important to aid productivity and growth of primary producers, larger inundated areas have been found to substantially drive groundwater recharge. In other words, the rate of infiltration to groundwater will be higher more inundation of the floodplain happens through flood.

The wetland hydrology and changes in floodplain inundation of the Gilbert catchment are largely driven by climate. However, water resource developments that lead to water diversion or harvesting have the potentials to disrupt productivity. As a result, there must be considerations regarding the role of these tropical rivers and the frequent floods, which sustain its productivity and also provide other functions including, groundwater recharge and improved water regulation. The use of remote sensing methods provides an opportunity to better understand the ecological implications of human activities as well as the influence of climate variability on wetland hydrology and floodplain productivity in the tropical rivers. This understanding directly feeds into policy frameworks and used in wetland conservation planning to support strategic water resource management

6.4 Concluding Remarks

Freshwater systems are highly valued because of the important roles they play in tourism and recreation, improvement of water quality, provision of shelter for fish and native wildlife, drought relief for wildlife, flood water storage, as well as support for a range of terrestrial and aquatic biodiversity. However, their monitoring using conventional approaches is limited and satellite technologies are filling such gaps. With the advent of Earth resources satellites, its monitoring and mapping as evolved and more sophisticated approaches to quantifying freshwater, including surface water dynamics have emerged. In large semi-arid domains and wet–dry tropics in northern Australia and Africa where this has been demonstrated, the potential of satellite observations to monitor changes in surface water extent provides a capability that underpins water resources management for sustainable development. This can aid understanding of climatic processes and several human actions that can limit water availability in such regions.

References

Agutu N, Awange J, Zerihun A, Ndehedehe C, Kuhn M, Fukuda Y (2017) Assessing multi-satellite remote sensing, reanalysis, and land surface models' products in characterizing agricultural drought in East Africa. Remote Sens Environ 194:287–302. https://doi.org/10.1016/j.rse.2017.03.041

Alsdorf DE, Lettenmaier DP (2003) Tracking fresh water from space. Science 301(5639):1491–1494. https://doi.org/10.1126/science.1089802

Alsdorf DE, Rodríguez E, Lettenmaier DP (2007) Measuring surface water from space. Rev Geophys 45(2):RG2002. https://doi.org/10.1029/2006RG000197

Amani M, Ghorbanian A, Ahmadi SA et al (2020) Google Earth Engine cloud computing platform for remote sensing big data applications: a comprehensive review. IEEE J Sel Top Appl Earth Obs Remote Sens 13:5326–5350. https://doi.org/10.1109/JSTARS.2020.3021052

Balcombe SR, Arthington AH (2009) Temporal changes in fish abundance in response to hydrological variability in a dryland floodplain river. Marine Freshw Res 60:146–159. https://doi.org/10.1071/MF08118

Ban Y, Zhang P, Nascetti A, Bevington AR, Wulder MA (2020) Near real-time wildfire progression monitoring with Sentinel-1 SAR time series and deep learning. Sci Rep 10. https://doi.org/10.1038/s41598-019-56967-x

Bazzaz FA (1990) The response of natural ecosystems to the rising global CO2 levels. Ann Rev Ecol Syst 21(1):167–196. https://doi.org/10.1146/annurev.es.21.110190.001123

Bernstein LS, Adler-Golden SM, Jin X, Gregor B, Sundberg RL (2012) Quick atmospheric correction (QUAC) code for vnir-swir spectral imagery: algorithm details. In: 2012 4th workshop on hyperspectral image and signal processing: evolution in remote sensing (WHISPERS), pp 1–4. https://doi.org/10.1109/WHISPERS.2012.6874311

Bernstein LS, Adler-Golden SM, Sundberg RL, Levine RY, Perkins TC, Berk A, Ratkowski AJ, Felde G, Hoke ML (2005) A new method for atmospheric correction and aerosol optical property retrieval for VIS-SWIR multi- and hyperspectral imaging sensors: QUAC (QUick atmospheric correction). In: Proceedings of the 2005 IEEE international geoscience and remote sensing symposium, 2005. IGARSS '05, vol 5, pp 3549–3552. https://doi.org/10.1109/IGARSS.2005.1526613

Bunn SE, Arthington AH (2002) Basic principles and ecological consequences of altered flow regimes for aquatic biodiversity. Environ Manag 30(4):492–507. https://doi.org/10.1007/s00267-002-2737-0

Bunn SE, Davies PM, Winning M (2003) Sources of organic carbon supporting the food web of an arid zone floodplain river. Freshw Biol 48(4):619–635. https://doi.org/10.1046/j.1365-2427.2003.01031.x

Bunn SE, Thoms MC, Hamilton SK, Capon SJ (2006) Flow variability in dryland rivers: boom, bust and the bits in between. River Res Appl 22(2):179–186. https://doi.org/10.1002/rra.904

Chander G, Markham BL, Helder DL (2009) Summary of current radiometric calibration coefficients for Landsat MSS, TM, ETM+, and EO-1 ALI sensors. Remote Sens Environ 113(5):893–903. https://doi.org/10.1016/j.rse.2009.01.007

Chen J, Famigliett JS, Scanlon BR, Rodell M (2016) Groundwater storage changes: present status from GRACE observations. Surv Geophys 37(2):397–417. https://doi.org/10.1007/s10712-015-9332-4

Chen T, de Jeu R, Liu Y, van der Werf G, Dolman A (2014) Using satellite based soil moisture to quantify the water driven variability in NDVI: a case study over Mainland Australia. Remote Sens Environ 140:330–338. https://doi.org/10.1016/j.rse.2013.08.022

Chen Y, Wang B, Pollino CA, Cuddy SM, Merrin LE, Huang C (2014) Estimate of flood inundation and retention on wetlands using remote sensing and GIS. Ecohydrology 7(5):1412–1420. https://doi.org/10.1002/eco.1467

Congalton R, Green K (2009) Assessing the accuracy of remotely sensed data: principles and practices. Taylor & Francis, 2nd

Davranche A, Lefebvre G, Poulin B (2010) Wetland monitoring using classification trees and SPOT-5seasonal time series. Remote Sens Environ 114(3):552–562. https://doi.org/10.1016/j.rse.2009.10.009

Davranche A, Poulin B, Lefebvre G (2013) Mapping flooding regimes in Camargue wetlands using seasonal multispectral data. Remote Sens Environ 138:165–171. https://doi.org/10.1016/j.rse.2013.07.015

Faggotter SJ, Webster IT, Burford MA (2013) Factors controlling primary productivity in a wet-dry tropical river. Marine Freshw Res 64:585–598. https://doi.org/10.1071/MF12299

Farahmand A, AghaKouchak A (2015) A generalized framework for deriving nonparametric standardized drought indicators. Adv Water Res 76:140–145. https://doi.org/10.1016/j.advwatres.2014.11.012

Feng M, Sexton JO, Channan S, Townshend JR (2016) A global, high-resolution (30-m) inland water body dataset for 2000: first results of a topographic-spectral classification algorithm. Int J Digit Earth 9(2):113–133. https://doi.org/10.1080/17538947.2015.1026420

Ferreira V, Montecino H, Ndehedehe C, Heck B, Gong Z, Westerhaus M, de Freitas S (2018) Space-based observations of crustal deflections for drought characterization in brazil. Sci Total Environ 644:256–273. https://doi.org/10.1016/j.scitotenv.2018.06.277

Feyisa GL, Meilby H, Fensholt R, Proud SR (2014) Automated water extraction index: a new technique for surface water mapping using landsat imagery. Remote Sens Environ 140:23–35. https://doi.org/10.1016/j.rse.2013.08.029

Filipponi F (2018) Bais2: burned area index for sentinel-2. Proceedings 2(7). https://doi.org/10.3390/ecrs-2-05177

Fisher A, Flood N, Danaher T (2016) Comparing landsat water index methods for automated water classification in eastern Australia. Remote Sens Environ 175:167–182. https://doi.org/10.1016/j.rse.2015.12.055

Gao B-C (1996) NDWI-a normalized difference water index for remote sensing of vegetation liquid water from space. Remote Sens Environ 58(3):257–266. https://doi.org/10.1016/S0034-4257(96)00067-3

Gidley SL (2009) Using high resolution satellite imagery to map aquatic macrophytes on multiple lakes in northern Indiana. Unpublished Msc thesis, Indiana University. https://core.ac.uk/download/pdf/46956355.pdf. Accessed 15th Apr 2019

Inman VL, Lyons MB (2020) Automated inundation mapping over large areas using landsat data and Google Earth Engine. Remote Sens 12(8). https://doi.org/10.3390/rs12081348

Karim F, Peña-Arancibia J, Ticehurst C, Marvanek S, Gallant J, Hughes J, Dutta D, Vaze J, Petheram C, Seo L, Kitson S (2018) Floodplain inundation mapping and modelling for the Fitzroy, Darwin and Mitchell catchments. A technical report to the Australian government from the CSIRO northern Australia water resource assessment, part of the national water infrastructure development fund: water resource assessments. CSIRO, Australia

Keddy PA, Fraser LH, Solomeshch AI, Junk WJ, Campbell DR, Arroyo MTK, Alho CJR (2009) Wet and wonderful: the world's largest wetlands are conservation priorities. BioScience 59(1):39–51. https://doi.org/10.1525/bio.2009.59.1.8

Khandelwal A, Karpatne A, Marlier ME, Kim J, Lettenmaier DP, Kumar V (2017) An approach for global monitoring of surface water extent variations in reservoirs using modis data. Remote Sens Environ 202:113–128. https://doi.org/10.1016/j.rse.2017.05.039

Kim D, Lee H, Laraque A, Tshimanga RM, Yuan T, Jung HC, Beighley E, Chang C-H (2017) Mapping spatio-temporal water level variations over the central Congo River using PALSAR ScanSAR and Envisat altimetry data. Int J Remote Sens 38(23):7021–7040. https://doi.org/10.1080/01431161.2017.1371867

Kingsford R, Watts R, Keohn J, Thompson R, Sims N (2014) Flow dependent ecological responses. A technical report from the ecological responses to altered flow regimes flagship research cluster (subproject 3). CSIRO water for a healthy country flagship, Australia

Krause CE, Newey V, Alger MJ, Lymburner L (2021) Mapping and monitoring the multi-decadal dynamics of Australia's open waterbodies using landsat. Remote Sens 13(8). https://doi.org/10.3390/rs13081437

Kruse F, Lefkoff A, Boardman J, Heidebrecht K, Shapiro A, Barloon P, Goetz A (1993) The spectral image processing system (SIPS)-interactive visualization and analysis of imaging spectrometer data. Remote Sens Environ 44(2):145–163. https://doi.org/10.1016/0034-4257(93)90013-N

Masocha M, Dube T, Makore M, Shekede MD, Funani J (2018) Surface water bodies mapping in Zimbabwe using landsat 8 OLI multispectral imagery: a comparison of multiple water indices. Phys Chem Earth Parts A/B/C 106:63–67. https://doi.org/10.1016/j.pce.2018.05.005

McFeeters SK (1996) The use of the Normalized Difference Water Index (NDWI) in the delineation of open water features. Int J Remote Sens 17(7):1425–1432. https://doi.org/10.1080/01431169608948714

Midwood JD, Chow-Fraser P (2010) Mapping floating and emergent aquatic vegetation in coastal wetlands of eastern Georgian bay, lake Huron, Canada. Wetlands 30(6):1141–1152. https://doi.org/10.1007/s13157-010-0105-z

Molinari B, Stewart-Koster B, Malthus TJ, Bunn SE (2021) Assessing spatial variation in algal productivity in a tropical river floodplain using satellite remote sensing. Remote Sens 13(9). https://doi.org/10.3390/rs13091710

Mueller N, Lewis A, Roberts D, Ring S et al (2016) Water observations from space: Mapping surface water from 25years of landsat imagery across Australia. Remote Sens Environ 174:341–352. https://doi.org/10.1016/j.rse.2015.11.003

Mulholland PJ, Best GR, Coutant CC, Hornberger GM, Meyer JL, Robinson PJ, Stenberg JR, Turner RE, Vera-Herrera F, Wetzel RG (1997) Effects of climate change on freshwater ecosystems of the south-eastern United States and the gulf coast of Mexico. Hydrol Process 11(8):949–970. https://doi.org/10.1002/(SICI)1099-1085(19970630)11:8<949::AID-HYP513>3.0.CO;2-G

Ndehedehe C (2020) Floodplain productivity of the Gilbert and Flinders catchments: Component 3 final report. Griffith University, Brisbane. https://research-repository.griffith.edu.au/handle/10072/394768on8thJuly2020

Ndehedehe CE (2019) The water resources of tropical West Africa: propblems, progress and prospect. Acta Geophysica 67(2):621–649. https://doi.org/10.1007/s11600-019-00260-y

Ndehedehe CE, Awange JL, Agutu NO, Okwuashi O (2018) Changes in hydro-meteorological conditions over tropical West Africa (1980–2015) and links to global climate. Glob Planet Change 162:321–341. https://doi.org/10.1016/j.gloplacha.2018.01.020

Ndehedehe CE, Burford MA, Stewart-Koster B, Bunn SE (2020) Satellite-derived changes in floodplain productivity and freshwater habitats in northern Australia (1991–2019). Ecol Indic 114:106320. https://doi.org/10.1016/j.ecolind.2020.106320

Ndehedehe CE, Ferreira VG, Agutu NO (2019) Hydrological controls on surface vegetation dynamics over West and Central Africa. Ecol Indic 103:494–508. https://doi.org/10.1016/j.ecolind.2019.04.032

Ndehedehe CE, Ferreira VG, Agutu NO, Onojeghuo AO, Okwuashi O, Kassahun HT, Dewan A (2021) What if the rains do not come? J Hydrol 595:126040. https://doi.org/10.1016/j.jhydrol.2021.126040

Ndehedehe CE, Ferreira VG, Onojeghuo AO, Agutu NO, Emengini E, Getirana A (2020) Influence of global climate on freshwater changes in Africa's largest endorheic basin using multi-scaled indicators. Sci Total Environ 737:139643. https://doi.org/10.1016/j.scitotenv.2020.139643

Ndehedehe CE, Onojeghuo AO, Stewart-Koster B, Bunn SE, Ferreira VG (2021) Upstream flows drive the productivity of floodplain ecosystems in tropical Queensland. Ecol Indic. https://doi.org/10.1016/j.ecolind.2021.107546

Ndehedehe CE, Stewart-Koster B, Burford MA, Bunn SE (2020) Predicting hot spots of aquatic plant biomass in a large floodplain river catchment in the Australian wet-dry tropics. Ecol Indic 117:106616. https://doi.org/10.1016/j.ecolind.2020.106616

Normandin C, Frappart F, Lubac B, Bélanger S, Marieu V, Blarel F, Robinet A, Guiastrennec-Faugas L (2018) Quantification of surface water volume changes in the Mackenzie Delta using satellite multi-mission data. Hydrol Earth Syst Sci 22(2):1543–1561. https://doi.org/10.5194/hess-22-1543-2018

Ojha C, Shirzaei M, Werth S, Argus DF, Farr TG (2018) Sustained groundwater loss in California's central valley exacerbated by intense drought periods. Water Res Res 54(7):4449–4460. https://doi.org/10.1029/2017WR022250

Ozesmi SL, Bauer ME (2002) Satellite remote sensing of wetlands. Wetlands Ecol Manag 10(5):381–402. https://doi.org/10.1023/A:1020908432489

Pekel J-F, Cottam A, Gorelick N, Belward AS (2016) High-resolution mapping of global surface water and its long-term changes. Nature 540:418–422. https://doi.org/10.1038/nature20584

Preisendorfer R (1988) Principal component analysis in meteorology and oceanography. Developments in atmospheric science, vol 17. Elsevier, Amsterdam

Rondeaux G, Steven M, Baret F (1996) Optimization of soil-adjusted vegetation indices. Remote Sens Environ 55(2):95–107. https://doi.org/10.1016/0034-4257(95)00186-7

Sakamoto T, Nguyen NV, Kotera A, Ohno H, Ishitsuka N, Yokozawa M (2007) Detecting temporal changes in the extent of annual flooding within the Cambodia and the Vietnamese Mekong Delta from MODIS time-series imagery. Remote Sens Environ 109(3):295–313. https://doi.org/10.1016/j.rse.2007.01.011

Seydi ST, Akhoondzadeh M, Amani M, Mahdavi S (2021) Wildfire damage assessment over Australia using sentinel-2 imagery and MODIS land cover product within the Google Earth Engine cloud platform. Remote Sens 13(2). https://doi.org/10.3390/rs13020220

Talbot C, Bennett E, Cassell K et al (2018) The impact of flooding on aquatic ecosystem services. Biogeochemistry 141:439–461. https://doi.org/10.1007/s10533-018-0449-7

Tapley B, Bettadpur S, Watkins M, Reigber C (2004) The gravity recovery and climate experiment: mission overview and early results. Geophys Res Lett 31:1–4. https://doi.org/10.1029/2004GL019920

Tockner K, Lorang MS, Stanford JA (2010) River flood plains are model ecosystems to test general hydrogeomorphic and ecological concepts. River Res Appl 26(1):76–86. https://doi.org/10.1002/rra.1328

Tulbure MG, Broich M (2019) Spatiotemporal patterns and effects of climate and land use on surface water extent dynamics in a dryland region with three decades of Landsat satellite data. Sci Total Environ 658:1574–1585. https://doi.org/10.1016/j.scitotenv.2018.11.390

Tulbure MG, Broich M, Stehman SV, Kommareddy A (2016) Surface water extent dynamics from three decades of seasonally continuous landsat time series at subcontinental scale in a semi-arid region. Remote Sens Environ 178:142–157. https://doi.org/10.1016/j.rse.2016.02.034

Twele A, Cao W, Plank S, Martinis S (2016) Sentinel-1-based flood mapping: a fully automated processing chain. Int J Remote Sens 37(13):2990–3004. https://doi.org/10.1080/01431161.2016.1192304

Ward D, Petty A, Setterfield S, Douglas M, Ferdinands K, Hamilton S, Phinn S (2014) Floodplain inundation and vegetation dynamics in the alligator rivers region (Kakadu) of northern Australia assessed using optical and radar remote sensing. Remote Sens Environ 147:43–55. https://doi.org/10.1016/j.rse.2014.02.009

Ward DP, Hamilton SK, Jardine TD, Pettit NE, Tews EK, Olley JM, Bunn SE (2013) Assessing the seasonal dynamics of inundation, turbidity, and aquatic vegetation in the Australian wet-dry tropics using optical remote sensing. Ecohydrology 6(2):312–323. https://doi.org/10.1002/eco.1270

Ward DP, Pettit NE, Adame M, Douglas MM, Setterfield SA, Bunn SE (2016) Seasonal spatial dynamics of floodplain macrophyte and periphyton abundance in the Alligator Rivers region (Kakadu) of northern Australia. Ecohydrology 9(8):1675–1686. https://doi.org/10.1002/eco.1757

Wolski P, Murray-Hudson M, Thito K, Cassidy L (2017) Keeping it simple: monitoring flood extent in large data-poor wetlands using MODIS SWIR data. Int J Appl Earth Obs Geoinform 57:224–234. https://doi.org/10.1016/j.jag.2017.01.005

Worden J, de Beurs KM (2020) Surface water detection in the Caucasus. Int J Appl Earth Obs Geoinform 91:102159. https://doi.org/10.1016/j.jag.2020.102159

Xu H (2006) Modification of normalised difference water index (NDWI) to enhance open water features in remotely sensed imagery. Int J Remote Sens 27(14):3025–3033. https://doi.org/10.1080/01431160600589179

Zhao D, Lv M, Jiang H, Cai Y, Xu D, An S (2013) Spatio-temporal variability of aquatic vegetation in Taihu Lake over the past 30 years. PLOS ONE 8(6):1–7. https://doi.org/10.1371/journal.pone.0066365

Chapter 7
Remote Sensing of Surface Vegetation

Vegetation is the basic instrument the creator uses to set all of nature in motion.
Antoine Lavoisier

Abstract Given the important role of vegetation to mankind and the numerous functions and services that forest ecosystems provide, monitoring changes in vegetation is critical. This chapter focuses on eco-hydrological systems through discussion of surface and aquatic vegetation and above-ground forest biomass. The technological progress and key science as well as scientific advancement surrounding satellite-based monitoring and assessments of vegetation and forest ecosystems are detailed. A case study to evaluate and compare the potential of rainfall, soil moisture and satellite-derived terrestrial water storage as ideal indicators to assess the impact of water availability on vegetation conditions is also discussed.

7.1 Introductory Remark

Healthy vegetation is an important source of food, energy, and provides essential services including, regulating the climate system, hosting biodiversity, and storing of carbon (e.g., Trumbore et al. 2015; Zeng et al. 1999). The effective delivery of these services and functions such as biodiversity richness can be affected when climate variability impacts on changes in surface vegetation and forest cover (Hansen et al. 2013). Understanding changes in surface vegetation and how they respond to climate variability and human activities is therefore important. On the one hand, the rise in human population through increased industrialization and urban sprawl, trigger substantial ecosystem disturbance, including changes in vegetation and forest cover. On the other hand, changes in climatic conditions impact on the spatio-temporal variability and distribution of surface vegetation. Humid regions such as coastal West Africa, Amazon, and Congo Basins have more vegetation and forest cover. This is in contrast to arid and semi-arid domains like the Sahel, Australia, and Middle East

C. Ndehedehe, *Satellite Remote Sensing of Terrestrial Hydrology*,
https://doi.org/10.1007/978-3-030-99577-5_7

where rainfall is limited. The typically arid nature of these regions results in limited vegetation and forest cover. However, due to urbanization and deforestation, forest cover and vegetation in humid regions of the world are in strong decline.

In 2013, change detection of global forest cover (2000–2012) using Earth resources satellite was undertaken by Hansen et al. (2013). They reported global forest loss of 2.3 million and a gain of 0.8 million Sq km with an increasing forest losses of about 2101 Sq km per year occurring in the tropics. While the loss of forest cover can be attributed to fire, deforestation also drives the consistent changes in the records of observed forest cover. In the Central African region, for instance, forest lost of 721.0 ha per year to deforestation between 2000 and 2010 has been reported and a rising trend of 0.3–0.5% per year is anticipated between 2020 and 2030 (Nogherotto et al. 2013). Given the important role of vegetation to mankind and the numerous functions and services that forest ecosystems provide, monitoring changes in vegetation is increasingly becoming crucial to improve our understanding of climate change and how they affect the environment and humanity. Some efforts have been made in this regard to highlight the constraints of water on terrestrial ecosystems and regional land surface phenology, by relying on soil moisture and rainfall products (e.g., Guan et al. 2014; Knauer et al. 2014; Chen et al. 2014; Zhou et al. 2014; Seghieri et al. 2012; Huber et al. 2011).

The use of rainfall as a indirect observation of water availability (Chen et al. 2014; Yang et al. 2014) is limited. Arguably, the uncertainties associated with the use of rainfall and water budget indicators like net precipitation (difference between rainfall and evapotranspiration) in some regions are other problems encountered in the attempt to understand how vegetation responds to the availability of water (e.g., Velicogna et al. 2015; Zhang et al. 2009). The impacts of climate change on ecosystem productivity across the globe could be highly varied depending on land surface conditions, human activities, climate, topography, and landscape among others. These factors are more likely to result in complex terrestrial ecosystem responses to change in hydrology and freshwater. For instance, in southwestern North America, surface vegetation greenness is perceived as dependent on dry season length and rainfall variability with the former acting as a strong indicator of vegetation growth (Zhang et al. 2010). Moreover, tree fractional covers in the African sub-regions are more impacted by the length of rainy season due to compound mechanisms of hydrological cycle, even though such impacts are considerably non-linear (Guan et al. 2014). But vegetation response to hydrological conditions in the dry tropical Sahelian areas of Africa are different. The study by Seghieri et al. (2012) showed that decreased temperature was the strongest predictor of both leafing and reproductive phenophases. This chapter focuses on eco-hydrological systems through discussion on surface and aquatic vegetation and above-ground forest biomass. The technological progress and key science as well as scientific advancement surrounding their monitoring and assessments are detailed.

7.2 Above Ground Biomass

The term biomass encapsulates the above-ground and belowground living and dead mass, including shrubs, trees, roots, vines, as well as fine and coarse litter associated with the soil (Liang and Wang 2020). Aboveground biomass is categorized as all living biomass above the soil. This can include stem, branches, stump, bark, seeds, and foliage. Liang and Wang (2020) defined Aboveground tree biomass as *'the weight of that portion of the tree found above the ground surface, when oven-dried until a constant weight is reached'*. Biomass is important because of its regulatory role in carbon emission, which could be released to the atmosphere due to deforestation. The impacts of changes in biomass on terrestrial ecosystems have warranted the delineation of biomass distribution on several spatial scales to aid improved understanding of carbon emission and sequestration, their roles in influencing soil fertility and land degradation, as well as several other environmental processes (e.g., soil nutrient allocations, carbon cycles, fuel accumulation, etc.). The role of above-ground forest biomass (AGFB) in regulating the global carbon cycle is natural and essential for the optimum functioning of the biosphere. Specifically, its ability in stabilizing atmospheric carbon dioxide in both regional and global scales is a key reason why the ravaging impacts of global warming and climate change over the years have not turned lethal. For instance, an approximated 2–4 GtC/year of atmospheric carbon is impounded by the AGFB annually (Qureshi et al. 2012; Lü et al. 2010). For sustainable and effective management of the AGFB, a series of reiterative and up-to-date monitoring of diverse forest parameters such as height, diameter-at-breast height (dbh), basal area, etc. is required (Joshi and Ghose 2014; Mutanga et al. 2012; Field et al. 1998). The studies undertaken by Bond and Keeley (2005) and Kalwij et al. (2014) provide insight to the factors contributing to the degradation of grassland ecosystems as well as its impacts on AGFB. These impacts spans across biodiversity conservation to farmers and residents of communities at large. Through adequate and repeated monitoring of the grassland ecosystems, such impacts can be mitigated.

The assessment and monitoring of above-ground biomass (AGB) in grassland ecosystems has been studied and still ongoing. In dryland rangelands of Australia, Mundava et al. (2015) argue that existing methods to quantify AGB are insufficient to provide adequate results given the vast landmass of areas designated as rangelands. Their optimized framework for the assessment of AGB combined satellite image with ground observation to improve the monitoring of AGB in rangelands. A large proportion of the Earth's land area is covered by natural rangelands. Their distribution cuts across regions that are typically arid and dry and characterized by highly variable climatic conditions with limited rainfall as well as complex vegetation composition and structure, topography, and soil type. They are usually found in woodlands, tropical savannahs, grasslands, and shrublands located mostly in the semi African Sahel and mainland Australia. Due to aridity, vegetation density and coverage is limited and usually comprises species with mixed composition. This vegetation setting and composition result in high spatio-temporal changes in AGB components (Mundava et al.

2015). The need to quantify AGB on spatial scales, which support better rangeland management decisions is thus increasingly becoming crucial. Firstly, to aid practices that improve management of feed supply, and lastly, to increased productivity that underpins producers to boost their income. The combination of ground-based observations and remote sensing techniques has been used as will be discussed in subsequent sections. However, due to improved capabilities in sensor technologies, accurate assessment of above-ground biomass is increasingly being the focus of most research to enhance the sustainable management of forests, especially in areas earmarked for protection and conservation.

7.2.1 Quantifying Above Ground Biomass

The traditional methods of AGFB monitoring rely on intensive field inventory, including taxonomical information as well as subsidiary data analysis and field surveys of forests and grasslands. The use of traditional techniques in monitoring AGB is difficult and challenging due to excessive labour, the amount of time, and cost associated with it. Some traditional forest AGB methods have been classified into direct or indirect and includes those that use biomass expansion factors, allometric equations, or conversion factors such as wood density. While the feasibility of traditional techniques is limited to smaller areas, it can sometimes be inapplicable because of inaccessibility due to difficult terrain (Timothy et al. 2016).

However, the advent of satellite remote sensing (SRS) technique has phased out the use of these conventional techniques and replaced them with more efficient remote sensing observation methods. Although Henry et al. (2011) pointed out the relative accuracy of the conventional technique over their SRS counterparts in estimating AGFB, the conventional technique is still considered to be environmentally unfriendly, time-exhausting, capital intensive, and is only largely applicable to a small sample area even without an actual reflection of the actual AGFB of the region (Segura and Kanninen 2005). SRS methods on the other hand provide better substitute for estimating AGFB in terms of large-scale repetitive coverage, data integration with other streaming services and open-source data packages, cost, and accessibility, especially in this era of technological advancements and big data.

7.2.2 Remote Sensing Optical Capabilities

The use of SRS in observing AGFB has gained considerable attention over the last few years (Straub et al. 2009; Mitchard et al. 2009; Li et al. 2009; Patenaude et al. 2005; Samimi and Kraus 2004; Zheng et al. 2004). Its feasibility is hinged on the fundamental eco-physiological discretion of the grasses functional types through its distinct photosynthetic pathways. The technological refinement of SRS techniques in AGFB monitoring has been significantly improved given the enhancement of the spectral, spatial, radiometric, and temporal resolutions of the system. The significance of this conventional monitoring technique is of utmost importance in adequate management of regional/global climate change, biodiversity, and food security. SRS provides an economical and efficient method of AGFB monitoring. Its proficiency in wider range assessment, varying spatial and spectral resolutions, canopy density stratification, as well as expediting forest types creates a competent and effectual platform, and also facilitates the management of a comprehensive database through the development of an adequate field inventory. The major distinction between the conventional AGFB monitoring techniques and the SRS method lies in the ability of the latter to provide repeated coverage offering an ample temporal data required for change detection analysis and studies, and also its ease in data transmission and interchange for further analysis especially through a geographic information system integration. The rapid developments in SRS are responsible for the availability of multi and hyper-spectral data, as well as light detection and radar (LiDAR), and radio detection and ranging (radar) information on the AGFB. This has been discussed further in the next section. The measurement of ground-feature reflectance in the visible, near, middle and far infrared portions of the electromagnetic spectrum is facilitated through observations from dedicated multispectral sensors which includes, Landsat 4–5 TM (Li et al. 2009; Samimi and Kraus 2004), Landsat 7 ETM+ (Zheng et al. 2004; Samimi and Kraus 2004), Landsat 8 (Dube and Mutanga 2015a, b; Dube et al. 2017), polarized synthetic aperture radar (Mitchard et al. 2009), airborne LiDAR (Ene et al. 2012; Popescu 2007), etc.

Apart from the use of satellite data acquired from multispectral sensors, hyperspectral data are also important source of satellite data to quantify AGB. Hyperspectral Image (HSI) obtain remotely sensed information relying on hundreds to thousands of spectral channels, thus making it a highly innovative remote sensing system. As oppose to the conventional panchromatic and multi-spectral images, the hundreds of contiguous narrow spectral channels allow a better discrimination of object classes. Even though some problems (e.g., data redundancy) associated with HSI have limited their applications, novel machine learning classifiers are emerging to improve feature extraction and image classification (Okwuashi and Ndehedehe 2020). As with the latter, the use of HSI in AGB estimation has shown some progress but the problem of data redundancy for some applications still exist (Timothy et al. 2016). However, on a local or catchment scale, satellite methods that combine Normalized Difference Vegetation Index (NDVI) from Landsat ETM+ and canopy height with field measurements acquired during growing seasons have also been valuable to the process of estimating AGB. Generally, SRS techniques, be it optical or radar techniques have provided better and improved estimations of the AGFB as opposed to conventional approaches, given their improved resolution, robust image processing techniques, and enhanced signal-to-noise ratio. In the next section, the use of radar methods as emerging technologies for AGB estimation is discussed in detail.

7.2.3 Radar Systems

AGFB estimation using radar technology encompasses the application of either interferometric techniques or backscattering parameters (Ghasemi et al. 2010). The most accurate and convenient method for AGFB estimations is the use of these backscattering parameters given the possibility of correlating them against field biomass estimations using regression analysis, especially in coniferous forests. Regardless of the accuracy level of radar in assessing AGFB, it is still limited and yet to be evaluated in some forest types. The limitations in SAR observations in AFGB estimation are largely due to

- polarization effects (e.g., HV, HH, VV),
- saturation levels at different wavelengths (P, L, and C bands),
- representation of vegetation stands including ground specifications.

These limitations, however, create opportunity for extra scrutiny, especially regarding ground specifications during AGFB radar assessment.

Machine learning applications and statistical methods using radar observations are innovative approaches to AGFB estimation. For instance, four effective AGFB modelling techniques, namely random forest (RF), support vector machine (SVR), and linear mixed effects (LME) regression were applied by Gleason and Im (2012) in a comparative assessment study that employed LiDAR data. An almost similar result was however, achieved from the models at individual tree level estimates (RMSE 506, 502, 457, and 505 kg tree-1). However, the study reported a more accurate biomass model using support vector machine (RMSE 671 kg per 380 m2 plot when modelled as a tree collection), and an improved AGFB when modelled at plot level. Zhao et al. (2009) estimated AGFB and assessed the potential of LiDAR technology in mapping biomass in varying scales. They achieved this by exploring the efficiency of canopy height quartile functions and canopy height distributions as LiDAR metrics. Their results showed that the models can precisely calculate AGFB and produce coherent predictive performances across a wide variety of scales among other fitted models. Zolkos et al. (2013) conducted a meta-analysis that includes more than 70 studies on AGFB, which employed LiDAR techniques. They reported improved accuracy of LiDAR-based method to radar-based method in quantifying AGFB, further highlighting the utility of radar systems in assessing AGFB.

Apart from LiDAR, the use of Synthetic Aperture Radar (SAR) is another evolving technology for AGFB monitoring. As observed in past studies, the accuracy of the SAR technique in monitoring is largely dependent on the number of images used for processing and the site conditions which include, moisture, wind speed, temperature, etc. However, atmospheric complexities, heterogeneous temporal, and spatial resolutions as well as saturation problems existing in the above technologies have led to the introduction of LiDAR sensor in AGFB assessment. The advent of LiDAR imaging technique integrated with the advanced statistical techniques between LiDAR-derived height metrics and AGFB to mathematical approaches including, machine learning, stochastic simulations, automated tree crown delineations have birthed several studies exploring LiDAR's potential in accurate biomass estimations (Gleason and Im 2012; Cho et al. 2012). Several studies have used hyperspectral remote sensing for AGFB monitoring (Clark et al. 2011; Goodenough et al. 2008). Its accuracy over the use of AVIRIS data was noted by Goodenough et al. (2008), as well as the significance of its measurements from microwave power or optical radiance modes (Treuhaft et al. 2003).

In the highly understudied savannahs and woodlands of African ecosystems, the use of satellite radar to quantify distribution and amount of above-ground woody biomass (AGB) is gradually emerging. Of particular note is the work by Bouvet et al. (2018), which relied on L-band PALSAR mosaic to generate a continental-wide map of the AGB of African savannahs and woodlands at a resolution of 25 m. As opposed to previous approaches and available AGB datasets at varying spatial scales (30 m to 1 km), AGB in the savannahs and woodlands of Africa based on radar remote sensing show improved outputs-based visual comparisons and other validation metrics. For instance, a cross-validation with ground observation and a validation with a LiDAR-derived AGB dataset showed a good accuracy between the L-band PALSAR-derived and others.

7.3 Aquatic Weeds

The increased pattern of water use, projected impacts of climate change as well as the invasion by lethal water weeds resulted in improved freshwater monitoring techniques. Using optical SRS techniques, an improved monitoring sequence of freshwater quality and management has been recorded over the past decades. Also, given the imminent freshwater threat from invasive aquatic weeds (Fig. 7.1), the need for a more diverse monitoring technique is justified. Studies have identified some of

Fig. 7.1 Illustration of aquatic weeds and aquatic plant biomass accumulation

Fig. 7.2 Aquatic plant biomass accumulation in the Mitchell catchment in northern Queensland, Australia. Source, Griffith University National Environmental Science Programme team

these threats as biological pollution, water quality deficiency, and significant alteration to aquatic and nutrient cycle of the ecosystem (Palmer et al. 2015; Giardino et al. 2015). Aquatic weeds can also cause considerable harm to the tourism and fishing industries which can result in financial loss (Parochetti et al. 2008; Shekede et al. 2008). However, aquatic plant biomass accumulation (Fig. 7.2) during extreme wet periods can contribute to a highly productive habitats that support higher order organisms such as fish, invertebrates, and waterfowl (Ndehedehe et al. 2020a). These aquatic plants can be characterized as emergent, floating, and submerged (Fig. 7.2). The proper assessment of aquatic weeds using optical-based monitoring techniques provides fundamental information that underpins adequate control and mitigation measures to conserve water quality as well as maintain optimum ecosystem functioning. For aquatic plant biomass accumulation in riverine, large water holes and intermittent floodplain water channels, their monitoring is required to predict the impacts of climate variability on floodplain wetland ecosystems (Ndehedehe et al. 2020a).

7.3.1 *Satellite-Derived Estimates of Aquatic Weeds*

Ground based observations provide the most accurate and direct distributions of aquatic weeds (Ritchie et al. 2003). However, the unsuitability of these methods in large sub-regional basis lies in its time consuming, labour-intensive, and spatially restricted nature (Ritchie et al. 2003). Also, it is very difficult to implement the conventional technique for time series analysis and change detection assessment. The availability of SRS monitoring, however, provides an ideal and appropriate spatial and temporal monitoring of aquatic weeds in a cost-effective and timely fashion (Giardino et al. 2015; Dube et al. 2014). The iterative coverage of the SRS methodology provides data for long-term monitoring, which is important in estimating the sensation of controlling techniques of aquatic weeds (Penatti et al. 2015). Also, it proves useful for a wide range of applications, ranging from the spatio-temporal analysis of the distribution to the modelling of their spread and potential environmental influence. This is very crucial to local residents, decision-makers as well as environmental management agencies.

Optical remote sensing (ORS) platforms utilize natural radiation from the Sun and present a 2-dimensional view of aquatic weeds, and vegetation as well as other Earth surface features. Its ease in accessibility and data acquisition is well detailed in recent studies (Basuki et al. 2013; Lu et al. 2012). ORS platforms exhibit distinct attributes such as local to global operational capabilities with sensors such as MODIS, AVHRR, Landsat TM providing frequent regional and global spatial data. The use of hyperspectral and optical datasets has dominated aquatic ecosystem research (i.e., water quality, wetland vegetation mapping, etc.), especially as a result of its high accuracy ranging from 75 to 95%. Andrew and Ustin (2008) achieved approximately 90% accuracy while mapping Lepidium using aggregated classification and regression tree models, while Dube et al. (2017) recorded a 92% accuracy in analyzing water hyacinth using Landsat 8 OLI data. Also, Siciliano et al. (2008) highlighted positive relationships between nitrogen levels and plant responses in nutrient enrichment estimation. The study also showed a significant correlation between Photochemical Reflectance Index (PRI) and Derivative Chlorophyll Index (DCI) with water nutrients. However, the study undertaken by Lu (2006) revealed poor estimation of biomass as a result of mixed pixels sizes while applying coarse spatial resolution optical sensors in biomass estimation.

7.4 Assessing Crop Conditions and Vegetation Health

Food insecurity a major global challenge, given the poor balance between food production and demand in the face of rising variations in climatic events, pests, and diseases, as well as the increasing impact of global warming, which directly affects crop health and overall yield. The impacts of pests and diseases on food production are alarming, since such impacts have the potential of affecting a wide range of crops, thereby resulting in substantial yield losses. This therefore results in a growing concern to frequently and accurately map and monitor farms as well as vegetation. This monitoring will aid the assessment of regional and global food security amidst surging crisis, especially in some parts of India and sub-Saharan Africa where poverty and hunger are immense. One of the severe impacts of climate change on crop yield is the periodic effects in the transpiration of diseases and pests. Emerging pests, including, stock borers and diseases, such as maize steak virus and grey leaf spot pose a significant amount of threat to regional and global food security. In some regions of Africa, e.g., Zimbabwe, Kenya, and South Africa, agricultural products like maize yields are lost either to pests or diseases. Despite these losses, a major challenge lies in the deficiency of appropriate techniques in monitoring the occurrence probability, direction, and spread of these diseases and pests, and the severity of their consequences on crops. Although, with the ongoing enhancement in remotely sensed data and satellite technology, a great mapping potential is imminent in monitoring cropland areas infected by diseases and pests, as well as potential susceptible zones over vast areas.

7.4.1 Remote Sensing Applications in Agriculture

Satellite remote sensing (SRS) techniques still remain the most viable, most efficient, and accurate method for obtaining critical crop information, like yield forecasts, levels of diseases and pests infestations, as well as other important crop information, such as areal statistics. Some merits of satellite method include its cost-effectiveness, frequency as well as synoptic coverage of regional and global scenes. The basic principle of crop monitoring using SRS technique lies in the ability of optical sensors on board these satellite systems to detect leaf colour, morphology, chlorosis, and changes in transpiration rates to aid the assessment of healthy and unhealthy crops. Several studies have used SRS method to monitor croplands and crop health, especially, for large-scale commercial farmlands (Thenkabail et al. 2010). Improvements in high-resolution remotely sensed data have shown great potential in assessing and monitoring fragmented crop lands, as well as striped regions infected or prone to diseases.

The need for the development of a reliable, cost-effective, and accurate means of monitoring perennial trees such as coffee requires urgent attention. This is particularly important given their tendency in growing over large areas as well as representing long term and higher levels of investment. With the use of remote sensing capabilities, it becomes easier to accurately monitor and identify farm areas experiencing poor crop growth, disease outbreaks, and pest infestations, thereby creating room for timely interventions. By relying on the spectral, temporal, and spatial characteristics of raw image bands and their potential to generate vegetation index, remotely sensed data provides accurate and indispensable opportunities for specific crop condition assessments. The robust nature of vegetation indices like the Normalized Difference Vegetation Index (NDVI) has made it one of the most popular and commonly applied remote sensing indices to assess plant conditions. Such eco-hydrological assessments include vegetation health and growth and crop yield (Glenn et al. 2008). The NDVI utilizes the contrasting effects of vegetation on the red and near infrared portion of the spectrum to present information on vegetation conditions. It has been widely applied in vegetative assessment of health and performance (Bento et al. 2018, 2020). It is a linear estimator of photosynthetically active radiation fraction diverted by vegetation and thus proficiently helpful in analysing net primary productivity patterns at varying spatial and temporal scales (Alcaraz-Segura et al. 2009).

However, one notable limitation associated with the use of NDVI in eco-hydrological assessment is its propensity to saturate under closed canopy and also its sensitivity to soil and atmospheric conditions (Xiao et al. 2003). In this regard, other vegetation indicators such as the Enhanced Vegetation Index (EVI) and Optimized Soil Adjusted Vegetation Index (OSAVI) were introduced (Rondeaux et al. 1996). Together with Land Surface Water Index, these vegetation indices have been employed in land cover monitoring and crop yield assessment. When combined with inundation metrics like modified normalized difference water index, OSAVI appears to be a suitable metric for quantifying green biomass to identify hotspots of aquatic primary producers in wet–dry areas unlike the NDVI (Ndehedehe et al. 2020c).

Dangwal et al. (2016) monitored water stress in wheat using Landsat TM. Their study using the SRS integrated with derived vegetation indices showed statistically significant results (RMSE = 0.12, r2 = 65%). This assessment was also corroborated by other studies attempting to show the importance of Landsat NDVI anomalies in land degradation identification as specifically applied over the hotspot of Eastern Tibetan Plateau, China (Fassnacht et al. 2015) and in Basilicata, Italy (Lanfredi et al. 2015). Comparatively, Ding et al. (2014) using 30 m spatial resolution Landsat data suggested that the use of NDVI parameters yields the highest observable heterogeneity in the developing stage of crop growth. Another study (Venteris et al. 2015) depicted the possibility of applying the 26-year Landsat data in anomalous patch identification and mapping of corn and soya bean at field level, based on their departure from the expected NDVI.

Additionally, Tatsumi et al. (2015) showed the ability of SRS capabilities in optimum (r2 = 81%, k = 0.71) agricultural crop mapping in homogeneous locations through the exploitation of spatial and temporal feature. The Synthetic aperture radar (SAR) data has been combined with optical Landsat in crop mapping, and is the latest of remote sensing applications in agriculture. This was demonstrated by Onojeghuo et al. (2017) who combined multi-temporal Sentinel-1A SAR data with NDVI images to map paddy rice fields. This assessment was enhanced by machine learning-based algorithms such as the support vector machines and random forest. But satellite technology, especially radar, and hyperspectral data certainly provide great opportunities in agricultural studies and analysis of land cover states (Okwuashi

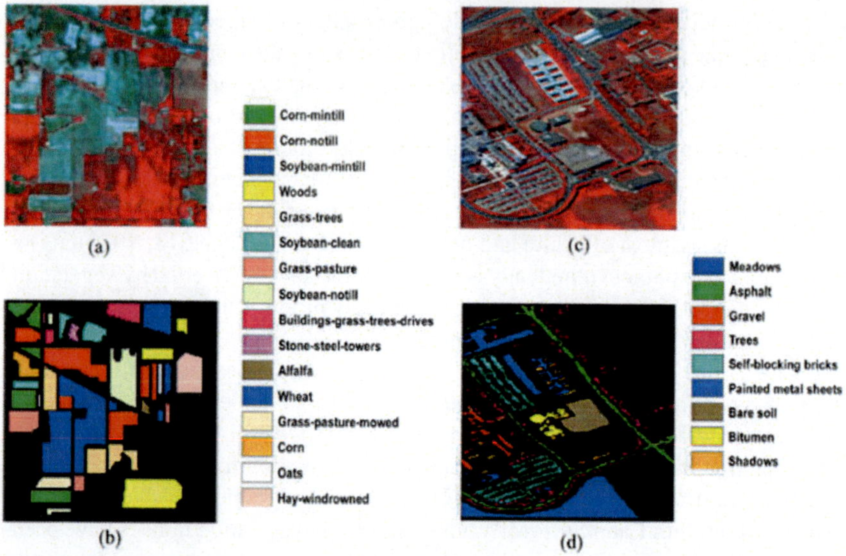

Fig. 7.3 Classification of crops using Hyperspectral data. **a** False colour composite for Indian Pines **b** Ground truth labels for Indian Pines **c** False colour for University of Pavia **d** Ground truth labels for University of Pavia. *Source* Okwuashi and Ndehedehe (2020)

et al. 2021; Okwuashi and Ndehedehe 2020). For instance, Fig. 7.3 shows the Hyperspectral data for the Indian Pines area in Northwest USA and the University of Pavia, Italy (Fig. 7.3a and c). The Indian Pines and University of Pavia data contain 16 and 9 ground-truth classes, respectively, (Fig. 7.3b and d). As opposed to the study undertaken by Onojeghuo et al. (2017) where popular machine learning enhanced the classification of paddy rice field, the use of novel deep machine learning methods like the Deep Support Vector Machine (DSVM) improved feature extraction (crops) from hyperspectral image regardless of regions (Figs. 7.4a–l and 7.5a–l). This classification was enhanced by the DSVM, which performed better than other several array of popular classifiers, including Support Vector Machine, Deep Neural Network, Gaussian Mixture Model, K Nearest Neighbour, and K Means in both regions (Figs. 7.4a–l and 7.5a–l). But the classification of Polarimetric Synthetic Aperture Radar (PolSAR) over agricultural or cultivated areas could be difficult. Arguably, this can be due to the presence of speckles in the image unlike hyperspectral data. However, the use of this novel machine learning (DSVM) has substantially improved land cover classification of PolSAR data (Okwuashi et al. 2021).

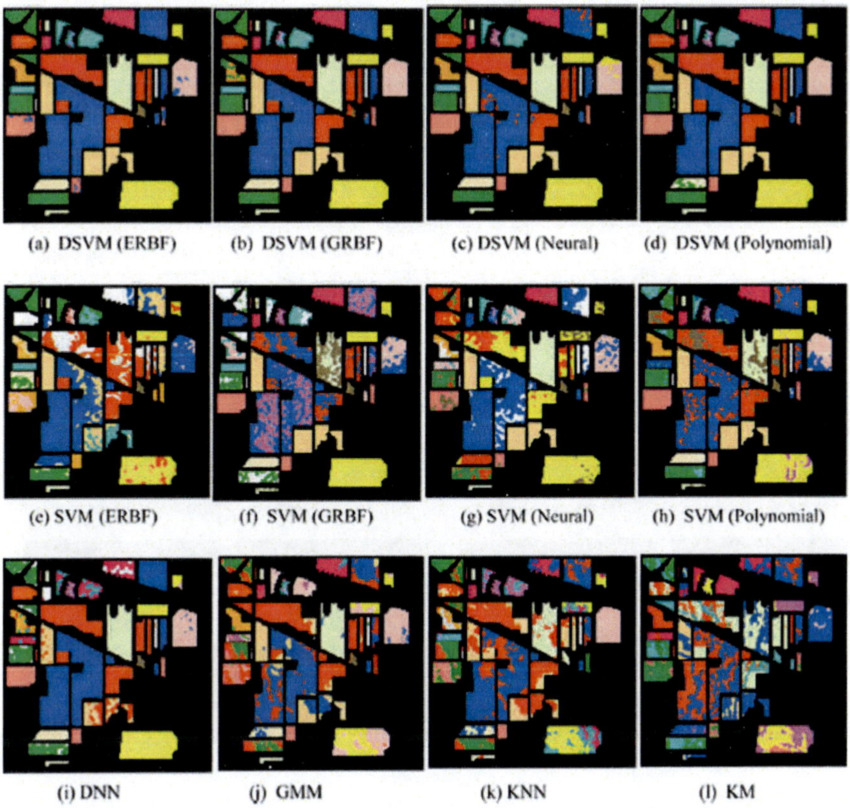

Fig. 7.4 Classified images of crops for Indian Pines using Deep Support Vector Machine (DSVM), Support Vector Machine (SVM), Deep Neural Network (DNN), Gaussian Mixture Model (GMM), K Nearest Neighbour (KNN), and K Means (KM). The legend is the same as indicated in Fig. 7.3. *Source*, Okwuashi and Ndehedehe (2020)

7.4.2 *State-of-the-Art and Emerging Technologies*

New multispectral space-borne Earth observation instruments (e.g., RapidEye, WorldView-2, Sentinel-2 multispectral image) have adopted narrow wavebands including, those in the red-edge region of the electromagnetic spectrum, which are unavailable in existing sensors such as ASTER, MODIS, and Landsat series. With the novel application of these new SRS-based techniques using machine learning algorithms, timely and accurate crop monitoring (yield estimation and health status) is now improved. The potential of the new Sentinel-2 multispectral image data in advancing satellite remote sensing applications has resulted in increasing interest in their broad use to assess environmental changes on different spatial and temporal scales (Ndehedehe et al. 2021, 2020a; Clevers and Gitelson 2013). This is as a result of its relatively spatial and temporal high resolution, strategically important spectral

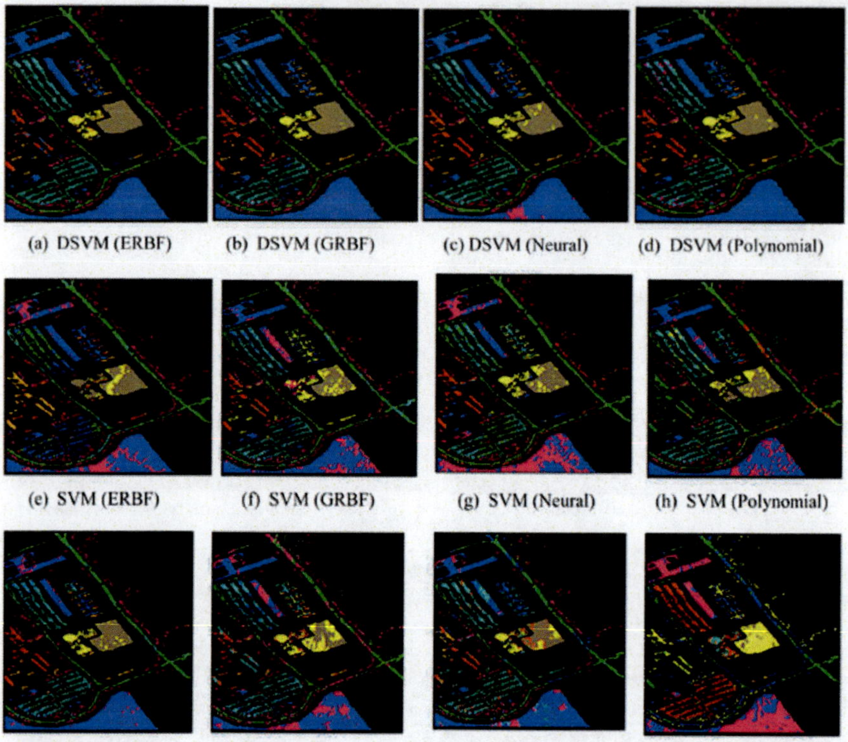

(a) DSVM (ERBF) (b) DSVM (GRBF) (c) DSVM (Neural) (d) DSVM (Polynomial)

(e) SVM (ERBF) (f) SVM (GRBF) (g) SVM (Neural) (h) SVM (Polynomial)

(i) DNN (j) GMM (k) KNN (l) KM

Fig. 7.5 Classified images of crops for University of Pavia using Deep Support Vector Machine (DSVM), Support Vector Machine (SVM), Deep Neural Network (DNN), Gaussian Mixture Model (GMM), K Nearest Neighbour(KNN), and K Means (KM). The legend is the same as shown in Fig. 7.3. *Source* Okwuashi and Ndehedehe (2020)

channels or bands, and free accessibility, thereby making it important for a range of applications, including crop yield and vegetation monitoring (Vincini et al. 2014). Another distinguishing feature of the Sentinel-2 multispectral satellite is its capitalization on the vast experience and technology of the Landsat and SPOT series over the past decades and yet maintaining relatively innovative capabilities. The Sentinel-2 multispectral imager is polar orbiting with a superior high quality spectral resolution having a huge swath-width of approximately 290 km with 13 distinct spectral bands. The spectral bands range from visible and near infrared to the shortwave infrared region of the electromagnetic spectrum. Among these bands, six provide data at 20 m spatial resolution, three at 60 m spatial resolution, while four provides at 10 m spatial resolution. With a temporal resolution of 15–30 days, providing 5-daily information about the Earth surface under cloud-free conditions, this mission captures a wide variety of crop assessment and feature analysis.

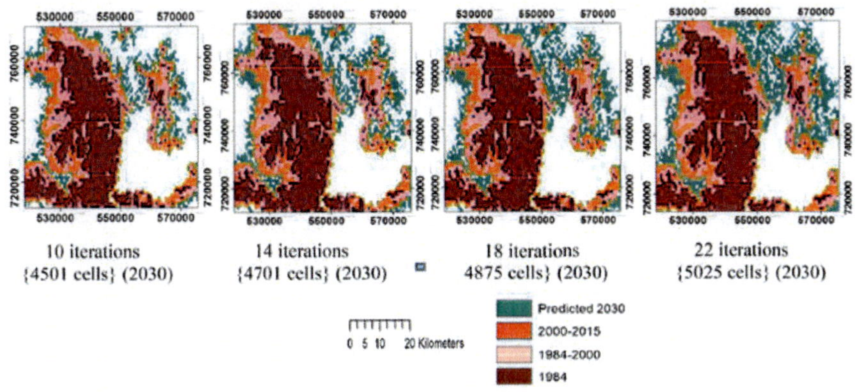

Fig. 7.6 Novel integration of Markov chain with support vector machines, and cellular automata models to predict urban states using Landsat data. *Source* Okwuashi and Ndehedehe (2021)

Novel satellite technologies with better capabilities are emerging. Their applications in various aspects of environmental monitoring remain the same and are well known. However, new machine learning algorithms have improved the use of data from these satellite platforms. For instance, the illustration in Fig. 7.6 shows how the innovative integration of Markov chain with support vector machines (SVM), and cellular automata models can be used to improve urban growth modelling based on Landsat data. The integration of novel satellite technologies with machine learning algorithms is aimed at exploring and exploiting the benefits of satellite Earth observation for informed management decision. One way this can be achieved is through the improvement of predictive models that use satellite data as key inputs. The study undertaken by Okwuashi and Ndehedehe (2021) highlighted the weakness of cellular automata models in urban growth simulation process. They argued that previous cellular automata calibrations (e.g., logistic regression and fractal-based calibrations) can be improved through the novel combination of Markov chain and cellular automata with machine learning algorithms like SVM. In this regard, the latter being a machine learning technique was employed by Okwuashi and Ndehedehe (2021) to mine the impact of several explanatory variables (e.g., distance to residential zones, industrial and commercial areas, major roads, as well as income and population potentials) that drive urban change. Urban transition probabilities between the various urban epochs were mined with Markov chain and the incremental discrete time steps based on neighbourhood interaction from an initial time to a future time were implemented using the CA model. The future urban state in 2030 as predicted based on previous urban states derived from Landsat data (1984–2000 and 2000–2015) was validated using several metrics including the McNemar's test (Fig. 7.6). The substantial fit between the predicted and reference data shows the interplay between state-of-the-art machine learning algorithms and satellite technologies, in an improved the robustness of urban change modelling urban.

7.4.3 Impact Assessment of Wildfire Events on Vegetation

With the rise in temperature caused by climate change, the occurrence of several wildfires in several regions of the world is now increasingly becoming more frequent. A wildfire is a fire that spreads rapidly and can also occur in wetlands and woodland areas. Drought periods create an ideal environment for the plenitude of dry biomass to be easily ignited, leading to fire outbreaks in vegetation and woodland areas. These wildfires destroy human life and wildlife and affect infrastructures (e.g., buildings), ecology (e.g., vegetation communities), and economy, causing damage that disrupts several natural and ecological processes on a large scale (Yeung and Li 2018). For instance, this may include increasing the probability of flood events and triggering a rise in global warming. Even though the opportunity for adaption and recovery could be restricted by other human activities and hydrological conditions like reduced rainfall and surface flow, vegetation communities can recover after a wildfire event. The use of satellite Earth observations like Landsat and Sentinel-2 is helping to provide opportunities to monitor this recovery and assess the impacts of wildfires on the environment. These satellite data are available and handy to help assess vegetation response to wildfire events under a changing climate to improve understanding of vegetation regrowth post wildfire recovery.

As with NDVI, EVI, OSAVI, and other remote sensing indicators developed to detect biophysical variables (vegetation and water) and assess productivity of ecosystems, areas affected by wildfire can be estimated by quantifying burn severity. To this end, the normalized burn ratio (NBR) is used and it is given as

$$NBR = b_{NIR} - b_{SWIR}/b_{NIR} + b_{SWIR} \tag{7.1}$$

where NIR is the near infrared band and $SWIR$ is the short wave infrared band of any optical satellite system (e.g., Landsat, Sentinel-2, etc.). The burn severity is estimated by computing NBR of the period before the fire and post-fire. The use of this indicator for burn severity assessment is based on the sensitivity of NIR to vegetation greenness and $SWIR$ to bare soil and charcoal (Yeung and Li 2018). Moreover, the difference between pre- and post-fire images from NBR can also be useful in burn severity assessment. This difference is called the Differenced Normalized Burn Ratio (i.e., $dNBR = NBR_{prefire} - NBR_{postfire}$). In the classification of NBR values (range from -1 to $+1$), higher values represent green vegetation while lower values represent areas affected by wildfire (i.e., burned areas), or bare soil, water or cloud. But in the dNBR (typically ranging from -2 to $+2$) classification, negative values indicate vegetation regrowth while values near zero (0) represent unburned areas (Yeung and Li 2018). Values greater than 0.1 in the dNBR indicate burned areas.

The example in Fig. 7.7a–b shows the temporal variations of fire events during the 2006–2020 period in the Macquarie Marshes in New South Wales, Australia based on the Fire Information for Resource Management System dataset (FIRMS). These

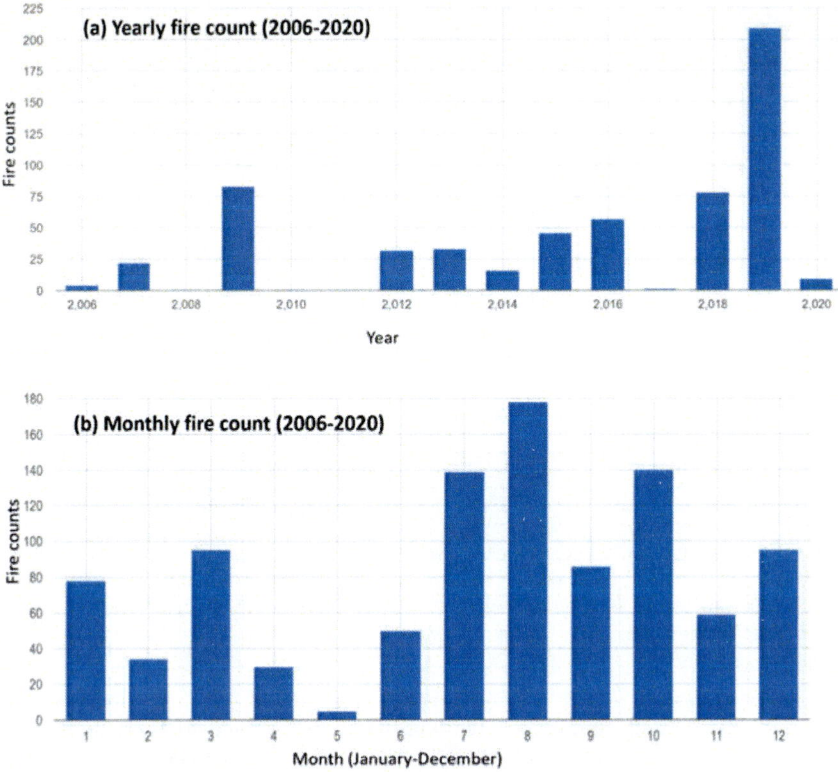

Fig. 7.7 Temporal patterns of fire events (2006–2020) in the Macquarie Marshes in New South Wales, Australia based on the Fire Information for Resource Management System dataset derived from MODIS MOD14/MYD14 times series. Fire extents can be estimated by multiplying the fire counts for each period by the area of each MODIS pixel (1000*1000 m), which has a spatial resolution of 1000 m)

data (https://earthdata.nasa.gov/firms) were retrieved from the Google Earth Engine platform (see Chap. 5) and show locations with near real-time active fire. These data are based on the standard MODIS MOD14/MYD14 Fire and Thermal Anomalies product. One of the biggest fire events in the Macquarie Marshes since 2006 is the 2019 wildfire (Fig. 7.7a). From the monthly occurrence of fire events, July, August, and October are notable months where these events peak (Fig. 7.7a). However, the Sentinel-2 data provides the opportunity to quantify estimated burn area on much higher spatial resolution and new remote indicators are facilitating the assessment of wildfires (Filipponi 2018). The false colour composite of the Macquarie Marshes pre-wildfire and during the wildfire which started in June 2019 shows some visible differences (Fig. 7.8a–b). The NBR presented in Fig. 7.8c is based on the novel Burned Area Index for Sentinel-2 (BAIS2) of Filipponi (2018). This index uses five multi-spectral Sentinel-2 bands (20 m spatial resolution) to map and identify areas

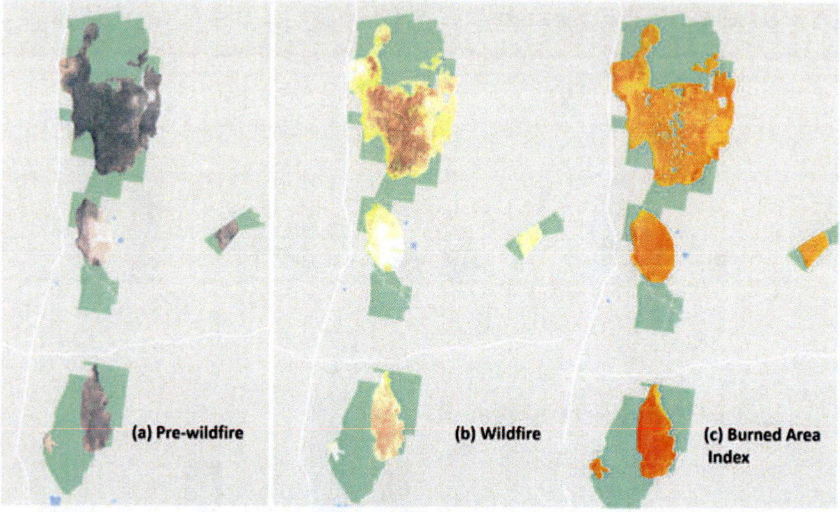

Fig. 7.8 Sentinel-2 false colour composites of the period **a** before and **b** during wildfire event and **c** spatial patterns for the 2019 fire events in the Macquarie Marshes in New South Wales, Australia. The burned area assessment is based on Sentinel-2 data using the Burned Area Index for Sentinel-2. Red pixels are burned areas

affected by fire. The spatial locations with red pixels are burned areas showing that much of southern areas of the marshes were affected by fire while the yellow pixels are unaffected or mildly affected locations (Fig. 7.8c).

7.4.4 *Water-Driven Changes in Crops and Vegetation*

The increasing population growth of major cities in the world has a significant contribution to food security and nature conservation efforts especially in the future. And given the significant variation in rainfall amount observed in some regions like West Africa, it can be argued that the pattern of crop growing which directly translates to food security is very unstable. Food insecurity is getting worse, especially in the Sudan and Sahelian regions where the high dependence of farmers on rain-fed agriculture makes them vulnerable to harsh climate, drought, and late rainfall. Given the need to sustain the growing demands of increasing human population, there must be a balance of climatic events such as rainfall and ET with crop growth and vegetation sustainability. The need to sustain population has caused several regions like West Africa to undergo significant medium-to-long term changes in the past three decades, with cropland extension, intensified grassland use, and variable rainfall patterns.

The basic results of water absorption are directly linked to water absorption occurring in the mid-infrared range, where they produce water mediated reduction in leaf reflectance in turgid plants, when compared to leaves with smaller amount of water content. While the ancillary results of water content on the leaves spectral reflectance is stimulated by the conduction of water occurring due to the leaf internal structure, cell size, and shape. This makes the ancillary method appealing in remote sensing. This is given the fact that it relies on the full width reflectance of the electromagnetic spectrum for identification and quantification of water deficit stress in crop managements and displays greater challenges in dimensionality, data handling, noise, and cost.

Remote sensing, particularly the use of hyperspectral sensors to understand water stress detection for crop management is very promising. Hyperspectral remote sensing data comprise several constricted contiguous spectral wavebands located from the visible, near infrared, mid-infrared, and thermal infrared portions of the spectrum. Many studies have confirmed the existence of strong water absorption bands in definite channels of the mid-infrared region of the electromagnetic spectrum specifically at around 1175, 1450, 1650, 1940, 2250, and 2500 nm (Jeger and Pautasso 2008). In addition to the mid-infrared region having basic bands associated with plant water content, the visible to near infrared (VIS/NIR) region (400e1300 nm) has ancillary water content related bands that can be used in water stress perception (Lin 2011). This is possible given that plant water content has several primary and secondary effects on leaf features, which in turn influences leaf, canopy, and top-of-the-atmosphere scale spectral reflectance.

Using ancillary effects of water absorption on leaf reflectance, Peñuelas et al. (1994) improved the water band index (WBI), which is built on the ratio between the reflectance at 900 nm and the water band at 970 nm and has been widely applied in plant water content evaluations with SRS techniques. For example, the WBI was the most significant parameter that uses reflectance from the optical domain in approximating plant water thickness (EWT) (Oumar and Mutanga 2014). Similarly, Eitel et al. (2006) found significant correlations between WBI and PWC ($r2 = 90\%$), and WBI and EWT ($r2 = 88\%$). This shows that ancillary effects of water absorption that occur in the VIS/NIR region of the spectrum could be beneficial in predicting plant water content and estimating plant water thickness in crop plants, without using the basic effects found beyond 1000 nm. This is critical because several notable wavebands as well as vegetation indices for water stress estimation that use primary effects of water absorption correlate with extreme atmospheric water vapour absorption (Kahn et al. 2011).

Several region and basin-specific studies have employed satellite geodetic missions like the Gravity Recovery and Climate Experiment (GRACE) as an important tool in hydrological research. These studies emphasized wide range of applications, e.g., in droughts, terrestrial water budget closure; hydrological features, sub-surface water storage, aquifer system procedures; and assessing the influence of global climate teleconnections on global and regional dynamics of terrestrial water storage (an integrated sum of changes in catchment stores, e.g., surface, canopy, and groundwater). Given that the observed correlation between long term soil moisture changes and fluctuations in some Sahelian vegetation was found to be unpredictable, more insightful studies using GRACE measurements might be useful to further evaluate water-related impacts on the region's terrestrial ecosystem (Ndehedehe et al. 2018a).

7.5 Understanding Changes in Surface Vegetation

The knowledge of vegetation water changes based on rainfall and soil moisture indicators is vital for informing the impacts of climate variability on crop production and ecosystem performance. However, established relationships of rainfall and soil moisture with changes in vegetation in some semi-arid and humid ecosystems of some regions suggest there is a need for further assessment. This is because these relationships are weak and complex, making our understanding of the response of vegetation to hydrological conditions limited. The complexity in these relationships has been linked to social (e.g., migration) and environmental (climate change, land use change, etc.) factors (Ndehedehe et al. 2019; Knauer et al. 2014). New satellite missions like the Gravity Recovery and Climate Experiment (GRACE) satellite observations have been used to improve understanding in this regard. Data from the GRACE mission have been used to measure the response of vegetation greenness to climate variability, highlighting the constraints of several environmental factors to the relationship between water availability and vegetation changes. Pioneering works along these lines (Ndehedehe et al. 2019; Yang et al. 2014) provide a foundation for monitoring the water-driven variability in surface vegetation using advanced satellite methods like GRACE. This section presents some findings and illustrations following the work of Ndehedehe et al. (2019) in the African sub-region (Fig. 7.9) where GRACE was used for the first time to assess changes in surface vegetation greenness.

7.5.1 Rainfall as Indicators of Changes in Surface Vegetation

Climate variability through changes in annual and seasonal rainfall is expected to result in surface vegetation greenness. However, the use of rainfall to assess the impact of water availability on vegetation conditions is difficult due to several factors, which include, complex landscapes, extreme rainfall variability, and complex hydrological processes among other factors (Knauer et al. 2014; Chen et al. 2014; Yang et al. 2014). Indeed, rainfall appears to be an important driver of changes in the spatial and temporal distribution of vegetation. But some studies have observed weak relationship between rainfall and vegetation indices (e.g., NDVI) in wetland areas (e.g., Georganos et al. 2017), suggesting that other factors drive changes in vegetation, and underpins the need for ideal indicators. As the relationship between vegetation conditions and rainfall in some regions could be complex, mounting evidence from several case-specific studies argue that rainfall has not adequately explained changes in surface vegetation greenness (Knauer et al. 2014). In West Africa, vegetation studies along this line have attributed this mostly to human influence and the impacts of climate change (e.g., Dardel et al. 2014; Jamali et al. 2014; Knauer et al. 2014; Boschetti et al. 2013; Olsson et al. 2005; Herrmann et al. 2005; Li et al. 2004).

Moreover, soil moisture derived from hydrological model is poorly associated with vegetation changes in the African Sahel (e.g., Huber et al. 2011). The complex nature and structure of the African rainforest restricts the use soil moisture observations from satellite in understanding how vegetation responds to water availability (e.g., Do et al. 2005). Notably, soil moisture observations from satellite remote sensing platforms may be limited and largely restricted in highly dense vegetation community (Dirmeyer et al. 2004). This is because plant tissues contain large amounts of water, which weaken the intensity of the microwave signal.

7.5.2 Soil Moisture as an Ideal Indicator for Surface Vegetation Greenness

Soil moisture is an ideal hydrological state variable that captures variability in surface vegetation greenness. However, the morphological and physiological adaptations of vegetation in some regions, especially arid domains like the Sahel and Australia can lead to complex water use by vegetation communities during the dry season (e.g., Guan et al. 2014; Seghieri et al. 2012; Huber et al. 2011). This means soil moisture may be restricted as a useful indicator to assess vegetation response to climate variability. Even though it is widely recognized that soil moisture is the most appropriate indicator that can improve our understanding of land surface phenology and the response of plants to climate variability (e.g., Chen et al. 2014; Huber et al. 2011), satellite soil moisture observations can be limited in landscapes where vegetation and plant cover, overtly depend on soil water from the saturated zone. This limitation, among other factors, has been associated with large retrieval errors (e.g., Högstrom et al. 2014) and the fact that observations of soil moisture from satellite platforms are restricted to the few centimetres of the top soil (e.g., Yang et al. 2014).

The Sahel is a good example of this. The root depth of vegetation in the Sahel, for instance, reaches the water table region, which lies up to 40 m below the surface during the dry season (Huber et al. 2011; Do et al. 2005). This makes some of the trees in the Sahel considerably less sensitive not only to inter-annual changes in rainfall but also to changes in soil moisture (be it satellite or model-derived). Additionally, environmental, pedological (e.g., soil characteristics), and changes in climatic conditions, among other factors contribute considerably to creating the diversity in local hydrology. This impact create uncertainties in our knowledge of how vegetation response to soil wetness or water availability. Under these conditions, both soil moisture and precipitation are expected to be restricted as indicators or controls for changes in vegetation in the African Sahel or semi-arid regions more generally.

7.5.3 Satellite Gravity Approach to Assessing Water-Driven Variations in Vegetation

Given the limitations of rainfall and soil moisture, terrestrial water storage (TWS) derived from satellite gravity observations such as the Gravity Recovery and Climate Experiment (GRACE, Tapley et al. 2004) has been introduced to further enhance our understanding of non-linear and spatially heterogeneous relationships between surface vegetation and hydrological indicators. TWS or equivalent water thickness (Chap. 4) as observed by GRACE is the sum of changes in soil moisture, groundwater, surface water, snow water equivalent, wetlands/canopy, etc.). TWS observed by the GRACE satellite has been used to study all of these water storage components. These components include groundwater, surface water, and outgoing water fluxes like evapotranspiration. For instance, GRACE estimates of changes in the Caspian Sea level were consistent with observations from altimeter and across several regions of the world, including the Central Valley and High Plain Aquifer and Murray–Darling Basin, well observations of groundwater level are consistent with groundwater predicted from GRACE (e.g., Chen et al. 2022). With the understanding of GRACE capability to capture changes in water storage components and hydrological stores (groundwater, surface water), arguably it can as well serve as a soil moisture surrogate, especially in arid or semi-arid regions, which are typically dry.

GRACE satellite data are now gradually emerging as an important alternative product to (i) evaluate temporal and spatial patterns in soil moisture and groundwater at watershed scale, and to assess the agro-ecological system of a semi-arid catchments (e.g., Ndehedehe et al. 2018a; Abiy and Melesse 2017; Famiglietti 2014; Wouters et al. 2014) (ii) study the impact of climate variability on vegetation water changes in arid ecosystems and Arctic biomes (see, Velicogna et al. 2015; Yang et al. 2014), and (iii) assess hydrological processes, water budget, drought variability, sub-surface water storage, and aquifer system processes (Ferreira et al. 2018; Ndehedehe et al. 2016a, 2017; Ferreira and Asiah 2015). However, its use in data deficient regions like Africa as a proxy for available plant/vegetation water is gradually emerging. Because of the restrictions of soil moisture in water-limited ecosystems as drivers of vegetation dynamics, and the fact that sub-surface water use during dry season exists in semi-arid regions (see, e.g., Guan et al. 2014; Huber et al. 2011), the potential use of GRACE-derived TWS as a driver of vegetation dynamics is therefore essential. This is critical and provides complementary perspectives on water availability as a driver of vegetation dynamics. This further advances our knowledge of wetland ecology and its biophysical systems, but much more extending the use of GRACE-derived TWS as an innovative biophysical tool that could support global terrestrial ecosystem assessment, especially in data sparse regions. In the next section, the use of GRACE satellite to provide hydrological control for observed changes in surface vegetation over sub-Sahara Africa based on temporal and spatial relationships is discussed. Surface vegetation is characterized by monthly time series of Normalized Difference Vegetation Index (NDVI). The goal is to (i) assess the temporal relationships of NDVI with GRACE-based TWS and rainfall, (ii) investigate the dominant spatio-temporal patterns of NDVI, and (iii) assess the spatial relationship of NDVI with TWS, rainfall, and groundwater at monthly and seasonal time scales.

7.6 Surface Vegetation Over West and Central Africa

7.6.1 West Africa

7.6.1.1 Climate

West Africa is located between latitudes 0 °N to 20 °N and longitudes 20 °W to 20 °E and comprises a group of 16 countries (Fig. 7.9) covering an area of approximately 7.5 Million km^2. The two major geographical zones are the countries of the gulf of Guinea (i.e., the area between latitudes 4 °N and 8 °N), and the Sahelian countries (Fig. 7.9). The boundary on the south is the Atlantic ocean; on the north is the Sahara desert, while the eastern boundary is flanked by the Cameroon mountains (Ndehedehe et al. 2016a). The climate of West Africa is regulated by the migration of the intertropical convergence zone (ITCZ). The location, position, and intensity (strength) of both

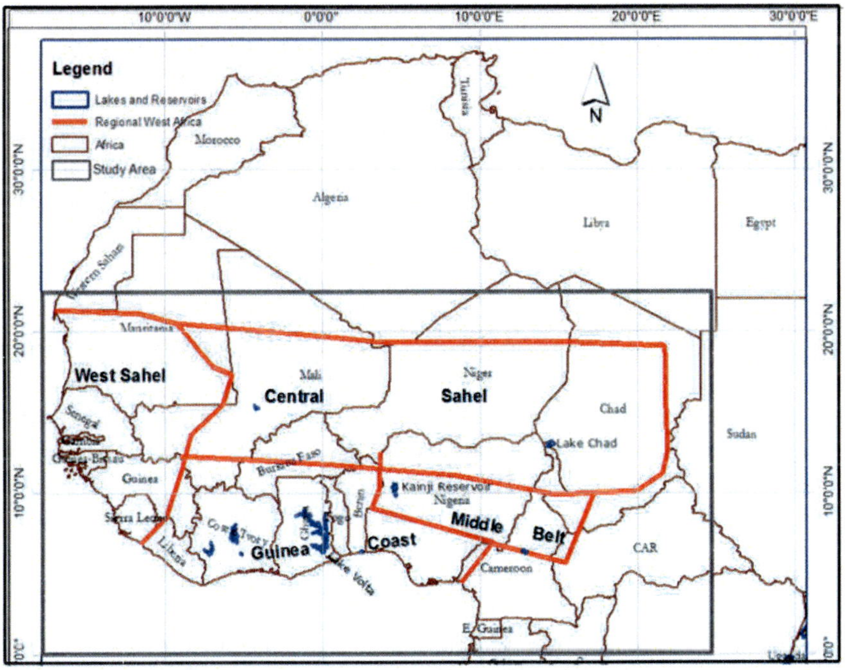

Fig. 7.9 The West African sub-regions showing **a** West Sahel, **b** Central Sahel, **c** Guinea Coast Middle belt. The blue polygons are freshwater bodies and lakes. *Source* Ndehedehe et al. (2016a)

the circulation features and ITCZ contribute to climate variability in the region and intra-annual rainfall distribution in the Sahel as well as the gulf of Guinea, which lies along the coast (Ndehedehe et al. 2016a). Rainfall ranges from less than 200 mm/yr in the dry Sahelian countries to over 2000 mm/yr in the gulf of Guinea or coastal West African countries. Rain seasons in the gulf of Guinea occur between April–June and July–September, with the wettest months being June and September or sometimes October (Nicholson et al. 2000). In the Sahel and fringe of the desert, the rain season is during the June–September period with maximum rainfall occurring in August.

7.6.1.2 Vegetation System

The coastal regions of West and Central Africa (WCA) (Fig. 1a) are heavily forested ecosystems largely due to high amounts of annual rainfall compared to other areas in the region (Ndehedehe et al. 2019). As shown in Fig. 7.10, there is a sharp contrast in the distribution of vegetation along the wet humid coastal regions West Africa compared to the typically dry semi-arid Sahel region in the North. This diversity in vegetation density in the region is largely driven by the migration of the tropical rainbelt and is discussed further in Sect. 7.6.2. This chapter showcases the results of GRACE hydrological monitoring of surface vegetation greenness over West and Central Africa (Fig. 7.9) based on published work of Ndehedehe et al. (2019). Here the temporal variations, seasonal, and mean annual cycles of Normalized Difference Vegetation Index (NDVI) are compared with those of terrestrial water storage (TWS) and rainfall. Specifically, the potential use of TWS as an alternate indicator to rainfall in understanding vegetation response to hydrological conditions in the region is being explored and discussed. The scope of this work is limited to countries in West and Central Africa (Fig. 7.9). However, the statistical relationships developed between NDVI and TWS/rainfall are valuable to exploring similar relationships in other tropical and semi-arid or arid ecosystems in the world.

7.6.2 Vegetation Distribution Along Climatic Gradients

The distribution of NDVI over West Africa shows that relatively higher NDVI values are observed in the coastal countries. This occurs during the January–March, April–June, and July–September periods (Fig. 7.10). Conversely, low NDVI values are observed during the September–November and October*–December periods (Fig. 7.10). Plant cover is usually expected to be higher in extremely tropical wet areas (e.g., the Guinea zone) irrespective of the season (i.e., dry or wet periods), leading to higher values in NDVI as opposed to semi-arid regions (Sahel zone) where vegetation and species richness respond gradually to changes in precipitation, which could be relatively low and highly varied. Changes in NDVI or vegetation activities still occur during dry seasons along the coastal countries of West and Central Africa (WCA). For example, this is notable in the January–March period in Liberia, Côte d'Ivoire,

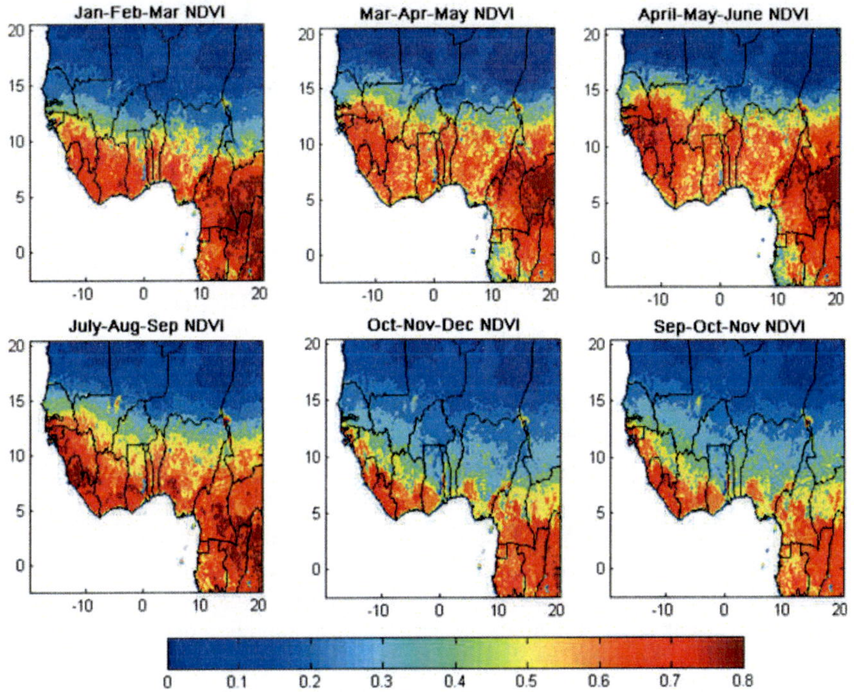

Fig. 7.10 Average seasonal NDVI distribution over West and Central Africa for the period between 2002 and 2013. The seasonal classifications are based on crop and rain seasons in WCA and are described as: January–March; March–May; April–June; July–September; September–November; October–December. *Source* Ndehedehe et al. (2019)

southern Ghana, and Cameroon where despite receiving relatively low rainfall during the period (dry season) indicate high NDVI values. This perhaps is due to the presence of permanent wetlands and huge evergreen broadleaf forests that thrives in water logged soils.

Consequently, as was earlier argued, rainfall would be inadequate in assessing vegetation conditions in such areas as the vegetation and forest compositions are mostly sensitive to soil moisture conditions that are in the root zones. Apart from soil type, climate variability play critical roles in the spatial and temporal distributions of surface vegetation in WCA. Hence, assessing the potential of GRACE-TWS as a hydrological indicator as opposed to other physical variables (precipitation, soil moisture) in providing useful information regarding our understanding of the spatial and temporal variations of vegetation dynamics is warranted. To this end, the spatial and temporal variations of monthly, annual, and residual NDVI anomalies over West Africa and selected catchments (Congo Basin, Mali, and Lake Chad Basin) are compared with those of TWS and precipitation. The selected catchments are tentative test beds employed to examine the relationship of NDVI with TWS and rainfall in a different climatic domain. In other words, they are so chosen to be representative of humid, semi-arid and arid characteristics of tropical WCA systems. Notably, exploring the water-driven variability of surface vegetation at catchment scale could be useful in advancing process-level insight into the vegetation or climate characteristics that modulate the couplings of NDVI with different hydrological indicators.

7.6.3 Inter-annual Changes in Vegetation Greenness

In this section, the relationship of annually aggregated NDVI cumulative departures with those of TWS and rainfall is compared. Comparing the cumulative annual departures (i.e., cumulative annual anomalies) of NDVI with TWS and rainfall masked over West Africa, the temporal patterns of NDVI appear to be consistent with those of rainfall, both showing a fall during the 2003–2005 period (Fig. 7.11a–b). NDVI showed a rise during the 2005–2013 period in their cumulative annual anomalies while the increase in TWS (2009–2013) only came after a steady condition (turning point) during the 2007–2009 period (Fig. 7.11a). The cumulative annual departure of TWS between 2007 and 2009 could be an hibernation period when the hydrology of the region was trying to reset owing to an improved rainfall condition (Fig. 7.11b). As opposed to TWS, the cumulative departure of rainfall (Fig. 7.11b) is characterized by several major turning points with rainfall plummeting in 2010.

Rainfall appears to be consistent with the temporal pattern of NDVI except for the anomalous years of 2009 and 2011 (Fig. 7.11b). Furthermore, the annual time series of cumulative annual anomalies (NDVI, TWS, rainfall) averaged in selected subregions or basins (south Mali, Lake Chad Basin, and Congo Basin) were also assessed given that cumulative annual anomalies over West Africa may be dominated by the tropical regions having very large NDVI, rainfall, and TWS values. The correlation results (Table 7.1) for these region-specific cases suggest that the temporal variations in surface vegetation greenness are more associated with TWS than with rainfall (Fig. 7.11c–h). The annual time series of rainfall and TWS in the southern Mali is significantly correlated with NDVI but over the Lake Chad and Congo Basins rainfall shows no statistically significant association with NDVI (Table 7.1). But the annual cumulative departures of NDVI and TWS in the Congo Basin and West Africa showed considerable and modest associations, respectively (Table 7.1).

The observed cumulative annual departures in Fig. 7.11 provide an overview of the eco-hydrological narrative of WCA during the 2003–2013 period. In particular, the role of land use change on surface runoff (e.g., Descroix et al. 2009; Favreau et al. 2009; Séguis et al. 2004). The impact of surface water schemes on eco-hydrological processes in West Africa (e.g., Ndehedehe et al. 2017) may also contribute to the relationship between NDVI and TWS, making it complex and difficult to understand. GRACE-derived TWS integrates precipitation over time, and for coastal areas of West Africa, the most variable components would mainly be surface waters (e.g., lakes and reservoirs), and soil moisture. The large changes in these components of GRACE-TWS along the coastal West African region could be inconsistent with

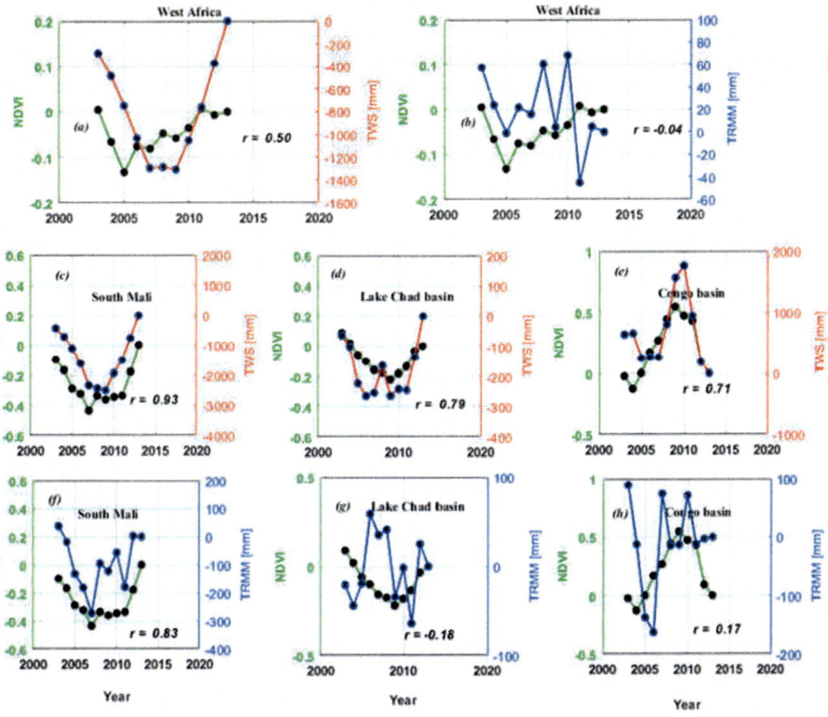

Fig. 7.11 Comparison of cumulative annual departures of NDVI with those of TWS and TRMM-based precipitation during the 2003–2013 period in WCA. **a**–**b** Cumulative annual departures (NDVI vs TWS and NDVI vs TRMM) over West Africa, **c**–**e** cumulative annual departures of NDVI with TWS, and **f**–**h** cumulative annual departures of NDVI with precipitation over the selected sub-regions (southern Mali, Lake Chad and Congo Basins). *Source* Ndehedehe et al. (2019)

NDVI patterns as variations in vegetation could be largely regulated by antecedent and complex conditions. For example, significant drops in rainfall may not necessarily translate to immediate loss of surface vegetation greenness. This is because the impact of a dry condition may also depend on the amount of water (or TWS) available in the hydrological stores of that catchment prior to rainfall deficits, as well as evapotranspiration rates and land surface conditions. This is one reason why observed spatial relationships of TWS in some coastal West African countries, while insignificant in some areas, are generally significant but considerably poor and non-linear in other locations.

But for other semi-arid ecosystems of West Africa, vegetation productivity is a function of soil moisture conditions among other factors. It is therefore expected that soil moisture in these areas will respond faster to severe water deficit conditions. Apparently, aridity is a typical climate feature of the Sahel region. Soil moisture and groundwater are therefore important hydrological stores that would be relatively the most variable water storage components of TWS as observed by the GRACE

Table 7.1 Temporal correlations (r) of NDVI with TRMM-based precipitation and TWS in southern Mali, Lake Chad Basin, Congo Basin, and West Africa during the 2002–2013 period. Correlation coefficients are significant at the 95% significant level using the Student t-test except for those marked with *

S/N	Temporal correlations	Southern Mali	Lake Chad Basin	Congo Basin	West Africa
1	NDVI versus TRMM (monthly)	0.72	0.73	0.70	0.68
2	NDVI versus TWS (monthly)	0.79	0.83	0.09*	0.74
3	NDVI versus TRMM (annual)	0.68	0.21*	0.09*	0.16*
4	NDVI versus TWS (annual)	0.79	0.58	0.42*	0.27
5	NDVI versus TRMM (annual cumulative)	0.83	−0.18*	0.17*	−0.04*
6	NDVI versus TWS (annual cumulative)	0.93	0.79	0.71	0.50

mission. Hence, the temporal association relationship between TWS and NDVI in the Sahel is therefore expected to be statistically significant and strong (Table 7.1). Nonetheless, the NDVI-TWS temporal relationship may result in significant time lags in some hydrological regions due to aquifer properties, the extent of inundated areas, and other human interventions (e.g., Ndehedehe et al. 2016a, 2017). Overall, unlike rainfall, TWS tend to mimic that of NDVI more closely in all cases and are modestly associated with each other for West Africa but shows considerably strong relationship in selected sub-regions (Fig. 7.11a, and c–e). TWS act as a buffer for water availability despite rainfall deficits, and such relationship is critical to aid the assessment of water availability (excess water or its lack thereof) for vegetation growth in terrestrial ecosystems.

7.6.3.1 Dominant Patterns in Surface Vegetation Changes

This section highlights the dominant or leading spatio-temporal patterns of NDVI over West and Central Africa (WCA) using independent component analysis (ICA Ndehedehe et al. 2016b; Cardoso 1999). ICA is an advanced multivariate technique used to decompose data grids into spatial maps and temporal patterns and is fully detailed in Chap. 13. In this case the NDVI grids over West Africa can be decomposed as

$$\mathbf{X}_{NDVI}(x, y, t) = \mathbf{TM}, \tag{7.2}$$

where the NDVI data matrix $\mathbf{X}$ consists of spatial maps $\mathbf{M}$, and temporal patterns $\mathbf{T}$, and (x, y) are grid or spatial locations, t is the time in months. $\mathbf{T}$ is unit-less because it has been normalized using its standard deviation. The corresponding spatial patterns **M** (i.e., associated with **T** have been scaled using the standard deviation of its independent components (i.e., **T**). As discussed further in Chap. 13, each ICA mode is a pair of the independent components and spatial patterns, which describes the combined variability in the data (temporal and spatial patterns). These are usually interpreted together, that is, the unit-less temporal pattern is multiplied with the corresponding spatial pattern to obtain the actual values of NDVI for a specific location. In Fig. 7.12 (right), the four leading temporal evolutions of explaining much of the observed variability in NDVI are obtained from the regionalization (or localization) of monthly NDVI grids over WCA using the ICA technique. These temporal series of NDVI correspond to spatial patterns (Fig. 7.12, left) that are distributed along different climatic domains. The first NDVI mode indicates variability localized over the Sudano-Sahelian band. The second NDVI mode on the other hand is exemplified by a pattern over the Central Sahel and Guinea Coast with an opposite phase to the Guinea Coast during the September period (Fig. 7.12, left). This pattern corresponds to an approximate phase shift of 0.5 years (or 6 months) and is likely caused by the latitudinal displacement of the rainbelt, leading to inter-annual variability of rainfall towards the north and south of approximately latitude 10 °N (Nicholson 2009).

With respect to the WCA region, changes in vegetation density and photosynthetic activity are mostly concentrated around the Sudano-Sahel belt (IC1, Fig. 7.12, left) and the Central Sahel (IC2, Fig. 7.12, left), both accounting for approximately 51.4% and 15.1% of the total variance, respectively. The observed temporal patterns of NDVI over the Sudano and Central Sahel suggest considerably strong annual and multi-annual variations, respectively (IC1 and IC2, Fig. 7.12, right). Overall, the leading NDVI mode of variability is consistent with the dominant patterns of rainfall and TWS in West Africa as shown in a recent study (Ndehedehe et al. 2016a). This presents a preliminary picture of both rainfall and TWS as indicators of surface vegetation over the Sudano-Sahel belt. As the main focus of this section is on the temporal assessment of the leading modes of NDVI with TWS, the relationship of the former with rainfall is also explored. Spatially averaged rainfall over West Africa is strongly associated with NDVI temporal evolutions in the Sudano-Sahel (IC1, Fig. 7.12 and Table 7.2) compared to the relationship of averaged rainfall over the Central Sahel with NDVI (IC2, Fig. 7.12 and Table 7.2).

This shows a strong response of vegetation to rainfall in the Sudano and semi-arid Sahel regions. Over the Central Sahel (IC2, Fig. 7.12), the temporal variation of TWS showed a relatively strong association with NDVI while in the Sudano-Sahel area (IC1, Fig. 7.12), TWS showed slightly less agreement with NDVI when compared to rainfall (Table 7.2). Their coefficient of determination (R^2) shows that TWS explains 47% while rainfall explains 59% of the annual signals in NDVI over the Sudano-Sahel belt (IC1, Fig. 7.12, Table 7.2). In the Central Sahel (IC2, Fig. 7.12, Table 7.2), TWS explains slightly higher portion of the variability in NDVI ($R^2 = 52\%$) than rainfall ($R^2 = 48\%$). The relationship between TWS and vegetation condition in the Central Sahel suggests that TWS imply a possibility of using TWS as a hydrological indicator for assessing vegetation response to changes in hydro-climatic conditions. This is true since it is a more direct measure of water availability in any given

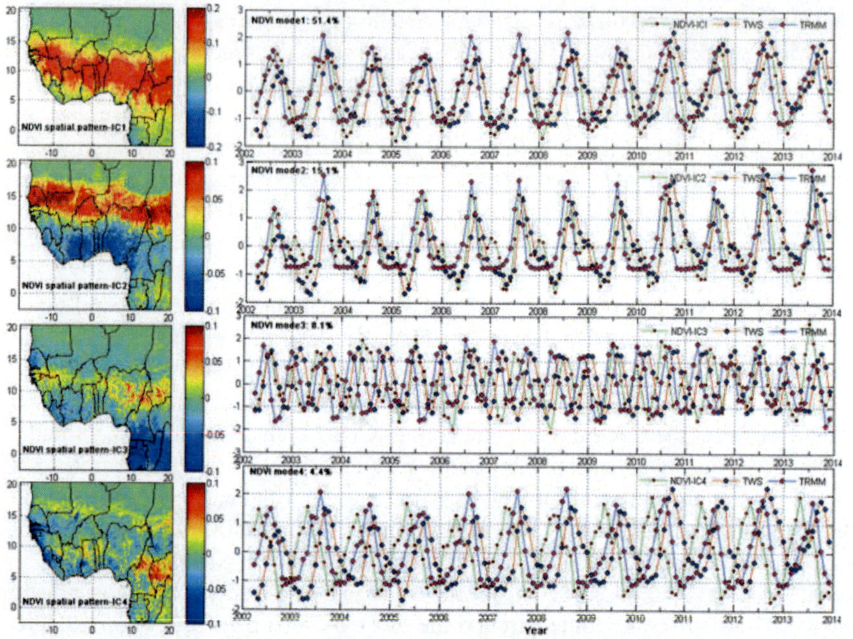

Fig. 7.12 Spatial and temporal variations of NDVI (2002–2013) over West and Central Africa using the Independent Component Analysis (ICA) technique. The temporal patterns (right) also known as independent components (ICs) have been standardized using their standard deviations. The NDVI spatial patterns (left) are scaled using the standard deviation of the computed ICs (i.e., temporal patterns). Averaged TWS and rainfall masked over West Africa and Central Sahel have been compared with the temporal evolutions of the first and second ICA modes of NDVI. Also, rainfall and TWS in the Congo Basin area (i.e., central Africa) and West Africa are compared with the third and fourth ICA modes (i.e., with their ICs), respectively. *Source* Ndehedehe et al. (2019)

Table 7.2 Temporal relationships (r/R^2) of NDVI with satellite-based precipitation (TRMM) and TWS in West and Central Africa as observed in the independent modes of variability in Fig. 7.12. Correlation coefficients are significant at the 95% significant level except for those marked with *. *Source* Ndehedehe et al. (2019)

S/N	Modes	NDVI versus TRMM		NDVI versus TWS	
		r	R^2	r	R^2
1	IC1	0.77	0.59	0.68	0.47
2	IC2	0.69	0.48	0.72	0.52
3	IC3	0.11*	0.00*	0.01	0.00*
4	IC4	0.32	0.10	−0.67	0.44

hydro-ecological system and complements precipitation, which provides an indirect observation of water availability.

The Central Sahel is obviously semi-arid, characterized by large grasslands, open shrublands, and croplands, rainfall is limited, ranging from 700 mm to less than 200 mm for most of the year (Ndehedehe et al. 2019). This region is also well known for its substantially high amount of land surface temperatures. Given its proximity to the Sahara Desert, which is typically hot and very arid, vegetation growth and the green-up of plant cover in this region will be largely restricted by soil wetness and pedological conditions such as infiltration and water holding capacity. High evapotranspiration rates caused by land surface temperatures decrease the amount of net precipitation or available water (precipitation minus evapotranspiration) and may contribute to the lower proportion of variability explained by rainfall. With such environmental conditions, a key component of TWS that is expected to be highly variable due to climate variability is soil wetness or soil moisture. GRACE-derived TWS changes would then be a better indicator and available surrogate for soil wetness or soil moisture. As was earlier argued in preceding sections of this chapter, the latter is an ideal state variable to assess the impacts of the climate system on vegetation changes. However, rainfall is still a more suitable indicator to track vegetation changes in the Sudano-Sahel region (IC1, Fig. 7.12). The impact of rainfall on temporal NDVI patterns in the Sudano-Sahel belt may be expected because it is largely an agricultural ecosystem and interestingly, heavily dependent on rainfall. Exceptions to this may include some of the selected catchments where observed temporal relationships (NDVI vs rainfall) based on annual aggregations show otherwise (Fig. 7.11c–h).

Furthermore, the third mode of NDVI variability, accounting for 8.1% of the total variability indicates multi-annual vegetation changes in the Congo Basin (IC3, Fig. 7.12). The fourth ICA mode represents NDVI annual variability in some coastal regions with an opposite phase in the southern areas of Central African Republic (CAR) and northern Cameroon (IC4, Fig. 7.12) and corresponds to 4.4% of the total variance. In the Congo Basin area in Central Africa (cf. Fig. 7.9), observed NDVI variability showed a rather poor relationship with TWS while averaged rainfall indicated a positive correlation (IC3, Fig. 7.12 and Table 7.2). TWS serves as a low-pass filter of precipitation, and this may also point towards a phase lag between TWS and NDVI (IC3, Fig. 7.12). But the last NDVI mode (IC4, Fig. 7.12) showed a statistically significant negative association with average TWS over West Africa and may be caused by a phase shift of 0.5 years. We also observed that average rainfall over the same area indicated a positive association (Table 7.2). Other complementary perspectives regarding complex environmental processes in the area exist. The influence of land use change, surface albedo alterations, uncontrolled deforestation and bush burning, among other factors have been linked to changes in hydrology and may also play important role in this relationship (e.g., Mahe et al. 2013; Paturel et al. 2003; Li et al. 2007; Favreau et al. 2009; Leduc et al. 1997). For example, from the land cover data analysed, the temporal aerial extents (km^2) of several land cover units declined (savanna, closed shrublands, mixed and evergreen forests). Others nonetheless, showed marked fluctuation during the 2002 and 2007 period. Whereas this is rather suggestive of land cover dynamics as important secondary drivers of temporal vegetation dynamics in the region, other reports Ndehedehe et al. (2019) support this notion. These impacts (land use change and climate variability) however, in the context of how they may primarily determine the trajectory of prospective future changes in the vegetation system of WCA is a subject for future considerations.

7.6.4 *Terrestrial Water Storage and Rainfall as Drivers of Changes in Surface Vegetation*

7.6.4.1 Temporal Variations

The temporal variations of terrestrial water storage (TWS) show a rather similar association with NDVI as with rainfall. TWS and rainfall indicate strong correlations of 0.74 and 0.68, respectively, with NDVI (Fig. 7.13a and b, Table 7.1). The residuals of TWS showed weak associations with those of NDVI unlike rainfall whose residuals indicated no association with NDVI (Fig. 7.13c and d). The MLRA results of Fig. 7.13a–b indicate that temporal series of TWS are more associated with NDVI compared to rainfall. In Figs. 7.13e-j, both TWS and rainfall account for substantial variability in observed NDVI. This can be seen over southern Mali, Lake Chad, and Congo Basins (except for TWS) with TWS in the Lake Chad Basin explaining the highest proportion of variability ($R^2 = 69\%$) in vegetation. At the annual scales, TWS and rainfall averaged over West Africa to some extent are less associated with NDVI. This is seen in their relatively low correlations with NDVI from Table 7.1. The landscapes here are very diverse, and factors like climate variability, ecosystem structure, may contribute to this poor annual associations. It is also possible this could result from data averaging and aggregation over the region. Similar analysis for selected catchments in WCA showed that TWS and rainfall are well associated with NDVI in southern Mali while only TWS is better correlated with NDVI in the Lake Chad Basin (Table 7.1). Also, the 'human-induced signal' such as expansion in cropping or cultivated areas, land use change, and a reduction in fallow period (Dardel et al. 2014), in our view, could significantly affect the temporal patterns of surface vegetation. This will lead to poor correlations of NDVI with TWS and rainfall.

The temporal variations of NDVI with those of TWS and rainfall over WCA shows that the monthly and inter-annual variability of NDVI are both explained by TWS and rainfall (Fig. 7.13a–j). Considering the somewhat strong correlations of TWS with NDVI, it can be argued that TWS appear to be more dependable than rainfall in its capacity to highlight the water-driven changes in the temporal patterns of surface vegetation in southern Mali and the Lake Chad Basin (Fig. 7.13g and i). This relationship is rather similar over West Africa as TWS accounts for about 55% or $r = 0.74$ (Fig. 7.13a) of the changes in monthly NDVI while rainfall explains about 46% or $r = 0.68$ (Fig. 7.13b and Table 7.1). This picture further accentuates the relative capacity and potential of TWS as a hydrological state variable for assessing influence of climate variability on the temporal variations in surface vegetation greenness over West Africa. This narrative may change nonetheless, when the association between

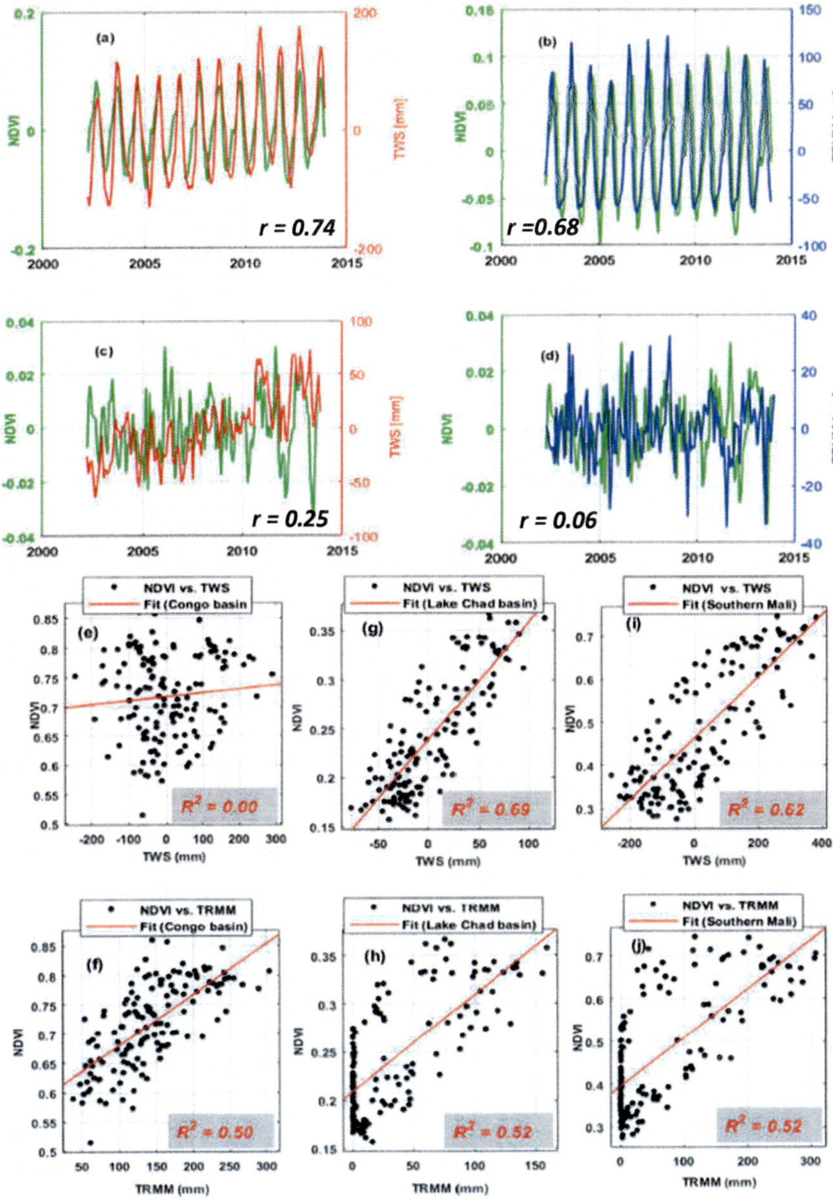

Fig. 7.13 Comparing the time series of temporal variations of NDVI, TWS, and rainfall averaged over West Africa and selected basins and sub-regions (Congo Basin, Lake Chad Basin, and southern Mali) after removing the mean of the period (2002–2013). **a–b** Temporal relationship of NDVI with TWS and precipitation (monthly) and **c–d** association of monthly residuals in NDVI with those of TWS and precipitation over West Africa. **e–j** Regression fits for NDVI with TWS and precipitation in the selected basins and sub-regions. The multi-linear regression analysis was used to estimate the residual parts of each data by removing the harmonic components (i.e., annual and semi-annual) from the data. *Source* Ndehedehe et al. (2019)

NDVI and TWS is assessed based on a grid by grid comparison. Such changes maybe subject to the complexity in ecosystem interactions and environmental factors (e.g., irrigation, use of chemical fertilizers, etc.). Also note that results presented in Fig. 7.13e–j are region-specific cases where TWS shows stronger association with NDVI. As the grid-based relationship in Sect. 7.6.4.2 indicates, rainfall considerably agrees with NDVI, showing higher correlations in much of WCA unlike TWS.

7.6.4.2 Spatial Relationships

NDVI shows relatively strong sensitivity to changes in terrestrial water storage (TWS) and rainfall mostly in the Sahel zone (i.e., Sudano and Sahel belt) (Fig. 7.14), consistent with the results in Sect. 7.6.3.1. Evaluating the spatial or grid-based relationship, rainfall shows statistically significant correlations ($\alpha = 0.05$) with NDVI in about 90% of WCA (includes non-linear response) during the 2002–2013 period compared to TWS (70% of the region) (Fig. 7.14a–b). From the spatial distribution of the correlation coefficients, rainfall has a much wider spread of positive correlations than TWS. Strong positive spatial correlations of TWS with NDVI are mostly found along the Sudano-Sahel belt (Fig. 7.14a). But rainfall showed significant correlation with NDVI beyond the Sahel, extending to the Sudano and Central African region (Fig. 7.14b). It is noted that regions characterized by savannas, croplands (natural vegetation), and grasslands are the dominant biomes in this Sahel zone where NDVI indicates a considerable spatial association with rainfall and TWS. Our contemporary understanding of the African eco-hydrological processes shows that much of the observed positive correlations of NDVI with TWS and rainfall are in the Sudano-Sahelian ecosystems. In this region (Sudano-Sahel), we earlier observed that NDVI showed the most variability (51.4%) (Fig. 7.12) based on the statistical decomposition of NDVI using ICA. But rainfall shows statistically significant positive correlations with NDVI in regions that encompass all the dominant modes of NDVI variability observed in Fig. 7.12.

Looking closely at Fig. 7.14a–b, TWS in some parts of the West African Sahel (e.g., Gambia, Guinea, Guinea Bissau, southern Mali) and some locations in Lake Chad Basin (parts of northern Nigeria, southern Niger, and southern Chad-Fig. 7.9) show higher correlations with NDVI compared to rainfall. In the northern flank of Niger, the positive correlations in Fig. 7.14a suggest the continuous oscillation of NDVI in some arid ecosystems and agree with TWS unlike the non-linear response indicated by rainfall (Fig. 7.14b). NDVI in other regions of WCA are better correlated with rainfall, and the suitability of GRACE-derived TWS as a hydrological indicator to assess ecosystem performance in these region-specific cases is also highlighted. Changes in soil moisture conditions in these ecosystems play an important role in the observed association of NDVI with TWS and rainfall. We also found relatively strong positive correlations of NDVI with rainfall in Gabon and some areas of the Congo Basin, where several vegetation types like the evergreen broadleaf forest are largely predominant. Other coastal countries in West Africa like Côte d'Ivoire, Liberia, and Sierra Leone however, show relatively poor and negative correlations. This is despite

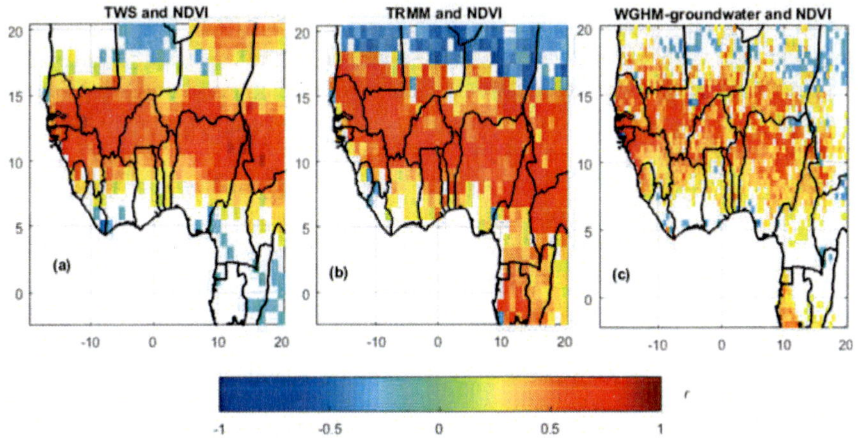

Fig. 7.14 Spatial correlations of monthly NDVI with **a** TWS (2002–2013), **b** TRMM-based precipitation (2002–2013), and **c** WaterGap model groundwater (2002–2009) over West and Central Africa. Only statistically significant correlations ($\alpha = 0.05$) are indicated. *Source* Ndehedehe et al. (2019)

these countries having almost similar climate and vegetation patterns as the countries in the Congo Basin.

Furthermore, the assessment of vegetation response to rainfall and TWS show some spatial limits and complexities. For example, be it rainfall or TWS, the correlations are either generally low or negative in humid (coastal areas) and extreme arid environments (Central Sahel). Apparently, high correlations are mostly present in regions of considerable seasonal change (cf. IC1, Fig. 7.12) rather than regions with little seasonal change such as the tropical coastal regions where NDVI-TWS is extremely low or non-existent in some cases (insignificant). The correlation of TWS and rainfall with NDVI are weak even in ecosystems with annual rainfall exceeding 2500 mm (areas with very little seasonal change as seen in IC1, Fig. 7.12). But those sub-regions with annual rainfall less than 1000 mm (areas with considerable seasonal change-IC1, Fig. 7.12) indicated strong association with NDVI. Arguably, different vegetation types and classes having different water critical points may also account for this.

The humid coastal countries of West Africa (e.g., Liberia, southern Côte d'Ivoire) have waterlogged soils and may suggest that vegetation growth is mostly a function of pre-existing soil conditions as argued earlier. As a result, rainfall is not primarily a factor for vegetation growth in such areas. This does not negate the role of rainfall in the region as an important source of water availability. Similar relationship is also observed with TWS in the coastal areas characterized by evergreen broadleaf forests (cf. Fig. 7.9) where there was no association with NDVI (Fig. 7.14a). This perhaps is due to the fact that TWS changes here are the result of a weak seasonal signal. But in the Sahel and Sudano belt regions, rainfall is an established indicator to assess changes in NDVI and is evidenced in their strong agreement (NDVI vs rainfall) (Fig. 7.14b). This observed relationship (rainfall vs NDVI) coincides with other reports for the Sudano-Sahel region but different for the wetland and irrigated areas, which produced weak correlations (Georganos et al. 2017). However, roots exploration, the adaptability, and tolerance of plants to arid conditions may impact such relationship (e.g., Chen et al. 2014; Zhou et al. 2014).

The correlation result in Fig. 7.14c indicates that model groundwater is well associated with NDVI in southern Mali and much of the Sahel and Sudano belt. Based on the response of NDVI to TWS and groundwater (e.g., southern Mali, Senegal, Gambia, south-west Guinea, northern Nigeria, etc.) in Fig. 7.14a and c, it is argued that surface vegetation in these ecosystems respond to water availability in all catchment stores, including groundwater. What this means is that there are groundwater dependent ecosystems in the region. The restriction of soil moisture in the southern Sahel regions as a driver of higher woody vegetation canopy was earlier reported (Huber et al. 2011). Several woody plants remain photosynthetically active during the year. This is very much unlike other plants, which are only green during rainy season (see, Brandt et al. 2016). With groundwater dependent ecosystems, it is more likely that TWS could be the most prominent indicator of vegetation changes in the region. This can happen in some parts of the semi-arid Sahel where photosynthetic activities can be seen even during periods of limited or no rainfall. The Sahel region is well known for its frequent droughts as well documented in the literature (e.g., Ndehedehe et al. 2020b, 2016b; Nicholson et al. 1998). Roots exploration and the adaptability of some plants to long dry periods validates such argument. Going by the response of NDVI to groundwater (e.g., Guinea Bissau, southern Mali, etc.) in Fig. 7.14c, TWS would be a substantive driver of vegetation dynamics. This is because it better represents the rooting depth of any vegetation, given that its water storage, including groundwater and soil moisture. The former extends to the unsaturated zone, meaning changes in vegetation or plants communities that rely on groundwater can be tracked using TWS.

More than 30% of global biodiversity benefits from groundwater (Fan et al. 2013) and extreme water requirements by some vegetation in water-limited ecosystems can be triggered by increased evapotranspiration, prolonged dry seasons, and rainfall deficits (Ndehedehe et al. 2018b). These factors can modify the structural properties of vegetation communities. This will make them even more dependent on groundwater and their water use more complex. The strong TWS-soil moisture relationship in a typical Sahelian catchment such as the Lake Chad Basin suggest that soil moisture is the principal driver of observed changes in the GRACE observed water column in the region. This presents TWS as a suitable soil moisture surrogate that may be considered as suitable and readily available hydrological control for surface vegetation in semi-arid regions.

7.6.4.3 Seasonal Relationships

It is also important to compare the seasonal variability of NDVI with terrestrial water storage (TWS) and rainfall. NDVI correlations with rainfall during dry sea-

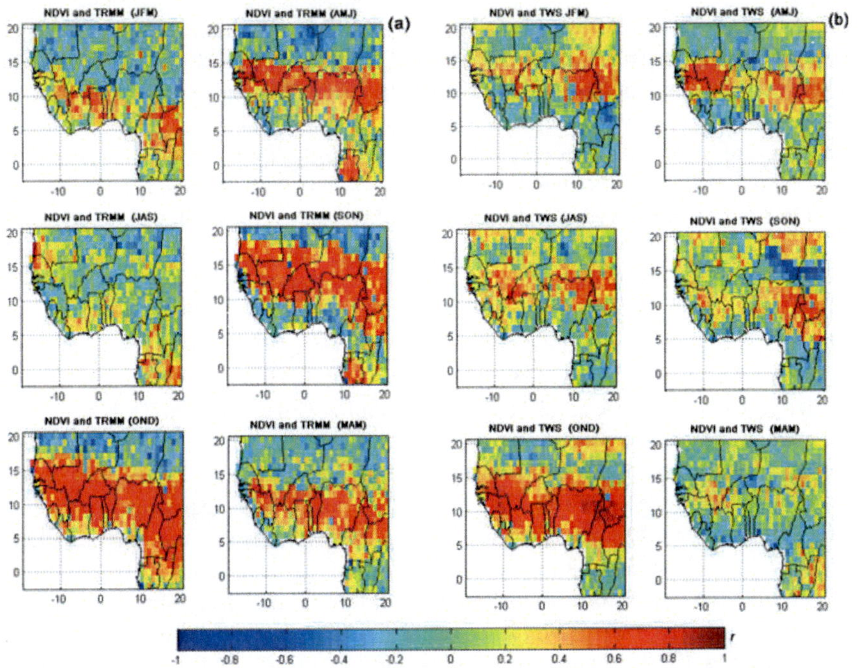

Fig. 7.15 Correlations of NDVI (2003–2013) with TRMM-based precipitation and TWS at pixel scale (spatial pattern) for seasonal periods (January, February, and March-JFM; March, April, and May-MAM; April, May, and June-AMJ; July, August, and September-JAS; September, October, and November-SON; October, November, and December-OND). **a** Correlations between seasonal NDVI and rainfall and **b** correlations between seasonal NDVI and TWS. *Source* Ndehedehe et al. (2019)

son introduces larger bias (Huber et al. 2011). Seasonal evaluations are therefore required to provide robustly unbiased assessment of how vegetation responds to climate variability. Correlations between NDVI and rainfall tend to be high in the Sahel zone as well as some coastal countries during the October–November, April–June, September–November, and March–May periods but very little correlation is found in the July-September and January-March periods (Fig. 7.15a). But TWS shows a higher relation in between October and November. Unlike this period, significant relationships with NDVI mostly in the Sahel zone except during the March–May period are obvious (Fig. 7.15b). Indeed, an overall assessment of the relation of the monthly and seasonal NDVI with those of rainfall and TWS (i.e., based on the respective grids) indicate that rainfall unarguably, provides a hydrological control on NDVI phenology in much of the sub-continent. A similar relationship (i.e., NDVI and rainfall) was observed by Nicholson et al. (1990) in the Sahel region, consistent with our result here.

However, rainfall seems to be restricted in the growing season (e.g., July–September) as an indicator of vegetation changes unlike TWS (Fig. 7.15b). The seemingly poor performance of rainfall as a predictor of vegetation greenness during growing season is consistent with the findings of Huber et al. (2011). They argued that the dry seasons introduce bias in the observed relationship (rainfall vs NDVI) because unlike NDVI, which still oscillates during such periods, rainfall does not. Rainfall as predictor of vegetation greenness shows wider spread of positive correlations during the October–November period (Fig. 7.15a). This is obviously different from TWS as indicated in Fig. 7.15b. However, TWS show linear relationships (positive correlations) in some locations (e.g., northern Niger) where the relationships between rainfall and NDVI are negative or does not exist. Even though TWS is also strongly associated with NDVI in much of WCA similar to rainfall, there are some locations and sub-regions where the relationship between NDVI and TWS is consistently strong. Be it monthly or seasonal time scales, NDVI demonstrates consistent sensitivity to TWS in these areas compared to rainfall (Figs. 7.14 and 7.15).

Presumably, these region-specific cases include southern Mali and Guinea in the Western Sahel and the Lake Chad Basin in north-east Nigeria and southern Chad. They are good examples of where TWS provides better hydrological controls for surface vegetation. In these cases, TWS can be used to complement rainfall in assessing or improving understanding of changes in hydrological conditions and how they impact on surface vegetation. Given the lack of in situ rainfall data to help with such assessment as well as the bureaucracies associated with its acquisition in the region, GRACE-derived TWS becomes very handy and helpful. Moreover, TWS in these landscapes would be more reliable than precipitation as soil moisture surrogate to assess water availability for the growth of surface vegetation. Technically, one would argue that GRACE-derived TWS changes are expected to be driven by soil moisture in the unsaturated soil moisture zone. Stressing this further, Yang et al. (2014) argued that these zones are more readily affected by changes in climate than the saturated zones. In the Lake Chad Basin, for example, changes in TWS would largely emanate from this catchment store (unsaturated zone and to some extent saturated zone during extreme drought periods). This is because there is no record of extensive groundwater abstractions and water loss is mostly through evapotranspiration (e.g., Ndehedehe et al. 2018b).

Furthermore, atmospheric conditions regulate the life span of canopy phenology. In Sahelian ecosystems, these factors also impact both leafing and reproductive phenophases (e.g., Seghieri et al. 2012; Do et al. 2005). In such a case, TWS would better account for plant water availability. Arguably, GRACE-TWS provides a water budget indicator and accounts for available freshwater changes in any hydrological system. This is because GRACE measure vertically integrated water storage changes (surface waters, soil moisture, and groundwater), regardless landscape characteristics (soil and vegetation types) and any other human influence (dams, water transfer, etc.) on terrestrial hydrology. Additionally, widespread drought tolerant species resulting from prolonged drought periods in the semi-arid catchments or regions could lead to transformative relationship and changes in the composition of species richness. This creates complexities in the way surface vegetation responds to hydrological conditions as well as mechanics of water use in such areas (e.g., Ndehedehe 2019; Yang et al. 2014; Huber et al. 2011). For example, this is evident in the Sahelian ecosystems, which have shown considerable relationship with groundwater (Fig. 7.14c). Thus, GRACE-derived TWS would be a more suitable hydrological indicator to assess the impact of climate variability on vegetation greenness in the basin and other ecosystems where similar relationship is observed.

7.7 Concluding Remark

Vegetation is an important source of food, energy, and provides several essential services including, regulating the climate system, hosting biodiversity, and storing of carbon. In this chapter, both surface and aquatic vegetation and above-ground forest biomass have been discussed in detail. The technological progress and key science as well as scientific advancement surrounding their monitoring and assessments have been detailed. Radar and optical remote sensing systems and satellite gravity missions like Gravity Recovery and Climate Experiment are now important tools to aid the assessment of surface and aquatic vegetation. The use of advance machine learning algorithms is also emerging to improve the applications of satellite technology, especially radar and hyperspectral data to provide great opportunities in agriculture and analysis of land cover states. The potential of GRACE-TWS as an important water budget indicator to account for the water-driven variability in surface vegetation in any eco-hydrological system has been demonstrated. GRACE satellites measure vertically integrated water storage changes (surface waters in lakes and reservoirs, snow water/ice, soil moisture, and groundwater). This chapter has demonstrated that GRACE data are better indicators for surface vegetation, especially in semi-arid areas compared to rainfall and state variables like soil moisture. They can therefore be used to complement rainfall in assessing or improving understanding of changes in hydrological conditions and how they impacts on surface vegetation. Given the lack of in situ rainfall data to help with such assessment, GRACE-derived TWS becomes very handy to assess water availability for the growth of surface vegetation. Moreover, GRACE satellites have been used to provide information on the depletion of groundwater aquifers caused by groundwater pumping for crop irrigation and several other industrial and domestic applications. There is evidence that it can be useful in assessing the productivity of groundwater-dependent ecosystems.

References

Abiy AZ, Melesse AM (2017) Evaluation of watershed scale changes in groundwater and soil moisture storage with the application of GRACE satellite imagery data. CATENA 153:50–60. https://doi.org/10.1016/j.catena.2017.01.036

Alcaraz-Segura D, Cabello J, Paruelo J et al (2009) Use of descriptors of ecosystem functioning for monitoring a national park network: a remote sensing approach. Environ Manag 43:38–48. https://doi.org/10.1007/s00267-008-9154-y

Andrew ME, Ustin SL (2008) The role of environmental context in mapping invasive plants with hyperspectral image data. Remote Sens Environ 112(12):4301–4317. https://doi.org/10.1016/j.rse.2008.07.016

Basuki TM, Skidmore AK, Hussin YA, Duren IV (2013) Estimating tropical forest biomass more accurately by integrating ALOS PALSAR and Landsat-7 ETM+ data. Int J Remote Sens 34(13):4871–4888. https://doi.org/10.1080/01431161.2013.777486

Bento VA, Gouveia CM, DaCamara CC, Libonati R, Trigo IF (2020) The roles of NDVI and land surface temperature when using the vegetation health index over dry regions. Global Planet Change 190:103198. https://doi.org/10.1016/j.gloplacha.2020.103198

Bento VA, Gouveia CM, DaCamara CC, Trigo IF (2018) A climatological assessment of drought impact on vegetation health index. Agric For Meteorol 259:286–295. https://doi.org/10.1016/j.agrformet.2018.05.014

Bond WJ, Keeley JE (2005) Fire as a global 'herbivore': the ecology and evolution of flammable ecosystems. Trends Ecology & Evolut 20(7):387–394. https://doi.org/10.1016/j.tree.2005.04.025

Boschetti M, Nutini F, Brivio PA, Bartholomé E, Stroppiana D, Hoscilo A (2013) Identification of environmental anomaly hot spots in West Africa from time series of NDVI and rainfall. ISPRS J Photogramm Remote Sens 78:26–40. https://doi.org/10.1016/j.isprsjprs.2013.01.003

Bouvet A, Mermoz S, Le Toan T, Villard L, Mathieu R, Naidoo L, Asner GP (2018) An above-ground biomass map of African savannahs and woodlands at 25 m resolution derived from ALOS PALSAR. Remote Sens Environ 206:156–173. https://doi.org/10.1016/j.rse.2017.12.030

Brandt M, Hiernaux P, Rasmussen K, Mbow C, Kergoat L, Tagesson T, Ibrahim YZ, Wélé A, Tucker CJ, Fensholt R (2016) Assessing woody vegetation trends in Sahelian drylands using MODIS based seasonal metrics. Remote Sens Environ 183:215–225. https://doi.org/10.1016/j.rse.2016.05.027

Cardoso JF (1999) High-Order contrasts for independent component analysis. Neural Comput 11:157–192

Chen T, de Jeu R, Liu Y, van der Werf G, Dolman A (2014) Using satellite based soil moisture to quantify the water driven variability in NDVI: a case study over mainland Australia. Remote Sens Environ 140:330–338. https://doi.org/10.1016/j.rse.2013.08.022

Chen J, Cazenave A, Dahle C et al (2022) Applications and challenges of GRACE and GRACE follow-on satellite gravimetry. Surv Geophys. https://doi.org/10.1007/s10712-021-09685-x

Cho MA, Mathieu R, Asner GP, Naidoo L, J v et al (2012) Mapping tree species composition in South African savannas using an integrated airborne spectral and LiDAR system. Remote Sens Environ 125:214–226. https://doi.org/10.1016/j.rse.2012.07.010

Clark ML, Roberts DA, Ewel JJ, Clark DB (2011) Estimation of tropical rain forest above-ground biomass with small-footprint lidar and hyperspectral sensors. Remote Sens Environ 115(11):2931–2942. https://doi.org/10.1016/j.rse.2010.08.029

Clevers J, Gitelson A (2013) Remote estimation of crop and grass chlorophyll and nitrogen content using red-edge bands on Sentinel-2 and -3. Int J Appl Earth Obs Geoinf 23:344–351. https://doi.org/10.1016/j.jag.2012.10.008

Dangwal N, Patel N, Kumari M, Saha S (2016) Monitoring of water stress in wheat using multi-spectral indices derived from Landsat-TM. Geocarto Int 31(6):682–693. https://doi.org/10.1080/10106049.2015.1073369

Dardel C, Kergoat L, Hiernaux P, Mougin E, Grippa M, Tucker C (2014) Re-greening Sahel: 30 years of remote sensing data and field observations (Mali, Niger). Remote Sens Environ 140:350–364. https://doi.org/10.1016/j.rse.2013.09.011

Descroix L, Mahé G, Lebel T, Favreau G, Galle S, Gautier E, J-C O et al (2009) Spatio-temporal variability of hydrological regimes around the boundaries between Sahelian and Sudanian areas of West Africa: a synthesis. J Hydrol 375(1–2):90–102. https://doi.org/10.1016/j.jhydrol.2008.12.012

Ding Y, Zhao K, Zheng X, Jiang T (2014) Temporal dynamics of spatial heterogeneity over cropland quantified by time-series NDVI, near infrared and red reflectance of Landsat 8 OLI imagery. Int J Appl Earth Obs Geoinf 30:139–145. https://doi.org/10.1016/j.jag.2014.01.009

Dirmeyer PA, Guo Z, Gao X (2004) Comparison, validation, and transferability of eight multiyear global soil wetness products. J Hydrometeorol 5:1011–1033. https://doi.org/10.1175/JHM-388.1

Do FC, Goudiaby VA, Gimenez O, Diagne AL, Diouf M, Rocheteau A, Akpo LE (2005) Environmental influence on canopy phenology in the dry tropics. For Ecol Manage 215(1–3):319–328. https://doi.org/10.1016/j.foreco.2005.05.022

Dube T, Gumindoga W, Chawira M (2014) Detection of land cover changes around Lake Mutirikwi, Zimbabwe, based on traditional remote sensing image classification techniques. Afr J Aquat Sci 39(1):89–95. https://doi.org/10.2989/16085914.2013.870068

Dube T, Mutanga O (2015a) Evaluating the utility of the medium-spatial resolution Landsat 8 multispectral sensor in quantifying aboveground biomass in uMgeni catchment, South Africa. ISPRS J Photogramm Remote Sens 101:36–46. https://doi.org/10.1016/j.isprsjprs.2014.11.001

Dube T, Mutanga O (2015b) Investigating the robustness of the new Landsat-8 Operational Land Imager derived texture metrics in estimating plantation forest aboveground biomass in resource constrained areas. ISPRS J Photogramm Remote Sens 108:12–32. https://doi.org/10.1016/j.isprsjprs.2015.06.002

Dube T, Mutanga O, Sibanda M, Bangamwabo V, Shoko C (2017) Testing the detection and discrimination potential of the new Landsat 8 satellite data on the challenging water hyacinth (Eichhornia crassipes) in freshwater ecosystems. Appl Geogr 84:11–22. https://doi.org/10.1016/j.apgeog.2017.04.005

Eitel JU, Gessler PE, Smith AM, Robberecht R (2006) Suitability of existing and novel spectral indices to remotely detect water stress in Populus spp. For Ecol Manage 229(1):170–182. https://doi.org/10.1016/j.foreco.2006.03.027

Ene LT, Næsset E, Gobakken T, Gregoire TG, Ståhl G, Nelson R (2012) Assessing the accuracy of regional LiDAR-based biomass estimation using a simulation approach. Remote Sens Environ 123:579–592. https://doi.org/10.1016/j.rse.2012.04.017

Famiglietti JS (2014) The global groundwater crisis. Nature 4:945–948. https://doi.org/10.1038/nclimate2425

Fan Y, Li H, Miguez-Macho G (2013) Global patterns of groundwater table depth. Science 339(6122):940–943. https://doi.org/10.1126/science.1229881

Fassnacht FE, Li L, Fritz A (2015) Mapping degraded grassland on the Eastern Tibetan Plateau with multi-temporal Landsat 8 data – where do the severely degraded areas occur? Int J Appl Earth Obs Geoinf 42:115–127. https://doi.org/10.1016/j.jag.2015.06.005

Favreau G, Cappelaere B, Massuel S, Leblanc M, Boucher M, Boulain N, Leduc C (2009) Land clearing, climate variability, and water resources increase in semiarid southwest Niger: a review. Water Resour Res 45:W00A16. https://doi.org/10.1029/2007WR006785.

Ferreira V, Asiah Z (2015) An investigation on the closure of the water budget methods over Volta Basin using multi-satellite data. Int Assoc Geod Symp. https://doi.org/10.1007/1345-2015-137

Ferreira V, Montecino H, Ndehedehe C, Heck B, Gong Z, Westerhaus M, de Freitas S (2018) Space-based observations of crustal deflections for drought characterization in brazil. Sci Total Environ 644:256–273. https://doi.org/10.1016/j.scitotenv.2018.06.277

Field CB, Behrenfeld MJ, Randerson JT, Falkowski P (1998) Primary production of the biosphere: integrating terrestrial and oceanic components. Science 281(5374):237–240. https://doi.org/10.1126/science.281.5374.237

Filipponi F (2018) Bais2: burned area index for sentinel-2. Proceedings 2(7). https://doi.org/10.3390/ecrs-2-05177

Georganos S, Abdi AM, Tenenbaum DE, Kalogirou S (2017) Examining the NDVI-rainfall relationship in the semi-arid Sahel using geographically weighted regression. J Arid Environ 146(Supplement C):64–74. https://doi.org/10.1016/j.jaridenv.2017.06.004

Ghasemi N, Sahebi MR, Mohammadzadeh A (2010) A review on biomass estimation methods using synthetic aperture radar data. Int J Geom Geosci 1(4):776–788

Giardino C, Bresciani M, Valentini E, Gasperini L, Bolpagni R, Brando VE (2015) Airborne hyperspectral data to assess suspended particulate matter and aquatic vegetation in a shallow and turbid lake. Remote Sens Environ 157:48–57. https://doi.org/10.1016/j.rse.2014.04.034

Gleason CJ, Im J (2012) Forest biomass estimation from airborne LiDAR data using machine learning approaches. Remote Sens Environ 125:80–91. https://doi.org/10.1016/j.rse.2012.07.006

Glenn EP, Huete AR, Nagler PL, Nelson SG (2008) Relationship between remotely-sensed vegetation indices, canopy attributes and plant physiological processes: what vegetation indices can and cannot tell us about the landscape. Sensors 8(4):2136–2160. https://doi.org/10.3390/s8042136

Goodenough DG, Niemann KO, Dyk A, Hobart G, Gordon P, Loisel M, Chen H (2008) Comparison of aviris and AISAairborne hyperspectral sensing for above-ground forest carbon mapping. IGARSS 2008 - 2008 IEEE international geoscience and remote sensing symposium, 2:II–129–II–132. https://doi.org/10.1109/IGARSS.2008.4778944

Guan K, Wood EF, Medvigy D, Kimball J, Pan M, Caylor KK, Sheffield J, Xu X, Jones MO (2014) Terrestrial hydrological controls on land surface phenology of African savannas and woodlands. J Geophys Res Biogeosci 119(8):1652–1669. https://doi.org/10.1002/2013JG002572

Hansen MC, Potapov PV, Moore R, Hancher M et al (2013) High-resolution global maps of 21st-century forest cover change. Science 342(6160):850–853. https://doi.org/10.1126/science.1244693

Henry C, Allen DM, Huang J (2011) Groundwater storage variability and annual recharge using well-hydrograph and GRACE satellite data. Hydrogeol J 19:741–755. https://doi.org/10.1007/s10040-011-0724-3

Herrmann SM, Anyamba A, Tucker CJ (2005) Recent trends in vegetation dynamics in the African Sahel and their relationship to climate. Glob Environ Chang 15(4):394–404. https://doi.org/10.1016/j.gloenvcha.2005.08.004

Högstrom E, Trofaier AM, Gouttevin I, Bartsch A (2014) Assessing seasonal backscatter variations with respect to uncertainties in soil moisture retrieval in Siberian Tundra regions. Remote Sens 6(9):8718–8738. https://doi.org/10.3390/rs6098718

Huber S, Fensholt R, Rasmussen K (2011) Water availability as the driver of vegetation dynamics in the African Sahel from 1982 to 2007. Global Planet Change 76(3–4):186–195. https://doi.org/10.1016/j.gloplacha.2011.01.006

Jamali S, Seaquist J, Eklundh L, Ardö J (2014) Automated mapping of vegetation trends with polynomials using NDVI imagery over the Sahel. Remote Sens Environ 141:79–89. https://doi.org/10.1016/j.rse.2013.10.019

Jeger MJ, Pautasso M (2008) Plant disease and global change - the importance of long-term data sets. New Phytol 177(1):8–11. https://doi.org/10.1111/j.1469-8137.2007.02312.x

Joshi H, Ghose M (2014) Community structure, species diversity, and aboveground biomass of the sundarbans mangrove swamps. Trop Ecol 55(3):283–303

Kahn PC, Molnar T, Zhang GG, Funk CR (2011) Investing in perennial crops to sustainably feed the world. Issues Sci Technol 27(4):75–81

Kalwij J, Steyn C, le Roux P (2014) Repeated monitoring as an effective early detection means: first records of naturalised Solidago gigantea Aiton (Asteraceae) in southern Africa. S Afr J Bot 93:204–206. https://doi.org/10.1016/j.sajb.2014.04.013

Knauer K, Gessner U, Dech S, Kuenzer C (2014) Remote sensing of vegetation dynamics in West Africa. Int J Remote Sens 35(17):6357–6396. https://doi.org/10.1080/01431161.2014.954062

Lanfredi M, Coppola R, Simoniello T, Coluzzi R, et al (2015) Early identification of land degradation hotspots in complex bio-geographic regions. Remote Sens 7(6):8154–8179. https://doi.org/10.3390/rs70608154

Leduc C, Bromley J, Schroeter P (1997) Water table fluctuation and recharge in semi-arid climate: some results of the HAPEX-Sahel hydrodynamic survey (Niger). J Hydrol 188(188–189):123–138. https://doi.org/10.1016/S0022-1694(96)03156-3

Liang S, Wang J (2004) Aboveground biomass. Adv Remote Sens 543–580. https://doi.org/10.1016/B978-0-12-815826-5.00014-3

Li J, Lewis J, Rowland J, Tappan G, Tieszen L (2004) Evaluation of land performance in Senegal using multi-temporal NDVI and rainfall series. J Arid Environ 59(3):463–480. https://doi.org/10.1016/j.jaridenv.2004.03.019

Li K, Coe M, Ramankutty N, Jong RD (2007) Modeling the hydrological impact of land-use change in West Africa. J Hydrol 337(3–4):258–268. https://doi.org/10.1016/j.jhydrol.2007.01.038

Li M, Qu JJ, Hao X (2009) Estimating aboveground biomass for different forest types based on Landsat TM measurements. In: 2009 17th international conference on geoinformatics, pp 1–6. https://doi.org/10.1109/GEOINFORMATICS.2009.5293408

Lin BB (2011) Resilience in agriculture through crop diversification: adaptive management for environmental change. Bioscience 61(3):183–193. https://doi.org/10.1525/bio.2011.61.3.4

Lu D (2006) The potential and challenge of remote sensing based biomass estimation. Int J Remote Sens 27(7):1297–1328. https://doi.org/10.1080/01431160500486732

Lu D, Chen Q, Wang G, Moran E et al (2012) Aboveground forest biomass estimation with landsat and LiDAR data and uncertainty analysis of the estimates. Int J For Res 2012:16. https://doi.org/10.1155/2012/436537

Lü X-T, Yin J-X, Jepsen MR, Tang J-W (2010) Ecosystem carbon storage and partitioning in a tropical seasonal forest in southwestern china. For Ecol Manage 260(10):1798–1803. https://doi.org/10.1016/j.foreco.2010.08.024

Mahe G, Lienou G, Descroix L, Bamba F, Paturel JE et al (2013) The rivers of Africa: witness of climate change and human impact on the environment. Hydrolog Proc 27(15):2105–2114. https://doi.org/10.1002/hyp.9813

Mitchard ETA, Saatchi SS, Woodhouse IH, Nangendo G, Ribeiro NS, Williams M et al (2009) Using satellite radar backscatter to predict above-ground woody biomass: a consistent relationship across four different African landscapes. Geophys Res Lett 36(23). https://doi.org/10.1029/2009GL040692

Mundava C, Schut A, Helmholz P, Stovold R, Donald G, Lamb D (2015) A novel protocol for assessment of aboveground biomass in rangeland environments. Rangel J 37:157–167. https://doi.org/10.1071/RJ14072

Mutanga O, Adam E, Cho MA (2012) High density biomass estimation for wetland vegetation using worldview-2 imagery and random forest regression algorithm. Int J Appl Earth Obs Geoinf 18:399–406. https://doi.org/10.1016/j.jag.2012.03.012

Ndehedehe C, Awange J, Agutu N, Kuhn M, Heck B (2016a) Understanding changes in terrestrial water storage over West Africa between 2002 and 2014. Adv Water Resour 88:211–230. https://doi.org/10.1016/j.advwatres.2015.12.009

Ndehedehe CE (2019) The water resources of tropical West Africa: problems, progress and prospect. Acta Geophys 67(2):621–649. https://doi.org/10.1007/s11600-019-00260-y

Ndehedehe CE, Agutu NO, Okwuashi O (2018) Is terrestrial water storage a useful indicator in assessing the impacts of climate variability on crop yield in semi-arid ecosystems? Ecol Ind 88C:51–62. https://doi.org/10.1016/j.ecolind.2018.01.026

Ndehedehe CE, Agutu NO, Okwuashi OH, Ferreira VG (2016b) Spatio-temporal variability of droughts and terrestrial water storage over Lake Chad Basin using independent component analysis. J Hydrol 540:106–128. https://doi.org/10.1016/j.jhydrol.2016.05.068

Ndehedehe CE, Awange J, Kuhn M, Agutu N, Fukuda Y (2017) Analysis of hydrological variability over the Volta river basin using in-situ data and satellite observations. J Hydrol: Reg Stud 12:88–110. https://doi.org/10.1016/j.ejrh.2017.04.005

Ndehedehe CE, Burford MA, Stewart-Koster B, Bunn SE (2020a) Satellite-derived changes in floodplain productivity and freshwater habitats in northern Australia (1991–2019). Ecol Ind 114:106320. https://doi.org/10.1016/j.ecolind.2020.106320

Ndehedehe CE, Ferreira VG, Agutu NO (2019) Hydrological controls on surface vegetation dynamics over West and Central Africa. Ecol Ind 103:494–508. https://doi.org/10.1016/j.ecolind.2019.04.032

Ndehedehe CE, Ferreira VG, Onojeghuo AO, Agutu NO, Emengini E, Getirana A (2020b) Influence of global climate on freshwater changes in Africa's largest endorheic basin using multi-scaled indicators. Sci Total Environ 737:139643. https://doi.org/10.1016/j.scitotenv.2020.139643

Ndehedehe CE, Okwuashi O, Ferreira VG, Agutu NO (2018b) Exploring evapotranspiration dynamics over Sub-Sahara Africa (2000 – 2014). Environ Monit Assess 190(7):400. https://doi.org/10.1007/s10661-018-6780-6

Ndehedehe CE, Onojeghuo AO, Stewart-Koster B, Bunn SE, Ferreira VG (2021) Upstream flows drive the productivity of floodplain ecosystems in tropical queensland. Ecol Ind. https://doi.org/10.1016/j.ecolind.2021.107546

Ndehedehe CE, Stewart-Koster B, Burford MA, Bunn SE (2020c) Predicting hot spots of aquatic plant biomass in a large floodplain river catchment in the Australian wet-dry tropics. Ecol Ind 117:106616. https://doi.org/10.1016/j.ecolind.2020.106616

Nicholson S (2009) A revised picture of the structure of the "monsoon" and land itcz over West Africa. Clim Dyn 32(7–8):1155–1171. https://doi.org/10.1007/s00382-008-0514-3

Nicholson SE, Davenport ML, Malo AR (1990) A comparison of the vegetation response to rainfall in the Sahel and East Africa, using normalized difference vegetation index from NOAA AVHRR. Clim Change 17(2):209–241. https://doi.org/10.1007/BF00138369

Nicholson SE, Some B, Kone B (2000) An analysis of recent rainfall conditions in West Africa, including the rainy seasons of the 1997 El Niño and the 1998 La Niña years. J Clim 13(14):2628–2640. https://doi.org/10.1175/1520-0442(2000)013<2628:AAORRC>2.0.CO;2

Nicholson SE, Tucker CJ, Ba MB (1998) Desertification, drought, and surface vegetation: an example from the West African Sahel. Bull Amer Meteor Soc 79(5):815–829. https://doi.org/10.1175/1520-0477(1998)079<0815:DDASVA>2.0.CO;2

Nogherotto R, Coppola E, Giorgi F, Mariotti L (2013) Impact of Congo basin deforestation on the African monsoon. Atmos Sci Lett 14(1):45–51. https://doi.org/10.1002/asl2.416

Okwuashi O, Ndehedehe CE (2020) Deep support vector machine for hyperspectral image classification. Pattern Recogn 103:107298. https://doi.org/10.1016/j.patcog.2020.107298

Okwuashi O, Ndehedehe CE (2021) Integrating machine learning with Markov chain and cellular automata models for modelling urban land use change. Remote Sens Appl: Soc Environ 21:100461. https://doi.org/10.1016/j.rsase.2020.100461

Okwuashi O, Ndehedehe CE, Olayinka DN, Eyoh A, Attai H (2021) Deep support vector machine for PolSAR image classification. Int J Remote Sens 42(17):6498–6536. https://doi.org/10.1080/01431161.2021.1939910

Olsson L, Eklundh L, Ardö J (2005) A recent greening of the Sahel-trends, patterns and potential causes. J Arid Environ 63(3):556–566. https://doi.org/10.1016/j.jaridenv.2005.03.008

Onojeghuo AO, Blackburn GA, Wang Q, Atkinson PM et al (2005) Mapping paddy rice fields by applying machine learning algorithms to multi-temporal Sentinel-1A and Landsat data. Int J Remote Sens 39(4):1042–1067. https://doi.org/10.1080/01431161.2017.1395969

Oumar Z, Mutanga O (2014) Predicting water stress induced by thaumastocoris peregrinus infestations in plantation forests using field spectroscopy and neural networks. J Spat Sci 59(1):79–90. https://doi.org/10.1080/14498596.2013.821679

Palmer SC, Kutser T, Hunter PD (2015) Remote sensing of inland waters: challenges, progress and future directions. Remote Sens Environ 157:1–8. https://doi.org/10.1016/j.rse.2014.09.021

Parochetti J, Arsenovic M, Getsinger K et al (2008) Addressing the need for herbicides for aquatic weeds in irrigation water in the US. Outlooks Pest Manag 19(3):112–116. https://doi.org/10.1564/19jun05

Patenaude G, Milne R, Dawson TP (2005) Synthesis of remote sensing approaches for forest carbon estimation: reporting to the Kyoto protocol. Environ Sci Policy 8(2):161–178. https://doi.org/10.1016/j.envsci.2004.12.010

Paturel J-E, Ouedraogo M, Servat E, Mahe G, Dezetter A, Boyer J-F (2003) The concept of rainfall and streamflow normals in West and Central Africa in a context of climatic variability. Hydrol Sci J 48(1):125–137. https://doi.org/10.1623/hysj.48.1.125.43479

Penatti NC, de Almeida TIR, Ferreira LG, Arantes AE, Coe MT (2015) Satellite-based hydrological dynamics of the world's largest continuous wetland. Remote Sens Environ 170:1–13. https://doi.org/10.1016/j.rse.2015.08.031

Peñuelas J, Gamon J, Fredeen A, Merino J, Field C (1994) Reflectance indices associated with physiological changes in nitrogen- and water-limited sunflower leaves. Remote Sens Environ 48(2):135–146. https://doi.org/10.1016/0034-4257(94)90136-8

Popescu SC (2007) Estimating biomass of individual pine trees using airborne Lidar. Biomass Bioenerg 31(9):646–655. https://doi.org/10.1016/j.biombioe.2007.06.022

Qureshi A, Pariva Badola R, Hussain SA (2012) A review of protocols used for assessment of carbon stock in forested landscapes. Environ Sci Policy 16:81–89. https://doi.org/10.1016/j.envsci.2011.11.001

Ritchie JC, Zimba PV, Everitt JH (2003) Remote sensing techniques to assess water quality. Photogramm Eng Remote Sens 69(6):695–704. https://doi.org/10.14358/PERS.69.6.695

Rondeaux G, Steven M, Baret F (1996) Optimization of soil-adjusted vegetation indices. Remote Sens Environ 55(2):95–107. https://doi.org/10.1016/0034-4257(95)00186-7

Samimi C, Kraus T (2004) Biomass estimation using Landsat-TM and -ETM+. towards a regional model for Southern Africa. GeoJournal 59:177–187. https://doi.org/10.1023/B:GEJO.0000026688.74589.58

Seghieri J, Carreau J, Boulain N, De Rosnay P, Arjounin M, Timouk F (2012) Is water availability really the main environmental factor controlling the phenology of woody vegetation in the central Sahel?? Plant Ecol 213(5):861–870. https://doi.org/10.1007/s11258-012-0048-y

Séguis L, Cappelaere B, Milési G, Peugeot C, Massuel S, Favreau G (2004) Simulated impacts of climate change and land-clearing on runoff from a small Sahelian catchment. Hydrol Process 18(17):3401–3413. https://doi.org/10.1002/hyp.1503

Segura M, Kanninen M (2005) Allometric models for tree volume and total aboveground biomass in a tropical humid forest in Costa Rica1. Biotropica 37(1):2–8. https://doi.org/10.1111/j.1744-7429.2005.02027.x

Shekede M, Kusangaya S, Schmidt K (2008) Spatio-temporal variations of aquatic weeds abundance and coverage in Lake Chivero, Zimbabwe. Phys Chem Earth, Parts A/B/C 33(8):714–721. https://doi.org/10.1016/j.pce.2008.06.052

Siciliano D, Wasson K, Potts DC, Olsen R (2008) Evaluating hyperspectral imaging of wetland vegetation as a tool for detecting estuarine nutrient enrichment. Remote Sens Environ 112(11):4020–4033. https://doi.org/10.1016/j.rse.2008.05.019

Straub C, Dees M, Weinacker H, Koch B (2009) Using airborne laser scanner data and CIR orthophotos to estimate the stem volume of forest stands. Photogrammetrie - Fernerkundung - Geoinformation 2009(3):277–287. https://doi.org/10.1127/0935-1221/2009/0022

Tapley B, Bettadpur S, Watkins M, Reigber C (2004) The gravity recovery and climate experiment: mission overview and early results. Geophys Res Lett 31:1–4. https://doi.org/10.1029/2004GL019920

Tatsumi K, Yamashiki Y, Canales Torres MA, Taipe CLR (2015) Crop classification of upland fields using random forest of time-series landsat 7 etm+ data. Comput Electron Agric 115:171–179. https://doi.org/10.1016/j.compag.2015.05.001

Thenkabail PS, Hanjra MA, Dheeravath V, Gumma M (2010) A holistic view of global croplands and their water use for ensuring global food security in the 21st century through advanced remote sensing and non-remote sensing approaches. Remote Sens 2(1):211–261. https://doi.org/10.3390/rs2010211

Timothy D, Onisimo M, Riyad I (2016) Quantifying aboveground biomass in African environments: a review of the trade-offs between sensor estimation accuracy and costs. Trop Ecol 57(3):393–405

Treuhaft RN, Asner GP, Law BE (2003) Structure-based forest biomass from fusion of radar and hyperspectral observations. Geophys Res Lett 30(9). https://doi.org/10.1029/2002GL016857

Trumbore S, Brando P, Hartmann H (2015) Forest health and global change. Science 349(6250):814–818. https://doi.org/10.1126/science.aac6759

Velicogna I, Kimball JS, Kim Y (2015) Impact of changes in GRACE derived terrestrial water storage on vegetation growth in Eurasia. Environ Res Lett 10(12):124024

Venteris E, Tagestad J, Downs J, Murray C (2015) Detection of anomalous crop condition and soil variability mapping using a 26 year Landsat record and the Palmer crop moisture index. Int J Appl Earth Obs Geoinf 39:160–170. https://doi.org/10.1016/j.jag.2015.03.008

Vincini M, Amaducci S, Frazzi E (2014) Empirical estimation of leaf chlorophyll density in winter wheat canopies using Sentinel-2 spectral resolution. IEEE Trans Geosci Remote Sens 52(6):3220–3235. https://doi.org/10.1109/TGRS.2013.2271813

Wouters B, Bonin JA, Chambers DP, Riva REM, Sasgen I, Wahr J (2014) GRACE, time-varying gravity, Earth system dynamics and climate change. Rep Prog Phys 77(11):116801. https://doi.org/10.1088/0034-4885/77/11/116801

Xiao X, Braswell B, Zhang Q et al (2003) Sensitivity of vegetation indices to atmospheric aerosols: continental-scale observations in Northern Asia. Remote Sens Environ 84(3):385–392. https://doi.org/10.1016/S0034-4257(02)00129-3

Yang Y, Long D, Guan H, Scanlon BR, Simmons CT, Jiang L, Xu X (2014) GRACE satellite observed hydrological controls on interannual and seasonal variability in surface greenness over mainland Australia. J Geophys Res Biogeosci 119(12):2245–2260. https://doi.org/10.1002/2014JG002670

Yeung CA, Li R (2018) Comparison of vegetation regeneration after wildfire between Mediterranean and tundra ecosystems by using Landsatimages. Ann GIS 24(2):99–112. https://doi.org/10.1080/19475683.2018.1424740

Zeng N, Neelin JD, Lau K-M, Tucker CJ (1999) Enhancement of interdecadal climate variability in the Sahel by vegetation interaction. Science 286(5444):1537–1540. https://doi.org/10.1126/science.286.5444.1537

Zhang Q, Xu C-Y, Becker S, Zhang ZX, Chen YD, Coulibaly M (2009) Trends and abrupt changes of precipitation maxima in the pearl river basin, china. Atmos Sci Lett 10(2):132–144. https://doi.org/10.1002/asl.221

Zhang X, Goldberg M, Tarpley D, Friedl MA, Morisette J, Kogan F, Yu Y (2010) Drought-induced vegetation stress in southwestern North America. Environ Res Lett 5(2):024008

Zhao K, Popescu S, Nelson R (2009) Lidar remote sensing of forest biomass: a scale-invariant estimation approach using airborne lasers. Remote Sens Environ 113(1):182–196. https://doi.org/10.1016/j.rse.2008.09.009

Zheng D, Rademacher J, Chen J et al (2004) Estimating aboveground biomass using Landsat 7 ETM+ data across a managed landscape in northern Wisconsin, USA. Remote Sens Environ 93(3):402–411. https://doi.org/10.1016/j.rse.2004.08.008

Zhou L, Tian Y, Myneni RB, Ciais P et al (2014) Widespread decline of congo rainforest greenness in the past decade. Nature 509(7498):86–90. https://doi.org/10.1038/nature13265

Zolkos S, Goetz S, Dubayah R (2013) A meta-analysis of terrestrial aboveground biomass estimation using lidar remote sensing. Remote Sens Environ 128:289–298. https://doi.org/10.1016/j.rse.2012.10.017

Chapter 8
Remotely Piloted Aircraft Systems

Abstract Advancements in technological capabilities of unmanned aircraft or close-range remote sensing systems are underpinned by technological advances in navigation and remote-control systems, making several service delivery and applications possible. This includes, reconnaissance, surveillance, high precision agriculture, localization and mapping, among others. These remotely piloted aircraft systems are gradually emerging as more efficient monitoring tools, especially due to their flexibility in altitude while monitoring objects on a relatively large scale. In addition to being relatively low cost, they provide better operational functions, enhanced imaging sensors, refined spectral and spatial resolutions, and with improved options for positioning and navigation. Their growing multi-disciplinary applications and benefits (e.g., in planetary science, environmental science, etc.) make them a technology for the future. In this chapter, their concepts and operations, classification, configurations, flight and data collection capabilities, as well as applications in environmental monitoring, including water resources are discussed.

8.1 Unmanned Aerial Vehicles—Close-Range Remote Sensing

Recall that in Chap. 3 we discussed about optical sensors and their platforms. The latter can be either ground based, space based, or airborne. Airborne platforms such as the use of aircrafts and airplanes are usually manned and typically relatively long-range remote sensing, operating in altitudes around 20 km. However, a new generation of remotely operated aircrafts which are unmanned operates on a much lower altitudes unlike manned aerial platforms. In this chapter, we will focus on the new generation of platforms popularly known as unmanned aerial vehicles (UAV). They are also referred to as remotely piloted aircraft systems, drones, or unmanned aerial systems. The Unmanned Aerial Vehicle (UAV), which is also often known as 'drone' is a unique aircraft specially modelled to operate without the control of

C. Ndehedehe, *Satellite Remote Sensing of Terrestrial Hydrology*,
https://doi.org/10.1007/978-3-030-99577-5_8

any human being on board but in relatively short or close range. These terms (UAV, drones, remotely piloted aircraft systems (RPAS)) will be interchangeably used in this chapter to represent close-range remote sensing platforms or unmanned aerial vehicles.

Even though the UAV technology was developed for use in the geomatics discipline, they are typically being used by the public and military for various applications. This has resulted in increased applications in numerous areas and disciplines (Fig. 8.1). Apart from its military applications in surveillance and monitoring capabilities, short-range remote sensing methods are in high demand and their use is growing in areas such as agriculture; forestry; and mapping, search, and rescue missions. Given its generic and dynamic application in several operations such as field and environmental research (Canas and Irwin 1986), its nomenclature varies. For instance, Remotely Piloted Aircraft System (RPAS), Medium Altitude Long Endurance (MALE) UAV, Unmanned Combat Air Vehicle (UCAV), the Remotely Operated Aircraft (ROA), etc. Nevertheless, the system still maintains a generic nomenclature, Unmanned Aerial System (UAS), which takes into cognizance the comprehension of the entire system, including the ground control stations and the aerial vehicle. The drone industry is growing and projected net value of total earnings could reach 15 billion USD by 2022 (Kazim et al. 2021). The broad range multi-disciplinary applications and benefits of drones in engineering, computer science, earth and planetary science, environmental science, etc. (Fig. 8.1) make it a technology for the future.

Generally, unmanned aerial systems or vehicles are flying robots designed with different capabilities based on purpose and application. These capabilities are underpinned by technological advances in navigation and remote control systems, making

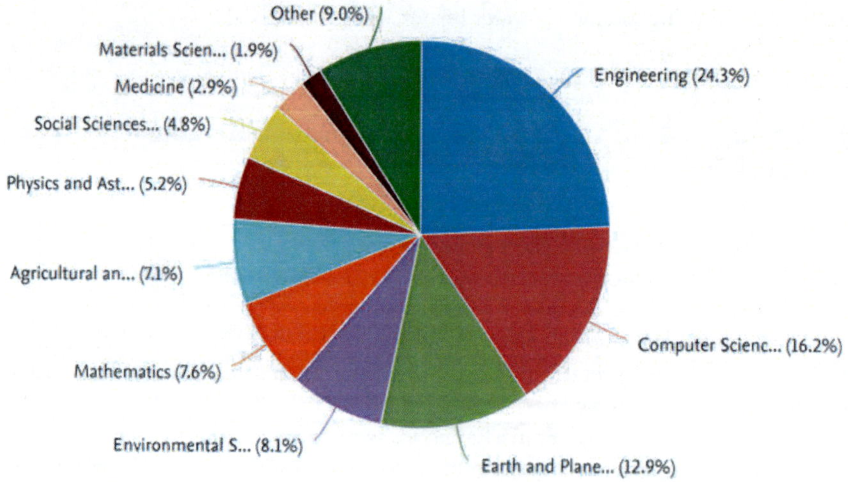

Fig. 8.1 Cross-disciplinary applications of UAVs/drones based on research outputs in Scopus, 4 September 2021

several service delivery and applications possible, including reconnaissance, surveillance, transportation of loads, high-precision agriculture, localization and mapping, and aerology (Kazim et al. 2021; Hassanalian and Abdelkefi 2017). Additionally, UAVs are now utilized as tools for enforcements, as well as behaviour modification in a managerial framework (Rees et al. 2018; Mulero-Pázmány et al. 2014). The plenitude of emerging cross-disciplinary applications of drones in the last 10 years highlights the important areas and value that this technology brings (Fig. 8.1). There are several types of drones and they are built around advances in sensor technology, navigation, and remote control systems. UAVs and drones can be classified in terms of size, range/endurance, and applications. In terms of sizes, they can be grouped as very small, small, medium, and large. Other classifications have focused on wing configurations, e.g., fixed wing, rotary wing, and a hybrid of fixed- and rotary-wing configurations (Kazim et al. 2021). The configuration of the fixed wing enables longer endurance flights, providing high-speed motion, as well as maintaining high payloads (Kazim et al. 2021). The classification based on range and how long they can last in the air (endurance) include

- Low-cost close range: These are low-cost and small-sized body drones that have a range of 5 km. They last between 20 and 45 min.
- Close range: These types have a range of 50 km and can last between 1 and 6 h. Mainly used for surveillance and reconnaissance activities.
- Short range: They have longer range of 150 km or more. Endurance times range from 8 to 12 h. Applications include reconnaissance and surveillance.
- Mid-range: These have super high speed compared to the mid-range and have working radius of 650 km. They are used to collect meteorological data. However, they can also be useful in reconnaissance and surveillance operations.
- Long range: These have the highest endurance time of about 36 h and a working radius of 300 km. These have reconnaissance and surveillance capabilities.

It should be noted that drones vary widely in their configurations and design. This may depend on their range and the purpose for which they were designed. Consequently, a more advanced classifications of drones can be obtained based on different parameters. These classifications have been extensively detailed in the literature and include drone categorization based on several parameters and characteristics, such as size, flight endurance, and capabilities for scientific and military applications (Hassanalian and Abdelkefi 2017). Figure 8.2a–i is an illustration of UAV classification based on several performance characteristics and design parameters. This can be weight, wing loading, wing span, range, maximum altitude, speed, endurance, and production costs. For instance, tailplane forward and flying wing UAVs have the propulsion systems at the back of the fuselage or at the front side of the UAV (Fig. 8.2a) as opposed to the fixed wing which uses a vertical propulsion system at the front of their fuselage, and have cross wings (Fig. 8.2b). The latter has the capability for vertical take-off and landing without the need for runway (Hassanalian and Abdelkefi 2017). Although there are relatively smaller UAVs which also do not require a runway for take-off, the drone classification in Fig. 8.2a–i is all based on their mission capabilities. This includes horizontal take-off landing (HTOL), vertical

Fig. 8.2 Drone classification based on platforms and a range of other design parameters. *Source* Hassanalian and Abdelkefi (2017)

take-off landing (VTOL), tilt wing, tilt rotor, tilt body, ducted fan, helicopter, heli-wing, and unconventional type drones (Fig. 8.2a–i, respectively). More generally, the guidance, navigation, and control of drones are usually undertaken through the use of a radio-controlled system, autopilot, and video base (Hassanalian and Abdelkefi 2017).

UAVs can also be categorized into military and domestic. Although this is not a well-known classification, the use of UAVs falls into either of this category. Military drones have different purposes, including surveillance and guiding air strikes of enemy targets. Surveillance drones (Fig. 8.3) are highly sophisticated military drones developed to collect still images, video recordings, or live video of specific targets. These could be images and pictures of individuals, vehicles, or specific areas of interest. They have been widely used in recent times by governments, militaries, law enforcement, or commercial organizations to collect information and intelligence that underpins important decision-making. As with typical optical sensing systems, these surveillance drones can provide access to areas and hotspots that are rather difficult to reach by humans due to complex terrain or topography.

Fig. 8.3 A fixed-wing United Kingdom manufactured surveillance drone. *Source* https://www.unmannedsystemstechnology.com/category/supplier-directory/platforms/surveillance-drones/

However, access to some potentially dangerous areas including crime scenes and battle fields for information have been made possible through the use of military-type UAVs (or drones) as shown in Fig. 8.3. Domestic UAVs are those designed with non-military capabilities in terms of range, body weight, and several other configurations. These types of drones are designed to improve understanding of environmental change and help with change detection and impact assessment through monitoring on a local scale. Some examples include assessing disaster areas, fighting wildfires, providing communication support during disaster, transportation, coastal ecosystem change evaluation, assessing crop water status, and monitoring crop health, among several other applications. This is discussed in more detail in subsequent sections.

The primary frames for the UAV basically applied in geomatics and photogrammetry are usually copters, fixed wings, multi-copter, and unpowered platforms, each having its own merit and demerits to suit their respective applications. They also vary in other aspects such as price range, payload and automation proficiency, flight autonomy, etc., for instance, a total autonomous flight is not allowed on low-cost solutions, and usually the action of take-off and landing is controlled by a human being. Another noteworthy cost-sensitive component is the engine type. Internal combustion engines are known to have longer tolerance and warrant higher payload when compared to the most used electrical ones. Although their build have higher costs and entails more pre-flight controls and maintenance. The specific characteristics of the three basic drone frameworks applied mostly in geomatics are outlined below

(Kazim et al. 2021; Hassanalian and Abdelkefi 2017; Ventura et al. 2017; Achille et al. 2015):

- Copters—This is produced to mimic real helicopters, but on a smaller scale. However, the bigger models are made up of gasoline engines, which enables it to carry heavier payloads and at the same time function autonomously. The elasticity of this type of UAV is well known. Another distinct feature of the big copters is that it is made up of strong resistivity and constancy to operate in extremely bad wind states. The need for flight preparations, repairs, and maintenance of this system is one of its major limitations in contrast with ICE motors (Gago et al. 2015; Archer et al. 2004).
- Multi-copters—These are known to comprise more than three propellers rigidly attached to a structure (e.g., Hassanalian and Abdelkefi 2017). With an adequate handling of each of the propeller's differential engine rotation, it becomes possible to navigate the systems position and motion. Given the direct propeller-payload proportion, this system has the potential of carrying more loads. However, its major limitation lies in the battery, which does not support very long autonomy (estimated 15–20 min). This therefore results in the changing of batteries for any long flight work sessions. This system, which is always characterized by its small dimension, diminutive deployment sessions, high stability, and extreme maneuverability, is usually preferred for the execution of terrestrial photogrammetric cultural heritage surveys (Achille et al. 2015). Moreover, this system has a very swift development phase, which is often emphasized on enhancing the flight time, sensors to be used, and most importantly the accuracy and precision of the entire on-board navigation systems.
- Fixed wing—The movement of this system is usually linear and is argued to be as a result of its propelling system, which is usually between one and three. Its operations are similar to traditional aircraft but in a fully autonomous mode with an adequate pre-planned flight plan. Its frequent use in aero-photogrammetry is due to its structural configurations, which encompasses a high autonomy and possibilities of wider range capture when compared to a single or multi-rotor configuration. For increased performance, autonomy, portability, and applications, the drone is designed with very light materials of less than 1 kg take-off weight. However, it is noticed that this solution exhibits an extremely poor wind resistance, which impedes its performance in bad wind conditions and limits the entire payload structure.

The developments and initial purpose of UAV systems date back to its application in military operations. They were being used specifically for the assessment of unfavourable regions, as well as for surveillance purposes. However, with the advancements in digital camera systems and the Global Navigation Satellite Systems (GNSS), its application was extended to civilians, primarily for geodetic operations, system navigation, and geo-referencing operations (Remondino et al. 2014). The introduction of UAV in this day and age has served as an alternative to the usual airborne photogrammetric procedure for monitoring large spatial extents even in 3D, thus extending its use in several domestic and scientific areas. Even though

UAV cannot be used to replace the usual airborne photogrammetric procedure in the mapping of large areas, its efficiency and resolution in mapping small areas and large-scale surveys are extremely superior. Its operation and application can support validation programmes as well as leading to quick explanation of some key phenomena, like the monitoring of vertical surfaces and pavements such as tall towers, using high-resolution procedures.

8.1.1 Concepts and Operations

Adequate overview of the historical advancement of UAV shows it has various country-specific regulations and a wide application range to remote sensing and photogrammetry. The UAV used by civilians maintains a diverse range of surveying and mapping applications. For instance, this can include their use in archaeology (Rinaudo et al. 2012), fire-fighting (Rufino and Moccia 2005), forestry (Puliti et al. 2015), mining (Nex and Remondino 2014), agricultural watershed analysis (Ouédraogo et al. 2014), geological resource monitoring (Johnson et al. 2014) as well as a wide range of sensor deployment tasks. The suitability of UAVs in airborne sensor deployment is due to their potential for 3D spatial coverage. For example, the studies carried out by Erman et al. (2008), Kerle et al. (2008), and Quaritsch et al. (2010) highlighted its applicability and performance in disaster management and ecological research. Various sensors for UAV assessment exist; however, the most common types are cameras-sensitive to a wide range of light spectra (Sorensen et al. 2017). The prevalence of these camera-based sensors over other sensor types is largely as result of their proficiency in automated and manual inspection, which is often times complemented by computer vision to enable easy depiction of even the most decipherable diminutive ground object. It is possible to program single drones to follow a specific pathway while covering the region and objects of interest. However, it can also be flown manually in either a line of sight or directed by Global Positioning Systems (GPS) or other sensory mechanisms. It is also possible to deploy multiple drones in swarms, for instance, to form a series of grid networks capable of relaying data over large distances, or mapping and tracking pollution, heat or other continuous events (Sorensen et al. 2017).

The first recorded use of the unmanned aircraft systems was documented long before the first powered flight of the Wright brothers. It was in Australia on 22 August 1849, when 200 pilotless balloons were launched with attached explosives and time fuse weighs 33 pounds, against the Venice City. The Aerodrome No. 5 designed by Samuel P. Langley was recorded to have flown over the Potomac River close to Washington DC on 6 May 1896. Also, the extensive application of radio-controlled aircrafts during the World wars for aerial surveillance was used to train antiaircraft gunners, as well as to operate as aerial torpedoes. The application of the drone technology during the cold war was primarily to capture intelligence from opposition camps even in denied and restricted areas. The unmanned aircraft systems were first deployed for large-scale operations in the Vietnam War. The twenty-first century saw the most frequent use and operation of the unmanned aircraft systems, especially in military applications as a tool to combat terrorism more efficiently (Tsach et al. 2010). These historical perspectives have however created a perception that the system was majorly used for military operations. But nowadays, we can see its application extend to several civilian-based operations and function ranging from climate change assessment, natural disaster monitoring, ecological studies, photography, and filming (Ventura et al. 2017).

Photo interpretations from UAVs have become a crucial tool in the assessment of vast surfaces and have been known to play key roles in the monitoring of ecological habitats, as well as land-use change assessments. It is however important to note that UAV platform has disadvantages similar to other remote sensing techniques, i.e., satellite and airborne methods for data capturing. For instance, the limitation of remotely piloted aircrafts is often felt in its over-reliance on weather conditions, flight speed, and altitude. Weather conditions such as rainfall can negatively impact on electronics while strong winds and cold temperatures can potentially increase battery usage and reduce performance, respectively.

For satellite methods, cloud cover due to poor weather conditions is also a problem. Furthermore, the possibility of some humid biotopes and tropic coast being covered by persistent cloud cover is also a serious concern, as it often results in creating specific gaps in satellite repositories, especially for distinct time periods and locations. Moreover, the temporal resolution of both aircraft platforms and satellite-derived imageries is largely dependent on the aircrafts availability and orbital features of the satellite missions (Koh and Wich 2012). Also, the finest spatial resolutions that can be achieved from these platforms are in the range of 30–50 cm/pixel. Therefore, we can conclusively say that satellite observations are often limited, especially as a result of revisit time (Turner et al. 2012), in the case of assessing highly heterogenous and dynamic environment, or even for real-time land-use monitoring for sensitive habitats and ecological services. Despite the advent of new satellite sensors, such as GeoEye-1, WorldView-2, Pleiades-1A, IKONOS, and Quickbird, which have been extensively applied in ecological studies (Phua et al. 2016; Wu and Bauer 2012; Hall et al. 2012; Song et al. 2010), challenges such as temporal coverage, cloud contamination, and high cost per scene still remain (Anderson and Gaston 2013). Financial constraints and limited funding in emerging economies can impede the acquisition of available images for environmental assessments and research in those regions.

Modern advancements in technological capabilities of the unmanned aircraft systems have however brought us to this point of reliance on UAVs, which keep getting smaller with better operational functions, enhanced imaging sensors, refined spectral and spatial resolutions, and more technical options for positioning and navigation. The upsurge in demand for UAV is high in both professional and recreational activities. This therefore provides opportunity for the application of these emerging technologies in unprecedented scientific operations even in the most diverse science fields. Also, the UAV technology provides earth scientists and particularly ecologists with novel procedures for timely, cost-effective, and responsive environmental assessment procedures, thus exploring the possibility of studying organisms and their corresponding niche at very close intervals (Anderson and Gaston 2013).

8.1.2 Sensor Characteristics and Capabilities of Low-Cost Unmanned Aerial Vehicles

UAVs utilize a broad range of sensors, which enable them to collect information regardless of complex urban environments. This section will discuss this more extensively, focusing on their characteristics. These characteristics determine the several capabilities of these sensors and will be highlighted. Green et al. (2019) pointed out the following environmental applications of low-cost UAVs:

- Field drainage patterns.
- Snowmelt characteristics.
- Identification of various tree species as well as their spacing.
- Applications in reconnaissance and mapping.
- Measurements over tree nurseries and experimental agricultural farms.
- Observations of domesticated farm animals and their environment.
- River embankment and older site protection.
- Assessing flood conditions at river ice jams.
- Quality of rural highway decks.
- Perception of underground quantities.
- Assessing storm-induced beach deposition.
- Recording of engineering facilities and structures.
- Observation of harbour sites and ports.
- Assessing deposition characteristics and coastal erosion.
- Observation of ice conditions.

8.1.2.1 Sensors

The swift advancements in technology, digital, and microprocessors today are the mainstay of the wide range of low-cost and miniaturized sensors and digital cameras, which have been found suitable for mounting on UAV platforms. For example, LiDAR, thermal, hyperspectral, multispectral, colour and colour infrared (CIR), as well as panchromatic sensors. Also, this includes the availability of digital camera slots on board the UAV platforms, as well as attachments on mobile devices. Even though several sensors were not originally designed to be used by UAVs, the advancements in the UAV technology have increased the demand for a wider range of sensors to be used for different applications. Another factor determining the limitation of various sensors to different platforms is the lift capacity and payload of the aerial platform. The high availability and surging range of both active and passive sensors observing naturally occurring radiation emitted or reflected by their targets are also noted (Kim et al. 2006).

- Passive sensors: Passive optical sensors are known to estimate radiation levels in the infrared (0.7–14 μm) and visible (0.4–0.7 μm) portions of the electromagnetic spectrum. They are most suitable in the presence of sunlight, which serves as an illumination source to them. A wide range of factors limit their performance, and they include smoke, haze, and even clouds, which are known to be a major course of refraction when signals penetrate them. However, a surge in the number of UAV-suitable optical sensors are being noticed, for example, small cameras, which have the potential of videography and still photography (i.e., iLook and GoPro series cameras), to both large and small DSLR cameras, stereo cameras, hyperspectral and multispectral cameras, high-resolution cameras mounted on smartphones, or other low-cost inventions like the HackHD cameras. The major constrain noticed by the optical sensors is due to the variety of opportunities and issues influencing the digital frame cameras that can be arrayed on lightweight UAVs. Factors such as camera weight in relation to available UAV payload; the various nonmetric features of a wide range of the low-cost cameras; the very small format of the camera images; the camera resolutions and lenses types; the photographic gaps; the demand for short exposure times to help tackle the influences of platform unsteadiness (e.g., roll, speed, pitch, and yaw); the need for high framing rates arising from the velocity of the UAV system above ground, though from a minute altitude; and, finally, the enormous lateral and longitudinal overlaps that need to be engaged for mapping targets (Sedunov et al. 2018).

- Multispectral and near infrared: Imageries derived from multispectral platforms are achieved through an observation of reflected energy within severally defined bands of the electromagnetic spectrum. It usually consists of three or more various band measurements located in each pixel of the produced images. The visible green, visible red, and near-infrared spectrums are well-known bands located in these sensors. It is also possible to obtain information that can be automatically interpreted or perceived visually through the simultaneous observations of multiple spectral wavelengths. In a specific location, it can be very useful to detect environmental features and depict their corresponding characteristics using the algebraic integration of parameters in different spectral wave bands. For instance, the adequate identification and detection of plant stress, nutrients of water status, and diseases are best achieved using multispectral imaging of vegetation. Also, it becomes possible to determine the Normalized Difference Vegetation Index images, which adequately portray the information regarding plant status and conditions with the use of infrared camera images which is obtained at definite time intervals, overlapped and aligned, with an effect of some specially filtered software such as the Pixel Wrench (Curran et al. 1983; Wright et al. 2018).
- Shortwave infrared: These types of radiations, which typically exist around 0.9–1.7 μm in the electromagnetic spectrum are not visible to the human eye but can be perceived with the use of dedicated indium gallium arsenide (InGaAs) sensors (Wang 2007). Images filtered through the InGaAs camera look exactly like visible images in their detail and resolution, which makes objects to be easily discerned. This, however, is in contrast with thermal imageries. One of the outstanding advantages of the shortwave infrared is its minimal power consumption, given that it employs a thermoelectric cooler or no cooler especially when the dark current is presumed to be low, and at the same time delivering adequate imageries even in low-light conditions. The InGaAs sensors are also known to be extremely

sensitive as it provides individual counts for photons. Therefore, when designed as focal plane arrays with thousands or millions of tiny sensor pixels, the short-wave infrared camera has the capacity of functioning optimally even in low-light conditions. This technology, however, is not cheap, although a rarer one exists, which is known as the detector material, Indium gallium arsenide (InGaAs) (Teke 2013; Bioucas-Dias 2013).

- Hyperspectral: This type of imaging is known to utilize a large amount of bandwidth in the light spectrum so as to present an intense dataset and perceive objects of interest which are not known to the single-bandwidth imaging sensors. Given its wealth of a wide variety of fine spectral bandwidths, it is easier to identify explicit conditions and image features. The use of these type of images having hundreds of bands like 255 is increasingly having a wide range of applications in environmental, coastal, vegetation, and even water resources management. The providers of these imageries are also known to provide highly flexible and customizable preferences for the resolution range and number of spectral bands existing in both the visible and infrared spectrums to allow end-users the choice of making adequate selections depending on the application of each dataset. The integration of hyperspectral imaging with a standard element in an electro-optical and infrared sensor suite, on the other hand, has posed a significant challenge for engineering, especially given its potential application in UAV technology.
- Fluorescence: The use of florescence spectroscopy has been seen in a wide range of applications. With the use of multi- and hyperspectral radiance sensors, it is possible to make estimates of fluorescence, while utilizing the Fraunhofer line and decoupling fluorescence from the reflected flux. Furthermore, the reflectance sensed by multispectral sensors is responsible for the optical indices associated to fluorescence fluxes. Meroni et al. (2013) however pointed out the difficulty and existing complication in estimating fluorescence from air especially, as a result of its atmospheric absorption through the sensors, and various techniques in understanding atmospheric effects that are not fully formed. The author also proposed and recommended very high spectral resolution (0.05–0.1 nm) for adequate estimation of fluorescence fluxes, so as to accommodate resampling and application of estimation techniques from both multi- and hyperspectral platforms.
- Thermal: Due to temperature variations existing between different surfaces, the features and characteristics of some objects and surfaces have been found to be adequately depicted in thermal infrared imageries. Due to the capabilities and peculiarities of the UAV, lightweight thermal cameras have been adapted for its use. The FLIR Quark 640 is a well-known long wave infrared and thermal sensor, which has been designed with lenses (7–35 mm) and has a scene range of −400–670 C and 50 mK sensitivity, has weight of 3–135 g, which is largely a function of configuration and settings (Voogt and Oke 2003).

8.1.2.2 Active Sensors

Active sensors are known to radiate emissions and assess the reflected fraction of the target objects as well as the time difference noticed between radiation and reception (Hill and Zimmerman 2010). One of the requirements of active sensors is the source generated power, which helps in complimenting some notable weight to the entire aerial framework. This makes it less popular to be used in UAVs in comparison to passive equipment. The Radar Synthetic Aperture radar (SAR) is a popular radar applying comparative motion between antenna and its corresponding target environment so as to deliver unique coherent-signal fluctuations. The radio waves of the SAR pulse exist within the 0.002–1 m wavelength when directed towards a specific target. Variety of echo waveforms obtained successively at different antenna positions are clearly perceived and stored, before undergoing a combined post-processing phase to resolve elements in an image of the target region. The wide range of application of the SAR imageries has been captured in many conventional studies. It has proved to be useful in detecting oil spills in water bodies and oceans because the oil is known to alter the backscatter features of the water. Similar to thermal imagery, the adequate detection of oil leaks is a function of the differential imagery technique, which is a prerequisite in this context. However, a distinct feature of the radar technology is its day and night capacity for data collection, which makes it very suitable for the surveillance of inaccessible and difficult environments even in poor weather conditions (Fitzgerald 2010).

- Radar: The military has integrated radar systems into large UAV systems for their operation. However, there is still need for low-cost, small, high-resolution radar systems to be used for operations meant for small UAVs (Samczyński 2018). With the advent of the micro-SAR system by the Brigham Young University in the United States, especially designed for small UAVs, a significant trade-off between coverage and precision versus size and cost is clearly represented. The system is known for its consistent data recording, which can last for over 60 min on a pair of compact flash discs, which can be transferred to a laptop for data processing using well-known SAR image formation and autofocusing software. The flight weight is reduced to <1 kg due to the minimal enclosures dominant in the system. The transmitted shortwaves are received at different times by the micro-SAR transmission with a definite interval, while reception occurs simultaneously using the continuous wave linear frequency modulation, which in turn translates to a low-power operation of the entire system. The micro-SAR makes use of bi-static operation in its transmission and reception in order to optimize performance. However, these transmissions take place through different antennas (Ding et al. 2020).
- LiDAR: The popularity of the airborne light detection and ranging (LiDAR) surveys is as a result of its wide range of applications (Keckhut et al. 2011). Usually, LiDAR surveys using UAVs are carried out at altitudes of 1500 m or above. This kind of height makes them vulnerable to a wide array of atmospheric conditions and also contributes to their poor image resolution or minute image footprints. One of the major limitations of the LiDAR on UAV, however, is in the size and

weight of the sensor. The entire LiDAR system is heavier than the UAV imaging system when compared. This is as a result of (i) LiDAR sensors are not as light and small as their cameras and (ii) LiDAR systems can be determined through a precise inertial navigation system. Although the imaging system makes use of traditional photogrammetry techniques not requiring and inertial navigation system, the UAV imaging system can compensate by covering a large area in a very short time having less prior expense (Mandlburger et al. 2015).

8.2 Remotely Piloted Systems in Environmental Monitoring

8.2.1 Coastal Management

The task of coastal zone monitoring and coastline detection is very crucial for environmental protection and coastal management. This is particularly so because coastal zones around the world are vulnerable to continental and marine processes (e.g., storms, erosion, etc.) caused by the composite influence of climate change and human activities. For effective integrated coastal management operations that will cushion the impacts of these processes, there is need for updated and precise information, given that climate fluctuations contribute substantially to coastal erosion. The role of remote sensing technique in coastal monitoring and assessment is, however, very crucial, especially considering its efficiency and effectiveness (e.g., repetitive and meaningful coverage and high temporal frequency) in adequate coastline detections. Over the years, the application of satellite remote sensing has improved especially in terms of having finer spatial resolutions. However, the available data is still insufficient for centrimetric-level accuracy assessment of small-to-medium scale. Studies, however, have shown the high-resolution information at low-cost output of integrating UAVs with several algorithms (for example, the structure for motion algorithm), which is also known as a UAV-SfM pipeline (Pajares 2015; Yastıklı et al. 2013).

The enhancements of UAVs especially in terms of minimizing the sensor sizes and increasing their flight potentials and agility combined with their available graphic imagery tool have resulted to the production of a spatial data acquisition tool for the purpose of environmental assessment (Gonçalves and Henriques 2015; Brunier et al. 2016; Clapuyt et al. 2016). Although past studies have explored techniques and methodologies that measure variations in coastal morphology zones with the use of UAVs (Gonçalves and Henriques 2015; Casella et al. 2014), more advancements in the application of UAV for further data collection, especially in remote sensing for a wide range of applications, (includes agricultural, environmental, and scientific applications), have also been reported (Brunier et al. 2016; Gonçalves and Henriques 2015; Klemas 2015; Remondino et al. 2014; Westoby et al. 2012; Harwin and Lucieer 2012).

With the provisions of UAV and its capabilities, it is easier to obtain digital image with high spatial and temporal resolutions, which have the capacity of overcoming some of the various limitations and shortcomings of spatially acquired data from aircrafts and satellites. Conventionally, the aircraft and satellite platforms provide datasets of resolutions ranging from 20 to 50 cm pixel (Turner et al. 2012). The spatial resolution obtained from the satellite systems ranges from 50 cm above, while the manned aircraft ranges from 10 to 100 cm as shown in Harwin and Lucieer (2012). On the other hand, observations derived from UAV have a high spatial resolution ranging from 2 to 5 cm and produce more detailed characteristics and landscape variations when used for topographic assessments, including coastal zone and morphology monitoring, beach morphological features, which can also be mapped in both 2D and 3D spaces depending on the purpose of assessment (Topouzelis and Kitsiou 2015; Remondino et al. 2014; Yastıklı et al. 2013). This distinctively shows the clear advantage of the UAV system compared to the satellite or manned aircraft systems. However, these advantages span across spatial resolution and entail other relevant aspects such as ease of deployment, ease of survey repeatability, frequent mission that provide spatial datasets with a high temporal and spatial resolutions (<5 cm), as well as the low cost in operating the UAVs compared to the manned remote sensing technique (Topouzelis and Kitsiou 2015; Harwin and Lucieer 2012). Other techniques in depicting and measuring beach morphology also exist, such as the application of digital surface models (DSMs) and georeferenced orthophotos in both 2D and 3D, which further permits the monitoring of variations in the beach as a result of harsh wave phenomena (Brunier et al. 2016).

Beaches are predominantly situated in locations with very strong sea–land interface, which are also concentrated regions that receive substantial human activities and developments. These anthropogenic influences are a major cause of the frequent erosion events noticed in beaches and often result in significant topographic variations in a short time. Lou Shilong (2013) pointed out that an estimated 50% of sandy coastlines in China are experiencing harsh coastal erosion. In an attempt to accurately determine, monitor, and maintain these beaches, surveys have been conducted using aerial laser scanning (McKean and Roering 2004; Notebaert et al. 2009), terrestrial laser scanning (Fairley et al. 2016; Brasington et al. 2012), total stations (Fuller et al. 2003), and even differential or real-time kinematic global position system (Squarzoni et al. 2005). In as much as these techniques and instruments have been widely applied in geomorphology and have produced accepted observational results, they also have their own limitations.

In the past, GPS measurements were the most popular technique used for the assessment of topographic details and erosions in beaches (Chen et al. 2018). However, the major limitation in this technique was that it could only be used to capture discrete points and profile datasets, which cannot be used for the adequate presentation of the topographic spatial distribution of the area, especially in cases of very steep and rugged terrains which need detailed profiling. Although its terrestrial applications produced relatively accurate result, its need for regular arrangement of various scanning stations over several kilometres of restricted beaches as a result of its inadequate scanning range (>500 km) is a major limitation of the instrument.

Other remote sensing techniques such as the airborne LiDAR technique are quite capable of generating detailed spatial elevation information of large-scale coastal regions (White and Wang 2003). However, its high cost and failure to generate homogenous datasets with GPS and terrestrial laser scanner are major shortcomings of this technique. With the advent of the low-cost and lightweight UAV systems, integrated with the development of the computer vision algorithms like the structure-from-motion, the use of the UAV technique in assessing beach geomorphology to aid coastal management is dominating other monitoring protocols and methodologies. It is now possible to generate topographic details from distributed and randomly oriented photographs using the structure-from-motion algorithm (Fonstad et al. 2013; Westoby et al. 2012). On the other hand, UAV-based photogrammetry serves as a significant complement and alternative to satellite and manned aircrafts given its growing utility in the generation of high-resolution topography of various surface processes (Colomina and Molina 2014).

UAV maintains a distinct advantage over other remote sensing methods used in coastal management, especially due to its flexibility, capability, and maneuverability in the acquisition of images with very high spatial resolutions and temporal frequencies even with consumer-grade cameras. Several applications of the UAV technology in various dynamic environmental studies have been undertaken over the past years. They include coastal topographic assessment and understanding geomorphic variation (Clapuyt et al. 2016; Delacourt et al. 2009), river bank and chasm erosion (Prosdocimi et al. 2015), glacial and fluvial dynamics (Immerzeel et al. 2014; Javernick et al. 2014), landslides (Turner et al. 2015), gully erosion (Gómez-Gutiérrez et al. 2014), Antarctic moss beds (Turner et al. 2015), and vegetation assessment (Jaakkola et al. 2010; Dandois and Ellis 2010).

8.2.2 *Vegetation and Forestry*

The contribution and role of forestry in the Earth's ecosystem cannot be overemphasized, especially given their function in mitigating harsh and adverse effects of climate variations. Their role also encompasses the provision of suitable habitats for various species of plants and animals, characterizing a significant carbon sink, as well as supporting an efficient ecological and hydrological cycle. Also, forests contribute to the production of timber and non-wood materials by being a source for relevant raw materials (Potapov et al. 2008; Boyd and Danson 2005). The roles and functions of forests earlier espoused have necessitated an extensive debate and discussion among the elite members of the public, scientists, as well as politicians in the 1980s, to establish techniques that will enhance forest health as well as its sustainability to the regional and global ecosystem (de Vries et al. 2003).

Veettil et al. (2018) also gave insights as to the importance of mangrove forests in the provision of various ecosystem services as well as coastal regional protection for both tropical and subtropical coastlines over the globe. Its support for elevated floral and faunal biodiversity and the provision of wide array of goods to various coastal communities is widely known. However, it has been reported that the amount of mangrove forests has been on a constant decline, especially due to population growth, which has intensified anthropogenic influences around it. Over the past decades, a substantial amount of global mangrove forests have been lost (Alongi 2002), especially as a result of factors such as urban development (Romañach et al.

2018), overexploitation of forest resources for timber, changes in rainfall amount, patterns and intensity (Ellison 2000; Field 1995), natural occurrences like typhoons (Villamayor et al. 2016; Cahoon et al. 2003), rise in sea-level extents (Field 1995), and transformation to aquaculture (Farnsworth and Ellison 1998; Primavera 2005, 2000), among others.

Frequent assessment of mangrove extent so as to determine its historical and current interactions with various ecosystem services is thus crucial. Accurate spatial information of mangrove forest extent is also important for natural resource management and land-use planning. The most popular means of this assessment is through remote sensing, given the efficiency and accuracy expected, which allows for precise estimation and global estimations. It is known to provide access to a wide range of procedures, techniques, and algorithms especially for field-based observations which are mostly known to be available at local extents. Remote sensing techniques are relatively inexpensive and also allow researchers to monitor real and spatial extent fluctuations of mangrove forest in real time for a long duration (Baloloy et al. 2020).

The fact that the global population is constantly increasing puts significant pressure and poses a growing concern to the rate of global plant production, as this growth is known to be directly proportional to the depletion of natural resources. This therefore calls for the need for more accurate and frequent monitoring of our regional and global vegetation so as to make informed decisions. The role of vegetation and forests transcends biomass productions, as they equally serve as significant repositories for freshwater (canopy water), and also aid in regional latent heat transfers through evapotranspiration. Among prominent remote sensing techniques used for this monitoring, UAV procedures have shown great distinction, especially due to its capability for timely tracking and accurate assessment through several procedures highlighted below:

- The UAV technique has the capability of generating digital surface models in 3D of the region of interest through the application of highly overlapping images and the utilization of photo reconstruction techniques using the well-known structure-from-motion algorithm (Dandois and Ellis 2013).
- They have the ability to utilize various perception and spatial structures to assimilate a wide range of vegetation spectrum.
- Given the high flexibility of the UAV system, the flights can be rescheduled to correspond with the timely progress of the vegetation index.
- Their altitude can be easily changed, which means that there is higher potential of obtaining ultra-high spatial resolution to discern very fine vegetation details when flown at a very low altitude.

Therefore, the advent of UAV creates a more feasible and efficient method of determining ultra-high spatial resolution even in 3D of trees and plants (de Castro et al. 2020). This opportunity bridges the gap between ground-based instruments and other existing conventional remote sensing systems for vegetation measurements. Additionally, UAV-based system can also complement manually obtained data from field, especially given that such data could have errors and can be inconsistent due to the strenuous procedure involved. Due to these peculiarities, the UAV system is preferred and seems more efficient in agro-forestry applications (Dainelli et al. 2021). These applications include land management (Librán-Embid et al. 2020), environmental assessment (Manfreda et al. 2018), vegetation mapping and characterization (Maes and Steppe 2019) and analysing plant phenotypic classification in breeding courses (Yang et al. 2017) as well as addressing a wide range of agricultural objectives and practice goals (Maes and Steppe 2019). Nevertheless, the enormous amount of detailed data implanted in UAV systems needs the application of strong algorithms and models to extract important information relating to the biochemical and structural vegetation composition, which also provides insight to the corresponding relevant plant traits. Hassler and Baysal-Gurel (2019) pointed out some statistical models such as pixel-based image analysis, extrema estimators, spatial statistics, and linear classifiers that are usually used for this analysis.

Although with the advent of more advanced procedures and techniques such as computer vision algorithms, machine learning, and object-based image analysis techniques, there is a strong potential and indication of more accurate analysis (Rehman et al. 2019). Besides the potential accuracy index expected in using these procedures, it can also help in precise vegetation analysis and can also be of great help to advisors, farmers, as well as public administrators in making decision regarding forestry and agricultural management.

8.2.3 Agriculture and Wetlands

Agricultural sector has also profited immensely from the advancement and innovations in sensor technology for several years. Innovation in this context is defined as an essential technique of achieving more by exploiting scarce resources even with minimal effort (Ahmad et al. 2021). The process of enhancing the quality of raw materials through innovative procedures certifies production competence, and also influences significantly on food safety and security, as well as economic growth (Ozdogan et al. 2017). The rapid success being noted in agricultural sector is largely a function of the introduction of conventional techniques and methodologies, which include autonomous robots, air vehicles, satellites, Global Positioning System (GPS), and Geographic Information System (GIS) procedures. The implementation of these technologies and innovations has led to the creation of new expressions such as 'agriculture 4.0', 'digital farming', 'precision approach', 'precision farming', 'precision agriculture', and so on. The whole idea of precision agriculture is that the overall dataset and technique-based agricultural production structure is utilized for the proper evaluation, determination, and supervision of field components, which include temporal and spatial fluctuations so as to discern the optimal profit, environmental protection, and sustainability.

With the advent of precision agriculture, better and more efficient strategies are being made for irrigation, disease control, enhanced fertilization, harvest, and even pest control. It also spans across technological development for gathering and analysing data, which is expected to result in the implementation of sufficient solutions (Urbahs and Jonaite 2013). Remote sensing methods have long been established as the means of data acquisition without any physical contact between the observer and the object and have proved to be a significant aspect of precision agriculture and farming. Even though this technique was primarily associated with the science of photogrammetry with the application of balloons for aerial measurements (Colomina and Molina 2014), its direct dominance in the assessment of agriculture and wetlands has become more popular. The aerial platforms used for remote sensing applications can be classified under several platforms such as helicopters, drones, balloons, as well as planes. Satellite platforms, on the other hand, make use of dedicated sensors, which are known to observe and measure emitted or reflected electromagnetic radiation from the object of focus.

The application of drones and satellites in precision farming was made popular with the availability of several satellite earth observation data (Toth and Jóźków 2016), which pioneered a new era of remote sensing data acquisition. However, given the efficiency and recorded capabilities of drone, which is largely found in its imagery systems and explicitly integrated sensors (Ozdogan et al. 2017), its use has become more popular in the analysis of vegetation and wetlands over the globe. The incorporation and application of the complex dynamic automation system (drone) is a growing agricultural trend. This has shown great potential in achieving new economical and agricultural traits (Ahmad et al. 2021).

A key advantage of utilizing UAV in precision agriculture is the possibility of customizing the imaging sensor to meet the needed spatial and spectral requirements and to attain rapid utilization. In contrast to other remote sensing platforms, UAVs can fly at a very low altitude, which is a factor in acquiring very high-resolution datasets, and suggests extremely detailed spatial and spectral field description of the region of interest (Turner et al. 2012). Its application in active vegetation and wetland assessment allows the system to precisely determine the location of plants during the growing period so as to act with corrective amounts (Vega et al. 2015; Hunt et al. 2013). The temporal frequency in the case of crop dynamics assessment is also critical to help address several forms of agricultural practices, environmental concerns, degasses, as well as pest control that can ultimately improve production and crop yield (Hatfield and Prueger 2010; Warren and Metternicht 2005). UAV sensors integrated with other geospatial techniques also aid in the establishment of optimum fertilization, field changes, disease diagnosis, as well as pest control for enhanced yield.

The popularity and efficiency of UAVs in vegetation and wetland monitoring has also provided a platform for agronomists, growers, as well as companies to have unrestricted observation access and more technical analytic procedures in analysing high-quality images and data obtained from this platform. The use of UAV is also well known in other aspects of vegetation monitoring and assessment such as seed planting, fertilizer application, weather analysis and prediction, irrigation schedules planning, and soil health scanning. The noted advantages of the UAV over other remote sensing platforms for wetland assessment are as follows:

- Proficiency in obtaining multi-angular data, especially from camera snapshots.
- Faster data collection rate even at high resolution, range, and quality.
- Proficiency in operating multiple sensors homogenously.
- Proficiency in integrating orthophoto datasets with 3D canopy height information.
- The capability of modelling canopy heights from structure-from-motion point clouds.

8.2.4 Water Resources

The need for constant water monitoring and protection is due to its importance in human survival. External factors such as agricultural fertilizers and industrial wastes are well known to threaten regional water safety. The need to develop and sustain water quality monitoring programmes is to eliminate water contamination sources, while sustaining and preserving water quality. To assess the quality of existing freshwater in ponds, lakes, and rivers, parameters such as electrical conductivity, temperature, pH, and dissolved oxygen are evaluated and monitored given that they are the most popularly used impairment indicators (Xu et al. 2016). Objectionable pH or temperature, low concentration of dissolved oxygen, as well as large concentration of salinity causes poor water quality. For efficient water characterization and trend identification of water quality with respect to time, periodic analysis and sampling are needed. For instance, storm water may consist of pollutants, which may include heavy metals, oils, sediments, litter, nutrient, and bacteria (Thomas et al. 2001). Maps of water quality distribution over water bodies are achieved from water quality indicator datasets. Such maps however, have a wide range of application, especially for hydrologists who make use of them to comprehend water body circulation and also to make estimations and predictions (Kaizu et al. 2011).

Monitoring also helps to adequately implement definite pollution prevention programmes by aiding in the acquisition of data for this purpose. Understanding the properties and interactions of processes in shallow and deep water is very crucial when viewed from various aspects. For instance, fish farm planning, navigation safety, as well as adequate designs for near-shore infrastructure like ports and wind farms is largely dependent on accurate bathymetric datasets. Basically, depth data is required for all maritime spatial planning and marine activities. However, the instrumentation, procedures, and manpower required to execute a hydrographic survey are very expensive. There are also regions that cannot be accessed by ships, especially shallow areas or regions composed with considerable amount of moving sand dunes, rocks, and coral reefs. Remote locations such as islands are also potentially dangerous zones for ships and are not feasible due to their remoteness and associated costs in carrying out survey there. Moreover, locations such as dynamic deltas need frequent and recurrent surveys, and this therefore makes the task of having updated nautical charts very expensive. The use of remote sensing techniques is one sure method of achieving appropriately accurate bathymetric maps over any study region of interest. However, it is noticed that remote sensing-based depth maps do not usually meet the requirements of the International Maritime Organization standards to nautical charts, which are often regarded as being too strict. On the other hand, some already existing nautical charts are known to consist of satellite-derived bathymetry. For instance, the UK Hydrographic Office published their original SDB nautical chart (GB2066) of Antigua in 2014.

The extensive role of marine coastal habitats in the ecosystem includes carbon sequestration (Lee et al. 2014), coastal protection (Spalding et al. 2014), and food security (Hicks et al. 2011). In the coastal zones, two of the most prolific services are sea grass meadows and coral reefs, which are considered to be highly susceptible to climatic variations (Ramesh et al. 2019). The need for adequate marine spatial planning is to enhance the ecological pliability of the ecosystem, as well as its corresponding services. To assess the significance of marine habitats and estimate their spatial extent variations, fine-scale monitoring, which is largely regarded as an element of spatial planning is needed (Collin et al. 2018). The essence of adequate monitoring of water resources is to be able to plan appropriately and establish habitat-type inventories for easy supervision of habitat logs seascape evolution and fragmentation (Saul and Purkis 2015).

The maps created from these habitat assessments are also useful in management planning and inauguration of conservation in marine protected locations (Kobryn et al. 2013). Satellite remote sensing techniques have been extensively used in obtaining information about water resources, especially in high resolution (1–10 m), moderate (10–100 m), and aerial forms (Hedley et al. 2016). Despite the substantial improvements made in the spatial and temporal resolutions of these satellite missions over the years, they are still restricted to providing submeter accuracy, which is needed to precisely map moderate-to-fine scale coastal variations (Ventura et al. 2018). However, with the advent of technology, UAV and its consumer-grade cameras, higher spatial and temporal resolution capabilities are now possible to aid water resource assessment (Duffy et al. 2018a; Anderson and Gaston 2013). The application of UAVs in environmental assessments has been reported (Koh and Wich 2012). It is common knowledge that UAVs have the capability of obtaining imageries with pixel sizes in the order of centimetres (2–5 cm) (Ventura et al. 2018; Duffy et al. 2018b; Topouzelis et al. 2018; Papakonstantinou et al. 2016). The fine-scale heterogeneity of potential marine environments is achieved using high-resolution datasets, which are often derived with the help of UAV (Murfitt et al. 2017).

Another extensive and widely known procedure for accurate water quality monitoring is the use of autonomous underwater vehicles (AUVs) and autonomous surface vehicles (ASV) (Blaas and Kroeze 2016; Karimanzira et al. 2014). These autonomous vehicles being operated in water are efficient and able to achieve constant water quality assessment of the study region. However, these devices also have their own challenges and limitations. One of the major limitations of water quality assessment using the autonomous underwater vehicles is the precise point positioning of the vehicle, given that GPS signals are obstructed when they have to penetrate water. As a result of this challenge, the AUV must be equipped with extra navigational system or acoustic localization device. On the other hand, the ASV has the ability to automatically navigate predefined sample positions and assess water quality determining parameters such as nitrate content, chlorophyll concentration, water and sensor depth, temperature, turbidity, pH, etc. (Blaas and Kroeze 2016). However, its major challenge is noticed in its unstable motion and difficulty in operation, due to its noticeable and vigorous sway from side to side and the ambiguous engine-control frequencies. These devices (ASVs and AUVs) provide a very high spatio-temporal data resolution and adaptive sampling as a result of their potential to undergo consistent monitoring. Another notable limitation for these devices lies in the assemblage of biased observations as a result of contaminated and over-used sensors and boards.

8.2.5 Environmental Science

Drones furnished with chemical sensing payloads are becoming increasingly popular in a wide range of meteorology disciplines, including industrial emission monitoring, atmospheric chemistry, precision agriculture, and, of course, environmental law enforcement. One distinct advantage of a maximum take-off weight unmanned aircraft system, which weighs less than 25 kg, for environmental assessment, is its ease in operating in several counties without need for flight permit from the aviation authorities. When being used for atmospheric chemistry, the chemical sensing payload provides a wide range of potential activities for the drone which include investigating the chemical composition of the lower atmospheric space, assessing toxic gases in dangerous or inaccessible locations, such as volcanoes and chimneys, and emitting 3D air quality maps with very high spatial resolutions. The creation of small drones has also facilitated adequate environmental monitoring because the cost of these drones has significantly dropped, therefore making it affordable and available for a wider range of researches, companies, and institutions (Brinkmann et al. 2018).

Moreover, the growing social concern and strict regulations on global warming and air pollution also form a cogent reason as to why the studies using drone for environmental assessments have sky rocketed over the years. The increasing concern in the application of drone technology to gas sensing tasks is also obvious from the contemporary market emergence of gas detectors especially built for drone applications and its associated gas sensors. The need to constantly monitor, regulate, and control the increasing greenhouse gas emissions, as well as air pollutants from the environment, has coerced industry operators, governmental agencies, and institutions to search and come up with cheaper means of environmental monitoring (Brinkmann et al. 2018). The advantages of assimilating drone technology into their operations are being recognized by notable oil/petroleum industries in the world (Burgués and Marco 2020), whereas many drone manufacturers have also designed and developed drones consisting of methane sensors to serve the oil and gas sectors. The same case applies in the waste management industries, where operators of waste water management plants and solid waste landfills are already testing and experimenting with drones that are made up of gas detectors. The goal is to reduce cost, as well as the risk of omission during continuous data acquisition using hand-held detectors (Emran et al. 2017).

Furthermore, the police department in many advanced countries in the world today also employs drone technology to regulate and check compliance to air emission regulations set up by environmental agencies in tandem with the government of the region. For instance, Maritime authorities in countries like Hong Kong, Denmark, etc. use drones to investigate and observe ships that are not conforming to international regulations on the maximum amount of sulphur content existing in their marine fuel (Zhou et al. 2019). Countries like Poland have begun using drones to snuffle the exhaust pipes of chimneys over the region so as to determine polluters utilizing low-quality fuel for their domestic activities. The use of small drones in extensive atmospheric research activities is largely a function of their potential in providing a dense sample of regions which were initially hard to access using existing techniques and methods, which include but not limited to weather balloons, manned aircrafts, or satellite observations (Villa et al. 2016). Their use in an extensive determination and investigation of greenhouse gas concentrations and other pollutants in the lower atmospheric space are well founded. They also function as efficient means of analysing volcanic plume composition and to estimate the amount of greenhouse gas emissions from permafrost or wildfires. Also, the data on gases like CO2 and CH4 can prove very useful in agriculture, as they can serve as adequate means of assessing and monitoring crop health and livestock emissions which has the potential of spreading over a large space.

8.2.6 Digital Mapping

Over the past few years, the growing application of UAV technology for digital mapping has been noted. With the improvements in UAV technological capabilities, the range and number of its application also improved rapidly. Specific types of monitoring including coastal and marine environments require high-resolution local data for the assessment of multi-temporal change detection analysis. The use of UAVs, which consists of a wide range of sensors, especially RGB cameras, thermal, and hyperspectral sensors for these assessments has led to consistent accurate results. Some studies have shown that UAVs also possess the capacity to efficiently fill current measurement gaps, especially in the remote sensing of the environment and is also responsible for the generation of crucial datasets needed for coastal change studies (Klemas 2015). In recent years, a large number of marine/coastal applications have surfaced, which explains and takes advantage of the value of this conventional technology for use by coastal managers and environmental researchers. The choice of UAV applications is proportional to the requirement of its corresponding job type. For instance, the GoPro and DSLR cameras, which are small off-the-shelf sport cameras can be integrated with more expensive sensors such as LiDAR and hyperspectral sensors for adequate coastal and bathymetric assessments.

Some of the applications of UAV in digital mapping are based on observations of the coastal environment and its habitats which include saltmarsh, mangroves, and sea grass. The digital mapping procedure involves the assembling of orthophotos and the application of video transects for the mapping of inter-tidal bathymetric habitats and to monitor inaccessible regions. It also enhances repeated surveys, evaluates the coverage and extent of vegetation, and the examination of dredging activities and plume extents. Its application also extends to the creation of 3D point clouds using stereo photography, routine marine fauna observation techniques, or even LiDAR. And also for minor oil spill contingency tracking, vegetation classification using hyperspectral imageries, designated engineering assessment of pipelines, offshore structures, thermal imaging, and safety intervention exercises. The opportunity to assess the coastal environmental changes has been one of its key applications, especially following its ease in obtaining multi-temporal photographic imageries to aid mapping of coastline changes. This is in terms of deposition and erosion, and measuring the rate of change.

The study undertaken by Green et al. (2019) highlighted a report that applied the popular DJI phantom 3 to assess and map coastal changes as a result of erosion relative to the coastal mitigation structures in Ghana over the Volta Delta shoreline. The study generated aerial photographs of the surroundings and structures using a basic platform and sensor configuration. They also generated high-resolution orthophotos and DEMs for the perception, detection, and analysis of both volumetric and planimetric changes. The study also highlighted the practicability of the technique for constant surveying and monitoring. Other studies have also implemented UAV to acquire and map spatial ecology and coastal water quality (Anderson and Gaston 2013). The primary advantage noticed here is that the system can be launched, accessed, operated, and analysed in a very short time while including the systems that can be easily and cheaply utilized by coastal environmental managers and aquaculture business for consistent monitoring of near-shore water quality.

Another study undertaken by Guillot and Pouget (2015) applied specially designed UAVs for the coastal environment to assess and precisely map the dunes and dikes of the Oleron Island in France earlier and after tide storms. The modified DJI F550 coastal UAV applied in their study has the potential of operating in high wind conditions. They are also configured to be resilient against moisture and sand particles. They also utilize off-the-shelf rugged sports camera with a universal fisheye lens. The fisheye distortion was removed from the imagery during the post-processing phase. This equally helped to geocorrect the imagery utilizing the ground control points. Using Agisoft soft copy photogrammetric software, an orthomosaic was generated, and consequently a digital surface model and a 3D portable document file (PDF) to serve as input parameters for the already designed geographic information system (GIS). The time series of imageries used to detect change was generated using a difference of digital volumetric model (DDVM). Using values generated from the Digital Elevation Models (DEM), it becomes easier to develop a two-dimensional topographic profile in the GIS software. Factors such as the ease with which the project was executed, low cost, and the coastal specific platforms were highlighted and identified as key benefits of the study. Additional projects for coastal assessments exist and include the study of pebble and sand movements, waste, birds, and seaweed, which has the tendency of providing indispensable information to enhance ecosystem knowledge in the future.

Several researchers have shown that drones have a significant and distinct advantage over the manned vehicles and other remote sensing techniques (Green et al.

2019). Beyond its low cost, platform type, and the relatively large number of imageries acquired in a short time, the operation and maintenance are relatively cheap when compared to other platforms. Another distinct advantage of the drone technology is that it can be used for data acquisition at any time, except during harsh weather conditions. Given its size, it also promotes relative freedom for image acquisition, since it does not require any booking process for the flying time, or the risk of flight cancellation as a result of poor visibility due to harsh weather. The manned aircraft may be limited by their flight regulation which doesn't allow it to fly below 200 m. However, using drones, high-quality images at close range can also be achieved since it is not bound by any altitude regulation. It is also known to have full flexibility to capture data during both day and night times, which can be found to be a bit difficult in the case of operating a manned aircraft for the same purpose. The effect of atmospheric haze, which has a great influence on the clarity of the imageries is reduced when the craft is flown at lower altitudes, hence another altitude advantage of drones over the manned aircraft systems.

8.3 Conclusion

The application of small unmanned aerial systems also known as drones as a tool for environmental monitoring and assessment is growing. They exist in different forms based on purpose and design and the broad range multi-disciplinary applications and benefits of drones make it a technology for the future. Several studies have shown the relevance of this system in performing adequate, efficient, and accurate tasks even beyond the field of surveying and geoinformatics. Some sensitive monitoring which requires ultra-high spatial and temporal resolutions such as ecological assessments makes use of the UAV systems in analysis and predictions of ecological processes and interactions. The same process is noticed in coastal management, vegetation and forestry, agriculture and wetlands, water resources, and the entire environmental science. UAVs are gradually emerging as more efficient monitoring tools, especially due to the flexibility of its altitude while monitoring objects of large scale, timely as it can be done anytime except under harsh weather conditions, and relatively cheaper including its maintenance. Satellite remote sensing and advances in remotely piloted aircraft systems (RPAS) (known as unmanned aerial vehicles) are providing improved capability to monitor the spatial and temporal variability of water storage components, complementing the capabilities of satellite data. UAVs are gaining grounds in hydrological applications and their integration with satellite data like synthetic aperture radar and satellite gravity observations to improve understanding of water stress (e.g., depletion of groundwater) and evolution of water resources in semi-arid agrarian domains are interesting future research directions in hydrology and cognate disciplines. The application of UAVs is expected to proliferate greatly in the near future as more and more UAV-based technological advancements and research are going on, even in the areas of improving their sensor characteristics and capabilities for finer and more effective monitoring and mapping of our local environment.

References

Achille C, Adami A, Chiarini S et al (2015) UAV-based photogrammetry and integrated technologies for architectural applications-methodological strategies for the after-quake survey of vertical structures in Mantua (Italy). Sensors 15(7):15520–15539. https://doi.org/10.3390/s150715520

Ahmad A, Ordoñez J, Cartujo P, Martos V (2021) Remotely piloted aircraft (RPA) in agriculture: a pursuit of sustainability. Agronomy 11(1). https://doi.org/10.3390/agronomy11010007

Alongi DM (2002) Present state and future of the world's mangrove forests. Environ Conserv 29(3):331–349. https://doi.org/10.1017/S0376892902000231

Anderson K, Gaston KJ (2013) Lightweight unmanned aerial vehicles will revolutionize spatial ecology. Front Ecol Environ 11(3):138–146. https://doi.org/10.1890/120150

Baloloy AB, Blanco AC, Ana RR, Nadaoka K (2020) Development and application of a new mangrove vegetation index (MVI) for rapid and accurate mangrove mapping. ISPRS J Photogramm Remote Sens 166:95–117. https://doi.org/10.1016/j.isprsjprs.2020.06.001

Archer F, Shutko AM, Coleman TL, Haldin A, Novichikhin E, Sidorov I (2015) Introduction, overview, and status of the Microwave Autonomous Copter System (MACS). In: IGARSS 2004. 2004 IEEE international geoscience and remote sensing symposium, vol 5, pp 3574–3576. https://doi.org/10.1109/IGARSS.2004.1370483

Bioucas-Dias JM, Plaza A, Gustavo CV et al (2013) Hyperspectral remote sensing data analysis and future challenges. IEEE Geosci Remote Sens Mag 1(2):6–36. https://doi.org/10.1109/MGRS.2013.2244672

Blaas H, Kroeze C (2016) Excessive nitrogen and phosphorus in European rivers: 2000–2050. Ecol Ind 67:328–337. https://doi.org/10.1016/j.ecolind.2016.03.004

Boyd DS, Danson FM (2005) Satellite remote sensing of forest resources: three decades of research development. Prog Phys Geograp: Earth Environ 29(1):1–26. https://doi.org/10.1191/0309133305pp432ra

Brasington J, Vericat D, Rychkov I (2012) Modeling river bed morphology, roughness, and surface sedimentology using high resolution terrestrial laser scanning. Water Resour Res 48(11). https://doi.org/10.1029/2012WR012223

Brinkmann T, Both R, Scalet BM, Roudier S, Sancho LD (2018) JRC reference report on monitoring of emissions to air and water from IED installations. In: European IPPC Bureau, European Commission, Joint Research Centre. https://eippcb.jrc.ec.europa.eu/reference/monitoring-emissions-air-and-water-ied-installations-0. Accessed March 2021

Brunier G, Fleury J, Anthony E, Gardel A et al (2016) Close-range airborne structure-from-motion photogrammetry for high-resolution beach morphometric surveys: examples from an embayed rotating beach. Geomorphology 261:76–88. https://doi.org/10.1016/j.geomorph.2016.02.025

Burgués J, Marco S (2020) Environmental chemical sensing using small drones: a review. Sci Total Environ 748. https://doi.org/10.1016/j.scitotenv.2020.141172

Cahoon DR, Hensel P, Rybczyk J et al (2003) Mass tree mortality leads to mangrove peat collapse at bay islands, honduras after hurricane mitch. J Ecology 91(6):1093–1105. https://doi.org/10.1046/j.1365-2745.2003.00841.x

Canas AAD, Irwin DA (1986) Airborne remote sensing from remotely piloted aircraft. Int J Remote Sens 7(12):1623–1635. https://doi.org/10.1080/01431168608948957

Casella E, Rovere A, Pedroncini A et al (2016) Study of wave runup using numerical models and low-altitude aerial photogrammetry: a tool for coastal management Estuarine. Coastal Shelf Sci 149:160–167. https://doi.org/10.1016/j.ecss.2014.08.012

Chen G, Weng Q, Hay GJ, He Y (2018) Geographic object-based image analysis (GEOBIA): emerging trends and future opportunities. GISci Remote Sens 55(2):159–182. https://doi.org/10.1080/15481603.2018.1426092

Clapuyt F, Vanacker V, Van Oost K (2016) Reproducibility of UAV-based earth topography reconstructions based on Structure-from-Motion algorithms. Geomorphology 260:4–15. https://doi.org/10.1016/j.geomorph.2015.05.011

Collin A, Ramambason C, Pastol Y, Casella E et al (2018) Very high resolution mapping of coral reef state using airborne bathymetric LiDAR surface-intensity and drone imagery. Int J Remote Sens 39(17):5676–5688. https://doi.org/10.1080/01431161.2018.1500072

Colomina I, Molina P (2014) Unmanned aerial systems for photogrammetry and remote sensing: a review. ISPRS J Photogramm Remote Sens 92:79–97. https://doi.org/10.1016/j.isprsjprs.2014.02.013

Curran PJ, Steven MD, Houghton JT, Cook Arthur H, Charnock, H (1983) Multispectral remote sensing for the estimation of green leaf area index. Philosoph Trans R Soc Lond. Series A, Math Phys Sci 309(1508):257–270. https://doi.org/10.1098/rsta.1983.0039

Dainelli R, Toscano P, Di Gennaro SF, Matese A (2021) Recent advances in unmanned aerial vehicle forest remote sensing—a systematic review. Part i: a general framework. Forests 12(3). https://doi.org/10.3390/f12030327

Dandois JP, Ellis EC (2010) Remote sensing of vegetation structure using computer vision. Remote Sens 2(4):1157–1176. https://doi.org/10.3390/rs2041157

Dandois JP, Ellis EC (2013) High spatial resolution three-dimensional mapping of vegetation spectral dynamics using computer vision. Remote Sens Environ 136:295–276. https://doi.org/10.1016/j.rse.2013.04.005

de Castro AI, Peña JM et al (2020) Mapping Cynodon Dactylon infesting cover crops with an automatic decision tree-OBIA procedure and UAV imagery for precision viticulture. Remote Sens 12(1). https://doi.org/10.3390/rs12010056

Delacourt C, Allemand P, Jaud M, Grandjean P, Deschamps A, Ammann J, Cuq V, Suanez S (2009) Drelio: an unmanned helicopter for imaging coastal areas. J Coastal Res 1489–1493

Dennison PH, Brunelle AR, Vachel A (2010) Assessing canopy mortality during a mountain pine beetle outbreak using GeoEye-1 high spatial resolution satellite data. Remote Sens Environ 114(11):2431–2435. https://doi.org/10.1016/j.rse.2010.05.018

de Vries W, Vel E, Reinds GJ et al (2003) Intensive monitoring of forest ecosystems in Europe: 1. Objectives, set-up and evaluation strategy. Forest Ecology Manag 174(1):77–95. https://doi.org/10.1016/S0378-1127(02)00029-4

Ding Man-Lai, Ding Chi-Biao, Tang Li et al (2020) A W-Band 3-D integrated mini-SAR system with high imaging resolution on UAV platform. IEEE Access 8:113601–113609. https://doi.org/10.1109/ACCESS.2020.3003273

Duffy JP, Cunliffe AM, DeBell L, Sandbrook CE (2018a) Location, location, location: considerations when using lightweight drones in challenging environments. Remote Sens Ecology Conserv 4(1):7–19. https://doi.org/10.1002/rse2.58

Duffy JP, Pratt L, Anderson K, Land PE, Shutler JD (2018b) Spatial assessment of intertidal seagrass meadows using optical imaging systems and a lightweight drone. Estuar Coast Shelf Sci 200:169–180. https://doi.org/10.1016/j.ecss.2017.11.001

Ellison JC (2000) How south pacific mangroves may respond to predicted climate change and sea-level rise: In: Climate change in the South Pacific: impacts and responses in Australia, New Zealand, and Small Island States. Springer Netherlands, pp 289–300. https://doi.org/10.1007/0-306-47981-8_16

Erman AT, van Hoesel L, Havinga P, Wu J (2008) Enabling mobility in heterogeneous wireless sensor networks cooperating with uavs for mission-critical management. IEEE Wireless Commun 15(6):38–46. https://doi.org/10.1109/MWC.2008.4749746

Emran BJ, Tannant DD, Najjaran H (2017) Low-altitude aerial methane concentration mapping. Remote Sens 9(8). https://doi.org/10.3390/rs9080823

Fairley I, Thomas T, Phillips M, Reeve D (2016) Terrestrial laser scanner techniques for enhancement in understanding of coastal environments: in seafloor mapping along continental shelves: research and techniques for visualizing benthic environments. Springer International Publishing, pp 273–289. https://doi.org/10.1007/978-3-319-25121-9_11

Farnsworth E, Ellison A (1998) The global conservation status of mangroves. Oceanogr Lit Rev 45(1):138–139

Field CD (1995) Impact of expected climate change on mangroves. Asia-Pacific Symposium on Mangrove Ecosystems, pp 75–81

Fitzgerald GJ (2010) Characterizing vegetation indices derived from active and passive sensors. Int J Remote Sens 31(16):4335–4348. https://doi.org/10.1080/01431160903258217

Fonstad MA, Dietrich JT, Courville BC, Jensen JL, Carbonneau PE (2013) Topographic structure from motion: a new development in photogrammetric measurement. Earth Surf Proc Land 38(4):421–430. https://doi.org/10.1002/esp.3366

Fuller IC, Large AR, Milan DJ (2003) Quantifying channel development and sediment transfer following chute cutoff in a wandering gravel-bed river. Geomorphology 54(3):307–323. https://doi.org/10.1016/S0169-555X(02)00374-4

Gago J, Douthe C, Coopman RE, Gallego PP, Ribas-Carbo M et al (2015) UAVs challenge to assess water stress for sustainable agriculture. Agricul Water Manag 153:9–19. https://doi.org/10.1016/j.agwat.2015.01.020

Gonçalves, J. A and Henriques, R (2015). UAV photogrammetry for topographic monitoring of coastal areas. ISPRS J Photogramm Remote Sens 104:101– 111 https://doi.org/10.1016/j.isprsjprs.2015.02.009

Green DR, Hagon JJ, Gómez C, Gregory BJ (2019) Using low-cost uavs for environmental monitoring, mapping, and modelling: examples from the coastal zone. Coastal Manag 465–501. https://doi.org/10.1016/B978-0-12-810473-6.00022-4

Guillot B, Pouget F (2015) UAV application in coastal environment, example of the oleron island for dunes and dikes survey. In: The international archives of the photogrammetry, remote sensing and spatial information sciences, vol XL-3/W3. https://doi.org/10.5194/isprsarchives-XL-3-W3-321-2015

Gómez-Gutiérrez A, Schnabel S, Berenguer-Sempere F, Lavado-Contador F et al (2014) Using 3D photo-reconstruction methods to estimate gully headcut erosion. CATENA 120:91–101. https://doi.org/10.1016/j.catena.2014.04.004

Hall K, Reitalu T, Sykes MT, Prentice HC (2012) Spectral heterogeneity of quickbird satellite data is related to fine-scale plant species spatial turnover in semi-natural grasslands. Appl Veg Sci 15(1):145–157. https://doi.org/10.1111/j.1654-109X.2011.01143.x

Harwin S, Lucieer A (2012) Assessing the accuracy of georeferenced point clouds produced via multi-view stereopsis from unmanned aerial vehicle (UAV) imagery. Remote Sens 4(6):1573–1599. https://doi.org/10.3390/rs4061573

Hassanalian M, Abdelkefi A (2017) Classifications, applications, and design challenges of drones: a review. Prog Aerosp Sci 91:99–131. https://doi.org/10.1016/j.paerosci.2017.04.003

Hassler SC, Baysal-Gurel F (2019) Unmanned aircraft system (UAS) technology and applications in agriculture. Agronomy 9(10). https://doi.org/10.3390/agronomy9100618

Hatfield JL, Prueger JH (2010) Value of using different vegetative indices to quantify agricultural crop characteristics at different growth stages under varying management practices. Remote Sens 2(2):562–578. https://doi.org/10.3390/rs2020562

Hedley JD, Roelfsema CM et al (2016) Remote sensing of coral reefs for monitoring and management: a review. Remote Sens 8(2). https://doi.org/10.3390/rs8020118

Hicks CC, Cohen PJ, Graham NA, Nash KL et al (2011) Harnessing global fisheries to tackle micronutrient deficiencies. Nature 574(7776):95–98. https://doi.org/10.1038/s41586-019-1592-6

Hill VJ, Zimmerman RC (2010) Estimates of primary production by remote sensing in the Arctic Ocean: assessment of accuracy with passive and active sensors. Deep Sea Res Part I 57(10):1243–1254. https://doi.org/10.1016/j.dsr.2010.06.011

Hunt ER, Daughtry CS, Mirsky SB, Hively WD (2013) Remote sensing with unmanned aircraft systems for precision agriculture applications. In: 2013 Second international conference on agro-geoinformatics (Agro-Geoinformatics), pp 131–134. https://doi.org/10.1109/Argo-Geoinformatics.2013.6621894

Immerzeel WW, Kraaijenbrink PD, Shea JM et al (2014) High-resolution monitoring of Himalayan glacier dynamics using unmanned aerial vehicles. Remote Sens Environ 150:93–103. https://doi.org/10.1016/j.rse.2014.04.025

Jaakkola A, Hyyppä J, Kukko A et al (2010) A low-cost multi-sensoral mobile mapping system and its feasibility for tree measurements. ISPRS J Photogramm Remote Sens 65(6):514–522. https://doi.org/10.1016/j.isprsjprs.2010.08.002

Javernick L, Brasington J, Caruso B (2014) Modeling the topography of shallow braided rivers using Structure-from-Motion photogrammetry. Geomorphology 213:166–182. https://doi.org/10.1016/j.geomorph.2014.01.006

Johnson K, Nissen E, Saripalli S, Arrowsmith JR, McGarey P, Scharer K, Williams P, Blisniuk K (2014) Rapid mapping of ultrafine fault zone topography with structure from motion. Geosphere 10(5):969–986. https://doi.org/10.1130/GES01017.1

Kaizu Y, Iio M, Yamada H, Noguchi N (2011) Development of unmanned airboat for water-quality mapping. Biosyst Eng 109(4):338–347. https://doi.org/10.1016/j.biosystemseng.2011.04.013

Karimanzira D, Jacobi M, Pfuetzenreuter T et al (2014) First testing of an AUV mission planning and guidance system for water quality monitoring and fish behavior observation in net cage fish farming. Inf Process Agric 1(2):131–140. https://doi.org/10.1016/j.inpa.2014.12.001

Kazim M, Azar AT, Koubaa A, Ibrahim ZF et al (2021) Event-driven programming-based path planning and navigation of UAVs around a complex urban environment. In: Unmanned Aerial Systems. Academic, New York, pp 531–565. https://doi.org/10.1016/B978-0-12-820276-0.00028-5

Keckhut P, Randel WJ, Claud C, Leblanc T et al (2011) An evaluation of uncertainties in monitoring middle atmosphere temperatures with the ground-based lidar network in support of space observations. J Atmos Solar-Terrestrial Phy 73(5):627–642. https://doi.org/10.1016/j.jastp.2011.01.003

Kerle N, Heuel S, Pfeifer N (2008) Real-time data collection and information generation using airborne sensors. In: Geospatial information technology for emergency response 32

Kim J, Sukkarieh S, Wishart S (2006) Real-time navigation, guidance, and control of a uav using low-cost sensors. Field Serv Robot: Recent Adv Res Appl 299–309. https://doi.org/10.1007/10991459_29

Klemas VV (2015) Coastal and environmental remote sensing from unmanned aerial vehicles: an overview. J Coastal Res 31(5):1260–1267. https://doi.org/10.2112/JCOASTRES-D-15-00005.1

Kobryn HT, Wouters K, Beckley LE, Heege T (2013) Ningaloo reef: shallow marine habitats mapped using a hyperspectral sensor. PLoS ONE. https://doi.org/10.1371/journal.pone.0070105

Koh Lian Pin, Wich Serge A (2012) Dawn of drone ecology: low-cost autonomous aerial vehicles for conservation. Trop Conser Sci 5(2):121–132. https://doi.org/10.1177/194008291200500202

Lee SY, Primavera JH, Dahdouh-Guebas F, McKee K, Bosire JO et al (2014) Ecological role and services of tropical mangrove ecosystems: a reassessment. Glob Ecol Biogeogr 23(7):726–743. https://doi.org/10.1111/geb.12155

Librán-Embid F, Klaus F, Tscharntke T, Grass I (2020) Unmanned aerial vehicles for biodiversity-friendly agricultural landscapes - a systematic review. Sci Total Environ 732. https://doi.org/10.1016/j.scitotenv.2020.139204

Lou S, Cai F, Wang H (2013) Development of coastal erosion and management. Adv Earth Sci 28(11):1239

Maes WH, Steppe K (2019) Perspectives for remote sensing with unmanned aerial vehicles in precision agriculture. Trends Plant Sci 24(2):152–164. https://doi.org/10.1016/j.tplants.2018.11.007

Mandlburger G, Hauer C, Wieser M, Pfeifer N (2015) Topo-bathymetric LiDAR for monitoring river morphodynamics and instream habitats–a case study at the Pielach River. Remote Sens 7(5):6160–6195. https://doi.org/10.3390/rs70506160

Manfreda S, McCabe MF, Miller PE et al (2018) On the use of unmanned aerial systems for environmental monitoring. Remote Sens 10(4). https://doi.org/10.3390/rs10040641

McKean J, Roering J (2004) Objective landslide detection and surface morphology mapping using high-resolution airborne laser altimetry. Geomorphology 57(3):331–351. https://doi.org/10.1016/S0169-555X(03)00164-8

Meroni M, Rossini M, Guanter et al. (2013) Remote sensing of solar-induced chlorophyll fluorescence: review of methods and applications. Remote Sens Environ 113(10):2037–2051. https://doi.org/10.1016/j.rse.2009.05.003

Mulero-Pázmány M, Stolper R, Van Essen LD, Negro JJ, Sassen T (2014) Remotely piloted aircraft systems as a rhinoceros anti-poaching tool in Africa. PloS One 9(1). https://doi.org/10.1371/journal.pone.0083873

Murfitt SL, Allan BM, Bellgrove A, Rattray A et al (2017) Applications of unmanned aerial vehicles in intertidal reef monitoring. Sci Rep 7(1):1–11. https://doi.org/10.1038/s41598-017-10818-9

Nex F, Remondino F (2014) UAV for 3D mapping applications: a review. Appl Geom 6(1):1–35. https://doi.org/10.1007/s12518-013-0120-x

Notebaert B, Verstraeten G, Govers G, Poesen J (2009) Qualitative and quantitative applications of LiDAR imagery in fluvial geomorphology. Earth Surf Processes Landf 34(2):217–231. https://doi.org/10.1002/esp.1705

Ouédraogo MM, Degré A, Debouche C, Lisein J (2014) The evaluation of unmanned aerial system-based photogrammetry and terrestrial laser scanning to generate DEMs of agricultural watersheds. Geomorphology 214:339–355. https://doi.org/10.1016/j.geomorph.2014.02.016

Ozdogan B, Gacar A, Aktas H (2017) Digital agriculture practices in the context of agriculture 4.0. J Econ Finan Acc 4:186–193. https://doi.org/10.17261/Pressacademia.2017.448

Pajares G (2015) Overview and current status of remote sensing applications based on Unmanned Aerial Vehicles (UAVs). Photogramm Eng & Remote Sens 81(4):281–330. https://doi.org/10.14358/PERS.81.4.281

Papakonstantinou A, Topouzelis K, Pavlogeorgatos G (2016) Coastline zones identification and 3d coastal mapping using UAV spatial data. ISPRS Int J Geo-Inf 5(6). https://doi.org/10.3390/ijgi5060075

Phua MH, Chong CW, Ahmad AH, Hafidzi MN (2016) Predicting rat occurrence in oil-palm plantation using GIS and GeoEye data. Environ Eng & Manag J 15(11):8. https://doi.org/10.30638/eemj.2016.275

Potapov P, Yaroshenko A, Turubanova S, Dubinin M, Laestadius L et al (2008) Mapping the world's intact forest landscapes by remote sensing. Ecol Soc 13(2)

Primavera JH (2005) Mangroves, fishponds, and the quest for sustainability. Science 310(5745):57–59. https://doi.org/10.1126/science.1115179

Primavera JH (2000) Development and conservation of Philippine mangroves: institutional issues. Ecol Econ 35(1):91–106. https://doi.org/10.1016/S0921-8009(00)00170-1

Prosdocimi M, Calligaro S, Sofia G, Dalla Fontana G, Tarolli P (2015) Bank erosion in agricultural drainage networks: new challenges from structure-from-motion photogrammetry for post-event analysis. Earth Surf Proc Land 40(14):1891–1906. https://doi.org/10.1002/esp.3767

Puliti S, Ørka HO, Gobakken T, Næsset E (2015) Inventory of small forest areas using an unmanned aerial system. Remote Sens 7(8):9632–9654. https://doi.org/10.3390/rs70809632

Quaritsch M, Kruggl K, Wischounig-Strucl D, Bhattacharya S, Shah M, Rinner B (2010) Networked UAVs as aerial sensor network for disaster management applications I. Elektrotechnik und Informationstechnik 127(3):56–63. https://doi.org/10.1007/s00502-010-0717-2

Ramesh R, Banerjee K, Paneerselvam A, Raghuraman R, Purvaja R, Lakshmi A (2019) Chapter 14 - importance of seagrass management for effective mitigation of climate change. Coastal Manag 283–299. https://doi.org/10.1016/B978-0-12-810473-6.00015-7

Rees AF, Avens L, Ballorain K, Bevan E et al (2018) The potential of unmanned aerial systems for sea turtle research and conservation: a review and future directions. Endanger Species Res 35:81–100. https://doi.org/10.3354/esr00877

Rehman TU, Mahmud S, Chang YK, Jin J, Shin J (2019) Current and future applications of statistical machine learning algorithms for agricultural machine vision systems. Comput Electron Agric 156:585–605. https://doi.org/10.1016/j.compag.2018.12.006

Remondino F, Spera MG, Nocerino E, Menna F, Nex F (2014) State of the art in high density image matching. Photogram Rec 29(146):144–166. https://doi.org/10.1111/phor.12063

Rinaudo F, Chiabrando F, Lingua A, Spanò A (2012) Archaeological site monitoring: UAV photogrammetry can be an answer. Int Arch Photogramm Remote Sens Spat Inf Sci 39(85):583–588. https://doi.org/10.5194/isprsarchives-XXXIX-B5-583-2012

Romañach SS, DeAngelis DL, Koh HL et al (2018) Conservation and restoration of mangroves: Global status, perspectives, and prognosis. Ocean Coast Manag 154:72–82. https://doi.org/10.1016/j.ocecoaman.2018.01.009

Rufino G, Moccia A (2005) Integrated VIS-NIR hyperspectral / thermal-ir electro-optical payload system for a mini-UAV. Infotech Aerosp. https://doi.org/10.2514/6.2005-7009

Samczyński P (2018) Applications of a modern micro-SAR system for small UAV operations (From active to passive SAR technology). In: 2018 19th international radar symposium (IRS), pp 1–8. https://doi.org/10.23919/IRS.2018.8448020

Sedunov A, Salloum H, Sutin A, Sedunov N, Tsyuryupa S (2018) UAV passive acoustic detection. In: IEEE International symposium on technologies for homeland security (HST), pp 1–6. https://doi.org/10.1109/THS.2018.8574129

Song C, Dickinson MB, Su L, Zhang S et al (2010) Estimating average tree crown size using spatial information from Ikonos and QuickBird images: across-sensor and across-site comparisons. Remote Sens Environ 114(5):1099–1107. https://doi.org/10.1016/j.rse.2009.12.022

Sørensen LY, Jacobsen LT, Hansen JP (2017) Low cost and flexible UAV deployment of sensors. Sensors 17(1):154. https://doi.org/10.3390/s17010154

Spalding MD, Ruffo S, Lacambra C et al (2014) The role of ecosystems in coastal protection: adapting to climate change and coastal hazards. Ocean Coast Manag 90:50–57. https://doi.org/10.1016/j.ocecoaman.2013.09.007

Toth C, Jóźków G (2016) Remote sensing platforms and sensors: a survey. ISPRS J Photogramm Remote Sens 115:22–36. https://doi.org/10.1016/j.isprsjprs.2015.10.004

Tsach S, Tatievsky A, London L (2010) Unmanned aerial vehicles (UAVs). Encycl Aerosp Eng. https://doi.org/10.1002/9780470686652.eae385

Saul S, Purkis S (2015) Semi-automated object-based classification of coral reef habitat using discrete choice models. Remote Sens 7(12):15894–15916. https://doi.org/10.3390/rs71215810

Squarzoni C, Delacourt C, Allemand P (2005) Differential single-frequency GPS monitoring of the La Valette landslide (French Alps). Eng Geol 79(3):215–229. https://doi.org/10.1016/j.enggeo.2005.01.015

Teke M, Deveci HS, Haliloğlu O et al (2013) A short survey of hyperspectral remote sensing applications in agriculture. In: 2013 6th international conference on recent advances in space technologies (RAST), pp 171–176. https://doi.org/10.1109/RAST.2013.6581194

Thomas KV, Hurst MR, Matthiessen P et al (2001) Toxicity characterisation of organic contaminants in stormwaters from an agricultural headwater stream in South East England. Water Res 35(10):2411–2416. https://doi.org/10.1016/S0043-1354(00)00535-2

Topouzelis K, Kitsiou D (2015) Detection and classification of mesoscale atmospheric phenomena above sea in SAR imagery. Remote Sens Environ 160:263–272. https://doi.org/10.1016/j.rse.2015.02.006

Topouzelis K, Papakonstantinou A, Doukari M, Stamatis P, Makri D, Katsanevakis S (2018) Coastal habitat mapping in the Aegean Sea using high resolution orthophoto maps. In: Fifth international conference on remote sensing and geoinformation of the environment (RSCy2017), p 10444. https://doi.org/10.1117/12.2279140

Turner D, Lucieer A, Watson C (2012) An automated technique for generating georectified mosaics from ultra-high resolution unmanned aerial vehicle (UAV) imagery, based on structure from motion (sfm) point clouds. Remote Sens 4(5):1392–1410. https://doi.org/10.3390/rs4051392

Turner D, Lucieer A, De-Jong SM (2015) Time series analysis of landslide dynamics using an unmanned aerial vehicle (UAV). Remote Sens 7(2):1736–1757. https://doi.org/10.3390/rs70201736

Urbahs A, Jonaite I (2013) Features of the use of unmanned aerial vehicles for agriculture applications. Aviation 17(4):170–175. https://doi.org/10.3846/16487788.2013.861224

Veettil BK, Pereira SFR, Quang NX (2018) Rapidly diminishing mangrove forests in Myanmar (Burma): a review. Hydrobiologia 822(1):19–35. https://doi.org/10.1007/s10750-018-3673-1

Vega FA, Ramírez FC, Saiz MP, Rosúa FO (2015) Multi-temporal imaging using an unmanned aerial vehicle for monitoring a sunflower crop. Biosys Eng 132:19–27. https://doi.org/10.1016/j.biosystemseng.2015.01.008

Ventura D, Bonifazi A, Gravina MF, Ardizzone GD (2017) Unmanned aerial systems (UASs) for environmental monitoring: a review with applications in coastal habitats. Aerial Robots-Aerodyn Control Appl 165–184. https://doi.org/10.1371/journal.pone.0083873

Ventura D, Bonifazi A, Gravina MF et al (2018) Mapping and classification of ecologically sensitive marine habitats using unmanned aerial vehicle (UAV) imagery and object-based image analysis (OBIA). Remote Sens 10(9). https://doi.org/10.3390/rs10091331

Villa TF, Gonzalez F, Miljievic B, Ristovski ZD, Morawska L (2016) An overview of small unmanned aerial vehicles for air quality measurements: present applications and future prospectives. Sensors 16(7). https://doi.org/10.3390/s16071072

Villamayor BM, Rollon RN, Samson MS et al (2016) Impact of Haiyan on Philippine mangroves: Implications to the fate of the widespread monospecific Rhizophora plantations against strong typhoons. Ocean Coast Manag 132:1–14. https://doi.org/10.1016/j.ocecoaman.2016.07.011

Voogt JA, Oke TR (2013) Thermal remote sensing of urban climates. Remote Sens Environ 86(3):370–384. https://doi.org/10.1016/S0034-4257(03)00079-8

Wang Menghua (2007) Remote sensing of the ocean contributions from ultraviolet to near-infrared using the shortwave infrared bands: simulations. Appl Opt 46(9):1535–1547. https://doi.org/10.1364/AO.46.001535

Warren G, Metternicht G (2005) Agricultural applications of high-resolution digital multispectral imagery. Photogramm Eng & Remote Sens 71(5):559–602. https://doi.org/10.14358/PERS.71.5.595

White SA, Wang Y (2003) Utilizing DEMs derived from LIDAR data to analyze morphologic change in the North Carolina coastline. Remote Sens Environ 85(1):39–47. https://doi.org/10.1016/S0034-4257(02)00185-2

Westoby MJ, Brasington J, Glasser NF, Hambrey MJ, Reynolds JM (2012) Structure-from-Motion photogrammetry: a low-cost, effective tool for geoscience applications. Geomorphology 179:300–314. https://doi.org/10.1016/j.geomorph.2012.08.021

Wright GG, Matthews KB, Cadell WM, Milne R (2018) Reducing the cost of multi-spectral remote sensing: combining near-infrared video imagery with colour aerial photography. Comput Electron Agric 38(3):175–198. https://doi.org/10.1016/S0168-1699(02)00141-2

Wu J, Bauer ME (2012) Estimating net primary production of turfgrass in an urban-suburban landscape with quickbird imagery. Remote Sens 4(4):849–866. https://doi.org/10.3390/rs4040849

Xu Z, Dong Q, Otieno B, Liu Y et al (2016) Real-time in situ sensing of multiple water quality related parameters using micro-electrode array (MEA) fabricated by inkjet-printing technology (IPT). Sens Actuators, B Chem 237:1108–1119. https://doi.org/10.1016/j.snb.2016.09.040

Yang G, Liu J, Zhao C, Li Z, Huang Y, et al (2017) Unmanned aerial vehicle remote sensing for field-based crop phenotyping: current status and perspectives. Front Plant Sci 8:1111. https://doi.org/10.3389/fpls.2017.01111

Yastıklı N, Bagci I, Beser C (2013) The processing of image data collected by light UAV systems for GIS data capture and updating. ISPRS Arch 7(W2). https://doi.org/10.5194/isprsarchives-XL-7-W2-267-2013

Zhou F, Pan S, Chen W, Ni X, An B (2019) Monitoring of compliance with fuel sulfur content regulations through unmanned aerial vehicle (UAV) measurements of ship emissions. Atmos Meas Tech 12(11):6113–6124. https://doi.org/10.5194/amt-12-6113-2019

Part V
Groundwater

Chapter 9
Groundwater from Space

Aquifer depletion is a largely invisible threat, but that does not make it any less real.
Lester R. Brown

Abstract More than half of the world's largest aquifers are showing rapid declines in groundwater storage. This loss in groundwater can jeopardize global food and water security. Yet, groundwater is difficult to monitor and manage largely due to lack of data and limited monitoring infrastructure and the nature of aquifers, which are vast and unseen. This chapter provides useful insight on how satellite methods like the Gravity Recovery and Climate Experiment mission are helping with assessing variability in groundwater on regional and water management scales. Other quantitative and empirical approaches to groundwater estimation are also discussed. Additionally, contemporary understanding of changes in groundwater systems and their response to human and climate factors is highlighted. An important outcome of this chapter is to introduce hydrologists, environmentalists, and researchers across the broad spectrum of geoscience and water resources to the use of satellite methods in assessing the status of groundwater to improve knowledge on groundwater systems.

9.1 Introductory Remarks

Groundwater is an important source of freshwater that is vulnerable to the impacts of a changing climate and unsustainable rates of extraction. It is an important source of global freshwater resource for more than two billion people, providing much of the world's drinking water supply and supporting food production on a large scale through irrigation water supply (Rodell et al. 2018; Famiglietti 2014). In arid and semi-arid regions (e.g., Australia) where surface water is limited, dependence on groundwater is astronomical and surpasses other humid regions like the Amazon and Congo Basins and coastal West Africa. For instance, water used for irrigation agriculture in Australia represents about 60% of annual groundwater withdrawals, while

C. Ndehedehe, *Satellite Remote Sensing of Terrestrial Hydrology*,
https://doi.org/10.1007/978-3-030-99577-5_9

ecosystem services, mining, manufacturing, and household water supply in other Australian territories depend absolutely on groundwater. During droughts, groundwater remains an important source of water for agriculture and other industrial applications as was evidenced during the Millennium drought in Australia and much of US and the arid northeastern Brazil (Ndehedehe et al. 2020b, 2021; Ojha et al. 2018; Famiglietti 2014; Leblanc et al. 2012). Substantial groundwater depletion rates in the last 15 years in India, northern Africa, California Central Valley, Middle East, North China Plain China, Canning Basin Australia, and southern America (Chen et al. 2016a, b; Famiglietti 2014), among several other regions confirm the threats posed by a changing climate and human water extraction. With increasing frequency of drought intensity and duration across the globe (Ndehedehe et al. 2020a, b; Spinoni et al. 2014), groundwater extraction will continue unabated as this is a natural human response to drought. This condition could even be exacerbated by the rise in human population, especially in regions that are typically dry with limited availability of surface water like in the middle east.

Yet, groundwater is difficult to monitor and manage largely due to the nature of aquifers, which are vast and unseen. In developed and some emerging economies, the limited monitoring infrastructure and inconsistent national reporting procedures in several regions of the world have hampered the availability of accurate statistics on groundwater abstraction and withdrawal. Data sparsity and lack of framework are key challenges to assessing aquifer systems and develop early warning indicators to intervene on impacts associated with extraction. These constraints also preclude diagnosis of important drivers (e.g., rainfall and runoff characteristics) of groundwater variations. This makes it difficult to predict its response to droughts, wildfires, and water use. For example, well data may not reflect true drought conditions because of withdrawals for irrigation and industrial purposes, which also impact the water table. Groundwater monitoring from satellites offers new national capability that can be integrated to national groundwater database architecture for freshwater accounting and improving water allocation and sharing. Addressing the lack and scarcity of in situ hydrological data by leveraging on satellite-derived freshwater changes, including the estimation of groundwater from satellite is therefore a key foundation to improving knowledge in groundwater hydrology, and for proactive water resources planning. Following the work of Ndehedehe et al. (2021), this chapter therefore provides useful insight on how satellite gravity methods like GRACE can be employed to assess variability in groundwater on a regional and water management scales. Changes in groundwater systems and their response to human and climate factors are highlighted. A key outcome of this chapter is to introduce hydrologists, environmentalists, and researchers across the broad spectrum of geoscience and water resources to the use of satellite methods in assessing the status of groundwater and improving knowledge on groundwater systems.

9.2 Groundwater Hydrology

9.2.1 *Quantifying Changes in Groundwater from Satellite*

Groundwater resources pose some general governance challenges because of its invisibility and lack of robust monitoring architecture. Unlike surface water, groundwater processes are more difficult to observe and quantify because of insufficient in situ measurements (well or ground observations). Even though ground observations are accurate and valuable in understanding groundwater variability and its response to climate variability or withdrawal, the distribution and availability of gauge observations across large spatial domains in several regions are usually limited. Several methods exist in monitoring and collecting information about groundwater change. They are usually direct approaches and consist of well observations or head measurements using piezometers. With several spatial domains not covered with observational approach, the knowledge of groundwater variability across large spatial domains is limited. In some regions, observational networks have been discontinued due to high cost of gauge installations along with their maintenance, which is also expensive and labour intensive. Moreover, age dating and hydro-geophysical methods, including field gravity surveys, have been used to detect changes in groundwater storage (Green et al. 2011; Pool and Eychaner 1995). This land-based gravity measurements rely on traditional surveying tools such as gravimeter and barometric level.

Gravity Recovery and Climate Experiment (GRACE) satellite observations have been used in the estimation of not only changes in terrestrial water storage but also in quantifying groundwater variations across several regions of the world (Tapley et al. 2019). Large-scale assessments of global and regional hydrology have benefited from the use of this satellite data for the quantitative estimates of monthly changes in terrestrial water storage, including variations in groundwater. Large aquifer systems have been tracked using GRACE satellites after the first results of GRACE were reported in 2004. Interestingly, these aquifer systems show considerable groundwater depletion that have been attributed to intensive groundwater abstraction and change in climatic condition (Ferreira et al. 2020; Ojha et al. 2019; Chen et al. 2016b; Famiglietti 2014). So, to overcome the challenge of data availability, the impacts of climate and water abstraction on changes in groundwater are now being increasingly monitored using advanced satellite methods like GRACE (Thomas and Famiglietti 2019a; Ojha et al. 2018; Chen et al. 2016b). Groundwater monitoring from satellites offers new capability in groundwater assessment. By integrating this satellite-based monitoring framework with national groundwater database architecture, this can lead to a better understanding of groundwater hydrology and freshwater accounting. GRACE satellites have enabled the monitoring of freshwater changes globally, including the assessment of important contributions to our understanding of global groundwater and its responses to changes in climate (Ndehedehe and Ferreira 2020; Thomas and Famiglietti 2019b; Rodell et al. 2018). Since the advent of GRACE, there are three methods or approaches that have been employed to quantify groundwater using GRACE satellite. This include the water budget approach, data assimilation, and statistical methods, which involves the reconstruction of groundwater by incorporating other auxiliary data from hydrological models and satellite. However, the big data fusion and inversion approach have only been recently employed to estimate groundwater variations. These approaches are discussed in detail in the following sub-sections.

9.2.2 Water Budget Approach

Groundwater storage estimation from GRACE using the water budget approach relies on auxiliary data to permit isolation of a groundwater storage signal from terrestrial water storage. This approach assumes that GRACE observed changes in terrestrial water storage (**T**) comprising changes in groundwater (**G**), soil moisture (**S**), ice and snow water as (**I**), inland surface water/reservoir storage (**R**), groundwater (**G**), and canopy water storage (**C**) as,

$$\mathbf{T}_{i,j} = \mathbf{G}_{i,j} + \mathbf{S}_{i,j} + \mathbf{I}_{i,j} + \mathbf{R}_{i,j} + \mathbf{C}_{i,j}, \tag{9.1}$$

where the sub-scripts i and j are the number of rows (time) and columns (spatial locations), respectively. Groundwater storage anomalies can be estimated by rearranging Eq. 9.1, which is referred to as the TWS-budget equation (e.g., Ferreira et al. 2020; Thomas and Famiglietti 2019b). Soil moisture, surface water, canopy water storage, and snow are usually auxiliary data derived from model observations such as GLDAS-Noah or other outputs from land surface models. These outputs include canopy water storage, soil moisture, and snow water equivalents or ice. Apart from soil moisture, using any of these components to subtract from terrestrial water storage to estimate groundwater will depend on regions. For example, Australia and the Middle East regions are arid environments where $\mathbf{I}_{i,j}$ and $\mathbf{C}_{i,j}$ are lacking or do not have significant changes as would be observed in the GRACE water column. In much of these regions, $\mathbf{R}_{i,j}$ is limited and its variations are significantly less and somewhat negligible in what GRACE observes as TWS. This means that only hydrological stores like soil moisture and groundwater undergo significant fluctuations or changes. In other words, groundwater changes will be limited to the difference between $\mathbf{T}_{i,j}$ and $\mathbf{S}_{i,j}$ (Eq. 9.1) in much of the region. However, in regions like the Lake Victoria and Lake Volta Basins where the presence of major lakes and water bodies contribute to changes in stored water, soil moisture, and surface will have to be removed from TWS to obtain accurate estimates of groundwater. Similarly in regions with the strong presence of snow, then **I** will have to be subtracted from $\mathbf{T}_{i,j}$, in addition to **S** and **R** if present. The water budget method is a widely used in the estimation of groundwater from GRACE satellites (Fig. 9.1). Across many regions of the world, including India, Central California Valley in the USA, the North China Plain, the Middle East, and the Murray–Darling Basin, groundwater estimates from GRACE compare well with

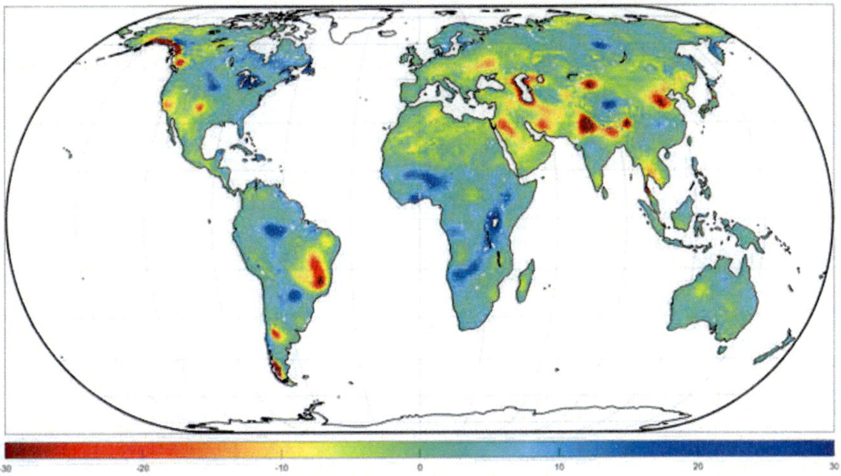

Fig. 9.1 Changes in global groundwater (mm/yr) estimated from GRACE satellite (April 2002–August 2021) using the water budget approach (Eq. 9.1). Areas with large surface water bodies have been masked out

well observations (Ojha et al. 2018; Chen et al. 2016b; Famiglietti 2014; Leblanc et al. 2009). As illustrated in Fig. 9.1, some of these regions, including northeast Brazil, north America, and Asia show significant groundwater depletion based on a more recent data (April 2002–August 2021), consistent with gauged observations.

9.2.3 *Signal Separation*

In the second approach, groundwater can be estimated through a signal separation that uses multi-resolution remotely sensed and outputs from hydrological models (Ferreira et al. 2020). In this regard, Eq. 9.1 is reduced to

$$\mathbf{T}_{ij} \approx \mathbf{G}_{ij} + \mathbf{S}_{ij} = \mathbf{c}_{\mathrm{G}}\mathbf{V}_{\mathrm{G}}^{\mathrm{T}} + \mathbf{c}_{\mathrm{S}}\mathbf{V}_{\mathrm{S}}^{\mathrm{T}}, \tag{9.2}$$

where $\mathbf{c}$ and $\mathbf{V}$ are matrices describing the series coefficients representing the temporal evolution (i.e., the principal components—PCs) and the orthonormal spatial base functions representing the spatial patterns (i.e., the empirical orthogonal functions—EOFs), respectively. The coefficients $\mathbf{c}_{\mathrm{G}}$ and $\mathbf{c}_{\mathrm{S}}$ (Eq. 9.2) are estimated by the least squares taken GRACE-TWS as the observations. With the estimated coefficients for groundwater and soil moisture compartments, now denoted as $\hat{\mathbf{c}}_{\mathrm{G}}$ and $\hat{\mathbf{c}}_{\mathrm{S}}$ (Eq. 9.1), groundwater can be estimated as

$$\hat{\mathbf{G}} = \hat{\mathbf{c}}_{\mathrm{G}}\mathbf{V}_{\mathrm{G}}^{\mathrm{T}}. \tag{9.3}$$

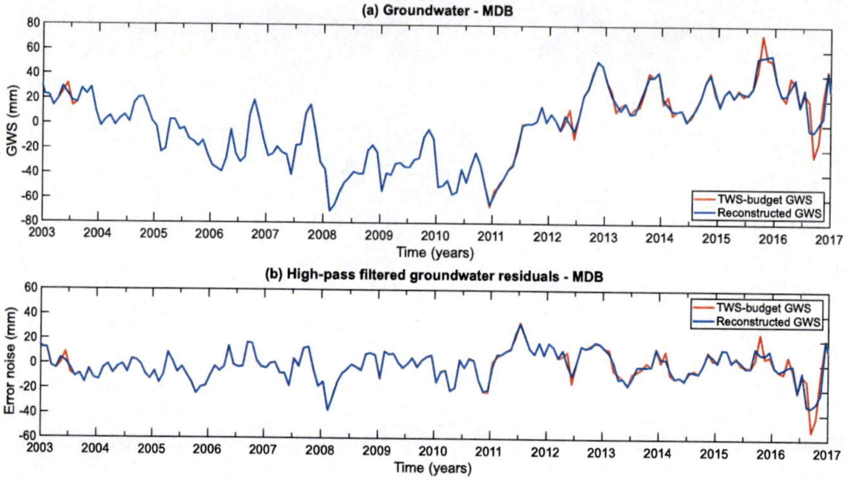

Fig. 9.2 Averaged groundwater series for MDB. **a** Comparison of the two approaches to GWS estimation and **b** the error noise between both estimates. *Source* Ndehedehe et al. (2021)

The same applies to soil moisture. The temporal similarity of GWS estimates quantified from reconstruction using geostatistical methods and water budget approaches confirms the validity of this method. An example of the error noise assessment of reconstructed water budget analysis derived groundwater over Murray–Darling Basin using GRACE observation is indicated in Fig. 9.2. The inversion of GWS based on the reconstruction using the estimated PCs ($\hat{\mathbf{c}}_G$), as per Eq. 9.3, offers an improvement regarding the direct application of the TWS-budget equation (Eq. 9.1). Nevertheless, as terrestrial water budget is supposed to be due to the hydrological cycle, any difference concerning the seasonal cycle of terrestrial water storage could give an estimation of the error. Such an approach has been used in GRACE monthly solutions by considering seasonal variations (see, e.g., Sakumura et al. 2014; Wahr et al. 2004). Thus, multi-linear regression analysis can be employed to assess the temporal variabilities of GWS series. In this example, an overall improvement of approximately 4% (i.e., estimated standard deviations for the respective series are 13.15 and 12.67 mm for GWS based on TWS budget and reconstruction, respectively) was found in the use of this method (Fig. 9.2). Error noise assessment of groundwater over Murray–Darling Basin estimated from both water budget and signal separation has been studied. The goal is to assess whether the inversion of GWS based on the reconstruction using the estimated PCs ($\hat{\mathbf{c}}_G$), as per Eq. 9.3, offers an improvement regarding the direct application of the TWS-budget equation (Eq. 9.1). In the absence of in situ groundwater data as is usually the case, this approach becomes handy to evaluate time series of GWS (see, e.g., Sakumura et al. 2014; Wahr et al. 2004). For example, consider a given time series, the model can be represented as

$$y(t) = a_0 + a_1 t + a_2 \frac{t^2}{2} + \sum_{i=1}^{3} A_i \cos\left(\frac{2\pi}{T_i} t - \varphi_i\right). \tag{9.4}$$

That is, Eq. (9.4) takes into account constant (a_0), linear (a_i), annual amplitude (A_1), and phase (φ_i) terms to describe the original input series y at an epoch t (given in years). The parameters were estimated through the least-squares fitting procedure with their corresponding accuracies. For GRACE-derived TWS changes, it is also necessary to consider the 161-day corresponding to the S_2 alias in Eq. (9.4) due to the insufficient ocean tide correction (Steffen and Petrovic 2009), which might not be significant in low-latitude regions, which encompasses the MDB. The areal weighted GWS time series for MDB shown in Fig. 9.2a are almost similar, whereas differences are seen during the middle 2002 and 2003 as well as 2012 afterward. Specifically, GWS derived from the TWS-budget equation (Eq. 14.1) shows generally higher amplitudes compared to the reconstructed GWS series from Eq. 9.3 between October 2015 and September 2016. In the characterization of the hydro-geological regime of Yangtze River Basin using remotely sensed and modelled products, Ferreira et al. (2020), the averaged time series of the reconstructed groundwater was well correlated with the 'observed', including the observed trend in both data series.

Simultaneous fitting of a time series taking into the functional model given by Eq. (9.4) was carried out for the respective GWS solutions shown in Fig. 9.2a considering the period of April 2002 to June 2017. Following this, the individual residual time series were computed after accounting for the de-trended or long-term and periodic components (de-seasoned) by removing the model given by Eq. 9.4. The remaining residuals of the model in Eq. 9.4 still contain both error effects and un-modelled signals such as those at inter-annual scale. To reduce the contribution of these signals still present, they were smoothed using a 13-month moving average, and the differences between both (original and filtered residuals) shown in Fig. 9.2a are the remaining high-frequency residuals. These high-frequency residuals present low correlation (autocorrelation) at which the respective autocorrelation functions cut off after the fourth lag. This indicates that more noise dominates the high-pass filtered residuals of both series in Fig. 9.2.

9.2.4 Data Assimilation

Even though the use of remote sensing makes it possible to comprehensively observe important water storage components of any hydrologic system, it cannot provide information on the entire system. This is because the measurements represent only a snap shot in time. The use of land surface hydrology process models has been handy and used to predict the spatio-temporal variations in hydrological state variables (soil moisture) and monitoring of changes in surface and sub-surface hydrology such as groundwater. However, a widely reported challenge with the use of hydrological models is that predictions are often poor, largely due to parameter and forcing errors, model initialization, and inadequate model physics and/or resolution as discussed in Sect. 9.2.5. All these factors impact on the faithfulness of these models and their ability to realistically represent land surface conditions, especially when anthropogenic influence such as land-use change, water resources development, and water harvesting contribute to hydrologic variations.

In data assimilation, the observational data are combined with outputs from numerical model to generate the true or optimal estimate of the evolving state of the system. In other words, past forecasts are combined with recent measurements (e.g., new river discharge data) to update and improve the model itself. This idea was pioneered by Charney et al. (1969), who first demonstrated it by combining incomplete historical data to infer the present state of the atmosphere. The goal of data assimilation is to find the best estimates of system states **X** from (noisy) observations **Z**, given a (noisy) model of the system dynamics. This problem can be solved from either the direct observer (i.e., Kalman filter) or dynamic observer (i.e., variational through time) techniques. The direct observer techniques use the difference between observation **Z** and model predicted observation Z to sequentially update the model forecast. The dynamic observer techniques, on the other hand, find the best fit between the forecast model state and the observations. This is subject to the initial state vector uncertainty and observation uncertainty, which is achieved by minimizing over space and time an objective function. This concept has fully developed into what is now known as data assimilation with successful applications in operational weather forecasts and oceanography. But recently, hydrological data assimilation to improve soil moisture and terrestrial water storage estimate is gradually increasing and gaining attention. Moreover, data assimilation techniques are emerging as important alternative approach to improve the accuracy of groundwater estimates on several scales. This enabled the understanding of land surface interaction with the climate system as well as assessing the response of water resources to climate variability. Data assimilation using GRACE satellite has played an important role in optimizing the accuracy of groundwater estimates at local and regional scales (e.g., Tangdamrongsub et al. 2020; Li et al. 2019). The key benefits of data assimilation include quality control, filling in data poor, regions combination of data, errors in data and in model, designing observational systems, maintaining consistency, and estimating unobserved quantities, among others (Rood et al. 1994).

9.2.5 Global Hydrological Models

Hydrological models have been used to synthesize groundwater. The WaterGAP hydrological model developed by Alcamo et al. (2003) accounts for the effect of structural and technological changes on water use, and an effect of climate on irrigation water requirements. A key ingredient of this model is that estimates surface runoff and groundwater recharge using daily water balances of the soil and canopy.

With the advent of GRACE satellite observations, the WaterGAP model is now optimized to simulate net groundwater abstractions and groundwater recharge from soils, in addition to groundwater recharge from surface water bodies in dry regions (Döll et al. 2014). The model uses a simple water balance approach and inputs that include monthly rainfall, temperature, cloud cover, and the number of wet days. There are two important data layers in the water gap groundwater data. The first is groundwater data where water use has been accounted for, i.e., deficit irrigation at only 70% of optimal irrigation with groundwater recharge from surface bodies. The other component is that which has no water use accounted for, i.e., no water use is assumed with groundwater recharge from surface bodies. This new global hydrological model (WaterGAP 2.2a) (see Döll et al. 2014) has been used to study groundwater changes globally (especially in data-deficient regions) and with observed groundwater depletion in such areas being consistent with other reports. This is because the model takes into account groundwater recharge from surface water bodies in semi-arid and arid regions and groundwater depletion. The data is freely available at a spatial resolution of $0.5° \times 0.5°$ and can be downloaded online.[1] Global hydrological models (GHMs) forced by global climate models have great potentials in global water assessment and in the simulation of groundwater changes (e.g., Prudhomme et al. 2014; Schewe et al. 2013). However, some inherent problems with these models include the following:

(i) Bias and conceptual model uncertainties (e.g., Todd et al. 2011; Schuol and Abbaspour 2006).
(ii) Dependence on computational estimates (e.g., Koster et al. 2004).
(iii) Lack of feedback processes involving anthropogenic impacts (e.g., Piao et al. 2010).
(iv) Model physics and choice of parameterizations (e.g., Oettli et al. 2011).

These problems have led to several optimization and reanalysis targeted at improving primary aspects of global water fluxes through simulations and data assimilation systems. The integration of gauged observations with satellite observations has significantly reduced uncertainties in key global hydrological fluxes from these models and reanalysis product.

9.2.6 Big Data Fusion and Inversion Approach

As already mentioned in preceding sections, satellite methods of estimating groundwater using GRACE observations have been largely studied across different regions of the world. These studies, including those that employed data assimilation techniques, showed declining trends that coincided with well observations. Apart from trends, inter-annual variability in the observed GRACE-derived time series is also well correlated with observations of groundwater from in situ networks. The use of

[1] http://www.uni-frankfurt.de/49903932/7_GWdepletion?

this method has been found useful in improving knowledge of groundwater hydrology, having made important contribution to understanding the role climate variability and anthropogenic influence on groundwater. However, it is still limited as a monitoring tool to inform policies on a water management scale. Apparently, this is due to the coarse spatial resolution of GRACE satellites. Groundwater water management agencies require much more detailed information on water availability to inform policies than what is currently provided by the GRACE mission. Although the use of high-resolution satellite radar observations provides some potentials in generating time series of groundwater loss, they usually require extensive and often unavailable lithological data (Castellazzi et al. 2018). Valencia et al. (2020) argued that combining higher resolution data with modelling is required to sufficiently provide detailed information on groundwater abstraction at the field scale.

To circumvent this problem, the scale difference between GRACE and the typical water management scales have to be reduced. To this end, the data fusion and inversion approach was employed over central Mexico where groundwater depletion of approximately 5000 Million Cubic Metres per year (MCM/yr) was reported by Castellazzi et al. (2018). Their GRACE and satellite radar (synthetic aperture radar) combination identified sub-regions associated with the reported substantial groundwater depletion. This inversion framework which combines GRACE and radar data to remotely assess groundwater storage loss is useful in water budgets and assessing groundwater changes in irrigated agricultural areas and urban centres where water abstraction is substantial (Castellazzi et al. 2016). New application has also shown that big data, inversion, and a satellite-model fusion can also be employed to quantify groundwater abstraction on a local and water management scale to improve water accounting.

Building on the capability of satellite geodetic systems (e.g., GRACE) and Synthetic Aperture Radar data, a new framework for water accounting has also been introduced (Valencia et al. 2020). Unlike the use of GRACE and radar data, this framework combines information from optical satellite systems (Landsat, NDVI Leaf Area Index, Albedo, etc.) with output from land surface hydrology and weather prediction models to quantify changes in irrigated crop water use and groundwater abstraction rates (Fig. 9.3). This approach was tested in two agricultural sub-regions in Saudi Arabia, and total groundwater abstraction was estimated at approximately 5.5 billion cubic metres for the year 2015. Although the big data fusion and inversion approach provide comprehensive details for water management planning at the local scale, they are not easily reproducible as they appear rather complex (Fig. 9.3) and not as simple as the GRACE water budget approach.

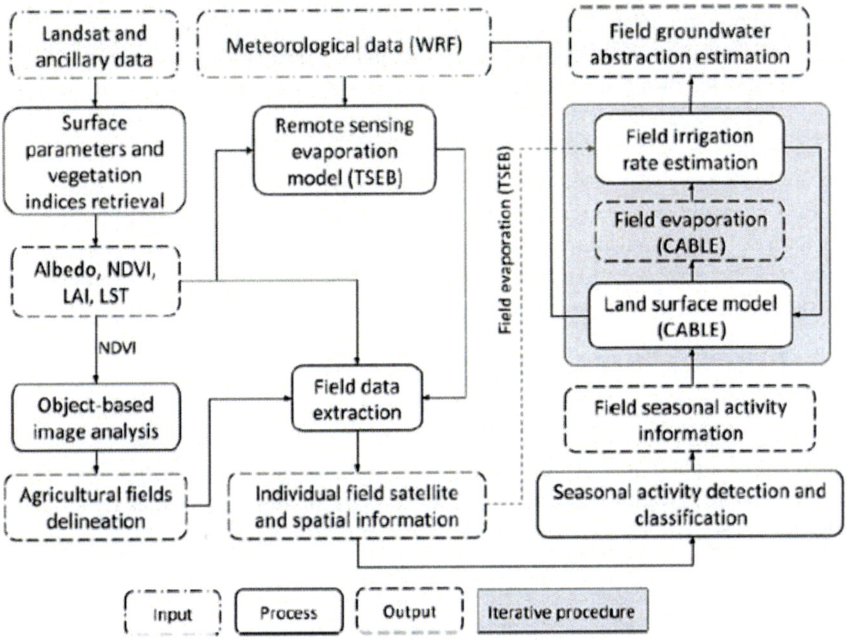

Fig. 9.3 The big data fusion and inversion approach for groundwater monitoring. *Source* Valencia et al. (2020)

9.3 Understanding Groundwater Systems and Drivers

A broad range of environmental impacts associated with droughts, rising temperature, wildfires, and intensive water abstraction contribute to changes in groundwater and impact groundwater processes and feedbacks. The composite influence of sustained groundwater pumping and frequent extreme climate events (e.g., droughts) on changes in groundwater could result in imminent risk of water scarcity and threat to water security.

9.3.1 Impacts of Climate Change and Groundwater Pumping

Drought frequency, long-term groundwater extraction, and low rates of natural recharge, which lowers groundwater levels within aquifers will result in groundwater depletion and eventually result in groundwater drought. That is, a long-term decline in groundwater levels caused by reduced rainfall and recharge, low well yields, and long-term extraction. Large aquifer systems like the Central California Valley have shown that prolonged groundwater extraction affects groundwater storage (Ojha et al. 2020). Groundwater depletion is caused by long-term water extraction and prolonged extreme drought. Under prolonged extreme droughts, extensive groundwater use increases the risk of groundwater loss. This depletion has implications for groundwater droughts, leading to permanent loss of aquifer storage. Observed groundwater depletion in several global hotspots, including North China Plain, Middle East, India, Australia, as measured by satellite measurements has been linked to persistent human water extraction (e.g., Chen et al. 2016a), most of which are the fall-outs of prolonged drought conditions and climate change and/or variability.

Changes in the intensity of droughts also drive groundwater trends. For example, groundwater abstractions temporarily exceed recharge in some areas with intensification of groundwater droughts while long-term annual average groundwater abstractions are smaller than recharge in hotspots with shorter drought events (Wendt et al. 2020). Furthermore, groundwater loss can be exacerbated by the combined influence of climate change and water extraction. For instance, increased water extraction will disrupt recharge, contributing to groundwater loss in time and space during prolonged drought events. In California's Central Valley aquifer system, observed groundwater loss during periods of severe drought increased the rates of groundwater decline. Extensive groundwater extraction and low rates of natural recharge during such period were exacerbated by droughts, causing a permanent damage to aquifer-system storage capacity (Ojha et al. 2018).

Local geology and soil, topography, vegetation, surface water hydrology, coastal flooding, and various forms of land use affect groundwater recharge. These factors result in the varied effects of surface and climatic conditions on different aquifer systems in several locations. In the West African Sahel, for example, severe drought conditions, well observations from an extensive network showed a strong rise in groundwater resources across several hydrological hotspots. This is a hydrological paradox and was attributed to changes in land use as well as the hydraulic conductivity of the soil and its infiltration capacity (e.g., Gal et al. 2017; Favreau et al. 2009; Leduc et al. 2001). Similar paradox has been reported in southeast Australia and southwest US, where human-induced changes in the ecosystem resulted in a long-term rise in water tables and increase in recharge (Scanlon et al. 2005; Allison et al. 1990).

9.3.2 Hydrological Response to Climate and Water Extraction: Example from the Murray–Darling Basin

9.3.2.1 Quantile Function of Storage and Cumulative Storage Deficit

The response of hydrological stores (TWS, SMS, and GWS) to climate variability (e.g., long-term declines or increase in rainfall) can be established using indicators that measure statistical dependence between two variables like linear regression, cross correlograms, and mutual information. But the quantile function of storage (QFS) and cumulative storage deficit are important and more effective methods to assess this response (Ferreira et al. 2020). The QFS is a plot (expressed in percentages) used to define the relationship between water storage and the time that storage is likely to equal or exceed some specified threshold (Ferreira et al. 2020). Past studies (e.g., Ndehedehe et al. 2021) have demonstrated the value of this plot in understanding hydrological response of some river basins (e.g., Yangtze River Basin) to climate variability and water use. For example, the MDB (Murray–Darling Basin) is one of those basins that was selected as a tentative test bed to measure the response of river basins to climatic conditions and human water use (e.g., Ndehedehe et al. 2021). This is because the MDB is the most significant agricultural hub and Australia's most extensive river system. This system is largely characterized by the composite influence of climate variability and human water management (van Dijk et al. 2013). As an illustration of the QFS curve (Fig. 9.4a), the exceedance at the zero net shows that MDB gained soil water only 40% of the time but lost it for 60% of the time. The MDB also lost its stored groundwater approximately 44% of the time during the entire period (2002–2017). However the QFS curve confirms that soil moisture responds faster to water deficit and climatic conditions than TWS and GWS. As a well-known agricultural hotspot, state variables like soil moisture are critical for the MDB. The curve in Fig. 9.4a can be useful to help assess this condition as it also shows that the probability of maximum soil water loss of about −83 mm is approximately 99%. In contrast, the exceedance probability for maximum soil water gain of about 123 mm is 1% (Fig. 9.4a). Furthermore, the slope of the QFS between the storage at 33% (33rd percentile) and 66% (66th percentile) is −1.24, −0.89, and −0.90 for TWS, SMS, and GWS, respectively. The slopes show that GWS and SMS have slightly less variable storage regimes. This means they have a more damped response than TWS over MDB, likely due to the contribution of SWS to the TWS. Furthermore, TWS presented less storage potential within MDB in comparison with SMS and GWS during the 2002–2017 period.

Moreover, patterns of water storage changes were identified using the cumulative SD curve (Fig. 9.4b). This is another approach to assess the response of hydrological stores like groundwater and surface water to climatic variability. An increasing trend represents wet and favourable hydrological conditions while decreasing trends indicate dry conditions. From the terrestrial water budget formulation, rainfall is primarily an important source of water supply to the MDB and hydrological stores (GWS, SMS TWS) respond accordingly to changes in precipitation, be it short or long term. The SD curve shows that soil moisture responded to the big wet period of 2010/2011 with very little or no lag (Fig. 9.4b). However, illustration in Fig. 9.4b suggests there is a delay between TWS (about 6 months) and GWS (approximately 2 years). GWS in the MDB maintains an opposite phase with rainfall because of this considerable lag time. The increase in GWS during the 2002–2005 period when rainfall was negative is another good example to emphasize this argument (Fig. 9.4b). Arguably, the rise in GWS increase during this period can be linked to the impact of preceding wet conditions in the basin during the 1999/2000 period.

There was a delayed response of GWS (approximately 2 years) to successive La-Niña events. These events were characterized by anomalous rainfall over much of Australia, evidenced in one of the biggest floods since European settlements. This

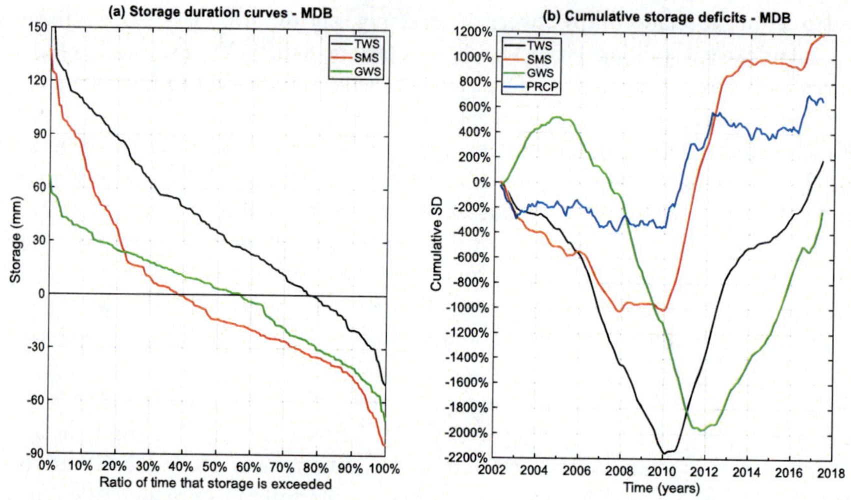

Fig. 9.4 Response of hydrological stores to droughts in the Murray–Darling basin **a** Quantile function of storage for TWS and GWS and **b** Cumulative storage deficit (SD) for TWS, precipitation, soil moisture, and GWS. *Source* Ndehedehe et al. (2021)

delay was evident in the considerable lag time observed between rainfall and groundwater. There is obvious decrease in river basin (MDB) response time to below-normal precipitation. Thus, creating a non-linear relationship that results in a diametrically opposite response of GWS to prolonged drought conditions as was observed during the Millennium Drought. Drought propagates very slowly to groundwater as opposed to soil moisture. For semi-arid river basins like the MDB, recharge might be negatively affected when surface water becomes limited under extreme drought conditions. This is because the changing properties of rainfall (intensity, quantity, frequency), flow, and higher evaporation rates during drought conditions can affect groundwater recharge. Droughts limit the availability of surface water, and in irrigated landscapes this will trigger changes in groundwater use and cause more depletion of groundwater resources (Ndehedehe et al. 2021).

Similar to drought, the propagation of extreme wet conditions is also slow as opposed to surface water and soil moisture. Unlike GWS, SMS and TWS showed a gradual decline during the 2002–2005 period in response to rainfall conditions (Fig. 9.4a). For the MDB, the QFS curve shows groundwater loss for approximately 44% during the 2002–2017 period. This loss and that of soil moisture (60% of the time) were responses to prolonged drought and land surface conditions. The observed hydrological signal or land water storage from GRACE along the tropical wet-dry areas of northern Australia provides a clue that groundwater response to climatic fluctuations will be significantly different as observed in the MDB. The MDB is the Australia's food bowl where surface and groundwater are heavily in use for agriculture. The QFS curve can be useful in providing insight to surface–moisture interaction with groundwater in such areas. For example, considerable changes in soil moisture and/or surface water are pointers to groundwater recharge or discharge. In arid regions, groundwater recharge can result from a small number of large rainfall events (Leblanc et al. 2012). This can make the impacts of climate variability on groundwater substantially different from those of surface water resources.

9.3.2.2 Human Actions Versus Water Abstraction

Human actions (e.g., groundwater withdrawal) and climate change affect the spatio-temporal distribution of surface water and groundwater availability. Past studies (e.g., Ndehedehe et al. 2020c; Tulbure and Broich 2019; Chen et al. 2016b) that quantified groundwater variations and surface water distribution, including freshwater habitats along floodplains, showed the effects of drought on hydro-ecological systems. In the MDB, for example, there was a depletion of surface water resources, primarily due to prolonged drought. With large reductions in surface water and increased application for irrigation and domestic use resulting from drought, there was an increase in groundwater extraction (e.g., Tulbure and Broich 2019; van Dijk et al. 2013; Leblanc et al. 2009). In other Australian territories, the loss of groundwater during the Millennium Drought has been reaffirmed further corroborating the constraints of climate change on groundwater (e.g., Ndehedehe et al. 2021; Yin et al. 2020). In all of these circumstances during which groundwater extraction increased due to the extreme drought, the below-normal rainfall apparently reduced recharge across the region resulting in limited groundwater supplies. Changes in recharge are heavily reliant on rainfall conditions. This is the effect of natural or climate variability. However, McCallum et al. (2010) increasing temperature and changes in the intensity of rainfall can drive substantial changes in recharge, especially in arid domains like Australia. This is because the water budget components such as precipitation and evaporation in these locations are key drivers of freshwater availability (or net precipitation). As a result, a change in their characteristics is best captured with meteorological and agricultural drought indicators.

Even though there is heavy reliance on groundwater for agriculture and domestic purposes due to limited surface water in Australia, the increased groundwater pumping during drought episodes as was observed during the Millennium drought supports the notion of combined influence of climate and anthropogenic activities (water extraction) on groundwater. A considerable rise in the value of groundwater anticipated the coming years largely due to the consistent upsurge in groundwater extraction for domestic, industrial, and agricultural applications (Taylor et al. 2013). Therefore, to guide expectations about drought events in the future and increase adaptation under climate change, a service application that combines hydrological drought indicators with GRACE observations will be required to consistently and frequently monitor groundwater drought. When designed after the similitude of the US Drought Monitor, this service application could help identify locations vulnerable to impacts and environmental hazards associated with groundwater drought and long-term extraction of groundwater. This will improve understanding in continental groundwater hydrology, especially how changes in recharge can be affected by variations in the characteristics of rainfall and land surface conditions, thus providing an integrated approach to water resources management and allocation.

9.3.3 Surface–Groundwater Interaction

Changes in the characteristic of rainfall can affect the availability of surface water resources and processes. Rainfall and evapotranspiration directly affect groundwater recharge and indirectly affect groundwater extraction or discharge. This means small changes in precipitation could trigger large changes in recharge in some semi-arid and arid regions (Green et al. 2011). Apparently, changes in climate through variability in rainfall and temperature regulate rates of recharge. Moreover, surface water has been identified as providing control to groundwater recharge as much of surface runoff not stored in reservoirs end up as groundwater or flows to oceans. For example, the rise in surface water extent during the 'big wet period' coincided with the rise in groundwater changes over Australia. The most obvious location where groundwater and surface water showed considerable response to strong positive anomalies in rainfall is the Murray–Darling Basin (Figs. 9.4b and 9.2). Recharge processes could be complicated in some aquifer systems such as those in Greater Horn of Africa where limited impact of climate variability and/or change on the groundwater level changes have been reported (Agutu et al. 2019). Notably, some factors such as rainfall intensity, flow, and temperature regulate water balance and drive changes in the amount or timing of groundwater recharge (Taylor et al. 2013; McCallum et al. 2010). This further highlights the need to improve knowledge of recharge rates and mechanisms that underpin a sustainable balance in groundwater use.

Moreover, surface water distribution in rivers and floodplains or in the form of flood events are beneficial to groundwater processes. They have positive impacts on groundwater recharge. Surface inundation can be an important predictor of groundwater recharge. Past studies (e.g., Talbot et al. 2018; Kazama et al. 2007) have argued that aerial coverage of surface inundation determines the quantity of flood water that infiltrates groundwater stores. Together with higher inundation extents will result in extensive groundwater recharge. There are also indications that water diversion projects (from a donor basin) that lead to substantial surface water availability in the recipient basin or region have slowed down groundwater depletion rates. Independent case studies of groundwater analysis in the north China Plain, for instance, have attributed the slowed depletion rates in groundwater to the mega water transfer from the Yangtze Basin to north China Plain (e.g., Liu et al. 2020; Zhang et al. 2020; Long et al. 2020). This phenomenon and the gradual depletion of groundwater rates observed in some areas of the north China Plain were caused by water diversion project. This is a positive feedback from human-driven changes in hydrology and emphasizes the impact of surface water in the recharge process. Although increased precipitation contributed to 30% of observed groundwater recovery in Beijing (in north China Plain), water diversion to the region accounted for substantial (40%) groundwater storage recovery (Long et al. 2020).

References

Agutu N, Awange J, Ndehedehe C, Kirimi F, Kuhn M (2019) GRACE-derived groundwater changes over Greater Horn of Africa: temporal variability and the potential for irrigated agriculture. Sci Total Environ 693:133467. https://doi.org/10.1016/j.scitotenv.2019.07.273

Alcamo J, Döll P, Henrichs T, Kaspar F, BL et al (2003) Development and testing of the WaterGAP 2 global model of water use and availability. Hydrolog Sci J 48(3):317–337. https://doi.org/10.1623/hysj.48.3.317.45290

Allison G, Cook P, Barnett S, Walker G, Jolly I, Hughes M (1990) Land clearance and river salinisation in the western Murray Basin Australia. J Hydrol 119(1):1–20. https://doi.org/10.1016/0022-1694(90)90030-2

Castellazzi P, Longuevergne L, Martel R, Rivera A, Brouard C, Chaussard E (2018) Quantitative mapping of groundwater depletion at the water management scale using a combined GRACE/InSAR approach. Remote Sens Environ 205:408–418. https://doi.org/10.1016/j.rse.2017.11.025

Castellazzi P, Martel R, Rivera A et al (2016) Groundwater depletion in Central Mexico: use of GRACE and InSAR to support water resources management. Water Resour Res 52(8):5985–6003. https://doi.org/10.1002/2015WR018211

Charney J, Halem M, Jastrow R (1969) Use of incomplete historical data to infer the present state of the atmosphere. J Atmos Sci 26(5):1160–1163. https://doi.org/10.1175/1520-0469(1969)026<1160:UOIHDT>2.0.CO;2

Chen J, Famigliett JS, Scanlon BR, Rodell M (2016a) Groundwater storage changes: Present status from GRACE observations. Surv Geophys 37(2):397–417. https://doi.org/10.1007/s10712-015-9332-4

Chen J, Wilson C, Tapley B, Scanlon B, Güntner A (2016b) Long-term groundwater storage change in Victoria, Australia from satellite gravity and in situ observations. Global Planet Change 139:56–65. https://doi.org/10.1016/j.gloplacha.2016.01.002

Döll P, Müller Schmied H, Schuh C, Portmann FT, Eicker A (2014) Global-scale assessment of groundwater depletion and related groundwater abstractions: combining hydrological modeling with information from well observations and GRACE satellites. Water Resour Res 50(7):5698–5720. https://doi.org/10.1002/2014WR015595

Famiglietti JS (2014) The global groundwater crisis. Nature 4:945–948. https://doi.org/10.1038/nclimate2425

Favreau G, Cappelaere B, Massuel S, Leblanc M, Boucher M, Boulain N, Leduc C (2009) Land clearing, climate variability, and water resources increase in semiarid southwest Niger: a review. Water Res Res 45:W00A16. https://doi.org/10.1029/2007WR006785

Ferreira V, Yong B, Tourian M, Ndehedehe C, Shen Z, Seitz K, Dannouf R (2020) Characterization of the hydro-geological regime of yangtze river basin using remotely-sensed and modeled products. Sci Total Environ 718:137354. https://doi.org/10.1016/j.scitotenv.2020.137354

Gal L, Grippa M, Hiernaux P, Pons L, Kergoat L (2017) The paradoxical evolution of runoff in the pastoral Sahel: analysis of the hydrological changes over the Agoufou watershed (Mali) using the KINEROS-2 model. Hydrol Earth Syst Sci 21(9):4591–4613. https://doi.org/10.5194/hess-21-4591-2017

Green TR, Taniguchi M, Kooi H, Gurdak JJ, Allen DM, Hiscock KM, Treidel H, Aureli A (2011) Beneath the surface of global change: impacts of climate change on groundwater. J Hydrol 405(3):532–560. https://doi.org/10.1016/j.jhydrol.2011.05.002

Kazama S, Hagiwara T, Ranjan P, Sawamoto M (2007) Evaluation of groundwater resources in wide inundation areas of the Mekong River basin. J Hydrol 340(3):233–243. https://doi.org/10.1016/j.jhydrol.2007.04.017

Koster RD, Dirmeyer PA, Guo Z, Bonan G, Chan E, Cox P, Gordon CT, Kanae S, Kowalczyk E, Lawrence D, Liu P, Lu C-H, Malyshev S, McAvaney B, Mitchell K, Mocko D, Oki T, Oleson K, Pitman A, Sud YC, Taylor CM, Verseghy D, Vasic R, Xue Y, Yamada T (2004) Regions of strong coupling between soil moisture and precipitation. Science 305(5687):1138–1140. https://doi.org/10.1126/science.1100217

Leblanc M, Tweed S, Van Dijk A, Timbal B (2012) A review of historic and future hydrological changes in the Murray-Darling Basin. Global Planet Change 80–81:226–246. https://doi.org/10.1016/j.gloplacha.2011.10.012

Leblanc MJ, Tregoning P, Ramillien G, Tweed SO, Fakes A (2009) Basin-scale, integrated observations of the early 21st century multiyear drought in southeast Australia. Water Resour Res 45(4):W04408. https://doi.org/10.1029/2008WR007333

Leduc C, Favreau G, Schroeter P (2001) Long-term rise in a Sahelian water-table: the continental terminal in south-west Niger. J Hydrol, 243(1–"2):43–54. https://doi.org/10.1016/S0022-1694(00)00403-0

Li B, Rodell M, Kumar S, Beaudoing H, Getirana A, Zaitchik B et al (2019) Global grace data assimilation for groundwater and drought monitoring: advances and challenges. Water Resour Res 55(9):7564–7586. https://doi.org/10.1029/2018WR024618

Liu X, Feng X, Ciais P, Fu B (2020) Widespread decline in terrestrial water storage and its link to teleconnections across Aasia and Eastern Europe. Hydrol Earth Syst Sci 24(7):3663–3676. https://doi.org/10.5194/hess-24-3663-2020

Long D, Yang W, Scanlon B et al (2020) South-to-north water diversion stabilizing Beijing's groundwater levels. Nat Commun 11(3665). https://doi.org/10.1038/s41467-020-17428-6

McCallum J, Crosbie R, Walker G et al (2010) Impacts of climate change on groundwater in Australia: a sensitivity analysis of recharge. Hydrogeol J 18:1625–1638. https://doi.org/10.1007/s10040-010-0624-y

Ndehedehe CE, Ferreira VG (2020) Assessing land water storage dynamics over Southern America. J Hydrol 580:124339. https://doi.org/10.1016/j.jhydrol.2019.124339

Ndehedehe CE, Ferreira VG, Agutu NO, Onojeghuo AO, Okwuashi O, Kassahun HT, Dewan A (2021) What if the rains do not come? J Hydrol 595:126040. https://doi.org/10.1016/j.jhydrol.2021.126040

Ndehedehe CE, Ferreira VG, Onojeghuo AO, Agutu NO, Emengini E, Getirana A (2020a) Influence of global climate on freshwater changes in Africa's largest endorheic basin using multi-scaled indicators. Sci Total Environ 737:139643. https://doi.org/10.1016/j.scitotenv.2020.139643

Ndehedehe CE, Haile GG, Agutu NO, Ferreira VG, Getirana A, Okwuashi O (2020b) Hydrological hotspots of climatic influence in Brazil: a two-step regularization approach. Atmos Res 246:105116. https://doi.org/10.1016/j.atmosres.2020.105116

Ndehedehe CE, Stewart-Koster B, Burford MA, Bunn SE (2020c) Predicting hot spots of aquatic plant biomass in a large floodplain river catchment in the Australian wet-dry tropics. Ecol Ind 117:106616. https://doi.org/10.1016/j.ecolind.2020.106616

Oettli P, Sultan B, Baron C, Vrac M (2011) Are regional climate models relevant for crop yield prediction in West Africa? Environ Res Lett 6(1):014008. https://doi.org/10.1088/1748-9326/6/1/014008

Ojha C, Shirzaei M, Werth S, Argus DF, Farr TG (2018) Sustained groundwater loss in california's central valley exacerbated by intense drought periods. Water Resour Res 54(7):4449–4460. https://doi.org/10.1029/2017WR022250

Ojha C, Werth S, Shirzaei M (2019) Groundwater loss and aquifer system compaction in San Joaquin valley during 2012–2015 drought. J Geophys Res: Solid Earth 124(3):3127–3143. https://doi.org/10.1029/2018JB016083

Ojha C, Werth S, Shirzaei M (2020) Recovery of aquifer-systems in Southwest US following 2012–2015 drought: Evidence from InSAR, GRACE and groundwater level data. J Hydrol 587. https://doi.org/10.1016/j.jhydrol.2020.124943

Piao S, Ciais P, Huang Y, Shen Z, Peng S, Li J, Zhou L, Liu H, YuecunMa Ding Y, Friedlingstein P, Liu C, Tan K, Yu Y, Zhang T, Fang J (2010) The impacts of climate change on water resources and agriculture in China. Nature 467:43–51. https://doi.org/10.1038/nature09364

Pool DR, Eychaner JH (1995) Measurements of aquifer-storage change and specific yield using gravity surveys. Groundwater 33(3):425–432. https://doi.org/10.1111/j.1745-6584.1995.tb00299.x

Prudhomme C, Giuntoli I, Robinson EL, Clark DB, Arnell NW, Dankers R, Fekete BM, Franssen W, Gerten D, Gosling SN, Hagemann S, Hannah DM, Kim H, Masaki Y, Satoh Y, Stacke T, Wada Y, Wisser D (2014) Hydrological droughts in the 21st century, hotspots and uncertainties from a global multimodel ensemble experiment. Proc Natl Acad Sci 111(9):3262–3267. https://doi.org/10.1073/pnas.1222473110

Rodell M, Famiglietti JS, Wiese DN, Reager JT, Beaudoing HK, Landerer FW, Lo M-H (2018) Emerging trends in global freshwater availability. Nature 557:651–659. https://doi.org/10.1038/s41586-018-0123-1

Rood RB, Cohn SE, Coy L (1994) Data assimilation for EOS: the value of assimilated data, Part 1. Earth Obs 6(1):23–25

Sakumura C, Bettadpur S, Bruinsma S (2014) Ensemble prediction and intercomparison analysis of grace time-variable gravity field models. Geophys Res Lett 41:1389–1397. https://doi.org/10.1002/2013GL058632

Scanlon BR, Reedy RC, Stonestrom DA, Prudic DE, Dennehy KF (2005) Impact of land use and land cover change on groundwater recharge and quality in the southwestern US. Glob Change Biol 11(10):1577–1593. https://doi.org/10.1111/j.1365-2486.2005.01026.x

Schewe J, Heinke J, Gerten D, Haddeland I, NWA et al (2013) Multimodel assessment of water scarcity under climate change. PNAS 111(9):3245–3250. https://doi.org/10.1073/pnas.1222460110

Schuol J, Abbaspour KC (2006) Calibration and uncertainty issues of a hydrological model (SWAT) applied to West Africa. Adv Geosci 9:137–143. https://www.adv-geosci.net/9/137/2006

Spinoni J, Naumann G, Carrao H, Barbosa P, Vogt J (2014) World drought frequency, duration, and severity for 1951–2010. Int J Climatol 34(8):2792–2804. https://doi.org/10.1002/joc.3875

Steffen H, Petrovic S et al (2009) Significance of secular trends of mass variations determined from GRACE solutions. J Geodynam 48(3–5):157–165. https://doi.org/10.1016/j.jog.2009.09.029

Talbot C, Bennett E, Cassell K et al (2018) The impact of flooding on aquatic ecosystem services. Biogeochemistry 141:439–461. https://doi.org/10.1007/s10533-018-0449-7

Tangdamrongsub N, Han S-C, Yeo I-Y, Dong J, Steele-Dunne SC, Willgoose G, Walker JP (2020) Multivariate data assimilation of GRACE, SMOS, SMAP measurements for improved regional soil moisture and groundwater storage estimates. Adv Water Resour 135:103477. https://doi.org/10.1016/j.advwatres.2019.103477

Tapley B, Watkins M, Flechtner F et al (2019) Contributions of GRACE to understanding climate change. Nat Climate Change 9:358-369. https://doi.org/10.1038/s41558-019-0456-2

Taylor R, Scanlon B, Döll P et al (2013) Ground water and climate change. Nat Clim Chang 3:322–329. https://doi.org/10.1038/nclimate1744

Thomas B, Famiglietti J (2019a) Identifying climate-induced groundwater depletion in GRACE. Sci Rep 9:4124. https://doi.org/10.1038/s41598-019-40155-y

Thomas B, Famiglietti J (2019b) Identifying climate-induced groundwater depletion in GRACE observations. Sci Rep 9:4124. https://doi.org/10.1038/s41598-019-40155-y

Todd M, Andersson L, Ambrosino C, Hughes D, Kniveton DR, Mileham L, Murray-Hudson M, Raghavan S, Taylor R, Wolski P (2011) Climate change impacts on hydrology in africa: Case studies of river basin water resources. In: African climate and climate change: physical, social and political perspectives, pp 123–153. https://doi.org/10.1007/978-90-481-3842-5_5

Tulbure MG, Broich M (2019) Spatiotemporal patterns and effects of climate and land use on surface water extent dynamics in a dryland region with three decades of Landsat satellite data. Sci Total Environ 658:1574–1585. https://doi.org/10.1016/j.scitotenv.2018.11.390

Valencia LOM, Johansen K, Solorio BJA et al (2020) Mapping groundwater abstractions from irrigated agriculture: big data, inverse modeling, and a satellite-model fusion approach. Hydrol Earth Syst Sci 24(11):5251–5277. https://doi.org/10.5194/hess-24-5251-2020

van Dijk AIJM, Beck HE, Crosbie RS, de Jeu RAM, Liu YY, Podger GM, Timbal B, Viney NR (2013) The millennium drought in southeast Australia (2001–2009): natural and human causes and implications for water resources, ecosystems, economy, and society. Water Resour Res 49(2):1040–1057. https://doi.org/10.1002/wrcr.20123

Wahr J, Swenson S, Zlotnicki V, Velicogna I (2004) Time-variable gravity from GRACE: first results. Geophys Res Lett 31(11):L11501. https://doi.org/10.1029/2004GL019779

Wendt DE, Van Loon AF, Bloomfield JP, Hannah DM (2020) Asymmetric impact of groundwater use on groundwater droughts. Hydrol Earth Syst Sci 24(10):4853–4868. https://doi.org/10.5194/hess-24-4853-2020

Yin W, Li T, Zheng W et al (2020) Improving regional groundwater storage estimates from GRACE and global hydrological models over Tasmania, Australia. Hydrogeol J 28:1809–1825. https://doi.org/10.1007/s10040-020-02157-3

Zhang C, Duan Q, Yeh PJ-F, Pan Y et al (2020) The effectiveness of the South-to-North Water Diversion Middle Route Project on Water delivery and groundwater recovery in North China Plain. Water Resour Res 56(10):e2019WR026759. https://doi.org/10.1029/2019WR026759

Part VI
Environmental Hazards

Chapter 10
Climate Change and Hydrological Systems

Abstract Improving knowledge of climate change, especially in the context of droughts and hydrology is required to predict environmental hazards, increase resilience to climate change, and ensure sustainable management of water resources. Climate is changing, and the rise in drought frequency and severity and their intensities would mean increased risk of food and water security, wildfires, and other environmental and life-threatening hazards. This understanding is therefore important not only to guide our expectations about future droughts, but increase opportunities for adaptation and drought preparedness (Ndehedehe et al. 2021).

10.1 Changes in the Global Climate System

Discussions on greenhouse effect and its implications continue to predominate the world's science and policy agenda on global change because of the growing evidence of warming in the climate system and the human influence that exacerbates it (Ndehedehe 2017). Despite an increasing number of climate change mitigation policies, anthropogenic emissions of greenhouse gases continue to rise (IPCC 2015). These greenhouse gases alter the radiative equilibrium of the atmosphere, causing increase in temperature and strong changes in other components of the hydrological cycle (e.g., Gleick 1989). Greenhouse gas emissions are the principal drivers of global warming, which, in turn, affects freshwater availability through an increased acceleration of the water cycle (Ndehedehe 2019). Warming of the climate system through anthropogenic greenhouse gas emissions affects natural systems, leading to accelerations and alterations in the global hydrological cycle. Global hydrological models forced by greenhouse gases concentration and climate models reveal a considerable rise in regional and global water scarcity (Schewe et al. 2013). Prolonged droughts, for example, have contributed substantially to the loss of surface water resources in some regions. Additionally, groundwater availability can also be impacted by climate change and droughts (e.g., Green et al. 2011). During the Millennium drought in Australia, there was a substantial decline in groundwater variations in the Murray Darling basin (Chen et al. 2016b). The increase in global severity of hydrological drought towards the end of the twenty-first century has been

C. Ndehedehe, *Satellite Remote Sensing of Terrestrial Hydrology*,
https://doi.org/10.1007/978-3-030-99577-5_10

linked to the impacts of climate change (Prudhomme et al. 2014). With growing evidence from considerable case studies (e.g., Bader and Latif 2011; Joly and Voldoire 2010; Losada et al. 2010), changes in climate through perturbations of the tropical oceans, mesoscale convective systems, indices of climate variability (e.g., El Niño Southern Oscillation-ENSO) and their impacts on precipitation patterns are increasingly becoming less debatable. This influence on precipitation patterns results in considerable changes in increased acceleration of the hydrological cycle (see, e.g., Ndehedehe et al. 2021; Todd et al. 2011). Tropical South America, for example, is one of the global climatic hotspots where the influence of ENSO and Pacific Decadal Oscillations on terrestrial water storage and drought events have been well documented (Ndehedehe and Ferreira 2020; Ndehedehe et al. 2020b; Linage et al. 2013; Phillips et al. 2012). These climate teleconnections and other atmospheric circulation patterns drive the water cycle creating several multi-year and decadal cycles of extreme floods and drought conditions. For instance, the recent historic extreme flood events, which occurred in late summer of 2021 in New South Washes, Australia were observed in areas previously affected by extreme drought and wildfire in the summer of 2020 (https://earthobservatory.nasa.gov/images/148093/historic-floods-in-new-south-wales). Furthermore, the notorious wet period in 2010/2011 period in Australia was preceded by the most extreme and prolonged drought since European settlement which is now popularly known as the 'Millennium drought' (Ndehedehe et al. 2021).

ENSO is an ocean–atmospheric climate phenomenon that occurs across the tropical Pacific Ocean and is characterized by somewhat periodic variations of 3–5 years in the sea surface temperature of the tropical eastern Pacific Ocean (Ndehedehe et al. 2017). ENSO has two opposite phases, the El Niño and La Niña. The latter is below-average sea surface temperature in the central and eastern tropical Pacific Ocean. El Niño, on the other hand is the reverse of La Niña in which a warming of the ocean surface in the central and eastern tropical Pacific Ocean occurs. These events are a natural part of the global climate system and they cause floods, droughts, and bush fires in many regions of the world, particularly those around the Pacific Ocean. There is also evidence on climate change impacts on annual runoff with some regions experiencing an increase particularly at higher latitudes, and other regions a decrease. The last half of the twentieth century has been replete with a rise in extreme events such as wild fires, droughts as well as floods. The anticipated increase in drought frequency and intensities across the globe under the IPCC-AR4 future climate projections (e.g., van der Molen et al. 2011) further stresses the increasing vulnerabilities of societies to the impacts of climate change (Ndehedehe et al. 2019).

10.1.1 The Earth System

The Earth system plays an important role in regulating global climate. It is categorized into two parts: the Geosphere and Biosphere. The latter is a global ecosystem that comprise living organisms and non-living factors from which they derive energy and nutrients. In other words it is an ecological system integrating all living and non-living things and their interactions. The Geosphere is the Earth itself and contains the Earth's interior, mantle, rocks and minerals, oceans, and gases. It is a term that collectively describes the Lithosphere, Atmosphere, Hydrosphere, and Cryosphere. The geosphere is an active player in the earth's systems, affecting the atmosphere and the oceans, as well as important processes such as the water and biogeochemical cycles. For example, the Earth'atmosphere contains various gaseous combinations, which are important for life in the human planet as well as protecting living organisms from solar ultraviolet radiation, solar wind and cosmic rays. These gases include, Nitrogen (78%), Oxygen (21%), Argon (0.93%), Carbon dioxide (0.04%), among others. The atmosphere has five different layers, extending all the way up to more than 400 km. These layers are, Exosphere, Thermosphere, Mesosphere, Stratosphere, and Troposphere, which is the region where the Earth's terrestrial weather is developed. The distance of these layers from the Earth is illustrated in Fig. 10.1. The troposphere is the closest (0–10 km) to the Earth and is an important layer of the atmosphere, given its role in regulating global climatology. It contains 75% of the atmosphere's total mass and 99% of its water vapour and aerosols (http://www.scientificaviation.com/what-is-the-boundary-layer/). The hydrosphere is another example that underpins the importance of the geosphere in key processes such as the water cycle. It consists of ocean, seas, lakes, wetlands, streams, groundwater, water vapour, and soil humidity. In other words, it's the sum of all water present on the planet. The water cycle is the result of difference between precipitation and evapotranspiration from water bodies, ground, and land surface. This exchange of moisture between land surface and the atmosphere can be affected by climate change and anthropogenic actions, resulting in global climate change. This change, in turn, affects the redistribution of continental land water storage. For example, deforestation can increase albedo

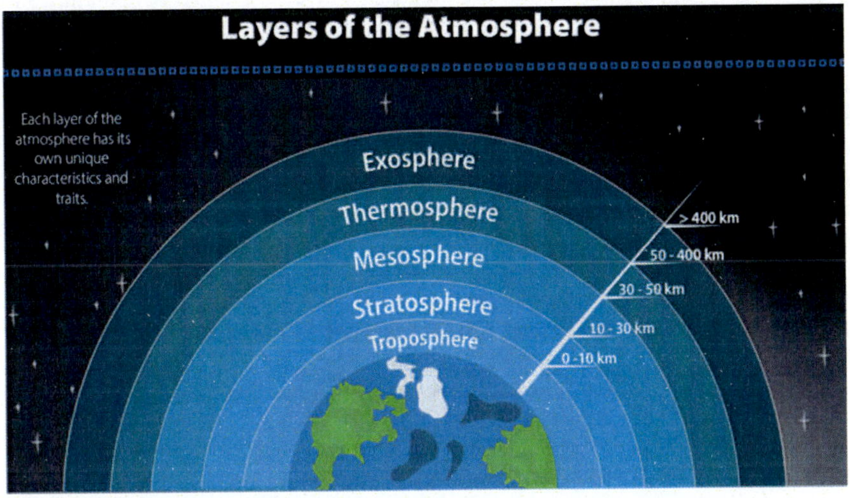

Fig. 10.1 Layers of the atmosphere

leading to cooler and drier climate conditions. In rainforest zones such as the Congo basin, the total atmospheric moisture contribution to local precipitation is substantial (Bell et al. 2015).

10.1.2 Understanding Climate Change

When solar radiation reaches the Earth's atmosphere after passing through space, much of it are absorbed by the Earth. This process warms the Earth, providing the energy for survival. However, some of the solar radiation is reflected and released from the Earth's atmosphere without being absorbed. After warming, the Earth emits infrared radiation, some of which pass through the atmosphere, while others are absorbed by greenhouse gases in the atmosphere and reflected back to the Earth. Greenhouse gases involved in this phenomenon include water vapor, carbon dioxide, nitrogen oxides, methane, ozone, chlorofluorocarbons, hydrofluorocarbons, and perfluorocarbons. These gases are considered as direct radiative forcing, which participates in radiation absorption directly. However, there are indirect greenhouse gases, which after chemical transformation contributes to global warming. Among these greenhouse gases, carbon dioxide as well as methane have greater effect on heat retention in the atmosphere compared to others. The exchange of energy creates an equilibrium in the Earth's temperature such that if there were no greenhouse effect in the atmosphere, the Earth's temperature would be on average about 15.5 °C lower than the status quo. This kind of temperature would make the human planet uninhabitable. But if the concentration of greenhouse gases inside the atmosphere surpasses the normal level, the Earth's energy balance will be disturbed. This will result in increased heat energy within the Earth's atmosphere, and will lead to an unprecedented warming of the Earth surface. The rise in average temperature in the atmosphere and oceans is mainly due to increased concentrations of greenhouse gases and heat energy trapped in the atmosphere. This is called global warming. This phenomenon is a key driver of global climate change. Greenhouse gases can be the result of human activities, such as thermal power plants, industrial activities, fossil fuels, among others. As part of the effects of climate change, glaciers in both the northern and southern hemispheres are shrinking as well as considerable acceleration in global water cycle.

10.2 Hydro-meteorological Changes and Links to Global Climate

As mentioned in Ndehedehe et al. (2018), global interest in climate change is growing because of its anticipated impacts on agriculture, water security, and economic growth. Impacts of climate change are expected to have direct and profound negative effects on freshwater availability (see, e.g., Tall et al. 2016; Prudhomme et al. 2014; Schewe et al. 2013). As a result, the focus on changes in hydro-meteorological conditions and water resources is receiving increasing attention (e.g., Ndehedehe et al. 2016a; Hall et al. 2014; Shiferaw et al. 2014; Conway et al. 2009; Descroix et al. 2009; Bekoe and Logah 2013), especially with the perceived risk and vulnerability of future losses and socio-economic problems (e.g., migration, famine, etc.) resulting from the acceleration of the water cycle. Extreme hydro-meteorological conditions and strong hydrological variability are unpredictable outcomes of changes in global climate that impacts on socio-economic systems of the world. In Thailand, for example, about 59 billion dollars was lost to the 2011 flood while economic growth was down by 38% due to hydrological variability in Ethiopia (see, Hall et al. 2014). Whereas the productive seasons of the year are restricted in monsoonal and tropical climates of the world due to strong seasonal and inter-annual rainfall variability (see, Hall et al. 2014), the preponderance of evidence from considerable case studies in the African sub-region (see, e.g., Ndehedehe et al. 2016b; Nicholson 2013; Mohino et al. 2011; Bader and Latif 2011; Losada et al. 2010; Giannini et al. 2008) confirm that atmospheric circulation features, warming of the tropical oceans, mesoscale convective systems, and climate teleconnections, amongst others have large impacts on meteorological processes and induce extreme climatic conditions. Such impacts, teeming up with other low-frequency variability that are connected to slow oceanic and climate signals from global sea surface temperature (SST) anomaly (e.g., Diaz et al. 2001; Enfield and Mestas-Nuñez 1999; Latif and Barnett 1996), may have profound influence on hydrological changes and water resources.

Studies of changes in global climate and how they impact on meteorological and hydrological processes, are without doubt, emerging as active research. So far, our understanding of global climate has improved due to significant progress and advances made in global and regional climate models (i.e., GCMs and RCMs) (see, e.g., Tall et al. 2016; Erfanian et al. 2016; Prudhomme et al. 2014; Schewe et al. 2013). As with rainfall, the changes in land water storage (terrestrial water storage-TWS) and fluxes such as river discharge are expected to be influenced by ENSO-related Pacific SST fluctuations and other triggers of ENSO, for example, SST anomalies of the north tropical Atlantic (see, e.g., Ham et al. 2013). These relationships are still being studied and are significant to understanding global aspects of ENSO effects on surface hydrology. Global climate is also affected by tropospherically connected ENSO signals in other global oceans (see, e.g., Enfield and Mestas-Nuñez 1999; Latif and Barnett 1996) and the link between long term changes in land water storage and anomalies in sea surface temperature of the Pacific, Indian, and Atlantic Oceans has been reported (e.g., Ndehedehe et al. 2018; Andam-Akorful et al. 2017; Phillips et al. 2012). However, there is still research gap in some regions and more studies are required to (i) enhance the skills of hydrological models and (ii) understand key climate variability indices that make substantial contributions to hydrological changes across different regions.

10.2.1 Rainfall Variability

Rainfall is a key source of water availability and is required for environmental functions, supporting the productivity of agricultural and hydro-ecological systems. Precipitation is a key component of the water cycle that shows an increasing acceleration due to climate change. When the climate warms, the hydrological cycle accelerates, resulting in changes in distribution of rainfall and also in the duration and intensity of extreme events such as droughts and floods. The increase in extreme weather events has implications (e.g., loss of biodiversity, increased risk of flood, soil erosion, etc.) that will affect the sustainability of water resources, hydrological systems, human health, management of storm runoff, and agriculture (Huntington 2006). In water-limited ecosystems, precipitation deficits and drought frequencies are more likely to reduce forest productivity and ecosystem services.

Several factors influence rainfall across the globe. For example, the West African climate has been linked to the West African Monsoon system, a significant element of the regional climate that brings about 70% of the annual rainfall (e.g., Sultan and Gaetani 2016; Janicot 1992). Apart from the West African Monsoon system, sea surface temperature forcings, land surface conditions, quasi-periodic phenomena (e.g., ENSO), and other low frequency global ocean signals, however, play key roles in the strong inter-annual and multi-decadal variability of West African rainfall. These climate teleconnection indices and low-frequency oscillations, e.g., Pacific Decadal Oscillations and Atlantic Multi-decadal Oscillation are also relevant competing physical mechanisms that drive meteorological processes and the variability in extreme wet conditions. In the Congo basin and much of Central Africa, sea surface temperature, atmospheric circulation features and mesoscale convective systems, regulate rainfall through their control on the rain belt (e.g., Farnsworth et al. 2011; Balas et al. 2007). In other regions, for example, topography also play key roles in the variability of annual rainfall.

10.2.2 Temperature

Considerable statistical, observational, and analytical evidence of climate change through rise in surface temperature have been documented. Apart from rainfall, temperature is a key climate component that affects the hydrologic cycle and biologic system, either directly or indirectly. Since the start of the nineteenth century, there has been a significant rise in global average temperature with the last three decades consistently warmer than the previous decades (IPCC 2015). Global warming produces an enhanced spring–summer melting of glaciers. This rise in global temperature results in the melting and shrinking of ice sheets in Greenland and Antarctic. Global warming of 1.5°–4.5°, for example, can raise sea levels by 20–140 cm, respectively. There has been about a 0.7 °C rise in global temperature in the past century. Considering the global climate, this value is much larger than any of the climatic changes experienced during the past centuries years (www.climatechange.gov.au). Due to additional increases in greenhouse gas concentrations, in which globally averaged surface temperature is projected to increase by 1.1–6.4 °C from 1990 to 2100, a warming climate is anticipated in the twenty-first century (www.climatechange.gov.au). The impacts of this warming may result in more heat waves and fewer frosts. Australia is climatic hotspot where the projected average warming of 0.4–2.0 °C by the year 2030 would lead to 10–80% decrease in frosts, among other deleterious effects of rising temperatures. Generally, these increasing temperatures are expected to result in reduced precipitation, and a rise in rainfall intensity will mean having more rainfall in a relatively shorter time as opposed to the past. This will trigger increased drying of the land surface in some areas, with the effect of an increased incidence and severity of droughts. Moreover, this rise in surface temperature has caused harsh drought conditions in the Sahel region of West Africa and in some humid regions like the Congo basin (e.g., Ndehedehe et al. 2019, 2020a; Zhou et al. 2014) while other regions have experience opposite trends in climatic conditions.

10.2.3 Extreme Climatic Events

The impacts of climate change, including long term variations in precipitation, rise in temperature and sea level, have a broad range of consequences for the availability of freshwater around the world. A warming climate will lead to an acceleration of the hydrological cycle, which in turn results in increased spatio-temporal variability of precipitation and also in the duration and intensity of extreme events such as droughts, storms, and floods (e.g., Ndehedehe 2019). This variability has impacts on the hydrological system, contributing to water scarcity during periods with prolonged below normal rainfall (drought). However, periods with large rainfall amounts may lead to flood, which, on the one hand, causes disaster and environmental damages. On the other hand, flood events could be very beneficial to freshwater ecosystems such as supporting extensive groundwater recharge (e.g., Talbot et al. 2018; Kazama et al. 2007). Climate change increases the intensity and frequency of extreme events such as floods and droughts, which both account for about 90% of natural disasters (Eichelter 2019). These events could be deleterious for most regions and families causing freshwater scarcity (Fig. 10.2), displacement, and even death. The loss of freshwater is accompanied by poor harvest, famine, hunger, and poverty.

10.2.3.1 Droughts

Extreme events such as droughts (prolonged below-normal rainfall) affect the availability of freshwater (groundwater, surface water in streams, rivers, lakes). Highly variable climatic conditions pose severe threats to water and ecological resources. Unprecedented climatic thresholds and extreme climatic events (droughts and floods) have been reported in several regions across the globe. The drying up of large

Fig. 10.2 Drought in Lake Chad. *Source* Eichelter (2019)

endorheic lakes and freshwater bodies such as the Lake Chad in West Africa (Fig. 10.2) and Lake Urmia in Iran are great examples of the influence of changing climate on freshwater availability. The iconic and dramatic decline of Lake Chad's surface area was caused by persistent and long drought episodes of the 1980s (Ndehedehe 2019). The disappearance of Lake Mead (Fig. 10.3), which forms the border between Nevada and Arizona, 30 miles from Las Vegas is another example of the impact of climate change on freshwater. The Lake, which was last full in 1998 has fallen by about 60%, and the water level dropped by more than a hundred feet (Fig. 10.3). Drought has significant impact on groundwater resources. For example, sustained groundwater Loss in California's Central Valley was linked to extreme drought (Ojha et al. 2018). Observed groundwater declines in several regions including, North China Plain, Middle East, India, and Australia have been associated with persistent human water extraction caused by prolonged drought conditions and climate change (e.g., Chen et al. 2016a).

Changes in the global climate change, especially increasing frequency and occurrence of droughts will fundamentally continue to amplify water scarcity. In large rainfed agricultural systems in Africa, changes in the distribution of rainfall patterns have been the main cause of fluctuation in food production, food scarcity, famine, and loss of biodiversity (Agutu et al. 2017; Ndehedehe et al. 2016c; Zhou et al. 2014; Shiferaw et al. 2014; Kasei et al. 2010). Regions that rely heavily on rainfall for hydro-power suffer from shortage of electricity with severe consequences on national income. The Lake Volta and Kianji dam, for example, are substantially used for hydro-power generation and during periods of limited rainfall, their hydro-power capacity are considerably diminished. As freshwater is limited during prolonged drought, there is tendency for agitations, inter-state tensions and increased competition for the available water resources (Ndehedehe 2019). Distant populations who indirectly depend on the water resources of such regions could be affected through trade disruption and increase in the prices of groceries (Ndehedehe et al. 2020a).

Fig. 10.3 The drying up of Lake Mead. *Source The New Yorker, May 18 2015*

10.2.3.2 Flood Events

Flood events also have consequences on the environment. They are naturally occurring events that results in land and coastal areas being inundated due to an increase in water level. Globally, floods occur more frequently and they have devastating impacts on mankind and the environment. Several floods across the globe (US, India, Bangladesh, Nepal, and Nigeria) have been highlighted as having devastating and far reaching consequences. The epic flood caused by Hurricane Harvey in the US, Texas for instance, received about 19 trillion gallons of rainfall in four days, the highest from a tropical system (http://mashable.com/2017/08/29/harvey-houston-flood-by-the-numbers-worst-flood/). Flood waters caused by Tropical Storm Harvey on August 30, 2017 in Northwest Houston left several houses in Houston submerged (Fig. 10.4) and a total of 25 million gallons of rainfall was nonetheless received in Texas when the tropical storm dissipated. In addition to more than 17,000 people being displaced, this flood also led to power outages and other catastrophic footprint of physical damages in the region. Unlike the US where such devastating flood seldom occurs, floods of this magnitude are somewhat frequent phenomena and even more catastrophic in some regions of West Africa. For example, it was mentioned in Basu (2009) that the 2007 severe flood caused by a sequence of above normal precipitation affected about 800,000 people in West Africa. A climatological diagnostics of this anomalous flood by Paeth et al. (2012) indicated that it was caused by La-Niña event in the Pacific and a heightened activity of African easterly waves, amongst other

Fig. 10.4 Flood caused by Hurricane Harvey in Texas in 2017. *Source NBC News, August 31, 2017*

factors. In 2009, torrential rains and floods affected an estimated 600,000 people in 16 West African countries, killing an estimated 159 people (Basu 2009).

In the Sahel, wet conditions and flood phenomenon are becoming more frequent and severe due to climate change (Ndehedehe et al. 2016b; Nka et al. 2015). Severe floods ravaged West Africa in 2012 and 2013, destroying human lives, farmlands, properties; and causing outbreaks of cholera and other diseases (WMO 2013, 2015). Mudslides and torrential flooding in Freetown, the capital city of Sierra Leone (http://abcnews.go.com/International/wireStory/official-200-dead-sierra-leone-floods-49204542) killed 400 people while atleast 3000 people were rendered homeless as a result of this torrential flood and mudslides. Intense rainfall, deforestation, and other forms of human activities were identified as immediate and preliminary factors for this deadly disaster. The city of Dhakar in Bangladesh is another hotspot of frequent devastating flood where increased flood vulnerability, have been attributed to poor drainage systems and urban expansion in flood-prone areas, primarily caused by rising population (Dewan 2012). In considerably humid regions with monsoon rainfall and influence of climate teleconnections such as coastal West Africa, India, Congo and Amazon basins, extensive surface water areas and flooded urban centres are very common during the peak of annual rainfall. While this phenomenon is largely caused by climatic factors, in Nigeria human activities such as increasing residential developments in floodplains are apparently major causes of floods. Globally, including semi-arid arid regions, there is a rise in flood frequency. The 'big wet' period in Australia in 2010–2011 period was one of the wettest periods due to extreme floods. While the gulf catchments in the Australia are usually flooded during summer the recent flood in New South Wales along the east coast as another major flood event in the region. Persistent, heavy rain occurring for

Fig. 10.5 The New South Wales floods in Australia during late summer (March) of 2021. *Source NASA Earth Observatory, 30 March, 2021*

several days in late summer (March 2021) in New South Wales, Australia, increased water levels along key rivers in the region. This resulted in more than 40,000 people being displaced across the sate of South Wales while several agricultural and live-stock livestock losses have been reported (https://earthobservatory.nasa.gov/images/148093/historic-floods-in-new-south-wales). As shown in Fig. 10.5, the capability of satellite technologies in environmental monitoring is again highlighted showing the floods in the Hawkesbury-Nepean River system, and extending towards the. The change detection showing flooded areas in light and dark brown along Hawkesbury-Nepean River system and Windsor (right, Fig. 10.5) is based on a natural-colour image of Landsat 8 for April 2020 and March 2021.

10.2.4 Changes in Freshwater

Few indicators of freshwater scarcity include drying of rivers, desiccation of lakes, and groundwater loss. Several important large endorheic lakes and freshwater bodies such as the Aral sea and lakes Urmia and Chad have been severely impacted by climate change resulting in substantial declines and permanent loss of water storage. In some regions, the rapid desiccation of these freshwater resources due to climate change and unmitigated water extraction for agriculture has increased their salinity, impacting on bio-diversity and the environment (Hassani et al. 2020). Changes in river flow regimes will affect hydrological systems (e.g., rivers, estuaries, water holes, freshwater habitats, and watersheds), navigation, hydro-power generation, and groundwater recharge among others. Glaciers are important freshwater resources in the arid and semi-arid regions of western China, and they are considerably vulnerable to climate change (Sun et al. 2018). In the arid Northwest China, snowmelt is an important water source for the growth and survival of oases outside the mountains, which sustains the fragile ecological balance in the region (Wang and Qin 2017).

Climate is changing, and the rise in the frequency and severity of droughts (periods with prolonged below normal rainfall) and their intensities would mean increased risk of water security (e.g., Ndehedehe et al. 2020b). In freshwater ecosystems, rainfall variability directly translates to flow variability, which in turn alters the ecology of wetlands, degrading aquatic biota and disrupting physical processes that sustain different levels of organisms (Bunn et al. 2006; Bunn and Arthington 2002). Furthermore, a rise in the rate of evapotranspiration will also affect water availability in irrigated agricultural lands. For this reason, information on the impacts of climate variability on the dynamics of evapotranspiration is profitable in developing water management adaptation strategies.

10.2.5 Groundwater

Groundwater is an important component of the global water cycle. It is regulated by the balance between inflow into the groundwater body, or aquifer, from the soil or surface water bodies (recharge) and outflow from groundwater to surface water systems (discharge) or groundwater abstractions (Chen et al. 2016b). Changes in river runoff (i.e., in rivers, floodplains, and reservoirs) affect surface water availability, which in turn impacts on groundwater recharge. Groundwater recharge is sensitive to and affected by changes in rainfall intensity, flow, and temperature (Taylor et al. 2013; McCallum et al. 2010). These factors regulate water balance and drive changes in the amount or timing of groundwater recharge. Using the GRACE satellite observations, groundwater declines in several global hotspots including, North China Plain, Middle East, India, Australia have been attributed to prolonged drought conditions and climate change and or variability (e.g., Chen et al. 2016a). The limits of groundwater sustainability in most advanced economies have been reached through development. As a result, small changes in recharge, discharge, or groundwater storage could have economic or environmental consequences (e.g., Green et al. 2011). Understanding the response of hydrological stores (e.g., groundwater, soil moisture) to extreme droughts is thus crucial in evaluating implications of limited rainfall on threats to water and food security (Ndehedehe et al. 2021).

References

Agutu N, Awange J, Zerihun A, Ndehedehe C, Kuhn M, Fukuda Y (2017) Assessing multi-satellite remote sensing, reanalysis, and land surface models' products in characterizing agricultural drought in East Africa. Remote Sens Environ 194:287–302. https://doi.org/10.1016/j.rse.2017.03.041

Andam-Akorful S, Ferreira V, Ndehedehe CE, Quaye-Ballard J (2017) An investigation into the freshwater variability in West Africa during 1979 – 2010. Int J Climatol 37(S1):333–349. https://doi.org/10.1002/joc.5006

Bader J, Latif M (2011) The 1983 drought in the West Sahel: a case study. Clim Dyn 36(3–4):463–472. https://doi.org/10.1007/s00382-009-0700-y

Balas N, Nicholson SE, Klotter D (2007) The relationship of rainfall variability in West Central Africa to sea-surface temperature fluctuations. Int J Climatol 27(10):1335–1349. https://doi.org/10.1002/joc.1456

Basu M (2009) West Africa flooding affects 600,000, U.N. reports. CNN. http://edition.cnn.com/2009/WORLD/africa/09/08/west.africa.flooding/index.html. Accessed 30 Aug 2017

Bekoe EO, Logah FY (2013) The impact of droughts and climate change on electricity generation in Ghana. Environ Sci 1(1):13–24

Bell JP, Tompkins AM, Bouka-Biona C, Sanda IS (2015) A process-based investigation into the impact of the Congo basin deforestation on surface climate. J Geophys Res: Atmos 120(12):5721–5739. https://doi.org/10.1002/2014JD022586

Bunn SE, Arthington AH (2002) Basic principles and ecological consequences of altered flow regimes for aquatic biodiversity. Environ Manage 30(4):492–507. https://doi.org/10.1007/s00267-002-2737-0

Bunn SE, Thoms MC, Hamilton SK, Capon SJ (2006) Flow variability in dryland rivers: boom, bust and the bits in between. River Res Appl 22(2):179–186. https://doi.org/10.1002/rra.904

Chen J, Famigliett JS, Scanlon BR, Rodell M (2016a) Groundwater storage changes: present status from GRACE observations. Surv Geophys 37(2):397–417. https://doi.org/10.1007/s10712-015-9332-4

Chen J, Wilson C, Tapley B, Scanlon B, Güntner A (2016b) Long-term groundwater storage change in Victoria, Australia from satellite gravity and in situ observations. Global Planet Change 139:56–65. https://doi.org/10.1016/j.gloplacha.2016.01.002

Conway D, Persechino A, Ardoin-Bardin S, Hamandawana H, Dieulin C, Mahé G (2009) Rainfall and water resources variability in Sub-Saharan Africa during the twentieth century. J Hydrometeorol 10(1):41–59. https://doi.org/10.1175/2008JHM1004.1

Descroix L, Mahé G, Lebel T, Favreau G, Galle S, Gautier E, O J-C et al (2009) Spatio-temporal variability of hydrological regimes around the boundaries between Sahelian and Sudanian areas of West Africa: a synthesis. J Hydrol 375(1–2):90–102. https://doi.org/10.1016/j.jhydrol.2008.12.012

Dewan AM (2012) Vulnerability of a megacity to flood: a case study of dhaka. In: Floods in a megacity: geospatial techniques in assessing hazards risk and vulnerability. Springer geography. https://doi.org/10.1007/978-94-007-5875-9_3

Diaz HF, Hoerling MP, Eischeid JK (2001) Enso variability, teleconnections and climate change. Int J Climatol 21(15):1845–1862. https://doi.org/10.1002/joc.631

Eichelter L (2019) Three solutions against climate change, simply explained. World Food Programme. https://medium.com/world-food-programme-insight/three-solutions-against-climate-change-simply-explained-bb3aecb0055d. Accessed 12 March 2021

Enfield D, Mestas-Nuñez A (1999) Multiscale variabilities in global sea surface temperatures and their relationships with tropospheric climate patterns. J Clim 12(9):2719–2733. https://doi.org/10.1175/1520-0442(1999)012<2719:MVIGSS>2.0.CO;2

Erfanian A, Wang G, Yu M, Anyah R (2016) Multimodel ensemble simulations of present and future climates over West Africa: impacts of vegetation dynamics. J Adv Model Earth Syst 8. https://doi.org/10.1002/2016MS000660

Farnsworth A, White E, Williams CJ, Black E, Kniveton DR (2011) Understanding the large scale driving mechanisms of rainfall variability over Central Africa. In: African climate and climate change: physical, social and political perspectives. African Climate and Climate Change: Physical, Social and Political Perspectives, pp 101–122. https://doi.org/10.1007/978-90-481-3842-5_5

Giannini A, Biasutti M, Held IM, Sobel AH (2008) A global perspective on African climate. Clim Change 90(7):359–383. https://doi.org/10.1007/s10584-008-9396-y

Gleick PH (1989) Climate change, hydrology, and water resources. Rev Geophys 27(3):329–344. https://doi.org/10.1029/RG027i003p00329

Green TR, Taniguchi M, Kooi H, Gurdak JJ, Allen DM, Hiscock KM, Treidel H, Aureli A (2011) Beneath the surface of global change: impacts of climate change on groundwater. J Hydrol 405(3):532–560. https://doi.org/10.1016/j.jhydrol.2011.05.002

Hall JW, Grey D, Garrick D, Fung F, Brown C, Dadson SJ, Sadoff CW (2014) Coping with the curse of freshwater variability. Science 346(6208):429–430. https://doi.org/10.1126/science.1257890

Ham Y-G, Kug J-S, Park J-Y, Jin F-F (2013) Sea surface temperature in the north tropical atlantic as a trigger for el nino/southern oscillation events. Nat Geosci 6(2):112–116. https://doi.org/10.1038/ngeo1686

Hassani A, Azapagic A, D'Odorico P, Keshmiri A, Shokri N (2020) Desiccation crisis of saline lakes: A new decision-support framework for building resilience to climate change. Sci Total Environ 703:134718. https://doi.org/10.1016/j.scitotenv.2019.134718

Huntington TG (2006) Evidence for intensification of the global water cycle: review and synthesis. J Hydrol 319(1–4):83–95. https://doi.org/10.1016/j.jhydrol.2005.07.003

IPCC (2015) Synthesis report. Contribution of working groups i, ii and iii to the fifth assessment report of the Intergovernmental Panel on Climate Change [core writing team, r.k. pachauri and l.a. meyer (eds)]. Intergovernmental panel on climate change, Switzerland, p 151. http://ar5-syr.ipcc.ch/ipcc/ipcc/resources/pdf/IPCC_SynthesisReport.pdf. Accessed 16 Jan 2017

Janicot S (1992) Spatiotemporal variability of West African rainfall. Part i: Regionalizations and typings. J Clim 5(5):489–497. https://doi.org/10.1175/1520-0442(1992)005<0489:SVOWAR>2.0.CO;2

Joly M, Voldoire A (2010) Role of the Gulf of Guinea in the inter-annual variability of the West African monsoon: what do we learn from CMIP3 coupled simulations? Int J Climatol 30(12):1843–1856. https://doi.org/10.1002/joc.2026

Kasei R, Diekkrüger B, Leemhuis C (2010) Drought frequency in the Volta Basin of West Africa. Sustain Sci 5(1):89–97. https://doi.org/10.1007/s11625-009-0101-5

Kazama S, Hagiwara T, Ranjan P, Sawamoto M (2007) Evaluation of groundwater resources in wide inundation areas of the Mekong River basin. J Hydrol 340(3):233–243. https://doi.org/10.1016/j.jhydrol.2007.04.017

Latif M, Barnett T (1996) Decadal climate variability over the North Pacific and North America: dynamics and predictability. J Clim 9(10):2407–2423. https://doi.org/10.1175/1520-0442(1996)009<2407:DCVOTN>2.0.CO;2

Linage C, Kim H, Famiglietti JS, Yu J-Y (2013) Impact of pacific and atlantic sea surface temperatures on interannual and decadal variations of GRACE land water storage in tropical South America. J Geophys Res: Atmos 118(19):10,811–10,829. https://doi.org/10.1002/jgrd.50820

Losada T, Rodríguez-Fonseca B, Janicot S, Gervois S, Chauvin F, Ruti P (2010) A multi-model approach to the Atlantic Equatorial mode: impact on the West African monsoon. Clim Dyn 35(1):29–43. https://doi.org/10.1007/s00382-009-0625-5

McCallum J, Crosbie R, Walker G et al (2010) Impacts of climate change on groundwater in Australia: a sensitivity analysis of recharge. Hydrogeol J 18:1625–1638. https://doi.org/10.1007/s10040-010-0624-y

Mohino E, Janicot S, Bader J (2011) Sahel rainfall and decadal to multi-decadal sea surface temperature variability. Clim Dyn 37:419–440. https://doi.org/10.1007/s00382-010-0867-2

Ndehedehe C, Awange J, Agutu N, Kuhn M, Heck B (2016a) Understanding changes in terrestrial water storage over West Africa between 2002 and 2014. Adv Water Resour 88:211–230. https://doi.org/10.1016/j.advwatres.2015.12.009

Ndehedehe CE (2017) Remote sensing of West Africa's water resources using multi-satellites and models. PhD thesis, Curtin University, Bentley, Perth, Western Australia. http://hdl.handle.net/20.500.11937/59637. Accessed 12 Jan 2018

Ndehedehe CE (2019) The water resources of tropical West Africa: propblems, progress and prospect. Acta Geophys 67(2):621–649. https://doi.org/10.1007/s11600-019-00260-y

Ndehedehe CE, Agutu N, Ferreira VG, Getirana A (2020a) Evolutionary drought patterns over the Sahel and their teleconnections with low frequency climate oscillations. Atmos Res 233:104700. https://doi.org/10.1016/j.atmosres.2019.104700

Ndehedehe CE, Agutu NO, Okwuashi OH, Ferreira VG (2016b) Spatio-temporal variability of droughts and terrestrial water storage over Lake Chad Basin using independent component analysis. J Hydrol 540:106–128. https://doi.org/10.1016/j.jhydrol.2016.05.068

Ndehedehe CE, Anyah RO, Alsdorf D, Agutu NO, Ferreira VG (2019) Modelling the impacts of global multi-scale climatic drivers on hydro-climatic extremes (1901–2014) over the Congo basin. Sci Total Environ 651:1569–1587. https://doi.org/10.1016/j.scitotenv.2018.09.203

Ndehedehe CE, Awange J, Kuhn M, Agutu N, Fukuda Y (2017) Climate teleconnections influence on West Africa's terrestrial water storage. Hydrol Process 31(18):3206–3224. https://doi.org/10.1002/hyp.11237

Ndehedehe CE, Awange JL, Agutu NO, Okwuashi O (2018) Changes in hydro-meteorological conditions over tropical West Africa (1980—2015) and links to global climate. Global Planet Change 162:321–341. https://doi.org/10.1016/j.gloplacha.2018.01.020

Ndehedehe CE, Awange JL, Corner R, Kuhn M, Okwuashi O (2016) On the potentials of multiple climate variables in assessing the spatio-temporal characteristics of hydrological droughts over the Volta Basin. Sci Total Environ 557–558:819–837. https://doi.org/10.1016/j.scitotenv.2016.03.004

Ndehedehe CE, Ferreira VG (2020) Identifying the footprints of global climate modes in time-variable gravity hydrological signals. Clim Change 159:481–502. https://doi.org/10.1007/s10584-019-02588-2

Ndehedehe CE, Ferreira VG, Agutu NO, Onojeghuo AO, Okwuashi O, Kassahun HT, Dewan A (2021) What if the rains do not come? J Hydrol 595:126040. https://doi.org/10.1016/j.jhydrol.2021.126040

Ndehedehe CE, Haile GG, Agutu NO, Ferreira VG, Getirana A, Okwuashi O (2020b) Hydrological hotspots of climatic influence in Brazil: a two-step regularization approach. Atmos Res 246:105116. https://doi.org/10.1016/j.atmosres.2020.105116

Nicholson S (2013) The West African Sahel: a review of recent studies on the rainfall regime and its interannual variability. ISRN Meteorol 2013(453521):1–32. https://doi.org/10.1155/2013/453521

Nka BN, Oudin L, Karambiri H, Paturel JE, Ribstein P (2015) Trends in floods in West Africa: analysis based on 11 catchments in the region. Hydrol Earth Syst Sci 19(11):4707–4719. https://doi.org/10.5194/hess-19-4707-2015

Ojha C, Shirzaei M, Werth S, Argus DF, Farr TG (2018) Sustained groundwater loss in california's central valley exacerbated by intense drought periods. Water Resour Res 54(7):4449–4460. https://doi.org/10.1029/2017WR022250

Paeth H, Fink A, Pohle S, Keis F, Machel H, Samimi C (2012) Meteorological characteristics and potential causes of the 2007 flood in sub-Saharan Africa. Int J Climatol 31:1908–1926. https://doi.org/10.1002/Joc.2199

Phillips T, Nerem RS, Fox-Kemper B, Famiglietti JS, Rajagopalan B (2012) The influence of ENSO on global terrestrial water storage using GRACE. Geophys Res Lett 39(L16705):2012. https://doi.org/10.1029/2012GL052495

Prudhomme C, Giuntoli I, Robinson EL, Clark DB, Arnell NW, Dankers R, Fekete BM, Franssen W, Gerten D, Gosling SN, Hagemann S, Hannah DM, Kim H, Masaki Y, Satoh Y, Stacke T, Wada Y, Wisser D (2014) Hydrological droughts in the 21st century, hotspots and uncertainties from a global multimodel ensemble experiment. Proc Natl Acad Sci 111(9):3262–3267. https://doi.org/10.1073/pnas.1222473110

Schewe J, Heinke J, Gerten D, Haddeland I, A NW et al (2013) Multimodel assessment of water scarcity under climate change. PNAS 111(9):3245–3250. https://doi.org/10.1073/pnas.1222460110

Shiferaw B, Tesfaye K, Kassie M, Abate T, Prasanna B, Menkir A (2014) Managing vulnerability to drought and enhancing livelihood resilience in sub-Saharan Africa: technological, institutional and policy options. Weather Clim Extremes 3:67–79. https://doi.org/10.1016/j.wace.2014.04.004

Sultan B, Gaetani M (2016) Agriculture in West Africa in the twenty-first century: climate change and impacts scenarios, and potential for adaptation. Front Plant Sci 7:1262. https://doi.org/10.3389/fpls.2016.01262

Sun M, Liu S, Yao X et al (2018) Glacier changes in the Qilian mountains in the past half-century: Based on the revised first and second Chinese Glacier Inventory. J Geograph Sci 28:206–220. https://doi.org/10.1007/s11442-018-1468-y

Talbot C, Bennett E, Cassell K et al (2018) The impact of flooding on aquatic ecosystem services. Biogeochemistry 141:439–461. https://doi.org/10.1007/s10533-018-0449-7

Tall M, Sylla MB, Diallo I, Pal JS, Faye A, Mbaye ML, Gaye AT (2016) Projected impact of climate change in the hydroclimatology of Senegal with a focus over the Lake of Guiers for the twenty-first century. Theor Appl Climatol 1–11. https://doi.org/10.1007/s00704-016-1805-y

Taylor R, Scanlon B, Döll P et al (2013) Ground water and climate change. Nat Clim Chang 3:322–329. https://doi.org/10.1038/nclimate1744

Todd M, Andersson L, Ambrosino C, Hughes D, Kniveton DR, Mileham L, Murray-Hudson M, Raghavan S, Taylor R, Wolski P (2011) Climate change impacts on hydrology in Africa: case studies of river basin water resources. In: African climate and climate change: physical, social and political perspectives. African climate and climate change: physical, social and political perspectives, pp 123–153. https://doi.org/10.1007/978-90-481-3842-5_5

van der Molen M, Dolman A, Ciais P, Eglin T, Gobron N, Law B, Meir P, Peters W, Phillips O, Reichstein M, Chen T, Dekker S, Doubková M, Friedl M, Jung M, van den Hurk B, de Jeu R, Kruijt B, Ohta T, Rebel K, Plummer S, Seneviratne S, Sitch S, Teuling A, van der Werf G, Wang G (2011) Drought and ecosystem carbon cycling. Agric For Meteorol 151(7):765–773. https://doi.org/10.1016/j.agrformet.2011.01.018

Wang Y-J, Qin D-H (2017) Influence of climate change and human activity on water resources in arid region of Northwest China: an overview. Adv Clim Chang Res 8(4):268–278. https://doi.org/10.1016/j.accre.2017.08.004

WMO (2013) WMO statement on the status of the global climate in 2012. World Meteorological Organization, WMO No.1108. www.wmo.int/pages/prog/wcp/wcdmp/documents/WMO1108.pdf Accessed 15 June 2015

WMO (2015) The climate in Africa:2013. World Meteorological Organization, WMO No.1147. http://www.wmo.int/pages/prog/wcp/wcdmp/documents/1147_EN.pdf Accessed 26 Jan 2017

Zhou L, Tian Y, Myneni RB, Ciais P et al (2014) Widespread decline of congo rainforest greenness in the past decade. Nature 509(7498):86–90. https://doi.org/10.1038/nature13265

Chapter 11
Drought Events

Abstract Multiple lines of evidence confirm that the influence of global climate on the frequency and magnitude of extreme events like drought across several regions of the world will continue to increase. Diagnostic studies that underpin drought preparedness and aid the optimization of existing templates on risk mitigation in drought-prone areas are therefore still emerging. But a non-trivial challenge in these studies is identifying ideal drought metrics that capture more effectively, the slow but dynamic large-scale climate processes (or global climate modes) that drive changes in drought characteristics. The need for an ideal metrics to explore the impacts of climate variability drivers on drought intensity and characteristics is explored in this chapter. The importance of such metrics and indicators to characterize and assess the influence of climate variability indices on hydrological systems is articulated. The concepts and complexities of droughts and extreme events, their diagnosis, and novel framework to assess their impacts on freshwater ecosystems are also discussed in detail.

11.1 Introductory Remarks

Drought is a natural hazard that can affect water availability, food supplies, the productivity of freshwater ecosystems, and, more generally, human lives. Compared to other natural hazards, drought is different in many ways. This can either be in terms of its creeping nature, not having a precise, generally accepted definition or the fact that its impacts are nonstructural and can affect much larger area than anticipated compared to other natural hazards (Wilhite et al. 2014). For instance, the impacts of drought on crop production in a particular country will affect its national income through diminished agricultural exports. On the other hand, the regions that benefit from the water resources of such countries through agricultural imports may experience increased prices and negative impacts on its gross domestic product, resulting from drought on the other country. As the impacts of droughts are far reaching, quantifying its impacts to provide relief is therefore complicated unlike other natural hazards like floods and bushfires. While drought can be a temporary climate condition, its impacts can be catastrophic, resulting in socio-economic challenges and

C. Ndehedehe, *Satellite Remote Sensing of Terrestrial Hydrology*,
https://doi.org/10.1007/978-3-030-99577-5_11

even hardship. Interestingly, drought is not just an event limited to regions where the climate is typically arid. Even though exposure to various drought intensity and frequency can vary spatially, the most humid regions of the world are not drought-proofed or exempted from its occurrence and impacts.

Early warning systems and new drought monitoring frameworks are therefore required to optimize risk-management solutions and strengthen national drought policies and drought preparedness (Ndehedehe et al. 2020a; Wilhite et al. 2014). These solutions underpin mitigation strategies and complementary interventions such as insurance programmes that could help cushion the impacts of drought risk. They can also aid opportunities for increase resilience and adaptation. There is an increased awareness of the cascading impacts of drought, and several countries around the world are beginning to respond to the deleterious impacts of droughts through the implementation of national drought strategies and programmes that foster the development of a comprehensive drought monitoring system (Hayes et al. 1999). These all have the potential that can support drought early warning, be it onset, or other characteristics such as severity and intensity. This information facilitates measures to cushion the impacts of droughts on socio-economic systems and improve water-related decisions. Diagnostic studies as well as those focused on the development of indicators to capture dynamic large-scale climate processes are emerging. These diagnostic studies (Ndehedehe et al. 2021a, 2020a; van Dijk et al. 2013; Spinoni et al. 2014) are some efforts to increase drought preparedness as well as aid the optimization of existing templates on risk mitigation in areas that are drought-prone. The concepts and complexities of droughts, their diagnosis, and a framework to assess their impacts on freshwater ecosystems are discussed in this chapter. The importance of drought metrics and indicators employed in drought characterization and assessing influence of global climate and climate variability indices are also discussed here.

11.2 Drought Events: Concepts, Causes, and Consequences

Drought is a prolonged below-normal rainfall that leads to water deficit and abnormal soil moisture deficiency, and limited water in streams, lakes, reservoirs, and groundwater stores. It could be slow and sometimes difficult to predict. This is due to the overarching complexities linked to its causes, magnitude, duration, and timing on different spatial and temporal scales (e.g., Kiem et al. 2016; van Dijk et al. 2013). People have died due to the impacts of droughts such as famine and lack of water. In addition to agitations within local communities and inter-state tensions created by limited and shared water resources, there are armed conflicts in some regions like the Lake Chad Basin. These conflicts and agitations will amplify the vulnerability of riparian agro-communities to poverty and other impacts resulting from water scarcity and droughts (Ndehedehe 2019). Distant populations who indirectly depend on the water resources of such regions could be victims of far-reaching impacts of limited alimentation caused by water deficit through trade disruption and price volatility (FAO 2016). The knowledge of drought characteristics as well as the climatic processes associated with the variability of extreme drought events are important to support future expectations of droughts. Such knowledge will underpin accurate assessment of the cascading impacts of various drought intensities on a broad range of ecosystem services.

11.2.1 Drought Classification Based on Concepts and Causes

There are five classification of droughts, namely, meteorological, agricultural, hydrological, socio-economic, and groundwater. The first three classifications are based on the variables of the hydrological cycle, precipitation, soil moisture, and streamflow, respectively (e.g., Loon and Laaha 2014). Socio-economic drought is associated with the impacts resulting from meteorological, agricultural, and hydrological drought events. But groundwater drought is a unique type of drought and may not be well known. It is the result of decrease in groundwater recharge or when groundwater wells run dry (water table reduces) due to prolonged drought and human water abstraction (Thomas et al. 2017; Goodarzi et al. 2016; Mishra and Singh 2010). These classifications, including anthropogenic drought, are further summarized below.

- Meteorological drought is a long-term precipitation deficit over a specific region. Although it is strictly based on long-term departure from normal precipitation, it can also be combined with increased potential evapotranspiration for an extensive period of time. Consequently, precipitation or its combination with evapotranspiration can be useful to analyse or quantify meteorological drought. Meteorological droughts can propagate through the terrestrial hydrological cycle and result in water loss in hydrological stores such as soil moisture, rivers, and groundwater (e.g., Uddameri et al. 2019; Loon and Laaha 2014). All drought categories begin with meteorological drought as rainfall is the main driver of the hydrological cycle.
- Soil moisture drought is a prolonged deficit of soil moisture. Hence, it is also called agricultural drought and refers to a period associated with declining or insufficient soil moisture at a particular time. This condition leads to crop failure or poor agricultural yield. Notably, soil moisture deficits are usually preceded by a meteorological drought and can also have impacts on natural ecosystems, groundwater recharge, and infrastructure. A plethora of drought indicators that combine precipitation, temperature, and soil moisture have been used to study this kind of drought (Van Loon 2015; Mishra and Singh 2010).
- Hydrological drought is the period associated with insufficient surface and subsurface water resources for various human requirements. This can be in the form of abnormally low water levels, in lakes, rivers, and reservoir levels. Apart from precipitation, society usually depends upon usable water from various sources, such as streamflow, lakes, and reservoir storage for agriculture, ecosystem functioning, domestic, and industrial purposes. As a result, several impacts of drought are related to hydrological drought unlike other forms of drought, e.g., meteorological, soil moisture, and droughts, among others (Ndehedehe et al. 2016a; Loon and Laaha 2014).

- Socio-economic drought is the failure of water resources systems to meet human water demands. Essentially, this type of drought is about drought impacts on socio-economic systems. That is, it relates the impacts of all meteorological, agricultural, or hydrological drought conditions on supply and demand of some economic goods, e.g., milk, vegetables, grains, and meat. Water availability is an important factor that drives global patterns of agricultural production and trade (Odorico et al. 2020). Prolonged and frequent drought events will therefore have considerable effects on the prices of economic goods due to limited supply. This limited supplies happen due to weather-related deficit in water supply.
- Groundwater drought is a long-term decline in groundwater levels caused by reduced rainfall and recharge, low well yields, and long-term extraction. It is the result of decreasing groundwater recharge and groundwater discharge (e.g., Goodarzi et al. 2016). The effects of droughts on aquifers affect the availability of groundwater (Ndehedehe et al. 2021a). Both direct and indirect consequences of human actions through intensive withdrawal also increase the vulnerability of groundwater drought (Uddameri et al. 2019; Thomas et al. 2017). It is a complex type of drought because of the non-linear response of aquifers to meteorological droughts, and the nature of surface water–groundwater interactions in some regions. Apart from climate change, groundwater droughts are also driven by extensive groundwater withdrawals, which makes it even more difficult to characterize and assess. For this reason, metrics and indicators to assess various characteristics of groundwater drought, including onset, duration, magnitude, and variability are still emerging (Uddameri et al. 2019; Van Loon et al. 2017; Thomas et al. 2017).
- Anthropogenic drought is a concept coined to reflect the various processes, e.g., land–atmosphere interactions, water, and energy balance that involve climate change and human actions that contribute to water-deficit conditions. Pioneering studies on this type of drought (e.g., AghaKouchak et al. 2021; Madani and Shafiee-Jood 2020) acknowledged the representation of human actions and climate in modelling human water systems and argue that the interplay of changes in natural systems, climate, anthropogenic activities, and local climatic conditions interact to drive water bankruptcy. In other words, increasing demand of freshwater for several applications under limited water availability scenario caused by prolonged drought could lead to anthropogenic drought. The permanent surface water loss in the Lakes Chad and Urmia as discussed in Chap. 12 is an example of anthropogenic drought. In these cases, water bankruptcy is the aftermath of extensive human water extraction largely triggered by both climate change and human actions. The composite role of unmitigated humans actions and climate on freshwater systems, including groundwater and surface water storage (Ndehedehe et al. 2021a), warrants an understanding of drought from multi-dimensional processes involving humans and natural systems (AghaKouchak et al. 2021; Madani and Shafiee-Jood 2020). This is anthropogenic drought and it can provide additional information that underpins planning and meaningful decisions on the management of water resources.

11.3 Characterization and Diagnoses

11.3.1 Drought Metrics

A drought index is an important variable used for the assessment of the impacts of droughts. It can also be used to define drought properties such as severity, duration, intensity, and spatial extent (Mishra and Singh 2010). Given the strong land–atmosphere interactions of semi-arid tropical areas (e.g., Koster et al. 2004), a non-trivial challenge is on identifying a suitable drought indicator that captures large-scale processes more effectively. For example, the long-term decline of rainfall in the North China Plain was linked to observed decline in evapotranspiration. Given that evapotranspiration can be used to support the management of water resources in irrigated agricultural landscapes, it has been recommended as a key variable in assessing agricultural drought and an indicator for early warning of food insecurity during drought periods (Ndehedehe et al. 2019b; Tadesse et al. 2015; Cao et al. 2014). Drought indicators can therefore be developed using water budget variables, like evapotranspiration, rainfall, river discharge, etc.

Indeed, indicators are now at the centre of numerous assessments, and this include those used to designate human development and environmental performance. The changing in local precipitation, land surface, and catchment characteristics in some areas have triggered the need to improve upon drought metrics used in assessing and evaluating drought characteristics (e.g., severity, duration, and intensity, among others). There are several drought indices used to measure how much precipitation for a given period of time has deviated from long-term records. They include the standardized precipitation index, percent of normal precipitation, standardized evapotranspiration index, surface water supply index, reclamation drought index, deciles, and Palmer drought severity index, among others (Hayes 2007). But, in this section, a brief description of three key indicators used to characterize various drought intensities is provided. However, emerging indicators from time-variable satellite observations are also discussed.

11.3.1.1 Standardized Precipitation Index

The standardized precipitation index (SPI, McKee et al. 1993) is computed based on long-term precipitation record for a given period because the assumption is that drought is primarily caused by prolonged below-normal precipitation or rainfall deficits (e.g., Spinoni et al. 2014). A probability distribution is then fitted to this long-term precipitation record, which is then transformed to a normal distribution so that the mean SPI for a given location and desired period is zero (Mishra and Singh 2010; McKee et al. 1993). The SPI is a very popular drought index. It is normalized or standardized so that wetter and drier climates can be represented in a similar way and can be calculated for other hydrological quantities or units such as snowpack, reservoir, streamflow, soil moisture, and groundwater (McKee et al. 1993).

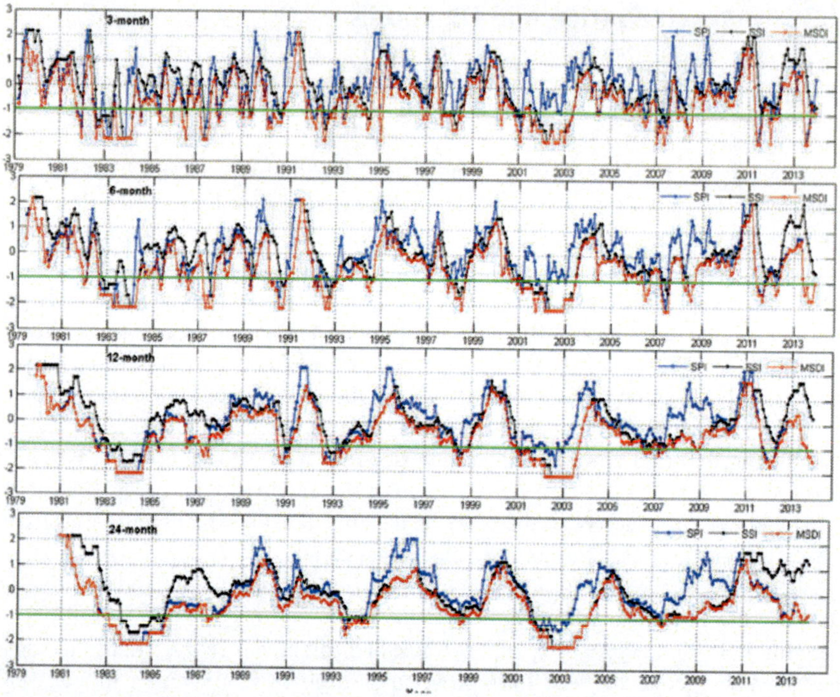

Fig. 11.1 Time series of different drought indicators over the Volta Basin in West Africa. These indicators are standardized precipitation index (SPI), standardized soil moisture index (SSI), and multivariate standardized drought index (MSDI). The MSDI is derived through the combination of precipitation and soil moisture. The green solid line is the adopted drought threshold based on the classification in Table 11.1. *Source* Ndehedehe et al. (2016b)

Participants of the *Lincoln Declaration on Drought Indices* unanimously adopted SPI and recommend same to WMO as a standard index to track meteorological drought events (Hayes et al. 2011). Following the Lincoln workshop, the National Meteorological and Hydrological Services (NMHSs) around the world have been encouraged to use the SPI to characterize meteorological droughts in addition to other indices currently in use.

Furthermore, because the SPI can be calculated for different time scales (e.g., 3, 6 12, 24 months), they can be used to monitor short (meteorological)-, medium (soil)-, and long (groundwater, rivers and reservoirs)-term changes in water availability. However, one of the limitations of SPI is that it is calculated with only precipitation data and does not consider other important climate variables such as evapotranspiration, which can have substantial influence on the nature of drought (Vicente-Serrano et al. 2010a). Different forms of droughts can be measured using either soil moisture, rainfall, or river discharge to construct drought indices on different time scales (e.g., 3, 6 12, 24 months) similar to the way SPI is computed (Fig. 11.1). As demonstrated in Fig. 11.1 over the Volta Basin in West Africa, SPI and standardized soil moisture index (SSI) show moderately strong correlation (i.e., r = 0.63 at 12 month scale at 95% confidence level). In other semi-arid regions like the Lake Chad Basin, this relationship between SPI versus SSI can be much higher ranging from 0.7 to 1.0 (Ndehedehe et al. 2016a). Additionally, these indicators can also be integrated to develop a multivariate standardized indicator (MSDI). The importance of this type of indicator (MSDI) is discussed further in the latter part of this chapter. MSDI assesses drought using multiple variables, linking single drought indicators (i.e., SPI, SSI, or hydrological indicator based on river discharge) into a composite model (Ndehedehe et al. 2016b; Farahmand and AghaKouchak 2015). This multivariate approach provides a more robust assessment of droughts. This is particularly so given that the MSDI improves the skill to feature drought onset and persistence (Fig. 11.1).

Despite the suitability of SPI in most regions, its use could be limited in arid regions. This is because of its inability to account for outgoing water fluxes like evapotranspiration, which is major source of water loss from farmlands. Moreover, it is a critical climate variable that also measures water availability. Arguably, for reasons that are not unconnected to climate change-induced extreme high temperatures, which are typical climate characteristics of dry and arid regions, the use of SPI may be restricted. For instance, the leafing and reproductive phenophases in arid climates such as the Sahel are more likely to be affected by changes in land surface temperature unlike rainfall (e.g., Seghieri et al. 2012). Arguably, rainfall is not a direct indicator of freshwater availability (e.g., Ndehedehe et al. 2019b; Yang et al. 2014; Chen et al. 2014). For this reason, the joint use of composite evapotranspiration-based index and other hydrological units to assess climate-driven influence on water availability was recommended (Ndehedehe et al. 2018).

11.3.1.2 Standardized Evapotranspiration Index

The standardized precipitation evapotranspiration index (SPEI) combines precipitation and temperature data in a water balance framework as detailed in the pioneering works of Vicente-Serrano et al. (2010a, b). The SPEI is multi-scalar in nature character, incorporating the effects of variations in temperature as well as rainfall deficits on drought assessment. This makes it an important drought metric to assess the impacts of drought under a global warming scenario. In addition to the multi-temporal nature of the SPEI, it is sensitive to changes in evaporation demand, which is caused by temperature fluctuations. The multi-scalar nature of SPEI enhances its potential in identifying several drought episodes (Vicente-Serrano et al. 2010b). In other words, it is crucial for evaluating the cascading impacts of droughts on agricultural (crops and vegetation), hydrological (groundwater, rivers, streams), and socio-economic systems (GDP and national income). The SPEI data is freely available online (http://digital.csic.es/handle/10261/153475). Drought metrics derived from multiple climate variables are increasing being used in drought assessment if not the most preferred approach to drought characterization. It is one of the reasons for the growing application of SPEI in drought characterization (e.g., Nguvava et al.

2019; Shiru et al. 2019; Asong et al. 2018). Although the multi-scalar nature of SPEI enhances its potential in identifying several drought episodes (Vicente-Serrano et al. 2010b), this index also agrees substantially with SPI in some regions (Ndehedehe et al. 2020a).

11.3.1.3 Palmer Drought Severity Index

The Palmer drought severity index is one of the most widely used drought indicators (Palmer 1965). It is based on a climatic water balance index, which incorporates precipitation and evapotranspiration anomalies and soil water-holding capacity. (Hayes 2007) describes it as a meteorological drought index that responds to abnormally dry or wet weather conditions. When there is a change in climatic condition, the PDSI-based indicator does not take into account streamflow, lake and reservoir levels, and other longer term hydrologic impacts. But is quite effective in measuring impacts sensitive to soil moisture conditions, such as agriculture. This index (http://www.cpc.ncep.noaa.gov/products/analysis_monitoring-/cdus/palmer_drought/) has been used as an important tool in evaluating the scope, severity, and frequency of prolonged periods of drought. The U.S. Department of Agriculture has employed PDSI to provide financial support to relevant agencies as part of emergency drought relief programmes (Hayes 2007). Other widely known applications of the PDSI include, delineating disaster areas and indicating water scarcity, amount of stock water, and potential intensity of forest fires. Although the PDSI has some resemblance to standardized evapotranspiration index because of the inclusion of temperature and rainfall, it is not multi-scalar in nature. This means that contrary to SPI and SPEI, which have no fixed temporal scale, relating different drought types such as meteorological, hydrological, and agricultural to specific water resources components such as runoff, snowpack, reservoir storage, etc. is difficult. As drought is usually multi-scalar, occurring at the same time across different time scales, this becomes a key limitation in PDSI (Vicente-Serrano et al. 2010b).

11.3.2 Drought Statistics and Characteristics

When using the aforementioned drought indicators, especially the SPEI and SPI, drought is identified when either of these metrics drops below a defined threshold value (i.e., –1) as indicated in Table 11.1. In the use of these drought metrics, four categories of drought or conditions (moderate, severe, extreme) that correspond to different thresholds are listed and can be used for interpretation (Table 11.1). The drought thresholds in Table 11.1 are the classification system used in drought characterization studies (McKee et al. 1993). There are several drought statistics that have been used in drought studies. These include frequency, duration, intensity, and proportion of areas affected by drought (see, e.g., Diaz et al. 2019; Wang et al. 2018; Spinoni et al. 2014). These properties have been very useful in understanding drought

Table 11.1 Drought thresholds based on McKee et al. (1993) classification system

Description	Threshold
Extreme wet	+2.0 and above
Very wet	+1.5 to +1.99
Moderately wet	+1.0 to +1.49
Near normal	–0.99 to +0.99
Moderate drought	–1.0 to –1.49
Severe drought	–1.5 to –1.99
Extreme drought	–2.0 or less

attributes and its impacts on hydro-ecological systems. Quantifying the proportion of areas under drought allows the assessment of drought impacts, focusing on drought-affected areas and the visualization of outcomes in space and time (Ndehedehe et al. 2020c). The integration of drought time series (SPEI, SPI, etc.) events into aggregated areas of drought, and the subsequent estimation of the percentage of drought area (PDA) for each time scale and step is calculated as (Diaz et al. 2019)

$$PDA = \frac{100}{A_r}\sum_{c=1}^{N}(d_s(t) * A), \tag{11.1}$$

where A is the area of the cell c and A_r is the region area, t is time step, d_s is drought state and the magnitude of drought is the PDA value. Drought duration is defined as the time length (i.e., the number of months) between the drought onset and termination (e.g., Mao et al. 2017; Spinoni et al. 2014). It is usually estimated using the sum of durations for all drought events divided by the number of drought events (e.g., Ndehedehe et al. 2020b; Xu et al. 2019) as

$$D = \frac{\sum_{i=1}^{n} d_i}{n}, \tag{11.2}$$

where D is the drought duration (months), d_i is duration of ith drought event, and n is the total number of drought events. Drought frequency is the number of drought occurrences in a given time period (Yu et al. 2014). It is estimated as the ratio between the number of drought months and the total number of months in the time series (e.g., Ndehedehe et al. 2020c; Spinoni et al. 2014) as

$$F = \frac{n_m}{N_m} \times 100, \tag{11.3}$$

where F is drought frequency (%), n_m is the number of drought months, and N_m is the total number of months. In other words, this can be given as the number of drought events in the period under consideration. Drought intensity measures drought severity per its duration and is useful to inform the strength of droughts (Zhang et al. 2015). It is the absolute value of the average of accumulated SPEI values during the drought events and is computed as (e.g., Haile et al. 2020; Wang et al. 2018)

$$I = \left| \frac{1}{n} \sum_{i=1}^{n} SPI_i \right|, \tag{11.4}$$

where I is drought intensity (–), n = number of drought occurrences in months, and SPI_i is SPI or any drought indicator (e.g., SPEI) value below the threshold (–1). For each duration of the drought event, drought severity measures the cumulative deficit below the truncation level to quantify drought intensity (Zhang et al. 2015). Drought intensity is also particularly useful in identifying how intensified droughts can result in drying of land surface, including increased evaporation.

11.3.2.1 Satellite-Based Crustal Deflections

There is an increasing interest in investigating the magnitude and distribution of crustal deformations and their relevance in different fields of study, such as hydrology, geodesy, geophysics, and climate change. But this is usually challenged because of poor distribution of hydrological gauge stations. Given the poor density of gauge networks and limited ground observations, hydro-geodetic research is therefore expected to benefit more from independent space observations and multi-resolution data. The techniques for measuring crustal deformations, e.g., the Global Positioning System and the combination of other space-borne geodetic sensors and surface-loading models provide opportunities to study crustal deformations (Ferreira et al. 2019a), and more importantly how drought-induced deflections can be monitored (Ferreira et al. 2018). The potential of Global Position Systems, satellite gravity-derived crustal deformations in drought studies are therefore articulated in this section.

Given the complexity of drought either as a phenomenon or a physical process, drought metrics that capture dynamics resulting from these processes are ongoing areas of research (Ndehedehe et al. 2020a, c). Satellite-based approaches to characterize drought and hydrological conditions are also gradually emerging. GRACE satellites, for example, have been used as a proxy for terrestrial water storage and hydrological indicator (e.g., Ferreira et al. 2019b; Fu et al. 2015) and to characterize hydrological and groundwater droughts (Ferreira et al. 2018; Thomas et al. 2017; Agutu et al. 2017; Yirdaw et al. 2008). The potential of other sensors such as the Global Positioning System (GPS) has been assessed in hydrological monitoring and quantifying droughts. In one of the pioneering studies over Washington and Oregon in the US (Fu et al. 2015), drought events between 2008 and 2010 and the recovery period (2011–2012) were detected using GPS-derived time-variable monthly water mass series. These applications are increasing as GPS-based crustal deflection are now being complemented by GRACE satellites to depict and capture unprecedented like the 2012–2015 drought over Brazil (Ferreira et al. 2018). As opposed to other

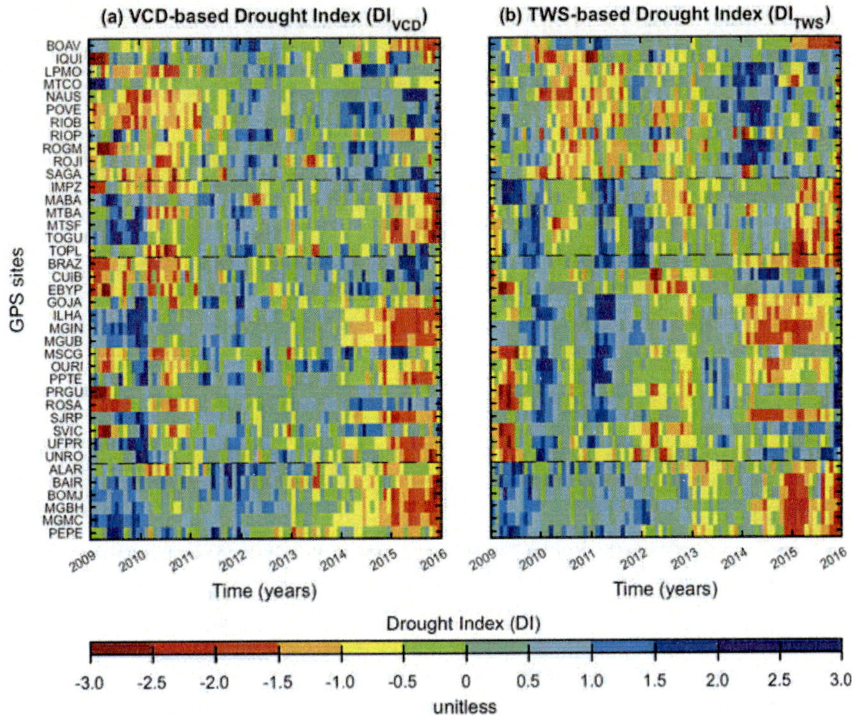

Fig. 11.2 Drought indices based on observations of global positioning system and gravity recovery and climate experiment-derived terrestrial water storage. The GPS sites are locations in river basins within Brazil and includes the São Francisco, Tocantins, La Plata, and Amazon river basins. *Source* Ferreira et al. (2018)

drought indicators, either as composite (SPEI) or single (SPI), the utility of GPS in drought monitoring comes with benefits that complement other drought indicators. The GPS stations over Brazil, for example, detected the changes in the Earth's shape as a response to the water mass loading and unloading (Fig. 11.2). With an improved accuracy better than 3 mm, this approach offers useful alternative in monitoring changes in extreme events (droughts and floods) of other humid regions like the Congo Basin and coastal West Africa, or regions were drought is endemic and severe. The frequency (daily and/or weekly basis) of such monitoring is even optimized by the use of GPS observations due to their considerably high sampling rate and availability in real time (Ferreira et al. 2018; Fu et al. 2015).

11.3.3 Assessing Drought Characteristics and Variability

Rainfall is highly variable in time and space because of the several mechanisms that drives it variability. With considerable spatial variability of rainfall at annual scale across regions, the use of a single mean SPI computed from averaged rainfall data hides the underlying spatial variability of the index (e.g., Ndehedehe et al. 2016a; Ali and Lebel 2009). This could result in poor characterization of wet/dry regimes. Specifically, it could lead to generations of extreme events like drought and might impact on modelling capabilities and the ability to effectively plan water resources (Ndehedehe et al. 2017). The influence of multiple ocean-atmosphere systems through the so-called climate teleconnections can also complicate understanding of drought evolutions and the water footprints in some regions. Drought analysis from a spatio-temporal point of view therefore provides opportunity to improve understanding of its occurrences and persistence. Several statistical methods have been employed to study the spatial and temporal variations of drought. These methods range from third- to fourth-order statistical methods like the principal component analysis (PCA) and independent component analysis (ICA), respectively (e.g., Jolliffe 2002; Cardoso 1999). Multivariate methods like the PCA have been employed in spatio-temporal drought analysis thereby addressing some of these problems (see, e.g., Ndehedehe et al. 2020a). The use of PCA to characterize drought evolutions was pioneered by Karl and Koscielny (1982). This has since resulted in the growing application of PCA in assessing spatio-temporal drought patterns globally (e.g., Agutu et al. 2017; Ivits et al. 2014; Bazrafshan et al. 2014; Santos et al. 2010). These methods, including ICA, wavelet analysis, canonical correlation analysis, and partial least-squares regression, have been employed to diagnose the influence of global climate on droughts and are fully discussed in the next chapter.

11.4 When the Rains Do Not Come

Rainfall is a major source of freshwater availability, supporting the sustainability of hydrological, ecological, and agricultural systems. The recent work of Ndehedehe et al. (2021a) explored this role by addressing a key hypothesis ('What if the rains do not come?') to aid understanding of water deficits on these systems and the socio-economic value of water. As they argued, risk-management solutions that increase opportunities for adaptation plans and resilience to climate change require an understanding of climate change processes and the cascading impact of extreme events (droughts and floods) on hydrological and socio-economic systems.

The large-scale variability of hydrological stores such as groundwater and surface water and how they respond to drought events in large semi-arid domains in Africa and Australia have been studied. Using multi-scale drought indicators, these studies (Ndehedehe et al. 2020a, 2021a; Agutu et al. 2017, 2019) highlighted the potential and suitability of different indicators (Sect. 11.2) to guide future expectations about droughts and its impacts. These indicators can be synthesized from any of the following variables:

- In situ or satellite-based rainfall based on standardized precipitation index (McKee et al. 1993, 1995).
- The combination of rainfall and evapotranspiration (Vicente-Serrano et al. 2010a, b).
- Hydrological state variables-soil moisture (Agutu et al. 2017; Sheffield and Wood 2008).
- Satellite observations of groundwater storage, surface water extent, and terrestrial water storage (Ndehedehe et al. 2021a; Agutu et al. 2019; Thomas et al. 2017).
- Water fluxes, e.g., river discharge, altimetry, etc. (Ndehedehe et al. 2019a, 2020b; Shukla and Wood 2008).
- Global positioning system-based crustal deflection (Ferreira et al. 2018).
- Surface vegetation using Normalized Difference Vegetation Index (NDVI) or enhanced vegetation index (Agutu et al. 2020, 2017).

In arid countries like Australia and the African Sahel where surface water is considerably limited, these indicators could be very handy to explore the impacts of droughts and rainfall variability on its hydrological systems with the aim of improving understanding of land water storage responses to drought conditions. The limits of groundwater sustainability in most advanced economies have been reached through development, largely caused by overexploitation of groundwater resources. Hence, small changes in recharge, discharge, or ground water storage could have substantial economic or environmental consequences (e.g., Green et al. 2011). It is therefore critical to understand the response of hydrological stores like groundwater and soil moisture to extreme events such as droughts. This will help reveal the implications of reduced or limited rainfall on threats to water and food security. This is particularly important in semi-arid regions where rainfall is highly variable and very limited.

Adaptation plans that underpin impact assessment from climate change and cushion its effects on mankind must include not only the prudent management and allocation of water resources, but the development of indicators and metrics to aid the assessment and evaluation of meaningful or significant changes in rainfall on surface hydrology. The development of indicators to measure water use efficiency as well as the level of water stress to assess progress towards the 2030 Agenda for Sustainable Development Goals was emphasized by Vanham et al. (2018). The use of *Water Vulnerability Index* was earlier advocated as key to supporting effective water management through governance (Sullivan 2011). In addition to the broad use of indicators in other sectors, a plethora of hydrological indicators have been introduced to capture the complexities of climate change, droughts, and land surface conditions, as well as oceanic processes on hydrological systems (e.g., Ndehedehe et al. 2021a). In the next section, we will discuss this further.

11.4.1 Impacts on Agricultural Systems

- **Australia**: In large semi-arid domains like the Murray Darling Basin in Australia, hydrological drought indicators were better predictors of hydrological stores such as groundwater and land water storage. Soil moisture and groundwater losses occurred 60% and about 44% of the time, respectively (2002–2017). This period (2002–2017) over Australia was characterized by prolonged drought (Millennium drought) and a big wet period (2010–2012). Furthermore, satellite observations of surface water extents confirmed that extreme events (droughts and floods) affect the distribution of surface water. Apart from hydrological systems, droughts also affect ecological and agricultural systems. As illustrated in (Fig. 11.3) drought indicators can capture impacts of drought events on food supply chain using annual crop production data. In the Murray Darling Basin, the relationship between climate data (drought indicators) and crop production was assessed using partial least square regression model (Ndehedehe et al. 2021a). As shown in Fig. 11.3, meteorological drought indicators were found to be excellent metrics that captured the variability in crop production. It is therefore important to carefully choose indicators, ensuring that they reflect the response of freshwater ecosystems (including groundwater) to climatic and anthropogenic constraints. This is part of a key process to guide investments in drought resilience and sustainable management of water resources.
- **East Africa**: Several combination of remotely sensed, in situ, and model products have been employed to characterize drought and assess the effectiveness of various drought indicators in topographically complex regions of Upper Greater Horn of Africa. The study by (Agutu et al. 2020) used multi-resolution data including precipitation (CHIRPS, GPCC, and CHIRP), soil moisture (ERA-Interim, MERRA-2, CPC, GLDAS, and FLDAS), vegetation condition index (VCI), and terrestrial water storage products from satellites and model outputs (GRACE and MERRA-2). Unlike the Murray Darling Basin where meteorological drought indicator showed the strongest relationship with crop production data, groundwater-based indicator performed exceptionally well, explaining between 92 and 99% of the variability in national annual crop production (Fig. 11.4). Soil moisture and vegetation-based products (FLDASN, FLDASV, CPC, VCI) and rainfall also accounted for high variabilities. There is an obvious evidence of good correlation between crop production (wheat and maize) and monthly rainfall anomalies in the region.

Drought impacts on crop yields will result in the loss of national income. This is typical for regions and countries that export large quantities of agricultural products or those where agricultural contributes substantially to their gross domestic product. In Australia, farms and freshwater ecosystems are increasingly being affected by frequent drought events. The adverse effects of such events and changes in climatic conditions on freshwater availability have reduced average profits by 35% for a typical cropping farm in Australia. This has resulted in national average loss in revenue (1.1 billion dollars per year based on 2015–2018 production levels) in the cropping industry (Hughes et al. 2019). The unprecedented impact of the Millennium

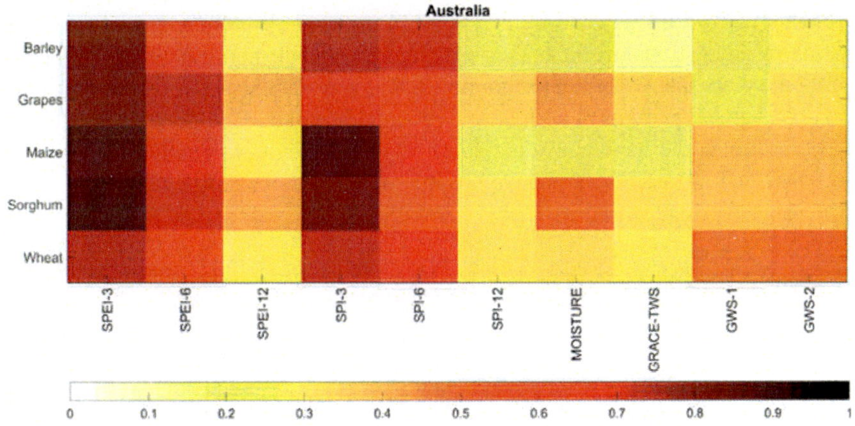

Fig. 11.3 Proportions of variability in national annual crop production data (barley, grapes, sorghum, maize, and wheat) (R^2) explained by various drought indices. The y-axis indicates crop while the x-axis represents the different indicators used in the partial least square regression model. *Source* Ndehedehe et al. (2021a)

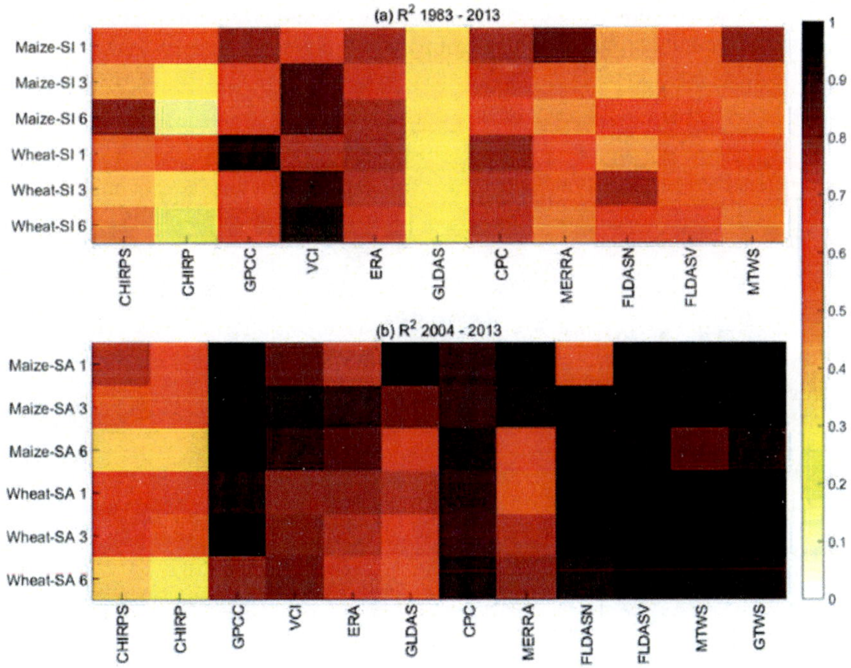

Fig. 11.4 Proportions of variability in national annual crop production data (maize and wheat) (R^2) explained by various drought indices (VCI, GPCC, ERA, CPC, FLDASN, and FLDASV). The y-axis indicates the crop (maize/wheat), standardized index-SI (for **a**), and standardized anomalies-SA (for **b**). The standardization time scales are 1, 3, and 6 for all indicators on the x-axis. *Source* Agutu et al. (2020)

Drought on Australia's economy is a good illustration of the relationship between hydrology and national income. This prolonged drought event resulted in more than 8 billion dollars spent on relief programmes. Water availability is an important factor that drives global patterns of agricultural production and trade (Odorico et al. 2020). Its variability could have considerable effects on agricultural goods and services. However, such variability can be well captured with various drought metrics.

11.4.2 Response of Hydrological Systems to Droughts

Studies have compared terrestrial water storage (TWS) and groundwater storage (GWS) with all cumulation scales of standardized precipitation index (SPI) and standardized precipitation evapotranspiration index (SPEI) (Fig. 11.5). This was undertaken over Australia as part of an assessment to understand the response of hydrological stores to climate variability. TWS shows moderately strong association with increasing SPEI aggregations on the east coast of Australia (Figs. 11.5a–c). SPEI-12 shows significant relationship with TWS in Queensland with correlation values mostly around 0.70, that is, excluding its extreme northern section (Fig. 11.5a). All SPEI and SPI cumulations were also compared with groundwater as with TWS. In

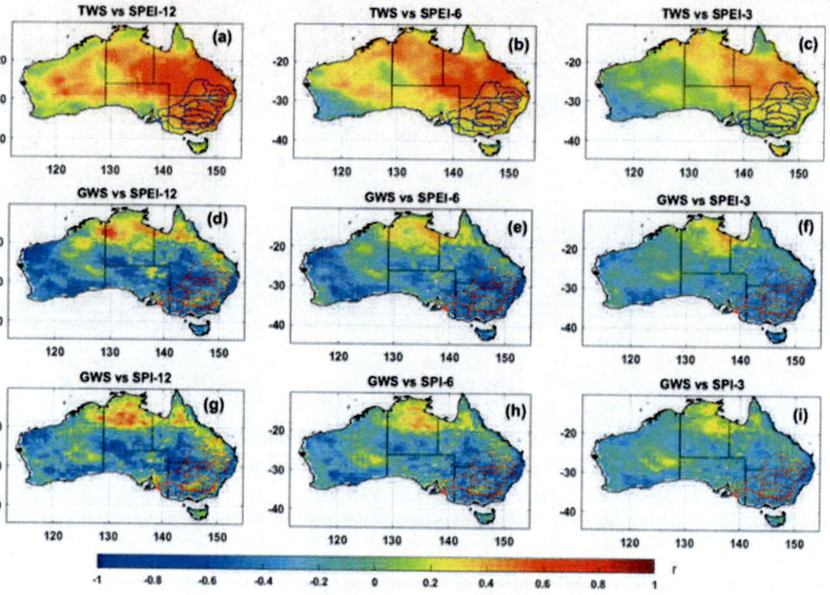

Fig. 11.5 Relationship of TWS and GWS with drought indicators (SPI and SPEI) based on Pearson's correlation. Spatial distribution of correlation coefficients for **a–c** TWS versus SPEI cumulations, **d–f** GWS versus SPEI cumulations, and **g–i** GWS versus SPI cumulations. *Source* Ndehedehe et al. (2021a)

the northern Australian territory, observed correlations between drought indicators and GWS are positive and negative in much of the region (Fig. 11.5d–i). The strong negative correlations emphasize the presence of considerable lags and non-linear relationships between these indicators and GWS. There is heavy reliance of groundwater in much of Australia. Superimposing this astronomical use of groundwater on the aridity experienced in the region is expected to result in non-linear hydrological processes as seen here. In other words, water use can be complex.

Groundwater-dependent ecosystems exist and water extraction for irrigation and those from resource extraction, e.g., mining activities are other forms of water use. Under prolonged drought, this usage will be exacerbated and recharge and discharge to streams and springs will be restricted. Moreover, it is noted that the relationship of GWS with drought indicators is stronger for those at higher cumulation scales (Fig. 11.5d and g). In practice, SPI-12/SPEI-12 are usually used to characterize hydrological drought, and in some cases groundwater drought. The observed relationship between groundwater and climate data (drought indicators) was linear in some sections of the wet-dry tropics of northern Australia. However, groundwater is strongly correlated with drought indicators (but this relationship is negative) in much of Australia where dependence on groundwater is astronomical and includes western, central, and the east coast. This negative correlation and opposite phase relationship (rainfall and groundwater) may be expected and is caused by recharge patterns, aquifer properties, and evapotranspiration rates.

11.4.3 Drought Impact on Food Security

Freshwater plays a key role in the sustenance of life on earth because of its uniqueness and irreplaceability. Humans make use of freshwater for several activities, including drinking, agriculture, and other economic activities. Among these, water use for agricultural purposes accounts for the most, claiming over 80% of human water consumption (Odorico et al. 2020). The economic significance of water in food security, crop production, and rural livelihood is often undervalued because most times it remains underpriced and users don't pay for its real value like oil. Additionally, most farmers pay for associated cost accrued in groundwater withdrawal and delivery. This uncontrolled use of freshwater resources, especially in crop production, is a major contributing factor to increased drought extents in some hydrological regions. There exists a strong interrelationship of food security with nutritional class and mental health. In cases of food insecurity, humans tend to consume very little foods, which may have very poor nutritional values. Studies have shown a strong relationship between food insecurity and psychiatric disorders, poor child development, as well as psychological distress (Friel et al. 2014).

Drought impact on food security. In rainfed-dependent systems like East Africa, the impacts of drought trigger famine cycles, resulting in lack of food and poverty (Agutu et al. 2017). Moreover, the drought of 2012 in the United States was characterized as multi-billion dollar agricultural disaster. Over 65% of the entire US region was affected by this drought, and using the Palmer Drought Index, studies showed it was the worst in the history of several drought events. A substantial reduction in some crops, such as grain corn (26–27%), sorghum (24%), soybeans (9%), was recorded by the U.S Department of Agriculture. This drought caused the loss of 87 million bushels of sorghum, 4 billion bushels of corn, and 170 million bushels of soybeans (Rippey 2015). Without the preparations for alternative water supply, especially for agriculture purposes in times of extreme drought, both terrestrial and aquatic habitats can suffer critical losses during such period. In the sections below, two region-specific examples of drought impacts on agricultural products (sugarcane and cocoa) are highlighted.

11.4.3.1 Impacts on Sugarcane

In 2015, the Food and Agricultural Organization of the United Nations reported a significant deterioration in agricultural produce over low-latitude tropical regions as a result of climatic fluctuations (FAO 2015). The Saccharum app (i.e., sugarcane) is a well-known subtropical plant having a long growing season. The productivity of sugarcane is significantly influenced by climatic fluctuations because it remains under the soil throughout the year (Francisco et al. 2014). In some countries, sugarcane is one of the main crops that support economic development, job creation as well as income generation. For example, Brazil extracts 35.6 million tonnes of sugar and 28.5 thousand cubic metres of ethanol from a total of 633.9 million tonnes of sugarcane. This therefore characterizes the region as the second largest ethanol producing country in the world, and also the largest sugarcane producer. The Brazilian centro-sul region accounts for 91.9% ethanol and 90.2% sugar production, as well as 90.4% sugarcane production, while the remainder of the production is accounted for by the north-northeastern region (de Medeiros Silva et al. 2019). According to the Brazilian Sugarcane Industry Association, the production of 6.7 million tonnes of sugarcane in the state of Paraiba alone generated 0.4 thousand cubic metres of ethanol and 0.1 million tonnes of sugar in the 2014/2015 harvest (UNICA 2015). Sugarcane, however, has shown significant susceptibility to climatic effects, which have hampered its growth and development (Alves and Marengo 2010). Drought and low temperatures affect sugarcane performance and even compromise the integrity of the final product (IPCC 2014).

11.4.3.2 Impacts on Cocoa Production

Cocoa farming is essential to the livelihood of an approximate 40–50 million people around the globe, including 5–6 million smallholder cocoa farmers who grow these crops. Cocoa farming is predominant in the West African region, where roughly 70% of the world's cocoa production originates (Schroth et al. 2016). The Benin republic is the only country not producing cocoa along the Guinea coast of West Africa. In 2011, West African countries like Sierra Leone, Cameroon, Ghana, Nigeria, and Ivory

Coast recorded cocoa beans as the most essential agricultural export by value, the second and third most essential for Guinea and Liberia, as well as Togo, respectively. Cocoa tree was imported from Brazil between the nineteenth and twentieth centuries into West Africa. Ever since, it has been mostly cultivated by small-scale farmers and is today regarded as a prototypal smallholder West African crop, which is different from the specie grown in Brazil and other Latin American regions where large cocoa farms are equally common (Clarence and Ruf 1996).

The West African region, however, is characterized by significant climatic perturbations due to notable fluctuations in the region's climate. These climatic impacts affect cocoa production and yield (Schroth et al. 2016), and can lead to little or no income generated from the cultivated cocoa. Given the poor income gotten from these cash crops, it becomes hard for farmers to afford the bills of external inputs from different experts, as well as technical help from fellow cocoa farmers in other regions, like South America regarding the management of climate change as regards to cocoa yield (Läderach et al. 2013). Majority of the West African cocoa regions are characterized by long dry seasons compared to majority of other cocoa producing regions (Schroth et al. 2016). Kotir (2011) reported a 30% decrease in annual rainfall of the West African savanna as a result of extreme drying of the climate over the second half of the twentieth century. Beside crops or vegetation, the impact of climate change also affected the forest zones (Ruf et al. 2015). During this period, some essential cocoa farms in the eastern forest belt of Ivory Coast became incongruous for the cultivation of cocoa (Ruf et al. 2015). Even though the dynamic nature of climatic conditions has brought about a positive change in the climatic condition of the West African region in the past decade (Ruf et al. 2015), other fluxes such as evapotranspiration rate, plant water demand, groundwater stress, and increased temperature changes are projected to result in an proliferated drought stress during the dry periods, thus leading to further deterioration of climatic conditions for cocoa in West Africa (Schroth et al. 2016).

11.4.4 Drought Impacts in Australia: An Example of the Millennium Drought

The millennium drought in Australia also known as the 'big dry' began in 1997 and lasted for about 12–13 years. The 'big dry' period was terminated by heavy rains in 2009 and 2010. These rains were ENSO-driven, causing large floods, which negatively affected properties and crops. During the millennium drought period, runoff and rainfall were restricted, leading to an increased occurrence of bush fires, dust storms, and widespread livestock and crop losses. The 'big dry' brought about several changes in the environment, affecting national income due to losses in agricultural products. There were also changes to water management framework of the continent, as well as protocols aimed at modernizing its water laws and institutions. The farmers were among the worst hit during this drought period. The agricultural sector contributes 2.5% of the total nation's economy and uses 66% of the total water supply. Prior to the end of the twentieth century, policies that supported irrigation practices were drafted to encourage and support settlements in sparsely settled and arid locations. As part of a distinctive approach to reform the economic sector in the 1990s, the Australian government started creating awareness to enlighten farmers about the inability of government to proceed with drought relief policies, which have been in existence. The 'big dry', however, brought about a change in these policies. The Australian government began an intervention programme where thousands of farmers were given incentives to cushion the effect of drought on farmers.

The effects of drought brought substantial changes to the Australian region over the past decade. For example, the sheep population in Australia once outnumbered the human population by a 10:1 ratio. However, this number has been substantially reduced to more than half over the past 10 years. Other products from the agricultural sectors such as rice and cotton production were also substantially affected. With the reoccurring policies on water restrictions, city dwellers developed creative means of reusing water, which birthed new industries aimed at water conservation using different techniques, such as building large desalination plants, and national aqueduct. Drought frequency also enhanced the sensitization of the average Australian on delicate nature of the country's ecosystem, and the dynamism of climate change over the entire sub-region. Also, drought effects were also considerably pronounced on the Australian water courses and landscapes. This resulted in drying of river beds and lakes. The deficits in soil moisture over this period as well as the lowering of water tables also contributed to the discernable effect on the continent's terrestrial and aquatic ecosystems (Ndehedehe et al. 2021a). It also contributed to the increase in dust storms and bushfires were more frequent over the past years. Moreover, the productivity of floodplain ecosystems in the wet-dry tropical areas of the Gulf regions was affected due to reduced upstream flow (Ndehedehe et al. 2021b, a).

The effect of the 'big dry' had a lasting effect on Australia's surface water, groundwater, and food production. For example, due to drought and over extraction of surface water in the Murray Darling Basin during the 'big dry', flows to the sea were limited in 2002. This was a significant problem for the entire region as it resulted in poor health of the river, which then prompted calls for restoration through better water management policies. The drought period equally affected groundwater reservoir as an unprecedented amount of groundwater withdrawal was noticed and recorded during this period (van Dijk et al. 2013; Leblanc et al. 2012). This affected both the availability and quality of freshwater resources in the region. It also had a profound effect on rural economies, and invigorated changes in the agricultural sector too. However, the drought period also brought about novel measures in tackling freshwater allocation such as the building spree of desalination plants, which increased the awareness of Australians on water conservation techniques and strategies. These numerous lessons derived from the recent years of the region's drought regime could prove to be of interest to other parts of the world having challenges with water management and effective allocation.

11.5 Understanding Droughts and the Impacts of Global Climate

We live in the era of complex interaction in landscape processes. Non-climatic factors such as large-scale human water groundwater extraction can interfere with the propagation process of drought characteristics and its intensity, e.g., from meteorological to hydrological (Kubiak-Wójcicka and Bąk 2018). Arguably, land-use change and increased water abstraction play critical role in future changes to drought. But climate change is also expected to drive large-scale ocean-atmospheric processes that will trigger large changes in rainfall patterns (e.g., Kiem et al. 2016). Climatic factors and human activities increase the frequency, variability, and even duration of drought intensity. Improving contemporary understanding of the processes that drives drought variability on a large-scale is thus an ongoing discussion. With the strong land–atmosphere interactions of semi-arid tropical areas (e.g., Koster et al. 2004), other non-trivial challenges in drought studies are those related to identifying a suitable drought indicator that captures large-scale processes more effectively. Indicators and metrics are obviously at the centre of numerous assessments. This includes those used to designate human development and environmental performance. The changing patterns of local precipitation, land surface, and the different characteristics of catchment in some regions have therefore warranted the need for ideal metrics to explore drought intensity. Following the work of Ndehedehe et al. (2020a) over the Sahel, this section explores some of these issues, discussing the complexities of droughts and their indicators in drought assessment and diagnosing the influence of global climate and climate variability indices.

11.5.1 The Complexities of Droughts

From preceding sections, we established the concept and causes of drought-temporary decrease in water availability caused by prolonged rainfall deficit (e.g., Van Loon et al. 2016; Spinoni et al. 2014). However, droughts can result from complex interactions due to prevailing land surface conditions, human activities, and even population growth can lead to increased water demand that makes drought impact complex. Contemporary understanding of these interactions, particularly those related to multi-scale global climate phenomena and how they contribute to drought evolutions, requires functional drought metrics that capture these processes. This is particularly essential because the propagation of drought and how it develops can be complex based on the factors mentioned earlier.

Apparently, when rainfall deficits in arid climates are prolonged, this would restrict surface water availability as well as freshwater stocks (Ndehedehe et al. 2021a). However, increase in evapotranspiration, aquifer type, changes in land use, soil type, temperature, and intensive groundwater abstraction will further drive the characteristics of drought and alter its propagation process (e.g., Cook et al. 2018; Ndehedehe et al. 2018; Kiem et al. 2016; Van Loon 2015; Van Loon et al. 2014). In the case of groundwater, the factors mentioned above interfere with drought characteristics and propagation through processes like infiltration, reduced discharge to springs, and flow dynamics, among others. In humid regions, catchment stores, e.g., reservoirs, soil column, and groundwater, can cause a delay in the propagation of drought signals or even its absence due to prolonged reservoir memory in hydrology (Ndehedehe et al. 2019a). In the case of typically dry or arid regions where rainfall is considerably limited, increasing temperature could impact on drought intensity. It is therefore important to measure droughts using multi-scale indicators or metrics incorporating water budget variables. It is for this reason that multiple climatic variables have been embedded in drought metrics like the Palmer Drought Severity Index (PDSI, Palmer 1965). The PDSI, SPI, and SPEI have been widely used along with other drought metrics. But the need to better account for actual water demand in irrigated landscapes resulted in the substitution of reference evapotranspiration in the SPEI with crop evapotranspiration Moorhead et al. (2015). And the use of NDVI and other vegetation variants like the enhanced vegetation index as drought indicators are also suitable for agricultural drought monitoring. Moreover, an earlier review of the Sahelian droughts (Epule et al. 2014) argued that metrics and predictive models incorporating the composite influence of climate and human actions are essential in effective characterization of droughts.

Precipitation deficits are obviously the primary cause of droughts, which in turn affects other hydrological stores (e.g., groundwater and surface water). Nevertheless, strong variations in temperature and changes in land cover states can interact with these deficits in complex ways that leads to different outcomes (e.g., identifying onset and drought duration) when these indicators (SPEI and SPI) are used to measure drought in specific locations. For instance, evaluation of different indicators in East Africa using satellite, reanalysis, and model data for agricultural drought assessment by Agutu et al. (2017) supports this notion. Their study which was undertaken in East Africa highlighted the suitability of these products to capture agricultural drought and emphasized the dissimilarity between meteorological-based indicators and those from vegetation and state variables like soil moisture, in capturing drought onset. The feedback processes of desertification, land-use change, and the complex water use mechanisms in vegetation in arid regions during the dry season (e.g., Ndehedehe et al. 2019b; Oguntoyinbo 1986) could contribute to drought complexity.

Even topography can also affect the capability of these indicators in measuring drought. In the characterization of drought in the Upper Greater Horn of Africa, the performance of various drought indicators (precipitation, soil moisture, and vegetation indices) was largely dependent on changes in topography (Agutu et al. 2020). In other words, uncertainties in the use of these indicators to measure drought characteristics and impact (areas under drought) were significantly different depending on the elevation of the said region. However, the uncertainties in some data can also limit their performance in drought characterization. For instance, can model-reanalysis soil moisture products in high-elevation areas be trusted? Agutu et al. (2021) evaluated the uncertainty of several of such products using the Greater Horn of Africa as a test bed. They argued that the magnitude of uncertainties in these products was higher in high-elevation areas of the Ethiopian highlands compared to low-elevation areas in Kenya, Uganda, and Tanzania. However, the magnitude of uncertainties in some soil moisture products was also linked to the quality of precipitation data used in their synthesis as well as difference in model physical dynamics. On the one hand, this outcome is important to inform choices of soil moisture products in agricultural drought assessment and improve knowledge on interactions between land surface and atmosphere. On the other hand, drought complexity in regions where limited in situ data are available for the calibration of model simulations and reanalysis data could even be higher.

Generally, drought events across the globe are perceived as a function of changes in global climate. The ENSO phenomenon has been linked to several extreme events such as floods and droughts on a variety of spatial scales (e.g., Ndehedehe et al. 2019a; Kiem et al. 2016; Rippey 2015; van Dijk et al. 2013; Sheffield and Wood 2008; Kumar et al. 2013; White et al. 2004). Major drought events in North America, West Africa, and East Asia have been linked to strong anomalies in tropical sea surface temperatures and other oceanic hot spots (e.g., Dai 2011). The perception of droughts in Australia confirm that the factors driving drought can be multi-scaled and its impacts could vary across regions (van Dijk et al. 2013). Several indices of oceanic variability trigger drought and the impacts and severity of drought on hydro-ecological systems can differ across Australian territories (Ndehedehe et al. 2021a). Climate teleconnection patterns, such as ENSO, Interdecadal Pacific Oscillation, and Indian Ocean Dipole (IOD), are connected to droughts in Australia, Africa, and Asia. This happens through direct causes such as higher temperatures, precipitation deficits, and the combined influence of precipitation and evaporation on hydrological stores such as soil moisture and groundwater (Ndehedehe et al. 2020a; Ferreira et al. 2018; Agutu et al. 2017; Kiem et al. 2016).

To assess drought impacts more effectively under a rising global warming scenario, it is pertinent to rely on multi-scale drought metrics derived from a combination of water budget and biophysical variables (e.g., vegetation, evapotranspiration, etc.). This is essential to enhance our understanding of evolutionary drought patterns. This is in terms of its variability, propagation, and impacts on water availability and ecosystem services. Moreover, the inclusion of multiple climate variables in drought analysis provides opportunity to inform more accurately on the role of various human activities (e.g., land-use change) in exacerbating drought events (Ndehedehe et al. 2019a). This could be helpful in optimizing predictive models, especially those that seek to account for the composite influence of natural variability and anthropogenic climate change on drought characteristics (Kiem et al. 2016). Notably, the performance of drought indicators is location-specific, depending on climate, topography, and human actions. Hence, the more reason to assess the suitability of multi-scaled indicators (synthesized from multiple climate variables) to improve perspective on drought events, their characteristics, and complexity.

11.5.2 Optimizing Drought Indicators

Some studies (e.g., Ndehedehe et al. 2016a; Agnew and Chappell 1999) have challenged the use of normalized rainfall deviations in drought assessment. This is mostly because the spatial aggregation of rainfall departures hides the underlying key local variation and drivers. The other challenge is that the averaging period for this data restricts drought characterization and interpretation. With this in mind, drought indicators that leverage multiple climate variables like the SPEI have been advocated (e.g., Ndehedehe et al. 2016a; AghaKouchak 2015; Farahmand and AghaKouchak 2015). The argument around this emphasizes drought metrics based on single climate variable (rainfall) do not possess the capacity to adequately describe or capture key characteristics of drought such as onset, duration, severity, frequency, and intensity (e.g., Ndehedehe et al. 2016b; Hao and AghaKouchak 2014). The limitations in the use of meteorological-based indicators like rainfall alone have been stressed (Nguvava et al. 2019). The effectiveness of SPEI in measuring drought severity and frequency under increased global warming has been reported (Shiru et al. 2019). The illustration in Fig. 11.1 further emphasizes the importance and robustness of multivariate or multi-scaled drought indicators (Sect. 11.3.1.1).

SPI is a well-known drought metric and it relies on long-term record of monthly precipitation. It's computationally simple, thus enabling its usage in a broad range of applications, including water management to provide interventions through emergency relief funds. However, there are contrasting outcomes as to the relationship between SPI and the SPEI in humid and arid regions (Ndehedehe et al. 2020a; Labudová and Takáč 2017). These outcomes take various forms. This can be in terms of temporal variations in predicted extents of various drought intensities as shown in Fig. 11.6. It is crucial to use drought metrics that capture land surface conditions and their interactions more effectively in arid regions where temperature is considerably high. On the other hand, the differences between SPI and the SPEI could also suggest evapotranspiration rates or both rainfall and temperature are important climate variables to improve our understanding of the variability or intensity of drought events in semi-arid domains. For example, in the West African Sahel, SPEI indicated that the notorious drought of the early 1980s was more extreme in central Niger (between 1984 and 1985) unlike SPI (Ndehedehe et al. 2020a). Even though estimated extents of droughts based on both indicators could be similar in some semi-arid catchments, the illustration in Fig. 11.6 over the Sahel region shows substantial difference in aerial extents of different drought intensities (moderate, severe, and extreme) for some periods. The argument that the multiscalar nature of SPEI enhances its utility in the characterization of various drought intensities (Vicente-Serrano et al. 2010b) is indirectly supported by the fact that sugarcane production in semi-arid region of Brazil showed significant relationship with temperature and rainfall (de Medeiros Silva et al. 2019).

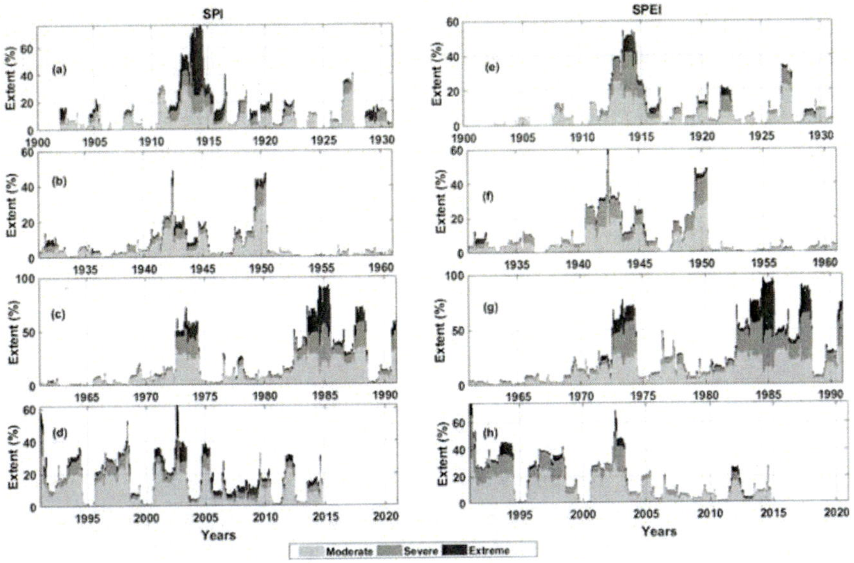

Fig. 11.6 Drought extents over the West African Sahel between 1901 and 2014. The estimated drought extents for **a–d** SPI and **e–h** SPEI are described in terms of their intensity (moderate, severe, and extreme). *Source* Ndehedehe et al. (2020a)

11.5.3 Coupled Variability of Climate Modes and Droughts

Some earlier studies have identified the coupled relationship of several teleconnections, e.g., ENSO and IOD, AMO, with rainfall (e.g., Ndehedehe et al. 2019a; Diatta and Fink 2014; Okonkwo 2014; Paeth et al. 2012). The influence of these climate modes, including that of ENSO on rainfall variability in East and other African subregion, has been widely received (e.g., Gizaw and Gan 2017; Andam-Akorful et al. 2017; Ujeneza and Abiodun 2015). This is because precipitation extremes are usually associated with circulation features and climate teleconnection indices of the Pacific and the Atlantic, land-ocean contrast among others (see, e.g., Diatta and Fink 2014; Nicholson and Grist 2001). The links between drought metrics with indices of oceanic variability and tropical oceans are further evidence that confirm the roles of low-frequency oscillations in inducing drought. The North Atlantic SSTs, for instance, are known for their roles in decadal and centennial timescale circulation and droughts in North America (Feng et al. 2011). On the other hand, observed variability in monsoon droughts in India has been linked to tropical SST (Kumar et al. 2013). The influence of multi-scale teleconnections (e.g., quasi-biennial, interannual, and inter-decadal, etc.) associated with global SST and sea-level pressure (SLP) have influence on the occurrence of drought episodes in Australia White et al. (2004). In East Africa, local circulations, such as the tropical easterly jet, low-level Westerlies, monsoons, and the Turkana jet, have induced the occurrence of droughts (Haile et al. 2019). The interactions of AMM, ENSO, and QBO with drought patterns in other regions re-emphasizes these teleconnections as important physical mechanisms that drives meteorological processes and drought evolutions. Meaning drought events can be predicted by the synthesis of these global multi-scale climate teleconnection elements. The mechanistic understanding of this multi-scale climate influence nonetheless, and how they drive precipitation extremes are interesting research directions.

11.6 Conclusion

Drought as a natural hazard affects water availability, food supplies, the productivity of freshwater ecosystems, and, more generally, human lives. It is different in many ways to other natural hazards, especially given it capability to affect larger area than anticipated compared to other natural hazards. Drought is not just an event limited to regions where the climate is typically arid. Arguably, exposure to various drought intensity and frequency can vary spatially. However, the most humid regions of the world are not drought-proofed or exempted from its occurrence and impacts. Rainfall is a major source of freshwater availability, supporting the sustainability of hydrological, ecological, and agricultural systems. But what happens when the rains do not come? Understanding the large-scale variability of hydrological stores such as groundwater and surface water and how they respond to drought events is therefore critical. Early warning systems and new drought monitoring frameworks are required to optimize risk-management solutions and strengthen national drought policies and drought preparedness. These solutions underpin mitigation strategies and complementary interventions such as insurance programmes that could help cushion the impacts of drought risk. They can also aid opportunities for increase in resilience and adaptation. In this chapter, drought concepts, classifications, and complexities, their diagnosis, and a framework to assess their impacts on freshwater ecosystems have been detailed. The importance of drought metrics and the capability of several indicators employed in drought characterization and assessing influence of global climate and climate variability indices are also discussed and demonstrated.

References

AghaKouchak A (2015) A multivariate approach for persistence-based drought prediction: application to the 2010–2011 East Africa drought. J Hydrol 526:127–135. https://doi.org/10.1016/j.jhydrol.2014.09.063

AghaKouchak A, Mirchi A, Madani K, et al (2021) Anthropogenic drought: definition, challenges and opportunities. Rev Geophys e2019RG000683. https://doi.org/10.1029/2019RG000683

Agnew C, Chappell A (1999) Drought in the sahel. GeoJournal 48(4):299–311. https://doi.org/10.1023/A:1007059403077

Agutu N, Awange J, Ndehedehe C, Kirimi F, Kuhn M (2019) GRACE-derived groundwater changes over Greater Horn of Africa: temporal variability and the potential for irrigated agriculture. Sci Total Environ 693:133467. https://doi.org/10.1016/j.scitotenv.2019.07.273

Agutu N, Awange J, Ndehedehe C, Kirimi F, Mwaniki M (2021) Understanding uncertainty of model-reanalysis soil moisture within Greater Horn of Africa (1982–2014). J Hydrol 603:127169. https://doi.org/10.1016/j.jhydrol.2021.127169

Agutu N, Awange J, Ndehedehe C, Mwaniki M (2020) Consistency of agricultural drought characterization over Upper Greater Horn of Africa (1982–2013): topographical, gauge density, and model forcing influence. Sci Total Environ 709:135149. https://doi.org/10.1016/j.scitotenv.2019.135149

Agutu N, Awange J, Zerihun A, Ndehedehe C, Kuhn M, Fukuda Y (2017) Assessing multi-satellite remote sensing, reanalysis, and land surface models' products in characterizing agricultural drought in East Africa. Remote Sens Environ 194:287–302. https://doi.org/10.1016/j.rse.2017.03.041

Alves LM, Marengo J (2010) Assessment of regional seasonal predictability using the PRECIS regional climate modeling system over South America. Theoret Appl Climatol 100(3):337–350. https://doi.org/10.1007/s00704-009-0165-2

Ali A, Lebel T (2009) The Sahelian standardized rainfall index revisited. Int J Climatol 29:1705–1714. https://doi.org/10.1002/joc.1832

Andam-Akorful S, Ferreira V, Ndehedehe CE, Quaye-Ballard J (2017) An investigation into the freshwater variability in West Africa during 1979–2010. Int J Climatol 37(S1):333–349. https://doi.org/10.1002/joc.5006

Asong ZE, Wheater HS, Bonsal B, Razavi S, Kurkute S (2018) Historical drought patterns over Canada and their teleconnections with large-scale climate signals. Hydrol Earth Syst Sci 22(6):3105–3124. https://doi.org/10.5194/hess-22-3105-2018

Bazrafshan J, Hejabi S, Rahimi J (2014) Drought monitoring using the multivariate standardized precipitation index (MSPI). Water Resour Mangagement 28:1045–1060. https://doi.org/10.1007/s11269-014-0533-2

Cao G, Han D, Song X (2014) Evaluating actual evapotranspiration and impacts of groundwater storage change in the North China Plain. Hydrol Process 28(4):1797–1808. https://doi.org/10.1002/hyp.9732

Cardoso JF (1999) High-order contrasts for independent component analysis. Neural Comput 11:157–192

Chen T, de Jeu R, Liu Y, van der Werf G, Dolman A (2014) Using satellite based soil moisture to quantify the water driven variability in NDVI: a case study over mainland Australia. Remote Sens Environ 140:330–338. https://doi.org/10.1016/j.rse.2013.08.022

Clarence-Smith WG, Ruf F (2018) Cocoa pioneer fronts: the historical determinants: In: Cocoa pioneer fronts since 1800: the role of smallholders, planters and merchants. Palgrave Macmillan UK, pp 1–22. https://doi.org/10.1007/978-1-349-24901-5_1

Cook BI, Mankin JS, Anchukaitis KJ (2018) Climate change and drought: from past to future. Curr Clim Chang Rep 4(2):164–179. https://doi.org/10.1007/s40641-018-0093-2

Dai A (2011) Drought under global warming: a review. Wiley Interdiscip Rev: Clim Chang 2(1):45–65. https://doi.org/10.1002/wcc.81

de Medeiros Silva WK, de Freitas GP, Coelho Junior LM, de Almeida Pinto PAL, Abrahão R (2019) Effects of climate change on sugarcane production in the state of Paraíba (Brazil): a panel data approach (1990–2015). Clim Change 154(1):195–209. https://doi.org/10.1007/s10584-019-02424-7

Diatta S, Fink AH (2014) Statistical relationship between remote climate indices and West African monsoon variability. Int J Climatol 34(12):3348–3367. https://doi.org/10.1002/joc.3912

Diaz V, Corzo G, Lanen HAV, Solomatine DP (2019) Spatiotemporal drought analysis at country scale through the application of the STAND Toolbox. In: Spatiotemporal Analysis of Extreme Hydrological Events, pp 77–93. https://doi.org/10.1016/B978-0-12-811689-0.00004-5

Epule ET, Peng C, Lepage L, Chen Z (2014) The causes, effects and challenges of Sahelian droughts: a critical review. Reg Environ Change 14(1):145–156. https://doi.org/10.1007/s10113-013-0473-z

FAO (2015) Food and agriculture organization of the united nations. Climate change and food systems: global assessments and implications for food security and trade. Food and Agricultural Organisation of the United Nations. https://www.fao.org/3/i4332e/i4332e.pdf 20 Sept 2020

FAO (2016) Climate change and food security: risks and responses. Food and Agricultural Organisation of the United Nations. http://www.fao.org/3/a-i5188e.pdf 20 Sept 2016

Farahmand A, AghaKouchak A (2015) A generalized framework for deriving nonparametric standardized drought indicators. Adv Water Resour 76:140–145. https://doi.org/10.1016/j.advwatres.2014.11.012

Feng L, Hu C, Chen X, Li R, Tian L, Murch B (2011) MODIS observations of the bottom topography and its inter-annual variability of Poyang Lake. Remote Sens Environ 115(10):2729–2741. https://doi.org/10.1016/j.rse.2011.06.013

Ferreira V, Montecino H, Ndehedehe C, Heck B, Gong Z, Westerhaus M, de Freitas S (2018) Space-based observations of crustal deflections for drought characterization in brazil. Sci Total Environ 644:256–273. https://doi.org/10.1016/j.scitotenv.2018.06.277

Ferreira VG, Montecino HD, Ndehedehe CE, del Rio RA, Cuevas A, de Freitas SRC (2019) Determining seasonal displacements of Earth's crust in South America using observations from space-borne geodetic sensors and surface-loading models. Earth, Planets and Space 71(1):84. https://doi.org/10.1186/s40623-019-1062-2

Ferreira VG, Ndehedehe CE, Montecino HC, Yong B, Yuan P, Abdalla A, Mohammed AS (2019b) Prospects for imaging terrestrial water storage in South America using daily GPS observations. Remote Sens 11(6). https://doi.org/10.3390/rs11060679

Francisco PR, Santos D, de Lima ER, de Moraes Neto JM, da Trindade HC (2014) Mapping of pedological potential of soils for cultivation of the sugar cane (Saccharum Spp) in Paraíba state using geotechnologies. J Hyperspectral Remote Sens 7(5):306–314

Friel S, Berry H, Dinh H, O'Brien L, Walls HL (2014) The impact of drought on the association between food security and mental health in a nationally representative Australian sample. BMC Public Health 14(1102):1–11. https://doi.org/10.1186/1471-2458-14-1102

Fu Y, Argus DF, Landerer FW (2015) GPS as an independent measurement to estimate terrestrial water storage variations in Washington and Oregon. J Geophys Res: Solid Earth 120(1):552–566. https://doi.org/10.1002/2014JB011415

Gizaw MS, Gan TY (2017) Impact of climate change and El Niño episodes on droughts in sub-Saharan Africa. Clim Dyn 49(1):665–682. https://doi.org/10.1007/s00382-016-3366-2

Goodarzi M, Abedi-Koupai J, Heidarpour M, Safavi HR (2016) Development of a new drought index for groundwater and its application in sustainable groundwater extraction. J Water Resour Plan Manag 142(9):04016032. https://doi.org/10.1061/(ASCE)WR.1943-5452.0000673

Green TR, Taniguchi M, Kooi H, Gurdak JJ, Allen DM, Hiscock KM, Treidel H, Aureli A (2011) Beneath the surface of global change: impacts of climate change on groundwater. J Hydrol 405(3):532–560. https://doi.org/10.1016/j.jhydrol.2011.05.002

Haile GG, Tang Q, Leng G, Jia G, Wang J, Cai D, Sun S, Baniya B, Zhang Q (2020) Long-term spatiotemporal variation of drought patterns over the Greater Horn of Africa. Sci Total Environ 704:135299. https://doi.org/10.1016/j.scitotenv.2019.135299

Haile GG, Tang Q, Sun S, Huang Z, Zhang X, Liu X (2019) Droughts in East Africa: causes, impacts and resilience. Earth Sci Rev 193:146–161. https://doi.org/10.1016/j.earscirev.2019.04.015

Hao Z, AghaKouchak A (2014) A nonparametric multivariate multi-index drought monitoring framework. J Hydrometeorol 15(1):89–101. https://doi.org/10.1175/JHM-D-12-0160.1

Hayes M, Svoboda M, Wall N, Widhalm M (2011) The lincoln declaration on drought indices: universal meteorological drought index recommended. Drought Mitigation Center Faculty Publications. http://digitalcommons.unl.edu/droughtfacpub/14. 9th June 2020

Hayes MJ (2007) Drought indices. Intermountain west climate summary. www.colorado.edu/climate/iwcs/archive/IWCS-2007-July-feature.pdf. 9 Dec 2015

Hayes MJ, Svoboda MD, Wilhite DA, Vanyarkho OV (1999) Monitoring the 1996 drought using the standardized precipitation index. Bull Am Meteorol Soc 80:429–438. https://doi.org/10.1175/1520-0477(1999)080$<$0429:MTDUTS$>$2.0.CO;2

Hughes N, Galeano D, Hattfield-Dodds (2019) The effects of drought and climate variability on Australian farms. Australian Bureau of Agricultural and Resource Economics and Sciences, Canberra. http://doi.org/10.25814/5de84714f6e08. 16th Oct 2020

IPCC (2014) Climate change 2014 synthesis report summary chapter for policymakers. Intergovernmental Panel on Climate Change. https://www.ipcc.ch/site/assets/uploads/2018/05/SYR_AR5_FINAL_full_wcover.pdf. 12th Aug 2021

Ivits E, Horion S, Fensholt R, Cherlet M (2014) Drought footprint on European ecosystems between 1999 and 2010 assessed by remotely sensed vegetation phenology and productivity. Glob Change Biol 20(2):581–593. https://doi.org/10.1111/gcb.12393

Jolliffe IT (2002) Principal component analysis, 2nd edn. Springer series in statistics. Springer, New York

Karl TR, Koscielny AJ (1982) Drought in the United States: 1895–1981. J Climatol 2(4):313–329. https://doi.org/10.1002/joc.3370020402

Kiem AS, Johnson F, Westra S, van Dijk A, Evans JP, O'Donnell A, Rouillard A, Barr C, Tyler J, Thyer M, Jakob D, Woldemeskel F, Sivakumar B, Mehrotra R (2016) Natural hazards in Australia: droughts. Clim Change 139(1):37–54. https://doi.org/10.1007/s10584-016-1798-7

Koster RD, Dirmeyer PA, Guo Z, Bonan G, Chan E, Cox P et al (2004) Regions of strong coupling between soil moisture and precipitation. Science 305(5687):1138–1140. https://doi.org/10.1126/science.1100217

Kotir JH (2011) Climate change and variability in Sub-Saharan Africa: a review of current and future trends and impacts on agriculture and food security. Environ Dev Sustain 13(3):587–605. https://doi.org/10.1007/s10668-010-9278-0

Kubiak-Wójcicka K, Bąk B (2018) Monitoring of meteorological and hydrological droughts in the Vistula basin (Poland). Environ Monit Assess 190(11):691. https://doi.org/10.1007/s10661-018-7058-8

Kumar KN, Rajeevan M, Pai D, Srivastava A, Preethi B (2013) On the observed variability of monsoon droughts over India. Weather Clim Extrem 1:42–50. https://doi.org/10.1016/j.wace.2013.07.006

Labudová L LM, Takáč J (2017) Comparison of SPI and SPEI applicability for drought impact assessment on crop production in the Danubian Lowland and the East Slovakian lowland. Theor Appl Climatol 128:491–506. https://doi.org/10.1007/s00704-016-1870-2

Läderach P, Martinez-Valle A, Schroth G, Castro N (2013) Predicting the future climatic suitability for cocoa farming of the world's leading producer countries, Ghana and Côte d'Ivoire. Clim Chang 119(3):841–854. https://doi.org/10.1007/s10584-013-0774-8

Leblanc M, Tweed S, Van Dijk A, Timbal B (2012) A review of historic and future hydrological changes in the Murray-Darling Basin. Glob Planet Chang 80–81:226–246. https://doi.org/10.1016/j.gloplacha.2011.10.012

Loon AFV, Laaha G (2014) Hydrological drought severity explained by climate and catchment characteristics. J Hydrol. https://doi.org/10.1016/j.jhydrol.2014.10.059

Madani K, Shafiee-Jood M (2020) Socio-hydrology: a new understanding to unite or a new science to divide? Water 12(7). https://doi.org/10.3390/w12071941

Mao Y, Wu Z, He H, Lu G, Xu H, Lin Q (2017) Spatio-temporal analysis of drought in a typical plain region based on the soil moisture anomaly percentage index. Sci Total Environ 576:752–765. https://doi.org/10.1016/j.scitotenv.2016.10.116

McKee TB, Doeskin NJ, Kieist J (1993) The relationship of drought frequency and duration to time scales. In: Conference on applied climatology, american meteorological society, Boston, Massachusetts, pp 179–184. www.ccc.atmos.colostate.edu/relationshipofdroughtfrequency.pdf. 27 June 2014

McKee TB, Doeskin NJ, Kieist J (1995) Drought monitoring with multiple time scales. In: Conference on Applied Climatology, American Meteorological Society, Boston, Massachusetts, pp 233–236. www.southwestclimatechange.org/node/911. 13 July 2014

Mishra AK, Singh VP (2010) A review of drought concepts. J Hydrol 391:202–216. https://doi.org/10.1016/j.jhydrol.2010.07.012

Moorhead JE, Gowda PH, Singh VP, Porter DO, Marek TH, Howell TA, Stewart B (2015) Identifying and evaluating a suitable index for agricultural drought monitoring in the Texas high plains. JAWRA J Am Water Resour Assoc 51(3):807–820. https://doi.org/10.1111/jawr.12275

Ndehedehe CE (2019) The water resources of tropical West Africa: problems, progress and prospect. Acta Geophys 67(2):621–649. https://doi.org/10.1007/s11600-019-00260-y

Ndehedehe CE, Agutu N, Ferreira VG, Getirana A (2020) Evolutionary drought patterns over the Sahel and their teleconnections with low frequency climate oscillations. Atmos Res 233:104700. https://doi.org/10.1016/j.atmosres.2019.104700

Ndehedehe CE, Agutu NO, Okwuashi OH, Ferreira VG (2016) Spatio-temporal variability of droughts and terrestrial water storage over Lake Chad Basin using independent component analysis. J Hydrol 540:106–128. https://doi.org/10.1016/j.jhydrol.2016.05.068

Ndehedehe CE, Anyah RO, Alsdorf D, Agutu NO, Ferreira VG (2019) Modelling the impacts of global multi-scale climatic drivers on hydro-climatic extremes (1901–2014) over the Congo basin. Sci Total Environ 651:1569–1587. https://doi.org/10.1016/j.scitotenv.2018.09.203

Ndehedehe CE, Awange J, Kuhn M, Agutu N, Fukuda Y (2017) Analysis of hydrological variability over the Volta river basin using in-situ data and satellite observations. J Hydrol: Reg Stud 12:88–110. https://doi.org/10.1016/j.ejrh.2017.04.005

Ndehedehe CE, Awange JL, Corner R, Kuhn M, Okwuashi O (2016) On the potentials of multiple climate variables in assessing the spatio-temporal characteristics of hydrological droughts over the Volta Basin. Sci Total Environ 557–558:819–837. https://doi.org/10.1016/j.scitotenv.2016.03.004

Ndehedehe CE, Ferreira VG, Agutu NO (2019) Hydrological controls on surface vegetation dynamics over West and Central Africa. Ecol Ind 103:494–508. https://doi.org/10.1016/j.ecolind.2019.04.032

Ndehedehe CE, Ferreira VG, Agutu NO, Onojeghuo AO, Okwuashi O, Kassahun HT, Dewan A (2021) What if the rains do not come? J Hydrol 595:126040. https://doi.org/10.1016/j.jhydrol.2021.126040

Ndehedehe CE, Ferreira VG, Onojeghuo AO, Agutu NO, Emengini E, Getirana A (2020) Influence of global climate on freshwater changes in Africa's largest endorheic basin using multi-scaled indicators. Sci Total Environ 737:139643. https://doi.org/10.1016/j.scitotenv.2020.139643

Ndehedehe CE, Haile GG, Agutu NO, Ferreira VG, Getirana A, Okwuashi O (2020) Hydrological hotspots of climatic influence in Brazil: a two-step regularization approach. Atmos Res 246:105116. https://doi.org/10.1016/j.atmosres.2020.105116

Ndehedehe CE, Okwuashi O, Ferreira VG, Agutu NO (2018) Exploring evapotranspiration dynamics over Sub-Sahara Africa (2000–2014). Environ Monit Assess 190(7):400. https://doi.org/10.1007/s10661-018-6780-6

Ndehedehe CE, Onojeghuo AO, Stewart-Koster B, Bunn SE, Ferreira VG (2021) Upstream flows drive the productivity of floodplain ecosystems in tropical queensland. Ecol Ind. https://doi.org/10.1016/j.ecolind.2021.107546

Nguvava M, Abiodun BJ, Otieno F (2019) Projecting drought characteristics over East African basins at specific global warming levels. Atmos Res 228:41–54. https://doi.org/10.1016/j.atmosres.2019.05.008

Nicholson S, Grist J (2001) A conceptual model for understanding rainfall variability in the West African Sahel on interannual and interdecadal timescales. Int J Climatol 21:1733–1757. https://doi.org/10.1002/joc.648

Odorico P, Chiarelli DD, Rosa Lorenzo EA (2020) The global value of water in agriculture. In: Proceedings of the national academy of sciences 117(36):21985–21993. https://doi.org/10.1073/pnas.2005835117

Oguntoyinbo JS (1986) Drought prediction. Clim Chang 9(1):79–90. https://doi.org/10.1007/BF00140527

Okonkwo C (2014) An advanced review of the relationships between Sahel precipitation and climate indices: a wavelet approach. Int J Atmos Sci 2014:11. https://doi.org/10.1155/2014/759067

Paeth H, Fink A, Pohle S, Keis F, Machel H, Samimi C (2012) Meteorological characteristics and potential causes of the 2007 flood in sub-Saharan Africa. Int J Climatol 31:1908–1926. https://doi.org/10.1002/Joc.2199

Palmer W (1965) Meteorological drought. Weather. Bureau, U.S. Department of Commerce, Washington. Research paper 45. D. C. (58 p)

Rippey BR (2015) The U.S. drought of 2012. Weather Clim Extrem 10(Part A):57–64. https://doi.org/10.1016/j.wace.2015.10.004

Ruf F, Schroth G, Doffangui K (2015) Climate change, cocoa migrations and deforestation in West Africa: what does the past tell us about the future? Sustain Sci 10(1):101–111. https://doi.org/10.1007/s11625-014-0282-4

Ruf F, Schroth G (2015) Introduction-Economic and ecological aspects of diversification of tropical tree crops. Economics and ecology of diversification. Springer, Dordrecht, pp 1–40

Santos JAF, Pulido-Calvo I, Portela MM (2010) Spatial and temporal variability of droughts in Portugal. Water Resour Res 46(3):W03503. https://doi.org/10.1029/2009WR008071

Schroth G, Läderach P, Martinez-Valle AI, Bunn C, Jassogn L (2016) Vulnerability to climate change of cocoa in West Africa: patterns, opportunities and limits to adaptation. Sci Total Environ 556:231–241. https://doi.org/10.1016/j.scitotenv.2016.03.024

Seghieri J, Carreau J, Boulain N, De Rosnay P, Arjounin M, Timouk F (2012) Is water availability really the main environmental factor controlling the phenology of woody vegetation in the central Sahel? Plant Ecol 213(5):861–870. https://doi.org/10.1007/s11258-012-0048-y

Sheffield J, Wood EF (2008) Global trends and variability in soil moisture and drought characteristics, 1950–2000, from observation-driven simulations of the terrestrial hydrologic cycle. J Clim 21(3):432–458. https://doi.org/10.1175/2007JCLI1822.1

Shiru MS, Shahid S, Chung E-S, Alias N (2019) Changing characteristics of meteorological droughts in Nigeria during 1901–2010. Atmos Res 223:60–73. https://doi.org/10.1016/j.atmosres.2019.03.010

Shukla S, Wood AW (2008) Use of a standardized runoff index for characterizing hydrologic drought. Geophys Res Lett 35(2):L02405. https://doi.org/10.1029/2007GL032487

Spinoni J, Naumann G, Carrao H, Barbosa P, Vogt J (2014) World drought frequency, duration, and severity for 1951–2010. Int J Climatol 34(8):2792–2804. https://doi.org/10.1002/joc.3875

Sullivan CA (2011) Quantifying water vulnerability: a multi-dimensional approach. Stoch Env Res Risk Assess 25(4):627–640. https://doi.org/10.1007/s00477-010-0426-8

Tadesse, T., Senay, G. B., Berhan, G., Regassa, T., and Beyene, S. (2015). Evaluating a satellite-based seasonal evapotranspiration product and identifying its relationship with other satellite-derived products and crop yield: A case study for Ethiopia. Int J Appl Earth Obs Geoinformation 40(Supplement C):39–54. https://doi.org/10.1016/j.jag.2015.03.006

Thomas BF, Famiglietti JS, Landerer FW, Wiese DN, Molotch NP, Argus DF (2017) GRACE groundwater drought index: evaluation of California central Valley groundwater drought. Remote Sens Environ 198(Supplement C):384–392. https://doi.org/10.1016/j.rse.2017.06.026

Uddameri V, Singaraju S, Hernandez E (2019) Is Standardized Precipitation Index (SPI) a useful indicator to forecast groundwater droughts?—insights from a Karst aquifer. JAWRA J Am Water Resour Assoc 55(1):70–88. https://doi.org/10.1111/1752-1688.12698

Ujeneza EL, Abiodun BJ (2015) Drought regimes in southern Africa and how well GCMS simulate them. Clim Dyn 44:1595–1609. https://doi.org/10.1007/s00382-014-2325-z

UNICA (2015) Brazilian sugarcane industry association. Crop production (Produção por safra). UNICA. https://unica.com.br/en/sugarcane-sector/sugar/

van Dijk AIJM, Beck HE, Crosbie RS, de Jeu RAM, Liu YY, Podger GM, Timbal B, Viney NR (2013) The millennium drought in southeast Australia (2001–2009): natural and human causes and implications for water resources, ecosystems, economy, and society. Water Resour Res 49(2):1040–1057. https://doi.org/10.1002/wrcr.20123

Van Loon AF (2015) Hydrological drought explained: WIREs. Water 2:359–392. https://doi.org/10.1002/wat2.1085

Van Loon AF, Kumar R, Mishra V (2017) Testing the use of standardised indices and GRACE satellite data to estimate the European 2015 groundwater drought in near-real time. Hydrol Earth Syst Sci 21(4):1947–1971. https://doi.org/10.5194/hess-21-1947-2017

Van Loon AF, Stahl K, Di Baldassarre G, Clark J, Rangecroft S, Wanders N, Gleeson T, Van Dijk AIJM, Tallaksen LM, Hannaford J, (2016) Drought in a human-modified world: reframing drought definitions, understanding, and analysis approaches. Hydrol Earth Syst Sci 20(9):3631–3650. https://doi.org/10.5194/hess-20-3631-2016

Van Loon AF, Tijdeman E, Wanders N, Van Lanen HA, Teuling AJ, Uijlenhoet R (2014) How climate seasonality modifies drought duration and deficit. J Geophys Res: Atmos 119(8):4640–4656. https://doi.org/10.1002/2013JD020383

Vanham D, Hoekstra A, Wada Y, Bouraoui F, de Roo A, Mekonnen M, van de Bund W, Batelaan O, Pavelic P, Bastiaanssen W, Kummu M, Rockström J, Liu J, Bisselink B, Ronco P, Pistocchi A, Bidoglio G (2018) Physical water scarcity metrics for monitoring progress towards SDG target 6.4: an evaluation of indicator 6.4.2 "level of water stress". Sci Total Environ 613–614:218–232. https://doi.org/10.1016/j.scitotenv.2017.09.056

Vicente-Serrano SM, Beguería S, López-Moreno JI (2010) A multiscalar drought index sensitive to global warming: the standardized precipitation evapotranspiration index. J Clim 23(7):1696–1718. https://doi.org/10.1175/2009JCLI2909.1

Vicente-Serrano SM, Beguería S, López-Moreno JI, Angulo M, El Kenawy A (2010b) A new global 0.5° gridded dataset (1901–2006) of a multiscalar drought index: comparison with current drought index datasets based on the palmer drought severity index. J Hydrometeorol 11(4):1033–1043. https://doi.org/10.1175/2010JHM1224.1

Wang F, Wang Z, Yang H, Zhao Y (2018) Study of the temporal and spatial patterns of drought in the Yellow River basin based on SPEI. Sci China Earth Sci 61(8):1098–1111. https://doi.org/10.1007/s11430-017-9198-2

White WB, Gershunov A, Annis JL, McKeon G, Syktus J (2004) Forecasting Australian drought using Southern Hemisphere modes of sea-surface temperature variability. Int J Climatol 24(15):1911–1927. https://doi.org/10.1002/joc.1091

Wilhite DA, Mannava Sivakumar MV, Pulwarty R (2014) Managing drought risk in a changing climate: the role of national drought policy. Weather Clim Extrem 3:4–13. https://doi.org/10.1016/j.wace.2014.01.002

Xu L, Chen N, Zhang X (2019) Global drought trends under 1.5 and 2 °c warming. Int J Climatol 39(4):2375–2385. https://doi.org/10.1002/joc.5958

Yang Y, Long D, Guan H, Scanlon BR, Simmons CT, Jiang L, Xu X (2014) GRACE satellite observed hydrological controls on interannual and seasonal variability in surface greenness over mainland Australia. J Geophys Res Biogeosci 119(12):2245–2260. https://doi.org/10.1002/2014JG002670

Yirdaw SZ, Snelgrove K, Agboma C (2008) GRACE satellite observations of terrestrial moisture changes for drought characterization in the Canadian Prairie. J Hydrol 356:84–92

Yu M, Li Q, Hayes MJ, Svoboda MD, Heim RR (2014) Are droughts becoming more frequent or severe in China based on the standardized precipitation evapotranspiration index: 1951–2010? Int J Climatol 34(3):545–558. https://doi.org/10.1002/joc.3701

Zhang Q, Qi T, Singh VP, Chen YD, Xiao M (2015) Regional frequency analysis of droughts in china: a multivariate perspective. Water Resour Manage 29(6):1767–1787. https://doi.org/10.1007/s11269-014-0910-x

Part VII
Human Interventions on Hydrology

Chapter 12
Anthropogenic Influence on Terrestrial Hydrology

Abstract Global water demand has increased at more than twice the rate of human population. The impacts resulting from increased water demands are now surpassing the effects of climate change. Across several regions of the world, much of the observed surface water depletion and groundwater loss are driven by human water withdrawals. For instance, even though the contraction of the Lake Chad's surface area was primarily caused by strong declines in precipitation, water withdrawals from the Lake for irrigation purposes exacerbated the effects of these declines, resulting in about 90% decline of the Lake's surface area. Increasing human population comes with additional pressure on existing water resources, and if freshwater is considered a finite resource, the rise in human population could be a major problem to its availability, especially in arid and semi-arid regions of the world. This chapter identifies some of these challenges, including complex hydrological processes caused by human water management efforts. Additionally, this chapter discusses some large reservoir systems where the impacts of human activities like dam constructions and climate change have negatively affected surface water hydrology and its variability. Specific examples of anthropogenic influence like, development of water infrastructures, water diversion projects, and groundwater extraction on terrestrial hydrology and hydro-ecological systems are discussed.

12.1 Introductory Remarks

Anthropogenic activities such as land cover change, water transfer, groundwater extraction, and damming of rivers have contributed substantially to freshwater variability (e.g., Getirana et al. 2020; Ndehedehe et al. 2017; Moore and Williams 2014; Margariti et al. 2019). These factors (i.e., climate and anthropogenic activities) drive terrestrial hydrology, contributing to major freshwater changes in unique ways (Ndehedehe 2019). For example, reservoirs and dams constructed along most of the world's major rivers have dramatically changed the seasonal flow rates and trends in stream flow rates (e.g., Dai et al. 2009; Yang et al. 2004; Lammers et al. 2001). Observed groundwater declines across several regions of the world have been linked to intensive human water extraction (e.g., Ojha et al. 2018; Chen et al. 2016).

C. Ndehedehe, *Satellite Remote Sensing of Terrestrial Hydrology*,
https://doi.org/10.1007/978-3-030-99577-5_12

From the influence of changing frequency of extreme events and unmitigated human actions to rise in human population, freshwater-related problems, especially its variability and predictability could be complex. For instance, a rise in human population could drive water scarcity in some regions while extreme events such as drought restricts agricultural production. In Africa where agriculture uses an estimated 88% of available water resources (Oyebande and Odunuga 2010), or in semi-arid countries where surface water is considerably limited, sustainable water resources development is thus crucial and is perceived as part of risk-based decisions and mitigation strategies to ensure food security. Such efforts, which are usually preceded by changes in water policies ultimately, leads to increased investments in hydraulic and water infrastructures, among others (Ndehedehe et al. 2020b; Vorosmarty et al. 2010). These management decisions and new policy directions can result in water supply disruption as was in Botswana and hydroelectric load shedding in Zambia (Siderius et al. 2018).

Moreover, hydrological changes in riverine ecosystems as well as the alteration of natural flow regimes, which provide valuable ecological services, such as water supply, hydro-power generation, flood control, recreation and navigation (Kennard et al. 2010) are the aftermaths of water governance decisions. The proposal to transfer water from the Congo basin to nourish the desiccated Lake Chad is another good example of a mitigation strategy to cushion the impact of drought. Apart from cultural and technological issues to be addressed in this kind of project, alteration of natural flow and potential hydro-ecological problems are likely. Under a changing climate, the attempt to ensure freshwater sustainability could inadvertently create a web of environmental problems that complicate knowledge on climate change and hydrological processes. On the one hand, the scientific framework that supports the development of environmental flow standards is thus crucial. On the other hand, improving knowledge on these processes and how human interventions interact with natural systems to induce hydro-climatic changes is important. Due to limited hydrologic data and lack of formal water allocation rights for both surface and groundwater in several emerging economies, water management is rather complicated. This chapter, therefore, focuses on novel contributions of remote sensing to hydrological research that supports meaningful decisions on water resources planning. As an illustration of anthropogenic influence on terrestrial hydrology, the impact of large reservoir systems on freshwater changes is discussed. This chapter identifies and discusses some hydrological systems where the impacts of human activities and climate change have negatively affected surface water hydrology and its variability.

12.2 Human Influence on Surface Water Hydrology

12.2.1 Lake Volta

As the world's largest man-made lake by surface area (8502 km^2), Lake Volta is a key source of freshwater, livelihood and hydropower, providing substantial economic benefits to approximately 20 million people within the Volta river basin in Western Africa (e.g., Ndehedehe et al. 2017). The Lake's freshwater also provides highly valued several ecosystem functions, including shelter for fish, flood water storage, drought relief, agriculture, recreation and tourism, and means of navigation within Ghana, its host country. The Lake Volta was built in 1964 and currently has an estimated capacity of 150 km^3 with an installed hydro-power generation capacity of about 912 MW (Owusu et al. 2008; Gyau-Boakye 2001). The construction of the Akosombo dam (Fig. 12.1) in the Volta river basin was accompanied by large floods that affected local communities, resulting in the creation of Lake Volta. The Lake is located at the lower Volta basin, a low elevation area that is naturally connected to the Volta river system and other small freshwater tributaries. The Volta river system comprises the Oti, Black Volta, and White Volta rivers and together with other smaller tributaries provide the perennial flows that sustain the Lake Volta. The natural connection of the Lake with the Volta river system leads to relatively strong

Fig. 12.1 The Akosombo Dam in Lake Volta, Ghana, West Africa

inter-annual variability in surface water hydrology around the precinct of the lower Volta (Ndehedehe et al. 2017).

The construction of the Akosombo dam (Fig. 12.1) came with a plenitude of hydrology-related environmental impacts. Prior to the construction of the dam, the Volta river was a key vehicle for sediments transport to the shoreline. However, Cheng (1980) reported that about 99.5% of the river drainage basin was blocked by the dam and with very limited supply of sand from the low-lying areas of coastal plains to the shoreline. Apart from sediment load changes, reservoir-induced seismicity, morphological and climatic changes have been reported as other resulting negative environmental impacts of the Akosombo dam (e.g., Gyau-Boakye 2001). For example, land cover transitions such as the conversion of vegetated areas to water bodies led to substantial changes in surface albedo and thereby altered the local heat budget. This and changes in wind speed induce changes in local weather conditions (e.g., Gyau-Boakye 2001) are few critical environmental challenges associated with the dam.

Moreover, despite having a surface area that corresponds to only 2% of the entire Volta basin, Lake Volta has a stronger hydrological signal (variations in surface water) compared to Lake Victoria with relatively larger surface area of approximately 68,800 km^2 (Moore and Williams 2014). This strong variation in surface water level is caused by water impoundment of the Akosombo dam and explains the considerable change in water levels over the Lake (Ndehedehe et al. 2016a). This contribution of the Lake to hydrological changes in the corresponding basin is very substantial even during drought periods. Contributions in the field of remote sensing hydrology to water resources development have shown how deliberate human interventions like the Akosombo dam can induce hydrological variability. Ndehedehe et al. (2017) found that water ponding by the Akosombo dam in the Lake Volta accounted for about 41% of the increasing trend in GRACE-TWS despite an apparent significant decline in precipitation between 2002 and 2014.

12.2.2 *Lake Victoria*

Lake Victoria is the second largest freshwater lake (68,000 km^2) in the world and similar to the Lake Volta is an important source of freshwater and livelihood for about 30 million people living within the riparian countries of the corresponding basin (e.g., Awange et al. 2013; Ogutu-Ohwayo et al. 1997). Lake Victoria is one of the African Great Lakes located in Eastern Africa, that provides ecological benefits, including shelter for unique and diverse fish faunas. (e.g., Ogutu-Ohwayo et al. 1997). For more than four decades now, Lake Victoria is well known for its hydro-power potential, providing an estimated 90% of Uganda's hydropower, and a significant fraction for Burundi and Rwanda (Getirana et al. 2020). Its socio-economic and ecological value and role in modulating local climate is well documented and unlike Lake Volta, has been widely studied (e.g., Getirana et al. 2020; Awange et al. 2013; Ogutu-Ohwayo et al. 1997). Recently, the study by Getirana et al. (2020) used computational models

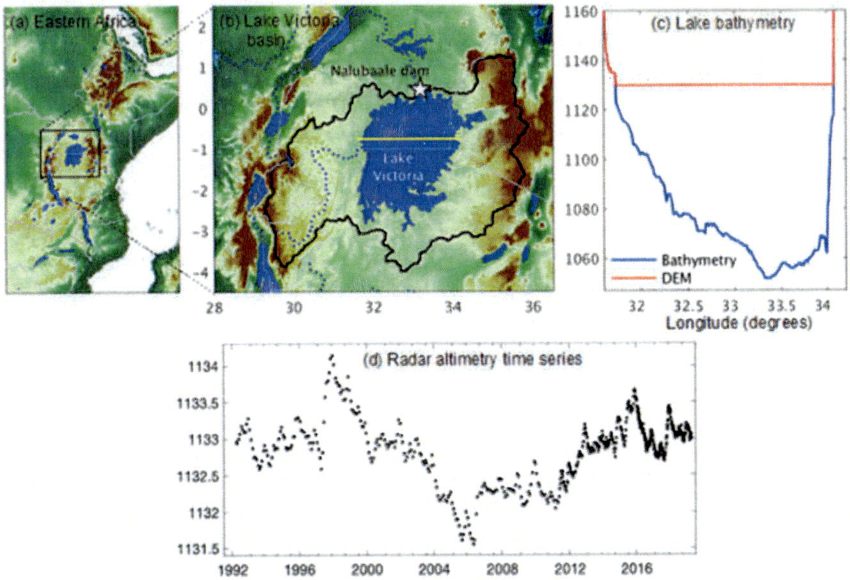

Fig. 12.2 Impact of hydro-power dam operation on Lake Victoria's freshwater storage variability. *Source* Getirana et al. (2020)

and a novel satellite-based reservoir operation scheme to quantify human impacts on Lake Victoria dynamics. Figures 12.2a–b show the location of Lake Victoria in Eastern Africa and the vicinity of the Nalubaale dam. The cross-sectional elevation profile in meters at 0.75° S latitude of the lake derived from satellite-based topography and ground-based bathymetry, and the lake's water level fluctuations from radar altimetry time series (meters) for 1992–2019 are indicated in Fig. 12.2c–d. Getirana et al. (2020) combined several satellite measurements from multiple sensors onboard Topex-Poseidon, Jason series, Envisat, GEOSTAT Follow-On, Saral and Sentinel 3A with river routing schemes to show that reservoir operation impact on water elevation, extent, storage and outflow, and controls changes in the lakes water storage. They showed that reservoir operation contributed substantially to water elevation, extent, storage and outflow, with lake water storage being 20% higher under dam control. Several transitions in the aerial extents of Lake Victoria have occurred in the last 30 years and both climatic and anthropogenic influence are arguably key drivers. The satellite altimetry-derived water level fluctuations show considerable and steady decline between 1998 and 2005 (Fig. 12.2d). While the lake can be sensitive to natural conditions (rainfall variability), human water management as seen in the operations of the Nalubaale dam contributes substantially to changes in the Lake's water storage. The footprint of this operation is evidenced in the strong hydrological changes observed over the Lake Victoria basin in which Lake Victoria contributes 64% to basin-wide variations in terrestrial water storage (Getirana et al. 2020).

12.2.3 Lake Urmia

The Lake Urmia in Iran is the largest inland saltwater body in the Middle East. It is an iconic permanent hypersaline lake in the world and well known for its vast area (613,253 Ha), rich aquatic ecosystems, and unique natural and ecological features (Zarghami 2011; Eimanifar and Mohebbi 2007). With an increasing human population (4.4 million) in the Urmia Lake basin, the Lake becomes a major source of freshwater availability for agriculture and domestic purposes (Delju 2013; Eimanifar and Mohebbi 2007). However, with intense agricultural development and rapid urbanization, there has been a substantial loss in the surface water extent of the Lake as well as decrease in groundwater level around the corresponding basin (Tourian et al. 2015; Zarghami 2011). Satellite measurements of the lake showed a loss of about 34 cm/yr in water levels during the 2002–2014 period and about 70% (220 km^2/yr) loss in its surface extent during the same period has been observed and largely attributed to the impacts of human activities (Tourian et al. 2015). The construction of causeway in the lake for transportation affected flow patterns and with the impacts of climate change, a rise in salinity of the lake was observed (Zarghami 2011; Eimanifar and Mohebbi 2007). A unanimous position on the key driver of desiccation of the Lake is the construction of a causeway in the middle of the Lake, segmenting it into two-northern and southern sections with only 1500 m gap for the movement of water in between the two sections (Eimanifar and Mohebbi 2007). However, other factors driving the drying of the lake, include increased groundwater abstraction for agriculture, dam construction and climate change (e.g., Tourian et al. 2015). Model simulation also indicated that the composite influence of excessive withdrawal of surface water resources and climate change, and the construction of four dams accounted for approximately 65 and 25% of the observed desiccation in Lake Urmia, respectively (Hassanzadeh et al. 2012).

The illustration in Fig. 12.3 shows a change detection of Lake Urmia as observed from the Thematic Mapper sensor aboard Landsat 5. Comparing the Landsat images of 1985 (left) and 2010 (right), there is a substantial change in the Lake's surface extent. Between 1985 and 2010 the Lake shrank by 60% primarily due to the combined impacts of dams on feeder streams, expanded use of groundwater, and prolonged drought. Along the edges of the 2010 image (right) are Light blue tones, which show shallow water and large salt deposits (Fig. 12.3), which has affected the refugia for migratory waterfowl and aquatic biodiversity. Moreover, in terms of hydrological challenges and key drivers, Lake Urmia bears some similarity with Lake Chad. First, they are both inland freshwater bodies located in arid regions and second, the composite influence of climate change and anthropogenic factors have contributed considerably to their dramatic decline. As with Lake Chad where inflows to the lake substantially reduced due to water resources development along the Chari-Logon river systems, (Delju 2013) reported that 35 dams were constructed on 21 rivers flowing to the Lake Urmia in response to increased water demand for agriculture and other applications. Moreover, the receding Lake Urmia has also

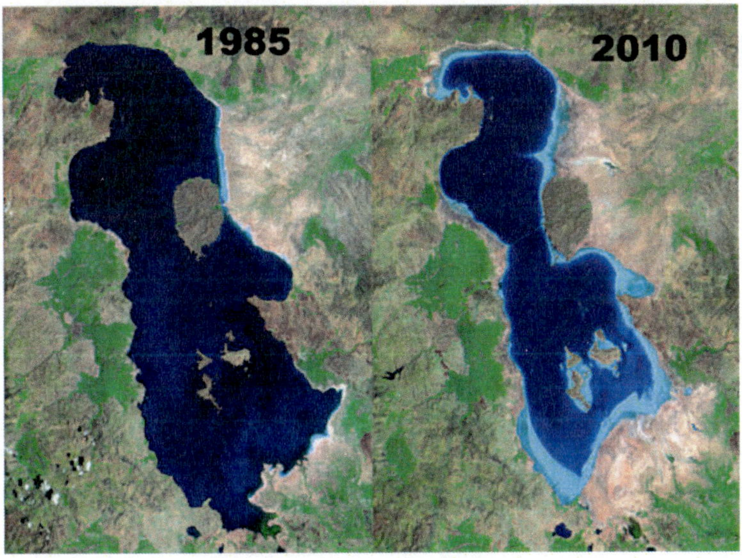

Fig. 12.3 Lake Urmia impacted by human water use. *Source* NASA/USGS Landsat missions

contributed to declines in groundwater resources and very much with permanent water loss conditions like the Lake Chad.

12.2.4 Lake Chad

The contraction of the Lake Chad's surface area was primarily caused by strong declines in regional precipitation patterns, restricting inflow from Chari-Logone river systems, which provide about 95% freshwater of the Lake. However, as indicated in several studies (see, e.g., Ndehedehe et al. 2016b; Birkett 2000; Coe and Foley 2001), water withdrawals from the Lake for irrigation purposes during water deficit periods compounded the effects of extreme droughts, resulting in about 90% decline of the Lake's surface area (Fig. 12.4). Apparently, this underscores the role of human activities is not only exacerbating the impacts of climate variability but in also reshaping the stability of Lake's ecosystem. The human-modified droughts of the Lake Chad basin are typical of the Anthropocene, where various forms of human activities impacts on catchment storage, soil properties, and hydrological processes, thereby modifying hydrological drought severity. Van Loon et al. (2016), for example, argued that urbanization impacts recharge and infiltration rates while soil moisture is influenced by deforestation, afforestation, desertification, and agricultural practices through evapotranspiration. In view of the aforementioned case studies, it can be reconciled that the combined effects of climate variations and human interventions conjugated with the well-known strong land-atmosphere coupling or interaction in the region (e.g.,

Fig. 12.4 The Lake Chad basin showing sections of the Lake Chad

Boone et al. 2009; Koster et al. 2004), will unarguably result in strong and profound influence on water resource systems and hydrological processes.

12.2.5 Large-Scale Water Diversion Projects

Inter-basin water transfer projects are other forms of human response to freshwater problems. This approach to water resources management are implemented to address insufficient water availability. A prominent water diversion project is the South-to-North Water Diversion project (SNWDP) in North China Plain and is of course the largest in China (Liu et al. 2020b; Zhang et al. 2020). The North China Plain is one of the world's hotspots of freshwater scarcity where groundwater depletion has been widely reported (Vorosmarty et al. 2010; Long et al. 2020). The water diversion from the south where the Yangtze River basin is located to the dry deficient North was part of key water management strategy to addressing water shortage in North China Plain where water used for agriculture and industrial applications is considerably high. Following the completion of this project in 2014, the litany of emerging studies has shown some positive feedback associated with the transition of surface water to North China Plain. For example, while water stress in Beijing and Tianjin was reduced (Liu et al. 2020b), the diversion also accounted for approximately 40% of total groundwater storage recovery during the 2006–2018 period in Beijing (Long et al. 2020). Apart from other perceived impacts on the ecosystem and environment (Guo et al. 2020), this water diversion project contributed to the recovery of groundwater levels in North China Plain (Zhang et al. 2020; Long et al. 2020). Similar ambitious water transfer project is being proposed for the desiccated Lake Chad in West Africa, which is expected to benefit from water transferred from the humid Congo basin. The Lake Chad basin is semi-arid with high evapotranspiration annual rates (Ndehedehe et al. 2018) and can we anticipate a similar response from groundwater?

12.2.6 Impacts on River Systems

Rivers contribute to human development and provide a chain of ecosystem services, such as recreational, tourism, commercial fisheries and support for a range of terrestrial and aquatic biodiversity. However, from rising population and demand for food to rapid urbanization, many rivers are currently under threat from numerous and intensifying anthropogenic influence. Irrigation, dam constructions, and inter water basin transfer are some anthropogenic activities stressing river ecosystems across the globe. Apart from contributing to the complexity of lake ecosystems (Ndehedehe 2019), such impacts are also known to have negatively affected river flow regimes, floodplain productivity, wetland hydrology and its connectivity and aquatic biodiversity (Ndehedehe et al. 2020c; Ndehedehe 2020). Maintaining regular flows and seasonal inundation of floodplains, which facilitates lateral connectivity, will ensure ecosystem productivity of instream ecosystems that underpins productive freshwater fisheries (e.g., Ndehedehe et al. 2020a; Balcombe and Arthington 2009; Thorp and Delong 1994; Hoeinghaus et al. 2009). This allows organisms to move into previously inaccessible habitats for feeding, foraging and reproduction (Ward and Stanford 1995). But human alteration of these natural flow regimes will lead to altered floodplain dynamics including declines in connectivity between floodplains and river channels (Górski et al. 2012; Bunn and Arthington 2002). Human threats on global rivers are well known (Vorosmarty et al. 2010), and in South and Southeast Asia, anthropogenic influence such as irrigation on rivers have been reported. For example, more than 50% of rivers were reported to have been affected by human activities with about 15% rise in human pressure during the 1990–2014 period, and observed hotspots included eastern Pakistan and northern India (Liu et al. 2020a). Moreover, the ecological impacts of damming rivers and development of water infrastructure have been studied and their threats to riverine ecosystems include impacts on the;

- movement and migration patterns of aquatic animals (Carpenter-Bundhoo et al. 2020),
- aquatic biodiversity in streams and riverine systems (e.g., Kingsford et al. 2014; Bunn and Arthington 2002),
- flow connectivity,
- commercial fish catch (Broadley et al. 2020),
- integrated physical processes that sustain different levels of organisms (e.g., primary and quaternary producers) (e.g., Bunn et al. 2006; Bunn and Arthington 2002).

12.3 Human Water Extraction

12.3.1 Groundwater and Irrigation

Water withdrawals in the form of groundwater extraction for irrigation and other applications also trigger significant changes in subsurface hydrology (Fig. 12.5). Over the Saharan aquifers, the extraction of fossil groundwater decreased terrestrial water storage (e.g., Ahmed et al. 2014). As revealed by satellite gravity observations, severe groundwater depletion in the last 15 years has been linked to extensive extraction for irrigation purposes (Famiglietti 2014). Arid regions of the world and Asia (e.g., India) are major hotspots of human water extraction. In the Middle East, satellite observations revealed an approximate loss of 17 mm/yr equivalent water height of groundwater (or 91 km^3 in volume) between January 2003 and December 2009 (Voss et al. 2013). Groundwater loss over India is alarming. The total estimated groundwater depletion in India ranges from 122 to 199 billion metre cube. The Northwest India is one of the sub-regions where observed groundwater losses remain very pronounced, with an estimated depletion rates of approximately 20.4 Gigatonne (Gt)/yr between January 2003 and December 2012 (Chen et al. 2014). Globally, there is broad reliance on groundwater for several purposes, including food security, eco-system functioning, and sustaining important natural and cultural values, as well as helping human adaptation to climate change (Taylor et al. 2013; Green et al. 2011). This over-dependence on groundwater has led to strong changes in groundwater in several regions of the world where considerable declines of up to a few hundred meters have been reported (Chen et al. 2014; Scanlon et al. 2012).

A good example is Australia where dependence on groundwater surpasses other developed economies and continues to increase due to limited availability of surface water. Water used for irrigation agriculture in Australia represents about 60% of Australia's annual groundwater withdrawals. Ecosystem services, mining, manufacturing, and household water supply in several Australian territories depend absolutely on groundwater. The intensity and increased need for groundwater withdrawals in the region resulted in substantial changes in groundwater (Leblanc et al. 2009). Given Australia's reliance on groundwater resources, there is a long history of groundwater management in several of its territories. From the Water Act of 1912 to water policies in the 1970 and 1980s, these were introduced to regulate human water use of groundwater (New South Wales Department of Planning, Industry and Environment, dpie.nsw.gov.au). For instance, the rapid expansion of the agriculture and increasing competition for water resources led to irrigation licences and allocation of water rights, among others. There is also a long history of groundwater development for pastoral use in the Great Artesian basin regions of New South Wales, which started in late 1870s shortly after the discovery of artesian groundwater at Kallara station west of Bourke. This groundwater development resulted in the Artesian Wells Act 1897, enabling groups of settlers to request government support to construct an artesian bore to serve their collective properties (dpie.nsw.gov.au). Some recent management of groundwater systems includes the Cap and Pipe the Bores program, which

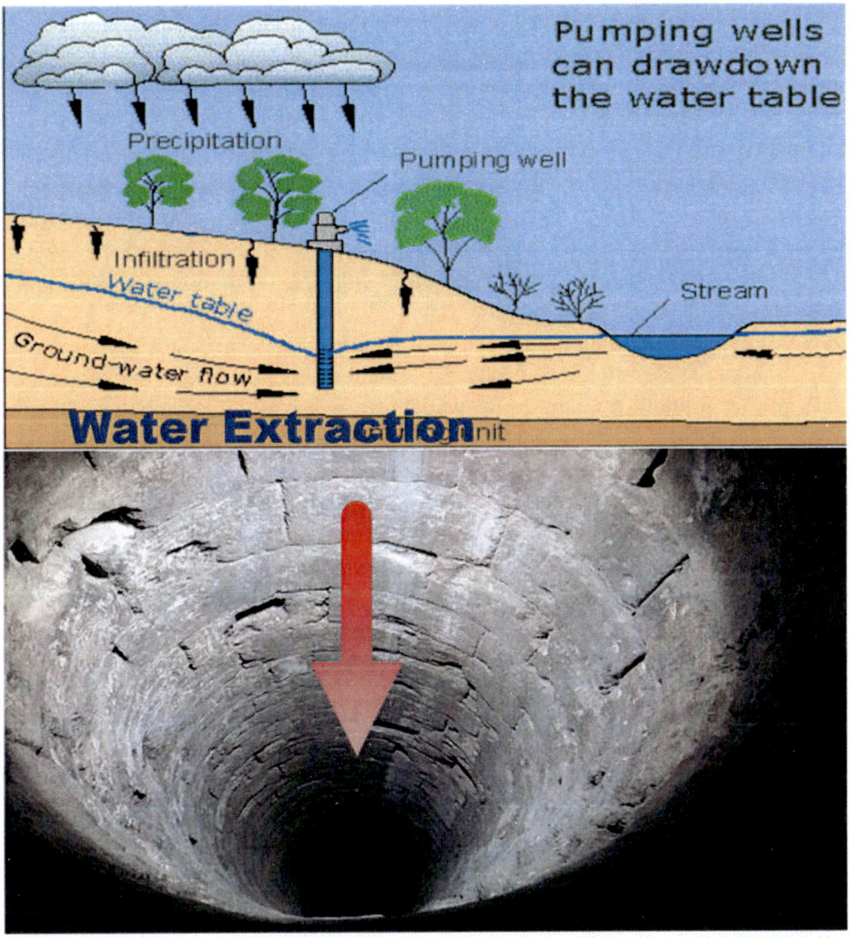

Fig. 12.5 Sustained groundwater extraction and changes in rainfall and surface flow influence spatial and temporal variations in groundwater

culminated in an estimated 80,328 ML/yr of water savings through the reduction of water extraction (dpie.nsw.gov.au). In California, rapid decline in groundwater levels across the Central Valley has been attributed to water extraction (Ojha et al. 2018). Generally, aquifer over-exploitation in the US results from about 60% irrigation water use, all of which emanates from groundwater (Scanlon et al. 2012) and as illustrated in Fig. 12.5, extensive groundwater pumping during drought or periods characterized by limited rainfall can result in drying out of wells.

Due to rise in global population and water demands, anthropogenic stress on freshwater is increasing. Global water demand has increased at more than twice the rate of population and the impacts of increased water demands are said to be larger than the effects of observed climate change (e.g., Ashraf 2017; Vörösmarty et al. 2000). Across several regions of the world, much of the observed groundwater loss observed from satellites are driven by human water withdrawals. For example, changes in selected aquifers in the United States and Iran are dominated by human activities and reduced human withdrawals have potentials to aid regime change in aquifers (Ashraf 2017). Aquifers respond to climate variability much more slowly than surface storage, and they provide a comfortable buffer that helps adaptation during such times. But with increased rainfall variability and evapotranspiration, climate change is likely to affect groundwater directly through a change in recharge and indirectly via increases in groundwater uses for agriculture, domestic, industry, and mining (OECD 2015).

Increasing human population comes with additional pressure on existing water resources, and if freshwater is considered a finite resource, the rise in human population could be a major problem to its availability, especially in arid and semi-arid regions of the world. Under a changing climate, the impacts of growing population on limited water resources could even be more complex. This could take the form of increased water demand for large agricultural, industrial and domestic water applications as the aftermath of both prolonged drought and population growth, and several water resources development. This creates a complicated freshwater system due to the interactions between climate and human actions. In the Zayandeh-Rud river basin of Iran, for example, the composite influence of human actions (extensive agricultural and industrial water use and rising human population and industry) and droughts has been a key challenge to the availability of water resources (Madani and Mariño 2009). Major management approaches and adaptation to this problem in the past and at least in the interim have been through inter-basin water transfer, and mining of fossil groundwater resources. In addition to these approaches, population control and regulating groundwater extraction have also been recommended as key policy initiatives for sustainability (Madani and Mariño 2009).

12.3.2 Hydrological Paradox

Despite severe drought conditions of the 1970 and 1980s, extensive network of well observations revealed that groundwater resources and water table in Niger, West Africa increased tremendously (Favreau et al. 2009; Leduc et al. 1997, 2001). This hydrological paradox was attributed to a change in land clearing and changes in land use, which caused an increase in runoff (Descroix et al. 2009; Favreau et al. 2009; Séguis et al. 2004). Although Descroix et al. (2009), in addition, argued from a hydrological perspective that this observed anomaly could possibly emanate from the hydraulic conductivity of the soil and its infiltration capacity, Li et al. (2007) confirmed the impacts of land-use change on the hydrological regimes of Niger and Lake Chad basins. They showed that complete deforestation increased simulated runoff ratio from 0.15 to 0.44 and annual stream flow by 35–65%. Human-induced changes in the ecosystem accompanied by long-term rise in water tables and increase in recharge were reported for south-east Australia and south-west United States (see, Scanlon et al. 2005; Allison et al. 1990). This increase in water table was observed after the transitions of natural ecosystems to rain-fed agriculture. In another study undertaken by Rost et al. (2008), land-use change reduced global evapotranspiration by 2.8% and increased discharge by 5.0% (1764 km^3/yr). It is also important to note that hydrological paradox can also be the aftermath of considerable human footprints.

References

Ahmed M, Sultan M, Wahr J, Yan E (2014) The use of GRACE data to monitor natural and anthropogenic induced variations in water availability across Africa. Earth-Sci Rev 136:289–300. https://doi.org/10.1016/j.earscirev.2014.05.009

Allison G, Cook P, Barnett S, Walker G, Jolly I, Hughes M (1990) Land clearance and river salinisation in the western Murray Basin, Australia. J Hydrol 119(1):1–20. https://doi.org/10.1016/0022-1694(90)90030-2

Ashraf B et al (2017) Quantifying anthropogenic stress on groundwater resources. Sci Rep 12910. https://doi.org/10.1038/s41598-017-12877-4

Awange J, Anyah R, Agola N, Forootan E, Omondi P (2013) Potential impacts of climate and environmental change on the stored water of Lake Victoria Basin and economic implications. Water Resour Res 49(0):8160–8173. https://doi.org/10.1002/2013WR014350

Balcombe SR, Arthington AH (2009) Temporal changes in fish abundance in response to hydrological variability in a dryland floodplain river. Mar Freshw Res 60:146–159. https://doi.org/10.1071/MF08118

Birkett CM (2000) Synergistic remote sensing of Lake Chad: variability of basin inundation. Remote Sens Environ 72:218–236. https://doi.org/10.1016/S0034-4257(99)00105-4

Boone A, Decharme B, Guichard F, Rosnay PD, Balsamo G, Belaars A, Chopin F, Orgeval T, Polcher J, Delire C, Ducharne A, Gascoin S, Grippa M, Jarlan L, Kergoat L, Mougin E, Gusev Y, Nasonova O, Harris P, Taylor C, Norgaard A, Sandholt I, Ottle C, Poccard-Leclercq I, Saux-Picart S, Xue Y (2009) The AMMA Land Surface Model Intercomparison Project (ALMIP). Bull Amer Meteorol Soc 90(12):1865–1880. https://doi.org/10.1175/2009BAMS27/86.1

Broadley A, Stewart-Koster B, Kenyon RA, Burford MA, Brown CJ (2020) Impact of water development on river flows and the catch of a commercial marine fishery. Ecosphere 11(7):e03194. https://doi.org/10.1002/ecs2.3194

Bunn SE, Arthington AH (2002) Basic principles and ecological consequences of altered flow regimes for aquatic biodiversity. Environ Manage 30(4):492–507. https://doi.org/10.1007/s00267-002-2737-0

Bunn SE, Thoms MC, Hamilton SK, Capon SJ (2006) Flow variability in dryland rivers: boom, bust and the bits in between. River Res Appl 22(2):179–186. https://doi.org/10.1002/rra.904

Carpenter-Bundhoo L, Butler G, Bond N et al (2020) Effects of a low-head weir on multi-scaled movement and behavior of three riverine fish species. Sci Rep 10:6817. https://doi.org/10.1038/s41598-020-63005-8

Chen J, Famigliett JS, Scanlon BR, Rodell M (2016) Groundwater storage changes: present status from GRACE observations. Surv Geophys 37(2):397–417. https://doi.org/10.1007/s10712-015-9332-4

Chen J, Li J, Zhang Z, Ni S (2014) Long-term groundwater variations in Northwest India from satellite gravity measurements. Glob Planet Chang 116:130–138. https://doi.org/10.1016/j.gloplacha.2014.02.007

Cheng KL (1980) The role of the Akosombo Dam on the Volta river in causing coastal erosion in central and eastern Ghana (West Africa). Mar Geol 37(3):323–332. https://doi.org/10.1016/0025-3227(80)90108-5

Coe MT, Foley JA (2001) Human and natural impacts on the water resources of the Lake Chad basin. J Geophys Res 106(D4):3349–3356. https://doi.org/10.1029/2000JD900587

Dai A, Qian T, Trenberth KE, Milliman JD (2009) Changes in continental freshwater discharge from 1948 to 2004. J Clim 22(10):2773–2792. https://doi.org/10.1175/2008JCLI2592.1

Delju AH et al (2013) Observed climate variability and change in Urmia lake basin, Iran. Theor Appl Climatol 111:285–296. https://doi.org/10.1007/s00704-012-0651-9

Descroix L, Mahé G, Lebel T, Favreau G, Galle S, Gautier E et al (2009) Spatio-temporal variability of hydrological regimes around the boundaries between Sahelian and Sudanian areas of West Africa: a synthesis. J Hydrol 375(1–2):90–102. https://doi.org/10.1016/j.jhydrol.2008.12.012

Eimanifar A, Mohebbi F (2007) Lake (Northwest Iran): a brief review 3(5). https://doi.org/10.1186/1746-1448-3-5

Famiglietti JS (2014) The global groundwater crisis. Nature 4:945–948. https://doi.org/10.1038/nclimate2425

Favreau G, Cappelaere B, Massuel S, Leblanc M, Boucher M, Boulain N, Leduc C (2009) Land clearing, climate variability, and water resources increase in semiarid southwest Niger: a review. Water Resour Res 45:W00A16. https://doi.org/10.1029/2007WR006785

Getirana A, Jung HC, Hoek JVD, Ndehedehe CE (2020) Hydropower dam operation strongly controls Lake Victoria's freshwater storage variability. Sci Total Environ 726:138343. https://doi.org/10.1016/j.scitotenv.2020.138343

Green TR, Taniguchi M, Kooi H, Gurdak JJ, Allen DM, Hiscock KM, Treidel H, Aureli A (2011) Beneath the surface of global change: impacts of climate change on groundwater. J Hydrol 405(3):532–560. https://doi.org/10.1016/j.jhydrol.2011.05.002

Guo C, Chen Y, Xia W, Qu X, Yuan H, Xie S, Lin L-S (2020) Eutrophication and heavy metal pollution patterns in the water suppling lakes of China's south-to-north water diversion project. Sci Total Environ 711:134543. https://doi.org/10.1016/j.scitotenv.2019.134543

Gyau-Boakye P (2001) Environmental impacts of the Akosombo dam and effects of climate change on the lake levels. Environ Dev Sustain 3(1):17–29. https://doi.org/10.1023/A:1011402116047

Górski K, van den Bosch LV, van de Wolfshaar KE, Middelkoop H et al (2012) Post-damming flow regime development in a large lowland river (Volga, Russian federation): implications for floodplain inundation and fisheries. River Res Appl 28(8):1121–1134. https://doi.org/10.1002/rra.1499

Hassanzadeh E, Zarghami M, Hassanzadeh Y (2012) Determining the main factors in declining the Urmia Lake level by using system dynamics modeling. Water Resour Manage 26:129–145. https://doi.org/10.1007/s11269-011-9909-8

Hoeinghaus DJ, Agostinho AA, Gomes LC, Pelicice FM, et al (2009) Effects of river impoundment on ecosystem services of large tropical rivers: embodied energy and market value of artisanal fisheries. Conserv Biol 23(5):1222–1231. https://doi.org/10.1111/j.1523-1739.2009.01248.x

Kennard MJ, Pusey Bradley J, Olden JD, Mackay SJ, Stein JL, Marsh N (2010) Classification of natural flow regimes in Australia to support environmental flow management. Freshw Biol 55(1):171–193. https://doi.org/10.1111/j.1365-2427.2009.02307.x

Kingsford R, Watts R, Keohn J, Thompson R, Sims N (2014) Flow dependent ecological responses: a technical report from the ecological responses to altered flow regimes flagship research cluster (subproject 3). CSIRO Water for a Healthy Country Flagship, Australia

Koster RD, Dirmeyer PA, Guo Z, Bonan G, Chan E, Cox P, Gordon CT, Kanae S, Kowalczyk E, Lawrence D, Liu P, Lu C-H, Malyshev S, McAvaney B, Mitchell K, Mocko D, Oki T, Oleson K, Pitman A, Sud YC, Taylor CM, Verseghy D, Vasic R, Xue Y, Yamada T (2004) Regions of strong coupling between soil moisture and precipitation. Science 305(5687):1138–1140. https://doi.org/10.1126/science.1100217

Lammers RB, Shiklomanov AI, Vörösmarty CJ, Fekete BM, Peterson BJ (2001) Assessment of contemporary Arctic river runoff based on observational discharge records. J Geophys Res: Atmos 106(D4):3321–3334. https://doi.org/10.1029/2000JD900444

Leblanc MJ, Tregoning P, Ramillien G, Tweed SO, Fakes A (2009) Basin-scale, integrated observations of the early 21st century multiyear drought in southeast Australia. Water Resour Res 45(4):W04408. https://doi.org/10.1029/2008WR007333

Leduc C, Bromley J, Schroeter P (1997) Water table fluctuation and recharge in semi-arid climate: some results of the HAPEX-Sahel hydrodynamic survey (Niger). J Hydrol 188(188–189):123–138. https://doi.org/10.1016/S0022-1694(96)03156-3

Leduc C, Favreau G, Schroeter P (2001) Long-term rise in a Sahelian water-table: the continental terminal in south-west Niger. J Hydrol 243(1–2):43–54. https://doi.org/10.1016/S0022-1694(00)00403-0

Li K, Coe M, Ramankutty N, Jong RD (2007) Modeling the hydrological impact of land-use change in West Africa. J Hydrol 337(3–4):258–268. https://doi.org/10.1016/j.jhydrol.2007.01.038

Liu C, Yang K, Bennett MM, Lu X, Guo Z, Li M (2020) Changes to anthropogenic pressures on reach-scale rivers in South and Southeast Asia from 1990 to 2014. Environ Res Lett 16(1):014025. https://doi.org/10.1088/1748-9326/abcf77

Liu X, Feng X, Ciais P, Fu B (2020) Widespread decline in terrestrial water storage and its link to teleconnections across Asia and eastern Europe. Hydrol Earth Syst Sci 24(7):3663–3676. https://doi.org/10.5194/hess-24-3663-2020

Long D, Yang W, Scanlon B et al (2020) South-to-north water diversion stabilizing Beijing's groundwater levels. Nat Commun 11(3665). https://doi.org/10.1038/s41467-020-17428-6

Madani K, Mariño MA (2009) System dynamics analysis for managing Iran's Zayandeh-Rud river basin. Water Resour Manage 23:2163–2187. https://doi.org/10.1007/s11269-008-9376-z

Margariti J, Rangecroft S, Parry S, Wendt DE, Loon AFV (2019) Anthropogenic activities alter drought termination. Elem Sci Anthropocene 7. https://doi.org/10.1525/elementa.365

Moore P, Williams SDP (2014) Integration of altimetry lake Lavels and GRACE gravimetry over Africa: inferences for terrestrial water storage change 2003–2011. Water Resour Res 50:9696–9720. https://doi.org/10.1002/2014WR015506

Ndehedehe C (2020) Floodplain productivity of the Gilbert and Flinders catchments: component 3 final report. Griffith University, Brisbane. https://research-repository.griffith.edu.au/handle/10072/394768. Accessed 8th July 2020

Ndehedehe C, Awange J, Agutu N, Kuhn M, Heck B (2016) Understanding changes in terrestrial water storage over West Africa between 2002 and 2014. Adv Water Resour 88:211–230. https://doi.org/10.1016/j.advwatres.2015.12.009

Ndehedehe CE (2019) The water resources of tropical West Africa: propblems, progress and prospect. Acta Geophys 67(2):621–649. https://doi.org/10.1007/s11600-019-00260-y

Ndehedehe CE, Agutu NO, Okwuashi OH, Ferreira VG (2016) Spatio-temporal variability of droughts and terrestrial water storage over Lake Chad Basin using independent component analysis. J Hydrol 540:106–128. https://doi.org/10.1016/j.jhydrol.2016.05.068

Ndehedehe CE, Awange J, Kuhn M, Agutu N, Fukuda Y (2017) Analysis of hydrological variability over the Volta river basin using in-situ data and satellite observations. J Hydrol: RegNal Stud 12:88–110. https://doi.org/10.1016/j.ejrh.2017.04.005

Ndehedehe CE, Burford MA, Stewart-Koster B, Bunn SE (2020) Satellite-derived changes in floodplain productivity and freshwater habitats in northern Australia (1991–2019). Ecol Ind 114:106320. https://doi.org/10.1016/j.ecolind.2020.106320

Ndehedehe CE, Ferreira VG, Onojeghuo AO, Agutu NO, Emengini E, Getirana A (2020) Influence of global climate on freshwater changes in Africa's largest endorheic basin using multi-scaled indicators. Sci Total Environ 737:139643. https://doi.org/10.1016/j.scitotenv.2020.139643

Ndehedehe CE, Okwuashi O, Ferreira VG, Agutu NO (2018) Exploring evapotranspiration dynamics over Sub-Sahara Africa (2000–2014). Environ Monit Assess 190(7):400. https://doi.org/10.1007/s10661-018-6780-6

Ndehedehe CE, Stewart-Koster B, Burford MA, Bunn SE (2020) Predicting hot spots of aquatic plant biomass in a large floodplain river catchment in the Australian wet-dry tropics. Ecol Ind 117:106616. https://doi.org/10.1016/j.ecolind.2020.106616

OECD (2015) Drying wells, rising stakes: towards sustainable agricultural groundwater use. OECD Studies on WaterOECD Publishing, Paris. https://doi.org/10.1787/9789264238701-en

Ogutu-Ohwayo R, Hecky R, Cohen A et al (1997) Human impacts on the African great lakes. Environ Biol Fishes 50:117–131. https://doi.org/10.1023/A:10073209323491997

Ojha C, Shirzaei M, Werth S, Argus DF, Farr TG (2018) Sustained groundwater loss in California's central valley exacerbated by intense drought periods. Water Resour Res 54(7):4449–4460. https://doi.org/10.1029/2017WR022250

Owusu K, Waylen P, Qiu Y (2008) Changing rainfall inputs in the Volta basin: implications for water sharing in Ghana. Geo J 71(4):201–210. https://doi.org/10.1007/s10708-008-9156-6
Oyebande L, Odunuga S (2010) Climate change impact on water resources at the transboundary level in West Africa: the cases of the Senegal, Niger and Volta Basins. Open Hydrol J 4:163–172. https://doi.org/10.2174/1874378101004010163
Rost S, Gerten D, Bondeau A, Lucht W, Rohwer J, Schaphoff S (2008) Agricultural green and blue water consumption and its influence on the global water system. Water Resour Res 44(9):W09405. https://doi.org/10.1029/2007WR006331
Scanlon BR, Faunt CC, Longuevergne L, Reedy RC, et al (2012) Groundwater depletion and sustainability of irrigation in the us high plains and central valley. Proc Natl Acad Sci 109(24):9320–9325. https://doi.org/10.1073/pnas.1200311109
Scanlon BR, Reedy RC, Stonestrom DA, Prudic DE, Dennehy KF (2005) Impact of land use and land cover change on groundwater recharge and quality in the southwestern US. Glob Change Biol 11(10):1577–1593. https://doi.org/10.1111/j.1365-2486.2005.01026.x
Séguis L, Cappelaere B, Milési G, Peugeot C, Massuel S, Favreau G (2004) Simulated impacts of climate change and land-clearing on runoff from a small Sahelian catchment. Hydrol Process 18(17):3401–3413. https://doi.org/10.1002/hyp.1503
Siderius C, Gannon KE, Ndiyoi M, Opere A, Batisani N, Olago D, Pardoe J, Conway D (2018) Hydrological response and complex impact pathways of the 2015/2016 El Niño in Eastern and Southern Africa. Earth's Future 6(1):2–22. https://doi.org/10.1002/2017EF000680
Taylor R, Scanlon B, Döll P et al (2013) Ground water and climate change. Nat Clim Chang 3:322–329. https://doi.org/10.1038/nclimate1744
Thorp JH, Delong MD (1994) The riverine productivity model: an heuristic view of carbon sources and organic processing in large river ecosystems. Oikos 70(2):305–308. https://doi.org/10.2307/3545642
Tourian M, Elmi O, Chen Q, Devaraju B, Roohi S, Sneeuw N (2015) A spaceborne multisensor approach to monitor the desiccation of Lake Urmia in Iran. Remote Sens Environ 156:349–360. https://doi.org/10.1016/j.rse.2014.10.006
Van Loon AF, Stahl K, Di Baldassarre G, Clark J, Rangecroft S, Wanders N, Gleeson T, Van Dijk AIJM, Tallaksen LM, Hannaford J, e. a. (2016) Drought in a human-modified world: reframing drought definitions, understanding, and analysis approaches. Hydrology and Earth System Sciences 20(9):3631–3650. https://doi.org/10.5194/hess-20-3631-2016
Vörösmarty CJ, Green P, Salisbury J, Lammers RB (2000) Global water resources: vulnerability from climate change and population growth. Science 289(5477):284–288. https://doi.org/10.1126/science.289.5477.284
Vorosmarty CJ, McIntyre PB, Gessner MO, Dudgeon D, Prusevich A, Green P, Glidden S, Bunn SE, Sullivan CA, Liermann CR, Davies PM (2010) Global threats to human water security and river biodiversity. Nature 467:555–555–561. https://doi.org/10.1038/nature09440
Voss KA, Famiglietti JS, Lo M, de Linage C, Rodell M, Swenson SC (2013) Groundwater depletion in the Middle East from GRACE with implications for transboundary water management in the Tigris-Euphrates-Western Iran region. Water Resour Res 49(2):904–914. https://doi.org/10.1002/wrcr.20078
Ward JV, Stanford JA (1995) Ecological connectivity in alluvial river ecosystems and its disruption by flow regulation. Regul Rivers: Res Manag 11(1):105–119. https://doi.org/10.1002/rrr.3450110109
Yang D, Ye B, Kane DL (2004) Streamflow changes over siberian yenisei river basin. J Hydrol 296(1–4):59–80. https://doi.org/10.1016/j.jhydrol.2004.03.017
Zarghami M (2011) Effective watershed management; case study of Urmia Lake, Iran. Lake Reserv Manag 27(1):87–94. https://doi.org/10.1080/07438141.2010.541327
Zhang C, Duan Q, Yeh PJ-F, Pan Y, et al (2020) The effectiveness of the South-to-North water diversion middle route project on water delivery and groundwater recovery in North China Plain. Water Resour Res 56(10):e2019WR026759. https://doi.org/10.1029/2019WR026759

Part VIII
Multivariate Methods in Hydrology

Chapter 13
Statistical Methods in Hydrology

Abstract With the growing applications of remote sensing observations in environmental monitoring, research efforts that focus on improving remote sensing classification algorithms are required to boost feature identification, optimization, and interpretation of remote sensing images. This has been the prime motivation of the use of advance machine learning (e.g., Deep Learning), multivariate methods, and fourth-order cumulant statistics in the identification of complex hydrological signals from satellite geodetic systems on a variety of spatial and temporal scales (Okwuashi and Ndehedehe 2020). Furthermore, the use of cumulant statistics in the decomposition of time-variable satellite gravity observations has proved to be robust in the identification of complex hydro-geodetic phenomena in regions where interactions between physical processes and climate fluctuations induce considerable large crustal displacements (Ndehedehe and Ferreira 2020). Improved and advanced algorithms in the analysis of remote sensing observations, be it from optical or geodetic systems are therefore crucial to effective feature identification and interpretation of earth observation (Okwuashi and Ndehedehe 2020). In this chapter, the suitability and efficacy of several multivariate statistics to characterize extreme events (e.g., drought and flood) and assess changes in terrestrial hydrology are explored.

13.1 Advanced Statistical Decomposition Methods

13.1.1 Principal Component Analysis

Principal component analysis (PCA) is an extraction technique that is used to reduce the dimension of large multivariate data, and at the same time account for the most dominant variations in the original dataset (e.g., Jolliffe 2002; Martinez and Martinez 2005; Westra et al. 2010; Preisendorfer 1988). This is achieved by seeking new directions of maximum variance in the dataset and then using these as new coordinate axes. A key benefit of replacing the variables by new sets of variables is that the sets are uncorrelated. Moreover, these new uncorrelated sets of variables are used to interpret the multivariate data (e.g., rainfall, temperature, soil moisture). Its mathematical simplicity and the ability to

C. Ndehedehe, *Satellite Remote Sensing of Terrestrial Hydrology*,
https://doi.org/10.1007/978-3-030-99577-5_13

explain the optimized variance, using a small number of principal components, has probably made this method the most widely used statistical data analysis tool in climate science (Westra et al. 2010). As a method that seeks to identify structures that explain the dominant patterns of variability in a two-dimensional dataset, some authors have referred to it as the Empirical Orthogonal Function analysis (Hannachi et al. 2007). Moreover, the Singular Value Decomposition (SVD) is another term used interchangeably with the PCA method, given that SVD is a term used for general decomposition of a matrix. In this sense, SVD can be used to decompose data matrices to simultaneously find both the principal components (PC's), which correspond to the EOF's or localized spatial patterns. The mathematical principle in PCA and SVD is similar. For instance, as with PCA, the application of SVD on the covariance matrix between two datasets will result in structures in each dataset that are best correlated with structures in the other dataset. These methods including factor analysis are classified as multivariate analyses. As mentioned earlier, they seek directions where the variance is maximum in the dataset and then apply these as new axes or a new orthogonal system. For instance, assuming three axes were required to describe a multivariate data (e.g., rainfall, temperature, groundwater) before it was transformed through statistical decomposition. The data dimension of this multivariate data can be reduced by dropping the second and third axes without losing much information from the data. This is because the first axis explains much of the total variability in the data. For instance, leading rainfall modes in catchments dominated by annual rainfall can be captured in the first axis as that will explain more than 95% of the maximum variance. The statistical principles and properties of these multivariate methods have been detailed in the literature (Martinez and Martinez 2005; Jolliffe 2002; Preisendorfer 1988). In this book, the term PCA will be used and principal components and empirical orthogonal functions are seen as the outcome of PCA or the result of statistical decomposition using the PCA technique.

13.1.1.1 Implementation

PCA reduces the dimensions of multivariate data by creating new variables that are linear functions of the original variables. Given k variables at a given time period i, the linear combinations for k principal components (PCs) as detailed in Ndehedehe et al. (2016a) are

$$\begin{bmatrix} y_{i,1} = p_{11}\mathbf{x}_{i,1} + p_{12}\mathbf{x}_{i,2} + p_{13}\mathbf{x}_{i,3} + \cdots + p_{1k}\mathbf{x}_{i,k} \\ y_{i,2} = p_{21}\mathbf{x}_{i,1} + p_{22}\mathbf{x}_{i,2} + p_{23}\mathbf{x}_{i,3} + \cdots + p_{2k}\mathbf{x}_{i,k} \\ \cdots \\ \cdots \\ y_{i,k} = p_{k1}\mathbf{x}_{i,1} + p_{k2}\mathbf{x}_{i,2} + p_{k3}\mathbf{x}_{i,3} + \cdots + p_{kk}\mathbf{x}_{i,k} \end{bmatrix} \quad i = 1, \ldots, n. \tag{13.1}$$

The data matrix $\mathbf{x}_{i,k}$ contains rows representing the time i in months and k, the variables, i.e., the observations, which, in our case, is TWS or rainfall. In the linear combinations above (Eq. 13.1), the y values are orthogonal and also the new uncorrelated variables called the PCs such that y_{i1} explains the highest variability while y_{i2} up to y_{ik} explain the remaining variance. Equation 13.1 can be summarized as

$$y_i = \sum_{k=1}^{n} p_k(i)x_k. \tag{13.2}$$

The coefficients of the linear combinations are called loadings (i.e., the eigenvectors) and they provide the weights of the original variables in the PCs. These loadings are spatially orthonormal while the PCs are orthogonal and linearly independent. The eigenvalues (i.e., the amount of covariance in time explained by each eigenvector) and eigenvectors, which are also referred to as empirical orthogonal functions (EOFs), can be derived from the sample covariance matrix or correlation matrix of the centred data matrix $\mathbf{x}_{i,k}$. Apart from its use being restricted, mostly to when observed variables are measured in different units, PCA based on the correlation matrix amounts to the analysis of the joint variance structure of the standardized data whereas PCA on the covariance matrix results in dominant patterns accounting for the highest variance in the data (e.g., Wilks 2011; Jolliffe 2002). The statistical decomposition can be based on either the correlation or the covariance matrix. The latter is more ideal in climate and hydrometeorlogical studies due to its ability to extract principle components that emphasize areas with very high temporal variability (e.g., Nicholson 2014; Jolliffe 2002). The EOF is the spatial distribution or the spatial patterns of rainfall or TWS while the EOF/PC pair is called the PCA mode. The eigenvectors (e.g., p_{11}, p_{12}, $p_{13} \cdots p_{kk}$) are normalized and usually called the loadings (EOFs). The eigenvalues (i.e., $\lambda_1 \geq \lambda_2 \geq \lambda_3 \cdots \lambda_k$) explain the fraction of the total variance explained by the these loadings (e.g., Santos et al. 2010).

PCA can be implemented for multivariate observations like climate and hydrological data or geophysical time series. For example, to analyze the spatial and temporal variations of rainfall over a region of interest, consider a matrix of rainfall (R) values as $\mathbf{X}_R = [x(p_k, t)]$ where p_k is space locations; k = 1, 2, ..., Nx, which is the number of spatial locations for $\mathbf{X}_R$, and t is the monthly time step. The temporal and spatial patterns of rainfall or the hydrological data will be decomposed as,

$$\mathbf{X}_R(t) = \sum_{k=1}^{N} a_{(k)}\mathbf{p}_k \tag{13.3}$$

where $a_{(k)}$(t) is the standardized score (or temporal patterns) and is also called expansion coefficient, $\mathbf{p}_k$ is the corresponding spatial pattern (or empirical orthogonal functions-EOF loading). The dominant modes are the first few pairs obtained, which usually explains the highest variability in the data and is usually an approximation of large-scale average while the remaining modes describe other variabilities, e.g., large-scale contrasts with dipoles and loadings (e.g., Montazerolghaem et al. 2016). Notably, the hydrological data in question must be continuous time series for it to be suitable for the PCA technique (Ndehedehe et al. 2016a). The use of this method to analyze the spatial and temporal variations of hydrological quantities (e.g., rainfall, terrestrial water storage) has been detailed in Chaps. 16 and 18 and reported in Ndehedehe et al. (2016a).

13.1.1.2 Identifying Statistically Significant Modes of Variability

The truncation of PCs, that is, the choice of dimensions or significant orthogonal modes to keep is an important decision in the use of statistical decomposition methods like the PCA. This process enables one to exclude redundant information from the original data matrix. However, the process is not a straightforward approach. In other words, there is no fixed criterion to determine the number of PCs that are best retained in a given situation but the choice of truncation can vary from available selection rules (e.g., rules based on hypothesis testing; size of cumulative variance; size of the last retained eigenvalue, etc.) to subjective choices (e.g., *scree plot*, *log-eigenvalue* diagram, and retaining PCs that represent a 'sufficient fraction' of the variance in the original data) depending on the data and the purpose of the PCA (see, e.g., Wilks 2011; Martinez and Martinez 2005; Jolliffe 2002). Approaches based on subjective criteria and size of the cumulative variance have been used. The *scree plot* analysis, for example (Fig. 13.1), uses a graph to determine the number of statistically significant PCs to retain (Jolliffe 2002). This scree graph is an eigenvalue spectrum that shows the eigenvalues in decreasing order as a function of their index. The truncation cutoff mark is identified at the point where the eigenvalues are very small, usually the PC number at which the separation occurs (e.g., Wilks 2011; Martinez and Martinez 2005). The illustration in Fig. 13.1a and b shows the scree plot analysis for terrestrial water storage and rainfall, respectively, over West Africa. The modes of variability at the point where the lines flatten or approaching zero (i.e., when eigenvalues are becoming very small) are not significant. Rainfall is a key driver of terrestrial water storage in the region and the first and second modes in the scree plot

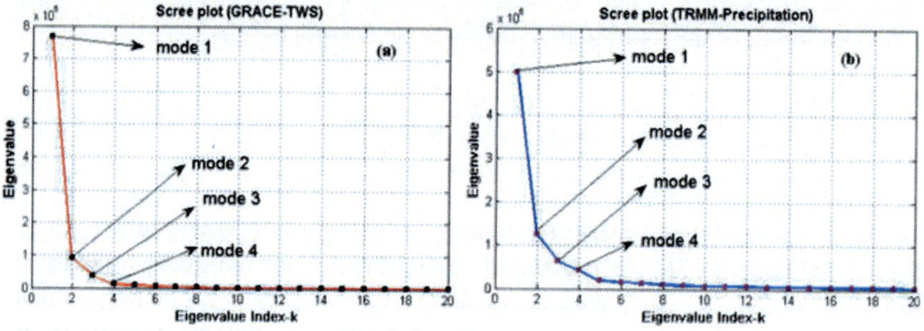

Fig. 13.1 Identifying statistically significant modes of variability using scree plot analysis. Significant modes of variability are identified for **a** terrestrial water storage obtained from Gravity Recovery and Climate Experiment and **b** satellite-based precipitation during the 2002–2014 period over West Africa

tend to have the same shape and have relatively large amount of eigenvalues compared to the third and fourth modes (Fig. 13.1a and b). This means the first two modes account for substantial variations in both data and represent the annual and multi-annual changes. While the third or fourth mode in rainfall could represent influence of climate oscillations, in GRACE data they could be the contributions of surface water (lakes and reservoirs) or a groundwater signal. The key issue is the ability to understand and interpret what PCA modes represent. As a further guide to selecting significant modes of variability, Jolliffe (2002) recommends that the total variance retained should be greater than some critical value (i.e., $70\% \leq V_{crit} \leq 90\%$) as in

$$\sum_{i=1}^{k} V_k \geq V_{crit}, \tag{13.4}$$

where V_k is the percentage of variance explained by each PCA mode from the total variation in the data (see Eq. 13.5) and V_{crit} the critical value. The cumulative percentage of variance explained (e.g., Martinez and Martinez 2005) can be used to determine the truncation point for the PCs whose joint variance describes the overall variability in the data. V_k above is estimated as

$$V_k = \frac{\lambda_i}{\sum_{i=1}^{k} \lambda_k} \times 100\%. \tag{13.5}$$

Other geometric and statistical properties of PCA can be found in the literature (Jolliffe 2002; Preisendorfer 1988).

13.1.2 Other Variants of Principal Component Analysis

The use of PCA in decomposing multivariate climate data into spatial and temporal patterns could be limited due to complexity in such data caused by atmospheric properties and other factors (e.g., Hannachi et al. 2007). In some regions, for example, the spatial patterns of PCA-derived rainfall might be difficult to interpret because of their geometric properties, including their orthogonality in space and time. Other variants of PCA, the rotated principal component analysis (RPCA, Richman 1986) was therefore introduced following the pioneering work of Kaiser (1958) to provide more spatially explicit and well localized spatial patterns. Other variants like the extended and complex PCA (CPCA) have also been introduced to address some limitations in RPCA such as the inability to deal with propagating features as they only rely on spatial correlation of the field (Hannachi et al. 2007). Despite the usefulness of the CPCA in identifying travelling and standing waves in geophysical observations, Horel (1984) argue that some of the drawbacks of the conventional PCA still remain and could even be more challenging when it comes to interpreting CPCA solutions.

Notably, the quest for optimized multivariate framework to improve the localization of hydrological signals has been the fall-out of limitations associated with the use of PCA in the analysis of climate data, e.g., rainfall (e.g., Ndehedehe et al. 2016b; Agutu et al. 2017). The RPCA is, however, considerably useful for this purpose and easier to implement and interpret compared to the CPCA. Consider a matrix of data values as $\mathbf{X}_{data} = [x(p_k, t)]$ where p_k is space locations; k = 1, 2, ..., jx, which is the number of spatial locations for $\mathbf{X}_{data}$, and t is the monthly time step. The PCA (e.g., Preisendorfer 1988) decomposes it as,

$$\mathbf{X}_{data}(t) = \sum_{k=1}^{j} a_{(k)}\mathbf{p}_k \tag{13.6}$$

where $a_{(k)}$(t) is the standardized score also called expansion coefficient, $\mathbf{p}_k$ is the corresponding empirical orthogonal functions (EOF). After this decomposition, the Log-eigenvalue (LEV) diagrams and other guidelines (Martinez and Martinez 2005; Jolliffe 2002) can be employed to identify and retain the significant orthogonal modes of variability. These modes can then be rotated using the RPCA scheme based on the varimax rotation (e.g., Forina et al. 1989; Kaiser 1958). The varimax rotation, which has been detailed in the pioneering work of Kaiser (1958), provides an improved spatially explicit and well-localized spatial patterns as opposed to the derived EOF from the PCA scheme. The spatial outputs from the varimax scheme are normalized using the standard deviation of their corresponding rotated principal components series while the latter is standardized using its standard deviation (e.g., Ndehedehe et al. 2021; Agutu et al. 2017).

13.1.3 Applications and Benefits of the PCA Technique

Exploring the spatial and temporal patterns of large multivariate climate data and those consisting of geophysical fields (e.g., Wilks 2011) have been significant strengths of PCA as an efficient tool in data analysis. The PCA method, for example, has shown unique capability in isolating both the inter-annual signals, and long-term periodic variations of geophysical fields (e.g., Rangelova et al. 2007), and extracting indices of climate variability (e.g., Compagnucci and Richman 2008). The use of a component extraction method such as PCA to decompose multivariate data help to address some questions such as what are the trends and dominant spatio-temporal patterns of observed variations in the hydrological data. It also provides a good knowledge of the spatio-temporal distribution of hydrological units, which is important for water resources planning and understanding the impacts of climate on the hydrological system. Not only is it useful in the analysis of meteorological and geophysical observations (e.g., Ndehedehe et al. 2016a; Sanogo et al. 2015; Mohino et al. 2011), it is now an effective tool in diagnosing the impacts of large-scale oceanic atmospheric predictors, atmospheric blocking, and climate teleconnection patterns

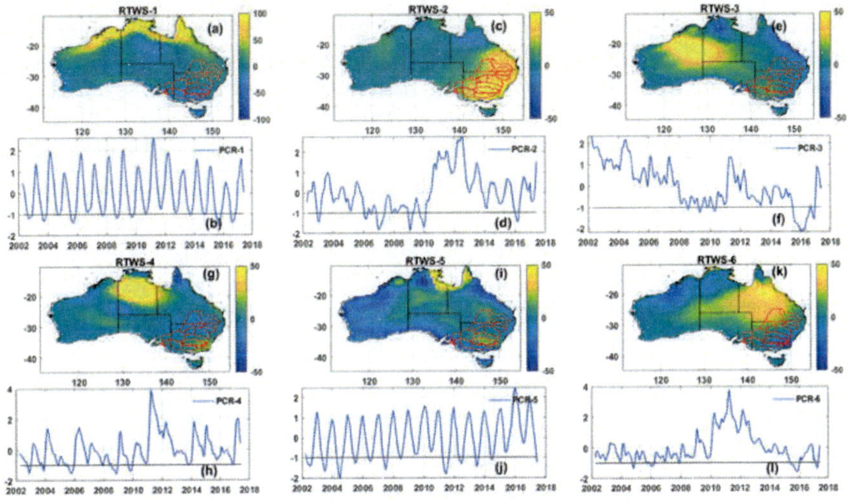

Fig. 13.2 Spatial and temporal patterns (2002–2017) of GRACE-derived terrestrial water storage over Australia. RTWS and PCR are the rotated spatial and temporal patterns of terrestrial water storage, respectively. *Source* Ndehedehe et al. (2021)

on hydro-meteorological patterns (e.g., Ndehedehe et al. 2019; Montazerolghaem et al. 2016; Kurnik et al. 2015; Nicholson 2014).

The PCA method has been used to analyze the spatial and temporal continental water storage changes. Over Australia, for example, the statistical decomposition of GRACE observations (Chap. 4) was undertaken to understand the temporal and spatial variability of terrestrial water storage (TWS) over the region (Ndehedehe et al. 2021). This is important to improve understanding on the key drivers of TWS in the region, especially in the context of response to extreme events (e.g., droughts and floods) and inter-annual variations to rainfall. Following a rotated PCA (Sect. 13.1.2) implementation of TWS, it was found that the northern region is Australia's water tower with strong spatial loadings and temporal patterns, which capture inter-annual variations (Fig. 13.2a–b). This is the first mode (RTWS-1/PCR-1) of variability and it accounts for the highest observed total variability (43.6%) while the total variability accounted for by the remaining five modes fluctuated between 12.5 and 8.5%. The second mode of TWS is localized over the Murray Darling basin and explains 8.6% of the total variability (RTWS-2/PCR-2, Fig. 13.2c–d) while the GRACE hydrological signal in the third mode shows a fall in TWS since 2002 (RTWS-3/PCR-3, Fig. 13.2e–f) around the Caning basin in Western Australia. This particular mode, the Caning basin represents the variations in groundwater and apparently shows substantial decline since 2002. However, between 2008 and 2011, the period coinciding with the La-Nina driven flood over Australia, this groundwater decline slowed down, further confirming the impact of flood on groundwater recharge. The GRACE-hydrological signal in the northern territory accounts for 12.5% of the total variability and its temporal pattern is also strongly annual (RTWS-4/PCR-4, Fig. 13.2g–h) but its spatial loadings are lower than what is observed in Fig. 13.2a. The localized GRACE hydrological signal around the northern part of the country also shows strong annual patterns as with TWS variability in mode 5 (RTWS-5/PCR-5, Fig. 13.2i–j). Similarly, the TWS signal in mode 6 (RTWS-6/PCR-6, Fig. 13.2k–l) appears to mimic the localized signal over the Murray Darling basin. The observed GRACE hydrological signals (the different modes of TWS variability localized over Australia) are largely driven by variations in annual rainfall. This is true and evident in the temporal patterns of TWS in the northern region and the Murray Darling basin (Fig. 13.2a–d, g–j). While hydrological signals from the 2010/2011 extreme wet periods are obvious in the temporal patterns of localized TWS over the region (Fig. 13.2b, d, f, h, j, l), there are also groundwater footprints in the observed changes in TWS.

13.1.4 Independent Component Analysis

To further improve the performance of the PCA method, advanced statistical methods like the Independent Component Analysis (ICA) have been used to localize and decomposed drought and TWS signals obtained from the PCA process, by rotating them towards statistical independence. As opposed to third-order statistical method like the PCA, the ICA (e.g., Cardoso and Souloumiac 1993; Cardoso 1999) is a fourth-order statistical technique and is usually referred to as blind source separation in signal processing (e.g., Cardoso 1999). Prior to implementing ICA, the PCA method is used to identify statistically significant modes of variability in the data. Subsequently, the derived significant dominant patterns are further explored through a classical rotation of the principle components, using the Joint Approximate Diagonalization of Eigen matrices (JADE) algorithm fully described in Cardoso and Souloumiac (1993) and Cardoso (1999). The JADE algorithm is statistical based and follows a Jacobi technique (so called because they seek to maximize measures of independence by a technique similar to the Jacobi method of diagonalization) in the process of orthogonal contrast optimization, different from other ICA online solutions that use the gradient techniques (Cardoso 1999). Fundamentally, ICA decomposes a data matrix $x_i(t)$, which consist of a number of statistically independent source signals $s_j(t)$ where t is the time index. This can be expressed as (e.g., Ziehe 2005)

$$x_i(t) = \sum A_{ij} s_j(t), \qquad (i = 1, \ldots, n, \;\; j = 1, \ldots, m), \tag{13.7}$$

x_i in Eq. (13.7) is the mixing model or the linear mixture and can be represented as a matrix for notational convenience as

$$\mathbf{X} = \mathbf{AS}, \tag{13.8}$$

where the entries of the data matrix $\mathbf{X}$ are samples of the $x_i(t)$ given in Eq. (13.7) leading to column vectors $x[t] = [x_1[t], \ldots, x_n[t]]^T$, the n x m matrix $\mathbf{A}$ has elements

$A_{i,j}$ and the source signals i.e., matrix **S**, similar to the construction of **X**, has column vectors $s[t] = [s_1[t], \ldots, s_m[t]]^T$. Two major challenges of ICA lie in the definition of a measure of independence and the choice of algorithms to find the change of basis (or separating matrix) that fully optimizes this measure (Cardoso 1999). These measures of independence are based on cumulant-based blind identification of the separation matrices. The fourth-order cumulant matrix is expressed as

$$\mathbf{C}_{i,j}(\mathbf{M}) = \sum \mathbf{M}_{u,v}\; cum(x_i, x_j, x_u, x_v), \tag{13.9}$$

where **M** is an arbitrary matrix (see, e.g., Ziehe 2005). The plane rotations are applied to these Cumalant-based matrices to estimate the independent components through a multiplicative update of an estimate of the separation matrix (Ziehe 2005; Theis et al. 2005; Cardoso 1999, 1991; Cardoso and Souloumiac 1993). The ICA algorithm is available at http://perso.telecom-paristech.fr/cardoso/Algo/Jade/jadeR.m and can be used to decompose hydrological data (Fig. 13.3) and drought indicators (SPI and SSI data). After decomposition, the independent components will have to be normalized using their standard deviations to be unitless. Each ICA mode is a combination of the temporal and spatial patterns and are usually interpreted together (i.e., the unitless temporal evolution is multiplied with the spatial pattern to obtain the actual values). As changes in TWS across the globe are driven by several climatic and human factors (Ndehedehe and Ferreira 2020; Rodell et al. 2018), the illustration in Fig. 13.3a–d highlights the importance of localizing hydrological signals like TWS. The spatial distribution of TWS trends in Fig. 13.3a shows considerable declines in the Patagonian region and increase along the Amazon floodplains, which is attributed to strong inter-annual changes in TWS observed over the Amazon basin region and

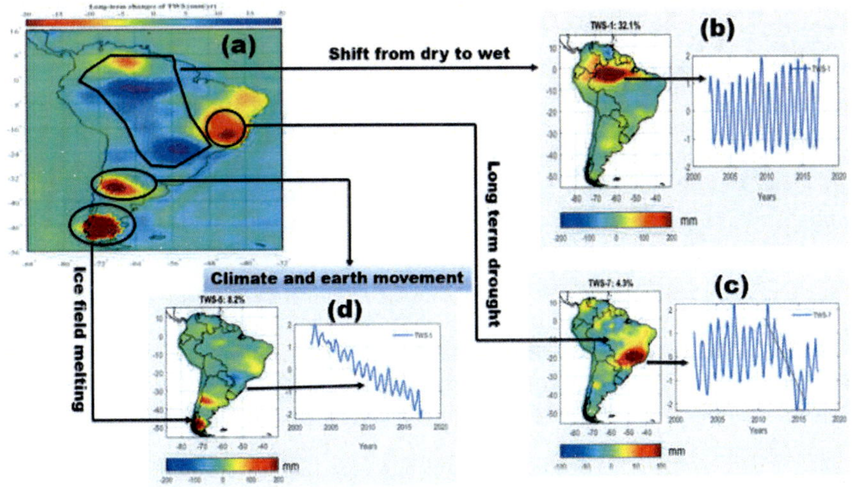

Fig. 13.3 Decomposition of GRACE-hydrological signals using independent component analysis

within its floodplain corridors. Whereas this is a major shift from dry to wet regimes in the basin (Fig. 13.3b), the strong mass loss in Patagonia ice-field is due to warming of the climate (Fig. 13.3d). The decline in north-east Brazil is linked to prolonged drought in the region (Fig. 13.3c). The statistical decomposition with ICA shows the temporal and spatial patterns of these signals localized over respective regions in South America.

13.2 Modelling and Understanding Relationships in Hydrology

13.2.1 Partial Least Square Regression

The partial least square regression (PLSR, e.g., Chen et al. 2018; de Jong 1993) model can be used to assess the coupled relationship of multi-scale climate modes with drought metrics like SPI. PLSR looks for latent vectors, which performs a simultaneous decomposition of independent variable, **X** and response variable **Y** (e.g., Lewis-Beck et al. 2003; Wold et al. 2001). These PLSR components are so determined to maximize the covariance between the two variables whilst complying with certain orthogonality and normalization constraints (de Jong 1993). Consider a simple formulation of the PLSR model (e.g., Chen et al. 2018) where the data elements $\mathbf{x}_i = [x_{i1}, x_{i2}, x_{i3} \ldots, x_{ip}]' \in \Re^p$ $(i = 1, 2, 3, \ldots, n)$ with n as the observation samples and $\mathbf{y}_i = [y_{i1}, y_{i2}, y_{i3} \ldots, y_{iq}]' \in \Re^q$ $(i = 1, 2, 3, \ldots, n)$ where n is the corresponding dependent samples. Then the independent variable, $\mathbf{X} = [\mathbf{x}_1, \mathbf{x}_2, \mathbf{x}_3 \ldots, \mathbf{x}_n]' \in \Re^{n\times p}$ and the response variable $\mathbf{Y} = [\mathbf{y}_1, \mathbf{y}_2, \mathbf{y}_3 \ldots, \mathbf{y}_n]'$ $\in \Re^{n\times q}$. The centred (i.e., removing the mean) data matrices **X** and **Y** are decomposed as (e.g., Chen et al. 2018; Wold et al. 2001),

$$\mathbf{X}_{n\times p} = \mathbf{t}_{n\times 1}\mathbf{p}'_{p\times 1} + \mathbf{E}_{n\times p}, \quad \mathbf{Y}_{n\times q} = \mathbf{u}_{n\times 1}\mathbf{q}'_{q\times 1} + \mathbf{F}_{n\times q}, \tag{13.10}$$

where **t** and **u** are latent vectors for the n observations, **p** and **q** are the loading vectors while **E** and **F** are the residual matrices. The assessment of climate influence on the evolutions of droughts based on the PLSR formulation can now be obtained as

$$\mathbf{Y} = \mathbf{X}\beta + \mathbf{F}, \tag{13.11}$$

where β is the PLSR coefficients, $\mathbf{X}_{Climate\ modes}$ and $\mathbf{Y}_{SPI/SPEI}$ are the predictor data matrix and response variables, respectively. The prediction of SPEI and SPI using these climate modes will be based on retained partial least square components.

13.2.2 Canonical Correlation Analysis

Canonical correlation analysis (CCA, see, Graham et al. 1987; Glahn 1968; Hotelling 1936) is another useful multivariate statistics that combines the Maximum Covariance Analysis with PCA to examine the interrelationship between two sets of multivariate data. In other words, it is 'double-barreled' PCA (e.g., Wilks 2011; Graham et al. 1987; Glahn 1968) due to its similarity to principal component analysis (PCA, e.g., Jolliffe 2002; Preisendorfer 1988) and multiple linear regression. Contrary to the PCA technique, the goal in CCA is to determine a linear combination of two different sets of variables such that the correlation between the two functions is a maximum. Key benefits of using this method include robustness in extracting relationships, diagnosing aspects of the coupled patterns in two multivariate fields, and limiting the probability of committing Type 1 error (e.g., Wilks 2011; Glahn 1968). CCA implements a simple bivariate correlation between a predictand and synthetic predictor variable, which is somewhat similar to Pearson correlation (Sherry and Henson 2005; Bretherton et al. 1992). That is, it seeks $\mathbf{f}_i$ and $\mathbf{q}_i$, such that observed correlation between individual and successive pairs of $\mathbf{f}_i$ $\mathbf{x}$ and $\mathbf{q}_i$ $\mathbf{y}$ are maximized. For instance, consider two multivariate datasets (Ndehedehe et al. 2019),

$$\mathbf{x}(t) = x_{i,j}, \quad i = 1, 2, 3 \ldots, n_x \quad j = 1, 2, 3 \ldots, n_h, \tag{13.12}$$

and

$$\mathbf{y}(t) = y_{i,j}, \quad i = 1, 2, 3 \ldots, n_y \quad j = 1, 2, 3 \ldots, n_h, \tag{13.13}$$

Assuming the means of $\mathbf{x}$ and $\mathbf{y}$ have been removed, then

$$u = \mathbf{f}'\mathbf{x}' \quad v = \mathbf{q}'\mathbf{y}. \tag{13.14}$$

The maximum correlation (r) between the canonical correlation coordinates (u and v) is desired by selecting $\mathbf{f}$ and $\mathbf{q}$ to maximize r. The maximization problem can be solved following a classical approach where the weight vectors are obtained using Lagrange multipliers or through a regularization approach where the original data is compressed through orthogonalization (pre-filtering) using the PCA concept (e.g., Bretherton et al. 1992; Barnett and Preisendorfer 1987). For climate data and hydrological time series, it is a standard practice to prefilter or regularize by applying PCA to identify statistical significant modes of variability in the data ($\mathbf{x}$ and $\mathbf{y}$) before applying CCA. This is because the large spatial fields in these multivariate data cause difficulty in inverting the matrices and in the eigenvalue problem, leading to instabilities in the CCA solution (see, e.g., Singh et al. 2012; Repelli and Nobre 2004; Yu et al. 1997; Shabbar and Barnston 1996). This is a potential drawback to CCA implementation for highly inter-correlated data fields, thus warranting prefiltering of the original data series ($\mathbf{x}$ and $\mathbf{y}$). This preprocessing operation was suggested by Barnett and Preisendorfer (1987) and is known to have several advantages. It eliminates noise, reduces the number of degrees of freedom in each field, invokes the principle of parsimony relevant to statistical modelling, and above all, it ensures that the resulting CCA modes are more stable with respect to sampling variability (e.g., Ndehedehe et al. 2018; Bretherton et al. 1992; Graham et al. 1987; Barnett and Preisendorfer 1987). Using Eq. (13.3) above to regularize (filter) $\mathbf{x}_{SPI}$ and $\mathbf{y}_{SST}$, the filtered components ($\acute{\mathbf{x}}$ and $\acute{\mathbf{y}}$) are given as

$$\acute{\mathbf{x}} = [a_1', \ldots, a_{m_x}']', \quad \acute{\mathbf{y}} = [b_1', \ldots, b_{m_y}']', \tag{13.15}$$

where the first few modes of m_x and m_y are retained as the leading PCA modes. The dimension reduction of these dataset (original data) through the truncation of their principal components prior can be obtained using a scree plot analysis (Martinez and Martinez 2005) but other guidelines and considerations are available in Livezey and Smith (1999). The CCA scheme can then be used to transform $\acute{\mathbf{x}}$ and $\acute{\mathbf{y}}$ into new sets of variables, the canonical variates (α_m and β_m) defined by the dot product as (e.g., Wilks 2011)

$$\alpha_m = \mathbf{d}_m' \acute{\mathbf{x}} = \sum_{i=1}^{n_x} d_{m,i}\, \acute{x}_i, \quad m = 1, \ldots, min(n_x, n_y) \tag{13.16}$$

$$\beta_m = \mathbf{h}_m' \acute{\mathbf{y}} = \sum_{i=1}^{n_y} h_{m,j}\, \acute{y}_j, \quad m = 1, \ldots, min(n_x, n_y) \tag{13.17}$$

where $\mathbf{d}_m'$ and $\mathbf{h}_m'$ are the canonical vectors (or canonical patterns) and n_x and n_y are elements of the concatenated vectors of the filtered data ($\acute{\mathbf{x}}$ and $\acute{\mathbf{y}}$). The grid points of $\acute{\mathbf{x}}$ and $\acute{\mathbf{y}}$ need not be the same in the canonical model as $\mathbf{d}_m'\acute{\mathbf{x}}$ and $\mathbf{h}_m'\acute{\mathbf{y}}$ have n_x and n_y elements, respectively. Consequently, the number of pairs, M, that can be obtained from the canonical variates (α_m and β_m) will depend on the smaller dimension between $\acute{\mathbf{x}}$ and $\acute{\mathbf{y}}$, i.e., M=$min(n_x, n_y)$. The correlation $c = [c_1, c_2, \ldots, c_n]$ is the canonical correlation between the canonical variates (predictand and predictor variables) for the retained CCA modes and indicates that each successive pairs of canonical variates exhibit smaller correlation than the previous pair. Also, note that the canonical vectors are chosen to maximize correlations with the resulting canonical variates (e.g., Wilks 2011). The association between the canonical vectors results in the canonical variates having the properties, $c_1 \geq c_2 \geq \ldots, \geq c_n$ and

$$c(\alpha_m, \beta_m) = \begin{cases} r_{Cm}, & k = m \\ 0, & k \neq m, \end{cases} \tag{13.18}$$

$$c(\alpha_k, \alpha_m) = c(\beta_k, \beta_m) = 0, \quad k \neq m. \tag{13.19}$$

Assuming the matrices $[S_x, S_y]$ and $[S_x, S_y]$ represent the covariances for different combinations of the elements of $\acute{\mathbf{x}}$ and $\acute{\mathbf{y}}$ and also related as $[S_x, S_y]$= $[S_y, S_x]'$, then the observed variance (ξ) for the canonical model is given as

$$\xi(\alpha_m) = \mathbf{d}'_m[S_x, S_x]\mathbf{d}_m = \xi(\beta_m) = \mathbf{h}'_m[S_y, S_y]\mathbf{h}_m = 1, \quad m = 1, \ldots, M. \tag{13.20}$$

As detailed in Ndehedehe et al. (2019), the canonical variates (α_m and β_m) are the projections of the regularized components ($\acute{\mathbf{x}}$ and $\acute{\mathbf{y}}$) on to the canonical vectors ($\mathbf{d}'_m$ and $\mathbf{h}'_m$). These canonical vectors are thereafter translated into corresponding spatial maps (i.e., for the predictands and predictors) to show ways in which the original variables contribute to the canonical variables. These spatial maps are referred to as $g - map$ (predictor map) and $h - map$ (predictand map) (e.g., Barnett and Preisendorfer 1987) and are used to investigate the regions of variability in the predictor fields that are strongly associated with the variability of the predictand fields (e.g., Repelli and Nobre 2004). The choice of the threshold for the number of statistically significant canonical modes can be determined using a number of test statistics. These tests include, the F statistics, chi square and Wilks's lambda criterion (Kshirsagar 2006).

13.2.3 Singular Spectral Analysis

Singular spectrum analysis (Ghil et al. 2002) is a nonparametric method that seeks to overcome the problems of finite sample length and noisiness of sampled time series by using a data-adaptive basis set. The method is useful in understanding nonlinear systems analysis. The method uses singular value decomposition (SVD) of the lagged covariance matrices and embeds a time series $\{R_d(t) : t = 1, \ldots, N\}$ in a vector space of dimension M. The embedding approach constructs a sequence of $\{\mathbf{R}_d(t)\}$ of M-dimensional vectors from the original time series (d) by using lagged copies of the scalar data $\{R_d(t) : 1 \leq t \leq N\}$ (see, Ghil et al. 2002; Ndehedehe et al. 2018),

$$\mathbf{R}_d(t) = (R(t), R(t+1, \ldots, R(t+M-1)), \tag{13.21}$$

the vectors $\mathbf{R}_d(t)$ are indexed by $t = 1, \ldots, N'$, where $N' = N - M + 1$. The SSA calculates the principal directions of extension of the sequence of augmented vector $\{\mathbf{R}_d(t) : t = 1, \ldots, N'\}$ in phase space using an eigenvalue-eigenvector decomposition of the $M \times M$ covariance matrix or simply through a SVD of the trajectory matrix (Ghil et al. 2002). From the SVD method, meaningful time series are reconstructed by means of diagonal averaging (e.g., Ndehedehe et al. 2018; Unnikrishnan and Jothiprakash 2015; Ghil et al. 2002).

13.2.4 Multiple Linear Regression Analysis

The Multiple Linear Regression Analysis (MLRA) method is a statistical technique used to model the relationships between a dependent variable and one or more independent variables. It uses a least square approach and has been widely applied in hydrology and climate science to explain the possible relationships between key variables (see, e.g., Diatta and Fink 2014; Seidou et al. 2007). In order to understand the seasonal and inter-annual variations in the data series (i.e., TWS and rainfall) at a given grid point, the regression model of the form (e.g., Rieser et al. 2010):

$$X(t) = \beta_0 + \beta_1 t + \beta_2 sin(2\pi t) + \beta_3 cos(2\pi t) + \beta_4 sin(4\pi t) + \beta_5 cos(4\pi t) + \varepsilon \quad (13.22)$$

can be fitted to the time series of the data. X(t) is time series of the data at time t, β_0 is a constant offset, β_1 is the linear trend, β_2 and β_3 account for the annual signal while β_4 and β_5 represent the semi-annual signals. The model bias ε is taken as the deviation between model outputs and observations. Least square fitting approach is used to estimate the regression coefficients, and the selected harmonic components (i.e., annual amplitude and semi-annual amplitude) are computed as:

$$Annual\ Amplitude = \sqrt{(\beta_2)^2 + (\beta_3)^2}, \quad (13.23)$$

$$Semi\ Annual\ Amplitude = \sqrt{(\beta_4)^2 + (\beta_5)^2}, \quad (13.24)$$

and the root mean square error was computed as:

$$RMSE = \sqrt{\frac{1}{n}\sum_{i=1}(x_{obs} - x_{sim})^2}, \quad (13.25)$$

where x_{obs} and x_{sim} are observations and simulated values from the regression model, respectively, for n months. In order to assess the model's fitness in simulating TWS and rainfall, the coefficient of multiple determinations was computed for each grid point of the data. Besides the root mean square errors (RMSEs) and the coefficients of multiple determinations, the bias (i.e., difference between the true value being estimated and the expected value of an estimator) can also be used to evaluate trend, annual and semi-annual components of hydrological data. The use of MLR to model trends, annual, and semi-annual amplitudes of precipitation is illustrated in Fig. 13.4. The spatial distribution of changes in decadal rainfall over the Volta basin between 1950 and 2010 shows large annual amplitudes between Latitudes 10 and 14 °C. The periods between 1950–1960 and 1960–1970 were relatively wetter in the Volta basin as the mean annual amplitude and mean monthly rainfall, respectively, were relatively stronger than in other decades. The observed increase in precipitation trends spreading mostly in the southern basin during 1980–1990 period appears to be the strongest increase since 1950 (Row 1, Fig. 13.4). While the annual rainfall patterns

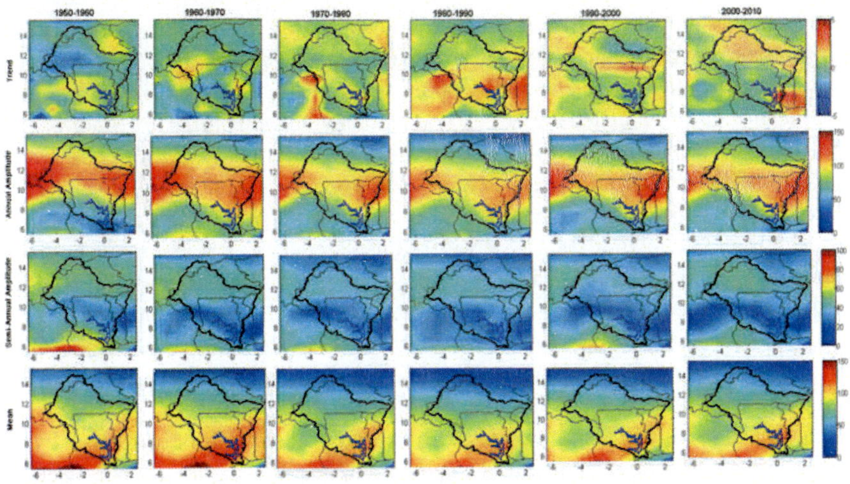

Fig. 13.4 Decadal rainfall (mm) patterns over the Volta basin (1950–2010). Row 1: Trends in precipitation. Row 2: Mean annual amplitude of precipitation. Row 3: Mean semi-annual amplitude of precipitation. Row 4: Decadal mean precipitation. *Source* Ndehedehe et al. (2016c)

in the 1980s were relatively weak due to the extreme drought of the 1983/1984, strong annual rainfall patterns similar to those of the 1950s are observed in northern Togo and Benin between 1990–2000 and 2000–2010 periods (Row 2, Fig. 13.4). This seems to be a recovery in the annual rainfall patterns of those countries (Row 2, Fig. 13.4). The semi-annual rainfall patterns seem rather weak in the last decades compared to the 1950s (Row 3, Fig. 13.4) while the mean rainfall shows a general decline at the southern basin as rainfall reduces from about 150 mm in the 1950–1960 period to 110 mm during 2000–2010 period (Row 3 and 4, Fig. 13.4).

13.2.5 Machine Learning

In this section, we will focus on the support vector machine regression model (SVMR, Vapnik 1995). The SVMR algorithm is a non-parametric machine learning system that implements the structural risk minimization inductive principle. The goal is to achieve an improved generalization with few learning patterns or training data (e.g., Smola and Schölkopf 2004). Although the support vector machine was originally introduced as a non-parametric machine learning classifier, Vapnik (1995) extended the algorithm for regression using an ε-insensitive loss function. The SVMR concept is based on the computation of a linear regression function in a high-dimensional feature space in which the input data (x_i) are mapped through a non-linear function (e.g., Okwuashi and Ndehedehe 2017). As the relationship between a multidimensional input vector x and the output y is unknown and could be non-linear (e.g.,

Wauters and Vanhoucke 2014), this mapping is warranted. Firstly, the model seeks a linear hyperplane that fits the multidimensional input vectors to output values, and finally, the SVMR predicts output values that are contained in a validation set (e.g., Okwuashi and Ndehedehe 2017; Wauters and Vanhoucke 2014; Smola and Schölkopf 2004; Vapnik 1995). The focus of the SVMR is not to minimize observed training error in any given dataset. However, it aims to minimize the generalization error bound with the overarching goal of achieving a generalized performance (e.g., Smola and Schölkopf 2004). SVMR is considered a non-parametric technique because it relies on kernel functions and its objective is to find a function, $f(x)$, with at most ε-deviation from the target y and at the same time is as flat as possible (e.g., Smola and Schölkopf 2004; Vapnik 1995). This is contrary to finding a function $f(x)$ that returns the best fit through minimizing the sum of squared errors as is the case in least squares.

The contributions of climate change to hydrological conditions or extreme events through climate teleconnection patterns can be diagnosed using the support vector machine regression model as was demonstrated in the Lake Chad and Murray Darling basins and Brazil (Ndehedehe et al. 2020a,b, 2021). In the use of SVRM in these regions (Brazil and Lake Chad and Murray Darling basins) to predict impacts of climate teleconnections on drought evolutions, it is noted that SVMR aims to seek a function $f(x)$ that deviates from $\mathbf{y}_i$ by a value no greater than ε for each training point in the data $\mathbf{x}_i$. Assuming these observations, i.e., the training data $[(\mathbf{x}_i\ \mathbf{y}_i), \ldots, (\mathbf{x}_n\ \mathbf{y}_n)] \subset \chi \times \Re$, where χ represent the spatial location of the input patterns. The linear SVMR function $f(x)$ will then be formulated as (e.g., Smola and Schölkopf 2004; Vapnik 1995)

$$f(x) = \langle \omega, x \rangle + b \; with \; \omega \in \chi, b \in \Re. \tag{13.26}$$

Minimizing the norm, i.e., $\parallel \omega \parallel = \langle \omega, \omega \rangle$ ensures flatness and can be written as a convex optimization problem:

$$\text{minimize } \tfrac{1}{2} \parallel \omega \parallel^2$$

$$subject\ to \begin{cases} yi - \langle \omega, x \rangle - b \leq \varepsilon \\ \langle \omega, x \rangle + b \leq -yi \leq \varepsilon. \end{cases} \tag{13.27}$$

It is possible that a function that approximates all pairs $(\mathbf{x}_i, \mathbf{y}_i)$ does not exist to satisfy these constraints for all points or that the convex optimization problem is not feasible. This is solved by introducing slack variables ξ_i, ξ_i^* to cope with otherwise infeasible constraints of the optimization problem in Eq. 13.27. With the inclusion of the slack variables, the objective function also called the primal formula as summarized in Vapnik (1995) is

$$\text{minimize } \frac{1}{2} \parallel \omega \parallel^2 + C \sum_{i=1}^{n} (\xi_i + \xi_i^*)$$

$$subject\ to \begin{cases} yi - \langle \omega, x \rangle - b \leq \varepsilon + \xi_i \\ \langle \omega, x \rangle + b \leq -yi \leq \varepsilon + \xi_i^* \\ \xi_i, \xi_i^* \ \geq 0 \end{cases} \tag{13.28}$$

C is a constant (a positive number) called the box constraint (or a pre-specified penalty value), which determines the trade-off between the flatness of $f(x)$ and the amount up to which deviations larger than ε are tolerated. In other words, we do not care about errors as long as they are less than ε. Essentially, C controls the penalty imposed on observations that lie outside the epsilon margin (ε) and helps to prevent regularization (Vapnik 1995). This is similar to dealing with the ε-insensitive loss function $\mid \xi_i \mid_\varepsilon$, which is described by

$$\mid \xi \mid_\varepsilon = \begin{cases} 0 & if \ \mid \xi \mid \leq \varepsilon \\ \mid \xi \mid - \varepsilon & otherwise. \end{cases} \tag{13.29}$$

The SVMR technique evaluates each run of the experiment by partitioning the data internally into training (i.e., 90% of the total data) and validation components. The remaining 10% of the observed data were, thereafter, used for forward prediction based on the hold-out method of cross-validation (e.g., Haley 2017). The stratified partitioning of the data using this approach ensures that each partition includes similar amount of observations from each group. The predicted and observed time series are then compared using Pearson correlation. In this kind of application, the SVM regression model is usually trained to fit the data by testing different kernel functions (linear, Gaussian, and polynomial) (e.g., Okwuashi and Ndehedehe 2017; Gunn 1998). As indicated in Fig. 13.5, three kernel functions were tested in the Murray Darling basin case and the Gaussian kernel improved the prediction of teleconnection patterns (Ndehedehe et al. 2021). The SVMR technique evaluates each run of the experiment using regression by partitioning the data into training and test sets components.

The illustration in Fig. 13.5 shows how rotated PCA was combined with SVMR to assess the constraints that could be posed by climatic variations on water availability. Climate teleconnection indices were used as predictors (independent variables) while the temporal series of drought indices obtained from the varimax process for all aggregation scales were designated as the response variables (dependent variable). Extreme events such as droughts and floods are triggered by global climate change, either through physical mechanisms, changes in atmospheric circulation patterns, or multi-scale oscillations like the ENSO, and other teleconnections (Ndehedehe et al. 2021). With this in mind, the first step to diagnosing the influence of climate variability indices in the study region was to identify a suitable support vector machine kernel function that will explore the complex non-linear interactions of these climate modes with drought indicators. Following the adoption of a suitable support vector machine kernel (Gaussian), the second step was to identify the cumulation scales that can be used to better predict drought evolutions based on the relationship between the observed and predicted values (drought indicators). The assumption here is that,

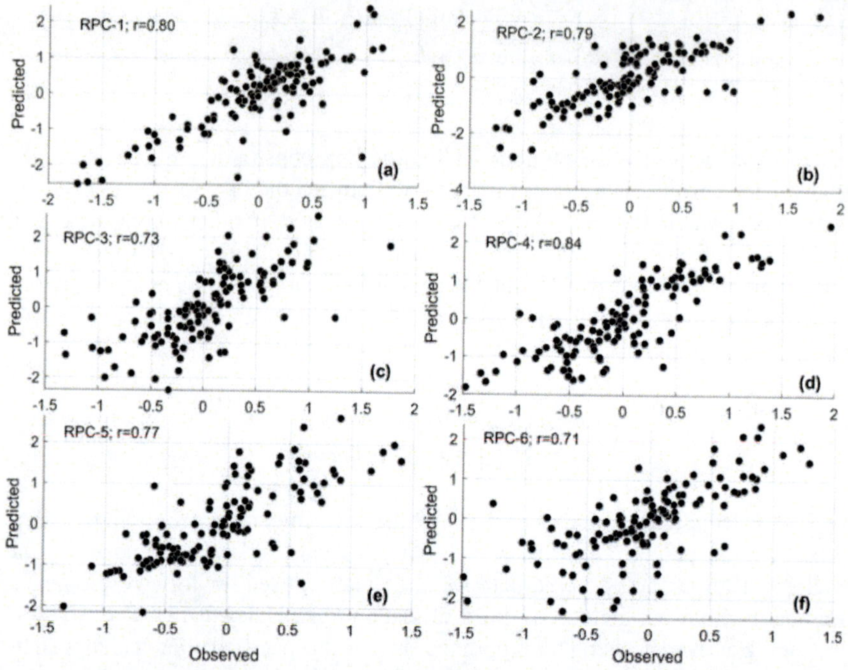

Fig. 13.5 Assessing the association of climate variability indices with climatic hotspots in Australia using support vector machine regression. RPC is rotated principal component derived from varimax rotation of the drought indicators. *Source* Ndehedehe et al. (2021)

if the latter is strongly correlated with the former, then the hydrological region in question is considerably vulnerable to the influence of global climate teleconnection patterns (Ndehedehe et al. 2021). The correlations between predicted and observed indicators (Fig. 13.5) show the efficacy of multivariate and machine learning methods in extracting the influence of climate modes on changes in hydro-climatic conditions.

13.2.6 Quantile Function of Hydrological Stores

The quantile function of storage (QFS) has been used to measure the response of hydrological systems to changes in climatic conditions (Ndehedehe et al. (2021); Ferreira et al. (2020)). QFS is a plot that represents the relationship between storage and the ratio of time (expressed in percentages) that storage in a river basin or region is likely to equal or exceed some specified value or amount (see, Ferreira et al. 2020). Studies have shown the usefulness of information generated from this plot to characterize the hydrological capacity of river basins to retain storage of various magnitudes or assess land water storage deficits through the use of QFS variants like terrestrial water storage duration curve and Flow Duration Curve (e.g., Ndehedehe et al. 2021; Tourian et al. 2013; Sawicz et al. 2011; Yirdaw et al. 2008). Recently, Ndehedehe et al. (2021) employed the QFS to measure the response of hydrological systems to changes in climatic conditions in the Murray Darling basin. Using the Weibull formula highlighted in Ferreira et al. (2020), the probability of exceedance is computed as,

$$P = \frac{M}{n+1}.100\% \tag{13.30}$$

where M is the ranked position on the listing starting with 1 for the largest monthly values as in the hydrological variables of interest (e.g., rainfall, terrestrial water storage, soil moisture, groundwater, etc.), which has n number of months. To quantify an index of variability for the hydrological indicators assessed, the slope of QFS (S_{QFS}) was computed as the interval between the 33rd (S33%) and 66th (S66%) storage percentiles and is given as (Sawicz et al. 2011),

$$S_{QFS} = \frac{S_{33\%} - S_{66\%}}{66 - 33}, \tag{13.31}$$

where Q33% and Q66% are the values at the 33rd and 66th percentiles, respectively. This means that a high S_{QFS} indicates a variable water storage regime. On the other hand, a low S_{QFS} implies a more damped response. Such response arguably is the aftermath of a combination of prolonged wet conditions or a wide-spread and year-round rainfall and/or the dominance of groundwater contribution among other factors (see, Sawicz et al. 2011). Ferreira et al. (2020) argued further, indicating that a plethora of factors such as hydro-geological setting, aquifer flow, storage types, storage potential, and recharge rates impact the S_{QFS} of groundwater storage. Moreover, the monthly percentage water storage deficit or its excess was computed using long-term median, maximum and minimum observations of the data as

$$D = \frac{S_i - median(S_j)}{max(S_j) - min(S_j)}.100\%, \tag{13.32}$$

where S_i is the monthly water storage anomaly and median(S_j), max(S_j), and min(S_j), are, respectively, the long-term median, maximum, and minimum monthly storage anomaly. For this anomaly, the mean removed from each monthly value of S has been remove and with i ranging from 1 to n for all monthly time steps in the data and j ranging from 1 to 12 for the 12 calender months of each year.

Recall that in Chap. 9, the QFS and cumulative storage deficit were illustrated for the Murray Darling basin using terrestrial water storage (TWS), groundwater storage (GWS) precipitation, and soil moisture. The QFS has been very valuable in understanding how hydrological systems respond to water climate change, be it long-term declines or increase in rainfall. Some testbeds implementing this method recently have demonstrated the importance of QFS (Ferreira et al. 2020; Ndehedehe et al. 2021), especially for agricultural hubs where the impacts of both climate variations

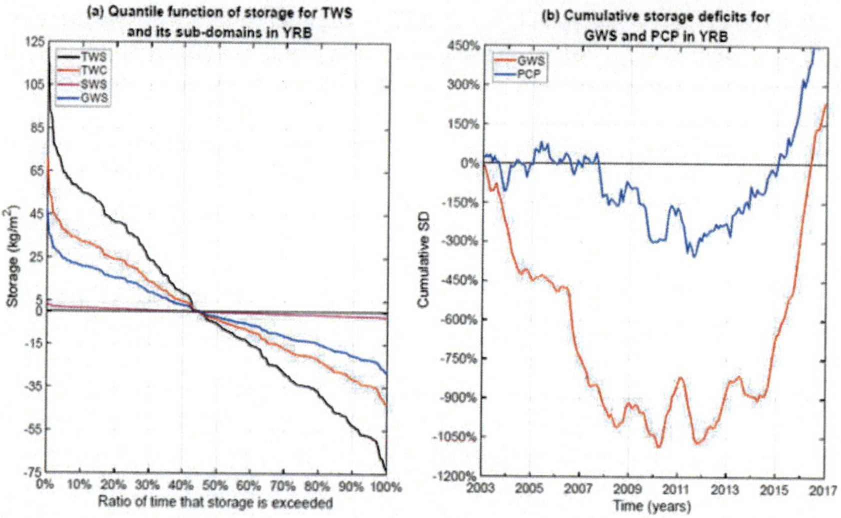

Fig. 13.6 Response of hydrological stores to droughts in the Yangtze River basin (YRB) **a** Quantile function of storage for terrestrial water storage (TWS), surface water storage (SWS), and groundwater storage (GWS) and **b** Cumulative storage deficit for GWS and precipitation (PRCP). *Source* Ferreira et al. (2020)

and human water management are profound. The QFS and cumulative storage deficit approach were also used in the characterization of the hydro-geological regime of Yangtze River basin. Figure 13.6a shows that during the period studied (2003–2016), aquifer storage loss occurred 55% of the time in the Yangtze River basin based on the quantile function of storage. This loss is also evident in the cumulative storage deficit for GWS, which shows a decrease between Jan 2003 until 2008 though some recovery happened after 2011 (Fig. 13.6b). The cumulative storage deficit curve also shows the rise in precipitation after the 2011 period when GWS recovered suggesting the role of rainfall in recharge and groundwater recovery (Fig. 13.6b).

Comparing the Yangtze basin with the Murray Darling basin, we see the importance and usefulness of the QFS and cumulative storage deficit approach. In particular, understanding response of hydrological systems to climate variability and emphasizing the intrinsic response of some regions to climate change and water use. In Fig. 9.4 (Chap. 9), for example, we showed that the Murray Darling basin gained soil water only 40% of the time and lost it for 60% of the time (2002–2017). Unlike terrestrial water storage, stored groundwater was lost approximately 44% of the time. However, the QFS highlights the vulnerability of soil moisture to climatic conditions compared to TWS and GWS, which respond much slower to drought conditions. The cumulative storage deficit, on the other hand, showed that soil moisture responded to the big wet period of 2010/2011 over Australia with rather little or no lag (Ndehedehe et al. 2021). Apparently, the curve in Fig. 9.4 (Chap. 9) also suggested there was a delayed response in TWS (about 6 months) and GWS (approximately

2 years). GWS showed an opposite phase with rainfall because of this lag time. The rise of GWS between 2002 and 2005 when rainfall was negative is another good example to emphasize this lag (Fig. 9.4). Ultimately, extreme climatic conditions like droughts are expected to propagate rather slowly to groundwater compared to soil moisture. Similarly, the propagation of extreme wet conditions caused by above-normal rainfall to nourishing hydrological stores (e.g., soil moisture, groundwater, surface water) is slow for groundwater compared to surface water and soil moisture (Ndehedehe et al. 2021).

13.2.7 *Quantifying Trends in Hydrological Data*

The spatial distribution of trends can be estimated using the least squares method. Furthermore, using a non-parametric method, trends in rainfall, soil moisture, groundwater and TWS were estimated for each grid cell. Based on the least squares estimation of the regression coefficient, MLRA can also be used to estimate trends as parameterized in Ndehedehe et al. (2016a). However, the Sen's slope (Sen 1968) estimator was used to estimate trends since it is robust and resistant to outliers. Sen slope (S_i) is the median overall values of the whole data and is estimated as

$$S_k = Median\left(\frac{P_j - P_i}{j - i}\right), \ for \ (1 \leq i < j \leq n), \tag{13.33}$$

where P_j and P_i represent data values at time j and i ($j > i$), respectively while n is the number of data. The slope can be positive indicating increasing trend or negative, indicating decreasing trend. The significance of observed trends ($\alpha = 0.05$) is usually tested using the Man–Kendall's test (Mann 1945; Kendall 1970) and is a widely used non-parametric method in testing the significance of trends. Parametric trend tests (e.g., Student t test, turning point, regression, inversion tests, etc.) can also be used to examine the statistical significance of trends but sometimes they violate normality, hence the popularity and preference of non-parametric methods such as the Man–Kendall, Hotelling–Pabst test, and Sen test amongst others (see, Machiwal and Jha 2012). The Mann-Kendall statistic (M) is a non-parametric method and is calculated as

$$M = \sum_{k=1}^{n-1} \sum_{j=k+1}^{n} sgn(P_j - P_i) \tag{13.34}$$

where n is the number of data locations. Machiwal and Jha (2012) suggests that even n values as low as 10 can be used in Man–Kendall's test provided there are no too many tied values. Supposing that $x = P_j - P_i$, then $sgn(x)$ is estimated as

$$sgn(x) = \begin{cases} 1, & if\ x > 0 \\ 0, & if\ x = 0 \\ -1, & if\ x < 0. \end{cases} \tag{13.35}$$

The M statistic represents the positive and negative differences for all data samples under consideration. The mean of the statistics under the null hypothesis is zero and is given as E[M] = 0 while its variance (σ) is given as

$$\sigma = \frac{n(n-1)(2n+5) - \sum_{k=1}^{n}(t_j - 1)(2t_j + 5)}{18}. \tag{13.36}$$

The Man–Kendall test statistics (M) is approximately normally distributed, subject to the following Z-transformation,

$$Z = \begin{cases} \frac{M-1}{[\sigma]^{1/2}}, & if\ M > 0 \\ 0, & if\ M = 0 \\ \frac{M+1}{[\sigma]^{1/2}}, & if\ M < 0. \end{cases} \tag{13.37}$$

The null hypothesis (no trend), H_0, which shows that a sample data ($\mathbf{X}_i, i = 1, 2 \ldots, n$) is independent and identically distributed was tested at $\alpha = 0.05$ (95% confidence level). If the computed absolute value of the test statistics (Z) is greater than the critical value of the standard normal distribution ($z_{\frac{\alpha}{2}}$), the hypothesis of negative or positive trend is rejected at the 95% confidence level and the alternative hypothesis (monotonic trend exist in $\mathbf{X}_i$), H_1, is accepted. Because of the presence of autocorrelation in some hydrological time series (e.g., rainfall and streamflow), which increases the probability of detecting trends (e.g., Yue et al. 2002), Hamed and Rao (1998) argued that the empirical significance levels of the original Mann-Kendall trend test are in great error vis-a-vis the nominal significance levels. To address this problem, a modified Mann-Kendall, which considers autocorrelation between the ranks of the data and provides an accurate empirical significance level was suggested by Hamed and Rao (1998) while Yue et al. (2002) proposed an alternative approach of *trend-free pre-whitening* procedure for detecting a significant trend in serially correlated series. The null hypothesis, H_0 is tested at $\alpha = 0.05$. Furthermore, based on the least square estimation of the regression coefficient, MLRA can also be used to estimate trends (Ndehedehe et al. 2016a).

13.2.8 *Correlation Analyses*

Several approaches appropriated for modelling the relationships between multivariate data measured at different times include autocorrelation, coherence analysis, dynamic factor analysis, cross-correlation, etc., (Boker et al. 2002). Pearson's correlation coefficient (r) and coefficient of determination can be used to examine the

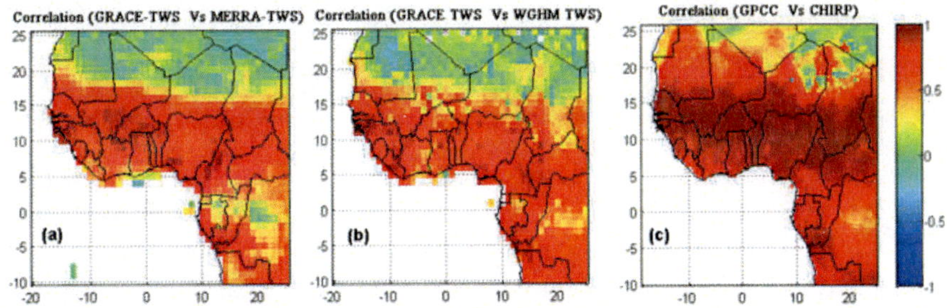

Fig. 13.7 Correlation of terrestrial water storage (TWS) products with two precipitation products (CHIRP and GPCC) over tropical West Africa. **a** GRACE and MERRA TWS (2002–2014) **b** GRACE and WGHM TWS (2002–2009) and **c** GPCC and CHIRP based precipitation (1980–2014). *Source* Ndehedehe et al. (2018)

strength of agreement between two different hydrological variables (e.g., rainfall and groundwater). Cross-correlation measures the similarity between two series as a function of the displacement of one relative to the other, and also determines the time lag between hydrological signals. Monthly grids of precipitation can be correlated with groundwater to understand the relationship between them, especially the response time of groundwater aquifers to rainfall. In the cross-correlation approach, the position of the peak value indicates the time offset when the peak association occurred (i.e., when the two signals are the most similar) in the hydrological time series. To understand the response time (i.e., the time lag with maximum correlation) of groundwater to rainfall of any region, cross-correlation method is crucial. The example in Fig. 13.7a–b shows that correlations are strong in much of the tropical West African region and suggest that the spatial and temporal dynamics of different terrestrial water storage products are consistent with GRACE-TWS. The CHIRPS precipitation data are considerably strong, consistent, and in agreement with GPCC-based precipitation in the region except in the extreme north-west Sahel (the Sahara Desert) (Fig. 13.7c). The GRACE-MERRA and GRACE-WGHM relationships in the Sahara Desert and the central Congo basin cuvette located in the upper Democratic Republic of Congo are considerably weak, and to some extent non-existent (Fig. 13.7a–b).

References

Agutu N, Awange J, Zerihun A, Ndehedehe C, Kuhn M, Fukuda Y (2017) Assessing multi-satellite remote sensing, reanalysis, and land surface models' products in characterizing agricultural drought in East Africa. Remote Sens Environ 194:287–302. https://doi.org/10.1016/j.rse.2017.03.041

Barnett TP, Preisendorfer R (1987) Origins and levels of monthly and seasonal forecast skill for United States surface air temperatures determined by canonical correlation analysis. Mon Weather Rev 115(9):1825–1850. https://doi.org/10.1175/1520-0493(1987)115<1825:OALOMA>2.0.CO;2

Boker SM, Xu M, Rotondo JL, King K (2002) Windowed cross correlation and peak picking for the analysis of variability in the association between behavioral time series. Psychol Methods 7(3):338–555. https://doi.org/10.1037/1082-989X.7.3.338

Bretherton CS, Smith C, Wallace JM (1992) An intercomparison of methods for finding coupled patterns in climate data. J Clim 5(6):541–560. https://doi.org/10.1175/1520-0442(1992)005<0541:AIOMFF>2.0.CO;2

Cardoso J-F (1991) Super-symmetric decomposition of the fourth-order cumulant tensor, blind identification of more sources than sensors. http://perso.telecom-paristech.fr/~cardoso/Papers.PDF/icassp91.pdf. 15 Jan 2016

Cardoso JF (1999) High-order contrasts for independent component analysis. Neural Comput 11:157–192

Cardoso JF, Souloumiac A (1993) Blind beamforming for non-gaussian signals. IEE Proc 140(6):362–370

Chen H, Sun Y, Gao J, Hu Y, Yin B (2018) Solving partial least squares regression via manifold optimization approaches. IEEE Trans Neural Netw Learn Syst 1–13. https://doi.org/10.1109/TNNLS.2018.2844866

Compagnucci RH, Richman MB (2008) Can principal component analysis provide atmospheric circulation or teleconnection patterns? Int J Climatol 28(6):703–726. https://doi.org/10.1002/joc.1574

de Jong S (1993) SIMPLS: an alternative approach to partial least squares regression. Chemom Intell Lab Syst 18(3):251–263. https://doi.org/10.1016/0169-7439(93)85002-x

Diatta S, Fink AH (2014) Statistical relationship between remote climate indices and West African monsoon variability. Int J Climatol 34(12):3348–3367. https://doi.org/10.1002/joc.3912

Ferreira V, Yong B, Tourian M, Ndehedehe C, Shen Z, Seitz K, Dannouf R (2020) Characterization of the hydro-geological regime of yangtze river basin using remotely-sensed and modeled products. Sci Total Environ 718:137354. https://doi.org/10.1016/j.scitotenv.2020.137354

Forina M, Armanino C, Lanteri S, Leardi R (1989) Methods of varimax rotation in factor analysis with applications in clinical and food chemistry. J Chemom 3(S1):115–125. https://doi.org/10.1002/cem.1180030504

Ghil M, Allen MR, Dettinger MD, Ide K, Kondrashov D, Mann ME, Robertson AW, Saunders A, Tian Y, Varadi F, Yiou P (2002) Advanced spectral methods for climatic time series. Rev Geophys 40(1):3–1–3–41. https://doi.org/10.1029/2000RG000092

Glahn HR (1968) Canonical correlation and its relationship to discriminant analysis and multiple regression. J Atmos Sci 25(1):23–31. https://doi.org/10.1175/1520-0469(1968)025<0023:CCAIRT>2.0.CO;2

Graham NE, Michaelsen J, Barnett TP (1987) An investigation of the el niño-southern oscillation cycle with statistical models: 1. predictor field characteristics. J Geophys Res: Ocean 92(C13):14251–14270. https://doi.org/10.1029/JC092iC13p14251

Gunn S (1998) Support vector machines for classification and regression, ISIS technical report. University of Southampton, Southampton. https://svms.org/tutorials/Gunn1998.pdf 10th Sept

Haley MR (2017) K-fold cross validation performance comparisons of six Naive portfolio selection rules: how Naive can you be and still have successful out-of-sample portfolio performance? Ann Finance 13(3):341–353. https://doi.org/10.1007/s10436-017-0301-4

Hamed KH, Rao AR (1998) A modified Mann-Kendall trend test for autocorrelated data. J Hydrol 204(1):182–196. https://doi.org/10.1016/S0022-1694(97)00125-X

Hannachi A, Jolliffe IT, Stephenson DB (2007) Empirical orthogonal functions and related techniques in atmospheric science: a review. Int J Climatol 27:1119–1152. https://doi.org/10.1029/96JC00922

Horel JD (1984) Complex principal component analysis: theory and examples. J Appl Meteorol Climatol 23(12):1660–1673. https://doi.org/10.1175/1520-0450(1984)023<1660:CPCATA>2.0.CO;2

Hotelling H (1936) Relations between two sets of variates. Biometrika 28(3–4):321–377. https://doi.org/10.2307/2333955

Jolliffe IT (2002) Principal component analysis, 2nd edn. Springer series in statistics. Springer, New York

Kaiser HF (1958) The varimax criterion for analytic rotation in factor analysis. Psychometrika 23(3):187–200. https://doi.org/10.1007/BF02289233

Kendall MG (1970) Rank correlation methods, 4th edn. Griffin, London, UK

Kshirsagar AM (2006) Wilks's lambda criterion. Encycl Stat Sci 15. https://doi.org/10.1002/0471667196.ess1388.pub2

Kurnik B, Kajfež-Bogataj L, Horion S (2015) An assessment of actual evapotranspiration and soil water deficit in agricultural regions in Europe. Int J Climatol 35(9):2451–2471. https://doi.org/10.1002/joc.4154

Lewis-Beck M, Bryman A, Futing T (2003) Encyclopedia of social sciences research methods. Thousand Oaks (CA), Sage. https://www.utdallas.edu/~herve/Abdi-PLS-pretty.pdf 4th Jan 2019

Livezey RE, Smith TM (1999) Considerations for use of the Barnett and Preisendorfer (1987) algorithm for canonical correlation analysis of climate variations. J Clim 12(1):303–305. https://doi.org/10.1175/1520-0442(1999)012<0303:CFUOTB>2.0.CO;2

Machiwal D, Jha MK (2012) Hydrological time series: theory and practice. Springer, India

Mann HB (1945) Nonparametric tests against trend. Econometrica 13(3):245–259. https://doi.org/10.2307/1907187

Martinez WL, Martinez AR (2005) Exploratory data analysis with MATLAB. Chapman and Hall/CRC Press LLC, UK. 1-58488-366-9

Mohino E, Rodríguez-Fonseca B, Losada T, Gervois S, Janicot S, Bader J, Ruti P, Chauvin F (2011) Changes in the interannual SST-forced signals on West African rainfall AGCM intercomparison. Clim Dyn 37(9–10):1707–1725. https://doi.org/10.1007/s00382-011-1093-2

Montazerolghaem M, Vervoort W, Minasny B, McBratney A (2016) Long-term variability of the leading seasonal modes of rainfall in south-eastern Australia. Weather Clim Extrem 13:1–14. https://doi.org/10.1016/j.wace.2016.04.001

Ndehedehe C, Awange J, Agutu N, Kuhn M, Heck B (2016) Understanding changes in terrestrial water storage over West Africa between 2002 and 2014. Adv Water Resour 88:211–230. https://doi.org/10.1016/j.advwatres.2015.12.009

Ndehedehe CE, Agutu NO, Okwuashi OH, Ferreira VG (2016) Spatio-temporal variability of droughts and terrestrial water storage over Lake Chad Basin using independent component analysis. J Hydrol 540:106–128. https://doi.org/10.1016/j.jhydrol.2016.05.068

Ndehedehe CE, Anyah RO, Alsdorf D, Agutu NO, Ferreira VG (2019) Modelling the impacts of global multi-scale climatic drivers on hydro-climatic extremes (1901–2014) over the Congo basin. Sci Total Environ 651:1569–1587. https://doi.org/10.1016/j.scitotenv.2018.09.203

Ndehedehe CE, Awange JL, Agutu NO, Okwuashi O (2018) Changes in hydro-meteorological conditions over tropical West Africa (1980–2015) and links to global climate. Glob Planet Chang 162:321–341. https://doi.org/10.1016/j.gloplacha.2018.01.020

Ndehedehe CE, Awange JL, Corner R, Kuhn M, Okwuashi O (2016) On the potentials of multiple climate variables in assessing the spatio-temporal characteristics of hydrological droughts over the Volta Basin. Sci Total Environ 557–558:819–837. https://doi.org/10.1016/j.scitotenv.2016.03.004

Ndehedehe CE, Ferreira VG (2020) Assessing land water storage dynamics over Southern America. J Hydrol 580:124339. https://doi.org/10.1016/j.jhydrol.2019.124339

Ndehedehe CE, Ferreira VG, Agutu NO, Onojeghuo AO, Okwuashi O, Kassahun HT, Dewan A (2021) What if the rains do not come? J Hydrol 595:126040. https://doi.org/10.1016/j.jhydrol.2021.126040

Ndehedehe CE, Ferreira VG, Onojeghuo AO, Agutu NO, Emengini E, Getirana A (2020) Influence of global climate on freshwater changes in Africa's largest endorheic basin using multi-scaled indicators. Sci Total Environ 737:139643. https://doi.org/10.1016/j.scitotenv.2020.139643

Ndehedehe CE, Haile GG, Agutu NO, Ferreira VG, Getirana A, Okwuashi O (2020) Hydrological hotspots of climatic influence in Brazil: a two-step regularization approach. Atmos Res 246:105116. https://doi.org/10.1016/j.atmosres.2020.105116

Nicholson SE (2014) Spatial teleconnections in African rainfall: a comparison of 19th and 20th century patterns. Holocene 24(12):1840–1848. https://doi.org/10.1177/0959683614551230

Okwuashi O, Ndehedehe C (2017) Tide modelling using support vector machine regression. J Spat Sci 62(1):29–46. https://doi.org/10.1080/14498596.2016.1215272

Okwuashi O, Ndehedehe CE (2020) Deep support vector machine for hyperspectral image classification. Pattern Recogn 103:107298. https://doi.org/10.1016/j.patcog.2020.107298

Preisendorfer R (1988) Principal component analysis in meteorology and oceanography. Dev Atmos Sci 17. Elsevier, Amsterdam

Rangelova E, van der Wal W, Braun A, Sideris MG, Wu P (2007) Analysis of gravity recovery and climate experiment time-variable mass redistribution signals over North America by means of principal component analysis. J Geophys Res Earth Surf 112(F3):2156–2202. https://doi.org/10.1029/2006JF000615

Repelli CA, Nobre P (2004) Statistical prediction of sea-surface temperature over the tropical Atlantic. Int J Climatol 24(1):45–55. https://doi.org/10.1002/joc.982

Richman MB (1986) Rotation of principal components. J Climatol 6(3):293–335. https://doi.org/10.1002/joc.3370060305

Rieser D, Kuhn M, Pail R, Anjasmara IM, Awange J (2010) Relation between GRACE-derived surface mass variations and precipitation over Australia. Aust J Earth Sci 57(7):887–900. https://doi.org/10.1080/08120099.2010.512645

Rodell M, Famiglietti JS, Wiese DN, Reager JT, Beaudoing HK, Landerer FW, Lo M-H (2018) Emerging trends in global freshwater availability. Nature 557:651–659. https://doi.org/10.1038/s41586-018-0123-1

Sanogo S, Fink AH, Omotosho JA, Ba A, Redl R, Ermert V (2015) Spatio-temporal characteristics of the recent rainfall recovery in West Africa. Int J Climatol 35(15):4589–4605. https://doi.org/10.1002/joc.4309

Santos JF, Pulido-Calvo I, Portela MM (2010) Spatial and temporal variability of droughts in Portugal. Water Resour Res 46(3):W03503. https://doi.org/10.1029/2009WR008071

Sawicz K, Wagener T, Sivapalan M, Troch PA, Carrillo G (2011) Catchment classification: empirical analysis of hydrologic similarity based on catchment function in the eastern USA. Hydrol Earth Syst Sci 15(9):2895–2911. https://doi.org/10.5194/hess-15-2895-201

Seidou O, Asselin JJ, Ouarda TBMJ (2007) Bayesian multivariate linear regression with application to change point models in hydrometeorological variables. Water Resour Res 43:W08401. https://doi.org/10.1029/2001jb000576

Sen PK (1968) Estimates of the regression coefficient based on Kendall's Tau. J Am Stat Assoc 63(324):1379–1389. https://doi.org/10.1080/01621459.1968.10480934

Shabbar A, Barnston AG (1996) Skill of seasonal climate forecasts in Canada using canonical correlation analysis. Mon Weather Rev 124(10):2370–2385. https://doi.org/10.1175/1520-0493(1996)124<2370:SOSCFI>2.0.CO;2

Sherry A, Henson RK (2005) Conducting and interpreting canonical correlation analysis in personality research: a user-friendly primer. J Pers Assess 84(1):37–48. https://doi.org/10.1207/s15327752jpa8401_09

Singh A, Kulkarni MA, Mohanty UC, Kar SC, Robertson AW, Mishra G (2012) Prediction of Indian summer monsoon rainfall (ISMR) using canonical correlation analysis of global circulation model products. Meteorol Appl 19(2):179–188. https://doi.org/10.1002/met.1333

Smola AJ, Schölkopf B (2004) A tutorial on support vector regression. Stat Comput 14(3):199–222. https://doi.org/10.1023/B:STCO.0000035301.49549.88

Theis FJ, Gruber P, Keck IR, Meyer-bäse A, Lang EW (2005) Spatiotemporal blind source separation using double-sided approximate joint diagonalization. In: Proc, EUSIPCO

Tourian MJ, Sneeuw N, Bárdossy A (2013) A quantile function approach to discharge estimation from satellite altimetry (ENVISAT). Water Resour Res 49(7):4174–4186. https://doi.org/10.1002/wrcr.20348

Unnikrishnan P, Jothiprakash V (2015) Extraction of nonlinear rainfall trends using singular spectrum analysis. J Hydrol Eng 20(12):1–14. https://doi.org/10.1061/(ASCE)HE.1943-5584.0001237

Vapnik V (1995) The nature of statistical learning theory. Springer, New York, NY

Wauters M, Vanhoucke M (2014) Support vector machine regression for project control forecasting. Autom Constr 47:92–106. https://doi.org/10.1016/j.autcon.2014.07.014

Westra S, Brown C, Lall U, Koch I, Sharma A (2010) Interpreting variability in global SST data using independent component analysis and principal component analysis. Int J Climatol 30(3):333–346. https://doi.org/10.1002/joc.1888

Wilks D (2011) Statistical methods in the atmospheric sciences, 3rd edn. Academic press, USA

Wold S, Sjöström M, Eriksson L (2001) PLS-regression: a basic tool of chemometrics. Chemom Intell Lab Syst 58(2):109–130. https://doi.org/10.1016/S0169-7439(01)00155-1

Yirdaw SZ, Snelgrove K, Agboma C (2008) GRACE satellite observations of terrestrial moisture changes for drought characterization in the Canadian Prairie. J Hydrol 356:84–92

Yu Z-P, Chu P-S, Schroeder T (1997) Predictive skills of seasonal to annual rainfall variations in the U.S. affiliated pacific islands: canonical correlation analysis and multivariate principal component regression approaches. J Clim 10(10):2586–2599. https://doi.org/10.1175/1520-0442(1997)010<2586:PSOSTA>2.0.CO;2

Yue S, Pilon P, Phinney B, Cavadias G (2002) The influence of autocorrelation on the ability to detect trend in hydrological series. Hydrol Process 16(9):1807–1829. https://doi.org/10.1002/hyp.1095

Ziehe A (2005) Blind source separation based on joint diagonalization of matrices with applications in biomedical signal processing. Ph.D. thesis, Universitat Potsdam. http://en.youscribe.com/catalogue/reports-and-theses/knowledge/blind-source-separation-based-on-joint-diagonalization-of-matrices-1424347. 15 May 2015

Part IX
Satellite Hydrology

Chapter 14
Satellite Observations of Terrestrial Water Storage

Abstract Global climate is changing and the impacts of such changes are evident in natural systems and acceleration of the water cycle. Satellite hydrology missions are helping with tracking the impacts of these changes on the global water cycle. The Gravity Recovery and Climate Experiment (GRACE) satellite is one of such missions that will continue to benefit studies of global mass transport and also help to understand the impacts of climate variability and anthropogenic influence on freshwater systems, including groundwater. This chapter elaborates on GRACE as an important tool that underpins remote sensing hydrology as an emerging discipline. Additionally, several innovative use of optical, radar, Global Navigation Satellite System, and other satellite geodetic missions in the context of quantitative and multidisciplinary aspects of remote sensing hydrology are also detailed. Scientific advancements in reconstructing hydrological quantities (e.g., water storage components, discharge, etc) from satellites and models using machine learning, data assimilation, geo-computational methods and advanced mathematical models are discussed. As a case study, terrestrial water storage changes over South America is assessed using GRACE data by developing a multivariate predictive framework to improve understanding of important drivers of continental water storage changes.

14.1 Introduction

Water is fundamental to environmental wellbeing, supporting a broad range of ecosystem services and the sustainability of hydro-ecological systems. But the usefulness of water is beyond its fundamental application to support food production, community health, renewable energy, and sanitation. This is because water is also essential for the nourishment of all forms of life in the human planet. The crucial role of freshwater (groundwater) in virtual water trade, water security, international policy, and sustainable development also makes it important (Jasechko and Perrone 2021). The United Nations (UN) has highlighted the importance of water in human development and sustainability in at least four of its sustainable development goals (SDGs United Nations 2015) – no poverty (Grey and Sadoff 2007), SDG2—zero hunger, SDG3—global health and wellbeing. The advancement in water resource

C. Ndehedehe, *Satellite Remote Sensing of Terrestrial Hydrology*,
https://doi.org/10.1007/978-3-030-99577-5_14

management has since shifted from the archaic simple storage, diversion and abstraction management techniques to the conventional reservoir storage, such as the grand Inga dam at Democratic Republic of Congo, as well as colossal engineering schemes for water transfer, like the transfer project between Northern and Southern China.

Regional water scarcity, however, is a serious hydrological concern, especially as a result of poorly managed demand for water and depletion of available water supplies (Veldkamp et al. 2015). Water supply is basically characterized as a function of precipitation average and variability, as well as its transformation into accessible water through groundwater recharge and surface water distribution. Potential water scarcity and crises are largely due to improper storage, poor reservoir management, and the over-dependency of regions and local communities on a specific water source (e.g., Weinthal et al. 2005). Water scarcity threat is also the result of ineffectiveness in water governance and regulatory agencies towards initializing adequate policies and capable institutional frameworks to sustainably and rightly balance striving needs with limited measures. It is expected that population increase, increased urbanization and growing demands from water-driven sectors such as energy and agriculture will lead to challenges for water quality and distribution, given that their impact may outweigh the effect of climate change (O'Connell 2017). However, regional changes in water resources, like varying rainfall and evapotranspiration patterns (Kalu et al. 2021), seasonal fluctuations evident in snowmelt associated with global warming, as well as increased potential for droughts and flood risks are inevitable given the growing impact of global climate change.

The increasing application of satellite remote sensing (SRS) for water resource management, especially with the use of Gravity Recovery and Climate Experiment (GRACE) mission since 2002 (Kalu et al. 2021; Ndehedehe et al. 2016a) has enabled our understanding of freshwater changes and impacts of climate change on continental freshwater. This innovation has therefore assisted with filling the gap of limited conventional in-situ technique in water resource management since SRS technique offers a wide range of observation even in sub-regional and continental scales (Ndehedehe and Ferreira 2020a; Sheffield et al. 2018). SRS techniques make direct and indirect observations of a large majority of hydrological components possible (McCabe et al. 2017; Zhang et al. 2016). This include the use of GRACE in Terrestrial Water Storage (TWS; i.e., surface water, soil moisture, groundwater, canopy water and snow) assessment (Ndehedehe et al. 2016a). The advancement in satellite technologies in the last few decades provides unprecedented opportunities in measurements and observations of key water budget (evapotranspiration, discharge etc.) that support and improve water resources management on several spatial scales (Sheffield et al. 2018). For instance, the GRACE mission has enabled the measurement of water storage changes across the globe by observing changes in gravity fields. Assessing the inherent variations of TWS components (surface water, soil moisture, groundwater, canopy water and snow) is important as it aids in establishing the basis of climate change. With such outcome, it is possible to make recommendation as to the required environmental response required to mitigate potential climatic hazards at basin or regional scale.

With the advent of satellite remote sensing and altimetry, a wide range of data collection and observation has been recorded in hydrology, especially in the assessment of water storage and fluxes in a changing world, thus enabling a global evaluation that spans political boundaries (Getirana et al. 2017; Famiglietti et al. 2015; Rodell and Famiglietti 2001). Several studies on the significance of SRS techniques in assessing changes in ecosystem functioning, vegetation conditions, as well as crop yield and health have been undertaken (Ndehedehe et al. 2016c; Zheng et al. 2004). Given its large coverage, SRS methods are also feasible in risk assessment and evaluations in the context of crop yield, water security, trade and repository functions (Dalin et al. 2017). Also, the large spatial resolution and high temporal coverage make it feasible for regional and global water resource monitoring in near real-time (Ndehedehe et al. 2016a). This chapter will focus on satellite hydrology applications as an integral aspect of remote sensing hydrology.

14.2 Satellite Gravity Missions: Potential in Global Freshwater Monitoring

Following the launch of GRACE satellite in 2002, monitoring changes in freshwater resources and complex hydrologic fluxes has become easier and more efficient. With the GRACE mission, one can apply monthly aggregated measurements of the earth's gravity field to track mass changes in the oceans, cryosphere and hydrosphere. Its measurement, however, is based on micron-level tracking of the satellite-to-satellite space, whose variance is subject to the individual gravitational attractions on the satellites as they move across the Earth's surface (Tapley et al. 2004a, b). This measurement is used to compute an estimated global spherical harmonic model of the earth's gravity field, which is subsequently used to determine the mass changes of the earth's surface. A major determinant of variation in mass change of the earth is changes in continental freshwater. GRACE data derived from gravity anomalies have been able to provide hydro-geodesists with new insights into variations in TWS that are difficult to observed from in-situ data. This, however, is dependent on our understanding of mass transport in the interior, and how their deviations are understood on human timescales. Given the mobility of water as well as its stability over rocks or other bodies capable of influencing the earth's gravity field, we understand that the gravity alteration caused by it can be detected on a wider range of time scales. Satellite remote sensing technique has made it possible to

- track ice-sheets and glacier ablations in the polar regions,
- monitor groundwater depletions, and changes in terrestrial stored water by collecting and archiving time variable monthly gravity fields (Tapley et al. 2004b),
- observe variations in the hydrological cycle,
- quantify the component of sea-level change due to the transmission of water between the continents and the ocean, and

- assess drought impacts and groundwater depletion caused by unsustainable irrigation water use and lack of government intervention.

The accuracy with which the GRACE mission deciphers monthly gravity fields ranges from an estimated 200,000 km^2 at low latitudes to approximately 90,000 km^2 near the poles (Swenson and Wahr 2002; Wahr et al. 1998).

The processing of GRACE data is undertaken in three science centres, which includes the Jet Propulsion Lab, Center for Space Research, and the German Research Center for Geosciences. The small difference among these three processing platforms, which can be overlooked, is as a result of maximum degree of the projected gravity harmonics, dissimilarity in data weighing and the periods over which the orbits are assimilated. Alternative approaches also exist. However, they are implemented by different institutions providing independent gravity solutions nowadays (Liu et al. 2008). Gravity is one of the basic forces of nature. The misconception of a uniform gravitational acceleration around the earth does not apply here, since we project that the gravity values around the earth are not consistent. The movement of earth (rotation and revolution) brings about little degrees of relatively large deviations from the average value, which can be extracted adequately from theory. Given that the earth's gravity field is a function of its mass redistribution, fluctuations in the earth's interior density and topography bring about further sub-regional fluctuations of tens of a millionth gravitational anomalies (Fig. 14.1). The basic computations implemented by the GRACE mission in assessing temporal gravimetric patterns are distributed as coefficients of spherical harmonic functions of degree l and order m. These can be evenly associated with water height fluctuations on the earth's surface. Given the various analysis and processing centres, the maximum harmonic degree is sufficiently based on the varying individual techniques. However, they are all appropriate to provide gravimetric estimations of approximately 200 km.

The Gravity field and steady-state Ocean Circulation Explorer (GOCE) is another satellite geodetic and geodynamic mission that came into limelight after its launch in 2009. GOCE measures the Earth's gravity field as well as model the geoid with an exceptionally high accuracy. The GOCE mission is owned by the European Space Agency and it combines gravity gradiometry with Global Positioning System to determine the Earth's mean gravity field at a global accuracy and spatial resolution of up to 80 km (Bouman et al. 2015). Although its mission objectives are somewhat different from the GRACE mission, GOCE was designed to quantify the thickness of the polar ice sheets, hence its potential in providing important information that underpin changes in global freshwater. This was achieved through a combination of bedrock topography with gravity, and ice sheet surface elevation acquired from altimetry. Moreover, unlike GRACE, which observed changes in the gravity fields of the earth, GOCE focused on the recovery of stationary (static) gravity fields of the earth, i.e., geoid and gravity anomalies. In other words, GRACE was designed to collect information on the time-variable gravity fields of the, especially on a large spatial scales of hundreds of kilometres. However, GOCE focused on the measurement of the static field but on the highest possible spatial resolution. Other mission objectives

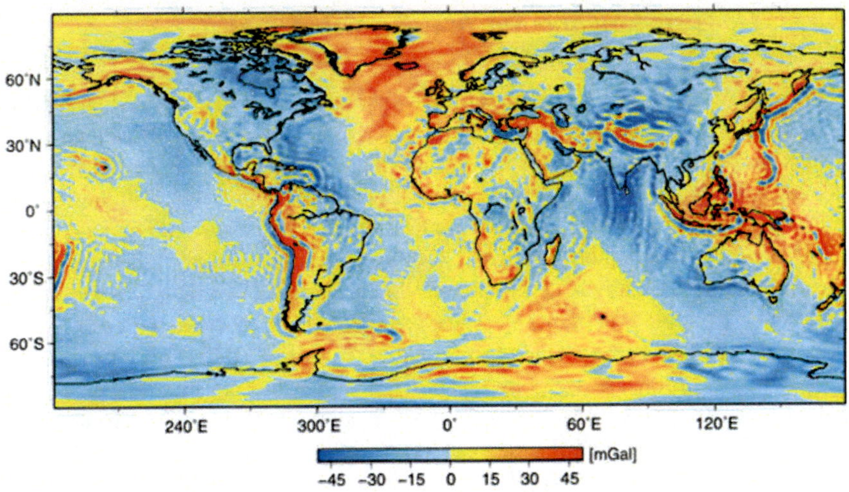

Fig. 14.1 Four-year observed static gravity anomalies of the GRACE mission; the sub-continental changes in gravity patterns are due to the varying density and topography of the regions. Units are milligal (1 mGal = 10^{-5} m s^{-2}). *Source*, Wouters et al. (2014)

of the GOCE mission as highlighted on the website (https://directory.eoportal.org/web/eoportal/satellite-missions/g/goce) include, to provide

- a new understanding of the physics of the Earth's interior such as the mantle composition lithosphere, and other physical process including geodynamic processes,
- an exceptionally high-accuracy global height reference system for datum operations and connection, which is also expected to serve as reference surfaces for land surface topography and evolution of ice sheets, and
- accurate estimate of the marine geoid to aid precise quantitative determination of ocean circulation and transport of mass.

Given its capability to acquire measurements of spatial variations of the Earth's gravitational field, GOCE has potential applications in the fields of geophysics, oceanography, climatology, and glaciology. For instance, the use of GOCE satellite observations to improve the modelling of the Earth's lithosphere was been reported by Bouman et al. (2015). Their study emphasized the usefulness of GOCE data in mapping crustal thickness and deep regional structures for frontier areas as well as improving understanding of Earth's dynamic processes. However, the GOCE mission was discontinued in November 2013.

14.2.1 Downscaling and Machine Learning

GRACE has enabled the monitoring of freshwater changes globally, making important contributions to understanding of global groundwater and its responses to changes in climate and human water management (Long et al. 2020; Ndehedehe et al. 2021; Rodell et al. 2018; Tapley et al. 2019; Famiglietti 2014). But the coarse spatial resolution of GRACE limits its application at practically water management scale. A downscaling method for GRACE-TWS and groundwater is gradually emerging as important alternatives to circumvent this problem (Li et al. 2019; Yin et al. 2018). There are some constraints regarding the effectiveness of downgraded GRACE products. For example, the method proposed by Yin et al. (2018) is only feasible in regions where groundwater is strongly correlated with evapotranspiration. However, new machine learning approaches and statistical methods have now been incorporated into GRACE downscaling scheme to improve the spatial resolution of GRACE-derived groundwater for hydrological applications (Jyolsna et al. 2021). A summary of recent studies on this shows that there are three levels at which GRACE downscaling can occur (Jyolsna et al. 2021; Li et al. 2019; Yin et al. 2018).

- The downscaling is spatial based. Here, the coarse-resolution of GRACE-derived TWS (110 km) is disaggregated into finer resolution for use on local to regional applications;
- The downscaling is temporal. The goal here is to further partition the monthly-averaged GRACE into diurnal variations;
- In this last stage, GRACE is being vertically downscaled through disaggregation into several hydrological stores that make up the GRACE water column. This includes canopy water storage, soil moisture, surface water storage, and groundwater storage

Relying on the relationship of GRACE with biophysical and hydro-climatic variables (e.g., vegetation, land surface temperature, precipitation, evapotranspiration) to predict new membership values of sub-grid changes in groundwater could also be a very valuable approach. This is because these variables play substantial role in observed changes in the water cycle. By relying on several techniques, including dynamic and statistically downscaling, TWS has been assimilated into a high-spatial resolution, land surface model to distribute GRACE data into finer spatial scales, thus improving the spatial resolution of GRACE data (Getirana et al. 2020; Li et al. 2019; Yin et al. 2018). Statistically downscaling methods employ these variables to aid understanding of the underlying physical dynamics of the system in a machine learning, non-parametric model functions (Yin et al. 2018). The benefits associated with statistical downscaling methods, e.g., flexibility, simplicity, and low computational cost have been echoed (Jyolsna et al. 2021). Although dynamic downscaling of GRACE data has been employed in several drought and water storage and flux simulation studies, some problems with GRACE statistical downscaling approach is that they are computationally complex and require substantial ground observation for validation (Jyolsna et al. 2021). The use of machine learning algorithms such as artificial neural networks and Random Forest to downscale GRACE data to improve understanding of changes in groundwater has been reported in the California's Central Valley (Chen et al. 2019).

14.2.2 Global Groundwater

The Earth's climatology is experiencing a rapid change, resulting in increased acceleration of the hydrological cycle. Substantial fluctuations in sub-regional and global TWS, evapotranspiration, and precipitation patterns (Ferreira et al. 2020; Rodell et al. 2018; Famiglietti et al. 2015; Kalu et al. 2021) emanate from this change. Severe drought and flooding episodes have occurred more frequently with increased intensity in some regions (Vanham et al. 2018) and predictions from climate model suggest severe drying incidents in the dry areas of the world and severe wetting episodes of wet areas. These extreme events impact on hydrological stores, including groundwater. The impacts of climate change on changes in groundwater have contributed to declining trends and loss of water storage in major aquifers around the world (Jasechko and Perrone 2021; Famiglietti et al. 2015). Some areas are expected to run out of water in the coming years due to the growing consumption of freshwater resources without adequate plans for re-injection and replenishment (Chen et al. 2016b).

Groundwater is a major source of freshwater, supplying water to an estimated population of 2.5 billion people in the world (Seo and Lee 2016). It is an important source of freshwater that maintains food and water security, sustains ecosystems, and enables human adaptation to climate variability and change (Green et al. 2011). However, groundwater suffers from poor recharge rate given the change in extreme events such as drought. Furthermore, there is a high groundwater consumption rate caused by urbanization, expansion of irrigated agriculture, and population growth, each having a direct and substantial impact on groundwater resources. This high dependency on groundwater resources has triggered groundwater draw down, groundwater and stream depletion, as well as seawater intrusion for several sub-regions (de Graaf et al. 2019; Scanlon et al. 2012). This is the case with China and United States, as studies have shown significant aquifer depletions in some regions (Feng et al. 2013; Scanlon et al. 2012). The major effect of this groundwater depletion is impact on national income and economic development. For example, in India, the government sponsored the cost for groundwater extraction by reducing electricity cost used for the exercise. This, however, was an attempt to enhance agricultural efficiency at the expense of reducing groundwater level (Shah et al. 2008). On the other hand, Dai et al. (2019) highlighted the effect of overloading groundwater resources carrying capacity (GRCC), stating the toll it will take on future socio-economic and developmental activities if adequate measures are not put in place to mitigate its outcome. The increasing pressure placed on groundwater resources due to urbanization and industrialization in many groundwater-dependent regions has limited the GRCC in those regions (Chen et al. 2010), thus the need to adequately monitor these GRCC for optimal groundwater resources management and sustainable economic development.

14.2.2.1 Can the Wells Run Dry?

Global groundwater wells are already at the risk of running dry (Jasechko and Perrone 2021). Climatic conditions in semi-arid domains like Australia and the Central California Valley impact on groundwater variability and the water table. For example, groundwater losses in the Murray Darling basin occurred 44% of the time during the 2002–2017 period, which for the most part was characterized by prolonged rainfall deficit (Ndehedehe et al. 2021). Droughts or lack of rainfall are occurring more frequently and impacts on surface water, soil moisture, and ecosystem functions have been studied (Ndehedehe et al. 2021; van Dijk et al. 2013). Changes in the characteristics of precipitation (intensity, quantity, frequency), flow, and evaporation affect groundwater recharge. They regulate water balance and drive changes in the amount or timing of groundwater recharge (Taylor et al. 2013; Green et al. 2011; McCallum et al. 2010). Droughts and increased evaporation rates limit the availability of surface water, triggering changes in groundwater use (Ndehedehe et al. 2021; Agutu et al. 2019). Excessive groundwater pumping draws down the water table and under prolonged rainfall deficit conditions and higher evaporation, groundwater depletion and low well yields will result in groundwater drought (when the wells run dry). Groundwater wells run dry because infiltration or recharge is unable to replace groundwater as quickly as pumping removes it. When the wells run dry, farmers are unable to irrigate and access to reliable and convenient drinking water is limited. The ability to map changes in groundwater will enable us to answer important questions like, how resilient are aquifers to drought and water extraction? Several observational methods for exploring groundwater storage including, age dating, chemical proxies, electromagnetic imaging and logging, and land surveying methods are limited and do not offer meaningful spatial coverage (Green et al. 2011). However, advanced satellite remote sensing methods like GRACE and big data streams now provide great opportunities to map groundwater availability and drought over very broad spatial domains and fine spatio-temporal resolution. The various approaches to quantifying groundwater from space, including the combination of GRACE with model outputs and other methods were detailed in Chap. 9.

14.2.2.2 Water Budget

In desiccated regions like the African Sahel and Central Asia, the majority of the populace depends exclusively on groundwater for consumption and agriculture, thereby leading to the current desertification and significant decrease in groundwater levels (Voss et al. 2013). As was mentioned in Chap. 9, techniques in assessing groundwater levels are constantly changing and evolving. Methods such as the base flow separation analysis, natural resources conservation service curve number, water budget analysis are conventional quantitative procedures in groundwater level monitoring. Studies have shown several challenges in the application of these techniques in analyzing groundwater levels due to variation of groundwater storage with several components, such as weather, topography and soil conditions. This, therefore, creates room for uncertainty in the unvalidated results obtained using these techniques. Despite these challenges, the introduction of time variable gravity technique in monitoring groundwater level has received increasing attention in the last few years.

With satellite measurement technique, a significantly higher accuracy index can be achieved compared to the ground-based point techniques (Adam 2002). Data obtained from the GRACE satellite mission have made this possible, particularly enabling important re-analysis of various hydrological components (e.g., evapotranspiration) and conditions across several regions. GRACE data have provided opportunities for integration with model outputs (land surface models) and in-situ data (ground observation data) through data assimilation to assess groundwater amounts and variations (Li et al. 2019). Furthermore, some studies have attempted to observe groundwater changes and surface water hydrology in key regions with increasing drought events and where groundwater use for irrigation is considerably substantial. This includes the Tigris-Euphrates river basin (Voss et al. 2013), the Lake Chad basin (Buma et al. 2016), Central Asia (Joodaki et al. 2014), India (Rodell et al. 2009), California valley (Scanlon et al. 2012), and Texas (Long et al. 2013), which is a well-known hotspot of water conflict. Also, a map depicting drought index parameters of groundwater fluctuations in the United States was developed using GRACE data (Houborg et al. 2012), further emphasizing the contribution of GRACE in hydrological studies and characterization of extreme events such as groundwater drought (Thomas et al. 2017).

14.2.2.3 Improving Observations and Models Through Assimilation

Data assimilation is a procedure that constrains the dynamic of a model with available observations to improve its estimates (Bertino et al. 2003). In hydrology, observations from satellite (e.g., GRACE) and in-situ sources (ground-based data) can be integrated with models (e.g., Catchment Land Surface Model) to improve the accuracy index of the model estimates (e.g., Crow and Wood 2003). The unavailability of in-situ groundwater data is not unconnected to the high cost of drilling and maintaining monitoring wells, as well as inaccessibility due to political considerations and public policies (Rodell et al. 2009). Also, it can be difficult to interpret semi-confined and confined aquifers for efficient groundwater assessment and storage evaluations.

The use of land surface models (LSMs) to monitor drought wetness conditions (Li et al. 2012; Houborg et al. 2012), and also to estimate groundwater storage changes are targeted efforts to improve understanding of the influence of climatic variations and anthropogenic influence on groundwater levels (Pokhrel et al. 2012). Advanced LSMs have been introduced to improve this understanding and can be used to assess spatial and temporal groundwater flow that captures climatic fluctuations (Fig. 14.2), which are appropriate for drought and water resource supervision when

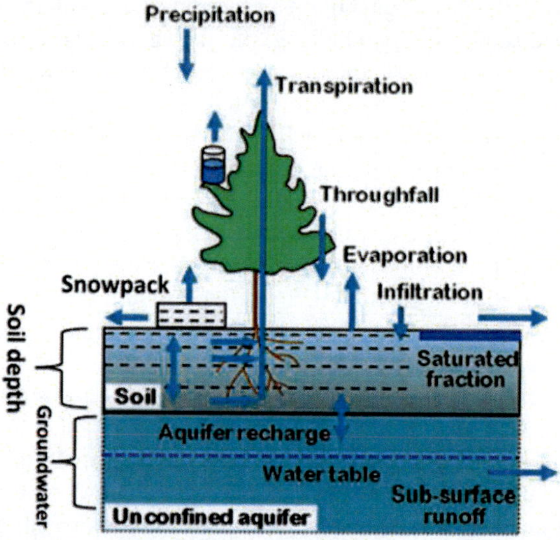

Fig. 14.2 A typical structure and water cycling process based on the Land Surface Models (LSMs). *Source*, Xia et al. (2017). Published 2017 by the American Meteorological Society. This Figure was originally adapted and modified from Lawrence et al. (2011)

forced by meteorological data. LSMs forced by both energy and water balances can provide substantially accurate evapotranspiration estimates (Mueller et al. 2011), which, if merged with enhanced precipitation and simulated drainage parameters can be used to simulate high-quality groundwater storage anomalies (Li et al. 2015). However, several case-specific studies exist where LSMs consist of highly simplified groundwater indices compared to dynamic groundwater flow models (Ducharne et al. 2000). The straightforwardness and poorly defined constraints on the temporal inconsistency of groundwater levels cause heterogeneity among groundwater change estimates from different LSMs, which may be greater than that of other hydrological parameters like runoff and precipitation patterns (Xia et al. 2017). Data assimilation of satellite with these models and bore monitoring data can, therefore, improve the simulation or prediction of groundwater on different spatial and temporal scales.

Observations from the GRACE mission integrated with auxiliary data has been the basis of several significant findings like the detection of groundwater depletion caused by excessive groundwater withdrawals for irrigation agriculture in China, India, United States, etc. (Rodell et al. 2018; Feng et al. 2013). Its relevance is also well founded in its attempt for uncertainty reduction in modelling anthropogenic effects on groundwater resources (Döll et al. 2014). The assimilation of satellite observations of terrestrial water storage with bore monitoring data is a key alternative to improve groundwater estimation (Li et al. 2019; Houborg et al. 2012). GRACE-groundwater estimates can be improved by assimilating satellite into land surface model and observational data. Such outputs from the data assimilation process can be useful in generating a groundwater drought indicator, which can be evaluated and validated using drought indicator derived from bore data. This will provide a reliable groundwater drought monitoring capability that can be incorporated into national groundwater information system. This innovative approach can be evaluated further to provide confidence in GRACE application in groundwater drought assessment.

Several techniques such as the Ensemble Kalman Filter (EnKF) have been employed in GRACE data assimilation into LSMs to force model state parameters while separating and downscaling GRACE data (Zaitchik et al. 2008). In downscaling groundwater level anomaly from GRACE anomalies, hydro-climate and LSM parameters are important, especially as a result of the temporal and effective spatial characteristic or resolution of GRACE observations (Save et al. 2012). Past studies have shown the integration of GRACE TWS anomalies over hydrological basins ranging from 100,000 to approximately 1,000,000 sq. km for assimilation into LSMs (Houborg et al. 2012; Forman et al. 2012; Zaitchik et al. 2008). These studies showed an existing homogeneity between the integrated parameters and the features of spherical harmonic-based data products whose effective spatial resolution is relatively coarse and demands filtering to minimize parallel uncertainties (Landerer and Swenson 2012). The application of newly developed GRACE products based on regional mass concentration (mascon) parameters with constraints to the transposition of satellite ranging measurements in a bid to conserve GRACE signals are also emerging (Save et al. 2012).

The usefulness of assimilated GRACE components spans across the conservation of measured data up to the creation of an enhanced alignment with prevailing land surface modelling infrastructures, which are hypothesized on systematic grids rather than disorganized basins (Kumar et al. 2016; van Dijk et al. 2014). One distinct benefit of satellite observations like GRACE for drought monitoring is its efficiency in detecting fluctuations in water storage occurring in the deeper subsurface, together with its increased susceptibility to water stress all year round (Li et al. 2012). This is very essential for recursive drought monitoring and understanding how droughts propagate through the water column or hydrological stores (soil moisture, surface water and groundwater). However, with the use of assimilated data, it is possible to achieve vertical disintegration combined with temporal and spatial depletion of GRACE-derived observations to efficiently assess discrete storage components such as groundwater at relatively higher scales than raw GRACE-derived observation (Li et al. 2019; Getirana 2016; Thomas et al. 2014). Also, Houborg et al. (2012) reported topical model estimates after forcing a data assimilated model to a near-real-time meteorological data. This up-to-date model is very useful for effective utilization aimed at filling observational gaps in the absence of raw (i.e., level 1 data) GRACE-derived observations and operations, like, weekly monitoring of drought.

14.2.3 Climate Change Impacts on Freshwater

The management of freshwater resources amidst series of incessant climatic episodes and thresholds can be very challenging. Hydrological quantities ranging from runoff, soil moisture, snowmelt, and precipitation are known to be variable under a changing climate over a range of time and spatial scales that could sometimes be difficult

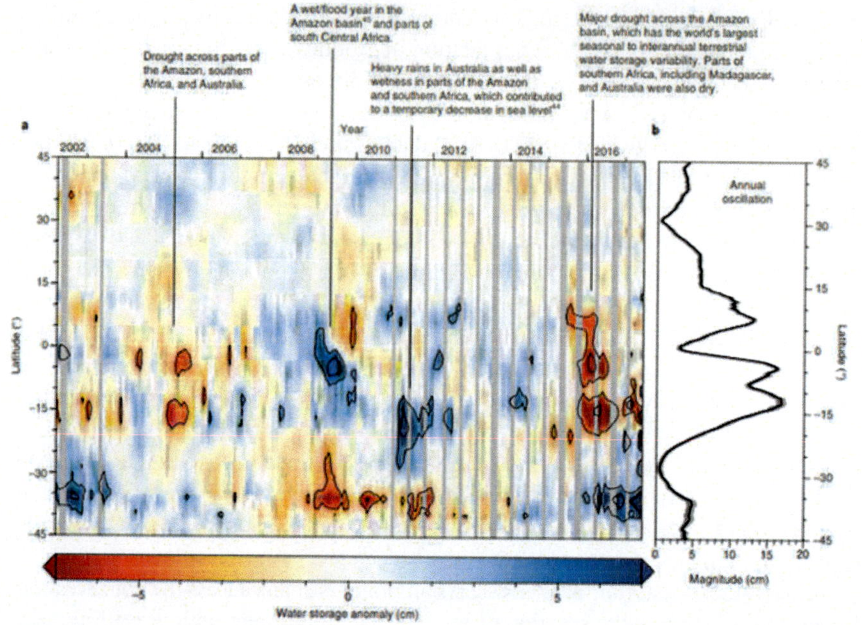

Fig. 14.3 TWS zonal mean of GRACE-derived observations from 2002 and 2017 **a** represents the anomaly of time series trend after the subtraction of annual periodic component, linear and offset parameters **b** the significance of annual oscillation from CSR RLO5M Mascons. *Source*, Tapley et al. (2019)

to be captured by climate and hydrological models. The most common cases of extreme events are floods and drought, though the effect of hydrologic variability is known to also have prolonged impacts. Among the most impressive contributions of the GRACE mission has been its usefulness in improving understanding of Earth's dynamic freshwater landscape, which play important role in food, and water security. Past studies have shown how GRACE-derived continental water mass change in relation to climate fluctuations has aid this understanding (Rodell et al. 2018; Reager et al. 2016). For example, in Fig. 14.3, we observe that the GRACE-derived zonal average of the annual TWS amplitude has a range of ± 17 cm equivalent water height. We also observed the relationship between climate-driven fluctuations of the annual TWS variation cycle with drought (Thomas et al. 2014; Long et al. 2013) and flood (Boening et al. 2012), episodes especially in low to mid scaled latitudes. The detection of ongoing and future hydro-climatic change can now be easily distinguished since the inception of GRACE-derived TWS dataset in monitoring the global water cycle (Rodell et al. 2015).

The integration of the annual climatic reports with the zonal mean water mass plots as indicators of groundwater and TWS anomalies has been in practice since 2011. The necessity of freshwater variability in the context of climatic events is facilitated by the idea that the global sea level record consists of substantial annual fluctuations around a fundamental secular trend. Natural freshwater fluctuations can be a notable basis or sink in the global ocean mass re-analysis, which has significantly contributed a minimum of 0.7 mm/year to sea-level rise in the GRACE period (Tapley et al. 2019). Specific ranges of TWS changes influence on sea level are

- interannual to decadal time range (Eicker et al. 2016),
- obscuring or intensifying the core trend (Reager et al. 2016), and
- transposing the sign and rate of sea level rise when more global freshwater is accumulated (Boening et al. 2012).

The combined effect of sea level and land-water storage is significant in clarifying the global sea-level fluctuations. The extreme impacts of climate change on freshwater resources have also been studied using conventional remote sensing techniques such as

1. Satellite and Radar altimetry to monitor lake fluctuations: Sea level is one of the indispensable variables recorded by the Global Climate Observing System (GCOS) in monitoring climate change. The application of satellite altimetry technique for adequate assessment of sea-level fluctuations from space has been in effect over the past two decades. However, the need to improve on the accuracy index of altimetry-based sea level profile for better monitoring periods is critical so as to answer some pertinent scientific questions like (a) Can sea level water budget be closed? (b) What is the rate of global sea level rise and how is it affecting the islands and coastal regions? (c) What are the factors influencing regional and inter-annual variability? (d) Can the internal/natural climate variability factors be easily separated from detected anthropogenic forcing signatures? Surface water bodies such as lakes, rivers and reservoirs are essential components of the terrestrial water cycle, which are useful for important water resource management. Water storage undergo rapid fluctuations, especially under the influence of climatic and anthropogenic factors. As a result of sparse in-situ observation networks, a comprehensive national-scale monitoring dataset of surface water bodies are unavailable for several regions in Africa, Asia, etc. The use of satellite altimetry for inland water assessment has been successfully undertaken in several of these regions to circumvent the lack of monitoring or in-situ data. However, the application of radar altimetry to monitor water level variations of large rivers, reservoirs and lakes has proved to be more accurate as well as complement the sparse in-situ monitoring networks for any specific region (Jiang et al. 2017).
2. Alternative remote sensing techniques: satellite remote sensing techniques have the potential to support prompt management, reiterative and cost-effective surveys of water resource over large regional and global scales. The assemblage of micro and nano-satellites, such as IceSat, Cubesat etc has generated lots of potential and data thus, revolutionizing techniques for high spatio-temporal hydrological assessment, which underpin tracking the pace of climate change, assessing surface water fluctuations (Cooley et al. 2017, 2019) and its impacts on regional and global freshwater resources (Sadeh et al. 2019). The repetitive observation mode of several Cubesats facilitates an adequate examination of river dynamics and

observations of various physical properties. Among the several Cubesats in space, the Cubesat assemblage provided by Plant Labs Inc. has the unique superiority of supplying a near-daily revisit time globally at 3 m ortho-rectified spatial resolution (Junqueira et al. 2021). Its use in identifying land and water areas using band variations of different multi-spectral images is well known (Acharya et al. 2018). This forms the basis of applying them in monitoring and assessing the impacts of climate change on surface flows/river discharge, lakes and reservoirs.

14.2.3.1 Satellite Hydrology of South and Central Africa and Climate Variability

The South and Central Africa (SCA) region is a notable climatic hotspot that plays a major role in the regulation of global climatology and hydrological fluxes. Characterized by several river basins such as, the Limpopo (415,000 km^2), Okavango (530,000 km^2), Orange (973,000 km^2), Zambezi (1.39 million km^2), and of course the popular Congo river basin (3.68 million km^2), which is the second largest river basin in the world. The rainfall climatology of the Congo River basin dominates global tropical rainfall patterns during transition seasons, which in turn leads to various regional and global climatic episodes (Ndehedehe et al. 2016a, 2018b, 2019). Studies have shown that the Congo basin GRACE hydrological signal is strongly associated with the multi-annual changes in the Congo River discharge ($r = 0.86$, $\rho < 0.05$) (Ndehedehe et al. 2018a). While the evidence from the relationship between the Congo River discharge and GRACE highlights the significant contribution of surface water changes in TWS, the three southern oceans (Atlantic, Indian, and Pacific) drive the hydrology of the basin and central Africa at large.

TWS as a basic constituent of the global water cycle is crucial for climate studies, agriculture, ecosystem experimentation, as well as water resource management. TWS as an indicator supports an enhanced understanding of the correlation between the various atmospheric phenomena and terrestrial water storage anomalies (Fig. 14.4), which could lead to an improved parameterization in climate models. The seasonal rainfall patterns existent in the Central Africa region are relatively stronger than that experienced in the Southern region. This could be due to the existence of massive river basins in the Central Africa region such as the Congo River basin. The influence of these basins brings about a multi-annual rainfall pattern in the region, which strongly contributes to changes in freshwater. The Central African region displays strong rainfall patterns all year round as opposed to the Southern region encompassing Namibia, Botswana, Lesotho, Swaziland, South Africa, Zimbabwe, Zambia, Malawi, and Mozambique, where rainfall is dominant only in the October-November-December period. The hydrological changes of the central African region is notable because of the surfeit of physical phenomena regulating its climate system. Local measurements from dedicated networks such as the Uncertainty Reduction in Models for Understanding Development Applications (UMFULA), Future Resilience for African Cities (FRACTAL), and Improving Model Processes for African Climate (IMPALA) networks have been used to advance understanding of scientific knowl-

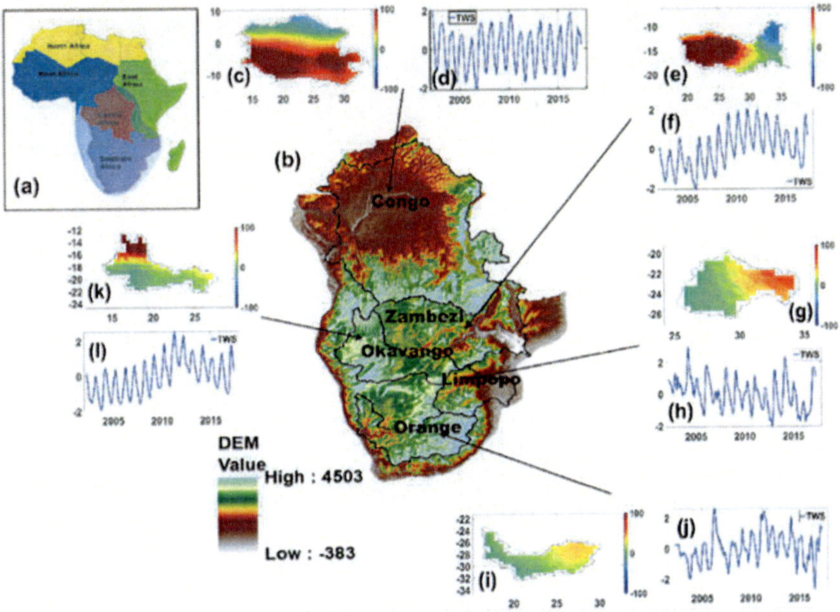

Fig. 14.4 TWS analysis of the South and Central Africa region **a**, **b** showing the spatial **c**, **e**, **g**, **I**, **k** and temporal **d**, **f**, **h**, **j**, **l** fluctuations of each river basin extent from GRACE-derived TWS datasets between April 2002 and December 2017. *Source*, Kalu et al. (2021)

edge about climate processes, regional and local climate trends, as well as develop the simulation of existing climatic models through a better understanding and representation of weather and climatic processes in the Central African region. However, hydrological studies over large spatially heterogeneous regions are still inadequate for accurate predictions due to poor coverage and spread in their respective territories of interest. As a result of the sporadic distribution of in-situ hydrological and meteorological monitoring infrastructure in the SCA region, our knowledge of the spatio-temporal patterns of TWS is substantially limited, thus warranting the large-scale assessment of water storage estimation undertaken in this contemporary studies.

Apart from TWS, evapotranspiration (ET) is another relevant flux but less understood constituent of the terrestrial water cycle. Its applicability in relating land and atmospheric subdivisions of the hydrological cycle by imposing heat and moisture exchange between them makes it indispensable in climatic studies and analysis (Ndehedehe et al. 2018a). It attempts to exploit an intense influence on the patterns of existing atmospheric water vapour transfer. Soil moisture, surface water, groundwater etc. in the SCA region (Fig. 14.4) are key components of the GRACE-derived TWS. In this region, the latter is largely driven by rainfall and ET rate and inten sity. The quantity of incoming and outgoing water in the hydrological budget in any sub-continent or regional scale is determined by rainfall and ET rates, which is con-

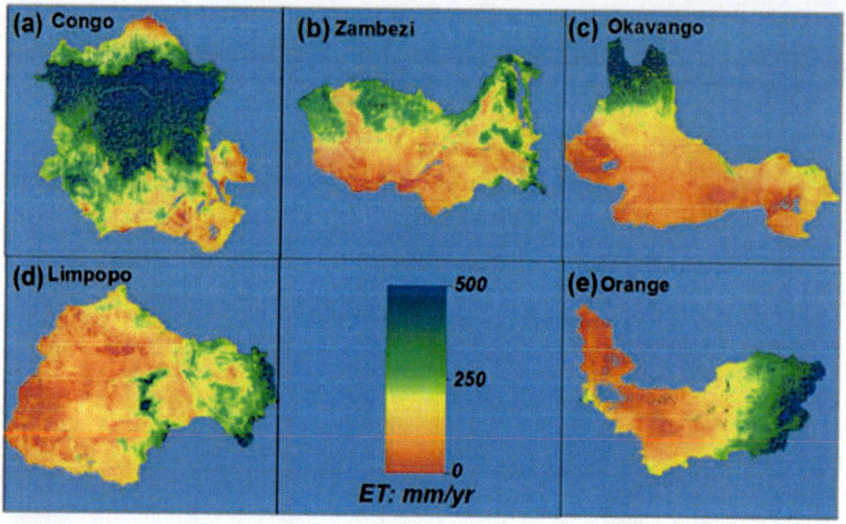

Fig. 14.5 Mean ET rates of major river basins in the SCA region from 2002 to 2017

sidered to have a net-zero loss. There also exists a direct variation between cloud evaporation and ET. The effect of this relationship results in humidity fluctuations, as well as rainfall variability in the region. The effect and presence of ET affect rainfall and circulation patterns.

As was earlier highlighted (Chap. 7), the rainfall pattern in the Congo River basin is distributed bi-modally in every annual cycle. This warrants the existence of two dry (June–August and December–February) and wet (March–May and September–October) seasons. The heart of the Congo River basin experienced higher ET rate (271.33 mm/yr) than the Northern and Eastern parts (Fig. 14.5), which recorded an ET rate of 239.75 mm/yr (Kalu et al. 2021). The relatively high ET rate in the region can be linked to (i) the optimal allocation of and movement of moisture within the soil in the SCA region, (ii) low vegetation transpiration as a result of the ongoing deforestation, (iii) increased rate of solar radiation, absorption and disposition. The effect of ET reduction over the Congo river basin was also highlighted as a major cause of the notable shift in rainfall pattern observed in the Central African region during the period. The study (Kalu et al. 2021) also pointed out the inconsistency of ET fluctuations over the SCA region, given that the lowest ET rate for the Congo, Limpopo, and Okavango river basins were recorded between June and August, while the lowest ET rate for the Zambezi and Orange River basins fluctuated between September and November and between July and August. The disparities in the seasonal ET frequencies of the main river basins in the SCA regions connote the presence of distinct factors that influence existing anomalies. Size also plays a key role in continental and regional climatic balance. Other determining factors significant to this changing fluctuation include, but not limited to, transpiration rate, latent energy flow, average surface temperature, and vegetation index. Since the dynamics and changes in ET influence energy balance, meaning that sunlight contributes to its overall effects, it follows that other major river basins in the world are expected to undergo this climatic process, which could lead to strong changes in TWS.

14.2.4 Global Navigation Satellite Systems

Given the ability of the Global Navigation Satellite System (GNSS) to serve as an alternative to estimating land water storage through precise observations of time-variable earth's gravity field in the absence of GRACE, its potential and applicability for TWS estimation are plausible (Ferreira et al. 2019b). Even though regions with sparse GNSS stations tend to suffer poor estimates of reversed land water storage fields, observed radial transpositions can be applied in weighing rivers and lakes in an attempt to improve the prediction of land water storage (e.g., Bevis et al. 2004, 2005), characterize hydrological drought patterns (Ferreira et al. 2018; Chew and Small 2014), and assess the period and timing of land water storage (Chew and Small 2014). GNSS-generated signals penetrate the atmosphere, providing subtle avenue for assessing water vapour information. These signals can equally be recovered from space or ground-based measurements (Bonafoni et al. 2019). As a result, several globally disseminated GNSS stations have become indispensable meteorological datasets for assessing water vapour, which plays a key role in weather structures (Jin et al. 2017).

The possibility of observing El-Nino southern oscillation (ENSO) evolution using GNSS-derived water vapour parameters was demonstrated by Wang et al. (2018). The idea of GNSS radio occultation which implements space-borne measurements has been severally applied in illustrating the atmosphere through vertical profiling. Missions such as GRACE, the Constellation of Observing Systems for Meteorology, Ionosphere, and Climate (COSMIC), and CHAllenging Minisatellite Payload (CHAMP) have been used for this illustration. ENSO-driven anomaly patterns together with temperature, water vapour, and COSMIC-derived specific humidity have been implemented to determine its suitability over continental scales (Llamendo et al. 2017). Another application of GNSS signals in hydrologic studies is in GNSS reflectometry, which applies satellite transmitted signals and ground reflected signals to assess and monitor the environment (Larson 2019).

14.2.4.1 Terrestrial Water Storage Observed from GNSS

Few techniques have been applied to estimate TWS. Some of these methods include, (i) sub-surface water dynamics assessment through ground-based observations using absolute gravimetries, and subsequently using the relative gravimetric observations to estimate TWS fluctuations (Hinderer et al. 2012; Naujoks et al. 2008), (ii) the application of GRACE on several spatial scales (Kalu et al. 2021; Ndehedehe and Ferreira 2020a; Tapley et al. 2019), (iii) the development of quantum sensors using a cold-atom interferometric technique for future TWS predictions (Müller et al. 2018). Estimating TWS via geodetic techniques is increasing, especially with the use of the Global Positioning System (GPS). Argus et al. (2014) suggested the reverse technique for TWS estimation using measurements of radial crustal distortion determined from GPS observations. Along these lines, studies have shown the possibility of inverting TWS with a 50 km horizontal scale for GPS datasets applied over California. GPS observations have been explored not only in hydrological applications as shown in some of these studies but also as independent observations to improve understanding of drought patterns. For example, Borsa et al. (2014) employed an array of GPS observations to understand drought pattern in United States by reversing the stimulated radial distortion gotten from GPS measurements. The reversed TWS showed total water reduction and correlated with the measured decline in rainfall patterns and streamflow. Using a checkboard test, an estimated spatial resolution of 200–300 km was achieved in the study. The checkerboard (also spike) test is widely used to illustrate the quality of inversion results as well as to assess the resolution of inversion. As further explained by Ferreira et al. (2019b), this is typically used in the field of inverse problems, especially in tomographic seismological studies.

Regions with wide distribution of GPS have benefited from the use of geodetic technique to quantify land water storage, including the prediction of water mass loadings associated by snowpack variations. In South America, the distributions of GPS stations as shown in Fig. 14.6 enabled a large-scale inversion of TWS from GPS observations of radial displacements. In Fig. 14.6, each GPS station is indicated with the colours of the circle number of years of GPS observations. This period of observations in each GPS location is overlaid on the spatial distribution of trends in TWS as observed by GRACE during the 2002–2017 (Fig. 14.6). As further detailed in Ferreira et al. (2019b), the data processing chain and experiments included an inversion based on checkerboard tests and closed-loop simulations of 'TWS' from hydrological models and then relying on radial displacements to image TWS on a daily time scale. The data processing protocols and workflow are illustrated in Fig. 14.7. Three different experiments were undertaken with the first experiment focusing on a checkerboard for three different spatial resolutions (Fig. 14.7). The second experiment focused on the root mean square analysis while the third considered the GPS-derived radial displacements in constructing images of TWS through an inverse modelling approach (Fig. 14.7). From the checkerboard test, the results of the inversion recovered the spatial patterns and amplitudes of the input model (Fig. 14.8). Inverted water storage distribution is slightly improved and the amplitude of the residual differences between the inputs (Fig. 14.8a) and inverted (Fig. 14.8c) model was at 1.10 kg/m2, and is almost negligible relative to all of South America. Even though the northeast region indicated relatively high uncertainties, due to low density distribution of GPS stations, recent progress in space-borne sensors provides opportunities to assess high temporal changes in freshwater resources.

These experiments demonstrate the feasibility of generating changes in land water storage on a daily temporal scale though with a rather course resolution of 300 km (Ferreira et al. 2019b). The high temporal frequency of daily time series of TWS obtained from GPS observations complements the monthly time scales of GRACE

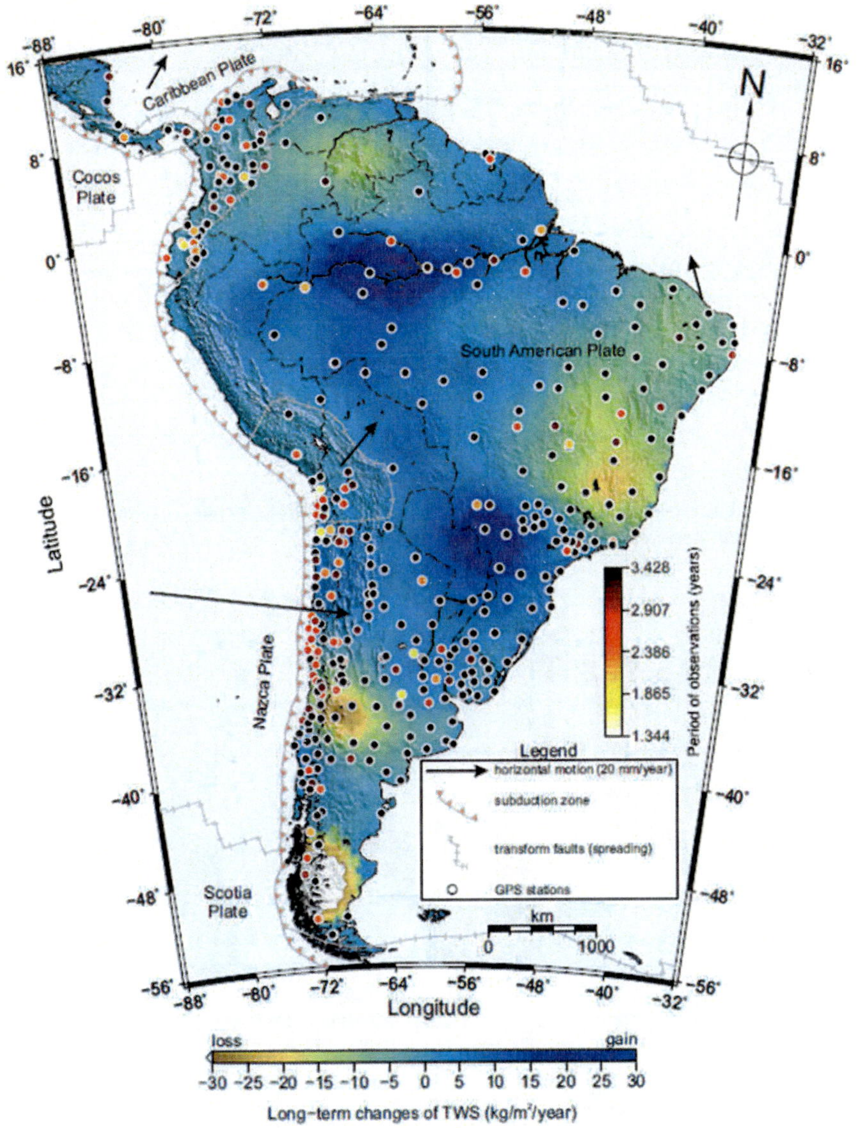

Fig. 14.6 Distribution of GPS stations (circles) over South America. The background image shows the long-term changes in terrestrial water storage based on GRACE data (2002–2017). *Source* Ferreira et al. (2019b)

data, thus extending the capabilities of geodetic sensors. Moreover, Fu et al. (2015) reported a significant correlation between the total TWS pattern and the physio graphic features of the study region by assessing TWS series using GPS-derived radial distortion technique, which also showed the spatial fluctuations of precipita-

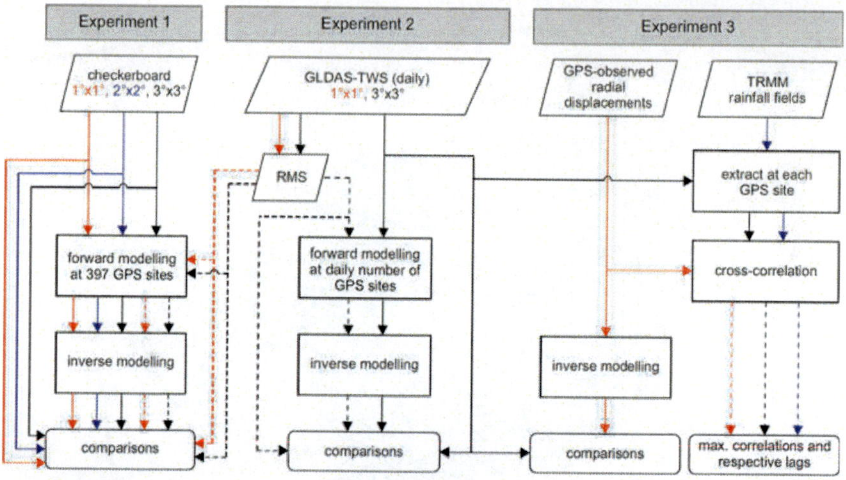

Fig. 14.7 The data processing protocols and work flow for analysis of an inversion of GPS observations over South America. *Source* Ferreira et al. (2019b)

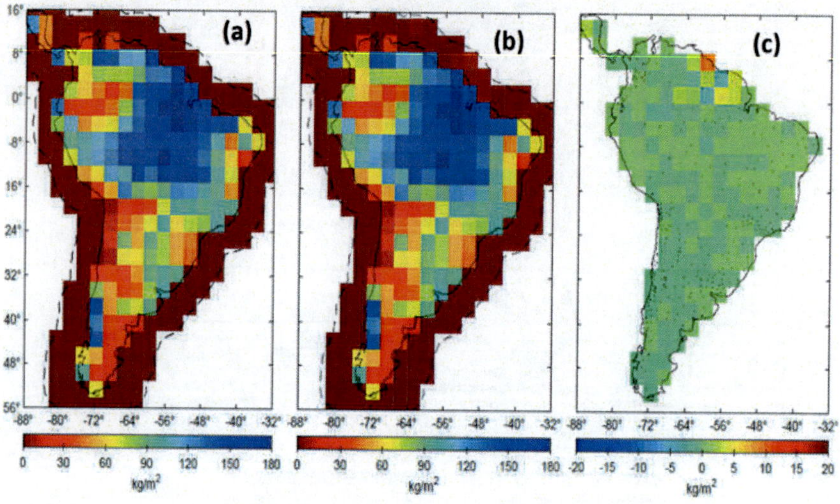

Fig. 14.8 GPS inversion results over South America on a spatial resolution of 3.0 arc degree. **a** Gridded mass loadings expressed by the RMS of the TWS series from hydrological model, **b** inverted TWS recovered from the 397 forward displacements, and **c** errors quantified as the difference between the input (**a**) and inverted loads (**b**). *Source* Ferreira et al. (2019b)

tion in the study area. A distinct advantage of GPS-derived TWS is the potential of widening the spatial resolution of TWS fields over GRACE measurements, while their products can serve as independent parameters to validate raw GRACE-derived TWS data.

14.2.5 Reconstructing GRACE-Derived Terrestrial Water Storage

The GRACE mission ended in October 2017, and the Follow On mission was launched in May 2018, that is, an approximately 1-year data interval. Since GRACE-derived observations are focused on providing estimates of terrestrial water storage anomalies (TWSA), it becomes pertinent to bridge the interval between these two missions. This, therefore warrants the need to fill the existing data gaps through reconstruction or the use of predictive modelling schemes. The reconstruction of TWSA offers a wide range of applicability especially in climate-related studies. This has enabled the reconstruction of climate-driven changes in terrestrial water storage. Climate-driven TWS has also been reconstructed using statistical models and water budget approach. For instance, the novel integration of independent component analysis with multiple linear regression to isolate climate teleconnection (e.g., ENSO)-induced TWS was achieved by reconstructing naturally driven TWS over South America (Ndehedehe and Ferreira 2020b). Satellite observations of terrestrial water storage from GRACE and surface water changes from altimetry with ancillary data from land surface models have been inverted to reconstruct groundwater storage changes (Fig. 14.9). This study, which was undertaken by Ferreira et al. (2020) over the Yangtze River basin, provides a transferrable reconstruction scheme to improve understanding of changes in all water storage components. For instance, the illustration in Fig. 14.9 shows that the reconstructed sub-domains (groundwater, surface water, and soil moisture and ice/snow) of TWS budget, specifically groundwater added approximately 38% to the observed changes in gain during the 2002–2017 period. But changes in soil moisture and ice/snow showed a much higher contribution to the rise in terrestrial water storage during the same period (Fig. 14.9). It is, therefore, easier to detect the hydrological stores (e.g., groundwater. snow/ice, soil moisture) driving water storage changes within a river basin or water planning areas, thus improving the assessment of climate change impacts on terrestrial hydrology.

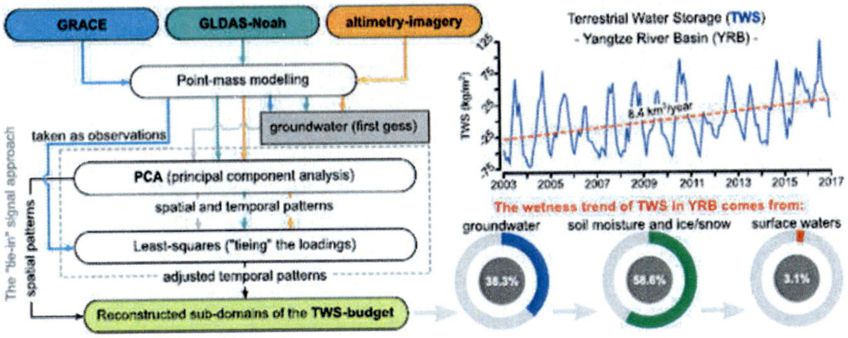

Fig. 14.9 Illustration of framework for reconstructing water storage components (groundwater, soil moisture and surface water) using satellites and models. *Source* Ferreira et al. (2020)

Further, several studies have attempted to provide alternate sensors and data processing techniques to obtain surface mass changes in the attempt to compensate for the interval between the two successive GRACE missions; none has been able to provide an acceptable accuracy or spatial resolution comparable to that of GRACE. For example, (i) the suggested swarm satellite method (Luck et al. 2018), (ii) the suggested global GNSS inversions technique (Rietbroek et al. 2014), (iii) the suggested satellite laser ranging (Talpe et al. 2017), etc. Other techniques aimed at reconstructing TWSA prior to the GRACE era have been highlighted in several studies. Most of them attempted to develop empirical relationships between GRACE-derived TWSA and associated climatic and hydrological products, such as sea surface temperature, rainfall, soil moisture, land surface temperature etc. The use of machine learning methods like the artificial neural network in reconstructing and forecasting water storage changes from GRACE time series has also been demonstrated (Ferreira et al. 2019). On the other hand, reconstruction of gridded TWSA using advance multivariate technique is also emerging. To achieve this, temporal modes and spatial patterns of the GRACE-derived TWSA were separated first, before the reconstruction of the TWSA fields by obtaining the correlation between GRACE temporal modes and correlated predictors (Forootan et al. 2014).

Advanced statistical methods have also been combined with other statistical techniques to reconstruct water storage components. The independent component analysis was used to separate the GRACE signals, while the autoregressive exogenous method was implemented to determine the association as well as replicate the reconstructions. Humphrey et al. (2016) adopted major climate pointers (temperature fields and rainfall) to reconstruct TWSA globally at each grid point. The decomposition of the gridded GRACE TWSA time series was undertaken first. Subsequently, this was followed by fitting temperature and rainfall parameters into a rectilinear trend, an inter-annual constituent, a seasonal constituent, and a high-frequency deposit constituent. Thereafter, the de-seasoned constituents of TWSA were reconstructed at a global scale through the derivation of interactions between the de-seasoned constituent of the GRACE-derived TWSA with temperature and rainfall. This methodology has been improved to engender global de-seasoned TWSA at a spatial resolution of 500 m $\times$ 500 m at monthly and daily extents using historical data from 1901 to date (Humphrey and Gudmundsson 2019).

Machine learning applications are also evolving as important tool to help reconstruct time series of TWS and other water storage components. For instance, Sun et al. (2020) applied a deep convolution neural network (CNN) to estimate spatial and temporal mismatch patterns between GRACE TWSA and water storage fluctuations as modelled by hydrological simulations and then correcting the simulated TWSA fields. The methodology entailed a spatial representation of disparity between GRACE TWSA and model parameters of the simulated TWSA using maps of the model parameters (temperature, rainfall, TWSA) implemented at each CNN-based epoch. Thereafter, the TWSA was reconstructed by subtracting the estimated mismatch from the TWSA model parameter map. Moreover, in the study undertaken by Sun et al. (2020), GRACE data were reconstructed for time beyond its mission period using machine learning-based algorithms while orienting to the GRACE RL06 dataset. The authors also implemented the multi-linear regression (MLR), deep neural network (DNN), and seasonal autoregressive integrated moving average (ARIMA) with exogenous parameters to estimate the correlation between the GRACE-derived TWSA and climatic fluxes or the past and present TWSA. Finally, it was possible to make predictions of the observed TWSA without the need for GRACE TWSA as input after training and validating the model parameters.

14.3 Satellite Remote Sensing for River Discharge Estimation

The major aim of for river flow and discharge measurements is for model calibration and hydrological data assimilation for flood forecast as well as other water resource management affairs. However, just like most hydrological products such as precipitation, runoff and evapotranspiration, in situ discharge observations are often very difficult to access for timely operations in disaster mitigation and flood forecasting or in most cases, are completely unavailable. This, therefore, has increased the relevance of microwave remote sensing information and altimetry-derived surface water elevations for flood inundation estimations, serving as alternate sources of surface water datasets used for hydrological operations (Alsdorf and Lettenmaier 2003; Alsdorf et al. 2001). Significant advances in water transport and storage techniques over the globe in the past decade have revolutionized hydrological science applications within the hydrological science community.

For example, using these altimetry datasets integrated with the observed discharges from the gauging stations located within the virtual stations, it is now easier to retrieve water discharges from stage–discharge relations. Rating curves are commonly used to represent these relations and allow for more flexibility in the prediction of water discharge from observed water levels, having the input data and flow regime as the accuracy index parameters. For instance, the estimation of river discharge from altimetry data has been reported for several tropical rivers of the world, e.g., Amazon River, Niger river, Chari river (Lamine et al. 2021; Getirana and Peters-Lidard 2013; Coe and Birkett 2004). Inasmuch as the discrepancies between the observed and predicted water discharges in most applications are relatively small and negligible, the applications of such techniques are often localized to virtual stations situated close to the gauging stations. Radar altimetry is also responsible for the enhanced projections and forecast of virtual stations at gauges downstream. Following the pioneering applications of this idea to forecast downstream discharges and levels in Lake Chad by Coe and Birkett (2004), other studies have been applied to forecast downstream discharges and water levels. This includes the work of Getirana et al. (2009) who assessed the likelihood of estimating discharges from radar altimetry datasets with the use of rating curves.

Given the relatively high demand in altimetry datasets, an unprecedented radar altimetry datasets at several hundred virtual stations over main rivers, lakes and tributaries have been recorded. They are often referred to as the Hydroweb datasets (Crétaux et al. 2011). Additionally, past studies have emphasized on the enhancement and development of a more accurate global modelling system, which can adequately simulate continental water fluxes (Decharme et al. 2010). The accuracy of river flow forecast downstream is largely a function of the upper-catchment satellite-based flow monitoring index, basically in the developing countries that still deal with insufficient ground-based river discharge measurements. Regardless of the potential accuracy index of in-situ observations, discharge estimations derived solely from remote sensing techniques can be seen to be the most sustainable solution to the inherent limitation in river discharge estimations.

14.3.1 Scientific Advancements in River Discharge Estimation

The advancements of new satellite hydrological missions such as radar altimetry have shown greater efficiency and feasibility in regional assessment of river discharge estimation as compared to in-situ observations. However, their proficiency in the direct estimation of river discharge for important river basins is still a challenge. New methods and algorithms that optimized the modelling and prediction of river discharge measurements have shown immense value in the advancement of knowledge on the nature of climate change influences on water resources. Recently, Lamine et al. (2021) obtained river section and river discharge using sole remote sensing products, i.e., satellite imagery and altimetry observations. Using the Alos Palsar RTC-DEM, the cross-sectional area as well as the average slope was estimated. The 'lateral-method' (LM), which directly provides the cross-sections, and the mixed method (MM), which incorporates a digital elevation model with bathymetry were introduced to aid the feasibility of estimating the cross-section of the Niger river. The lateral method illustrated in Fig. 14.10 shows the (a) position of the starting points Jn, right-side sections Rn and left-side sections Ln, (b) horizontal (imagery) and vertical (altimetry) evolution of water over time, (c) derived cross-section of the river obtained by combining the two sides sections (left and right) shown in panel (b).

In the mixed method approach, Fig. 14.11 shows the (a) extracted river banks' points (X, Y, Z) from the RTC-DEM at which the riverbed was masked out, (b) derived river bottom points (X,Y,Z) from satellite altimetry, (c) mixed points (X, Y, Z) from RTC-DEM (a) and satellite altimetry (b), and (d) resulting digital elevation model based on the mixed points, which provides the topography and bathymetry. The predicted discharge at four cross-sections at a river reach of 30 km at Niger river was used to observe downstream flow gains, suggesting the efficacy of this approach. It is however, important to note that the various techniques, and methodologies highlighted in studies that have used remote sensing data to estimate discharge (Lamine et al. 2021; Getirana and Peters-Lidard 2013; Coe and Birkett 2004) have varied accu-

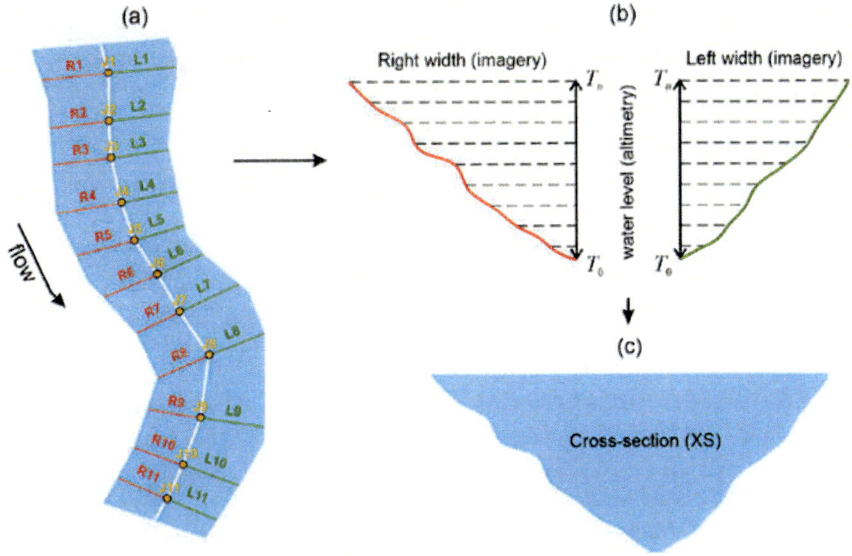

Fig. 14.10 Illustration of the lateral method. *Source* Lamine et al. (2021)

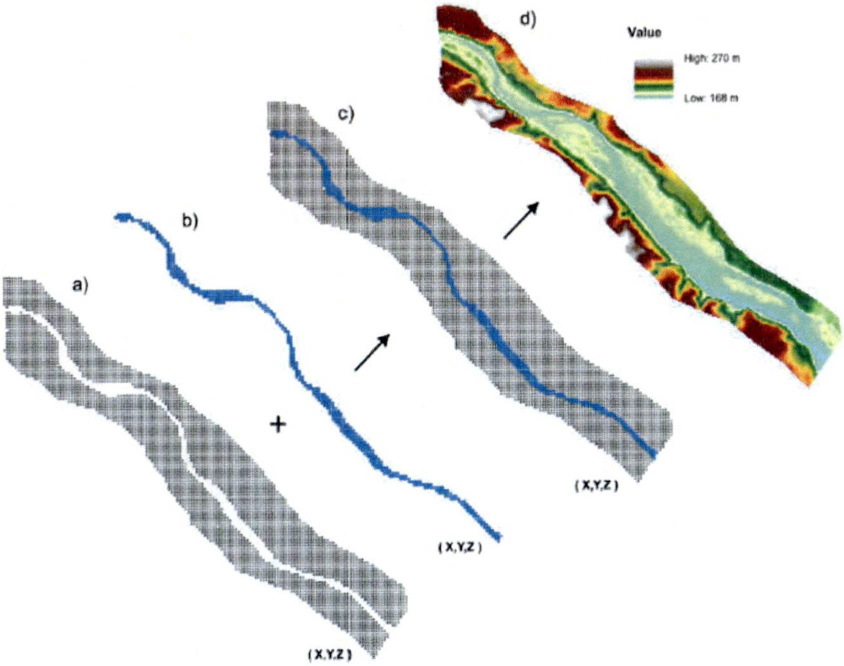

Fig. 14.11 Illustration of the mixed method. *Source* Lamine et al. (2021)

racy index, which is largely a function of their data-driven procedures and empirical approaches in problem-solving.

14.4 Understanding Changes in Land Water Storage

Understanding global freshwater response to important stressors such as human abstraction and climate is important to model threats to water and food security (Rodell et al. 2018; Vorosmarty et al. 2010). Inter-annual variations in surface water hydrology, including rivers, lakes, and floodplains drive changes in global terrestrial water storage. As with influence of multi-scaled climate variability indices such as the ENSO phenomenon on changes in rainfall patterns, freshwater can also be affected by these large-scale oceanic processes (ENSO). But substantial large scale changes in hydrological processes can also result from multiple strings of human activities, e.g., surface water developments and water diversion in other large watersheds, which also drive surface water hydrology (e.g., Ndehedehe et al. 2019). The underlying uncertainties in the forecast of changes in freshwater (groundwater, streams, rivers, etc.) in some regions can be induced by several unmitigated human actions, multi-scale climatic drivers, as well as large scale physical and dynamic processes (e.g., Rodell et al. 2018; Famiglietti 2014; Han et al. 2010; Chen et al. 2007b). These factors have enduring hydro-ecological effects on the environments and combine to limit our understanding of large-scale hydrological processes and impacts of climate on water availability.

It is true that the composite influence of various human actions and climate variability affects hydrological changes and hampers sustainable management of freshwater resources. However, accounting for human-induced influence (e.g., surface water developments and increased water abstraction) in predictive models is challenging and may require advance statistical approaches. By combining GRACE data with other multi-resolution data, there has been some progress in this regard but isolating the influence of climate variability and human activities on water storage is still a challenge, warranting further assessment. In the previous section, the various approaches to reconstruction of TWS were detailed. Building on these reconstruction approaches, Liu et al. (2021) combined a reconstruction technique with statistical-based models, using an indirect means that relies on a combination of GRACE satellite data with models to separate climate- and human-driven water storage anomalies. Their reconstruction of TWS was very consistent with satellite observations of TWS when compared with in situ observations of water use. This represents an important progress in freshwater studies seeking to quantify human-driven impacts on water availability. However, the complexity of hydrological processes in some areas resulting from a mix of physical processes like earth quakes, groundwater-induced subsidence and climate change, makes understanding hydrological processes more difficult. A case-specific study over South America (Ndehedehe and Ferreira 2020a) as will be discussed in the next section demonstrates how continental water storage can be driven by these multiple factors.

Assessing changes in freshwater on a global scale is difficult because of lack or limited ground observations. Satellite remote sensing, however, is making the assessment of freshwater possible thus providing opportunities to track changes in ice-sheet, glacier ablation, as well as quantifying variations in hydrological stores (groundwater, soil moisture, and surface water), and impacts of climate change on freshwater variations (Ndehedehe and Ferreira 2020a; Ndehedehe et al. 2021; Thomas and Famiglietti 2019; Rodell et al. 2018; Famiglietti and Rodell 2013). Specifically, satellite hydrology missions such as the Gravity Recovery and Climate Experiment (GRACE) provide an opportunity to monitor changes in continental stored water. In South America (SA), repeated hydrogeological and climatic disturbances have been reported during the last decade. A continent-wide assessment of terrestrial stored water that relies on GRACE observations is important to explore observed hydrogeodetic influence as well as those of climate. This chapter, therefore, presents the work of Ndehedehe and Ferreira (2020a), which employed a two-step regularization approach that combined the JADE (Joint Approximate Diagonalization of Eigen matrices) algorithm and PLSR (partial least squares regression) to assess satellite-derived changes in terrestrial water storage over SA, using GRACE observations during the 2002–2017 period.

14.5 Terrestrial Water Storage over South America

In the context of dynamic earth processes, South America is a well-known hotspot of crustal and lithospheric deformations, seismicity, and geo-hazards (see, e.g., Ruiz and Madariaga 2018; Ferreira et al. 2018; Montecino et al. 2017a, b; Jaramillo et al. 2017; Vigny et al. 2011; Han et al. 2010; Lomnitz 2004). The influence of climate and physical processes as well as geodetic disturbances could have implications on surface mass variations and the acceleration of the water cycle. In view of these perturbations on surface hydrology, time-variable geophysical signals observed by the Gravity Recovery and Climate Experiment (GRACE, Tapley et al. 2004a) are expected to be driven not only by climate oscillations and key processes of inter-annual variability (e.g., Ni et al. 2018; Ndehedehe et al. 2018a, 2017b; Dong et al. 2015; Linage et al. 2013; Phillips et al. 2012) but natural and other non-climatic elements, e.g., human water abstraction and deformations (e.g., Rodell et al. 2018; Chen et al. 2016a; Castellazzi et al. 2016; Famiglietti 2014; Chen et al. 2007b). This assumption is anchored on the fact that apart from the redistribution in continental water storage, other dynamic processes such as gravitational tide in the solid Earth, post-glacial rebound and variations in Antarctic and Greenland ice volumes cause significant changes in the Earth' s gravity (e.g., Swenson and Wahr 2002; Wahr et al. 1998). Whereas a plethora of scientific reports on freshwater dynamics are mostly focused on climate variability related changes (e.g., Rodell et al. 2018; Anyah et al. 2018; Reager et al. 2014; Thomas et al. 2014; Linage et al. 2013; Phillips et al. 2012), very little attention is paid to improving our understanding of the possible contributions of non-climatic factors, especially those not related to groundwater abstraction.

Changes in freshwater can be poorly understood in areas where these factors are frequently observed. Consequently, predicting freshwater systems at regional or continental scales requires large-scale assessment of TWS (terrestrial water storage) variations and an understanding of prominent drivers of surface hydrology. On the one hand, this knowledge provides foundation for an efficient modelling framework for water resources. On the other hand, limited ground observations and the lack of a suitable modelling approach to characterize key hydrological metrics, however, are some important constraints to such assessments. Further, the proliferation of dams and surface water developments for hydropower, agriculture and other relevant applications are gradually emerging as considerable drivers of TWS. This kind of impacts has been reported in Africa's biggest lakes like the lakes Volta and Victoria (see, Ndehedehe et al. 2017a; Moore and Williams 2014; Ahmed et al. 2014). The expansion of dams across several regions of the world and hydro-power stations are the results of management initiatives to substantially boost hydroelectricity capacity around the world (Zarfl et al. 2015). This, however, is also expected to result in considerable impact on hydrological changes. The Lake Volta (see, Ndehedehe et al. 2017a; Ndehedehe 2019), or the Amazon basin where dam constructions are impacting on ecosystems by modifying vital flood pulses are good examples of how surface water resources development drives hydrology.[1] This will be discussed in detail in the Chap. 15.

The GRACE satellite has been operational since 2002 and with its Follow-On mission, which was launched in 2018 has continuously provided information on changes in the Earth's gravity field and has enabled quantitative estimates of changes in TWS (sum of soil moisture, groundwater, surface water, wetlands, etc.) across the world. Several of past GRACE-hydrological studies in South America-SA (Fig. 14.12a, b) focused on the Amazon basin (e.g., Frappart et al. 2013a, 2012, 2011a; Alsdorf et al. 2010; Kim et al. 2009; Han et al. 2009) and new studies on the hydrogeological structures and impacts of climate on TWS dynamics are emerging (e.g., Ndehedehe and Ferreira 2020b; Tourian et al. 2018; de Linage et al. 2014) thus emphasizing the need to further assess the inter-annual variations of land water storage and the representation of dynamic processes in time-variable gravity observations.

To improve understanding of the spatial and temporal variations of extended GRACE-derived TWS (2002–2017) over SA, this study localizes GRACE-hydrological signals by rotating it towards statistical independence using the Joint Approximate Diagonalization of Eigen matrices (JADE) algorithm (e.g., Ndehedehe et al. 2016b; Ziehe 2005; Cardoso and Souloumiac 1993; Cardoso 1999; Common 1994; Cardoso 1991). Here, we provide findings following the work of Ndehedehe and Ferreira (2020a) who combined partial least squares regression (PLSR) model with JADE rotation and multi-linear regression to assess large-scale changes in TWS over SA. The specific aims of this study was to (i) assess localized spatial and tem-

[1] https://news.mongabay.com/2018/01/study-amazon-dams-are-disrupting-ecologically-vital-flood-pulses/.

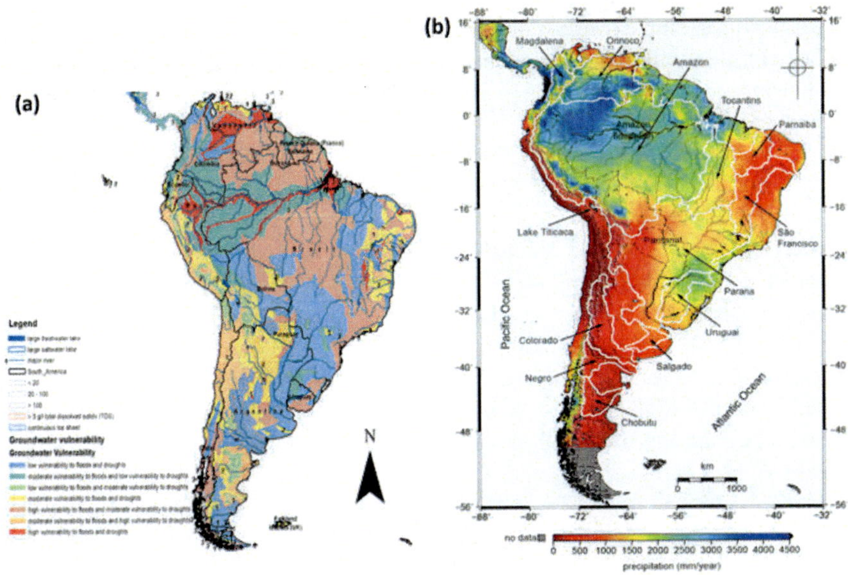

Fig. 14.12 Map showing countries, rivers, and groundwater aquifers in South America. **a** The hydrogeological map of South America indicating aquifers, rivers, lakes, and groundwater vulnerability areas. The classification of groundwater recharge (mm/a) rates is in numbers (e.g., 2(= 20 − 100)). The aquifer maps and hydrological units (river distribution networks and lakes) are those of World-wide Hydrogeological Mapping and Assessment Programme produced under the Global Groundwater Vulnerability to Floods and Droughts project (WHYMAP GWV © BGR & UNESCO 2015). **b** The prominent river basins in South America and the network of transboundary rivers in each wetland and the annual distribution of rainfall from the Tropical Rainfall Measuring Mission satellite-based precipitation (2003–2016)

poral variations of GRACE-derived TWS over SA through the rotation of prefiltered GRACE-hydrological signals, (ii) assess inter-annual variations in TWS in relation to precipitation, and (iii) predict temporal evolutions of GRACE-derived TWS over 12 prominent river basins in SA. To this end, sea surface temperature (SST) anomalies were statistically decomposed with the JADE method and used as input in a partial least square regression (PLSR) model. The concept or argument in this work was formulated around the assumption that strong ocean-land atmosphere interactions through SST (sea surface temperature) generate the systems that drive changes in precipitation around the region. Indeed, precipitation in several tropical systems is driven by these interactions, resulting in increasing acceleration of the water cycle. This large changes in rainfall in turn drive variations in continental stored water. This is generally the case as the influence and the feedback mechanism of anomalous warming of the surrounding oceans on inter-annual rainfall variations have been widely reported (e.g., Zou et al. 2018; Ndehedehe et al. 2018a; Odekunle and Eludoyin 2008; Nicholson 2013; Fontaine and Bigot 1993). In the sections that follow, a case-specific example over SA of how satellite and the use of multivariate methods can be helpful in improving our understanding of freshwater changes is detailed.

14.5.1 Data and Methods

Rainfall was based on the Global Precipitation Climatology Centre (GPCC, Schneider et al. 2014) data while Gravity Recovery and Climate Experiment (GRACE) mascons (Chap. 4) was used to estimate land water storage. The mascon data (2002–2017) are available in the CSR data portal (http://www2.csr.utexas.edu/grace/RL05_mascons.html). River basin estimates of TWS values $\overline{TWS}$ were recovered from the global mascon solutions by using the area weighted average approach, i.e., the approximated area of the basin (region) as (e.g., Ndehedehe et al. 2017a)

$$\overline{TWS}(t) = \frac{1}{A}\sum_{i=1}^{n} \overline{TWS}(\phi_i, \lambda_i, t)A_i, \tag{14.1}$$

where n is the number of cells within the basin, A_i is the area of each cell i, A is the total area of the basin and ϕ_i and λ_i are the corresponding latitudes and longitudes of the center of each grid cell, respectively. The Sea Surface Temperature (SST) data (monthly means from 2002 to 2017) used in this study were the NOAA' s Optimum Interpolation SST V2 and were downloaded from NOAA' s portal (https://www.esrl.noaa.gov/psd/data/gridded/data.noaa.oisst.v2.html). Except in cases of human water management (e.g., water transfers), SST is a key predictor of rainfall and water availability because de Linage et al. (2014) identified several studies that showed evidence of the interactions between SST and Intertropical Convergence Zone, which resulted in the severe droughts of 2005 and 2010 in the central and western Amazon.

14.5.2 Rotation of Terrestrial Water Storage Using Higher Order Statistical Algorithms

There are a range of higher order statistical decomposition methods some of which uses the fixed-point iteration scheme (Frappart et al. 2011b). This algorithm was previously used to filter satellite gravity signals observed by GRACE (Frappart et al. 2011b, 2010). Contrary to these algorithms, the JADE technique exploits the fourth-order cumulants of the data matrix based on remote properties that include non-stationarity, spectral non-Flatness, and non-Gaussianity (e.g., Ndehedehe et al. 2016b; Ziehe 2005; Cardoso 1999; Common 1994). Because of the statistical and numerical efficiency of the JADE approach in solving the optimization problem (e.g., Ziehe 2005; Cardoso and Souloumiac 1993), eight significant independent and well-localized spatio-temporal patterns (will be interchangeably designated as 'regions')

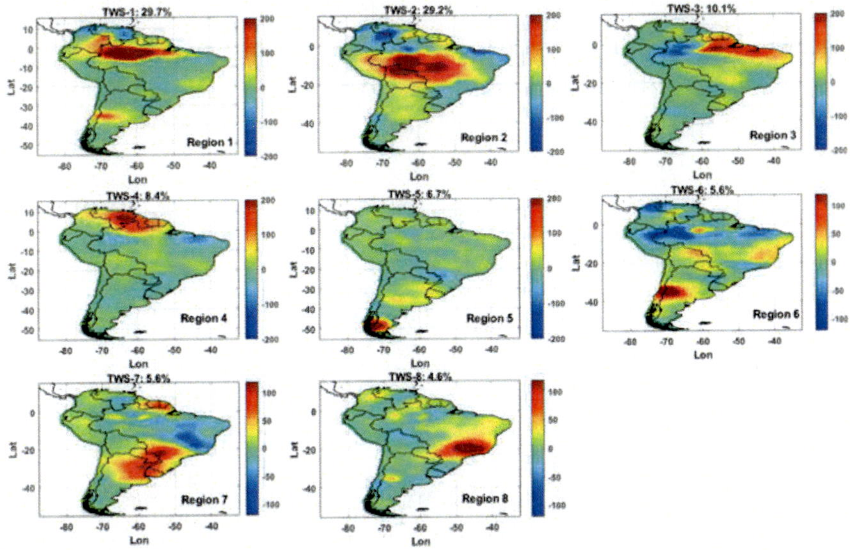

Fig. 14.13 Regionalization of TWS (2002–2017) based on pre-orthogonalization and cumulant decomposition methods. The spatial patterns are scaled using the standard deviation of the computed independent components, i.e., the temporal series in Fig. 14.14. They are also interpreted in conjunction with their corresponding temporal patterns (independent components). The axis labels, latitudes (Lat) and longitudes (Lon) are indicated

of TWS were recovered over South America (SA). These patterns highlight considerable inter-annual variability and regional GRACE-hydrological signals over the continent (Fig. 14.13). The surface water dynamics of the Amazon basin and much of Brazil dominate the independent patterns of TWS in SA (TWS-1 and TWS-2, Figs. 14.13 and 14.14) and together account for approximately 60% of the total variability. Region one is localized over the Amazon floodplain (cf. Fig. 14.12b) and the exchange of water fluxes within this domain is somewhat complex. Apart from the indication that the highest proportion of stored water on the Amazon floodplain comes from the mainstem river, Alsdorf et al. (2010) also found that the mainstem discharge was higher compared to the sum of annual water storage and that drained from the Amazon floodplain. Although variations in surface waters represent a considerable component of TWS as observed by GRACE (e.g., Frappart et al. 2011a; Kim et al. 2009; Han et al. 2009), the strong exchange of fluxes within the floodplain corridors explains why the dominant patterns of TWS over SA is observed over the Amazon floodplain (TWS-1, Figs. 14.13 and 14.14). This is consistent with the results of Frappart et al. (2013b), who found the strongest spatial loadings of TWS along the Solimões-Amazon corridor and includes the south of the Amazonian and the Negro basins.

In regions 3 and 4 (TWS-3 and TWS-4, Figs. 14.13 and 14.14), observed GRACE-hydrological signals are annual variations with probable contributions from multi-

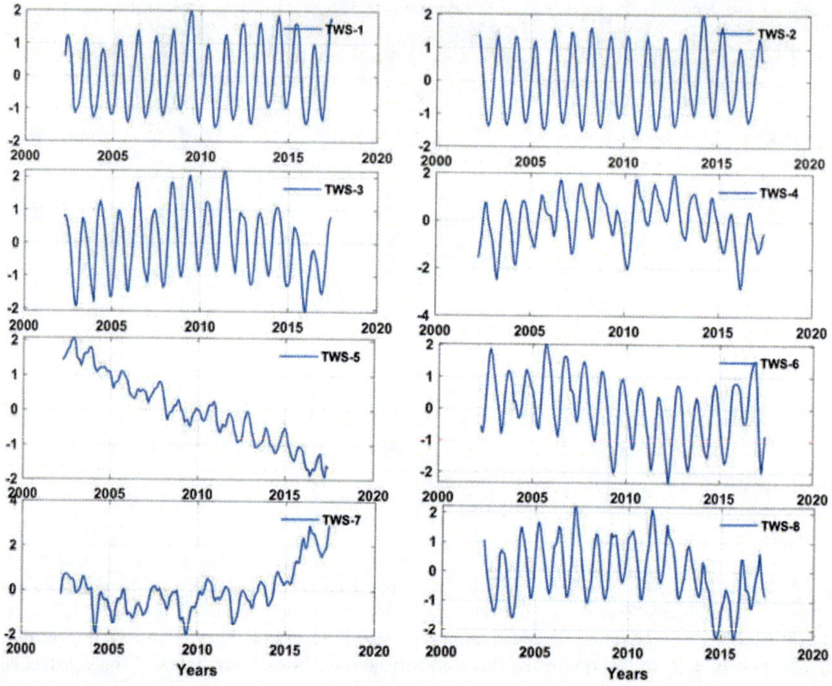

Fig. 14.14 Localized temporal patterns of TWS (2002–2017) based on the cumulant decomposition methods similar to Fig. 14.13. These temporal patterns are in standardized units (y-axis) and correspond to the spatial evolutions and hydrological regions shown in Fig. 14.13. Amplitudes of TWS for each region (e.g., TWS-1, TWS-2, etc.) are recovered by a joint interpretation of these temporal patterns (in standardized units) with the corresponding spatial patterns in Fig. 14.13

scale climate oscillations, e.g., El-Ninõ Southern Oscillation (ENSO)-related teleconnections. Some studies found the connections between ENSO and inter-annual TWS variations in these regions (e.g., Ni et al. 2018; Linage et al. 2013; Phillips et al. 2012). By further exploring the interplay between ENSO-teleconnections and GRACE-hydrological signals for regions three (northern Brazil) and four (Venezuela), we also found strong evidence that suggests climate teleconnection-driven influence on TWS dynamics. But it is not clear, which hydrological stores or component of GRACE TWS (e.g., surface water, soil moisture, aquifer, etc.) provides this response to climate. The GRACE-hydrological signal in region 5 (TWS-5, Figs. 14.13 and 14.14) is the melting of the Patagonia ice field. Consistent with this study, different methods, which includes forward modelling approach have been employed to highlight the continued mass loss in Patagonia (Rodell et al. 2018; Willis et al. 2012; Chen et al. 2007a). This extensive and unabated mass loss in Patagonia ice-field (TWS-5, Fig. 14.14) is caused by the warming of the climate system, and being the second largest ice body in the Southern Hemisphere, GRACE is a viable tool for the continued monitoring of the impact of rising temperature on Ice fields.

Note that similar to the JADE algorithm, estimated negative trends observed over the Patagonia ice field using the Sen' s slope and multi-linear regression model are consistent (Sects. 14.5.3.1 and 14.5.4.2).

Region six (TWS-6, Figs. 14.13 and 14.14) highlights the amplitudes of TWS in the Chile region but with most parts falling within Argentina while region 7 (TWS-7, Figs. 14.13 and 14.14) depicts multi-annual variations in TWS over south Brazil and the neighbouring Central Argentina with a weak opposite phase in north-east Brazil. Having experienced repeated earthquakes, deformations, and other forms of natural disturbances (e.g., Ruiz and Madariaga 2018; Jaramillo et al. 2017; Vigny et al. 2011; Lomnitz 2004), the geophysical signals in Chile and the neighbouring Argentina are expected to be dominated and driven by natural and climatic elements. The temporal pattern (TWS-6, Fig. 14.14) associated with this signal in the Chile/Argentina regions (TWS-6, Fig. 14.13) shows a falling amplitude between 2005 and 2011 while the period after 2011 shows a rising trend, consistent with the results in Sect. 14.5.4. Note that region 6 (TWS-6, Fig. 14.13) also indicates relatively strong opposite TWS anomalies in the central Amazon basin area and could be related to floodplain and wetland water storage gain between 2005 and 2010, i.e., when jointly interpreted from the corresponding temporal patterns (TWS-6, Fig. 14.14) and again coincides with numerical results in Sect. 14.5.4. The argument here is that elevation of the flood plains of Amazon basin (low elevation areas less than 200 m above mean sea level) coincides with the unique features of TWS for region 6 (e.g., the region with the most freshwater-Fig. 14.12) as depicted in its spatial patterns (TWS-6, Fig. 14.13). That is, the spatial patterns with positive loadings in Chile and environ (TWS-6, Fig. 14.13) are interpreted as wet conditions (or gain in surface mass) if multiplied with its corresponding positive amplitudes in Fig. 14.14. Similarly, the spatial patterns with negative loadings in Fig. 14.13 (Amazon region) are interpreted as wet conditions (or rise in surface mass) if multiplied with its corresponding negative amplitudes (TWS-6, Fig. 14.14). The low elevation areas of the Amazon and the strong annual fluctuations (TWS-6, Fig. 14.14) suggest that it appears to be a rather permanent wetland water storage induced by rainfall seasonality. As opposed to the high elevation areas of Chile (greater than 3500 m above mean sea level), which shows annual fluctuations in TWS, it can be argued that elevation plays a role in the Amazon flood plain wetland water storage. Notably, this Amazon signal (region 6) is different from regions 1 and 2 where river storage, exchange of fluxes between tributaries, including vertical movement of shallow groundwater could be major drivers of GRACE-derived TWS variations along the Amazon corridor.

Regarding the spatial patterns in region 7 (TWS-7, Fig. 14.13), the strongest signals with opposite phase occur within southern Brazil. The corresponding time series (TWS-7, Fig. 14.14) shows relatively strong negative anomalies in 2004 and 2009 before the strong rise that occurred after 2010. The peak negative anomalies in 2004 and 2009 appear to coincide with severe droughts in the region, especially the summer drought of 2004/2005. As highlighted further in Sect. 14.5.3.1, the short-term trends in TWS and rainfall during this period around this region (2012–2017) show consistency between 2012 and 2014 but with small dissimilarity during the 2010–2012 period and can be explored further in future studies that focus on this region.

Region 8 (TWS-8, Figs. 14.13 and 14.14) is the GRACE-hydrological signal native to north-east Brazil (e.g., Sun et al. 2016). The declining trend in TWS during the 2011–2015 period and the strongest negative anomaly observed between 2014 and 2015 (TWS-8, Fig. 14.14) coincide with the widely reported super extreme droughts that ravaged most eastern sections of Brazil during the same period, especially 2015 (e.g., Ferreira et al. 2018; Zou et al. 2018; Getirana 2016). This prolonged drought, which occurred in much of north-east Brazil during most of the 2010–2013 period, resulting in considerable declines in TWS (TWS-8, Fig. 14.14) triggered the need to improve drought-related policy and management strategies at various levels of government Brazil (see, Gutiérrez et al. 2014). Due to limited annual rainfall, north-east Brazil is generally a water deficit region and depends on surface water from the Amazon basin for irrigation and hydropower generation. The Sao Francisco river basin in Brazil depends on water transfers between river basins. Moreover, the National Water Agency of Brazil reported in 2015 that 79% of total water withdrawn was for irrigated agriculture. Hence, GRACE is a viable hydrological tool to support the monitoring of hydrological drought and its impacts on water availability and human abstraction (see, Ferreira et al. 2018).

14.5.3 Assessing Inter-Annual Variations and Changes in TWS

In this section, results for the region-specific (12 river basins) estimates of linear rates in TWS are highlighted. These river basins are among the well-known global river basins. Add to this, time series of TWS grids are compared to those of rainfall over SA based on cross-correlograms.

14.5.3.1 Recent Changes in TWS in South American River Basins

Observed temporal variations in the 15-year GRACE-derived TWS over 12 river basins are marked with strong fluctuations. With an estimated linear rate of -11.29 ± 1.71 mm/yr, Sao Francisco showed the strongest fall while Parana indicated the strongest rise (6.88 ± 1.13 mm/yr) in TWS during the 2002–2017 period (Fig. 14.15). The summary of Mann-Kendall's test statistics for all significant ($\alpha = 0.05$) linear trends observed in the 12 river basins for the two periods (2002–2017 and 2010–2017) analyzed are highlighted in Table 14.1. Considering the estimated linear rates between 2010 and 2017, the strongest decline (-38.48 ± 7.90 mm/yr) in TWS was found in Orinoco while Uruguay showed a considerable increase in TWS (28.28 ± 3.49 mm/yr) unlike other basins (Fig. 14.15 and Table 14.1). During the same period (2010–2017), significant increase in TWS was also observed in Parana (17.92 ± 3.44 mm/yr), Rio Colorado (23.99 ± 2.60 mm/yr), Negro (13.0 ± 3.27 mm/yr), and Rio Salado (25.17 ± 2.78 mm/yr) while Sao Francisco (-30.84 ± 4.17) and Parnaiba

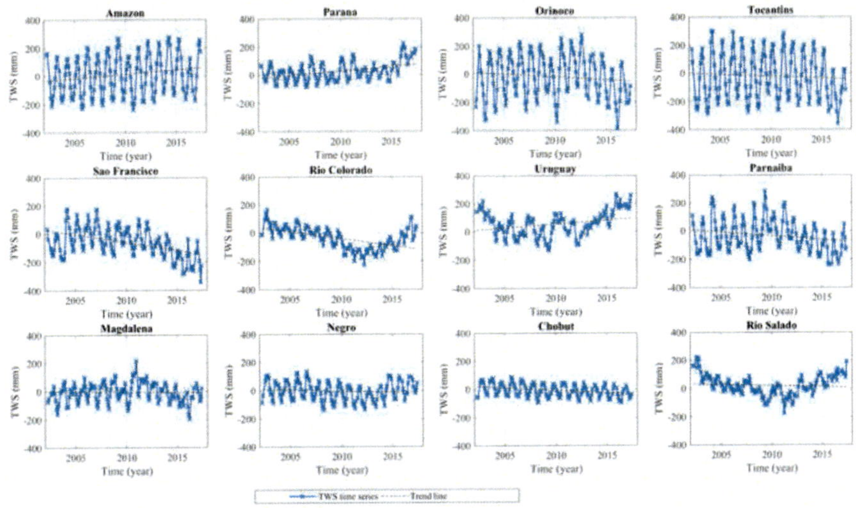

Fig. 14.15 Assessing inter-annual variations of TWS (2002–2017) for 12 prominent global river basins in Southern America

(−24.42 ± 4.95) indicated relatively strong negative trends in TWS (Fig. 14.15 and Table 14.1).

The spatial distribution of trends in TWS and precipitation over South America suggests complex hydrological structures and processes. For instance, the long-term declines in TWS over Chile/Argentina regions and melting of the Patagonia ice field are inconsistent with rainfall (Fig. 14.16a, d). This is because the factors driving the hydrology of these areas are beyond rainfall. Recall that these signals coincide with the spatial and temporal patterns in regions 5 and 6 (TWS-5 and TWS-6, Figs. 14.13 and 14.14) and further highlight the effectiveness of the JADE rotation in localizing geophysical signals. Further, the analyses of short-term trends (2010–2012 and 2012–2014) also highlight the remarkable difference in recent changes in TWS in relation to rainfall. Such inconsistencies are mostly observed over Venezuela, Brazil, and Argentina (Fig. 14.16b, c and e, f). However, the TWS trends in Northeast Brazil between 2012 and 2014 are consistent with those of rainfall (Fig. 14.16e, f). Whereas TWS in some regions in SA respond to changes in rainfall and climatic conditions as is the case in north east and southern Brazil (Fig. 14.16b, c and e, f), the complex hydrogeological structures in some areas may trigger interesting hydrological processes. To understand this process, the grid-based TWS-rainfall association and phase lags during the entire period are explored in the next section by implementing a cross-correlation.

Table 14.1 Estimated trends in temporal variations of TWS for 12 river basins in South America during the 2002–2017 and 2010–2017 periods. The null hypothesis (no significant trend), H_0, was tested at $\alpha = 0.05$ (95% confidence level) using the Mann-Kendall' s statistics. The p-value (probability threshold agreed for the significance level) and root mean square error are indicated. Note, the trends with asterisks (*) are not statistically significant

	River basins	2002/04–2017/06			2010/04–2017/06		
S/N		Trend (mm/yr)	p-value	RMSE	Trend (mm/yr)	p-value	RMSE
1	Amazon	6.54 ± 2.60	0.004	141.755	*11.06 ± 8.49	0.221	151.317
2	Parana	6.88 ± 1.13	0.000	61.441	17.92 ± 3.44	0.000	61.29
3	Orinoco	*−4.25 ± 2.89	0.096	157.598	−38.48 ± 7.90	0.000	142.128
4	Tocantins	*−2.60 ± 3.06	0.5186	166.345	*−16.65 ± 8.98	0.265	160.039
5	Sao Francisco	−11.29 ± 1.71	0.000	93.126	−30.84 ± 4.17	0.000	74.30
6	Rio Colorado	−9.91 ± 1.22	0.000	66.437	23.99 ± 2.60	0.000	46.377
7	Uruguay	5.33 ± 1.54	0.000	83.982	28.28 ± 3.49	0.000	62.13
8	Parnaiba	−5.38 ± 2.05	0.0108	111.733	−24.42 ± 4.95	0.000	88.151
9	Magdalena	*−0.44 ± 1.25	0.925	68.134	−18.76 ± 3.58	0.000	63.86
10	Negro	*−1.51 ± 1.20	0.1699	65.124	13.0 ± 3.27	0.000	58.268
11	Chobut	−3.74 ± 0.78	0.000	42.578	*−4.17 ± 2.30	0.011	41.068
12	Rio Salado	*−1.49 ± 1.31	0.1671	71.137	25.17 ± 2.78	0.000	49.611

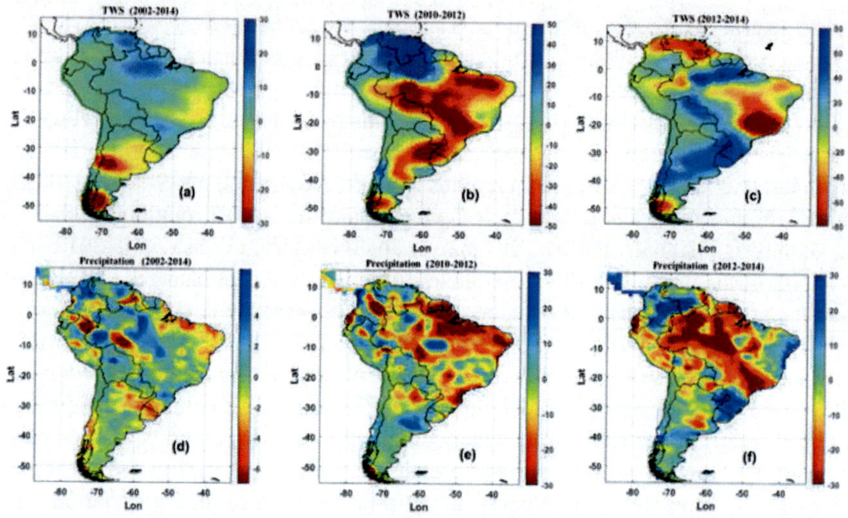

Fig. 14.16 Spatial distribution of trends in TWS (**a**–**c**) and precipitation (**d**–**f**) for three different periods (2002–2014, 2010–2012, and 2012–2014). All units are in mm/year

14.5.3.2 GRACE TWS Versus Rainfall

To explore the TWS–rainfall relationship, cross-correlation between detrended series of GRACE-TWS and GPCC-rainfall was analyzed. From the water budget equation, TWS changes in time. In other words, observed changes in TWS balance precipitation minus evaporation and runoff. Figure 14.17 shows the lags (presented as coloured surface) at which the two series present the maximum correlation coefficients (presented as contour lines). Overall, rainfall rates lead TWS in approximately 0 to 3 months for almost all of SA with correlation coefficient ranging from approximately 0.60 to 0.80 (Fig. 14.17). For example, considering the physiographic regions of SA, the Brazilian Highlands, the Amazon Rainforest, and the Gran Chaco regions (spreads across eastern Bolivia, western Paraguay, northern Argentina and sections of Brazil) all indicate relatively high correlations with rainfall leading TWS. Moreover, the Lianos and Guiana Highlands (both at northern SA), and the Andes Mountains

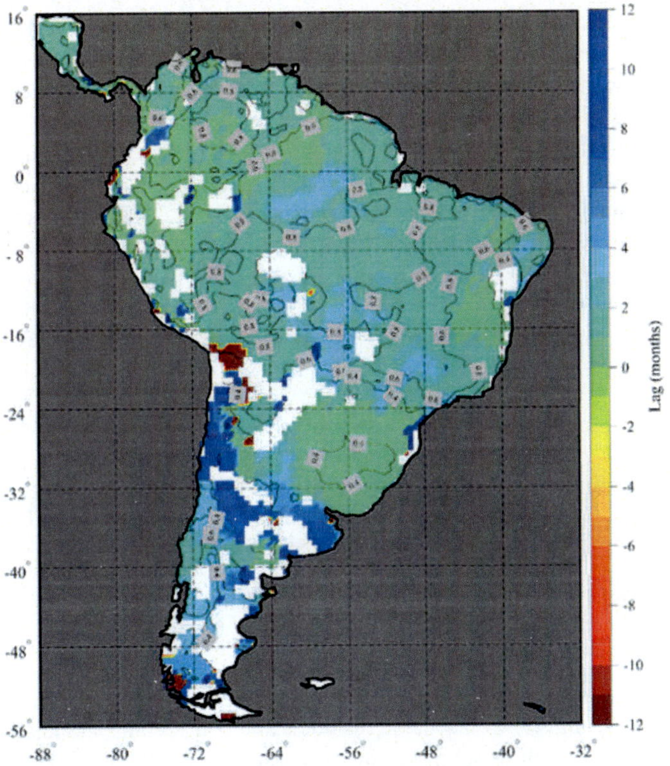

Fig. 14.17 Correlation analysis showing phase lags at which maximum correlation coefficients occurred for TWS versus rainfall during a common period (2002–2017). The contour lines show the correlation coefficients while the values depicted in the colour bar indicate the lags in months. Regarding lags, negative values mean TWS leads rainfall while positive values imply rainfall leads TWS. Only statistically significant correlations ($\alpha = 0.05$) are presented

(apart from the portion that surrounds the Altiplano and Atacama Desert) present good phase agreement with high correlation coefficients. While this shows good feedback of TWS on rainfall for almost all SA, the relationship between TWS and rainfall during the period is statistically not significant at the 95% confidence interval.

However, there are exceptions in the regions below latitude 16°, which are arid regions characterized by relatively low precipitation rates (Fig. 14.12b). The Chile coastline and the Patagonia are regions where correlations are modest ranging from 0.40 to 0.60 and rainfall generally leads TWS with a phase greater than four months (Fig. 14.17). There are also some hydrological hot spots (regions below latitude 16°) with TWS leading rainfall with a phase of 12 months but with high correlation only for the Lake Titicaca region (cf. Figs. 14.12 and 14.17). But around the Patagonian region and Chobut, Colorado and Salgado basins, correlation coefficients between the series are low ($r = 0.40$) indicating that the key driver of variations in TWS is beyond rainfall. For example, the Pampas of Argentina (approximately between latitudes 40 °S to 32 °S and longitudes 56 °W to 64 °W) is characterized by annual precipitations of 500–1000 mm (Fig. 14.12b). TWS shows relatively low annual peaks with negative linear trends of −5 to −15 mm/yr of mass loss. This could be attributed to changes in the land use due to the crop rotations or natural vegetation, which might decrease groundwater recharge and thus decrease TWS storages (e.g., Kroes et al. 2019). This is particularly true for Patagonia since it is characteristically a dry climate. As shown in region 6 (Figs. 14.13 and 14.14), the Patagonia ice fields are undergoing a contemporary melting and retreating of glacial ice and is consistent with previous studies (e.g., Willis et al. 2012; Chen et al. 2007a). The Andean Plateau (Altiplano) presents a phase lead greater than 12 months for rainfall relative to TWS series and includes part of the Pampas region. Furthermore, part of the Pampas region (southern) also shows a phase lead of about 12 months (TWS lags rainfall). In both cases, we found that correlation coefficients were low, mostly between around 0.10 and 0.20 and insignificant.

14.5.4 Modelling TWS Using Parameter Estimation Techniques

14.5.4.1 Partial Least Square Regression Prediction of River Basin TWS

Overall, the temporal variations of TWS are well predicted for most of the river basins (Fig. 14.18) and show reasonable residuals and considerable association with observed TWS based on three validation metrics used (Fig. 14.19a–b). In terms of the performance of PLSR model, time series of TWS in the Amazon, Orinoco Tocantins, and Chobut basins were well predicted and showed optimum skills based on these validation metrics (R^2, IA, and NSE). These basins indicated R^2, IA, and NSE coefficients above 0.83 (Fig. 14.19b) but with Chobut having the least residual and

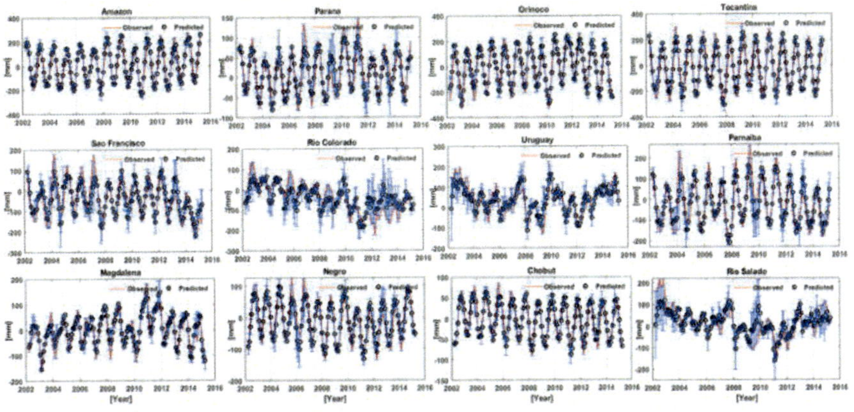

Fig. 14.18 Modelling GRACE-derived TWS (2002–2017) using the PLSR model. Temporal patterns of predicted and observed TWS for the 12 prominent river basins in Fig. 14.16 are compared using several validation metrics

RMSE (Fig. 14.19a). However, Rio Colorado, Rio Salado, Uruguay, Sao Francisco, and Parnaiba indicated relatively higher TWS residuals and RMSEs (Figs. 14.18 and 14.19a). The uncertainties (i.e., residuals and RMSEs) in predicted time series of TWS in these locations were also reflected in their observed validation metrics with Rio Colorado showing the lowest R^2 (0.49), IA (0.66), and NSE (0.49) (Fig. 14.19b). The obvious poor performance of the PLSR model in Rio Colorado, Rio Salado, and Sao Francisco (Fig. 14.19a, b) gives credence to the argument of the perceived contributions of non-climatic elements to observed inter-annual variations in TWS. Especially for the basins located in Brazil (e.g., Sao Francisco), Uruguay, and Argentina, the TWS–rainfall relationship as was highlighted in Sect. 14.5.3.2 indicated poor association with considerable phase lags in most catchments. This may not be unconnected with human water management as Getirana (2016), for example, found considerable correlations of monthly time series of TWSA and ground-based water storage observations with most reservoirs within southeastern Brazil. Given the observed residuals and the index of agreement (Figs. 14.18 and 14.19a), it seems the predictability of TWS based on hydro-climatic variables could be challenging in some river basins of SA. As indicated in Fig. 14.17, some of these regions (Andean Plateau, Pampas, etc.) showed low correlation values ranging from 0.00 to 0.20 and with more than 12 months in phase lag between.

14.5.4.2 Multi-linear Regression of TWS Grids

In this section, the results of simulated TWS and rainfall grids over SA using the MLRA technique are highlighted. The mean annual amplitudes of rainfall observed over the Amazon basin area are consistent with TWS (Fig. 14.20a, f) while the

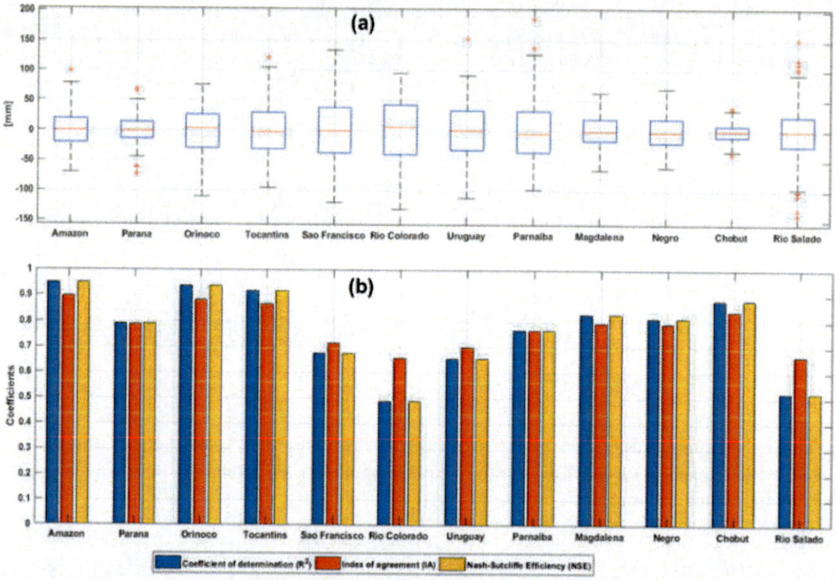

Fig. 14.19 Analysis of PLSR model outputs for TWS in South American river basins. **a** Uncertainties in the PLSR model prediction of TWS over the 12 river basins for 157 time steps (2002/04–2015/04). The TWS residuals (i.e., the difference between the predicted and observed obtained in mm) indicated here are for PLSR calibration based on nine significant components. **b** Performance of PLSR model in predicting TWS in the 12 river basins based on three skill metrics (coefficient of determination (R^2), Index of Agreement (IA), Nash-Sutcliffe Efficiency (NSE))

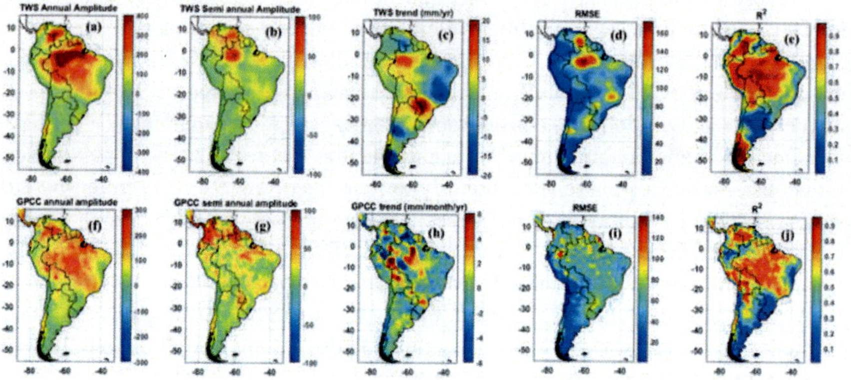

Fig. 14.20 Modelling **a–e** GRACE-derived TWS and **f–j** GPCC precipitation using the multi-linear regression formulation during the period 2002–2017

trends and mean semi-annual amplitudes of TWS show significant disparity and do not completely mimic the modelled patterns of rainfall (Fig. 14.20b, c and g, h). As opposed to rainfall, the only spot with considerable semi-annual signal is in the Brazil section of the Amazon basin (Fig. 14.20b, g). The dissimilarity in trends (TWS and rainfall), especially those observed in Brazil and Argentina (Fig. 14.20c, h) confirm the complex geophysical processes in the continent as highlighted in previous sections. Overall, the root mean square error (RMSE) is relatively low over the arid regions of SA for rainfall and TWS but somewhat higher for TWS in the central Amazon basin (Fig. 14.20d, i). Such relatively high uncertainties may result from the influence of important processes of inter-annual variability (e.g., Pacific Decadal Oscillation) or the complex properties of the Amazon floodplain as it relates to exchange of fluxes and water storage (e.g., Alsdorf et al. 2010). However, the coefficient of determination (R^2) suggests that the model predicts variations in TWS and rainfall quite well with R^2 values ranging from 1.00 to 0.80 (Fig. 14.20e, j). Except for R^2 values (i.e., TWS) over Patagonia (Fig. 14.20e), the regions that are poorly simulated in the two products are generally found below Latitude 20° (Fig. 14.20e, j). The low R^2 values in TWS, for example, could imply strong interactions with non-climatic factors and/or the multi-linear regression model is not suitable for those locations. Overall, the modelling of TWS and rainfall over the Amazon basin show a good fit and suggest it can be predicted quite well. This is justified by the result in Sect. 14.5.3.2 where the TWS–rainfall relationship around the Brazilian highlands and much of the Amazon rainforest is relatively high with rainfall leading TWS. But the plenitude of observed R^2 ranging from 0 to 0.30 in some regions of the continent could imply the predominant influence of other hydrological drivers, e.g., natural disturbances and perturbations of multi-scale global climate signals. The influence of the latter on TWS in tropical SA has been reported (e.g., Linage et al. 2013) while the contributions of the former to changes in TWS in the Chile/Argentina regions have been affirmed in some studies (e.g., Rodell et al. 2018; Han et al. 2010).

14.5.5 Assessing Drivers of Land Water Storage

Global climate is changing. The impacts of such changes, which have been attributed to rising anthropogenic emissions of greenhouse gases, amongst other things are evident in natural systems and acceleration of the water cycle, be it at regional or global scales. The rise in global sea levels, long-term declines in rainfall and TWS, and increased frequency in flood and drought events (e.g., Rodell et al. 2018; Spinoni et al. 2014; Nka et al. 2015; Boening et al. 2012; Paeth et al. 2012; van der Molen et al. 2011; Sheffield and Wood 2008) across the globe are some apparent indices of a changing global climate. As with other regions whose TWS variations and discharge are largely rainfall-dependent (e.g., Ndehedehe 2019; Ndehedehe et al. 2016a; Phillips et al. 2012; Conway et al. 2009), land water storage dynamics (includes river discharge and surface runoff) in SA are arguably driven by changes in climate and other key processes of inter-annual variability such as ENSO (e.g., Linage et al. 2013; de Linage et al. 2014; Frappart et al. 2013a, b). Incessant extreme droughts event in South America, for instance, have been linked to large scale variations in the tropical oceans (Pacific and Atlantic). This was recently echoed by Erfanian et al. (2017) who also found significant relationship between unparalleled drought events in South America and extreme anomalies in SST of the nearby oceans. Whereas inter-annual changes in rainfall over tropical ecosystems are significantly modulated by the tropical oceans (Ndehedehe et al. 2018a; Odekunle and Eludoyin 2008; Semazzi et al. 1988; Nicholson 2013), our results in Sect. 14.5.2 confirm that these drought episodes have considerable impacts on TWS. This is true for Brazil where the observed decline in TWS between 2012 and 2015 (TWS-8, Figs. 14.13, 14.14; and 14.16) coincided with the extreme drought that ravaged Brazil during the period (e.g., Ferreira et al. 2018). Apart from the continued mass loss in Patagonia (TWS-5, Figs. 14.13 and 14.14) caused by climate warming, the observed TWS–rainfall relationship (Sect. 14.5.3) also highlights the importance of climate variability in observed variations in TWS over most sub-regions of SA. Within the context of TWS response to climate variability induced rainfall, there are indications, however, that some hydrological signals are not unconnected with non-climatic influence.

In practical geodetic concepts as it relates to mass redistributions caused by changes in gravity fields, Chile (TWS-6, Figs. 14.13 and 14.14) is one of the most complex hydrological regions in SA (Ferreira et al. 2019a). Because of its hydro-geodetic characteristics and geophysical formation, the Chilean region and environ, for example, are vulnerable to vertical deformations, seismicity, and earthquakes. (see, e.g., Montecino et al. 2017a, b; Jaramillo et al. 2017). Some of these geodetic perturbations and disturbances as also found in other regions (e.g., Chen et al. 2007b), have enduring hydro-ecological effects on the environments (e.g., Jaramillo et al. 2017). Hydrologically, they leave behind extraneous geophysical signals that become available as an integral part of observed hydrological changes in such regions. These signals complicate our understanding of land water storage drivers. For example, in the Chile earthquake of 27 February 2010, Han et al. (2010) found a gravity anomaly of $-5\ \mu$Gal with a spatial scale of 500 km east of the epicentre after the earthquake. If we consider the relation between water storage change (Δh, in meters) and gravitational attraction due to the water mass (Δg_{TWS}, m/s^2) as: $\Delta g_{\mathrm{TWS}} \approx 4.2 \cdot s_y \cdot \Delta h$, the change in gravity due to the Maule event of -5×10^{-8} m/s^2 would be equivalent to an apparent change in TWS of about -39 mm considering s_y, the specific yield (dimensionless), equals 0.3. Correcting the impacts of this gravity shift on the overall surface mass changes around the region is a complex geodetic problem and this was not addressed in this study since it is a localized effect.

Since the inception of GRACE, there have been at least four earthquakes (e.g., 2010(Mw = 8.8), 2014(Mw = 8.2), 2015(Mw = 8.3), and 2016(Mw = 7.6)) and other geodetic disturbances, e.g., tsunamis and volcanism in Chile (e.g., Ruiz and Madariaga 2018; Jaramillo et al. 2017; Montecino et al. 2017a, b; Vigny et al. 2011; Chen et al. 2007a). While Han et al. (2010) found strong gravity shift after the 2010 Chile earthquake, which affected gravity anomalies in Argentina, the observed TWS trend in Central Argentina was partly attributed to this magnitude-8.8 earthquake, in addition to natural drivers such as rainfall (Rodell et al. 2018). It is obvious that the GRACE hydrological time series in Fig. 14.14 are mostly dominated by strong inter-annual variations. But when the trends, annual and semi-annual amplitudes of TWS were isolated using a multi-linear regression parameterization, region 6 (TWS-6, Fig. 14.14) showed a considerable rise in TWS after 2011 when it was statistically decomposed. This rise in TWS could be artificial jumps in surface mass variations related to post-seismic deformations or coseismic jump in the geoid (Montecino et al. 2017a).

Based on the PLSR model, TWS in Sao Francisco, Rio Colorado, and Rio Salado showed less association with climatic parameters. This association results probably from a combined influence of human water management practices and other complex environmental conditions and processes mentioned earlier. For instance, in the Sao Francisco river basin in South America, for example, about 44 million people depend on water transfers between river basins. Further, the National Water Agency of Brazil reported in 2015 that 79% of total water withdrawn was for irrigated agriculture. This kind of footprints and perhaps those caused by geodetic perturbations could leave behind surface mass variations that can be misconstrued or interpreted as those induced by rainfall seasonality. Also, water-limited regions in Brazil mostly coincide with semi-arid areas and those with fractured aquifers where there is heavy reliance on surface waters (ANA 2017). The linear trend observed in water-limited basins such as Sao Francisco (-30.84 ± 4.17 mm/yr) between 2010 and 2017 in Brazil shows a considerable fall (Table 14.1) and could result in a rise in water demand, especially in the light of unprecedented changes in rainfall. With a total of 19,361 man-made reservoirs as at 2016, Brazil apparently shows increased dependence on surface water resources for irrigation and hydropower with Sao Francisco being one of the major hotspots for surface water consumption (ANA 2017). Hence, this obvious anthropogenic footprint in the region is expected to impact the prediction of TWS variations.

14.5.6 Predicting Land Water Storage

For an optimal prediction of TWS (response or dependent variables) over each river basin, a two-step regularization approach that combined JADE method and PLSR was employed. Fourteen independent components of SST anomalies (i.e., seven independent components each from the Pacific and Atlantic oceans), i.e., (independent variables) obtained from the combined PCA-cumulant decomposition were used in the PLSR model to predict TWS. A subset of the latent variables based on nine PLSR components were retrieved and used to develop a prediction model. The observed PLSR model uncertainties (Fig. 14.19) in the simulation of river basin TWS using the leading independent SST modes from the Pacific and Atlantic oceans for some regions (e.g., Amazon, Orinoco, Tocantins, etc.) suggest the considerable role of climate variability in long-term changes in river basin TWS. However, the PLSR model was somewhat less effective in other river basins (Rio Salado, Sao Francisco, Rio

Colorado, and Uruguay) given their modest skills (R^2, IA, and NSE; Fig. 14.19b) and relatively high RSMEs (not shown). The Jarque-Bera statistical test for these basins ($p = 0.5$, 0.1, 0.2, 0.5, 0.39 for Sao Francisco, Rio Colorado, Uruguay, and Rio Salado, respectively) indicated the model' s estimated residuals were not normally distributed as their probability values were greater than the 0.05 confidence level. The modest predictive capacity of TWS using climate components as exhibited by the PLSR scheme in these basins merely imply that TWS are significantly driven by non-climatic factors as we have unpacked in previous sections. The increasing dynamics in global TWS owing to the combined influence of climate and other stressors restricts other conventional models (e.g., least squares) in predictive frameworks. Although it bears some similarity to principal component analysis, i.e., in terms of seeking a hyperplane that maximizes the variance of the input variable (Preisendorfer 1988), one key advantage of the PLSR model is that it mitigates the effect of multicollinearity (e.g., Biswas and Kundu 2018; Lewis-Beck et al. 2003; Wold et al. 2001) by ensuring that only significant components relevant to the response variable are retained for the regression. Although the orthogonality of the principal components apparently also solves this multi-collinearity problem, the choice of an optimum subset of predictors however, remains a key issue (e.g., Lewis-Beck et al. 2003; Wold et al. 2001). The PLSR scheme addresses this problem, thus making it more suitable in the forecast of hydrological quantities such as GRACE-derived TWS. This innate potential of the PLSR model is what we have explored in this study for the prediction of river basin TWS in South America for the first time and complements existing frameworks of other statistical predictive models deployed recently for routine analysis and modelling of climatic variables (Ndehedehe et al. 2019; Krepper et al. 2018; Biswas and Kundu 2018).

Considering the low R^2 values in modelled TWS, optimising the multi-linear regression model for improved freshwater prediction could imply an expansion of the independent variables. Unlike other regions where variations in rivers, lakes, and floodplains do not contribute to observed changes in GRACE-TWS (Getirana et al. 2017), TWS over the Amazon basin is considerably driven mostly by its surface waters (rivers and estuaries) along the floodplain. The dominant patterns of TWS observed over the Amazon basin in this study align with an earlier report by Kim et al. (2009) who found that river storage and sub-surface flow accounted for about 73% of TWS variations in Amazon. As opposed to the Amazon region, significant large-scale alteration of hydrological processes resulting from multiple strings of human activities, e.g., surface water developments and water diversion in other large watersheds are well-known drivers of surface water hydrology (e.g., Ndehedehe et al. 2019; Wada et al. 2017). Hence, independent variables, e.g., river discharge and evapotranspiration are critical water budget quantities expected to improve freshwater prediction for the Amazon basin, in addition to accounting for climate teleconnection-driven rainfall.

However, accounting for human-induced influence (e.g., surface water developments and increased water abstraction) in predictive models is challenging and may require advance statistical approaches. For instance, in the light of the observed strong gravimetric contributions of Lake Volta to GRACE-derived TWS over the

Volta basin, a weighted least square formulation of global spherical harmonic analysis was integrated with a fourth-order cumulant statistics to isolate non-climatic hydrological time series of surface water storage (see, Ndehedehe et al. 2017a). By recovering these surface water contributions caused by human water management strategies, TWS over the Volta basin can be predicted more accurately. Research problems that seek to quantify the impacts of anthropogenic contributions on regional or continental hydrology are new innovative directions that will support predictive frameworks and effective characterization of key hydrologic metrics.

14.6 Conclusion

The knowledge of global freshwater response to critical stressors, e.g., human water abstraction is crucial to improving predictive frameworks that support water governance and management schemes. Hence, an innovative approach that combined JADE (Joint Approximate Diagonalization of Eigen matrices) algorithm and partial least square regression was employed in this study to assess GRACE-derived terrestrial water storage (TWS) over South America (SA). The temporal evolutions of TWS over prominent river basins in the continent were also predicted using independent patterns of sea surface temperature (SST) anomalies of the nearby oceans. Satellite measurements from GRACE and other optical missions provide the opportunity to assess water storage changes that underpin freshwater accounting and allocation to support management.

References

Adam D (2002) Gravity measurement: amazing GRACE. Wetl Ecol Manag 416(6876):10–12. https://doi.org/10.1038/416010a

Agutu N, Awange J, Ndehedehe C, Kirimi F, Kuhn M (2019) GRACE-derived groundwater changes over Greater Horn of Africa: temporal variability and the potential for irrigated agriculture. Sci Total Environ 693:133467. https://doi.org/10.1016/j.scitotenv.2019.07.273

Ahmed M, Sultan M, Wahr J, Yan E (2014) The use of GRACE data to monitor natural and anthropogenic induced variations in water availability across Africa. Earth Sci Rev 136:289–300. https://doi.org/10.1016/j.earscirev.2014.05.009

Alsdorf D, Birkett C, Dunne T, Melack J, Hess L (2001) Water level changes in a large amazon lake measured with spaceborne radar interferometry and altimetry. Geophys Res Lett 28(14):2671–2674. https://doi.org/10.1029/2001GL012962

Alsdorf D, Han S-C, Bates P, Melack J (2010) Seasonal water storage on the amazon floodplain measured from satellites. Remote Sens Environ 114(11):2448–2456. https://doi.org/10.1016/j.rse.2010.05.020

Alsdorf DE, Lettenmaier DP (2003) Tracking fresh water from space. Science 301(5639):1491–1494. https://doi.org/10.1126/science.1089802

ANA (2017) Brazilian water resources report 2017. National Water Agency, Brazil

Anyah R, Forootan E, Awange J, Khaki M (2018) Understanding linkages between global climate indices and terrestrial water storage changes over Africa using GRACE products. Sci Total Environ 635:1405–1416. https://doi.org/10.1016/j.scitotenv.2018.04.159
Acharya TD, Subedi A, Lee DH (2018) Evaluation of water indices for surface water extraction in a landsat 8 scene of Nepal. Sensors 18(8). https://doi.org/10.3390/s18082580
Argus DF, Fu Y, Landerer FW (2014) Seasonal variation in total water storage in California inferred from GPS observations of vertical land motion. Geophys Res Lett 41(6):1971–1980. https://doi.org/10.1002/2014GL059570
Bertino L, Evensen G, Wackernagel H (2003) Sequential data assimilation techniques in oceanography. Int Stat Rev 71(2):223–241. https://doi.org/10.1111/j.1751-5823.2003.tb00194.x
Bevis M, Kendrick E, Cser A, Smalley R (2004) Geodetic measurement of the local elastic response to the changing mass of water in Lago Laja. Chile Phys Earth Planet Inter 141(2):71–78. https://doi.org/10.1016/j.pepi.2003.05.001
Bevis M, Alsdorf D, Kendrick E et al (2005) Seasonal fluctuations in the mass of the Amazon River system and Earth's elastic response. Geophys Res Lett 32(16). https://doi.org/10.1029/2005GL023491
Biswas HR, Kundu PK (2018) A principal component analysis based model to predict post-monsoon tropical cyclone activity in the Bay of Bengal using oceanic Niño index and dipole mode index. Int J Climatol 38(5):2415–2422. https://doi.org/10.1002/joc.5344
Boening C, Willis JK, Landerer FW, Nerem RS, Fasullo J (2012) The 2011 La Niña: so strong, the oceans fell. Geophys Res Lett 39(19):L19602. https://doi.org/10.1029/2012GL053055
Bonafoni S, Biondi R, Brenot H, Anthes R (2019) Radio occultation and ground-based GNSS products for observing, understanding and predicting extreme events: a review. Atmos Res 230. https://doi.org/10.1016/j.atmosres.2019.104624
Borsa AA, Agnew DC, Cayan DR (2014) Ongoing drought-induced uplift in the western United States. Science 345(6204):1587–1590. https://doi.org/10.1126/science.1260279
Bouman J, Ebbing J, S M et al (2015) GOCE gravity gradient data for lithospheric modeling. Int J Appl Earth Obs Geoinf 35:16–30. https://doi.org/10.1016/j.jag.2013.11.001
Buma WG, Lee S, Seo JY (2016) Hydrological evaluation of Lake Chad Basin using space borne and hydrological model observations. Water 8(5). https://doi.org/10.3390/w8050205
Cardoso J-F (1991) Super-symmetric decomposition of the fourth-order cumulant tensor, blind identification of more sources than sensors. Retrieved from: http://perso.telecom-paristech.fr/~cardoso/Papers.PDF/icassp91.pdf. Accessed 15 January 2016
Cardoso JF (1999) High-order contrasts for independent component analysis. Neural Comput 11:157–192
Cardoso JF, Souloumiac A (1993) Blind beamforming for non-gaussian signals. IEE Proc 140(6):362–370
Castellazzi P, Martel R, Galloway DL, Longuevergne L, Rivera A (2016) Assessing groundwater depletion and dynamics using GRACE and InSAR: potential and limitations. Groundwater 54(6):768–780. https://doi.org/10.1111/gwat.12453
Chen J, Famigliett JS, Scanlon BR, Rodell M (2016) Groundwater storage changes: present status from GRACE observations. Surv Geophys 37(2):397–417. https://doi.org/10.1007/s10712-015-9332-4
Chen J, Wilson C, Tapley B, Scanlon B, Güntner A (2016) Long-term groundwater storage change in Victoria, Australia from satellite gravity and in situ observations. Glob Planet Chang 139:56–65. https://doi.org/10.1016/j.gloplacha.2016.01.002
Chen JL, Wilson CR, Tapley BD, Blankenship DD, Ivins ER (2007) Patagonia icefield melting observed by gravity recovery and climate experiment (GRACE). Geophys Res Lett 34(22):L22501. https://doi.org/10.1029/2007GL031871
Chen JL, Wilson CR, Tapley BD, Grand S (2007) GRACE detects coseismic and postseismic deformation from the Sumatra-Andaman earthquake. Geophys Res Lett 34(13):L13302. https://doi.org/10.1029/2007GL030356

Chen L, He Q, Liu K, Li J, Jing C (2019) Downscaling of GRACE-derived groundwater storage based on the random forest model. Remote Sens 11(24). https://doi.org/10.3390/rs11242979

Chen N, Qu J, Xu J, Li Z, Yang Li (2010) Evaluation of groundwater environment carrying capacity based on catastrophe theory. In: 2010 4th international conference on bioinformatics and biomedical engineering, pp 1–4. https://doi.org/10.1109/ICBBE.2010.5515459

Chew C, Small E (2014) Terrestrial water storage response to the 2012 drought estimated from GPS vertical position anomalies. Geophys Res Lett 41(17):6145–6151. https://doi.org/10.1002/2014GL061206

Coe MT, Birkett CM (2004) Calculation of river discharge and prediction of lake height from satellite radar altimetry: example for the Lake Chad basin. Water Resour Res 40(10):W10205. https://doi.org/10.1029/2003WR002543

Common P (1994) Independent component analysis, a new concept? Signal Process 36:287–314

Cooley SW, Smith LC, Stepan L, Mascaro J (2017) Tracking dynamic northern surface water changes with high-frequency planet CubeSat imagery. Remote Sens 9(12). https://doi.org/10.3390/rs9121306

Cooley SW, Smith LC, Ryan JC, Pitcher LH, Pavelsky TM (2019) Arctic-boreal lake dynamics revealed using CubeSat imagery. Geophys Res Lett 46(4):2111–2120. https://doi.org/10.1029/2018GL081584

Conway D, Persechino A, Ardoin-Bardin S, Hamandawana H, Dieulin C, Mahé G (2009) Rainfall and water resources variability in Sub-Saharan Africa during the twentieth century. J Hydrometeorol 10(1):41–59. https://doi.org/10.1175/2008JHM1004.1

Crétaux J-F, Jelinski W, Calmant S, Kouraev A, Vuglinski V, Bergé-Nguyen M, Gennero M-C, Nino F, Rio RAD, Cazenave A, Maisongrande P (2011) Sols: a lake database to monitor in the near real time water level and storage variations from remote sensing data. Adv Space Res 47(9):1497–1507. https://doi.org/10.1016/j.asr.2011.01.004

Crow WT, Wood EF (2003) The assimilation of remotely sensed soil brightness temperature imagery into a land surface model using Ensemble Kalman filtering: a case study based on ESTAR measurements during SGP97. Adv Water Resour 26(2):137–149. https://doi.org/10.1016/S0309-1708(02)00088-X

Dai D, Sun M, Xu X, Lei K (2019) Assessment of the water resource carrying capacity based on the ecological footprint: a case study in Zhangjiakou City, North China. Environ Sci Pollut Res 26(11). https://doi.org/10.1007/s11356-019-04414-9

Dalin C, Wada Y, Kastner T, Puma MJ (2017) Groundwater depletion embedded in international food trade. Nature 543(7647):700–704. https://doi.org/10.1038/nature21403

Decharme B, Alkama R, Douville H, Becker M, Cazenave A (2010) Global evaluation of the ISBA-TRIP continental hydrological system. Part II: uncertainties in river routing simulation related to flow velocity and groundwater storage. Cover J Hydrometeorol 11(3):601–617. https://doi.org/10.1175/2010JHM1212.1

de Linage C, Famiglietti JS, Randerson JT (2014) Statistical prediction of terrestrial water storage changes in the Amazon Basin using tropical Pacific and North Atlantic sea surface temperature anomalies. Hydrol Earth Syst Sci 18(6):2089–2102. https://doi.org/10.5194/hess-18-2089-2014

de Graaf IE, Gleeson T, van Beek LR, Sutanudjaja EH, Bierkens MF (2019) Environmental flow limits to global groundwater pumping. Nature 574(7776):90–94. https://doi.org/10.1038/s41586-019-1594-4

Döll P, Schmied HM, Schuh C, Portmann FT, Eicker A (2014) Global-scale assessment of groundwater depletion and related groundwater abstractions: combining hydrological modeling with information from well observations and GRACE satellites. Water Resour Res 50(7):5698–5720. https://doi.org/10.1002/2014WR015595

Dong L, Shimada J, Kagabu M, Fu C (2015) Teleconnection and climatic oscillation in aquifer water level in Kumamoto plain, Japan. Hydrol Process 29(7):1687–1703. https://doi.org/10.1002/hyp.10291

Ducharne A, Koster RD, Suarez MJ, Stieglitz M, Kumar P (2000) A catchment-based approach to modeling land surface processes in a general circulation model: 2. Parameter estimation and model demonstration. J Geophys Res Atmos 105(D20):24823–24838. https://doi.org/10.1029/2000JD900328

Eicker A, Forootan E, Springer A, Longuevergne L, Kusche J (2016) Does GRACE see the terrestrial water cycle "intensifying"? J Geophys Res Atmos 121(2):733–745. https://doi.org/10.1002/2015JD023808

Erfanian A, Wang G, Fomenko L (2017) Unprecedented drought over tropical south america in 2016: significantly under-predicted by tropical sst. Sci Rep 7(5811). https://doi.org/10.1038/s41598-017-05373-2

Famiglietti JS (2014) The global groundwater crisis. Nature 4:945–948. https://doi.org/10.1038/nclimate2425

Famiglietti JS, Cazenave A, Eicker A, Reager JT, Rodell M, Velicogna I (2015) Satellites provide the big picture. Science 349(6249):684–685. https://doi.org/10.1126/science.aac9238

Famiglietti JS, Rodell M (2013) Water in the balance. Science 340(6138):1300–1301. https://doi.org/10.1126/science.1236460

Famiglietti J, Cazenave A, Eicker A, Reager JT, Rodell M, Velicogna I (2015) Satellites provide the big picture. Science (New York, NY) 349(6249):684–685. https://doi.org/10.1126/science.aac9238

Feng W, Zhong M, Lemoine J-M, Biancale R, Hsu H-T, Xia J (2013) Evaluation of groundwater depletion in North China using the Gravity Recovery and Climate Experiment (GRACE) data and ground-based measurements. Water Resour Res 49(4):2110–2118. https://doi.org/10.1002/wrcr.20192

Ferreira V, Montecino H, Ndehedehe C, Heck B, Gong Z, Westerhaus M, de Freitas S (2018) Space-based observations of crustal deflections for drought characterization in brazil. Science of The Total Environment 644:256–273. https://doi.org/10.1016/j.scitotenv.2018.06.277

Ferreira V, Yong B, Tourian M, Ndehedehe C, Shen Z, Seitz K, Dannouf R (2020) Characterization of the hydro-geological regime of yangtze river basin using remotely-sensed and modeled products. Sci Total Environ 718:137354. https://doi.org/10.1016/j.scitotenv.2020.137354

Ferreira VG, Montecino HD, Ndehedehe CE, del Rio RA, Cuevas A, de Freitas SRC (2019) Determining seasonal displacements of Earth's crust in South America using observations from space-borne geodetic sensors and surface-loading models. Earth Planets Space 71(1):84. https://doi.org/10.1186/s40623-019-1062-2

Ferreira VG, Ndehedehe CE, Montecino HC, Yong B, Yuan P, Abdalla A, Mohammed AS (2019b) Prospects for imaging terrestrial water storage in South America using daily GPS observations. Remote Sens 11(6). https://doi.org/10.3390/rs11060679

Ferreira VG, Andam-Akorful SA, Dannouf R, Adu-Afari E (2019) A multi-sourced data retrodiction of remotely sensed terrestrial water storage changes for West Africa. Water 11(2). https://doi.org/10.3390/w11020401

Fontaine B, Bigot S (1993) West African rainfall deficits and sea surface temperatures. Int J Climatol 13(3):271–285. https://doi.org/10.1002/joc.3370130304

Forman BA, Reichle RH, Rodell M (2012) Assimilation of terrestrial water storage from GRACE in a snow-dominated basin. Water Resour Res 48(1). https://doi.org/10.1029/2011WR011239

Forootan E, Kusche J, Loth I, Schuh WD et al (2014) Multivariate prediction of total water storage anomalies over West Africa from multi-satellite data. Surv Geophys 35(4):913–940. https://doi.org/10.1007/s10712-014-9292-0

Frappart F, Papa F, da Silva JS, G R et al (2012) Surface freshwater storage and dynamics in the Amazon basin during the 2005 exceptional drought. Environ Res Lett 7(4):044010

Frappart F, Papa F, Guntner A, Werth S, da Silva JS, Tomasella J, Seyler F, Prigent C, Rossow WB, Calmant S, Bonnet M-P (2011) Satellite-based estimates of groundwater storage variations in large drainage basins with extensive floodplains. Remote Sens Environ 115(6):1588–1594. https://doi.org/10.1016/j.rse.2011.02.003

Frappart F, Ramillien G, Leblanc M, Tweed SO, Bonnet M-P, Maisongrande P (2011) An independent component analysis filtering approach for estimating continental hydrology in the GRACE gravity data. Remote Sens Environ 115(1):187–204. https://doi.org/10.1016/j.rse.2010.08.017

Frappart F, Ramillien G, Maisongrande P, Bonnet M-P (2010) Denoising satellite gravity signals by independent component analysis. Geosci Remote Sens Lett, IEEE 7(3):421–425. https://doi.org/10.1109/LGRS.2009.2037837

Frappart F, Ramillien G, Ronchail J (2013) Changes in terrestrial water storage versus rainfall and discharges in the amazon basin. Int J Climatol 33:3029–3046. https://doi.org/10.1002/joc.3647

Frappart F, Seoane L, Ramillien G (2013) Validation of GRACE-derived terrestrial water storage from a regional approach over South America. Remote Sens Environ 137:69–83. https://doi.org/10.1016/j.rse.2013.06.008

Fu Y, Argus DF, Landerer FW (2015) GPS as an independent measurement to estimate terrestrial water storage variations in Washington and Oregon. J Geophys Res Solid Earth 120(1):552–566. https://doi.org/10.1002/2014JB011415

Getirana A (2016) Extreme water deficit in brazil detected from space. J Hydrometeorol 17(2):591–599. https://doi.org/10.1175/JHM-D-15-0096.1

Getirana A, Jung HC, Hoek JVD, Ndehedehe CE (2020) Hydropower dam operation strongly controls Lake Victoria's freshwater storage variability. Sci Total Environ 726:138343. https://doi.org/10.1016/j.scitotenv.2020.138343

Getirana ACV, Peters-Lidard C (2013) Estimating water discharge from large radar altimetry datasets. Hydrol Earth Syst Sci 17(3):923–933. https://doi.org/10.5194/hess-17-923-2013

Getirana ACV, Bonnet M et al (2009) Hydrological monitoring of poorly gauged basins based on rainfall-runoff modeling and spatial altimetry. J Hydrol 379(3):205–219. https://doi.org/10.1016/j.jhydrol.2009.09.049

Getirana A, Kumar S, Girotto M, Rodell M (2017) Rivers and floodplains as key components of global terrestrial water storage variability. Geophys Res Lett 44(20):10,359–10,368. https://doi.org/10.1002/2017GL074684

Green TR, Taniguchi M, Kooi H, Gurdak JJ, Allen DM, Hiscock KM, Treidel H, Aureli A (2011) Beneath the surface of global change: impacts of climate change on groundwater. J Hydrol 405(3):532–560. https://doi.org/10.1016/j.jhydrol.2011.05.002

Grey D, Sadoff CW (2007) Sink or swim? Water security for growth and development. Water Policy 9(6):545–571. https://doi.org/10.2166/wp.2007.021

Gutiérrez APA, Engle NL, Nys ED, Molejón C, Martins ES (2014) Drought preparedness in brazil. Weather Clim Extrem 3:95–106. https://doi.org/10.1016/j.wace.2013.12.001

Han S-C, Kim H, Yeo I-Y, Yeh P, Oki T, Seo K-W, Alsdorf D, Luthcke SB (2009) Dynamics of surface water storage in the Amazon inferred from measurements of inter-satellite distance change. Geophys Res Lett 36(9):L09403. https://doi.org/10.1029/2009GL037910

Han S-C, Sauber J, Luthcke S (2010) Regional gravity decrease after the 2010 Maule (Chile) earthquake indicates large-scale mass redistribution. Geophys Res Lett 37(23). https://doi.org/10.1029/2010GL045449

Hinderer J, Pfeffer J, Boucher M et al (2012) Land water storage changes from ground and space geodesy: first results from the GHYRAF (Gravity and Hydrology in Africa) experiment. Pure Appl Geophys 169(8):1391–1410. https://doi.org/10.1007/s00024-011-0417-9

Houborg R, Rodell M, Li B, Reichle R, Zaitchik BF (2012) Drought indicators based on model-assimilated Gravity Recovery and Climate Experiment (GRACE) terrestrial water storage observations. Water Resour Res 48(7):W07525. https://doi.org/10.1029/2011WR011291

Humphrey V, Gudmundsson L, Seneviratne SI (2016) Assessing global water storage variability from GRACE: trends, seasonal cycle, subseasonal anomalies and extremes. Surv Geophys 37(2):357–395. https://doi.org/10.1007/s10712-016-9367-1

Humphrey V, Gudmundsson L (2019) GRACE-REC:a reconstruction of climate-driven water storage changes over the last century. Earth Syst Sci Data 11(3):1153–1170. https://doi.org/10.5194/essd-11-1153-2019

Jaramillo E, Melnick D, Baez JC, Montecino H, Lagos NA, Acuña E, Manzano M, Camus PA (2017) Calibrating coseismic coastal land-level changes during the 2014 iquique (mw=8.2) earthquake (northern chile) with leveling, gps and intertidal biota. PLOS One 12(3):1–16. https://doi.org/10.1371/journal.pone.0174348

Jasechko S, Perrone D (2021) Global groundwater wells at risk of running dry. Science 372(6540):418–421. https://doi.org/10.1126/science.abc2755

Jiang L, Nielsen L, Andersen OB, Bauer-Gottwein P (2017) CryoSat-2 radar altimetry for monitoring freshwater resources of China. Remote Sens Environ 200:125–139. https://doi.org/10.1016/j.rse.2017.08.015

Jin S, Park J, Cho J, Park P (2017) Seasonal variability of GPS-derived zenith tropospheric delay (1994–2006) and climate implications. J Geophys Res Atmos 112(D9). https://doi.org/10.1029/2006JD007772

Joodaki G, Wahr J, Swenson S (2014) Estimating the human contribution to groundwater depletion in the Middle East, from GRACE data, land surface models, and well observations. Water Resour Res 50(3):2679–2692. https://doi.org/10.1002/2013WR014633

Junqueira AM, Mao F, Mendes TSG, Simões SJC et al (2021) Estimation of river flow using CubeSats remote sensing. Sci Total Environ 788. https://doi.org/10.1016/j.scitotenv.2021.147762

Jyolsna PJ, Kambhammettu BVNP, Gorugantula S (2021) Application of random forest and multi-linear regression methods in downscaling GRACE derived groundwater storage changes. Hydrol Sci J 66(5):874–887. https://doi.org/10.1080/02626667.2021.1896719

Kalu I, Ndehedehe CE, Okwuashi O, Eyoh AE (2021) Assessing freshwater changes over Southern and Central Africa (2002–2017). Remote Sens 13(13). https://doi.org/10.3390/rs13132543

Kim H, Yeh PJ-F, Oki T, Kanae S (2009) Role of rivers in the seasonal variations of terrestrial water storage over global basins. Geophys Res Lett 36(17):L17402. https://doi.org/10.1029/2009GL039006

Krepper G, Romeo F, de Sousa Fernandes DD, Diniz PHGD, de Araújo MCU, Nezio MSD, Pistonesi MF, Centurión ME (2018) Determination of fat content in chicken hamburgers using nir spectroscopy and the successive projections algorithm for interval selection in pls regression (ispa-pls). Spectrochim Acta Part A: Mol Biomol Spectrosc 189:300–306. https://doi.org/10.1016/j.saa.2017.08.046

Kroes J, van Dam J, Supit I, de Abelleyra D, Verón S, de Wit A, Boogaard H, Angelini M, Damiano F, Groenendijk P, Wesseling J, Veldhuizen A (2019) Agrohydrological analysis of groundwater recharge and land use changes in the Pampas of Argentina. Agric Water Manag 213:843–857. https://doi.org/10.1016/j.agwat.2018.12.008

Kumar SV, Zaitchik BF, Peters-Lidard CD, Rodell M et al (2016) Assimilation of Gridded GRACE terrestrial water storage estimates in the North American land data assimilation system. J Hydrometeorol 17(7):1951–1972. https://doi.org/10.1175/JHM-D-15-0157.1

Lamine BOM, Ferreira VG, Yang Y, Ndehedehe CE, He X (2021) Estimation of the Niger River cross-section and discharge from remotely-sensed products. J Hydrol Reg Stud 36. https://doi.org/10.1016/j.ejrh.2021.100862

Landerer FW, Swenson SC (2012) Accuracy of scaled GRACE terrestrial water storage estimates. Water Resour Res 48(4):W04531. https://doi.org/10.1029/2011WR011453

Larson KM (2019) Unanticipated uses of the global positioning system. Annu Rev Earth Planet Sci 47(1):19–40. https://doi.org/10.1146/annurev-earth-053018-060203

Lawrence DM et al (2011) Parameterization improvements and functional and structural advances in Version 4 of the Community Land Model. J Adv Mod Ear Syst 3(M03001). https://doi.org/10.1029/2011MS00045

Lewis-Beck M, Bryman A, Futing T (2003) Encyclopedia of social sciences research methods. Thousand Oaks (CA): Sage. Accessed from https://www.utdallas.edu/~herve/Abdi-PLS-pretty.pdf on 4th January 2019

Li B, Rodell M, Famiglietti JS (2015) Groundwater variability across temporal and spatial scales in the central and northeastern U.S. J Hydrol 525:769–780. https://doi.org/10.1016/j.jhydrol.2015.04.033

Li B, Rodell M, Zaitchik BF, Reichle RH, Koster RD, van Dam TM (2012) Assimilation of GRACE terrestrial water storage into a land surface model: evaluation and potential value for drought monitoring in western and central Europe. J Hydrol 446:103–115. https://doi.org/10.1016/j.jhydrol.2012.04.035

Li B, Rodell M, Kumar S, Beaudoing H, Getirana A, Zaitchik B et al (2019) Global GRACE data assimilation for groundwater and drought monitoring: advances and challenges. Water Resour Res 55(9):7564–7586. https://doi.org/10.1029/2018WR024618

Linage C, Kim H, Famiglietti JS, Yu J-Y (2013) Impact of pacific and atlantic sea surface temperatures on interannual and decadal variations of GRACE land water storage in tropical South America. J Geophys Res Atmos 118(19):10,811–10,829. https://doi.org/10.1002/jgrd.50820

Liu B, Zou X, Yi S, Sneeuw N, Cai J, Li J (2021) Identifying and separating climate- and human-driven water storage anomalies using GRACE satellite data. Remote Sens Environ 263:112559. https://doi.org/10.1016/j.rse.2021.112559

Liu M, Tian H, Chen G, Ren W, Zhang C, Liu J (2008) Effects of land-use and land-cover change on evapotranspiration and water yield in China during 1900–20001. J Am Water Resour Assoc 44(5):1193–1207. https://doi.org/10.1111/j.1752-1688.2008.00243.x

Lomnitz C (2004) Major earthquakes of Chile: a historical survey, 1535–1960. Seismol Res Lett 75(3):368–378. https://doi.org/10.1785/gssrl.75.3.368

Llamedo P, Hierro R, de la Torre A, Alexander P (2017) ENSO-related moisture and temperature anomalies over South America derived from GPS radio occultation profiles. Int J Climatol 37(1):268–275. https://doi.org/10.1002/joc.4702

Long D, Scanlon BR, Longuevergne L, Sun AY, Fernando DN, Save H (2013) GRACE satellite monitoring of large depletion in water storage in response to the 2011 drought in Texas. Geophys Res Lett 40(13):3395–3401. https://doi.org/10.1002/Grl.50655

Long D, Yang W, Scanlon B et al (2020) South-to-north water diversion stabilizing Beijing' s groundwater levels. Nat Commun 11(3665). https://doi.org/10.1038/s41467-020-17428-6

Lück C, Kusche J, Rietbroek R, Löcher A (2018) Time-variable gravity fields and ocean mass change from 37 months of kinematic Swarm orbits. Solid Earth 9(2):323–339. https://doi.org/10.5194/se-9-323-2018

McCabe MF, Rodell M, Alsdorf DE, Miralles DG, Uijlenhoet R, Wagner W, Lucieer A, Houborg R, Verhoest NEC, Franz TE, Shi J, Gao H, Wood EF (2017) The future of Earth observation in hydrology. Hydrol Earth Syst Sci 21(7):3879–3914. https://doi.org/10.5194/hess-21-3879-2017

McCallum J, Crosbie R, Walker G et al (2010) Impacts of climate change on groundwater in Australia: a sensitivity analysis of recharge. Hydrogeol J 18:1625–1638. https://doi.org/10.1007/s10040-010-0624-y

Montecino HD, de Freitas SR, Báez JC, Ferreira VG (2017) Effects on chilean vertical reference frame due to the maule earthquake co-seismic and post-seismic effects. J Geodyn 112:22–30. https://doi.org/10.1016/j.jog.2017.07.006

Montecino HDC, Ferreira VG, Cuevas A, Cabrera LC, Báez JCS, Freitas SRCD (2017) Vertical deformation and sea level changes in the coast of chile by satellite altimetry and tide gauges. Int J Remote Sens 38(24):7551–7565. https://doi.org/10.1080/01431161.2017.1288306

Moore P, Williams SDP (2014) Integration of altimetry lake lavels and GRACE gravimetry over Africa: inferences for terrestrial water storage change 2003–2011. Water Resour Res 50:9696–9720. https://doi.org/10.1002/2014WR015506

Mueller B, Seneviratne SI, Jimenez C, Corti T, Hirschi M, Balsamo G et al (2011) Evaluation of global observations-based evapotranspiration datasets and IPCC AR4 simulations. Geophys Res Lett 38(6). https://doi.org/10.1029/2010GL046230

Müller J, Dirkx D, Kopeikin SM et al (2018) High performance clocks and gravity field determination. Space Sci Rev 214(5). https://doi.org/10.1007/s11214-017-0431-z

Naujoks M, Weise A, Kroner C, Jahr T (2008) Detection of small hydrological variations in gravity by repeated observations with relative gravimeters. J Geod 82(9):543–553. https://doi.org/10.1007/s00190-007-0202-9

Ndehedehe C, Awange J, Agutu N, Kuhn M, Heck B (2016) Understanding changes in terrestrial water storage over West Africa between 2002 and 2014. Adv Water Resour 88:211–230. https://doi.org/10.1016/j.advwatres.2015.12.009

Ndehedehe CE (2019) The water resources of tropical West Africa: propblems, progress and prospect. Acta Geophys 67(2):621–649. https://doi.org/10.1007/s11600-019-00260-y

Ndehedehe CE, Agutu NO, Okwuashi OH, Ferreira VG (2016) Spatio-temporal variability of droughts and terrestrial water storage over Lake Chad Basin using independent component analysis. J Hydrol 540:106–128. https://doi.org/10.1016/j.jhydrol.2016.05.068

Ndehedehe CE, Anyah RO, Alsdorf D, Agutu NO, Ferreira VG (2019) Modelling the impacts of global multi-scale climatic drivers on hydro-climatic extremes (1901–2014) over the Congo basin. Sci Total Environ 651:1569–1587. https://doi.org/10.1016/j.scitotenv.2018.09.203

Ndehedehe CE, Awange J, Kuhn M, Agutu N, Fukuda Y (2017) Analysis of hydrological variability over the Volta river basin using in-situ data and satellite observations. J Hydrol Reg Stud 12:88–110. https://doi.org/10.1016/j.ejrh.2017.04.005

Ndehedehe CE, Awange J, Kuhn M, Agutu N, Fukuda Y (2017) Climate teleconnections influence on West Africa's terrestrial water storage. Hydrol Process 31(18):3206–3224. https://doi.org/10.1002/hyp.11237

Ndehedehe CE, Awange JL, Agutu NO, Okwuashi O (2018) Changes in hydro-meteorological conditions over tropical West Africa (1980 – 2015) and links to global climate. Glob Planet Change 162:321–341. https://doi.org/10.1016/j.gloplacha.2018.01.020

Ndehedehe CE, Awange JL, Corner R, Kuhn M, Okwuashi O (2016) On the potentials of multiple climate variables in assessing the spatio-temporal characteristics of hydrological droughts over the Volta Basin. Sci Total Environ 557–558:819–837. https://doi.org/10.1016/j.scitotenv.2016.03.004

Ndehedehe CE, Ferreira VG (2020) Assessing land water storage dynamics over Southern America. J Hydrol 580:124339. https://doi.org/10.1016/j.jhydrol.2019.124339

Ndehedehe CE, Ferreira VG (2020) Identifying the footprints of global climate modes in time-variable gravity hydrological signals. Clim Change 159:481–502. https://doi.org/10.1007/s10584-019-02588-2

Ndehedehe CE, Ferreira VG, Agutu NO, Onojeghuo AO, Okwuashi O, Kassahun HT, Dewan A (2021) What if the rains do not come? J Hydrol 595:126040. https://doi.org/10.1016/j.jhydrol.2021.126040

Ndehedehe CE, Okwuashi O, Ferreira VG, Agutu NO (2018) Exploring evapotranspiration dynamics over Sub-Sahara Africa (2000 – 2014). Environ Monit Assess 190(7):400. https://doi.org/10.1007/s10661-018-6780-6

Ni S, Chen J, Wilson CR, Li J, Hu X, Fu R (2018) Global terrestrial water storage changes and connections to ENSO events. Surv Geophys 39(1):1–22. https://doi.org/10.1007/s10712-017-9421-7

Nicholson S (2013) The West African Sahel: a review of recent studies on the rainfall regime and its interannual variability. ISRN Meteorol 2013(453521):1–32. https://doi.org/10.1155/2013/453521

Nka BN, Oudin L, Karambiri H, Paturel JE, Ribstein P (2015) Trends in floods in West Africa: analysis based on 11 catchments in the region. Hydrol Earth Syst Sci 19(11):4707–4719. https://doi.org/10.5194/hess-19-4707-2015

O'Connell E (2017) Towards adaptation of water resource systems to climatic and socio-economic change. Water Resour Manag 31(10):2965–2984. https://doi.org/10.1007/s11269-017-1734-2

Odekunle TO, Eludoyin AO (2008) Sea surface temperature patterns in the Gulf of Guinea: their implications for the spatio-temporal variability of precipitation in West Africa. Int J Climatol 28:1507–1517. https://doi.org/10.1002/joc.1656

Paeth H, Fink A, Pohle S, Keis F, Machel H, Samimi C (2012) Meteorological characteristics and potential causes of the 2007 flood in sub-Saharan Africa. Int J Climatol 31:1908–1926. https://doi.org/10.1002/Joc.2199

Pokhrel YN, Hanasaki N, Yeh PJ, Yamada TJ, Kanae S, Oki T (2012) Model estimates of sea-level change due to anthropogenic impacts on terrestrial water storage. Nat Geosci 5(6):389–392. https://doi.org/10.1038/ngeo1476

Phillips T, Nerem RS, Fox-Kemper B, Famiglietti JS, Rajagopalan B (2012) The influence of ENSO on global terrestrial water storage using GRACE. Geophys Res Lett 39(L16705):2012. https://doi.org/10.1029/2012GL052495

Preisendorfer R (1988) Principal component analysis in meteorology and oceanography. Developments in Atmospheric Science 17. Elsevier, Amsterdam

Reager JT, Thomas BF, Famiglietti JS (2014) River basin flood potential inferred using GRACE gravity observations at several months lead time. Nat Geosci 7(8):588–592. https://doi.org/10.1038/ngeo2203

Reager JT, Gardner AS, Famiglietti JS, Wiese DN, Eicker A, Lo M-H (2016) A decade of sea level rise slowed by climate-driven hydrology. Science 351(6274):699–703. https://doi.org/10.1126/science.aad8386

Rietbroek R, Fritsche M, Dahle C et al (2014) Can GPS-derived surface loading bridge a GRACE mission gap? Surv Geophys 35(6):1–17. https://doi.org/10.1007/s10712-013-9276-5

Rodell M, Beaudoing HK, Ecuyer TS, Olson WS, Famiglietti JS et al (2015) The observed state of the water cycle in the early twenty-first century. J Clim 28(21):8289–8318. https://doi.org/10.1175/JCLI-D-14-00555.1

Rodell M, Famiglietti JS, Wiese DN et al (2018) Emerging trends in global freshwater availability. Nature 557(7707):651–659. https://doi.org/10.1038/s41586-018-0123-1

Rodell M, Famiglietti JS, Wiese DN, Reager JT, Beaudoing HK, Landerer FW, Lo M-H (2018) Emerging trends in global freshwater availability. Nature 557:651–659. https://doi.org/10.1038/s41586-018-0123-1

Rodell M, Velicogna I, Famiglietti JS (2009) Satellite-based estimates of groundwater depletion in India. Nature 460(7258):999–1002. https://doi.org/10.1038/nature08238

Rodell M, Famiglietti JS (2001) An analysis of terrestrial water storage variations in Illinois with implications for the Gravity Recovery and Climate Experiment (GRACE). Water Resour Res 37(5):1327–1339. https://doi.org/10.1029/2000WR900306

Ruiz S, Madariaga R (2018) Historical and recent large megathrust earthquakes in Chile. Tectonophysics 733:37–56. https://doi.org/10.1016/j.tecto.2018.01.015

Sadeh Y, Zhu X, Chenu K, Dunkerley D (2019) Sowing date detection at the field scale using CubeSats remote sensing. Comput Electron Agric 157:568–580. https://doi.org/10.1016/j.compag.2019.01.042

Save H, Bettadpur S, Tapley BD (2012) Reducing errors in the GRACE gravity solutions using regularization. J Geod 86(9):695–711. https://doi.org/10.1007/s00190-012-0548-5

Scanlon BR, Faunt CC, Longuevergne L, Reedy RC et al (2012) Groundwater depletion and sustainability of irrigation in the us high plains and central valley. Proc Natl Acad Sci 109(24):9320–9325. https://doi.org/10.1073/pnas.1200311109

Schneider U, Becker A, Finger P, Meyer-Christoffer A, Ziese M, Rudolf B (2014) GPCC's new land surface precipitation climatology based on quality-controlled in situ data and its role in quantifying the global water cycle. Theor Appl Climatol 115(1–2):15–40. https://doi.org/10.1007/s00704-013-0860-x

Semazzi FHM, Mehta V, Sud Y (1988) An investigation of the relationship between sub-Saharan rainfall and global sea surface temperatures. Atmos-Ocean 26(1):118–138. https://doi.org/10.1080/07055900.1988.9649293

Seo JY, Lee S-I (2016) Integration of GRACE, ground observation, and land-surface models for groundwater storage variations in South Korea. Int J Remote Sens 37(24):5786–5801. https://doi.org/10.1080/01431161.2016.1249301

Shah T, Bhatt S, Shah RK, Talati J (2008) Groundwater governance through electricity supply management: assessing an innovative intervention in Gujarat, western India. Agric Water Manag 95(11):1233–1242. https://doi.org/10.1016/j.agwat.2008.04.006

Sheffield J, Wood EF (2008) Global trends and variability in soil moisture and drought characteristics, 1950–2000, from observation-driven simulations of the terrestrial hydrologic cycle. J Clim 21(3):432–458. https://doi.org/10.1175/2007JCLI1822.1

Sheffield J, Wood EF, Pan M, Beck H, Coccia G, Serrat-Capdevila A, Verbist K (2018) Satellite remote sensing for water resources management: potential for supporting sustainable development in data-poor regions. Water Resour Res 54(12):9724–9758. https://doi.org/10.1029/2017WR022437

Spinoni J, Naumann G, Carrao H, Barbosa P, Vogt J (2014) World drought frequency, duration, and severity for 1951–2010. Int J Climatol 34(8):2792–2804. https://doi.org/10.1002/joc.3875

Sun T, Ferreira VG, He X, Andam-Akorful SA (2016) Water availability of São Francisco river basin based on a space-borne geodetic sensor. Water 8(5):213. https://doi.org/10.3390/w8050213

Sun Z, Long D, Yang W, Li X, Pan Y (2020) Reconstruction of GRACE data on changes in total water storage over the global land surface and 60 basins. Water Resour Res 56(4). https://doi.org/10.1029/2019WR026250

Swenson S, Wahr J (2002) Methods for inferring regional surface-mass anomalies from Gravity Recovery and Climate Experiment (GRACE) measurements of time-variable gravity. J Geophys Res-Solid Earth 107(B9). https://doi.org/10.1029/2001jb000576

Tapley B, Bettadpur S, Watkins M, Reigber C (2004) The gravity recovery and climate experiment: mission overview and early results. Geophys Res Lett 31:1–4. https://doi.org/10.1029/2004GL019920

Tapley B, Watkins M, Flechtner F et al (2019) Contributions of GRACE to understanding climate change. Nat Clim Change 9:358–369. https://doi.org/10.1038/s41558-019-0456-2

Tapley BD, Bettadpur S, Ries JC, Thompson PF, Watkins MM (2004) Grace measurements of mass variability in the earth system. Science 305(5683):503–505. https://doi.org/10.1126/science.1099192

Tapley BD, Watkins MM, Flechtner F, Reigber C et al (2019) Contributions of GRACE to understanding climate change. Nat Clim Change 9(5):358–369. https://doi.org/10.1038/s41558-019-0456-2

Talpe MJ, Nerem RS, Forootan E, Schmidt M, Lemoine FG, Enderlin EM, Landerer FW (2017) Ice mass change in Greenland and Antarctica between 1993 and 2013 from satellite gravity measurements. J Geod 91(11):1283–1298. https://doi.org/10.1007/s00190-017-1025-y

Taylor R, Scanlon B, Döll P et al (2013) Ground water and climate change. Nat Clim Change 3:322–329. https://doi.org/10.1038/nclimate1744

Thomas AC, Reager JT, Famiglietti JS, Rodell M (2014) A GRACE-based water storage deficit approach for hydrological drought characterization. Geophys Res Lett 41(5):1537–1545. https://doi.org/10.1002/2014GL059323

Thomas B, Famiglietti J (2019) Identifying climate-induced groundwater depletion in GRACE observations. Sci Rep 9:4124. https://doi.org/10.1038/s41598-019-40155-y

Thomas BF, Famiglietti JS, Landerer FW, Wiese DN, Molotch NP, Argus DF (2017). GRACE groundwater drought index: evaluation of California central valley groundwater drought. Remote Sens Environ 198(Supplement C):384–392. https://doi.org/10.1016/j.rse.2017.06.026

Tourian MJ, Reager JT, Sneeuw N (2018) The total drainable water storage of the Amazon River Basin: a first estimate using GRACE. Water Resour Res 54(5):3290–3312. https://doi.org/10.1029/2017WR021674

United Nations (2015) Transforming our world: the 2030 agenda for sustainable development. United Nations

van der Molen M, Dolman A, Ciais P, Eglin T, Gobron N, Law B (2011) Drought and ecosystem carbon cycling. Agric For Meteorol 151(7):765–773. https://doi.org/10.1016/j.agrformet.2011.01.018

van Dijk AIJM, Renzullo LJ, Wada Y, Tregoning P (2014) A global water cycle reanalysis (2003–2012) merging satellite gravimetry and altimetry observations with a hydrological multi-model ensemble. Hydrol Earth Syst Sci 18(8):2955–2973. https://doi.org/10.5194/hess-18-2955-2014

van Dijk AIJM, Beck HE, Crosbie RS, de Jeu RAM, Liu YY, Podger GM, Timbal B, Viney NR (2013) The millennium drought in southeast Australia (2001–2009): Natural and human causes and implications for water resources, ecosystems, economy, and society. Water Resour Res 49(2):1040–1057. https://doi.org/10.1002/wrcr.20123

Vanham D, Hoekstra A, Wada Y, Bouraoui F, de Roo A, Mekonnen M, van de Bund W, Batelaan O, Pavelic P, Bastiaanssen W, Kummu M, Rockström J, Liu J, Bisselink B, Ronco P, Pistocchi A, Bidoglio G (2018) Physical water scarcity metrics for monitoring progress towards SDG target

6.4: an evaluation of indicator 6.4.2 "level of water stress". Sci Total Environ 613–614:218–232. https://doi.org/10.1016/j.scitotenv.2017.09.056

Veldkamp TI, Wada Y et al (2015) Changing mechanism of global water scarcity events: impacts of socioeconomic changes and inter-annual hydro-climatic variability. Glob Environ Change 32:18–29. https://doi.org/10.1016/j.gloenvcha.2015.02.011

Vigny C, Socquet A, Peyrat S, Ruegg J-C, Métois M, Madariaga R, Morvan S, Lancieri M, Lacassin R, Campos J, Carrizo D et al (2011) The 2010 mw 8.8 Maule megathrust earthquake of Central Chile, monitored by GPS. Science 332(6036):1417–1421. https://doi.org/10.1126/science.1204132

Vorosmarty CJ, McIntyre PB, Gessner MO, Dudgeon D, Prusevich A, Green P, Glidden S, Bunn SE, Sullivan CA, Liermann CR, Davies PM (2010) Global threats to human water security and river biodiversity. Nature 467:555–561. https://doi.org/10.1038/nature09440

Voss KA, Famiglietti JS, Lo M, de Linage C, Rodell M, Swenson SC (2013) Groundwater depletion in the Middle East from GRACE with implications for transboundary water management in the Tigris-Euphrates-Western Iran region. Water Resour Res 49(2):904–914. https://doi.org/10.1002/wrcr.20078

Wada Y, Bierkens MFP, de Roo A, Dirmeyer PA, Famiglietti JS, Hanasaki N, Konar M, Liu J, Müller Schmied H, Oki T et al (2017) Human-water interface in hydrological modelling: current status and future directions. Hydrol Earth Syst Sci 21(8):4169–4193. https://doi.org/10.5194/hess-21-4169-2017

Wahr J, Molenaar M, Bryan F (1998) Time variability of the Earth's gravity field: hydrological and oceanic effects and their possible detection using GRACE. J Geophys Res-Solid Earth 103(B12):30205–30229. https://doi.org/10.1029/98jb02844

Wang F, Wang Z, Yang H, Zhao Y (2018) Study of the temporal and spatial patterns of drought in the Yellow River basin based on SPEI. Sci China Earth Sci 61(8):1098–1111. https://doi.org/10.1007/s11430-017-9198-2

Weinthal E, Vengosh A, Marei A, Gutierrez A, Kloppmann W (2005) The water crisis in the gaza strip: prospects for resolution. Groundwater 43(5):653–660. https://doi.org/10.1111/j.1745-6584.2005.00064.x

Willis MJ, Melkonian AK, Pritchard ME, Ramage JM (2012) Ice loss rates at the Northern Patagonian icefield derived using a decade of satellite remote sensing. Remote Sens Environ 117:184–198. https://doi.org/10.1016/j.rse.2011.09.017

Wold S, Sjöström M, Eriksson L (2001) PLS-regression: a basic tool of chemometrics. Chemom Intell Lab Syst 58(2):109–130. https://doi.org/10.1016/S0169-7439(01)00155-1

Wouters B, Bonin JA, Chambers DP, Riva REM, Sasgen I, Wahr J (2014) GRACE, time-varying gravity, Earth system dynamics and climate change. Rep Prog Phys 77(11):116801. https://doi.org/10.1088/0034-4885/77/11/116801

Xia Y, Mocko D, Huang M et al (2017) Comparison and assessment of three advanced land surface models in simulating terrestrial water storage components over the United States. J Hydrometeorol 18(3):625–649. https://doi.org/10.1175/JHM-D-16-0112.1

Yin W, Hu L, Zhang M, Wang J, Han S-C (2018) Statistical downscaling of GRACE-derived groundwater storage using ET data in the North China plain. J Geophys Res Atmos 123(11):5973–5987. https://doi.org/10.1029/2017JD027468

Zaitchik BF, Rodell M, Reichle RH (2008) Assimilation of GRACE terrestrial water storage data into a land surface model: results for the Mississippi River Basin. J Hydrometeorol 9(3):535–548. https://doi.org/10.1175/2007JHM951.1

Zarfl C, Lumsdon AE, Berlekamp J, Tydecks L, Tockner K (2015) A global boom in hydropower dam construction. Aquat Sci 77:161–170. https://doi.org/10.1007/s00027-014-0377-0

Zhang Y, Pan M, Wood EF (2016) On creating global gridded terrestrial water budget estimates from satellite remote sensing. Remote Sens Water Resour 59–78. https://doi.org/10.1007/978-3-319-32449-4_4

Zheng D, Rademacher J, JC et al (2004) Estimating aboveground biomass using Landsat 7 ETM+ data across a managed landscape in northern Wisconsin, USA. Remote Sens Environ 93(3):402–411. https://doi.org/10.1016/j.rse.2004.08.008

Ziehe A (2005) Blind source separation based on joint diagonalization of matrices with applications in biomedical signal processing. PhD thesis, Universitat Potsdam. Retrieved from: http://en.youscribe.com/catalogue/reports-and-theses/knowledge/blind-source-separation-based-on-joint-diagonalization-of-matrices-1424347. Accessed 15 May 2015

Zou Y, Macau EEN, Sampaio G, Ramos AMT, Kurths J (2018) Characterizing the exceptional 2014 drought event in são paulo by drought period length. Clim Dyn 51(1):433–442. https://doi.org/10.1007/s00382-017-3932-2

Part X
Human Water Management on Hydrology

Chapter 15
Impacts of Water Resources Development on Hydrology

Abstract Changes in terrestrial water storage (e.g., groundwater, surface water) are largely driven by climate change and anthropogenic influence. The latter includes, construction of dams and reservoirs, water transfer, water harvesting, land-use change, groundwater overexploitation, among others. These human activities either directly or indirectly trigger alterations in natural water equilibrium of terrestrial ecosystems. For instance, the impacts of reservoir operations drive changes in discharge timings of rivers and can dominate changes in land water storage. Anthropogenic influence affect water balance and can amplify the impacts of climate change on hydrology. This influence can emanate from other indirect channels such as deforestation and expansion of agricultural areas in semi-arid domains. This chapter discusses these issues, exploring different monitoring perspectives and frameworks to help quantify such impacts to improve understanding of complex hydrological processes triggered by human actions. A case study following the work of Ndehedehe et al. (2017) who employed a two-step procedure based on spherical harmonic synthesis and statistical decomposition to support the understanding of the Lake Volta natural hydrology is also highlighted. The combination of satellite geodetic observations with advanced computational methods is emphasized as novel approaches embedded in remote sensing hydrology as an emerging discipline.

15.1 Contributions of Human Activities to Hydrology

15.1.1 Anthropogenic Influence on Stored Water

Freshwater resources reside in watersheds, lakes, aquifers, wetlands, or floodplains. These resources are susceptible to the impacts of climate change as well as anthropogenic effects (Usman et al. 2021; Ahmed et al. 2014; Zhang et al. 2008). The impacts of climate change on the environment can take various forms. This can be in terms considerable fluctuations in rainfall patterns, flood and drought episodes,

C. Ndehedehe, *Satellite Remote Sensing of Terrestrial Hydrology*,
https://doi.org/10.1007/978-3-030-99577-5_15

sea-level rise, as well as increasing temperature. Decline in global food security can result from incessant extreme events like frequent bushfires, drought and floods. However, mismanagement of available water resources can also impact negatively on food security (United Nations 2006). Arid regions are becoming increasingly drier as a result of climatic fluctuations being experienced on regional and global scales. Consequently, tropical regions like Africa and Asia have experienced an intense and continuous rainfall pattern and shift in climate, resulting in incessant flooding events, crop damage as well as soil erosions (Hulme et al. 2001). Natural variabilities comprise changes in temperature, rainfall, as well as runoff patterns. These are normally obfuscating, given the trans-boundary description of several continental freshwater resources in arid sub-region, such as Africa (Ahmed and Wiese 2019). Freshwater resources across several continents are increasingly being shared and human water use at local, regional, and global aspects are distinct and varies across a broad range of applications, e.g., irrigation, hydro-power, etc. (Sharma et al. 1996).

Anthropogenic impacts are caused by several activities such as, deforestation, surface water diversion projects, groundwater overexploitation and contamination, urbanization resulting from population growth. The increase in freshwater demands is usually as a result of population growth. The increasing demand for groundwater due to expansion of irrigated agriculture and rise in human population has resulted in intensive groundwater pumping and depletion (Bierkens and Wada 2019). These factors (climate change and increased human population) will impact on future water requirements, triggering increased reliance on groundwater resources. The result of these will be a continued increase in the use of non-renewable groundwater resources. This increasing consumption of groundwater resources will eventually translate to large aquifer depletion. Several human actions like aquifer dewatering also limits freshwater availability thereby increasing its cost in a regional scale. Approximately 1 in 4 billion people do not have access to fresh and safe-drinking water (United Nations 2011). Regional and global freshwater resources are envisaged to experience increased variabilities, especially as a result of population growth. The rate of this rise in human population is expected to create more demand for freshwater resources on a global scale. Creating channels for assessing freshwater resources will ultimately reinforce developing sustainable management policies and solutions that are vital to global economic growth, food security, poverty reduction and optimal ecosystem functioning.

Sub-regions and regional ecosystems are maintained as a result of available surface water resources (Gleick 2003). The volume of freshwater resources stored on the Earth's surface has been greatly facilitated through dam constructions and operations (Chao et al. 2008). The construction of these dams has contributed to flood reduction in several societies and regions. Through increased water storage capacity in reservoirs, the availability and accessibility of freshwater resources for agricultural and domestic use as well as applications in renewable energy are now improved. This technique of storing freshwater resources in reservoirs has also led to the amplification of seasonal variability in water storage, which can be extremely harmful to the environment largely due to several factors including

- intensified greenhouse gas emissions (Deemer et al. 2016),
- high evapotranspiration (ET) rates (Zhao and Gao 2019),
- decomposed water quality, poor ecosystem functioning (Gillespie et al. 2015), and
- increased downstream erosion (Pelicice et al. 2015).

Despite the creation of new dams, the amount of synthetic water confiscation of approximately 184,000 km^2 of original water bodies is alarming, causing global concerns, especially in emerging economies. This has also increased the need for adequate tracking of human modification to surface water storage, which is aimed at achieving sustainable ecosystem functioning and water resources globally. The task of tracking and quantifying anthropogenic influence on regional and global freshwater resources is extremely challenging, especially for relatively smaller water storages. Data depicting reservoir water level are usually inaccessible and often provided in formats that impede global scale analysis (Cooley et al. 2021).

Conversely, few natural water bodies are usually assessed, especially in less industrialized and remote locations (Lawford et al. 2013; Shiklomanov et al. 2002). The advent of radar altimeters has been very helpful in tracking reservoir storage in large regional and global scales. However, their coarse spatial resolutions constrain their applicability to a few hundreds of existing global freshwater storages (Gao et al. 2012). Studies attempting to identify anthropogenic influences on global freshwater storages have distinctively depended on model simulations (Zhou et al. 2016), or granular gridded satellite gravimetric measurements, which cannot isolate synthetic and natural variations in water level fluctuations (Rodell et al. 2018; Getirana et al. 2017). Therefore, global storage variability and its corresponding anthropogenic impacts are still unknown. Cooley et al. (2021) quantified the influence of human intervention on seasonal freshwater storage fluctuations by applying the Global Reservoirs and Dams database (GRanD, GRanD Lehner et al. 2011) and the Global Georeferenced Database of Dams (Mulligan et al. 2020). The study reported a 0.03 km^2 sized ICESat-2 observations of 8964 reservoirs a minimum of two times over the study region. Among these reservoirs, the median seasonal fluctuation for managed reservoirs was 0.86 m (0.75–0.97 m), against the non-managed which was 0.22 m (0.12–0.32 m). In total, they reported 12.8% (1151) showing seasonal fluctuations that exceeded 5 m, while about 4.5% (405) indicating seasonal fluctuations greater than 10 m. In contrast, it was discovered that about 0.3% of natural lakes showed inconsistencies greater than 5 m, some of which may be unknown. Finally, it can be deduced that due to anthropogenic influence and patterns, human-managed reservoirs drive the majority of the Earth's surface water variability (Fig. 15.1).

15.1.2 How Does Human Influence Affect Water Balance?

Changes in terrestrial water storage are largely influenced by two main factors-climate change and anthropogenic influence. The severity of the latter is driven by man-made activities (e.g., construction of dams, water transfer, land-use change,

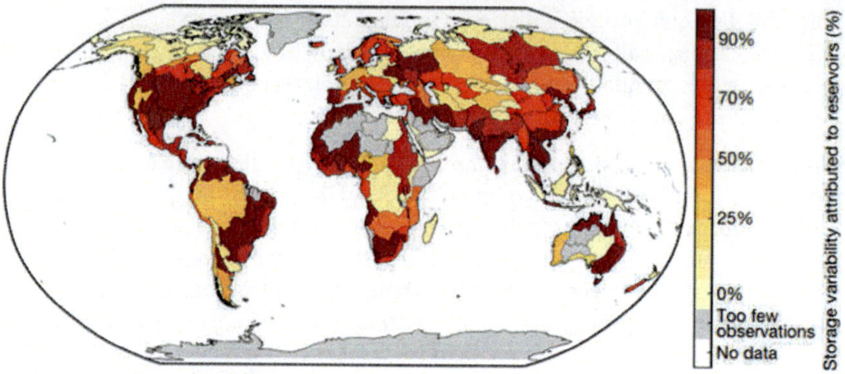

Fig. 15.1 Ratio of seasonal freshwater storage variability correlated with reservoirs by hydrologic basins from 2018 to 2020. Greater effect on human-managed reservoirs on freshwater storages is depicted using darker colours, while less effect is depicted using lighter colours. *Source* Cooley et al. (2021)

etc.) and population growth. The construction of dams for water storage aimed at generating hydro-power or irrigation agriculture coupled with water withdrawals for industrial and domestic purposes are all human-induced developments. These activities either directly or indirectly trigger alterations in the natural water equilibrium of terrestrial ecosystem. For instance, the impacts of operating reservoirs drive changes in discharge timings of rivers. From the use of the water balance model, Fekete et al. (2010) reported that the influence of human interventions on some river basins is either greater than or equal to the effect of anticipated climatic factors on annual runoff over a duration of 40 years. Humans also rely on groundwater reservoirs for irrigation, which account for the largest water-sector use of an estimated 70% of global water withdrawals and approximately 90% of consumptive water use.

Studies have shown that human-induced irrigation is estimated to amount to 1250 km^3 per annum in total groundwater withdrawn volumes (Hoff et al. 2010). These data show that apart from climate change, anthropogenic influence are also strong drivers of the imbalance in our terrestrial ecosystem. Anthropogenic water consumption can lead to an estimated 5% reduction in average annual runoff for most rivers as recorded during the period of 1971–2000. The annual runoff parameters can be significantly influenced by the surface water evaporation rate from man-made reservoir functionalities, which can result in seasonal regime shifts over multi-years. Heavily irrigated regions in some western parts of the United States and across much of Asia experienced substantial impact of this anthropogenic-induced runoff. Over 15% of median ensemble runoff decrease has been observed in some river basins in central Asia, the Middle East, and the Indian sub-continent. Other river basins such as the Orange, Colorado, Nile, Murray-Darling, among other basins in Asia have recorded between 5 and 15% ensemble median decrease as a result of excessive anthropogenic water consumption (Haddeland et al. 2014).

Apart from the direct impacts on freshwater depletion, anthropogenic influence also affect water balance. This influence can emanate from other indirect channels such as deforestation, expansion of agricultural areas and groundwater over exploitation. Moreover, forests play an important role in the terrestrial water cycle at both global and local scales. They transport large amounts or volumes of freshwater to the atmosphere. This of course happens through plant transpiration activities, which increase the total sum of evapotranspiration activities in the region. Increased evapotranspiration activities are known to replenish clouds and help generate rainfall amounts that sustain large rainforests such as the Amazon. A significant amount of water that falls in the tropical rainforests is recycled within the system. This, however, is mostly sustainable in very large rainforests. Impacts of lake, wet soil and river evaporation form a bulk of the recycling process. However, a significant amount of these impacts are fast-tracked by plants, especially trees. Given that the taproot of trees is able to penetrate the water table to get moisture even in dry periods, they play a great role in completing the water circulatory system by releasing moisture through their leaves through transpiration. The amount of water released depends on the extent of soil moisture, which is indirectly a function of precipitation rates in the region. The Amazon basin is one of the regions in the world, which maintains a minimum amount of 50–80% moisture at all in times in the terrestrial ecosystem water cycle.

15.1.3 Assessing Impacts of Human Activities on Water Resources

Water resource in a catchment is influenced by topographic, meteorological, geological, as well as land-use factors given that human activities maintain a significant role in the dynamics of basin-scale hydrological cycle. Meteorological factors regulate climatic and weather conditions. Topographic and geological factors, on the other hand, generally remain constant within the time scale of interest to water experts and hydrologists. Human activities generally control land-use factors. However, the type of natural vegetation could play a key role in forested and remote catchments of interest. Other factors caused by human activities are water infrastructure projects, which include reclamation tasks, diversion to/from catchments, dam and reservoir constructions as well as other actions that alter or interrupt natural drainage patterns. The changes in landscape of a catchment can be due to natural alterations of cultivated agricultural crops, vegetation, or other human-induced events such as afforestation and deforestation. The tendency of land-use/land cover changes resulting in varied flood frequencies has been captured by past literatures (Brath et al. 2006). Other factors such as mean annual runoff, and base-flow (Wang et al. 2006) also undergo significant changes as a result of land-use changes. Due to observed trend between land-use data and global river runoffs, studies have argued that the dynamics of land use plays a key role in regulating regional water resources, especially in the tropics.

Changes in land use/land cover drive surface runoff and water availability in general. This is exemplified in several studies (Usman et al. 2021; Ndehedehe 2019), including the work of Zhang et al. (2008) who reported a 50% reduction in annual runoff over the Loess Plateau in China. The role of human-induced afforestation and deforestation in assessing catchment hydrological responses has been captured by past studies (Li et al. 2012). These studies reported a notable decrease in catchment runoff after afforestation and a significant increase after deforestation activities. As will be demonstrated later in this chapter, water infrastructure and developmental projects such as dams and reservoir construction and diversion projects impact on runoff and eco-hydrological processes (Adel 2001). For instance, evapotranspiration rates from large water infrastructures or reservoirs can be compared to that of a forested land cover at an annual time scale, especially in humid sub-tropical climates (Degu et al. 2011).

15.1.3.1 Hydrological Modelling

Several existing hydrological modelling procedures have revealed the influence of climate change on large river basin discharge. Some of these procedures involved the application of a single hydrologic model forced with various climate projections (Arnell 2003). Global hydrological models demonstrate a variety of systematic biases in an attempt to stimulate primary attributes of an observed climate system (Hagemann et al. 2013). Regardless of the biases from these models, notable international bodies such as the Inter-governmental Panel on Climate Change (IPCC) conclude that there exists a substantial confidence interval that climate models provide valid quantitative predictions of future climate fluctuations. However, it is difficult to implement these models in hydrological impact studies without undergoing some bias corrections until they seamlessly reproduce the current climatic state. Without applying these corrections on the hydrological models, their resulting seasonal distributions and runoff amounts tend to vary substantially from in-situ observations especially when they are used as input to hydrological simulations. The corrections are, however, based on a fitted histogram equalization operation.

By applying them to the hydrological models, it becomes possible to conserve robustness properties, as well as eradicate large jumps at either monthly or seasonal time series. Inasmuch as bias corrections of climatic forcing models are an indispensable procedure in climatic impact simulations, several recent studies have also characterized pitfalls and limitations inherent in the process (Ehret et al. 2012). The study by (Hagemann et al. 2013) assessed hydrological projections from eight global hydrological models. The study highlighted a large spread in estimated runoff variations within the climate-hydrology modelling framework over several regions. However, a unique increase in runoff was noticed at high latitudes, while a strong signal of reduced runoff was shown over the mid-latitude regions. The authors also concluded that the choice of hydrological models contributes to the uncertainty for hydrological variations caused by the atmosphere-ocean general circulation models used. Models such as the WaterMIP and ISI-MIP have been extensively applied by several studies in assessing future water scarcity as well as analysing variations in naturalized runoff (Haddeland et al. 2014). With the use of these models, the extent of water withdrawals and consumptions from different sectors can be simulated to model important hydrological parameters accurately.

15.1.3.2 Satellite Remote Sensing Techniques

With the advent of satellite remote sensing techniques aimed at tracking terrestrial freshwater changes from space, i.e., using the GRACE mission, researchers and scientists have been able to improve understanding on the sustainability of most of the world's largest aquifers, noting that much of it have undergone significant deteriorations as a result of human influence (Rodell et al. 2018). Human population continues to grow unabated leading to rapid urbanization and rising food demand. This rise contributes to anthropogenic influences and pressures on global freshwater resources (Best 2019). It is, however, imperative to evaluate anthropogenic pressures on freshwater sources, especially in the areas of ascertaining their spatio-temporal dynamics and locating hotspots, i.e., areas experiencing the most intense anthropogenic pressures. This action will help in mitigating impact of anthropogenic pressures on freshwater sustainability (Elliott et al. 2014), low river connectivity, decreasing runoff, biodiversity loss as well as aquatic contamination (Mekonnen and Hoekstra 2018; Grill et al. 2019). Failure to address these issues will result in the inhibition of the UN's sixth Sustainable Development Goal; i.e., clean water and sanitation (United Nations 2015).

15.1.4 Hydropower Dam Operations and Repercussions on Hydrology

Over the past few decades, there has been an increase in the number of dam constructions. Over half of the world's largest river systems are regulated with a minimum of 50,000 dams constructed globally (ICOLD 2007). Factors contributing to this upsurge in construction include urbanization caused by population growth and socio-economic development. Obviously, the rise in dam constructions is driven by the need to meet the increasing demand for water, electricity and infrastructure. Dam construction on natural wetlands and biodiversity has an enormous negative effect on the expected functions of ecosystem. Some examples of these negative impacts include poor water quality, unstable flow patterns, and movement interception for fishes and other higher order consumers (Wu et al. 2019; Bunn et al. 2006). Increase in evaporation rates from extended water surfaces caused by river impoundments

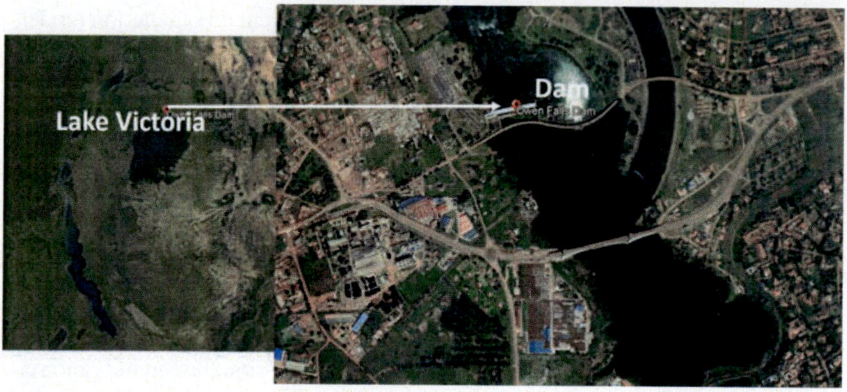

Fig. 15.2 Overview map of Lake Victoria and the Nalubaale Power Station (Owen Falls Dam)

also impacts on land-atmosphere relations, as well as regulating regional and global climatic fluxes (Pokhrel et al. 2012; Degu et al. 2011). It also aids in the advancement of the increasing greenhouse gas emission from immersed organic components (Deemer et al. 2016). Understanding and quantifying the influence of reservoir actions and river impoundment on the regional and global water cycle, especially from the perspective of resource management and planning are thus crucial.

Several sub-regional reservoirs and lakes around the world are largely dedicated income-generating hydro-power schemes, as a result of their intense economic potential. For example, Lake Victoria is a large freshwater body that contributes to the socio-economic prosperity and sustainability of the Lake Volta basin located in East Africa. This lake, however, has been influenced by the Nalubaale hydro-power dam, situated at its only channel on the White Nile river (Fig. 15.2). Much of the hydropower supplied to Uganda, Burundi, and Rwanda emanates from Lake Victoria. Lake Volta and the Kainji dam in West Africa are additional examples where reservoirs and lakes have very strong economic benefits. However, drought events and extreme climate-induced fluctuations are anticipated to impact on the level of income generated amidst other social factors, which include migration and inadequate power supply. Given the over-dependence of the global populace on freshwater resources for sustained livelihood, the impacts of climatic variability on surface water hydrology could lead to a disruption in global food production sequence (Getirana et al. 2020).

The Lake Victoria contributes to regulating the regional climate and possible influence of repetitive patterns in the sub-region, like the El-Nino/Southern Oscillation and Indian ocean dipole has been identified (Awange et al. 2013). Past reports highlighted the synopsis of Lake Victoria's hydrology attributing the cause of the water imbalance to a combination of climatic and anthropogenic factors. Case-specific studies highlighted the vulnerability of surface water hydrology to climatic fluctuations on a global scale (Awange et al. 2008; Oyebande 2001), leading to rise in water demand as a result of industrialization, as well as urbanization. It is, therefore, necessary to assimilate science with sustainable reservoir protocols and techniques as part of an assimilated approach to predict the impacts of future development of water structures to aid appropriate management.

Several studies have also examined global scale anthropogenic influence and their impacts focused only on streamflow, thereby neglecting other important hydrological variables (Yassin et al. 2019; Haddeland et al. 2006) such as the Nalubaale hydro-power dam effect. Doll et al. (2009) used a hydrological model in undertaking a global-scale analysis and reported that a regulated Lake Victoria improves low flow caused by temporal stream flow uniformity, giving room for hydro-power invention, food security or water supply. The study undertaken by Getirana et al. (2017) highlighted the relevance of surface water storage as a major constituent of terrestrial water storage (TWS) fluctuations in the tropics, along major rivers passing through arid region at high latitudes and elevations, especially as detected in the Nile and Amazon river basins, making up a total TWS of 28%. This would generally be the case in these sub-humid regions, given the substantial changes observed in rainfall and TWS patterns (Kalu et al. 2021; Ndehedehe et al. 2016a). The limitations underlined in the study by Getirana et al. (2017) hamper the suitable modelling of the global water cycle and an adequate decomposition of TWS components.

Significant progress have been made in the past two decades in the area of enhancing numerical representation of anthropogenic influence on the global water cycle (Yassin et al. 2019; Haddeland et al. 2006; Doll et al. 2009). Much of these studies are, however, limited to the influence of reservoir operation on river flow while disregarding spatio-temporal water storage fluctuations and backwater impacts. It therefore places emphasis on the importance of a global modelling system that takes into account both natural and anthropogenic processes, given its usefulness in optimizing assessment of varied freshwater systems.

Lake Victoria water storage under dam operation is estimated to be approximately 2369 km^3. This relates to about 20% lake water storage acceleration, when compared to a naturalized framework (1985 km^3) (Getirana et al. 2020) . Hamilton et al. (2016) used a bathymetric data to predict a 2424 km^3 water storage for Lake Victoria. The water volume obtained from this study differs slightly from that of Getirana et al. (2020) with a 2% increase. This was, however, linked to the means of data acquisition employed in the two studies. While Hamilton et al. (2016) used a static water altitude dataset, Getirana et al. (2020) on the other hand employed a 17-year average dataset. The dam influenced lake water storage fluctuation ($\pm$28 km^3) to a minimum of 2310 km^3 and a maximum of 2430 km^3, which was recorded in October 2005 and June 2015. The influence of dam operations in the lake is noticeable under natural conditions, as the lake water storage fluctuation is minimized to less than half ($\pm$12 km^3). When compared with GRACE observations at basin scale, it was noticed that surface water storage fluctuations under dam operation maintained a high correlation with TWS fluctuations at 72%. However, the fluctuation noticed in the variability ratio and correlation was not as high between surface water storage under a naturalized system and measured TWS (Getirana et al. 2020).

Global water problems can be described as multivariate in nature. Factors such as population growth and highly variable climatic conditions drive water scarcity in some region as well as restricts agricultural yield in other region based on their level of reliance on rainfall. The shortage of field-based hydrological measurements to monitor the various series of inter-annual fluctuations associated with drought coupled with the influences of drought frequency on regional livelihood and income complicate these problems (Ndehedehe et al. 2020a; Agutu et al. 2017). Given the dominance of Agriculture in continental water resource consumption (Oyebande 2001), sustainable water resource development in most regions can be adopted as part of risk-based decisions to ensure food security. Furthermore, climate-induced severe drying has triggered some management decisions around the world. For example, in the South African sub-region, these impacts have led to water supply disruption in Botswana and hydroelectric load sheltering in Zambia (Siderius et al. 2018).

Due to the scarcity of water infrastructures, the enhanced computational depiction of reservoir operation will influence highly on water scarcity assessment and early warning frameworks applied for global decision making (McNally et al. 2017; Sheffield et al. 2014). The integration of this scheme into an operational water assessment systems in the absence of surface water components, like the Global Land Data Assimilation (GLDAS Rodell et al. 2004) and North America Land Data Assimilation models will be of immense benefit to water managers and the scientific community at large, especially in the area of providing available and comprehensive datasets depicting spatial and temporal distribution of water fluxes. New methodological framework can be implemented in atmospheric and climatic modelling to give a detailed representation of human-induced impacts on surface water extent fluctuations on land-atmosphere interactions, which has influence on local and global climatology. Finally, its capacity in enhancing the workability of hydrological and climatic models as well as their capacity in realistic reproduction of hydro-climatologic and land surface conditions is basically a key outcome of this computational modelling approach.

15.1.5 Impacts of Dams on Rivers

Global energy demand has increased in the last few years. This is caused by rapid growth in human population and the need for economic development (United Nations 2011). A 72% increase in electricity production was reported between 1993 and 2010 and is expected to increase by an additional 56% by 2040 (The World Bank 2014). This, however, does not make up for the fact that over 1.4 billion people are disconnected from electricity supplies, especially in the rural communities of Sub-Saharan and South Asia (United Nations 2012). The need to secure the future energy demand and also close the electricity access gap is therefore of utmost importance, especially for the energy sector (United Nations 2011). Energy production concerns spans across reachability, as it is also one of the major causes of depleting fossil energy, greenhouse emission, and the over exploitation of uranium in the ecosystem

(United Nations 2012). The fact that they are unequally distributed on a global scale, however, provides reasons for country dependence. Political conflicts can impact negatively on such dependence (Asif and Muneer 2007). Renewable energy sources currently account for 20% of the global electricity production, with hydropower contributing 80% of the total share (The World Bank 2014). Globally, 37,600 dams have been designed whose heights exceed 15m. Out of these dams, about 8,600 projected for hydro-power generation are in operation (ICOLD 2011). According to the World Bank Database World Development Indicators, about 32 countries (e.g., Mozambique, Brazil, Norway, and Nepal) rely on hydropower to address more than 80% of their electricity needs (The World Bank 2014).

The creation of dams makes provisions for novel and artificial types of environment given their contributions to aquatic ecology, surface water hydrology, as well as water quality and quantity in aquatic habitats. Factors such as dam-induced current and increased temperature caused by dam can greatly affect delicate fishes, e.g., snow, trout, catfish, loaches etc. This impact on aquatic life can cause a reduction in fish populations. Another effect of impoundment induced from dam construction is that the water in reservoirs may become thermally stratified. This may, however, depend on the depth of reservoir and its water retention time. Since the water uptake for the turbines is usually in the hypolimnion, the downstream discharge of cold water can be made possible by the operation of the power station. The de-oxygenation of the hypolimnion coupled with the downstream discharge has a negative impact, especially on the aquatic fauna in the river below the dam.

Dam creation not only impacts rivers but also the residents of the location where it is cited. Its negative impacts on marine life can also trickle down marine fishery operations and can result in loss of income and migration (professional fishermen) to other locations in search of other opportunities. This may create more competition in their new destination. Although social interactions between the power station employees and local communities will have both negative and positive effects on the values and traditional cultures of the local communities, the public safety concern is of utmost importance, especially given the possibility of accidents from sudden peak release downstream during the dry seasons. Flooded landscapes resulting from the construction of new reservoirs also foster the need to relocate the fishing settlements as well as permanent loss of arable land, and by extension agricultural production in that sub-region. However, these settlements may be provided with constant electricity, and some other benefits coming from the presence of the dam and reservoir in that region such as the use of reservoir for fisheries, catering services, as well as the development of local tourism.

15.2 Water Diversion Projects

15.2.1 Impacts on Groundwater

The importance of groundwater as a critical freshwater resource with global significance has been highlighted in several studies (Long et al. 2020). Groundwater remains an important source of freshwater that supports global food and water security, constraining eco-hydrological systems, and enabling human adaptation to climate change (Taylor et al. 2013). The relevance of this vital resource is mostly felt in the arid and semi-arid regions where an approximated 26% global water withdrawal emanates from while 40% is used for irrigation (Margat and Van der Gun 2013; Siebert et al. 2010). Approximately 70% of global groundwater withdrawals are channelled to the agricultural sector (Margat and Van der Gun 2013). However, important factors such as urbanization caused by rise in human population, higher gross domestic products, among other factors, have led to an over exploitation of this crucial freshwater reserves, thereby leading to a substantial impact on food and water security (Dalin et al. 2017; Gleeson et al. 2012; Scanlon et al. 2012). Several studies have reported global groundwater depletion (Rodell et al. 2018; Long et al. 2017; Richey et al. 2015), for example, in U.S California's Central Valley Northwest India, US High Plains, Middle East, and the Northern China Plain (Voss et al. 2013; Feng et al. 2013; Famiglietti et al. 2011; Rodell et al. 2009). The groundwater depletion reported in the Northern China Plain is about the highest. A record depletion of 3.5 km^3/year during the 1983–1993 period based on data obtained from shallow groundwater level observation was reported. Further, approximately 8.3 km^3/year loss in groundwater is reported with data from 2003 to 2010 (from GRACE observation) and consists of mass changes of both deep and shallow groundwater reservoirs (Feng et al. 2013).

The threat to water security extends to many large cities globally. Cities like Mexico have reported land subsidence rate of up to 30 cm/year, which was caused by extensive groundwater pumping, especially for agricultural and municipal water use, diminishing water resources, and proliferating water adulteration and flood risk (Chaussard et al. 2014). Other cities like Chennai, Sao Paulo, Jakarta, Melbourne, and some regions in India have also encountered extreme water scarcity and marked global groundwater depletion. This led to endangered water supply for agricultural and municipal use and land subsidence and desiccation of rivers and wetlands (Famiglietti 2014; Scanlon et al. 2012; Doll et al. 2009; Rodell et al. 2009). Major contributing factors to groundwater depletion include rapid urbanization, severe agricultural irrigation, as well as climate change. In Beijing, China where large population growth is obvious (21.5 million in 2018), severe water shortages have been experienced over the past decade. The average groundwater depletion rate in the Beijing plane was 0.7 m/year. Given the groundwater depletion rates, Beijing water supply is projected to worsen in the coming years. This is particularly so due to fluctuating climatic events, changing characteristics of groundwater recharge through precipitation, as well as drought and heat waves in some regions (Kang and Eltahir 2018; Su et al. 2018; Trenberth et al. 2002).

With the growing concerns of severe drought and climatic extremes, it is important to improve the resilience of water supplies, especially to large cities where most of the human population reside. Several approaches can be employed to tackle issues of water management to improve resilience. The balance between water inputs and outputs is reflected in the water storage budget. A major cause of historical storage depletion is increased groundwater withdrawal (e.g., irrigation pumping), with a corresponding low recharge rate especially from rainfall or drought. Different approaches can be employed to replenish underground water storage, including regulating withdrawal rate, and improving recharge. In regulating withdrawal rate, the pumpage of water can be decreased and available terrestrial water can be recycled for reuse. In replenishing underground storage, water can be transported from more humid regions to the arid regions (e.g., Arizona and California) (Scanlon et al. 2012), and alternative water sources can also be developed (e.g., desalinating seawater) (Schiermeier 2008; Muller 2018).

Using surface water and groundwater mutually exclusively underpins effort to improve water management. The use of surface water during wet periods and the groundwater during dry/drought periods are some efforts to aid sustainability. Managed aquifer replenishment is also beneficial for storing surplus surface water from wet periods in diminished aquifers for use during drought periods, which helps in equal distribution of water (Muller 2018). With the use of data from the GRACE mission and its Follow-On, it is easier to track terrestrial water storage especially due to its contribution to the variations noticed in the earth's gravity field (Tapley et al. 2019, 2004). In places like Northern China where terrestrial water storage is decreasing, it is easier to detect this from GRACE signals, as well as regions with rising land water storage like Southern China (Long et al. 2017; Feng et al. 2013). One solution to ease regions of water shortages is by water diversion from the humid regions. This involves a dedicated national project where systems of pipelines are constructed through a canal connecting the humid region to the arid region. For example, the South-to-North Water Diversion (SNWD) project was launched in 2002 and was designed to move an approximated 45 km^3 of water every year from the Yangtze River through three canals and pipeline routine.

Studies have shown the impacts of climate fluctuations and water diversion on groundwater from humid to arid regions using groundwater flow models or coupled surface-ground water techniques to model variations in groundwater levels under varied climatic and water diversion schemes (Xiaet al. 2018; Li et al. 2017). However, there is a need to also understand the impact of policies that restrict groundwater withdrawal. Moreover, the lack of physical phenomenon of groundwater recharge modelling such as preferential flow and percolation may produce a misleading assessment of the impact of climatic fluctuations on groundwater. The need to understand the individual water diversion policies of each region to enhance groundwater recovery is crucial. Of particular note is that the climatic contributions of the areas involved in the project need to also be understood (Long et al. 2020). In water diversion projects, it is important to note that uneven spatio-temporal distribution of water resources has led to water scarcity in many densely populated regions, this calls for a need to investigate consequences associated with this kind of project (Long et al. 2020).

15.2.2 Effects on Water Quality and Ecosystem Services

The increasing water consumption needs and limited water availability has created an upsurge in the growing global concern for water quality and water scarcity. This has, however, led to the implementation of several physical water transfer projects as seen in Canada, Australia, India, China, United States, etc. (Ghassemi and White 2007). As was mentioned earlier, the South-to-North water project in China is one of the largest physical water transfer projects in the world. It includes the eastern, middle, and western routes, which attracts water from the upstream, mid-stream, and downstream of the Yangtze River, respectively (Zhang and Li 2014). The volume of water estimated to be diverted is substantially large (44.8 billion m^3/year) and is equivalent to the terrestrial water volume of the Yellow river. This diversion has been evaluated to benefit over 200 million people in the region (e.g., Yang et al. 2015). The increase in production trades is also another crucial reason why substantial water volumes have been transferred virtually among other regions, this is regarded as virtual water flow. Hoekstra and Mekonnen (2012) estimated the total volume of international water flows from industrial applications and trades in agriculture as $2.32 \times 1012\,m^3$. Using crops like rice, barley, maize, soya beans and wheat, Lee et al. (2017) estimated virtual water exports within and beyond Asia (2000–2012), comparing it to those of European Union. Further, studies depicting virtual water flows on irrigation borough scales, region, and national scales (Fu et al. 2018; Liu et al. 2018; Mohammed and Darwish 2018) have been reported. This is besides several important large-scale studies that have been undertaken.

Studies highlighting the analysis of water scarcity and water stress have emerged in recent years, drawing the attention of many researchers. Analysis on water quality and quantity (Gosling and Arnell 2016; Li et al. 2017) has been heavily discussed in the literature. In 2002, the use of the water footprint concept as an inclusive water appropriation indicator to show consumed and polluted water volumes was proposed by Hoekstra et al. (2011). In their study, they represented grey water footprints as the volume of freshwater needed to integrate the load of pollutants, taking into consideration the pollutant contents, natural contextual concentrations, and existing ambient water quality basics and the blue water footprints as the utilization of ground and surface water along the product supply chain. Some water scarcity indicators based on the blue water footprint were suggested and employed at varying scales (Liu et al. 2017; Hoekstra et al. 2011). Liu et al. (2020) later proposed the need for a widespread water scarcity indicator encompassing water consumption and pollution rate, bringing into perspective the need for adequate water scarcity assessment, including their interactions. In their study, the authors assessed water stress using the proportional relationship between water availability and regional water consumption, bearing in mind the impact of water quality and quantity.

15.3 Assessing Human Impacts on Hydrology Using Remote Sensing, Computational and Statistical Methods

As mentioned earlier, freshwater availability, including its variability can be affected by several forms of anthropogenic influence. These human footprints may include, land-use changes, construction of dams and other water storage infrastructures. For example, reservoir impacts and dam construction along several of the world's largest rivers have altered seasonal flow rates and the hydrological regimes of these rivers (Dai et al. 2009; Yang et al. 2004; Lammers et al. 2001). In south east Australia and south-west United States, human-induced alterations in the regional ecosystem combined with long-term escalation in the water tables and recharge increase were noticed (Allison et al. 1990; Scanlon et al. 2005). Similar cases have been highlighted in West Africa where observations from extensive well networks show a rising groundwater trend despite severe drought condition experienced for more than a decade (Favreau et al. 2009; Leduc et al. 1997). These complex hydrological processes were described as the 'Sahelian paradox'. In the West African Sahel, alterations in land use and land clearing have resulted in increased runoff rate (Favreau et al. 2009; Descroix et al. 2009; Séguis et al. 2004). This phenomenon has also been reported in other Sahelian countries of West Africa (Gal et al. 2017; Mahé and Paturel 2009; Mahé and Olivry 1999), and new studies providing more insight on these hydrologically complex processes are emerging (Gal et al. 2017). Descroix et al. (2009) argued that the observed changes could possibly have originated from the hydraulic transmission capability of the soil as well as its infiltration volume. Other studies (Li et al. 2007) confirmed the influence of land-use change on the hydrological regimes of Lake Chad and Niger basins. These studies together illustrate the positive impact of complete deforestation on simulated runoff ratio (0.15–0.44) and annual stream (35–65%).

The assertion that this phenomenon is also typical in the Sahel can be based on the preponderance of evidence from other reports. However, these Sahelian paradox studies were undertaken at relatively smaller watershed scales (e.g., 500 and 5000 km^2), and further studies can be undertaken to further assess this phenomenon on a large scale. Other reports tend to agree that the Sahel responds to the influence of droughts and land-use change in the same manner. But the combination of satellite observations like GRACE as well as model observations can improve understanding of large-scale groundwater assessment and its response to anthropogenic influence. This approach has the tendency of offering a wider perspective and perhaps improved understanding of the hydro-dynamics of the sub-region. Using higher-order statistical tools and framework for spatial and temporal analysis, the synopsis by Ndehedehe (2019) demonstrated the benefit of this computational technique to this assessment.

15.3.1 Statistical Methods

The impact of deliberate human influence on hydrological fluxes around many sub-regions has been recently captured using remote sensing hydrology. For instance, using advanced statistical methods like the independent component analysis, Ndehedehe et al. 2017 showed that water ponding by the Akosombo dam in the Lake Volta described about 41% of the escalating trend in GRACE-derived TWS regardless of a relative significant regression in rainfall from 2002 to 2014. Yet, the Volta basin, which recorded a notable decadal fall in rainfall rate, maintained a relatively stronger changes in terrestrial water storage even during this period (Ndehedehe et al. 2017; Moore and Williams 2014; Ahmed et al. 2014). While the case of the Volta basin has been detailed in subsequent sections of this chapter, this accentuates the impacts and implications of human water management schemes on hydrological variables. Hydrological operations (e.g., hydraulic setups) in Senegal triggered the evolution of water resources as opposed to the 1980s when hydrological specifications were rather geological (Ngom et al. 2016). A conceptual and statistical approach has also been employed to assess the drivers of water storage in the Lake Chad located at the intersection of the humid West African sub-region and the Sahara desert (Zhu et al. 2019).

15.3.2 Remote Sensing and Multi-resolution Data

One of the primary impacts of strong declines in rainfall patterns around the Lake Chad is the contraction noticed in its total surface area. This impact inhibits inflow from Chari-Logone river structure, which is responsible for the provision of approximately 95% of Lake freshwater. The effect of extreme droughts in the area was, however, compounded due to water withdrawals from the Lake for irrigation purposes and facilitated by water deficit periods (Ndehedehe et al. 2016a; Coe and Foley 2001). This led to an approximately 90% decline of the lake surface area (Fig. 15.3). The satellite assessment and monitoring of the evolutionary patterns in Lake Chad water resources as illustrated in Fig. 15.3 reveal the substantial loss of aerial extent of the lake since 1973 when it was full. The hydrological conditions in the Lake Chad basin led to several studies seeking to understand the key drivers of drought and the exceedingly high loss of water storage in the lake Chad. Given the lack of in situ data, satellite observations are increasingly being employed to assess the hydrological conditions and eco-hydrology of the Lake Chad basin (Ndehedehe et al. 2018, 2020b; Skaskevych et al. 2019). The combination of remotely sensed data from several optical missions with drought metrics to assess changes in freshwater over the Lake Chad basin has been perceived as valuable and a key approach that underpins contemporary understanding of key climate and human stressors in the region (Ndehedehe et al. 2020b). More details on the Lake Chad basin is discussed in Chap. 20.

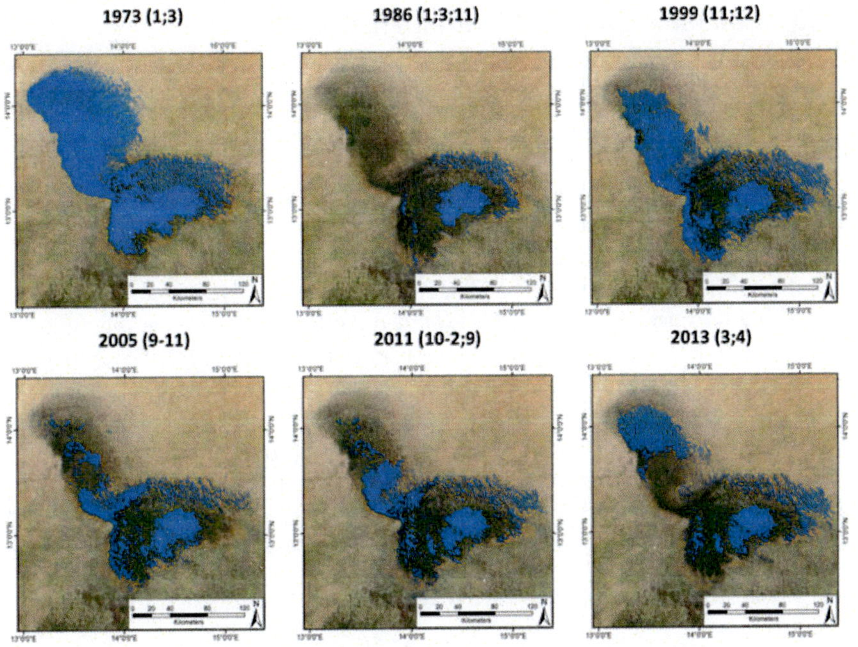

Fig. 15.3 Overview historical changes in the Lake Chad surface area between 1973 and 2013. The blue polygons are areas with water presence. *Source*, Guzinski et al. (2014)

The impact of human activities on the Lake's ecosystem transcends worsening the impact of climatic variability in the region but also extends to the readjustment of the Lake's stability. The human-modified droughts of the Lake Chad basin are characteristic of the Anthropocene era, especially given the impact of human activities on soil interactions, hydrological procedures, and catchment storage. These impacts exacerbate the intensity of hydrological drought over the sub-region. The impacts of urbanization on groundwater infiltration and recharge rates have been underlined in previous studies (Van Loon et al. 2016), whereas anthropogenic actions such as deforestation, afforestation, desertification, and agricultural practices through evapotranspiration have influence on soil moisture. The aforementioned studies have shown that the combined effects of climate fluctuations and human actions integrated with the well-known robust land-atmosphere coupling over the sub-region (Boone et al. 2009; Koster et al. 2004) will definitely produce a strong and profound effect on hydrological processes as well as water resource schemes.

15.3.3 Computational and Hydrological Models

Computational models have been combined with satellite data to study terrestrial water storage changes (reservoirs and lakes) and how human water management affects surface water hydrology. The study undertaken by Getirana et al. (2020) implemented a state-of-the-art model integrated with a proposed satellite-based reservoir operation system to assess the influence of dam construction over the Lake Victoria's storage fluctuations and water dynamics. The result obtained from the study indicates that surface water strongly impacts on the available freshwater resources and fluctuations within the basin. The study also extended our knowledge on the role of dam operations and its essence in accurate assessment of Lake Victoria's water height, water storage extent as well as its fluctuations, which has a meaningful contribution to Lake outflow and water accessibility in the sub-region. The proposed satellite-based reservoir operation structure used in their study proved useful in advancing the illustration of human water management operations on the global water cycle.

15.4 Human Water Use Resulting from Drought Impacts

The possibility for more intense, severe, and intense drought under future climate change is of serious global concern (Diffenbaugh et al. 2015). Studies have described the influence of drought threat on agriculture, as shown by the loss of crops and extreme water stress associated with drought events in many countries like Syria, Russia, Horn of Africa, U.S. Midwest, Southeast Asia, Cape Town, and California (Solander and Wilson 2018; Maxwell and Fitzpatrick 2012; Mallya et al. 2013), which has experienced a series of consecutive multi-year droughts (Medellín-Azuara et al. 2016). These incidents emphasize the ubiquity of drought, as well as its potentially disastrous effects, especially on regional and global scales. For example, in California, there were eight unusually dry years over two drought periods in the last decade; 2007–2009 and 2012–2016, with a period in 2010–2011 showing average weather conditions (Fig. 15.4).

The most severe drought was observed between 2013 and 2017 (Fig. 15.4). Other climatic thresholds reported include the most intense temperatures in over 1200 years, the most minimal snowpack in 500 years, and the duration with the most intense drought events (e.g., Belmecheri et al. 2016). The notable drought, which occurred from 2012 to 2016 affected a large extent of the State of California, with over 67% encountering exceptional or extreme drought conditions from 2014 to 2016 (Fig. 15.4). The Central Valley in California is one of the largest global and economically important regions with irrigated agriculture. With a production of over 50% fruits, vegetables, and nuts, the sub-region is among the predominant manufacturer of highly nutritious specialty crops, like pistachios, almonds, and grapes. The sub-region accounts for an estimated 75% of the irrigated land in California and 17% of the irrigated lands over the country (Medellín-Azuara et al. 2016; Faunt et al. 2009).

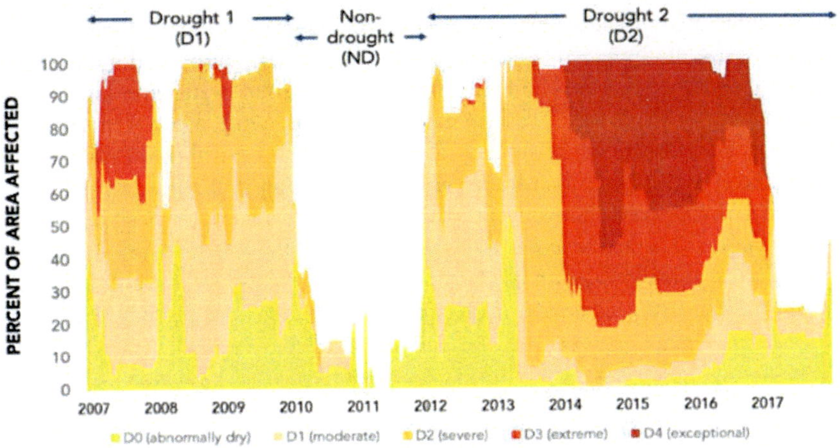

Fig. 15.4 Total area influenced by drought in California, United States. *Source*, Gebremichael et al. (2021)

It has also recorded the highest absorption of endangered species over the country (Hanak et al. 2019).

Another instance where drought frequency has contributed to increased development of water resources is evidenced in Brazil's increased water resources development. In light of several precarious drought events, as with other regions, the Brazilian government responded to several climatic thresholds through the construction of substantial water infrastructures. Brazil's National Water Agency (ANA) reported about 19,361 man-made reservoirs used for irrigation and hydropower existed as at 2016 (ANA 2017). In the drought analysis over Brazil (Chap. 20), there are two periods when drought-affected areas reached and exceeded 50%. Based on the MERRA indicator, the first was during the 1998/1999 period and the last occurred between 2012 and 2016. The former was the cause of the biggest energy crisis, which occurred in Brazil (Getirana 2016). Notably, the cascading impacts of the 2012–2015 extreme droughts in the region have been fully documented (Chap. 20). Some of these impacts were evidenced in freshwater crises, agricultural losses, groundwater depletion, depletion of moisture influx from the Amazon basin, among several other effects that have been reported in the literature. (e.g., Ferreira et al. 2018; Marengo et al. 2018; Awange et al. 2016; Getirana 2016).

From previous studies, the possibility of examining some aspects of land-use changes, its driving influences and impacts on underground water has been highlighted from 2007 to 2016 in Central Valley. During the study period, increase in fallow land regions and decrease in irrigated land were noticed (Medellín-Azuara et al. 2015, 2016; Cooley et al. 2015). Studies also reported an increase in acreage for fruit tree crops and nuts and a reduction in acreage for cereals, cotton, and alfalfa (Christian-Smith et al. 2011). The shift in cropping patterns was had significant surge in crop water use per hectare. The Central Valley recorded a reduction of

8 billion cubic metre of surface water, of which 6.2 billion cubic metre was replenished through groundwater pumping, resulting in an intense groundwater depletion (Medellín-Azuara et al. 2016). On the other hand, the groundwater overdraft led to higher pumping costs, reduced well yields, poorer water quality, and significant land subsidence that puts the entire sub-region, including its physical infrastructures at risk (Jeanne et al. 2019). Groundwater has been declining in the last few decades. Therefore without proper mitigation strategies and measures; a continuation of this decline is to be expected, even at a higher rate given the contribution of climate change and urbanization (Alam et al. 2019). This, therefore, led to recent studies focusing on different techniques of quantifying irrigation rates in current climate models in the affected regions, e.g., the Central Valley (Liu et al. 2018).

15.5 Hydrology of the Lake Volta

The Lake Volta is one of the highly esteemed projects from the period of Africa's decolonization. It is the largest man-made reservoir by surface area in the world, covering a total area of approximately 8500 km^2. This Lake was built in 1964, and it has an estimated capacity of 150 km^3. It has an installed hydro-power generation capacity of about 912 MW (e.g., Owusu et al. 2008; Gyau-Boakye 2001). Moreover, Lake Volta is one of the most important surface water in the West African sub-region in terms of size, economic value and the riparian countries it supports. It is formed by the Akosombo dam in Ghana and represents progress and a symbol of socio-economic importance in the region. It is still a critical source of livelihood and hydropower to the almost 20 million people whose livelihood depends on the Volta river basin (hereafter the Volta basin, Fig. 15.5). Along with the Kpong dam, which was constructed in 1982, Ghana is able to generate approximately 1060 MW of hydropower. As with numerous other nations in West Africa, which rely heavily on hydropower as a major source of energy, the Akosombo and Kpong dams provide an estimated 80% of the Ghana's electricity.

The rapid upsurge in human population and economic development comes with an increasing demand in energy. For emerging economies, which rely on hydropower, a rise in surface water developments is inevitable. Several West African countries have their rivers regulated through dams, and the Volta River is one among them, supplying huge sediments to the Volta River Delta (Boateng et al. 2012). However, not only was this development associated with substantial decline in fluvial sediment supply from the river, there was a far-reaching negative effects such as coastal erosion on the river delta as well as changes in hydrological regimes of the Volta River Delta (Amenuvor et al. 2020). As indicated in Fig. 15.6, there is an obvious alteration in the inter-annual variability of river discharge and sediment load of the Volta river in the pre- and post-Akosombo dam construction. Apparently, average annual discharge and sediment load in much of the period prior to the dam construction in 1964 were higher compared to the post-dam period after 1970. Maximum annual peaks in discharge in pre-dam era for most of the years ranged from greater than 1500 m^3/s to more

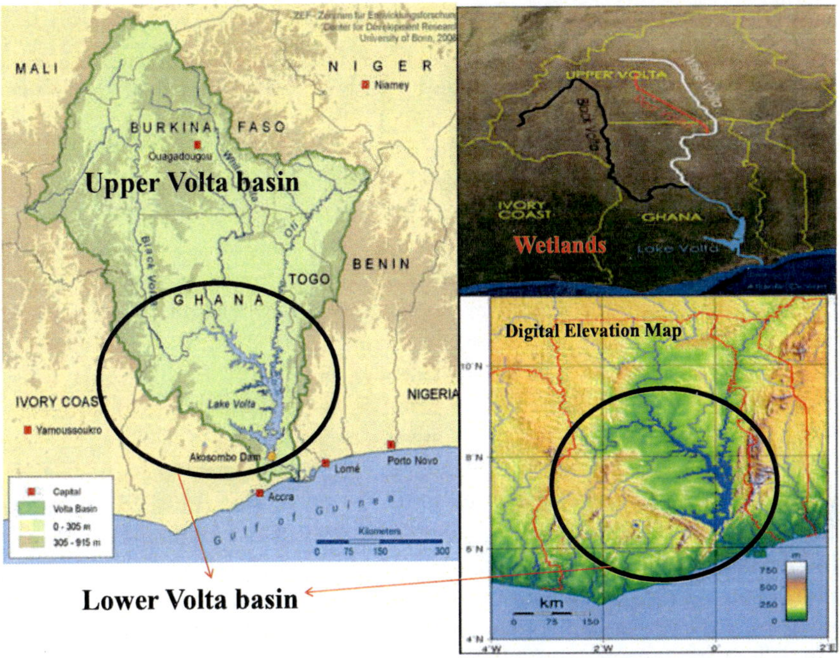

Fig. 15.5 Study area showing the Lake Volta river basin (LVRB). The riparian countries (Ghana, Togo, Benin, Burkina Faso, Mali, and Ivory Coast) of the basin are also indicated. The Lake Volta, the Volta river system, and the location of the Akosombo dam have also been indicated. The Volta river system comprises the Oti river and White and Black Volta rivers). *Source* Ndehedehe et al. (2017)

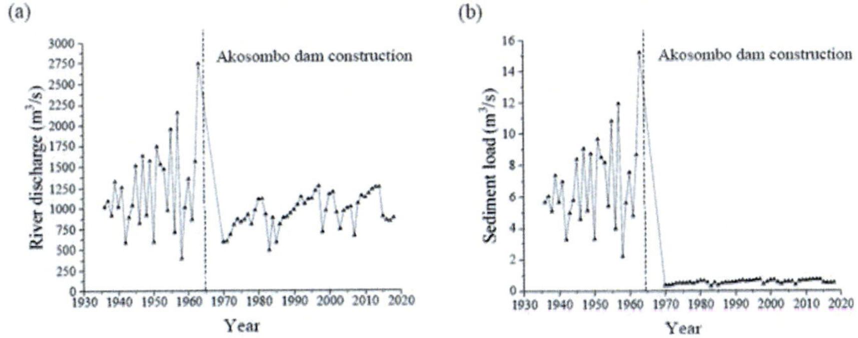

Fig. 15.6 Impact of dam construction on **a** surface water hydrology and **b** sediment load. *Source* Amenuvor et al. (2020)

than 2500 m^3/s while similar annual peaks in discharge in the post-dam construction period were mostly between 1000 and 750 m^3/s (Fig. 15.6a). Although post-dam-induced decline in annual river discharge was reasonably moderate, the effect of the Akosombo dam on sediment led to substantially high reduction (Fig. 15.6b). The effects of reduced sediment and flow alteration in the Volta river caused by dam construction on sediment load could potentially result in erosion of the downstream river channel and reduction of flood peaks (Amenuvor et al. 2020). Apart from effects on the morphological evolution of the Volta river delta, there are ecological impacts of water resources development of this nature on the environment and the ecological communities that rely on the ecosystem services provided by the Volta river systems. Some of these consequences have been covered in Chaps. 12 and 19 of this book.

This section presents the work of Ndehedehe et al. (2017) who employed a two-step procedure of weighted least squares formulation of global spherical harmonic analysis and statistical decomposition to improve understanding of the Volta basin's natural hydrology. It employs a multi-satellite approach to monitor and quantify inter-annual changes in satellite-derived terrestrial water storage (TWS) over the Volta basin (also includes the Lake Volta) and use is made of a cumulant based statistical method, the independent component analysis (e.g., Common 1994; Cardoso and Souloumiac 1993) to support the localization of hydrological signals that are obscured by leading signals caused by the impact of damming (Akosombo dam). To this end, the spatio-temporal variability of TWS over the Volta basin after removing the hydrological signal induced by Lake Volta water level changes (hereafter called Lake Volta induced TWS) was analyzed for the first time. The specific aims include (i) understanding hydrological changes, (ii) assessing the contributions of Lake Volta to TWS, and (iii) assessing the impact of climate variability on the freshwater systems of the Volta basin.

15.5.1 Data and Method

This study used GRACE (Tapley et al. 2004) Release-05 (RL05) spherical harmonic coefficients truncated at degree and order 60 from Center for Space Research (CSR), covering the period 2002–2014 (data files accessed at http://icgem.gfz-potsdam.de/ICGEM/shms/monthly/csr-rl05/). The processing chains of GRACE spherical harmonic coefficients as well as post-processing operations including, accounting for the geophysical signal loss caused during the pre-processing of GRACE data are as detailed in Chap. 4. Average TWS values (i.e., ΔW) over the basin were estimated using area weighted averaging (e.g., Sneeuw et al. 2014):

$$\Delta W(\chi; t) = \sum_{l=1}^{n} \Delta W(\psi_i, \lambda_i, t) \frac{A_i}{A_\chi}, \tag{15.1}$$

Table 15.1 Details of the stream flow records of the Volta river system (consists of the Oti, White, and Black Volta rivers) retrieved from the GRDC archives

River	GRDC No	Latitude	Longitude	Area (km^3)	Station	Duration
White Volta	1531450	9.70	−1.08	92950	Nawuni	1986 – 2007/02
Oti	1531800	9.28	0.23	58670	Sabari	1996–2007/02
Black Volta	1531050	10.63	−2.92	93820	Lawra	1990/03–2007/02

where χ is the basin index, n is the number of grid points in the basin, A_i is the area of the grid cell i in χ and A_χ is the total area of χ.

This study also used lake height variations computed from TOPEX/POSEIDON (T/P), Jason-1 and Jason-2/OSTM altimetry. These data are provided by the United States Department of Agriculture and can be accessed online. Stream flows of the Volta river system and the Akosombo dam (obtained from the Water Research Institute of Ghana) were also employed to assess the hydrological condition of the Lake Volta. Also, the river discharge data from the Global Runoff Data Centre (GRDC) (www.bafg.de/GRDC) for the Black Volta, White Volta and Oti rivers, which drain into Lake Volta have been retrieved and used as a complementary data to support the analysis of hydrological variability of the basin (Table 15.1. Global Precipitation Climatology Centre (GPCC) based precipitation (Schneider et al. 2014; Becker et al. 2013) product with a spatial resolution 0.5° × 0.5° was used for this study. The data covering the period from 1979 to 2014 period was used in the study to analyze seasonal rainfall variability in the region. The MODIS global terrestrial evapotranspiration (ET) products (Mu et al. 2011) were also used in this study. ET data were combined with precipitation to quantify net precipitation. The latter is a measure of available renewable freshwater resource. The data (spatial resolution of 0.5° x 0.5°) are available for download NASA's website (http://www.ntsg.umt.edu/project/mod16).

15.5.2 *Removing the Stored Water in Lake Volta from the Volta Basin Hydrology*

15.5.2.1 Lake Volta Induced TWS Changes

Observed trends in TWS in the Lake Volta river basin (LVRB) are dominated by TWS changes. These changes are primarily caused by hydrological changes through fluctuations in the Lake Volta water level and that of lower Volta where the Volta river system converges. Even though Lake Volta is approximately 2% of the basin its contribution to observed trends in TWS change is three times that of Lake Victoria despite having a surface area that is only one-eighth of Lake Victoria (Moore and Williams 2014). This hydrological signal from the water changes in the lake alone can result in substantial changes in observed TWS over the corresponding basin.

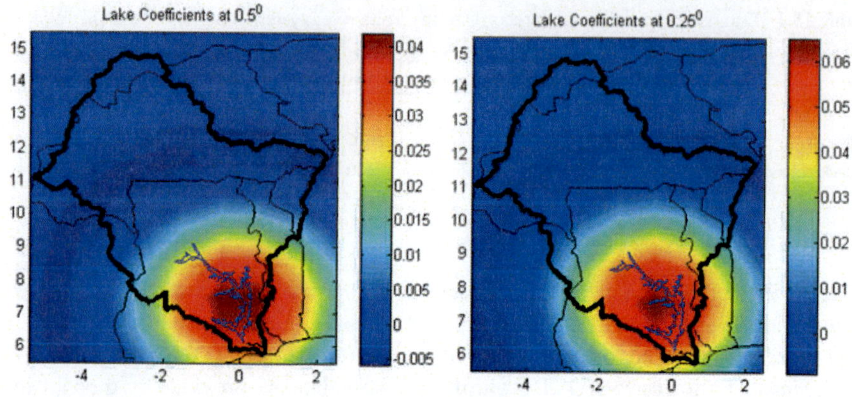

Fig. 15.7 Lake kernel functions for Lake Volta (blue polygon is the Lake's boundary) whose surface extent of 8500 km^2 was used for the synthesized mask. A grid mask constructed at grid intervals of 0.5° x 0.5° (left) was compared with a grid mask constructed at 0.25° x 0.25° (right). Uncertainties in the residual water storage over the Lake are somewhat reduced using a grid mask of 0.25° x 0.25°. The solid back polygon is the basin boundary. *Source* Ndehedehe et al. (2017)

The problem with this is that it can lead to a biased assessment of total water storage changes over the entire basin. This is because the water ponding by the lake will lead to pseudo trends in TWS over the basin. Separating the lake's water storage from the observed TWS over the basin is thus crucial to aid understanding of the natural variability of land water storage over the basin.

To assess the impact of climate variability on freshwater over the basin, it is critical to remove the Lake Volta signal (i.e., water storage changes due to variations in the Lake's surface) from the observed GRACE-derived TWS. To do this, a methodology was developed to convert the Lake Volta water level variations were to mass changes similar to how TWS obtained from GRACE was processed (e.g., Ndehedehe et al. 2017; Ferreira and Asiah 2015; Moore and Williams 2014). The method assumes a constant surface over the Lake given that impact of the tidal gravitational force on it can be assumed to be negligible. As a result, the surface water storage or the satellite altimetry water storage over the Lake would be an approximation. This is particularly because of the relatively small size, and the nature of the lake's north-south orientation (cf. Fig. 15.5). A global mask was constructed and used to define the lake kernel function f_{Lake} (Fig. 15.7) using the geodetic coordinates of the Lake extent as

$$f_{Lake}(\varphi, \lambda) = \begin{cases} 1, & \text{Over Lake Volta} \\ 0, & \text{Outside Lake Volta,} \end{cases} \quad (15.2)$$

where φ is geodetic latitude and λ the geodetic longitude. The global mask is constructed at grid intervals of 0.25° x 0.25° to adequately capture the lake's shape. A value of 1 was assigned within the Lake surface and 0 outside the Lake. These values correspond to 1 mm and 0 mm of EWH over the inland water and elsewhere, respectively. Using the global spherical harmonic analysis, a grid integrable function $f(\varphi, \lambda)$ on a sphere can be synthesized into harmonic coefficients $C_{l,m}$ and $S_{l,m}$ as (e.g., Sneeuw 1994)

$$\begin{Bmatrix} C_{l,m} \\ S_{l,m} \end{Bmatrix} = \frac{1}{4\pi} \int\int_{\sigma} f(\varphi, \lambda) P_{l,m}(\sin\varphi) \begin{Bmatrix} \cos m\lambda \\ \sin m\lambda \end{Bmatrix} d\sigma, \tag{15.3}$$

where $d\sigma = \sin\varphi \; d\varphi \; d\lambda$ and $P_{l,m}$ is the normalized associated Legendre function of degree l and order m. GRACE fields are snapshots of the static geoid, which are given in terms of spherical harmonic coefficients. The Lake's function $f_{Lake}(\varphi, \lambda)$ (Eq. 15.2) was therefore reconstructed (Eq. 15.3) and expressed through a spherical harmonic expansion as (e.g., Moore and Williams 2014)

$$f_{Lake}(\varphi, \lambda) = \frac{1}{4\pi} \sum_{l=1}^{l_{max}} \sum_{m=0}^{l} (C_{l,m}^{Lake} \cos m\lambda + S_{l,m}^{Lake} \sin m\lambda) P_{l,m}(\sin\varphi), \tag{15.4}$$

where $C_{l,m}^{Lake}$ and $S_{l,m}^{Lake}$ are the dimensionless lake spherical harmonic coefficients. This approach will help recover the time series of TWS over the basin not associated with variations surface in Lake Volta. As with CSR RL05 GRACE monthly fields, the spherical harmonic coefficients $C_{l,m}^{Lake}$ and $S_{l,m}^{Lake}$ were truncated at degree and order 60. These were then scaled using altimetry-derived water level variations. This was achieved by multiplying the changes in the altimetry height with the synthesized kernel function (Fig. 15.7). The difference between the Lake induced TWS and TWS changes over the entire basin will give the contribution of the Lake.

15.5.2.2 Multivariate Statistics

Recall that the independent component analysis (ICA) was earlier described in Chap. 13 of this book. It is a higher order multivariate technique and was used here to statistically decompose or separate TWS changes into spatial and temporal patterns. After the decomposition of TWS, the evolving spatio-temporal patterns will be localized over the region. The regionalization (signal localization) based on the ICA process employs the Joint Approximate Diagonalization of Eigen matrices (JADE). This was also discussed in more detail in Chap. 13 and further information are available in the pioneering works of Cardoso (1999) and Cardoso and Souloumiac (1993). GRACE TWS $\mathbf{X}_{TWS}$ was decomposed into spatial maps $\mathbf{S}$, and temporal variations $\mathbf{T}$ (i.e., after removing the water storage contributions from Lake Volta) as:

$$\mathbf{X}_{TWS}(x, y, t) = \mathbf{TS}, \tag{15.5}$$

where (x, y) are grid locations (e.g., geographic coordinates) and t is the time steps in months. $\mathbf{T}$ is unit-less because it has been normalized using its standard deviation.

The corresponding spatial patterns **S** are also scaled using the standard deviation of its independent components (i.e., **T**). As was mentioned in Chap. 13, **T** and **S** are jointly interpreted. That is, the spatial and temporal patterns are usually integrated to form what is traditionally called the ICA mode.

GRACE measures vertically integrated water storage changes. For the Volta basin, these changes are given in terms of soil moisture, groundwater and surface water as

$$\triangle TWS = \triangle SW + \triangle SM + \triangle GW + \triangle S_T, \tag{15.6}$$

where $\triangle SW$ is the change in total surface water storage (i.e., Lake Volta and the rivers), $\triangle SM$ is the change in soil moisture, $\triangle GW$ is the change in groundwater, and $\triangle S_T$ is the change in unquantified surface waters in wetlands and vegetation. After removing the contribution of $\triangle SW_{Lake\ Volta}$ (i.e., the Lake induced TWS signal), the residual water storage change component, $\triangle W_T$ is expressed as

$$\triangle W_T = \triangle TWS - \triangle SW_{Lake\ Volta}. \tag{15.7}$$

This quantity (i.e., $\triangle W_T$), which is the summation of $\triangle GW$, $\triangle SM$, and $\triangle S_T$ was statistically decomposed using the ICA method (Chap. 13) to understand its dominant spatio-temporal patterns over the basin. The Multiple Linear Regression Analysis (MLRA) was used to parameterize the trends, annual and semi-annual signals or harmonic components. The dataset $\mathbf{Y}_{Rainfall}$ as

$$\mathbf{Y}(i, j, t) = \beta_0 + \beta_1\mathbf{t} + \beta_2\mathbf{sin}(2\pi\mathbf{t}) + \beta_3\mathbf{cos}(2\pi\mathbf{t}) + \beta_4\mathbf{sin}(4\pi\mathbf{t}) + \beta_5\mathbf{cos}(4\pi\mathbf{t}) + \varepsilon(\mathbf{t}), \tag{15.8}$$

where (i, j) are the grid locations, t is the time in months, β_0 is the constant offset, β_2 and β_3 accounts for the annual signals while β_4 and β_5 represents the semi-annual signals and ε the error term. The deseasonalized data, $\mathbf{X}_{Rainfall}$, are characterized as

$$\mathbf{X}_{Rainfall} = \mathbf{Y} - [\beta_1 t + \beta_2 sin(2\pi t) + \beta_3 cos(2\pi t) + \beta_4 sin(4\pi t) + \beta_5 cos(4\pi t)]. \tag{15.9}$$

The strongest signals in rainfall variability over the Lake Volta basin emanate from the annual and semi-annual signals of the data. The MLRA was then used to isolate (deseasonalize) these components from rainfall and TWS. The deseasonalize rainfall data were then standardized using its standard deviation after removing the mean (Chap. 13). This was undertaken to help investigate the relationship between extreme rainfall conditions and the behaviour of discharge at Akosombo and water level variations over the Lake. A least square fit was employed in this study to estimate the trends in

- the time series of averaged residual TWS (after removing the Lake induced water storage) for the period during 2002–2014,
- the temporal patterns of independent modes derived from residual TWS decomposition using the ICA method,
- time series of the lake height variations after removing the long term mean, and

Table 15.2 The classification system of hydrological conditions of the Lake Volta based on the annual anomalies of lake water level and stream flow records of the Akosombo station, White Volta and Oti rivers

Description	Hydrologic condition class
Very wet	> 1.5
Wet	$+0.5$ to $+1.5$
Average	-0.5 to $+0.5$
Dry	-0.5 to -1.5
Very dry	< -1.5

- the precipitation, evapotranspiration and TWS (before and after removing the Lake induced water storage) datasets.

Furthermore, the annual anomalies of the Lake Volta time series, Akosombo and the White Volta river streamflow records were also assessed. The value of one standard deviation was used to set the limits for the groupings of the hydrological conditions (Table 15.2). This is similar to the range of variability approach (RVA) and standardized precipitation index (Genz and Luz 2012; McKee et al. 1993), which has been discussed in Chap. 11.

15.5.2.3 TWS Changes Induced by Lake Volta Water Level Changes

The Lake Volta TWS change was quantified using satellite altimetry water level changes via the method outlined in Sect. 15.5.2.1. This component was removed from the averaged TWS over the basin to obtain the residual TWS (i.e., $\triangle W_T$ in Eq. 15.7). Linear trends or rates in GRACE-TWS, (Fig. 15.8, top) and GRACE-TWS residual (Fig. 15.8, bottom) were compared. Note that the latter is the residual water storage changes. Assessing these trends will improve our understanding of the proportion of water storage contributed by the Lake to the total observed trend in land water storage over the basin. The water storage over the Volta basin (i.e., GRACE-TWS and GRACE-TWS residual) indicates an increasing trend during the period investigated (see top and bottom panel, Fig. 15.8). Specifically, GRACE-TWS (i.e., ΔTWS) shows a rising trend of 13.96 ± 2.02 mm. This resulted in about 5.83 km^3 of land water storage over the basin (2002–2014), while GRACE-TWS residual (i.e., $\triangle W_T$) shows a monthly increase of 8.15 ± 1.3 mm (i.e., 3.40 km^3 of water). The difference between the two trends provides an indication on the water storage contribution of the Lake. The estimated amount of this contribution is approximately 5.81 mm/month around the vicinity of the Lake. This contribution (i.e., the trend from the Lake's water storage) amounts to approximately 41.6% of the trend in water storage over the basin (2002 2014). Theoretically, the effort to represent the Lake's boundary in a spherical harmonics domain will impact on the amount of surface water storage that can be recovered from the Lake. Nonetheless, the empirical approach employed to

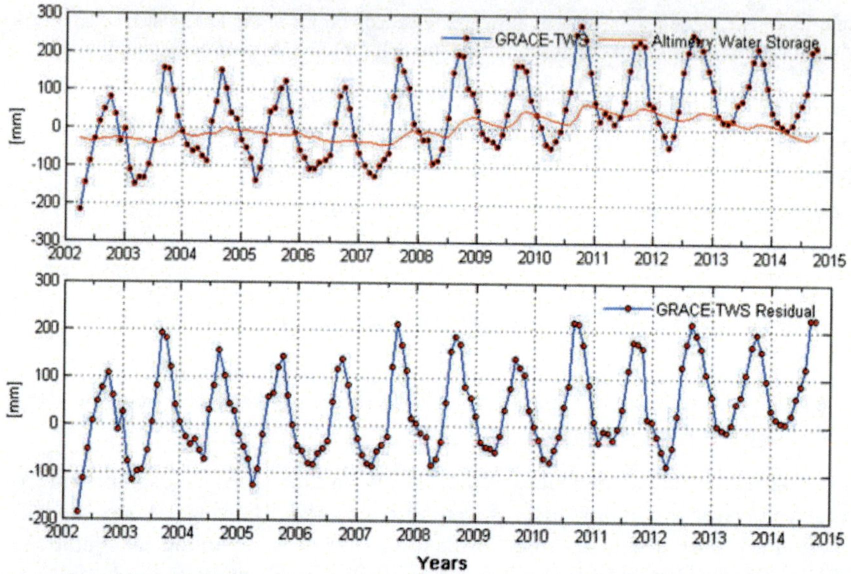

Fig. 15.8 Observed variations in terrestrial water storage (TWS) based on GRACE satellite data. The surface water variability caused by the Lake Volta as observed by satellite altimetry is also shown. Top panel: Temporal variations of GRACE-derived TWS over the Volta basin and altimetry-derived water storage changes over the Lake for the common period (2002–2014). Bottom panel: Temporal series of GRACE-derived TWS over the Volta basin after removing the surface water contributions of the Lake Volta. *Source* Ndehedehe et al. (2017)

removing the Lake induced water storage provides a clue to the natural hydrological variability or behaviour of the Volta basin.

Water storage changes over the Lake (i.e., altimetry water storage) declined during the 2002–2004 and 2006–2008 periods (Fig. 15.8, top). Drought events have been reported during these periods in the region (e.g., Ndehedehe et al. 2016b; Bekoe and Logah 2013). More than that, the residual TWS over the basin is instructive and shows the non-linear hydrological response of the basin (Fig. 15.8, bottom). Notably, decline in precipitation (2001–2002) affected flows, resulting in large reductions in stream flows of the Oti and Volta rivers (White and Black Volta rivers). This reduction in flows in turn affected the water level of the reservoir in 2003. As would be expected, the reduction in stream flows of the White and Black Volta rivers culminated in water deficit in the Lake (i.e., the surface water observed from satellite altimetry) between 2002 and 2004 (Fig. 15.8, top). However, rainfall condition improved in 2003. This impacted on the Volta basin resulting in a much higher amplitude of residual TWS (i.e., GRACE-TWS residual, Fig. 15.8, bottom).

15.5.3 Hydrological Characteristics Based on TWS Changes

Here, the results of applying independent component analyses (ICA) on TWS and residual TWS changes ΔW_T over the basin are discussed further. One of the main goal of these analyses is to improve understanding of the region's terrestrial hydrology. Another aim is to also find a scientific justification for the response of TWS changes to rainfall, especially during periods when rainfall restricted in the basin. After applying the ICA method on TWS, two statistically significant independent modes of TWS variability were identified. Note that the ICA technique was applied on TWS prior to and after removing Lake Volta signal, which is the surface water variability caused by the Akosombo dam. In the first mode, strong spatial variability of TWS over the entire basin was localized in areas like Ivory Coast, Ghana, Togo, and western Burkina Faso (IC1, Fig. 15.9), which is the source of the Black Volta. The second mode showed spatial variability around the Lake area (IC2 Fig. 15.9). IC1 and IC2 of Fig. 15.9 accounted for 96.5% and 1.9% variances, respectively, from the cumulated variance.

The first temporal pattern (IC1) is driven by inter-annual rainfall variability over the entire basin. But IC2 contains much of the hydrological signals around the Lake Volta. The Lake induced TWS (see Sect. 15.5.2.3) showed a correlation of 0.51 with IC2 of Fig. 15.9, indicating that at least 50% of the variability in the observed TWS over the Lake is explained by IC2. Their temporal variations (IC1 and IC2 of Fig. 15.9) indicate an increase of 26.0 ± 4 mm/yr and 2.25 ± 0.5 mm/yr for IC1 and IC2, respectively, when jointly derived from the spatial and temporal patterns. The observed trends in the temporal patterns of TWS (IC1 and IC2,

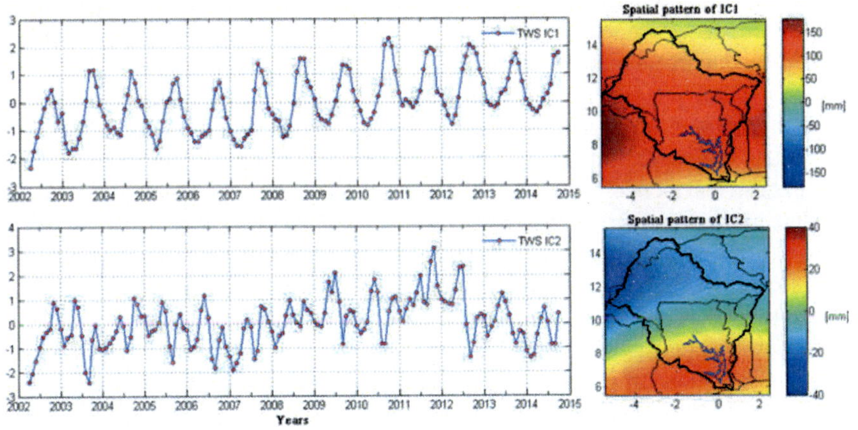

Fig. 15.9 Temporal variations of TWS over the Volta basin (i.e., before removing the Lake Volta signal) after applying the Independent Component Analysis. These temporal patterns are unit less since they have been standardized using their standard deviations. The spatial patterns (right) are scaled using the standard deviation of the computed temporal patterns (left). *Source* Ndehedehe et al. (2017)

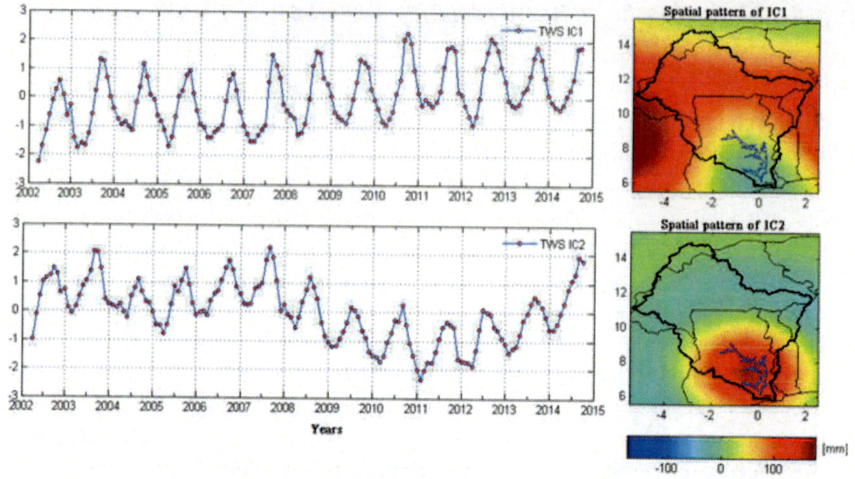

Fig. 15.10 Temporal variations of TWS over the Volta basin (i.e., after removing the Lake Volta induce TWS signal) using the Independent Component Analysis. These temporal patterns are unit less because they have been standardized using their standard deviations. The spatial patterns (right) are scaled using the standard deviation of the temporal patterns (left). *Source* Ndehedehe et al. (2017)

Fig. 15.9) also include the unquantified water storage in wetlands and smaller coastal rivers, which include Chi-Nakwa, Ochi Amissah, Ayensu, Densu and the Tordzie (see, Barry et al. 2005).

After removing the contribution of Lake Volta, the ICA results also show two statistically significant modes of TWS change over the basin. Here, the first independent mode shows spatial and temporal patterns of TWS localized over Ivory Coast, upper Black Volta catchment and the Oti river catchments. It accounts for about 76.2% of the total variability, and this includes Ghana and Burkina Faso. The second independent mode shows TWS signal localized in the lower Volta catchment and explains about 21.5% of the total variability, and is found in Ghana (Fig. 15.10). The first ICA mode indicates an increase of 24 ± 4 mm/yr in TWS over the basin (IC1, Fig. 15.10). A decline in water storage (59.5 ± 8.5 mm/yr) around the lower Volta between 2007 and 2011, which approximates to a loss of 14.19 km^3 of water in Ghana, is observed in the second ICA mode. This loss is consistent with the declines in precipitation in the lower Volta (Fig. 15.11a) during the same period (2002–2014). Evapotranspiration does not indicate declines within the vicinity of the Lake (Fig. 15.11b). But the trends in the spatial patterns of TWS (Fig. 15.11c) over the basin (statistically significant at 95% confidence interval) agrees with the observed trends in dominant patterns of TWS (IC1, Figs. 15.9 and 15.10). The spatial pattern of IC1 of Fig. 15.10 is a prototype of the estimated residual TWS trend shown in Fig. 15.11d. The lake-induced TWS represented about 20% of the TWS signal of the basin when Figs. 15.9 and 15.10) were compared.

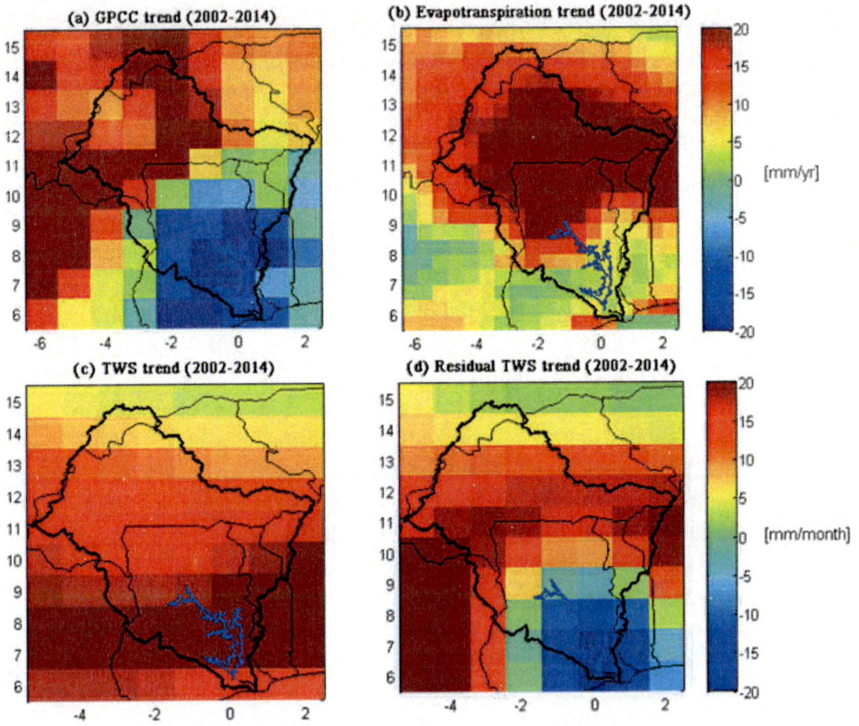

Fig. 15.11 Spatial distribution of trends in water fluxes over the Volta basin (2002–2014). **a** Precipitation, **b** evapotranspiration, **c** GRACE-TWS (i.e., prior to removing the lake induced-TWS), and **d** GRACE-TWS residual (i.e., after removing the lake induced-TWS). *Source* Ndehedehe et al. (2017)

Looking at IC2 of Fig. 15.10 closely, one would see that the TWS change over Lake Volta area revolves around latitudes 9°N and 6°N. This is where the Volta river system is mostly concentrated. Observed variabilities in TWS in the Lake Volta area (IC2, Fig. 15.10) are largely driven not only by rainfall variability, but also by the annual variability of the Black Volta River on the west, White Volta river to the east, and the Oti river on northwestern Benin and Togo, see Fig. 15.5) in Ghana. These rivers have high flows during heavy rainfall periods (e.g., June and July), which culminates in a higher runoff, are the main river systems that drain into Ghana. The loss of terrestrial stored water at the lower Volta basin, which is in Ghana during the 2007–2011 period (IC2, Fig. 15.10) can also be attributed to the loss of freshwater from these main river systems. This loss is the result of low stream flows and base flow recessions due to long term precipitation decline (2002–2014 period). During such period when rainfall is limited, lasting for almost a decade, human intervention like the development and expansion of small- and medium-scale reservoirs in the basin (e.g., Leemhuis et al. 2009) are usually a common response.

There is evidence of increased irrigation schemes and the development of surface water resources, be it small- or medium-scale reservoirs within the basin (e.g., Leemhuis et al. 2009; Andreini et al. 2002). Arguably, the use of surface water resources in riparian regions like Burkina Faso must have contributed to limiting inflow into the lower Volta basin in Ghana, resulting in the observed loss of stored water ΔW_T within that period. With the decline in Lake Volta water level heights since 2011, coupled with the observed loss of water storage over the lower Volta (IC2, Fig. 15.10), the impression is that the Volta basin may be tilting towards drier conditions.

However, it is unclear if TWS decline (2007–2011) around the lower Volta (IC1, Fig. 15.10) is strongly linked to the negative trends in rainfall during the 2002–2014 period (Fig. 15.11a). The hydrology of the lower Volta responds to rainfall conditions in the basin. However, about 1 year lag is noticed in this response. The results in previous sections show relatively drier conditions during the 2011–2013 period (Fig. 15.13) but a recent analysis of droughts over the Volta basin showed that the lower Volta area experienced severe drought conditions during the 2010–2013 period (see, Ndehedehe et al. 2016b). Apparently, rainfall conditions in the western Burkina Faso and the surrounding areas have improved (i.e., showing positive trends-Fig. 15.11a). Although this is rather speculative, the decline in rainfall patterns around the lower Volta (cf. Fig. 15.11a) may account for the decline of TWS indicated in IC2, Fig. 15.10. As will be highlighted further in Sect. 15.5.3.2, the losses in net precipitation over the basin coincides with a decrease in residual water storage of the lower Volta during the 2002–2014 period (IC2, Fig. 15.10). Notably, the hydro-meteorological cycles of the region changes and the loss in water storage changes of the tropical lower Volta can be recovered. This is because the hydrology naturally resets to reflect new conditions as more water becomes available through rainfall or upstream flows.

15.5.3.1 Spatial Patterns of Seasonal Precipitation and Stored Water

Rainfall is predominant in the July–September period (Fig. 15.12, row 1) towards the northern part of the basin (upper Volta). In this region, mean rainfall in northern Ghana and Burkina Faso is about 200 mm while it is relatively higher in Ivory Coast/Mali and Benin. Mean rainfall distribution is more than 200 mm in the upper Volta area only during the July–September period. But around the lower Volta, the distribution of rainfall for all seasons reaches 120 mm (except January–March). Areas with high rainfall (i.e., Fig. 15.12, row 1) coincide with relatively high elevation regions such as Ivory Coast and Benin (cf. Fig. 15.5). Moreover, the basin area shows amplitudes of about 150 mm in the spatial patterns of July–August period (Fig. 15.12, row 2). The spatial patterns of TWS in Côte d' Ivoire (Fig. 15.12, row 2) between July and September appear to be that of Lake Kossou. This is where the hydro-electric power plant is located 16 km upstream of the Bandama river and about 40 km away from Yamoussoukro the administrative capital.

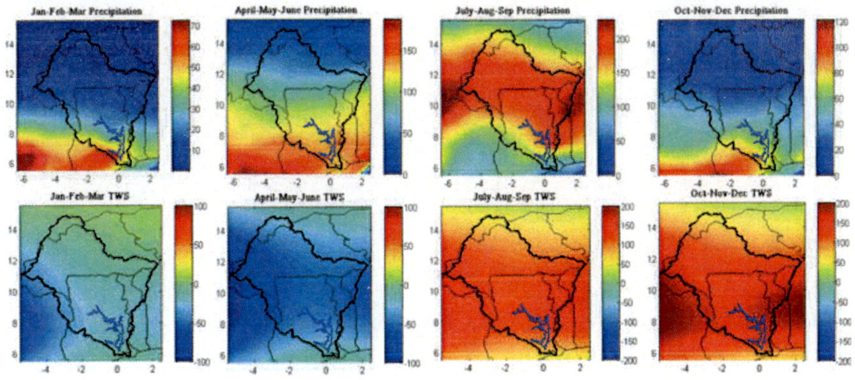

Fig. 15.12 Seasonal rainfall and TWS patterns over the Volta basin and its surroundings. First row: Spatial variability of seasonal precipitation (2002–2014). Second row: Spatial variability of seasonal TWS (2002–2014). *Source* Ndehedehe et al. (2017)

15.5.3.2 Precipitation Impacts on Observed Stream Flow

The confined aquifer system of the basin naturally links the Volta river system with the man-made reservoir (Lake Volta, cf. Fig. 15.5). As shown in Fig. 15.13 (top), relatively dry and wet years are obvious and changes in the lake water levels and flows responded to these wet and dry episodes except for 2003 and 2007 (Fig. 15.13 top). Even though 2003 was generally a wet year, observed flows at Akosombo and the altimetry water levels over Lake Volta indicate extreme negative anomalies, emphasizing a dry condition. Arguably, the impact extreme dry years of 2001 and 2002 prevailed upon the wet period of 2003. This resulted in strong negative anomalies in river discharge and water level variations as observed from satellite altimetry. Arguably, after a wet period, considerable stream flow from the Volta river system reaches the lake after substantial proportion of precipitation has been absorbed by the dry areas around the lake's catchment, creating a non-linear response.

The late response of reservoirs and surface water to extreme rainfall conditions was earlier reported by Ndehedehe et al. (2016b). They showed that drought metrics like standardized precipitation index (SPI) and standardized runoff index (SRI) were inconsistent in terms of their behaviour in observed wet years over the Volta basin. This presupposes a non-linear relationship that demonstrates the phase lag between long term river discharge and precipitation, especially after a previous extreme dry period. The decline in the lake water levels between 2014 and September 2015 can be seen as a direct response to the dry condition in late 2013. But the fall in the lake's water level between 2011 and 2013 could be due to precipitation deficits of 2011 (Fig. 15.13, top). A 12-month lag between rainfall and the flow of surface water from the upper Volta basin to the Lake is assumed. Even though the pronounced peaks of water storage changes (i.e., TWS and residual TWS) over the basin in some years (2003, late 2007, and 2010) coincide with those of rainfall anomalies (Fig. 15.13, middle and bottom panels), net precipitation does not capture these

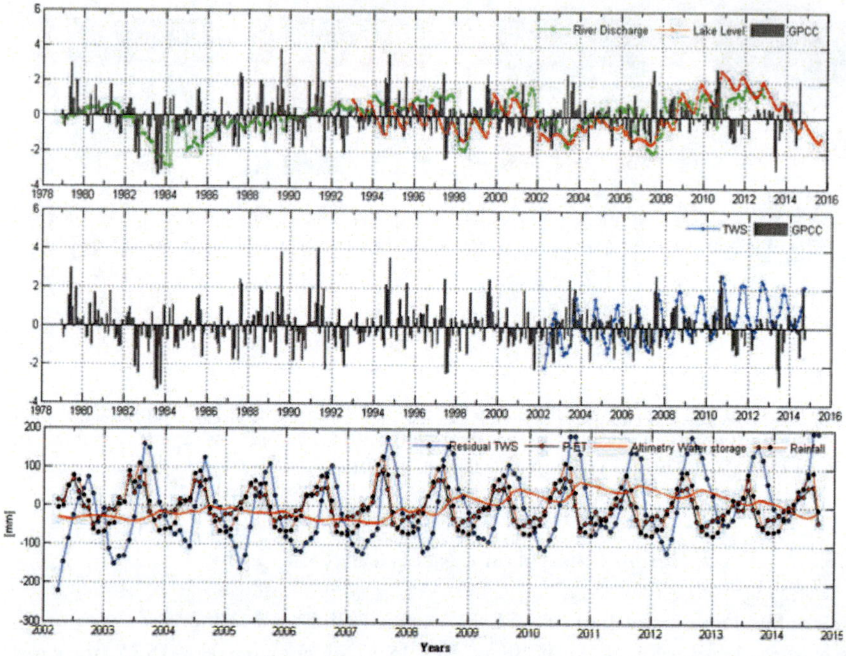

Fig. 15.13 Relationship between extreme precipitation conditions and in-situ river discharge and altimetry as observed from satellite. The annual and semi-annual components of precipitation have been removed and then standardized using its standard deviation. Top: The temporal relationship of deseasonalize rainfall (1979—2014) with standardized anomalies of river discharge (Akosombo, 1979—2012) and water level (1993—2016) from satellite altimetry. Middle: Temporal relationship between deseasonalize rainfall and standardized anomalies of TWS over the Volta basin. Bottom: Time series of net precipitation, residual TWS, altimetry water storage (lake-induced TWS) and rainfall anomalies during the GRACE period (2002–2014). *Source* Ndehedehe et al. (2017)

hydrological periods. Instead, a net precipitation (i.e., $P - ET$) decline of -15 mm/yr was observed during the period (Fig. 15.13, bottom). This coincided with the losses in residual water storage in the lower Volta area (2007–2011) (IC2, Fig. 15.10), as well as the negative trends in precipitation patterns reported in Sect. 15.5.3.

The hydro-power generation of 2003 was low compared to other years. In an earlier study (Bekoe and Logah 2013), this was linked to the impacts of a previous drought condition (i.e., in 2001) and below average rainfall of the following here (2002). The extreme low anomalies of river discharge and altimetry water levels in late 2007 (Fig. 15.13, top) are inconsistent with rainfall in the same period. This pattern demonstrates the general physical hydrological phenomenon within the basin. That is, wet periods are preceded by low water level in the Lake, due to limited rainfall. Overall, this behaviour can be seen in several wet years (1994, 1999, 2003, 2007, and 2010) in the basin. Notably, these years are preceded by dry periods (Fig. 15.13, top).

Moreover, TWS is largely driven by rainfall over the Volta basin as their relationship appears to be consistent (Fig. 15.13, middle). The drier conditions of 2013 reflect in the extreme negative anomalies of rainfall in the mid-year. But this tends to be inconsistent with observed TWS change (Fig. 15.13, middle). No potential cause regarding this inconsistency was identified but further discussions on the cumulative annual rainfall anomalies and TWS have been provided in Sect. 15.5.3.3.

15.5.3.3 Freshwater Systems

The natural variability of rivers in the Volta river system has been compared with flow records of Akosombo and Kpong dams (Fig. 15.14a). These dams manage the reservoir outflow for hydro-power operations. From visual inspection, the White Volta and the Oti rivers seem to have patterns that are more related to the flows from Akosombo and Kpong facilities (hereafter called dam flows), that is in terms of their lowest minimum observed peaks (lowest minimum peaks of the White Volta and Oti rivers impact on dam flows). There are periods (1983/1984, 1997/1998, and 2006/2007) in the basin, which were characterized by hydrological droughts and water shortage deficits. Water storage in the Lake was affected by these drought years and the resulting low flows (Fig. 15.14a) during the same periods are consistent with the observed precipitation anomalies (Fig. 15.14b).

The Volta river system is not that complicated as it shows a linear response to rainfall. The White Volta (31%) and Black Volta (34%) rivers account for higher variabilities in rainfall as opposed to the Oti river (Table 15.3). The Oti river, nonetheless, is well associated with rainfall over the basin at 2 months lag. The Black Volta is a far larger catchment nourishing Burkina Faso and other catchments outside Ghana.

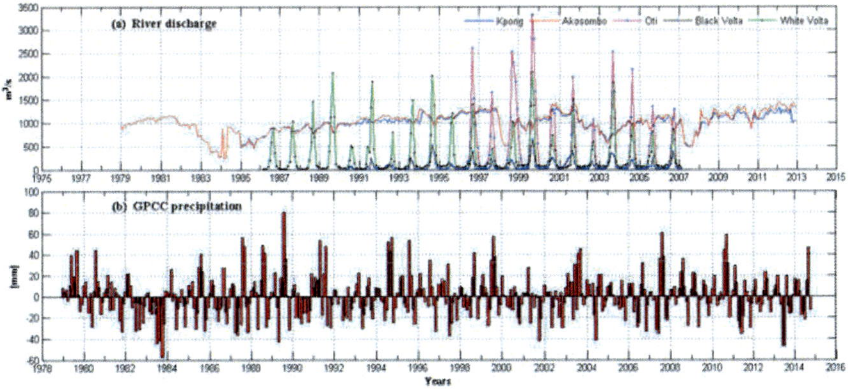

Fig. 15.14 Comparing precipitation patterns with stream flow records of the Volta river system to assess how rainfall influences surface water hydrology. **a** Time series of observed river discharge at the Kpong/Akosombo, Black Volta, White Volta and Oti rivers. **b** Time series of precipitation anomalies over the Volta basin. *Source* Ndehedehe et al. (2017)

Table 15.3 The relationship between precipitation in the Volta basin and the Volta river system (i.e., White Volta, Black Volta, and Oti rivers). Correlations (r) are statistically significant at 95% confidence interval. Trends marked with * are statistically insignificant (95% confidence interval). The maximum r is with respect to the lags indicated. *Source* Ndehedehe et al. (2017)

River	Correlation (r)	Maximum r	R^2	Lag	Period	Trends ($m^3/s/yr$)
White Volta	0.56	0.75	0.31	1 month	1986/03 – 2007/02	−6.96*
Oti	0.38	0.80	0.14	2 months	1996/03–2007/02	−380.24
Black Volta	0.58	0.80	0.34	1 month	1990/03 – 2007/02	27.6*

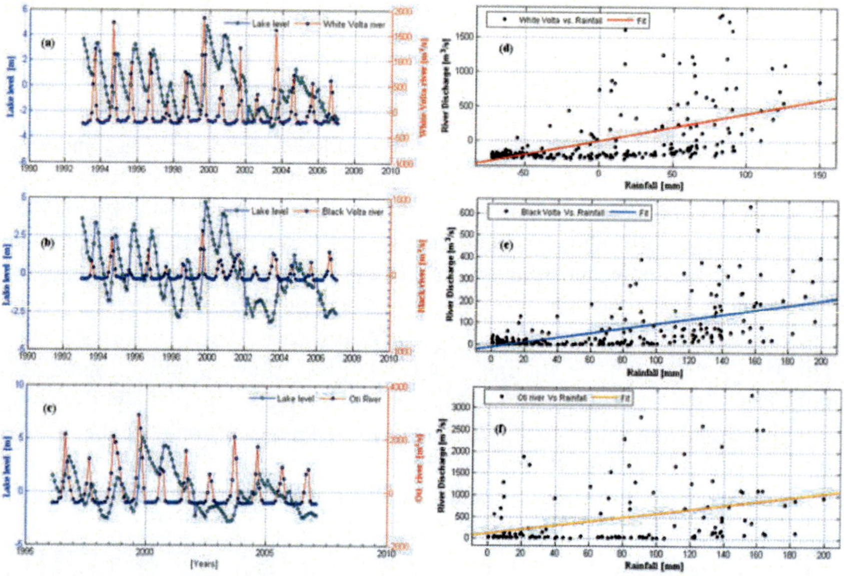

Fig. 15.15 Relationship between precipitation in the Volta basin and the Lake Volta/the Volta river system. **a–c** Temporal variations of Lake Volta water level anomalies after the mean has been removed and discharge anomalies of the White Volta (1993–2007), Black Volta (1990/03–2007/02), and Oti (1996/03–2007/02) rivers. **d–f** Relationship between rainfall with the stream flows of White Volta (1986–2007/02), Black Volta (1990/03–2007/02), and Oti (1996/03–2007/02) rivers. *Source*, Ndehedehe et al. (2017)

But the White Volta and Oti rivers tend to contribute higher proportion of freshwater to the Lake Volta (Figs. 15.14a and 15.15a–c). Of all the water flowing into Lake Volta, about 67% comes from outside Ghana through the Volta river system (Andreini et al. 2002). The Oti river provides the highest contribution (32%) to the Lake. This could explain the observed reduced nourishment and flows into the Lake during the 1996–2007 period (Sect. 15.5.4), and is consistent with the declines of 380 and 6.96 $m^3/s/yr$ in the Oti and White Volta river flows (Table 15.3), respectively. The Oti river accounted for only 14% of the changes in rainfall (i.e., $R^2 = 0.14$). This is very much like the White and Black Volta rivers which explained higher proportion of variabilities (see Table 15.3 and Fig. 15.15d–f). But stronger extremes in Oti river flow (e.g., 2003) are also sometimes inconsistent with observed water levels in Lake Volta (Fig. 15.15c). The results here show that the dam flow records are largely reflective of the Lake's water level, which, in turn, relies on the Volta river system.

15.5.4 *Hydropower Reservoir System of the Lake Volta and the Impact of Climate Variability*

The construction of dams and land-use change contributes to non-stationarity in surface water hydrology (e.g., Ngom et al. 2016; Descroix et al. 2009; Leblanc et al. 2008). But they also amplify natural climate changes, environmental conditions (e.g., increase in Hortonian runoff due to changes in infiltration capacity), and the impacts of atmospheric circulations and low-frequency internal variability. This is what was observed in the Volta basin, where the flow records of the Volta river downstream Akosombo during the 1936–1966 period indicated natural variability, consistent with precipitation changes in the pre-dam construction era. Comparatively, this is very much unlike the post dam era, where the generation of flows at Akosombo are the direct impacts of water management strategy. When the river at Akosombo was dammed, large changes in the hydrological regimes of the river were observed. But, the impact of small changes in precipitation also contributed to tremendous changes in runoff (Oguntunde et al. 2006; Friesen et al. 2005). There are several consequences of damming rivers. This includes decline in annual sediment flux and discharge, reservoir aggradations, and water quality degradation) (see, e.g., Fan et al. 2015; Gyau-Boakye 2001). However, climate variability impacts on water availability in the Lake and the corresponding basin (Ndehedehe et al. 2016b) and the nature of this impact is an important issue for consideration. This section will focus on the mechanisms of extremes in rainfall and how they impact on stream flow. Most importantly, the hydrological condition of Lake Volta is characterized and discussed.

15.5.4.1 Analysis of Extremes on the Volta River System and Lake Volta

The rising trend (1.66 $m^3/s/yr$; $\alpha = 0.05$) in minimum flow records of the White Volta coincides with the increase in rainfall (0.13 mm/yr) during 1986–2006 period (Fig. 15.16a). But the statistically significant observed decline in maximum flow (20.4 $m^3/s/yr$) of the White Volta is in contrast to the decrease observed in maximum rainfall (1.92 mm/yr) during the period (Fig. 15.16b). Comparing the White Volta and Lake Volta water levels (1992–2006), the trends in their minimum quantities (0.56 $m^3/s/yr$ and -0.1 m/yr, respectively) are not statistically significant (Fig. 15.16c). Only Lake Volta showed a negative trend, which was statistically significant (-0.25 m/yr) (Fig. 15.16d) when the trends in their maximum values were estimated. Rain-

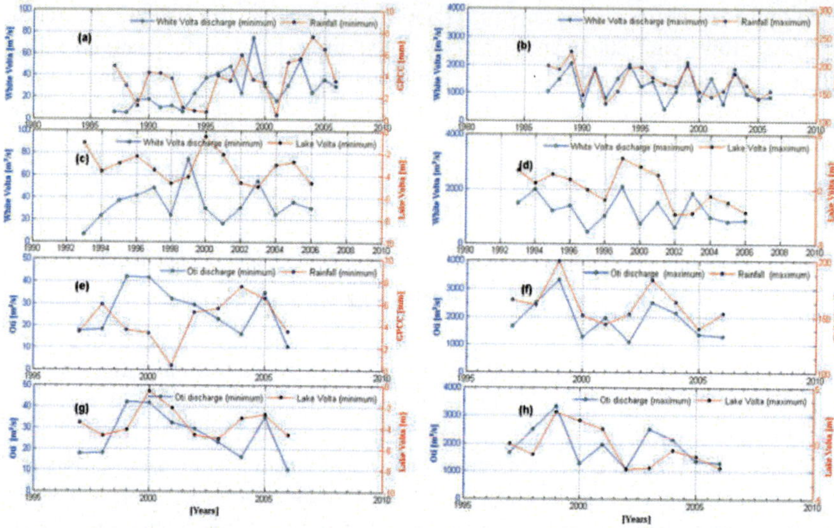

Fig. 15.16 Analysis of extreme annual rainfall (1986–2006), White Volta river discharge (1986–2006), Oti river discharge (1997–2006), and satellite altimetry derived water levels (1993–2006). **a–d** Maximum and minimum White Volta river discharge compared with those of Lake Volta water levels, and precipitation over the Volta basin. **e–h** Maximum and minimum flows from Oti river compared with those of water levels from Lake Volta, and precipitation over the Volta basin. *Source* Ndehedehe et al. (2017)

fall is expected to affect the generation of flows because it is a key input to any hydrological system.

Existing smaller water infrastructures in the basin contribute to water loss. However, evaporation from Lake Volta also contributes to water loss and is three times higher than losses emanating from small- and medium-scale water infrastructures in the basin (Leemhuis et al. 2009). Also, the impact of decline in rainfall and human intervention, which limits the amount of inflow into the Lake, when superimposed on the water loss through evaporation from the Lake's surface are possible reasons that may cause the decline (statistically significant at 95% confidence interval) in maximum lake water levels observed during the period. Changes in the characteristics of evaporation from lakes can substantially impact on the energy and water budgets of hydrological regions. For instance, a net precipitation deficit in the Lake Volta caused by water budget imbalance between precipitation and evaporation was reported in Giesen et al. (2001). Another study (Gyau-Boakye 2001) attributed decline in the Lake's water level to increasing temperatures and diminished runoff from the Volta river system. As shown earlier in Sect. 15.5.3.2, observed decline in net precipitation during the 2002–2014 period coincides with precipitation decline and loss of water storage in the Volta basin. However, the decline in maximum rainfall is consistent with those of White Volta (−41.9 $m^3/s/yr$, statistically insignificant) and Lake levels.

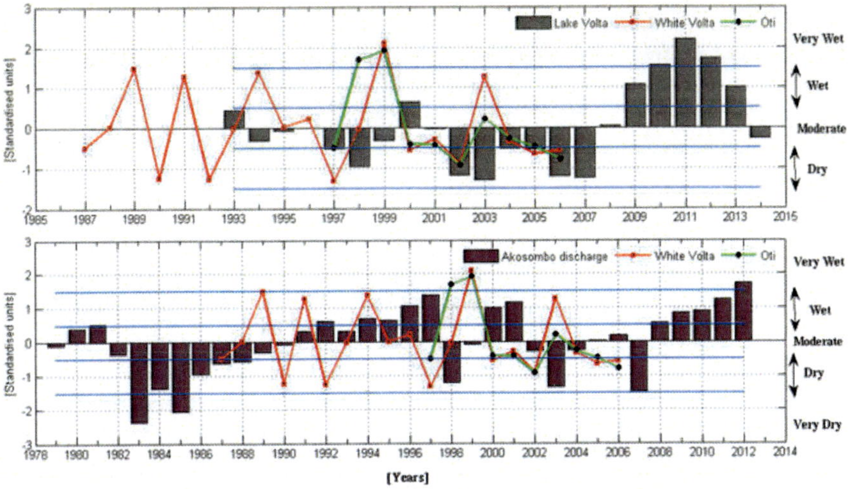

Fig. 15.17 Hydrologic condition of the Lake Volta. Top: Temporal patterns of annual Lake Volta water level changes (1993–2014), the White Volta (1987–2006), and Oti (1997–2006) rivers. Bottom: Temporal patterns of annual Akosombo discharge (1979–2012), White Volta and Oti (1997–2006) rivers. The blue lines are thresholds or limits of each hydrological condition class. *Source* Ndehedehe et al. (2017)

For the Oti river (Fig. 15.16e–h), changes in maximum and minimum river discharge (-88.1 m^3/s/yr and -0.82 m/yr, respectively) were consistent with those of Lake Volta water levels (-0.33 and -0.04 m/yr for maximum and minimum estimates, respectively) and rainfall (only the maximum estimate indicated a decline of -2.0 mm/yr while the minimum estimate was 0.22 mm/yr). Correlation ($r = -0.42$) between minimum White Volta river and the Lake Volta water level is negative (Fig. 15.16c) while the Oti river showed positive correlation of 0.53 with Lake Volta water level (Fig. 15.16g). The maximum flows of these two rivers, however, showed positive correlations of 0.38 and 0.38 with the Lake (Fig. 15.16d, h). These relationships coincide with their strong positive correlations ($r = 0.69$ White Volta river) and (0.81 Oti river) with rainfall (Fig. 15.16b, f) during the periods considered. Overall, maximum correlations ranging from 0.79−0.94 between the two rivers and rainfall show a one year lag and less (that is, based on their minimum and maximum quantities). This is consistent with the results presented in previous sections. This relationship is useful in assessing critical changes in hydrological regime resulting from climate variations. Furthermore, the Lake Volta appears to be consistently dry between 2001 and 2007 (Fig. 15.17, top) compared to the 2009–2013 period that is wet. The dry condition of the 2002–2007 period coincides with the water storage deficit over the Lake during the same period (Fig. 15.8, top). The Lake's water storage, which is marginally affected by small water infrastructures during long dry periods, naturally affects the generation of flows downstream Akosombo (Fig. 15.17, bottom).

15.6 Conclusions

Changes in terrestrial water storage are largely influenced by two main factors—climate change and anthropogenic influence. The severity of the latter is driven by man-made activities (e.g., construction of dams, water transfer, land-use change, etc.) and population growth. The construction of dams for water storage aimed at generating hydro-power or irrigation agriculture coupled with water withdrawals for industrial and domestic purposes are all human-induced developments that have impacted on natural variability of terrestrial hydrology. Lakes Volta and Victoria are good examples of where these activities directly or indirectly affect natural water balance in the surrounding terrestrial ecosystem. The impacts of large water projects like the Akosombo dam in the Volta river basin and other anthropogenic influences have contributed to complicating knowledge of natural hydrological variability and the influence of climate variations on the basin's freshwater systems. In this chapter, we have shown the potential of multi-satellite and in-situ data in assessing surface water hydrology. An important outcome of this chapter is demonstrating how advanced statistical methods like the independent component analysis can be combined with other computational methods and frameworks using a suite of these satellite observations to investigate hydrological variability and quantify the impacts of dams and water infrastructures on water storage changes. The Lake Volta basin was used as testbed in this regard. A two-step procedure that integrated weighted least square formulation of global spherical harmonic analysis and advanced statistical decomposition method was employed to support this assessment. Despite being relatively small in extent, the Lake Volta shows very large fluctuations in surface water, contributing almost half (41.6%) of the observed increase in water storage changes to the Volta basin hydrology. But the two-step procedure revealed the substantial loss of land water storage (59.5 ± 8.5 mm/yr) during the period studied after the contributions of Lake Volta water storage to the basin's hydrology were removed. Moreover, this chapter also identified several impacts resulting from large water resources development. For instance, alteration in the inter-annual variability of river discharge and sediment load of the Volta river in the pre- and post-Akosombo dam construction was noted. Even though post-dam-related decline in annual flows was reasonably moderate, the effect of the Akosombo dam on sediment led to substantially high reduction.

References

Adel MM (2001) Effect on water resources from upstream water diversion in the Ganges basin. J Environ Qual 30(2):356–368. https://doi.org/10.2134/jeq2001.302356x

Agutu N, Awange J, Zerihun A, Ndehedehe C, Kuhn M, Fukuda Y (2017) Assessing multi-satellite remote sensing, reanalysis, and land surface models' products in characterizing agricultural drought in East Africa. Remote Sens Environ 194:287–302. https://doi.org/10.1016/j.rse.2017.03.041

Ahmed M, Sultan M, Wahr J, Yan E (2014) The use of GRACE data to monitor natural and anthropogenic induced variations in water availability across Africa. Earth-Sci Rev 136:289–300. https://doi.org/10.1016/j.earscirev.2014.05.009

Ahmed M, Wiese DN (2019) Short-term trends in Africa's freshwater resources: Rates and drivers. Sci Total Environ 695:133843. https://doi.org/10.1016/j.scitotenv.2019.133843

Alam S, Gebremichael M, Li R, Dozier J, Lettenmaier DP (2019) Climate change impacts on groundwater storage in the central valley California. Clim Change 157(3):387–406. https://doi.org/10.1007/s10584-019-02585-5

Allison G, Cook P, Barnett S, Walker G, Jolly I, Hughes M (1990) Land clearance and river salinisation in the western Murray Basin Australia. J Hydrol 119(1):1–20. https://doi.org/10.1016/0022-1694(90)90030-2

Amenuvor M, Gao W, Li D, Shao D (2020) Effects of dam regulation on the hydrological alteration and morphological evolution of the Volta River Delta. Water 12(3). https://doi.org/10.3390/w12030646

ANA (2017). Brazilian water resources report 2017. National Water Agency, Brazil

Andreini M, Vlek P, Giesen NVD (2002). Water sharing in the Volta basin. FRIEND. In: Proceedings of the Fourth International FRIEND Conference, South Africa, pp 329–335

Arnell NW (2003) Effects of IPCC SRES* emissions scenarios on river runoff: a global perspective. Hydrol Earth Syst Sci 7(5):619–641 https://doi.org/10.5194/hess-7-619-2003

Asif M, Muneer T (2007) Energy supply, its demand and security issues for developed and emerging economies. Renew Sustain Energy Rev 11(7):1388–1413. https://doi.org/10.1016/j.rser.2005.12.004

Awange J, Anyah R, Agola N, Forootan E, Omondi P (2013) Potential impacts of climate and environmental change on the stored water of Lake Victoria Basin and economic implications. Water Res Res 49(0):8160–8173. https://doi.org/10.1002/2013WR014350

Awange J, Ogalo L, Bae K, e. a. (2008) Falling lake victoria water levels: is climate a contributing factor? Clim Change 89:281–297. https://doi.org/10.1007/s10584-008-9409-x

Awange JL, Mpelasoka F, Goncalves RM (2016) When every drop counts: analysis of droughts in Brazil for the 1901–2013 period. Sci Total Environ 566–567:1472–1488. https://doi.org/10.1016/j.scitotenv.2016.06.031

Barry B, Obuobie E, Andreini M, Andah W, Pluquet M (2005) Comprehensive assessment of water management in agriculture: Comparative study of river basin development and management. International Water Management Institute. Retrieved from www.iwmi.cgiar.org/assessment, Accessed 11 November, 2015

Becker A, Finger P, Meyer-Christoffer A, Rudolf B, Schamm K, Schneider U, Ziese M (2013) A description of the global land-surface precipitation data products of the Global Precipitation Climatology Centre with sample applications including centennial (trend) analysis from 1901 to present. Earth Syst Sci Data 5(1):71–99. https://doi.org/10.5194/essd-5-71-2013

Bekoe EO, Logah FY (2013) The impact of droughts and climate change on electricity generation in Ghana. Environ Sci 1(1):13–24

Belmecheri S, Babst F, Wahl ER, Stahle DW, Trouet V (2016) Multi-century evaluation of sierra Nevada snowpack. Nat Clim Chang 6(1):2–3. https://doi.org/10.1038/nclimate2809

Best J (2019). Anthropogenic stresses on the world's big rivers. Nat Geosci 12(1):7–21

Bierkens MFP, Wada Y (2019) Non-renewable groundwater use and groundwater depletion: a review. Environ Res Lett 16(4):063002. https://doi.org/10.1088/1748-9326/ab1a5f

Boateng I, Bray M, Hooke J (2012) Estimating the fluvial sediment input to the coastal sediment budget: a case study of Ghana. Geomorphology 138(1):100–110. https://doi.org/10.1016/j.geomorph.2011.08.028

Boone A, Decharme B, Guichard F et al (2009) The AMMA Land Surface Model Intercomparison Project (ALMIP). Bull Am Meteorol Soc 90(12):1865–1880. https://doi.org/10.1175/2009BAMS27/86.1

Brath A, Montanari A, Moretti G (2006) Assessing the effect on flood frequency of land use change via hydrological simulation (with uncertainty). J Hydrol 324(1):141–153. https://doi.org/10.1016/j.jhydrol.2005.10.001

Bunn SE, Thoms MC, Hamilton SK, Capon SJ (2006) Flow variability in dryland rivers: boom, bust and the bits in between. River Res Appl 22(2):179–186. https://doi.org/10.1002/rra.904

Cardoso JF (1999) High-order contrasts for independent component analysis. Neural Comput 11:157–192

Cardoso JF, Souloumiac A (1993) Blind beamforming for non-gaussian signals. IEE Proc 140(6):362–370

Chao BF, Wu YH, Li YS (2008) Impact of artificial reservoir water impoundment on global sea level. Science 320(5873):212–214. https://doi.org/10.1126/science.1154580

Chaussard E, Wdowinski S, Cabral-Cano E, Amelung F (2014) Land subsidence in central Mexico detected by ALOS InSAR time-series. Remote Sens Environ 140:94–106. https://doi.org/10.1016/j.rse.2013.08.038

Christian-Smith J, Levy M, Gleick PH, Ross N, Luu P (2011) Impacts of the California drought from 2007 to 2009. Oakland, CA: Pacific Institute, pp 59–68. Accessed from on 8th Aug 2021. https://pacinst.org/wpcontent/uploads/2013/02/ca_drought_impacts_full_report2.pdf

Coe MT, Foley JA (2001) Human and natural impacts on the water resources of the Lake Chad basin. J Geophys Res 106(D4):3349–3356. https://doi.org/10.1029/2000JD900587

Common P (1994) Independent component analysis, a new concept? Signal Process 36:287–314

Cooley SW, Ryan JC, Smith LC (2021) Human alteration of global surface water storage variability. Nature 591(7848):78–81. https://doi.org/10.1038/s41586-021-03262-3

Cooley H, Donnelly K, Phurisamban R, Subramanian M (2015) Impacts of California's ongoing drought: agriculture. Oakland, CA, USA, Pacific Institute

Dai A, Qian T, Trenberth KE, Milliman JD (2009) Changes in continental freshwater discharge from 1948 to 2004. J Clim 22(10):2773–2792. https://doi.org/10.1175/2008JCLI2592.1

Dalin C, Wada Y, Kastner T, Puma MJ (2017) Groundwater depletion embedded in international food trade. Nature 543(7647):700–704. https://doi.org/10.1007/s11356-019-04414-9

Deemer BR, Harrison JA, Li S, Beaulieu JJ et al (2016) Greenhouse gas emissions from reservoir water surfaces: a new global synthesis. Bioscience 66(11):949–964. https://doi.org/10.1093/biosci/biw117

Degu AM, Hossain F et al (2011) The influence of large dams on surrounding climate and precipitation patterns. Geophys Res Lett 38(4). https://doi.org/10.1029/2010GL046482

Descroix L, Mahé G, Lebel T, Favreau G, Galle S, Gautier E et al (2009) Spatio-temporal variability of hydrological regimes around the boundaries between Sahelian and Sudanian areas of West Africa: A synthesis. J Hydrol 375(1–2):90–102. https://doi.org/10.1016/j.jhydrol.2008.12.012

Diffenbaugh NS, Swain DL, Touma D (2015) Anthropogenic warming has increased drought risk in California. Proc Natl Acad Sci 112(13):3931–3936. https://doi.org/10.1073/pnas.1422385112

Doll P, Fiedler K, Zhang J (2009) Global-scale analysis of river flow alterations due to water withdrawals and reservoirs. Hydrol Earth Syst Sci 13(12):2413–2432. https://doi.org/10.5194/hess-13-2413-2009

Ehret U, Zehe E, Wulfmeyer V, Warrach-Sagi K, Liebert J (2012) HESS Opinions "Should we apply bias correction to global and regional climate model data? Hydrol Earth Syst Sci 16(9):3391–3404. https://doi.org/10.5194/hess-16-3391-2012

Elliott J, Deryng D et al (2014) Constraints and potentials of future irrigation water availability on agricultural production under climate change. Proc. Natl. Acad. Sci 111(9):3239–3244. https://doi.org/10.1073/pnas.1222474110

Famiglietti JS, Lo M, Ho SL, Bethune J et al (2011) Satellites measure recent rates of groundwater depletion in California's Central Valley. Geophys Res Lett 38(3). https://doi.org/10.1029/2010GL046442

Famiglietti JS (2014) The global groundwater crisis. Nature 4:945–948. https://doi.org/10.1038/nclimate2425

Fan H, He D, Wang H (2015) Environmental consequences of damming the mainstream Lancang-Mekong river: a review. Earth Sci Rev 146:77–91. https://doi.org/10.1016/j.earscirev.2015.03.007

Faunt CC, Hanson R, Belitz K (2009). Groundwater availability of the central valley aquifer, California, professional paper 1766. US Geological Survey. Accessed from on 8th August, 2021. https://www.researchgate.net/profile/Claudia-Faunt/publication/255565139_Groundwater_Availability_of_the_Central_Valley_Aquifer/links/549197390cf269b0486165f4/Groundwater-Availability-of-the-Central-Valley-Aquifer.pdf

Favreau G, Cappelaere B, Massuel S, Leblanc M, Boucher M, Boulain N, Leduc C (2009) Land clearing, climate variability, and water resources increase in semiarid southwest Niger: a review. Water Resour Res 45:W00A16. https://doi.org/10.1029/2007WR006785

Fekete BM, Wisser D, Kroeze C et al (2010) Millennium Ecosystem Assessment scenario drivers (1970–2050): climate and hydrological alterations. Global Biogeochem Cycles 24(4). https://doi.org/10.1029/2009GB003593

Feng W, Zhong M, Lemoine J et al (2013) Evaluation of groundwater depletion in North China using the Gravity Recovery and Climate Experiment (GRACE) data and ground-based measurements. Water Resour Res 49(4):2110–2118. https://doi.org/10.1002/wrcr.20192

Ferreira V, Asiah Z (2015) An investigation on the closure of the water budget methods over Volta Basin using multi-satellite data. In: International Association Geodesy Symposia. https://doi.org/10.1007/1345-2015-137

Ferreira V, Montecino H, Ndehedehe C, Heck B, Gong Z, Westerhaus M, de Freitas S (2018) Space-based observations of crustal deflections for drought characterization in brazil. Sci Total Environ 644:256–273. https://doi.org/10.1016/j.scitotenv.2018.06.277

Friesen J, Andreini M, Andah W, Amisigo MA, Giesen NVD (2005) Storage capacity and long-term water balance of the Volta Basin, West Africa. Proceedings of symposium S6 held during the Seventh IAHS Scientific Assembly, Brazil. Accessed 18 December 2015. http://iahs.info/uploads/dms/13203.2220138-14520Foz20S6-2-1420Friesen.pdf

Fu YC, Zhao J, Wang C, Peng W et al (2018) The virtual Water flow of crops between intraregional and interregional in mainland China. Agri Water Manag 208:204–213. https://doi.org/10.1016/j.agwat.2018.06.023

Gal L, Grippa M, Hiernaux P, Pons L, Kergoat L (2017) The paradoxical evolution of runoff in the pastoral Sahel: analysis of the hydrological changes over the Agoufou watershed (Mali) using the KINEROS-2 model. Hydrol Earth Syst Sci 21(9):4591–4613. https://doi.org/10.5194/hess-21-4591-2017

Gao H, Birkett C, Lettenmaier DP (2012) Global monitoring of large reservoir storage from satellite remote sensing. Water Resour Res 48(9). https://doi.org/10.1029/2012WR012063

Gebremichael M, Krishnamurthy PK, Ghebremichael LT, Alam S (2021) What drives crop land use change during multi-year droughts in California's Central Valley? prices or concern for water? Remote Sens 13(4):650. https://doi.org/10.3390/rs13040650

Genz F, Luz L (2012) Distinguishing the effects of climate on discharge in a tropical river highly impacted by large dams. Hydrol Sci J 57(5):1020–1034. https://doi.org/10.1080/02626667.2012.690880

Getirana A (2016) Extreme water deficit in brazil detected from space. J Hydrometeorol 17(2):591–599. https://doi.org/10.1175/JHM-D-15-0096.1

Getirana A, Jung HC, Hoek JVD, Ndehedehe CE (2020) Hydropower dam operation strongly controls Lake Victoria's freshwater storage variability. Sci Total Environ 726:138343. https://doi.org/10.1016/j.scitotenv.2020.138343

Getirana A, Kumar S, Girotto M, Rodell M (2017) Rivers and floodplains as key components of global terrestrial water storage variability. Geophys Res Lett 44(20):10,359–10,368. https://doi.org/10.1002/2017GL074684

Ghassemi F, White I (2007) Inter-basin water transfer: case studies from Australia, United States, Canada, China and India. Cambridge University Press

Gleick PH (2003) Global freshwater resources: soft-path solutions for the 21st century. Science 302(5650):1524–1528. https://doi.org/10.1126/science.1089967

Giesen NVD, Andreini M, Edig AV, Vlek P (2001) Competition for water resources of the Volta basin. Reg Manag Water Resour (268):199–205. Accessed 12 Nov 2014. http://www.zef.de/fileadmin/template/Glowa/Downloads/

Gillespie BR, Desmet S, Kay P, Tillotson MR, Brown LE (2015) A critical analysis of regulated river ecosystem responses to managed environmental flows from reservoirs. Freshw Biol 60(2):410–425. https://doi.org/10.1111/fwb.12506

Gleeson T, Wada Y, Bierkens MF, Van Beek LP (2012) Water balance of global aquifers revealed by groundwater footprint. Nature 488(7410):197–200. https://doi.org/10.1038/nature11295

Grill G, Lehner B, Thieme M, Geenen B, Tickner D et al (2019) Mapping the world's free-flowing rivers. Nature 569(7755):215–221. https://doi.org/10.1038/s41586-019-1111-9

Gosling SN, Arnell NW (2016) A global assessment of the impact of climate change on water scarcity. Clim Change 134(3):371–385. https://doi.org/10.1007/s10584-013-0853-x

Guzinski R, Kass S, Huber S, Bauer-Gottwein P, e. a. (2014) Enabling the use of earth observation data for integrated water resource management in Africa with the water observation and information system. Remote Sens 6(8):7819–7839. https://doi.org/10.3390/rs6087819

Gyau-Boakye P (2001) Environmental impacts of the Akosombo dam and effects of climate change on the lake levels. Environ Dev Sustain 3(1):17–29. https://doi.org/10.1023/A:1011402116047

Haddeland I, Skaugen T, Lettenmaier DP (2006) Anthropogenic impacts on continental surface water fluxes. Geophys Res Lett 33(8). https://doi.org/10.1029/2006GL026047

Haddeland I, Heinke J, Biemans H et al (2014) Global water resources affected by human interventions and climate change. Proc Natl Acad Sci 111(9):3251–3256. https://doi.org/10.1073/pnas.1222475110

Hamilton S, Munyaho AT, Krach N, Glaser S (2016) Lake Victoria bathymetry contours, V7. Climatic Change. https://doi.org/10.7910/DVN/ZSWZIA

Hagemann S, Chen C, Clark DB et al (2013) Climate change impact on available water resources obtained using multiple global climate and hydrology models. Earth Syst Dyn 4(1):129–144. https://doi.org/10.5194/esd-4-129-2013

Hanak E, Escriva-Bou A, Gray B, Green S, Harter T, Jezdimirovic J, Lund J e. a. (2019) Water and the future of the san joaquin valley. Public Policy Institute of California. Accessed from on 8th August 2021. https://www.ppic.org/wp-content/uploads/water-and-the-future-of-the-san-joaquin-valley-overview.pdf

Hoekstra AY, Mekonnen MM (2012) The water footprint of humanity. PNAS 109(9):3232–3237. https://doi.org/10.1073/pnas.1109936109

Hoekstra AY, Chapagain AK, Mekonnen MM, Aldaya MM (2011) The water footprint assessment manual: setting the global standard. Routledge

Hoff H, Falkenmark M, Gerten D, Gordon L, Karlberg L, Rockström J (2010) Greening the global water system. J Hydrol 384(3):177–186. https://doi.org/10.1016/j.jhydrol.2009.06.026

Huang S, Huang Q, Chang J, Zhu Y, Leng G, Xing L (2015) Drought structure based on a nonparametric multivariate standardized drought index across the Yellow River basin, China. J Hydrol 530:127–136. https://doi.org/10.1016/j.jhydrol.2015.09.042

Hulme M, Doherty R, Ngara T, New M, Lister D (2001) African climate change: 1900-2100. Climate Res 17(2). https://doi.org/10.3354/cr017145

International Commission on Large Dams (2011) World Register of Dams. International Commission on Large Dams

ICOLD (2007) World Register of Dams. International Commission on Large Dams

Jeanne P, Farr TG, Rutqvist J, Vasco DW (2019) Role of agricultural activity on land subsidence in the San Joaquin Valley, California. J Hydrol 569:462–469. https://doi.org/10.1016/j.jhydrol.2018.11.077

Kalu I, Ndehedehe CE, Okwuashi O, Eyoh AE (2021) Assessing freshwater changes over Southern and Central Africa (2002–2017). Remote Sens 13(13). https://doi.org/10.3390/rs13132543

Kang S, Eltahir EA (2018) North China Plain threatened by deadly heatwaves due to climate change and irrigation. Nat Commun 9(1):1–9. https://doi.org/10.1038/s41467-018-05252-y

Koster RD, Dirmeyer PA, Guo Z, Bonan G et al (2004) Regions of strong coupling between soil moisture and precipitation. Science 305(5687):1138–1140. https://doi.org/10.1126/science.1100217

Lammers RB, Shiklomanov AI, Vörösmarty CJ, Fekete BM, Peterson BJ (2001) Assessment of contemporary Arctic river runoff based on observational discharge records. J Geophys Res: Atmos 106(D4):3321–3334. https://doi.org/10.1029/2000JD900444

Lawford R, Bogardi J, Marx S et al (2013) Basin perspectives on the water-energy-food security Nexus. Curr Opin Environ Sustain 5(6):607–616. https://doi.org/10.1016/j.cosust.2013.11.005

Leblanc MJ, Favreau G, Massuel S, Tweed SO, Loireau M, Cappelaere B (2008) Land clearance and hydrological change in the Sahel: SW Niger. Global Planet Change 61:135–150. https://doi.org/10.1016/j.gloplacha.2007.08.011

Leduc C, Bromley J, Schroeter P (1997) Water table fluctuation and recharge in semi-arid climate: some results of the HAPEX-Sahel hydrodynamic survey (Niger). J Hydrol 188(188–189):123–138. https://doi.org/10.1016/S0022-1694(96)03156-3

Lehner B, Liermann CR, Revenga C et al (2011) High-resolution mapping of the world's reservoirs and dams for sustainable river-flow management. Front Ecol Environ 9(9):494–502. https://doi.org/10.1890/100125

Leemhuis C, Jung G, Kasei R, Liebe J (2009) The Volta Basin water allocation system: assessing the impact of small-scale reservoir development on the water resources of the Volta basin, West Africa. Adv Geosci 21:57–62. https://doi.org/10.5194/adgeo-21-57-2009

Lee SH, Yoo SH, Choi JY, Mohtar RH (2017) Evaluation of external virtual water export and dependency through crop trade: an Asian case study. Paddy Water Environ 15(3):525–539. https://doi.org/10.1007/s10333-016-0569-4

Liu J, Cao X, Li B, Yu Z (2018) Analysis of blue and green water consumption at the irrigation district scale. Sustainability 10(2). https://doi.org/10.3390/su10020305

Li H, Zhang Y, Vaze J, Wang B (2012) Separating effects of vegetation change and climate variability using hydrological modelling and sensitivity-based approaches. J Hydrol 420–421:403–418. https://doi.org/10.1016/j.jhydrol.2011.12.033

Li X, Ye SY, Wei AH, Zhou PP, Wang LH (2017) Modelling the response of shallow groundwater levels to combined climate and water-diversion scenarios in Beijing-Tianjin-Hebei Plain, China. Hydrogeol J 25(6):1733–1744. https://doi.org/10.1007/s10040-017-1574-4

Li K, Coe M, Ramankutty N, Jong RD (2007) Modeling the hydrological impact of land-use change in West Africa. J Hydrol 337(3–4):258–268. https://doi.org/10.1016/j.jhydrol.2007.01.038

Liu J, Wang Y et al (2017) A comprehensive analysis of blue water scarcity from the production, consumption, and water transfer perspectives. Ecol Ind 72:870–880. https://doi.org/10.1016/j.ecolind.2016.09.021

Liu J, Li M et al (2020) Influences of the south-to-north water diversion project and virtual water flows on regional water resources considering both water quantity and quality. J Clean Prod 244:118920. https://doi.org/10.1016/j.jclepro.2019.118920

Long D, Pan Y, Zhou J, Chen Y et al (2017) Global analysis of spatiotemporal variability in merged total water storage changes using multiple GRACE products and global hydrological models. Remote Sens Environ 192:198–216. https://doi.org/10.1016/j.rse.2017.02.011

Long D, Yang W, Scanlon B e. a. (2020) South-to-north water diversion stabilizing Beijing' s groundwater levels. Nat Commun 11(3665). https://doi.org/10.1038/s41467-020-17428-6

Mahé G, Olivry J-C (1999) Assessment of freshwater yields to the ocean along the intertropical Atlantic coast of Africa (1951–1989). C R Acad Sci Ser IIA Earth Planet Sci 328(9):621–626. https://doi.org/10.1016/S1251-8050(99)80159-1

Mallya G, Zhao L, Song XC, Niyogi D, Govindaraju RS (2013) 2012 midwest drought in the United States. J Hydrol Eng 18(7):737–745. https://doi.org/10.1061/(ASCE)HE.1943-5584.0000786

Mahé G, Paturel J-E (2009) 1896–2006 Sahelian annual rainfall variability and runoff increase of Sahelian Rivers. CR Geosci 341(7):538–546. https://doi.org/10.1016/j.crte.2009.05.002

Marengo JA, Alves LM, Alvala RC, Cunha AP, Brito S, Moraes OL (2018) Climatic characteristics of the 2010–2016 drought in the semiarid Northeast Brazil region. Anais da Academia Brasileira de Ciencias 90:1973–1985. https://doi.org/10.1590/0001-3765201720170206

Margat J, Van der Gun J (2013). Groundwater around the world: a geographic synopsis. CRC Press, p 376

Maxwell D, Fitzpatrick M (2012) The 2011 Somalia famine: Context, causes, and complications. Glob Food Sec 1(1):5–12. https://doi.org/10.1016/j.gfs.2012.07.002

McKee TB, Doeskin NJ, Kieist J (1993) The relationship of drought frequency and duration to time scales. In: Conference on applied climatology, American meteorological society, Boston, Massachusetts, pp. 179–184. Accessed 27 June 2014. www.ccc.atmos.colostate.edu/relationshipofdroughtfrequency.pdf

McNally A, Arsenault K, Kumar S, Shukla S et al (2017) A land data assimilation system for sub-Saharan Africa food and water security applications. Sci data 4(1):1–19. https://doi.org/10.1038/sdata.2017.12

Medellín-Azuara J, MacEwan D, Howitt RE, Koruakos G, Dogrul EC, Brush CF, TN, K., T, H., F, M., and Lund, J. R. (2015) Hydro-economic analysis of groundwater pumping for irrigated agriculture in California's Central Valley, USA. Hydrogeol J 23(6):1205–1216. https://doi.org/10.1007/s10040-015-1283-9

Medellín-Azuara J, MacEwan D, Howitt RE, Sumner DA, Lund JR, Scheer J, R, G., Q, H., ND, A., Arnold B (2016) Economic analysis of the 2016 California drought on agriculture. Center for Watershed Sci. Accessed from on 9th August, 2021. https://watershed.ucdavis.edu/files/biblio/DroughtReport_20160812.pdf)

Mekonnen MM, Hoekstra AY (2018) Global anthropogenic phosphorus loads to freshwater and associated grey water footprints and water pollution levels: a high-resolution global study. Water Resour Res 54(1):345–358. https://doi.org/10.1002/2017WR020448

Mohammed S, Darwish M (2018) Water footprint and virtual water trade in Qatar. Desalin Water Treat 117–132. Accessed from on 9th August 2021. https://gggi.org/site/assets/uploads/2019/06/Water_footprint_and_virtual_water_trade.pdf

Moore P, Williams SDP (2014) Integration of altimetry lake lavels and GRACE gravimetry over Africa: inferences for terrestrial water storage change 2003–2011. Water Resour Res 50:9696–9720. https://doi.org/10.1002/2014WR015506

Mulligan M, van Soesbergen A, Sáenz L (2020). GOODD, a global dataset of more than 38,000 georeferenced dams. Sci Data 7(1):1–8. https://doi.org/10.1038/s41597-020-0362-5

Mu Q, Zhao M, Running SW (2011) Improvements to a MODIS global terrestrial evapotranspiration algorithm. Remote Sens Environ 115(8):1781–1800. https://doi.org/10.1016/j.rse.2011.02.019

Muller M (2008) Lessons from Cape Town's drought. Nature 559:174–176

Ndehedehe C, Awange J, Agutu N, Kuhn M, Heck B (2016) Understanding changes in terrestrial water storage over West Africa between 2002 and 2014. Adv Water Resour 88:211–230. https://doi.org/10.1016/j.advwatres.2015.12.009

Ndehedehe CE (2019) The water resources of tropical West Africa: problems, progress and prospect. Acta Geophys 67(2):621–649. https://doi.org/10.1007/s11600-019-00260-y

Ndehedehe CE, Agutu N, Ferreira VG, Getirana A (2020) Evolutionary drought patterns over the Sahel and their teleconnections with low frequency climate oscillations. Atmos Res 233:104700. https://doi.org/10.1016/j.atmosres.2019.104700

Ndehedehe CE, Agutu NO, Okwuashi O (2018) Is terrestrial water storage a useful indicator in assessing the impacts of climate variability on crop yield in semi-arid ecosystems? Ecol Ind 88C:51–62. https://doi.org/10.1016/j.ecolind.2018.01.026

Ndehedehe CE, Awange J, Kuhn M, Agutu N, Fukuda Y (2017) Analysis of hydrological variability over the Volta river basin using in-situ data and satellite observations. J Hydrol: Reg Stud 12:88–110. https://doi.org/10.1016/j.ejrh.2017.04.005

Ndehedehe CE, Awange JL, Corner R, Kuhn M, Okwuashi O (2016) On the potentials of multiple climate variables in assessing the spatio-temporal characteristics of hydrological droughts over the Volta Basin. Sci Total Environ 557–558:819–837. https://doi.org/10.1016/j.scitotenv.2016.03.004

Ndehedehe CE, Ferreira VG, Onojeghuo AO, Agutu NO, Emengini E, Getirana A (2020) Influence of global climate on freshwater changes in Africa's largest endorheic basin using multi-scaled indicators. Sci Total Environ 737:139643. https://doi.org/10.1016/j.scitotenv.2020.139643

Ngom F, Tweed S, Bader J-C, Saos J-L, Malou R, Leduc C, Leblanc M (2016) Rapid evolution of water resources in the Senegal delta. Global Planet Change 144:34–47. https://doi.org/10.1016/j.gloplacha.2016.07.002

Oguntunde PG, Friesen J, van de Giesen N, Savenije HH (2006) Hydroclimatology of the Volta river basin in West Africa: trends and variability from 1901 to 2002. Phys Chem Earth, Parts A/B/C 31(18):1180–1188. https://doi.org/10.1016/j.pce.2006.02.062

Owusu K, Waylen P, Qiu Y (2008) Changing rainfall inputs in the Volta basin: implications for water sharing in Ghana. GeoJournal 71(4):201–210. https://doi.org/10.1007/s10708-008-9156-6

Oyebande L (2001) Water problems in Africa–how can the sciences help? Hydrol Sci J 46(6):947–962. https://doi.org/10.1080/02626660109492888

Pelicice FM, Pompeu PS, Agostinho AA (2015) Large reservoirs as ecological barriers to downstream movements of Neotropical migratory fish. Fish Fish 16(4):697–715. https://doi.org/10.1111/faf.12089

Pokhrel YN, Hanasaki N, Yeh PJ, Yamada TJ, Kanae S, Oki T (2012) Model estimates of sea-level change due to anthropogenic impacts on terrestrial water storage. Nat Geosci 5(6):389–392. https://doi.org/10.1038/ngeo1476

Richey AS, Thomas BF, Lo M-H, Reager JT, Famiglietti JS et al (2015) Quantifying renewable groundwater stress with GRACE. Water Resour Res 51(7):5217–5237. https://doi.org/10.1002/2015WR017349

Rodell M, Famiglietti JS, Wiese DN, Reager JT, Beaudoing HK, Landerer FW, Lo M-H (2018) Emerging trends in global freshwater availability. Nature 557:651–659. https://doi.org/10.1038/s41586-018-0123-1

Rodell M, Velicogna I, Famiglietti JS (2009) Satellite-based estimates of groundwater depletion in India. Nature 460(7258):999–1002. https://doi.org/10.1038/nature08238

Rodell M, Houser PR, Jambor U, Gottschalck J, Mitchell K, K. M. et al (2004) The global land data assimilation system. Bull Am Meteorol Soc 85(3):381–394. https://doi.org/10.1175/BAMS-85-3-381.R

Rodell M, Famiglietti JS, Wiese DN, Reager JT et al (2018) Emerging trends in global freshwater availability. Nature 557(7707):651–659. https://doi.org/10.1038/s41586-018-0123-1

Scanlon BR, Faunt CC, Longuevergne L, Reedy RC, e. a. (2012) Groundwater depletion and sustainability of irrigation in the us high plains and central valley. Proc Natl Acad Sci 109(24):9320–9325. https://doi.org/10.1073/pnas.1200311109

Scanlon BR, Reedy RC, Stonestrom DA, Prudic DE, Dennehy KF (2005) Impact of land use and land cover change on groundwater recharge and quality in the southwestern US. Glob Change Biol 11(10):1577–1593. https://doi.org/10.1111/j.1365-2486.2005.01026.x

Schiermeier Q (2008) Water: purification with a pinch of salt. Nature 452:260–261. https://doi.org/10.1038/452260a

Sharma NP, Damhaug T, Gilgan-Hunt E, Grey D, Okaru V, Rothberg D (1996) African water resources; challenges and opportunities for sustainable development 331

Schneider U, Becker A, Finger P, Meyer-Christoffer A, Ziese M, Rudolf B (2014) GPCC' s new land surface precipitation climatology based on quality-controlled in situ data and its role in quantifying the global water cycle. Theoret Appl Climatol 115(1–2):15–40. https://doi.org/10.1007/s00704-013-0860-x

Séguis L, Cappelaere B, Milési G, Peugeot C, Massuel S, Favreau G (2004) Simulated impacts of climate change and land-clearing on runoff from a small Sahelian catchment. Hydrol Process 18(17):3401–3413. https://doi.org/10.1002/hyp.1503

Sheffield J, Wood EF, Chaney N, Guan K, Sadri S, Yuan X, Olang L, Amani A, Ali A, Demuth S, Ogallo L (2014) A drought monitoring and forecasting system for Sub-Sahara African water resources and food security. Bull Am Meteor Soc 95(6):861–882. https://doi.org/10.1175/BAMS-D-12-00124.1

Siebert S, Burke J, Faures JM, Frenken K, Hoogeveen J et al (2010) Groundwater use for irrigation - a global inventory. Hydrol Earth Syst Sci 14(10):1863–2010. https://doi.org/10.5194/hess-14-1863-2010

Shiklomanov AI, Lammers RB, Vörösmarty CJ (2002) Widespread decline in hydrological monitoring threatens Pan-Arctic Research. EOS Trans Am Geophys Union 83(2):13–17. https://doi.org/10.1029/2002EO000007

Siderius C, Gannon KE, Ndiyoi M, Opere A, Batisani N, Olago D, Pardoe J, Conway D (2018) Hydrological response and complex impact pathways of the 2015/2016 El Niño in Eastern and Southern Africa. Earth's Future 6(1):2–22. https://doi.org/10.1002/2017EF000680

Skaskevych A, Lee J, Jung HC, Bolten J, David JL et al (2019) Application of grace to the estimation of groundwater storage change in a data poor region: a case study of Ngadda catchment in the Lake Chad Basin. Hydrolog Process 0(ja). https://doi.org/10.1002/hyp.13613

Sneeuw N (1994) Global spherical harmonic analysis by least-squares and numerical quadrature methods in historical perspective. Geophys J Int 118(3):707–716. https://doi.org/10.1111/j.1365-246X.1994.tb03995.x

Sneeuw N, Lorenz C, Devaraju B, Tourian M, Riegger J, Kunstmann H, Bárdossy A (2014) Estimating runoff using hydro-geodetic approaches. Surv Geophys 35(6):1333–1359. https://doi.org/10.1007/s10712-014-9300-4

Solander KC, Wilson CJ (2018) The cape town drought: what is happening and will it happen again? Los Alamos National Lab. Accessed from 9th August 2021. https://permalink.lanl.gov/object/tr?what=info:lanl-repo/lareport/LA-UR-18-21239

Su B, Huang J, Fischer T et al (2018) Drought losses in China might double between the 1.5C and 2.0C warming. Proc Natl Acad Sci 115(42): 10600–10605. https://doi.org/10.1073/pnas.1802129115

Tapley B, Bettadpur S, Watkins M, Reigber C (2004) The gravity recovery and climate experiment: mission overview and early results. Geophys Res Lett 31:1–4. https://doi.org/10.1029/2004GL019920

Tapley B, Watkins M, Flechtner F e. a. (2019) Contributions of GRACE to understanding climate change. Nat Climate Change 9:358–369. https://doi.org/10.1038/s41558-019-0456-2

Taylor R, Scanlon B, Döll P et al (2013) Ground water and climate change. Nat Clim Chang 3:322–329. https://doi.org/10.1038/nclimate1744

Trenberth KE, Caron JM, Stepaniak DP, Worley S (2002) Evolution of El-Niño–Southern oscillation and global atmospheric surface temperatures. J Geophys Res: Atmos 107(D8). https://doi.org/10.1029/2000JD000298

United Nations (2015) Transforming our world: the 2030 agenda for sustainable development. United Nations

United Nations (2006) African Water Development Report 2006. United Nations

United Nations (2011) Sustainability and equity: a better future for all (2011 human development report. United Nations

United Nations (2012) The Emissions Gap Report 2012. United Nations. United Nations Environment Programme (UNEP)

United Nations (2011) The energy challenge for achieving the millennium development goals. United Nations 47(1-2):123–138

UN (2015) Transforming our world: the 2030 Agenda for Sustainable Development. New York, USA. United Nations

Usman M, Ndehedehe CE, Farah H, Manzanas R (2021) Impacts of climate change on the streamflow of a large river basin in the Australian tropics using optimally selected climate model outputs. J Clean Prod 315. https://doi.org/10.1016/j.jclepro.2021.128091

Van Loon AF, Stahl K, Di Baldassarre G, Clark J, Rangecroft S, Wanders N, Gleeson T, e. a. (2016) Drought in a human-modified world: reframing drought definitions, understanding, and analysis approaches. Hydrol Earth Syst Sci 20(9):3631–3650. https://doi.org/10.5194/hess-20-3631-2016

Voss KA, Famiglietti JS, Lo M, de Linage C, Rodell M, Swenson SC (2013) Groundwater depletion in the Middle East from GRACE with implications for transboundary water management in the Tigris-Euphrates-Western Iran region. Water Resour Res 49(2):904–914. https://doi.org/10.1002/wrcr.20078

Wang G, Zhang Y, Liu G, Chen L (2006) Impact of land-use change on hydrological processes in the Maying River basin, China. Sci China Ser D: Earth Sci 49(10):1098–1110

The World Bank (2014). Database World Development Indicators 'Electricity production (kWh)'. https://data.worldbank.org/indicator/EG.USE.ELEC.KH.PC

Wu H, Chen J, Xu J, Zeng G et al (2019) Effects of dam construction on biodiversity: a review. J Clean Prod 221:480–489. https://doi.org/10.1016/j.jclepro.2019.03.001

Yang Y, Yin L, Zhang Q (2015) Quantity versus quality in China's South-to-North water diversion project: a system dynamics analysis. Water 7(5):2142–2160. https://doi.org/10.3390/w7052142

Xia J, Wang Q, Zhang X, Wang R, She D (2018) Assessing the influence of climate change and inter-basin water diversion on Haihe River basin, eastern China: a coupled model approach. Hydrogeol J 26(5):1455–1473. https://doi.org/10.1007/s10040-018-1773-7

Yang D, Ye B, Kane DL (2004) Streamflow changes over Siberian Yenisei river basin. J Hydrol 296(1–4):59–80. https://doi.org/10.1016/j.jhydrol.2004.03.017

Yassin F, Razavi S, Elshamy M, Davison B, Sapriza-Azuri G, Wheater H (2019) Representation and improved parameterization of reservoir operation in hydrological and land-surface models. Hydrol Earth Syst Sci 23(9):3735–3764. https://doi.org/10.5194/hess-23-3735-2019

Zhao G, Gao H (2019) Estimating reservoir evaporation losses for the United States: fusing remote sensing and modeling approaches. Remote Sens Environ 226:109–124. https://doi.org/10.1016/j.rse.2019.03.015

Zhang X, Zhang L, Zhao J, Rustomji P, Hairsine P (2008). Responses of streamflow to changes in climate and land use/cover in the Loess Plateau, China. Water Res Res 44(7). https://doi.org/10.1029/2007WR006711

Zhang Y, Li GM (2014) Influence of south-to-north water diversion on major cones of depression in North China Plain. Environ Earth Sci 71(9):3845–3853. https://doi.org/10.1007/s12665-013-2771-7

Zhou T, Nijssen B, Gao H, Lettenmaier DP (2016) The contribution of reservoirs to global land surface water storage variations. J Hydrometeorol 17(1):309–325. https://doi.org/10.1175/JHM-D-15-0002.1

Zhu W, Jia S, Lall U, Cao Q, Mahmood R (2019) Relative contribution of climate variability and human activities on the water loss of the Chari/Logone River discharge into Lake Chad: A conceptual and statistical approach. J Hydrol 569:519–531. https://doi.org/10.1016/j.jhydrol.2018.12.015

Part XI
Impacts of Global Climate Teleconnections

Chapter 16
Droughts and Impacts of Climate Teleconnections

Abstract Apart from inducing unusually high atmospheric CO_2 growth rates in productive regions, and its other far-reaching long-term impacts on ecophysiology and ecosystem dynamics, drought events are the world's most costly natural disasters. The United States, for example, suffered agricultural losses estimated at 40 and 30 billion due to the 1988 and 2012 extreme drought events, respectively. The overarching complexities of drought make it difficult to predict and accurately assess its causes, intensity, duration, magnitude and impacts (e.g., extents) on several spatio-temporal scales. This chapter explores a satellite remote sensing monitoring perspective and a range of multivariate framework and geospatial tools that underpin accurate assessment of the role of global climate teleconnection patterns.

16.1 Understanding Droughts and Its Variability

Drought is difficult to predict, especially with the overarching complexities of several climate and anthropogenic phenomena linked to its causes and magnitude, on several spatio-temporal scales (Ndehedehe et al. 2020a; Kiem et al. 2016). Anthropogenic factors such as large-scale human water withdrawals interfere with the propagation process of drought characteristics and intensity (meteorological to hydrological) (Kubiak-Wójcicka and Bąk 2018). These factors can be complicated by the influence of oceanic phenomena on local precipitation patterns, making drought impacts even more deleterious. To assess and measure drought impacts, various drought monitoring approaches have been implemented for the detection and characterization of historic drought episodes based on several water budgets or hydrological state variables. Some of these variables include soil moisture, rainfall, stream flow, as well as other hydrological variables, and integration of oceanic phenomena and climate teleconnections (Kiem et al. 2016; Sheffield and Wood 2007; McKee et al. 1993). The use of remote sensing techniques, soil moisture from observations, and model simulations have been implemented as drought variables in past studies (Hunt et al. 2009; Sheffield et al. 2009). Generally, the soil surface layer of less than 30 cm responds faster to short-term meteorological episodes like evaporation, snow, and rainfall, while the soil surface layer beyond 30 cm is driven principally by long-term

C. Ndehedehe, *Satellite Remote Sensing of Terrestrial Hydrology*,
https://doi.org/10.1007/978-3-030-99577-5_16

drivers such as moisture reallocation and evapotranspiration seasonal cycle (Hunt et al. 2009).

Drought impacts (e.g., reducing agricultural yield and income, restricting hydropower, etc.) on the environment have been emphasized as a major disaster in our ecosystem and have attracted the attention of a wide range of environmental scientists, especially geologists, meteorologists, environmentalists, ecologists, hydrologists, and agricultural scientists. The occurrence of drought in almost all climatic zones including high and low rainfall areas has been highlighted in past studies (Mishra and Singh 2010), and is majorly related to the reduction in the amount of rainfall received in a specific region over a particular study period. Several factors, which include low relative humidity, temperature fluctuations, high winds, timing and characteristics of rainfall, including the distribution and timing of rainy days during crop growing seasons, duration, and intensity, as well as inception and termination of rainfall, play a significant role in regional and global drought occurrence. Aridity has been characterized as a distinctive and permanent component of climate and is restricted to typically low rainfall areas like the West African Sahel, Middle East, and Australia. This is different from drought, which is a temporary rainfall anomaly or period of rainfall deficits. There is also a distinction between heat wave and drought. A conventional time scale coupled with a heat wave is in the order of a week, whereas a typical drought has the potential of lasting for months or even years. In an event where drought and heat wave effects are combined, it results in severe socio-economic consequences for the region.

Anthropogenic influence in today's society is one direct cause of water scarcity as noticed in several sub-regions. This is noticed in the increased urbanization rate, which is directly linked to population growth, agricultural expansion, and development of the industrial and energy sectors. Other factors which include climate change and water pollution have further promoted sub-regional water scarcity in certain regions. In recent years, drought and flood episodes have shown more severe levels and peaks. The duration between two consecutive water disasters is gradually closing in as pointed out by past studies (Aswathanarayana 2001). Drought impacts on both surface and groundwater resources will result in diminished water supply, low crop yield, depreciating water quality, deteriorated power generation, reduced range productivity, muddled riparian habitats, and interrupted recreation activities, as well as social and economic events in any specific region (Riebsame et al. 2019). Drought impacts also extend to water quality; especially as moderate climatic variations change, the hydrologic systems have a significant effect on lake chemistry (Webster et al. 1996). Organic components, sediment, and nutrients are conveyed to surface waters by runoff—a corridor that is interposed during drought episodes.

16.1.1 *Remote Sensing Applications in Drought Monitoring and Assessment*

Identifying drought propensity, estimation, and probability assessment are imperative for optimal earth science monitoring. With the means to ascertain drought occurrences, it is possible to implement programmes aimed at improving food security. Adequate food-aid delivery and risk management programmes involve proper knowledge of drought occurrence probability. Also, drought information at administrative stages is very crucial for better understanding of the potential effects of regional and global climate change.

One of the major problems of continental drought risk monitoring and assessment is the unavailability of reliable rainfall data, especially in regions like Africa (Ndehedehe et al. 2020a). The coverage of operational weather stations in most regions like Africa is insufficient with enormous spatial gaps, and distinct stations often provide discontinuous data. As a result of these, rainfall observations are usually substituted by data obtained from remotely sensed platforms and atmospheric circulation models. The potential of these observations, including reanalysis, global positioning systems, and several outputs from hydrological models (soil moisture, rainfall) in the characterization of several drought intensities and extreme events have been assessed (Agutu et al. 2020; Ferreira et al. 2018; Agutu et al. 2017; Yirdaw et al. 2008). Models of rainfall datasets for assessing food security include rainfall estimate (RFE) developed by the Climate Prediction Center (CPC) of the National Oceanic and Atmospheric Administration (NOAA), precipitation prognosis of the European Center for Medium Range Weather Forecasts (ECMWF), and the Food and Agricultural Organization rainfall estimates.

16.1.1.1 Satellite Precipitation

Satellite-derived observations from the Tropical Rainfall Measuring Mission (TRMM) and the Tropical Applications of Meteorology using SATellite products developed at Reading University are also popular in assessing sub-regional and continental rainfall schemes. However, regions with deficient rain gauge stations like Africa have been identified as hotspots with substantial amount of errors and deviations in the parameters from the above-mentioned rainfall estimates (e.g., Dinku et al. 2007). Moreover, authentication or better calibration of these rainfall estimates is impeded by the unavailability of rain gauge observations. Validating these products is challenging because of this constraint on the availability and quality of rainfall dataset. As several of these regions are largely rain-fed dependent, lack of in situ observations limits application in food security assessment (Verdin et al. 2005). Meteorological drought assessment to support drought impact on food security and freshwater ecosystems requires access to long-term records of precipitation data on a sufficient spatial scale. Satellite observations of precipitation from NASA hosted in the agroclimatology archive provide the capability for the use of globally available data to undertake such assessment. Such opportunities with a particular focus on evaluating the reliability of this data in detecting meteorological drought using the standardized precipitation index have been explored in Jordan (Al-Kilani 2021). The underlying challenges, e.g., overestimating drought intensity (moderate, mild, and severe droughts) as well as their poor performance in detecting extremely wet condition were observed over Jordan in the study undertaken by Al-Kilani (2021). With the poor representation of land surface conditions and hydrological state variables in some reanalysis data that used satellite precipitation products, the continued evaluation of earth observations and space-borne measurements of rainfall and global climatology is crucial.

16.1.1.2 Vegetation Indices

Given that changes in vegetation are considerably tied to rainfall variability, vegetation indices like the Normalized Difference Vegetation Index (NDVI) or its variant and Enhanced Vegetation Index can be implemented for drought assessment (Ndehedehe et al. 2019b; Agutu et al. 2017). Field studies and experiments from airborne scanner showed that the spectral reflectance properties of vegetation canopies and distinct integrations of red and near-infrared reflectance, often referred to as vegetation indices, are very crucial for assessing green vegetation. Among the various vegetation indices based on these two spectral channels, the NDVI (Deering 1978) is well known and serves as an indicator for assessing vegetation health and crop yield. Agricultural drought has been characterized using vegetation indices. In East Africa where this was demonstrated, the vegetation condition index accounted for high proportions of observed variabilities in annual crop production in Kenya, Tanzania, and Uganda (Agutu et al. 2017). However, be it vegetation, model-derived soil moisture, or meteorological-based indicators, topography influences the performance of these metrics in characterizing the spatial and temporal variability of drought (Agutu et al. 2020). The inconsistencies arising from this characterization in high elevation areas of Upper Greater Horn of Africa have been linked not only to the direct effect of topographical changes but also to gauge density and forcing parameters for model-derived data (Agutu et al. 2020).

However, NDVI has some shortcomings and could be restricted. For instance, the effects of surface anisotropy and soil moisture could be a challenge in the use of NDVI to measure droughts impacts and variability. Besides target anisotropy and soil moisture, other influences such as angular geometry of the sensor radiance and the time of observation are also known to influence NDVI parameters. This is particularly crucial for AVHRR-derived data given that the orbit of the NOAA platforms was susceptible to periodic drift and the 25-year time series is made up of data from various sensors (Tucker et al. 2005). Presently, various time series of satellite-derived vegetation indices at the regional and global scale integrated with a high temporal frequency are accessible. These include Satellite Pour l'Observation de la Terre (SPOT), Advanced Very High Resolution Radiometer (AVHRR), and Moderate Resolution Imaging Spectroradiometer (MODIS). Among these missions,

AVHRR has the longest time series and is also known to provide data from thermal channels. Its sensor has been flown on different NOAA satellites and is responsible for a large number of derived products, and is still capturing even to date.

16.1.1.3 Multi-sensor Remotely Sensed Drought Indicators

Quantifying variations in water cycle constituents and their climatological average parameters is important to monitor and assess drought. Such assessment underpins our understanding of integrated driving factors and consequences of drought and its impacts across several domains. Apart from in situ or gauged observations and hydrological model simulations of key water cycle parameters (rainfall, evapotranspiration, surface runoff, etc.), remotely sensed drought indicators, such as soil moisture observations, and satellite-based meteorological and hydrological drought indices have significantly enhanced the capabilities for large-scale and cheaper drought monitoring and assessment techniques (e.g., Ndehedehe et al. 2020d; Agutu et al. 2017; AghaKouchak 2015). For instance, satellite observations of precipitation have aided the feasibility of operational assessment of global meteorological drought. The use of MODIS satellite observations to create a drought severity index is aimed at integrating vegetation greenness and evapotranspiration fluctuations for drought monitoring and assessment. Drought severity index determined from optical-infrared remote sensing facilitates near real-time drought assessment. However, its applicability is restricted by noise effects from atmospheric aerosols, clouds, or deficient solar radiance and is also limited for catchment-scale drought impacts due to its coarse spatial resolution (500 m). Some of the capabilities of multi-sensor remotely sensed drought monitoring capability are highlighted briefly below.

1. Advancements in microwave remote sensing of regional and global soil moisture over the past decade have improved the capacity for satellite assessment of agricultural drought. Particularly, soil moisture products acquired from the Advanced Microwave Scanning Radiometer for the Earth Observation System (AMSR-E) as well as other space-borne microwave sensors have been implemented for operational drought monitoring and crop yield assessment (Yin et al. 2018). Contemporary satellite missions for global soil moisture assessment include the National Aeronautics and Space Administration Soil Moisture Active Passive (SMAP) missions and the European Space Agency Soil Moisture Ocean Salinity (SMOS), which have the ability to provide lower frequency (L-band) microwave measurements. These observations have capabilities in drought assessments, including characterizing agricultural drought on multi-temporal scales.
2. For hydrological drought, satellite gravimetric observations of terrestrial water storage anomalies derived from the Gravity Recovery and Climate Experiment (GRACE) as well as the integrated drought indices developed from reanalysis and models have been successfully implemented in assessing severe drought episodes (Ndehedehe et al. 2020c, 2021; Zhao et al. 2017; Thomas et al. 2017; Agutu et al. 2017; Thomas et al. 2014). The fundamental idea of the GRACE mission is to assess the integrated water storage fluctuations over the earth's water column.

This assessment includes freshwater resources (groundwater and surface water), soil moisture, canopy, and ice. The use of this novel mission in drought monitoring and assessment has improved the understanding of drought propagation (Zhang et al. 2017; Awange et al. 2016; Thomas et al. 2014; Yirdaw et al. 2008). Thomas et al. (2014). The negative eccentricities from monthly meteorological data in drought analysis warranted the implementation of GRACE-derived terrestrial water storage anomalies dataset as an unconstrained Global Drought Severity Index (Thomas et al. 2014). The integration of GRACE-derived terrestrial water storage data into the well-known catchment land surface model to improve drought predictions is an obvious benefit of the GRACE mission in improving the understanding of climate change impacts on water availability. This has resulted in the development of an optimal GRACE-based drought indicator (Li et al. 2019; Thomas et al. 2017; Houborg et al. 2012). The study undertaken by Yirdaw et al. (2008) is an exemplary case study highlighting the importance of exploring GRACE for use in continuous drought assessment over the Canadian Prairie sub-region. Several regions around the world, e.g., Texas, Amazon, and China, among others (e.g., Getirana 2016; Long et al. 2013; Chen et al. 2010), have used GRACE to monitor and assess drought frequency. However, the direct implementation of GRACE on drought backed with a generally accepted globally gridded model is yet to be developed (Gerdener et al. 2020; Li et al. 2019). The GRACE-based drought index is popular due to its ability to operate without the use of weather estimates and their integrated uncertainties. Using GRACE, it is now possible to estimate the amount of water loss from the ecosystem (Fig. 16.1a), even with a clear definition of the start and end of any drought episode over a 30-day period. One of the distinct advantages of remotely sensed method of drought estimation is its ease in quantifying the water volume needed in storage for a drought-infected region to return to its normal and average condition (Fig. 16.1b). This is achieved by GRACE through its detection of unequivocal hydrological drought salvaging targets.

3. Groundwater drought is difficult to measure and characterize because of limited groundwater bore and the impact of human water extraction. But satellite-based groundwater drought assessment is emerging and will provide an evidence-based insight into the role of climate variability and water extraction on groundwater systems. GRACE data has been used to explore the possibility of characterizing groundwater drought in the United States and Europe (Thomas et al. 2017; Van Loon et al. 2017a). The use of GRACE in groundwater drought assessment can provide insight that will enable a predictive understanding of the influence of climate variability on groundwater availability. This is crucial for sustained decisions on water resources management, and to increase resilience to climate change. Moreover, the ability to quantify groundwater withdrawals using multi-temporal satellite data (Majumdar et al. 2020) is significant progress that could aid the optimization of future modelling scenarios of water use. This is particularly crucial for areas where resource extraction increases the risks of changes in interconnectivity of groundwater systems and disruption to groundwater flow patterns.

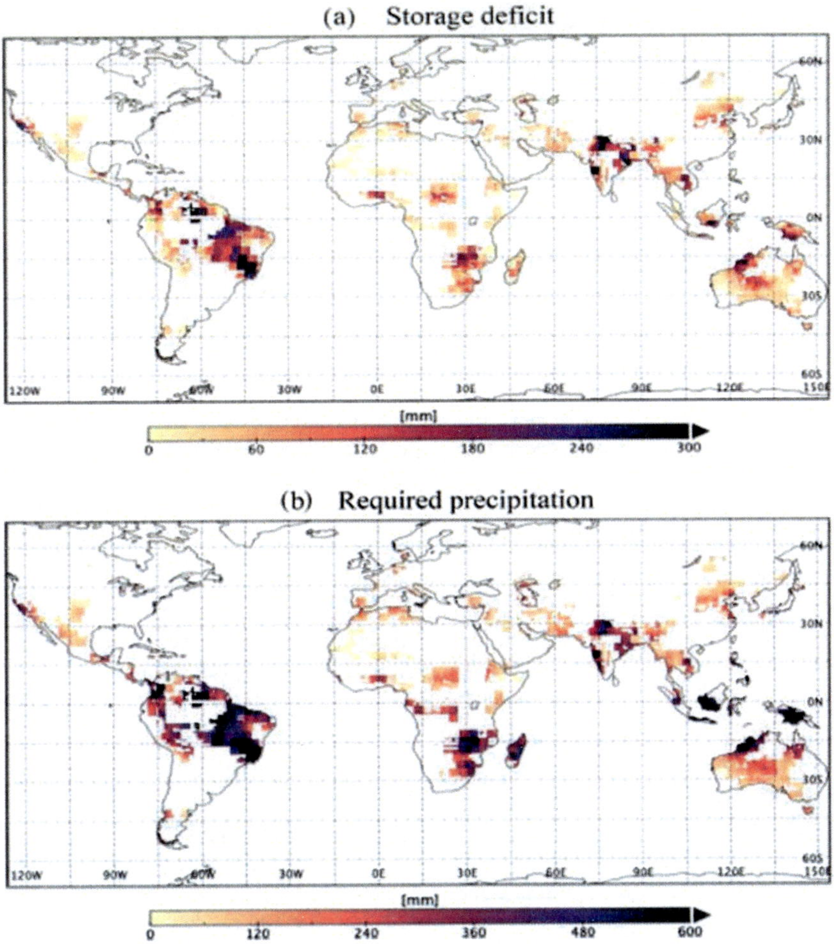

Fig. 16.1 GRACE-satellite observation showing **a** global storage deficit for January, 2016 and **b** the volume of rainfall needed to compensate the storage deficit. *Source* Gerdener et al. (2020)

16.1.2 Drought Characterization: Principles and Methods

Droughts are classified as one of the most expensive natural disasters in today's world (WMO and GWP 2017), having substantial impacts on regional and global ecosystems and human environments (Ndehedehe et al. 2021, 2019a; Sheffield et al. 2014; Cooley et al. 2015). It is however expected that climate change will greatly impact drought frequency and severity, especially in arid and semi-arid regions which are already under water stress (Seager et al. 2013). Enhancing our knowledge of drought characterization, dynamics and influence in the context of climate change is a crucial, especially due to devastating droughts in several regions of the world (Ndehedehe

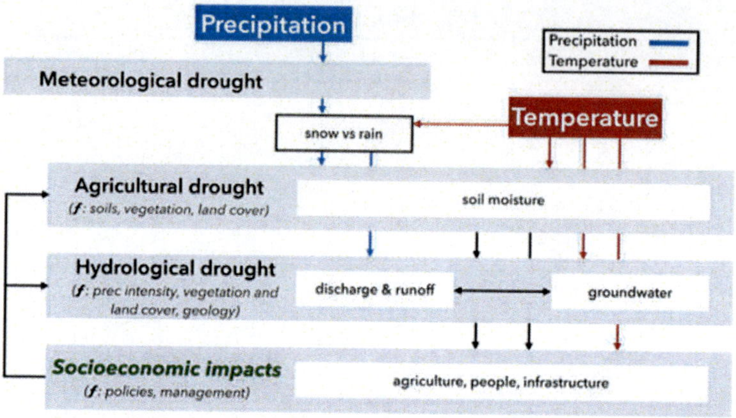

Fig. 16.2 Drought characterization depicting its associated processes Cook et al. (2018)

et al. 2021; He et al. 2016; Sheffield et al. 2014; Spinoni et al. 2014). Comprehensive characterization of drought and climate change could be difficult, however, the nature of drought events allows it to be categorized as cross-disciplinary, encompassing a wide variety of fields such as hydrology, climatology, meteorology, ecology, sociology, agronomy, economics, and anthropology. Droughts are basically classified into four categories: hydrological, meteorological, agricultural socio-economic, and anthropogenic drought. These were detailed in Chap. 11.

However, groundwater drought is another type of drought (Chap. 11). Groundwater drought is a long-term decline in groundwater levels caused by reduced rainfall and recharge or when the water table drops below normal levels due to intensive water extraction This type of drought is the result of low rainfall affecting aquifer recharge rates and can sometimes be induced by extensive human water abstraction resulting in low well yields or even a drying well. This usually occurs on a time scale of months to years. The process of quantifying the total amount of water available in groundwater systems is an extremely complicated activity. Even if it is quantified, in many groundwater systems, adverse effects of storage deterioration can be seen long before the total storage is depleted. This understanding informed the definition of groundwater drought as the decrease in groundwater levels (Mishra and Singh 2010). However, it is possible to use groundwater storage or recharge to characterize or quantify a groundwater drought (Peters et al. 2001).

Drought is majorly characterized as a typical moisture dearth relative to some usual baseline or thresholds. They usually commence as rainfall deficits (i.e., meteorological droughts), cascading over time, usually from a few months to a few years, through the hydrologic cycle to influence soil moisture (agricultural drought) (Fig. 16.2). These deficits extend to other hydrological units, which include runoff, stream flow, and surface water in lakes, springs, wetlands, and reservoirs (hydrological drought), and storage in aquifers (Fig. 16.2). Additional processes at the land surface and the land–atmosphere interface have the ability to deepen or ameliorate rainfall-forced drought anomalies, especially for agricultural and drought indices. For example, increased land surface temperatures can increase evaporation and evapotranspiration demands over the atmosphere, thereby lessening available moisture content from the Earth's surface. It also improves rainfall fraction falling as rain rather than snow, and enhances the timing of the snow melt period noticed in the spring. Land surface constituents (e.g., topography, soil type, etc.) and vegetation components (e.g., land cover, phenology, etc.) can potentially influence drought impacts, especially through their soil water-holding ability, runoff, and the segregation of energy and moisture variability on the atmosphere. Recently, the effects of human influence on both societal susceptibilities and drought variabilities have generated attention (Van Loon et al. 2016). The cost of droughts to communities and regions is also dependent on groundwater extractions, irrigation agriculture, as well as water conservation policies. They also influence other parameters such as the rate of drought propagation through the hydrologic cycle, the severity of impacts on ecosystem services and agriculture, and how quickly natural systems can be revitalized (He et al. 2016; Van Loon 2015).

The relationship of drought processes with climate change events like rainfall (Knutti 2013) is still saddled with large uncertainties. This is, however, necessary, given the importance of climate fluctuations in modulating drought variability. Simulated projections can also be complicated, given the quantitative evaluations across models, historical measurements, and the paleo-climate record which are usually complex, and also due to the dearth of direct long-term measurements, for several drought indicators, like runoff and soil moisture. Additionally, the extent to which forced climatic trends and pointers in drought episodes can be distinguished is strongly influenced by natural climate fluctuations, which in the short term can intensify or inhibit forced trends and pointers. Reuniting across these incongruent perspectives and uncertainties is challenging and in the past often resulted in contradictory conclusions about the influence of climate change on drought, even for the same event. The extent to which the recent drought in California has been attributed to climate change, for instance, depends largely on whether the studies in question focused on soil moisture on rainfall as their assessment indices (e.g., Diffenbaugh et al. 2015; Griffin and Anchukaitis 2014).

16.2 Drought Impacts

16.2.1 Impacts on Humans

The sustainability of global ecosystems is largely a function of freshwater availability. Situations that warrant the unavailability of freshwater resources are a cause for serious concern as it heavily impacts regional ecology, causing incessant fire outbreaks, poor crop yield, land subsidence, as well as death of aquatic and human lives. The complex and hazardous effects of drought are a huge challenge for local communities, as well as the global societies. Frequent drought episodes have been reported in most sub-regions in the last three decades (Mishra and Singh 2010). Drought-prone regions like Africa have been devastated by droughts, and impacts such as livestock death, crop failures, and human losses are rampant, especially due to frequent drought cases reported in some regions, for example, eastern and central parts of Africa (Agutu et al. 2019, 2020; Ndehedehe et al. 2020a; Agutu et al. 2017). Cases of drought-induced deaths in Ethiopia and Sudan between 1984 and 1985 were estimated to be 450,000, over 325,000 deaths in the Sahel (1974–1975), and more than 100,000 deaths in Mozambique (Haile et al. 2019; Vicente-Serrano et al. 2012; Guha-Sapir and Hargitt 2004). Some studies also suggested that drought impacts could have been made worse due to the results of anthropogenic influences, like deforestation, land degradation, excessive groundwater withdrawal, and climate change caused by global warming (Trenberth et al. 2014; Sheffield et al. 2012).

Drought incidence, heat waves, high wind speed episodes as well as their integrated influences contribute to enhancing fire dangers in most fire-prone regions in the world (Bowman et al. 2017; Molina-Terrén and Cardil 2016). These drought events are triggered by strong changes in sea surface temperature (SST) modes, which include the pacific decadal oscillations (PDO), El Nino southern oscillation (ENSO), or the Atlantic multi decadal oscillation (AMO), including its corresponding climatic teleconnections, from inter-annual to multi-decadal periods (Molina-Terrén and Cardil 2016). Climate teleconnections affect the atmosphere, stimulating surging impressions on local weather patterns around the world. They indirectly influence inter-annual fluctuations in biomass creation, fuel moisture, land water storage, and vegetation phenological cycles (Kitzberger et al. 2017; Ni et al. 2017; Li et al. 2016; Ndehedehe et al. 2017). Just like tropical cyclones, drought is one of the most prolonged climate phenomena and is one of the most costly natural disasters (Ndehedehe et al. 2019a). For instance, severe droughts in the United States in 1998 and 2012 resulted in losses valued at 40 and 30 billion dollars in agricultural products and other services (Rippey 2015).

16.2.2 Water Resources and Agriculture

Due to the integrated effect of population growth, climate change, and anthropogenic influence, water resources are facing an unprecedented pressure (Oelkers et al. 2011). The rate of terrestrial water extraction over the past two decades increased by six-fold so as to meet food needs as well as to improve the economic wellbeings of nations and the ever-increasing world population (Cosgrove 2014). Moreover, increasing variability in climatic conditions continues to impact decadal land surface temperature, resulting in a rise of approximately 0.850C (IPCC 2014). This change in global climate has also triggered a redistribution of rainfall patterns and increased frequency of drought episodes (Trenberth 2011). Water deficits resulting from improved water extraction integrated with climate change are common. In 2005, an estimated 35% of the global population were living in areas with lingering water shortage, whereas 80% are exposed to water safety problems, and approximately 50% of the global population are projected to live under serious water stress between 2071 and 2100 (Hanasaki et al. 2018; Oelkers et al. 2011; Vorosmarty et al. 2010). Addressing the growing demand for water is thus crucial to aid the implementation of an action water-policy plan that underpins sustainable allocation of water resources.

Another concern worthy of note is the effects of drought on vegetation and agriculture. For example, one of the first viewpoints on the reasons for the Sahelian droughts was based on the Charney hypothesis (Charney et al. 1977; Fuller 2002). In the Charney hypothesis, droughts in the Sahel were due to deteriorating vegetation cover. This model proposed that deterioration in vegetation cover was the result of overgrazing, as well as the translation of woodlands into agricultural lands which resulted in increased reflectivity of the landscape and its albedo-like properties (e.g., Zeng 2003; Fuller 2002). This was evident in reduced moisture release into the atmosphere, and less absorption of sunlight, thereby resulting in less convection for rainfall (Zeng 2003). There exist an antithetical relationship between the coefficient of reflectance and average annual rainfall, as well as a direct relationship between albedo and mean latitude in the Sahel (Epule et al. 2014).

We also understand that a significant northward reduction in the Sahel was observed as a result of the decrease in evapotranspiration rates (Nicholson et al. 1998). Although it is still unclear if the magnitude of land-use change is adequate to explain the rate of observed drought noticed in the Sahel, an evaluation of population dynamics and land-use history in the region over the past four decades reported only an average variation in land use. Therefore, the relationship between the cause (anthropogenic influence) and effect (drought) is still complicated at best (Zeng 2003). Although some studies suggested that the region witnessed significant land degradation and deforestation, the relationship between the cause and effect is plausible (Lamprey 1988).

16.3 Global Climate Teleconnection Patterns and Impacts on Droughts

Droughts have been listed as one of the highest among other natural hazards, causing extensive destruction to humans and their environment, as well as to the economy. For instance, drought in South Africa and California costs billions of dollars to mitigate its effect on human lives (Wolski et al. 2017; Diffenbaugh et al. 2015). Incessant drought causes overwhelming destruction to humans, and the impacts resulting from these extreme events are estimated to amplify in arid and warmer regions that will likely experience a more severe and frequent drought episode (Su et al. 2018). Past studies have highlighted notable variations in droughts as well as their climate related causes (Gibson et al. 2002). The prolonged duration in anomalously dry periods in some regions is caused by anticyclonic flow or atmospheric blocking (Wolf et al. 2018). This type of blocking usually leads to long-term drought events and is usually modulated by remotely forced teleconnections, like large-scale climatic oscillations, and anomalous sea surface temperatures (Swain et al. 2017). Past studies conducted over large sub-regions like Australia, Amazon, the United States, and even Africa suggest that apart from the influence of the popular El-Nino Southern Oscillation (ENSO) mode of climate fluctuations, Pacific Decadal Oscillation (PDO), North Atlantic Oscillation (NAO), Pacific-North America (PNA) pattern, and the Atlantic Multi-decadal Oscillation (AMO) are also notable low-frequency climatic oscillations that directly or indirectly impact on extreme events such as drought (Ndehedehe et al. 2019a; Dai 2011; Giannini et al. 2003; Diaz et al. 2001; Zeng et al. 1999).

The AMO diminishes rainfall rate and air temperature in the North Pacific, and exhibits a direct relationship with humidity reduction in the area. But ENSO in its frosty phase, i.e., La Nina, causes a shortage in rainfall over the South American equatorial zone by moderating the sea surface temperature of the region. The PDO, on the other hand, correlates with the cold sea surface temperature as well as decline in humidity, especially in its negative stage (NJ and SR. 2002). As a climatic pattern, the PDO possesses both cold and warm phases. During the cold phases, the eastern pacific displays below-average temperatures, while the western and central Pacific Ocean have warm anomalies in the sea surface temperatures of the region. The reverse is noticed during the warm phases. The warm phase of the PDO consists of two varieties of pressure anomalies, which include, (i) high pressure in the subtropical Pacific north directing a clockwise flow of wind and (ii) low pressure in the north Pacific directing an anti-clockwise flow of wind (Guha-Sapir and Hargitt 2004; NJ and SR. 2002).

The dry seasons observed over Korea, Japan, Australia, and the eastern part of Russian correspond with the PDO's warm phase. Similarly, this phase also corresponds with the noticeable wet seasons over southeast Brazil, south-central South America, Western Australia, Southwestern Mexico and the United States, as well as the coast of the Gulf of Alaska (NJ and SR. 2002). There is a relationship between the PDO and ENSO, which exerts substantial influence on sea surface temperature anomalies over the Pacific. The relationship between the PDO's warm phase and the ENSO is as a result of the weakness in trade wind, as well as the existence of warmer waters, whereas in the cold La Nina phase, the reverse is the case, as a reduction in the sea surface temperature is noticed and trade winds are reinforced (Quintana and Aceituno 2012). Further, the AMO is usually referred to as the first mode of sea surface temperature fluctuations over the North Atlantic having a duration of 20–40 years. It is a global scale fluctuation mode that efficiently substitutes cold and warm phases, especially in the Northern Hemisphere. Its role in driving changes in hydro-climatic extremes in humid and semi arid-domains have been documented (e.g., Ndehedehe et al. 2020c; Ndehedehe and Ferreira 2020).

The influence of the PDO and AMO are also felt in the South American region. Studies undertaken in the region suggest that the PDO influences the northern part of Chile (Quintana and Aceituno 2012), whereas the climatic fluctuations in the northern and southern parts of the region is connected to the AMO. The aforementioned study argued that during the early 1970s, a dry condition which was associated with a negative PDO phase was observed. Other regions have received direct climatic influence due to the predominant oscillations fostered by the Tropical Pacific Ocean, and to a smaller degree by other oceanic disturbances from the Atlantic Ocean. The ENSO, on the other hand, is initiated from the Pacific, exhibiting an atmosphere–ocean oscillation structure between warm and cold conditions, especially in the Pacific. It contains numerous and varied weather conditions, with a significant relationship with drought and flood episodes noticed over the Pacific.

The global environmental and economic impact of ENSO is largely based on its prevalent inter-annual mode of climatic fluctuations over a wide range of sub-regions (Mcphaden et al. 2006). As a fundamental aspect of the global climatology that draws attention to the composite influences of oceanic and atmospheric phenomena, ENSO has a substantial impact on the interaction and notable large-scale variations over the tropical latitudes of the Pacific Ocean, especially in sea levels, winds, rainfall patterns, sea current, and atmospheric pressure (Williams and Hanan 2011; White et al. 2004). The influence of tropospherically connected ENSO signals on global climate has been identified (Enfield and Mestas-Nuñez 1999; Latif and Barnett 1996) and their links with prolonged drought events have been diagnosed in sub-Sahara Africa (Ndehedehe et al. 2018a; Giannini et al. 2003). The impact of ENSO as an important driver of drought events in several climatic hotspots around the world, like north-east Brazil, emphasizes the important role of the Pacific Ocean in hydro-meteorological processes. For Brazil, this role has been widely recognized because of the well-known influence of ENSO on drought occurrence (Ndehedehe et al. 2020d; Marengo et al. 2018).

The opposite phases of ENSO are adequately represented by the La Nina and El Nino oscillations (McPhaden 2004). The trade wind moves from east to west along the Equator with the southern equatorial current, especially during neutral intervals (Trenberth et al. 2002). The intense tropical sun which is warmed by the warm surface water current is accumulated in the western Pacific. The vertical gradient in temperature that isolates the warm surface stratum of the ocean from the cold stratum deeper rises in the east and is directed downward in the west (Mcphaden et al. 2006). The transpiration of cold water is enabled by the relative exteriority of the vertical gradient, especially over the eastern Pacific. The heterogeneous temperature noticed over the east–west surface flows below the air pressure descent in the sea surface near the Equator especially as it boosts the eastern trade winds. The trade wind also collects humidity and heat from the ocean because it moves from east to west (McPhaden et al. 2015). Given the cluster production of heavy rainfall and clouds formed through deep convection, by extension, the humid and warm mass becomes less dense and escalates above the western Pacific warm pool. Wolter and Timlin (2011), however, pointed out the possibility of the mass of air that escalates to this region to revert to the east above troposphere ranks, before descending to the cold waters of the eastern Pacific.

An enervated trade wind in the Equator is noticed with respect to the intensifying atmospheric pressure over the western Pacific, while it reduces over the eastern Pacific during the El-Nino phase. The western Pacific warm pool migrates eastward as a result of the weaker trade wind, which also causes the thermocline to intensify along the pool, thereby allowing for the deep cloud accretion and heavy rains associated with the escalation of the walker circulation branch (McPhaden 2004). Drought probability in several sub-regions like Indonesia, the Philippines, and Australia has increased as a result of deep convection towards the east. Islands along the western coasts of South America as well as the central Pacific Ocean on the other hand experience an increase in torrential rainfall rates. Changes in global wind are fueled by energy from the heat released in the upper middle zone of the troposphere due to the tropical rainfall pattern. Variations in these rainfall patterns tend to influence the changes in global weather patterns, especially as a result of the El-Nino atmospheric teleconnection (Trenberth 1997).

16.3.1 Drought Recovery Forecast

Understanding water loss from a freshwater ecosystem is important to aid the characterization of the cascading impacts of drought. This is especially helpful in setting a clear target and expectations for optimal drought recovery. To achieve this, observational data capture changes in the water column like terrestrial water storage change, which can only be obtained from the GRACE mission. It can be difficult to understand drought recovery time, especially in terms of depicting the proposed ecosystem duration to revert to its original pre-drought operational state (CR et al. 2017). However, the importance of this process is well founded in the increase in the number of drought events (Cook et al. 2018), since ecosystem recovery before the subsequent drought is essential. Expected drought recovery estimates provide good and technical information to water management authorities to help them in water supply regulation pending the drought recovery phase. Drought recovery techniques are not very popular. However, historical drought assessments using various predictors (climatic parameters and teleconnections) have been reported (e.g., Maity et al. 2016). Studies attempting to characterize the beginning and ending phases of drought patterns and causes have been undertaken (e.g., Maxwell and Fitzpatrick 2012), though with little or no focus on the statistical evolution of drought recovery. Drought impact can persist until all hydrological storage compartments (i.e., groundwater and deep soil moisture) are restored. This is despite the intensity and characteristics of such drought on the hydro-ecological system, e.g., surface water, soil moisture, and vegetation. Drought prediction and forecast, as well as its onset and recovery, can therefore be improved by integrating GRACE-derived terrestrial water storage (TWS) over all the storage compartments. Drought recovery forecasts can be facilitated by the application of GRACE-derived TWS. This is because TWS as measured by GRACE can easily provide a complete profile of the water column or hydrological stores like groundwater and terrestrial stored water to augment the sparse in situ groundwater observations datasets (Fig. 16.3).

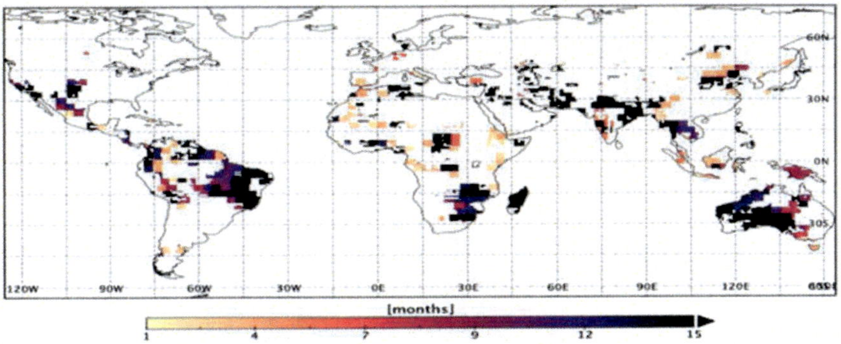

Fig. 16.3 Plot showing an estimated required rainfall rate for drought recovery from the January 2016 drought observed using the GRACE mission. *Source* Singh et al. (2021)

16.4 Droughts in the Congo Basin and Impacts of Climate Teleconnections

The Congo basin (Fig. 16.4) is the third largest basin in the world (approximate size of 3.7 million km^2), and plays substantial roles in global climate (e.g., Washington et al. 2013). Yet, it is an understudied region when it comes to drought-related investigations. Despite the seeming opportunities for hydrology-related research, it is apparently less documented unlike the Amazon basin where a plethora of such studies exist (Alsdorf et al. 2016). Some hydro-meteorological-focused investigations have shown evidence of drying in some sections of the basin (e.g., Ndehedehe et al. 2018a; Hua et al. 2016; Zhou et al. 2014). But climatological extremes and historical evolutions of drought episodes in the basin are, however, conspicuously undocumented. The strong ocean–land atmosphere interaction, the impact of nearby oceans on precipitation patterns, and large-scale moisture flux convergence, which show considerable influence on rainfall seasonality in the Congo basin (Dyer et al. 2017) provide a basis to diagnose the contributions of oceanic factors on drought patterns in the region. Drought events are becoming more unprecedented and complex due to the combined effects of unmitigated climate change and perceived human factors (e.g., Thomas et al. 2017; Van Loon et al. 2016). Droughts influence the equilibrium in terrestrial carbon by changing the rates of carbon uptake by photosynthesis and release by total ecosystem respiration, as well as their coupling (see van der Molen et al. 2011). This could be a source of concern for the ecosystem carbon cycle of the Congo basin. Even though the Congo basin is one of the wettest regions in the world, droughts are increasingly becoming frequent. The rise in severe droughts in other humid regions like the Amazon basin clearly suggests the non-inviolability of the Congo basin to several drought risks and its cascading impacts. This is already happening following an extended severe drought events in the Congo basin. The impacts of prolonged and extreme droughts on the tropical rainforest systems of Central Africa were stressed by Zhou et al. (2014). They argued that the observed

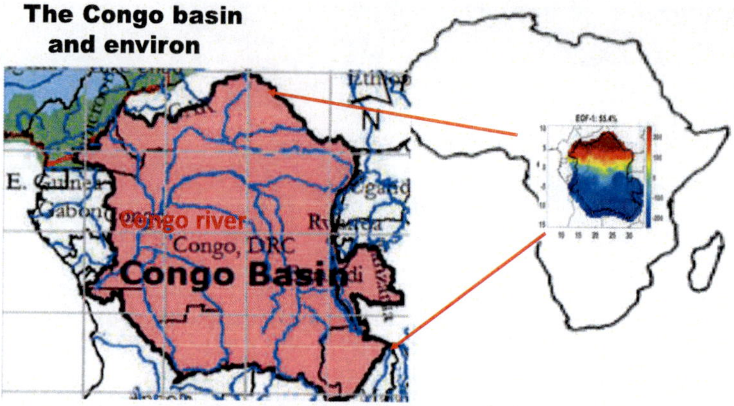

Fig. 16.4 Study area showing the Congo basin (i.e., black solid line on the left). The Congo river and its numerous tributaries (i.e., blue lines on the left) are also indicated. The insert on the right is the spatial patterns of the leading precipitation mode (data obtained from Global Precipitation Climatology Centre) over the Congo basin during the 1991 – 2014 period and accounts for a total variability of 55.4%. The dominant patterns of rainfall (annual variations) increased during the 1931–1960 (56.3%) and 1961–1990 (57.3%) periods as opposed to start of the century 1901–1930 (51.3%). *Source* Ndehedehe et al. (2019a)

persistent drying in the region could lead to substantial changes in the composition and structure of the Congolese forest.

The Congo basin contains some of the largest areas of the world's tropical forests and wetlands. These wetlands are considerably important to global carbon and methane cycle (O'Loughlin et al. 2013; Achard et al. 2002). With its numerous water resources and being one of the largest rainforests after the Amazon basin, the Congo basin is well positioned to provide several essential services to the ecological communities who depend on the basin. These services include regulating the nutrient cycle, water supply, recreation and tourism, as well as support for a range of aquatic biodiversity (Ndehedehe et al. 2020b; Tockner et al. 2010; Gidley 2009). In this era of the Anthropocene where climate change and multiple strings of anthropogenic activities (e.g., deforestation) are dominant drivers of global environmental change, regions like the Congo basin are expected to be more vulnerable to climatic disturbance as well as the influence of human activities. But the impact of climate change induced by the tropical oceans, physical mechanisms, and climate teleconnections is expected to trigger more hydrological changes in the basin. This is because these factors regulate precipitation and the transport of moisture in the region. Process-based understanding of the cascading impacts of extreme events such as drought on the Congo basin hydrology is thus important and can directly feed into management and policy frameworks.

This section presents results and analyses from the work of Ndehedehe et al. (2019a) over the Congo basin (Fig. 16.4) who developed statistical relationships between observed standardized precipitation index (SPI) and global sea surface temperature using principal component analysis as a regularization tool prior to the implementation of a canonical correlation scheme. Specifically, this chapter will look at (i) spatio-temporal drought patterns and their coupled variability with sea surface temperature (SST) anomalies and (ii) the influence of large-scale global climate signals on drought evolutions. The commonly used WMO (World Meteorological Organisation) recommended drought index, the standardized precipitation index (SPI, McKee et al. 1993), and standardized runoff index (SRI) are employed as drought indicators. SPI and global sea surface temperature are then integrated in a statistical predictive framework that combines principal component analysis (PCA, e.g., Preisendorfer 1988), canonical correlation analysis (CCA, e.g., Glahn 1968; Hotelling 1936), and non-parametric autoregressive framework to examine the role of global climate on drought evolutions in the Congo basin.

16.4.1 Data

This study used the Global Precipitation Climatology Centre (GPCC, Schneider et al. 2014; Becker et al. 2013) to analyse historical hydrological drought over the Congo basin. The Global Runoff Data Centre (GRDC)-based (www.bafg.de/GRDC) river discharge data observed at the Congo Kinshasa station was used to study hydrological response of the Congo river to drought conditions. GRDC has a number of gauges distributed across the basin. The global sea surface temperature (SST) data (Reynolds et al. 2002) used in this study (1982–2014) was accessed from NOAA's official earth system research laboratory portal (http://www.esrl.noaa.gov/psd/data/gridded/data.noaa.oisst.v2.html). Several prominent indices of oceanic variability associated with the climate variabilities in the region were used in this study to further examine the role of global climate on droughts in the Congo basin. They include El-Niño Southern Oscillation (ENSO), Indian Ocean Dipole, Atlantic Multi-decadal Oscillation, Atlantic Meridional Mode, Pacific Decadal Oscillation, Quasi-Biennial Oscillation, and Maiden Julian Oscillation (MJO).

16.4.2 Method

16.4.2.1 Multivariate Methods

Drought phenomenon is described using indices, which are usually based on a single variable such as rainfall and soil moisture. The standardized precipitation index (SPI, McKee et al. 1993) is employed in this study. Similar to the SPI values of McKee et al. (1993), we assume that a drought condition occurs when the drought index is consistently negative and reaches a value of -1. The methodology and workflow used in this study is illustrated in Fig. 16.5. The principal component analysis (PCA, e.g., Jolliffe 2002) method was used to statistically decompose time

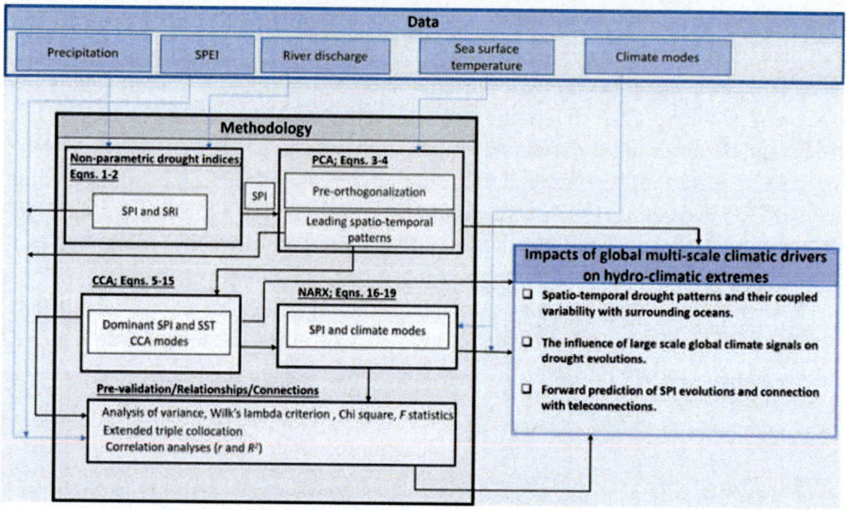

Fig. 16.5 Schematic representation of the methodology and workflow used in the study undertaken by Ndehedehe et al. (2019a). Note that a number of statistically significant climate modes that explained a considerable proportion of the variability of observed global climate teleconnections were also identified in the canonical scheme implemented by Ndehedehe et al. (2019a). This is one of the basis for the coupled variability between SPI and the surrounding oceans reported in this section

series of computed SPI ($\mathbf{X}_{SPI}$) at 12-month scale into leading orthogonal modes of variability (i.e., temporal and spatial patterns). Again, the PCA method and the rules for implementation and analysis have been described in Chap. 13. For the purpose of physical interpretability, this study focused on three spatio-temporal modes of SPI evolutions over the Congo basin. However, in the pre-implementation of the canonical correlation analysis scheme (Chap. 13, Sect. 13.2.2), which attempts to connect SPI evolutions with oceanic processes, the PCA technique was rather employed as a regularization tool to minimize the noise in the data (see details in Sect. 13.2.2). On the spatio-temporal consistency of SPI and SPEI between 1901 and 2013, this was assessed using simple correlation while GPCC precipitation was compared to those of Climate Research Unit based rainfall and TRMM-satellite precipitation using the extended triple collocation method and analysis of variance (Fig. 16.5). To examine the inter-relationship between global SST anomalies and SPI over the Congo basin, canonical correlation analysis (CCA, e.g., Graham et al. 1987; Bretherton et al. 1992; Glahn 1968; Hotelling 1936) was employed. The implementation of CCA and its application in this study is as detailed in Chap. 13.

16.4.2.2 Machine Learning

The connections of leading SPI modes to indices of oceanic variability are diagnosed by employing the CCA technique. Based on this diagnosis, a standard ANN (Artificial Neural Network) autoregressive framework that incorporates these indices as predictors is introduced to predict future SPI temporal patterns. To this end, our inputs to the neural network framework were dominant SPI mode and the large-scale climatic drivers identified from the main portion of the canonical model that showed a significant relationship with well-known climate oscillations (cf. Fig. 16.5). Some of these climatic drivers, e.g., ENSO, IOD, and AMO are primary drivers that also describe the properties of rainfall variability in the African sub-region. For example, the pervasive influence of the equatorial Pacific and the Atlantic oceans on interannual variability of rainfall in Central Africa was reported by Balas et al. (2007). And even more recently, Dyer et al. (2017) drew attention to the fact that large-scale circulation patterns over the Indian Ocean are significant dominant moisture sources for the Congo basin. Hence, these climate modes and those whose perceived influence in the African sub-region have been reported (Sect. 16.4.1) and are included in the predictive framework. The association between inputs ($y_1, y_2, y_3 ... y_n$) and outputs (y_t) in a standard neural network model can be represented mathematically as (e.g., Khashei and Bijari 2010)

$$y_t = w_0 + \sum_{j=1}^{q} w_j . g(w_{0,j} + \sum_{i=1}^{n} w_{i,j} . y_{t-i}) + \epsilon_t, \tag{16.1}$$

where w_j $(j = 0, 1, 2, 3, ..., q)$ and $w_{i,j}$ $(i = 0, 1, 2, 3, ..., n,\ j = 1, 2, 3, ..., q)$ are the model parameters referred to as the connection weights; n and q are number of input nodes and hidden nodes, respectively. Other implementation details and choice of activation functions can be found in Zhang et al. (1998) and Khashei and Bijari (2010). In a typical ANN forecasting problem, the inputs are usually the predictor (independent) variables while the outputs are dependent variables similar to a non-linear regression model. This relationship as modelled in Eq. 16.1 can be further simplified as (e.g., Zhang et al. 1998).

$$\gamma = f(\zeta_1, \zeta_2, \zeta_3, ..., \zeta_n), \tag{16.2}$$

where γ is the dependent variable and $\zeta_1, \zeta_2, \zeta_3, ..., \zeta_n$ are n predictor variables. But when the ANN is a prototype of the non-linear autoregressive model, the inputs variables are the past observations of the time series while outputs are the future values. The functional mapping is

$$\gamma_{t+1} = f(\gamma_t, \gamma_{t-1}, \gamma_{t-2}, ..., \gamma_{t-n}), \tag{16.3}$$

where γ_t is the observation for time t. As opposed to the non-linear autoregressive neural network, this study employed the non-linear autoregressive with exogenous inputs (NARX) detailed in Lin et al. (1996). The NARX belongs to a prominent class of discrete-time non-linear systems that derive future values of the time series based on previous data points of the original time series, in addition to an external data (e.g., Ruiz et al. 2016; Lin et al. 1996). For example, NARX predicts time series β_{SPI} at a time (t) using the past values (p) of series β_{SPI} and another external series $\mu_{Indices}$. In other words, the model takes in one input time series for SPI at time $t-1$, $\beta(t-1)$, and another input with exogenous components (climate modes) at time $t-1$, $\mu(t-1)$. This results in a single forward predicted output series β_{SPI}. This is modelled in Eq. 16.4 as

$$\beta_{SPI}(t) = h(\mu(t-1), \mu(t-2), ..., \mu(t-k), \beta(t-1), \beta(t-2), ..., \beta(t-p) + \varepsilon(t)), \quad (16.4)$$

where the function h is unknown in advance, $\varepsilon(t)$ is the error term, and the objective of the neural network training is to approximate the function through network optimization of the weights and neuron bias (e.g., Ruiz et al. 2016). It should be noted that the neural network system evaluates each run of the experiment using regression by splitting the first components of the data internally into training, validation, and testing components (i.e., 70–30% of the total data). The other part of the data is thereafter used for the forward prediction based on the hold-out method of cross-validation (e.g., Haley 2017). In other words, both the SPI and climate modes used for simulation and training were 372 time steps, i.e., from 1981 to 2012 different from the remaining part used for prediction in a closed-loop mode. To further examine the teleconnections and potential influence on the evolutions of SPI, the predicted and validated SPI evolutions are compared with the estimated climate modes described in Sect. 16.4.1 using Pearson correlation and coefficient of determination (R^2). This is different from the analysis in Sect. 16.4.2.1 where the canonical scheme was used to formulate the relationship between SPI patterns and observed climate modes from SST anomalies of the three oceans.

16.4.3 Spatial and Temporal Variability of Droughts During the 1901–2014 Period

16.4.3.1 Precipitation and Drought Indicators

1. *(a) Precipitation*: The uncertainties and ambiguities surrounding the spatial and temporal distribution of rainfall over the Congo basin (e.g., Washington et al. 2013) have remained an area of disagreement that warrants further investigation. So three rainfall products (GPCC, CRU, and TRMM) were evaluated prior to their use in this study. The evaluation metrics used in this study show that the GPCC and CRU ($F = 1.71/p = 0.1915$; $F = 1.99/p = 0.1589$; $F = 0.86/p = 0.351$; and $F = 0.11/p = 0.736$ for the 1901−1930, 1931−1960, 1961−1990, 1991−2013 periods, respectively) are consistent as there are no significant difference in the their population means at the 95% confidence level

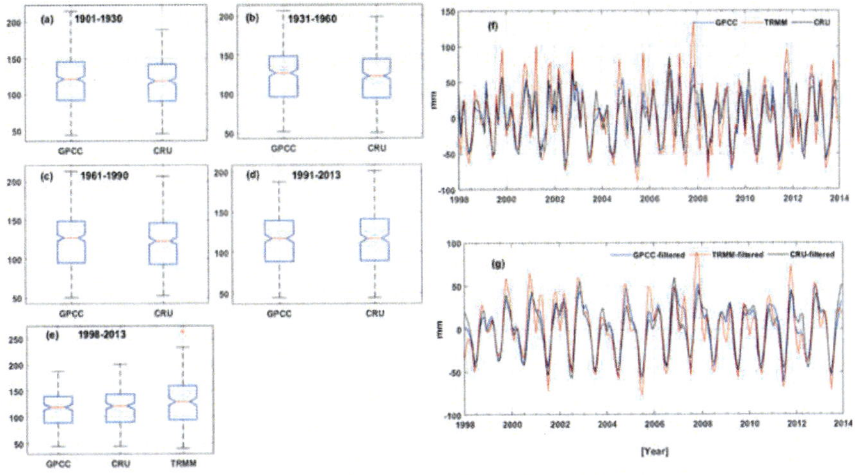

Fig. 16.6 Relationship of satellite precipitation (TRMM) with gauge and observational reference (GPCC and CRU) over the Congo basin. **a–d** Analyses of variance for GPCC and CRU data in different climatological period between 1901 and 2013 and **e** for all products (GPCC, TRMM, and CRU) during a common climatological window (1998–2013). **f** Precipitation anomalies of TRMM, GPCC, and CRU products with respect to the mean of the 1998–2013 period (i.e., after removing the mean) and **g** filtered precipitation anomalies of the three precipitation products using the Savitzky–Golay filter (Savitzky and Golay 1964)

(Fig. 16.6a–d). However, their comparison with TRMM for the common period shows a statistical significant difference ($F = 8.04/p = 0.0004$ at $\alpha = 0.05$) (Fig. 16.6e). The temporal relationships for the common period for all precipitation products, i.e., after removing the mean of the period are strong ($r = 0.90$ for both GPCC vs. TRMM and GPCC vs. CRU, Fig. 16.6f) and relatively better after applying the Savitzky–Golay filter (Savitzky and Golay 1964), which helps to cushion the effects of strong seasonal amplitudes and noise ($r = 0.92$ and 0.94 for both GPCC vs. TRMM and GPCC vs. CRU, respectively, Fig. 16.6g). The extended triple collocation (ETC) results indicate GPCC data has the least error variance (4.25) with a considerably high correlation coefficient ($r = 0.99$) as opposed to TRMM and CRU ($r = 0.81$ for both). Although TRMM shows some strong amplitudes that coincide with ENSO events (e.g., 2007), it is generally noisier unlike GPCC and CRU as it shows the highest error variance. But their correlation value (i.e., TRMM and CRU) in the ETC scheme suggests both data have a higher signal to noise ratio and are thus sensitive to the signal similar to GPCC. During seasonal periods when precipitation peaks are relatively stronger (e.g., December–February and March–May), the relationship of GPCC with CRU and TRMM are largely consistent ($r = 0.78$ and 0.73 (for GPCC vs. TRMM and GPCC vs. CRU, respectively, in DJF); $r = 0.85$ (MAM)) as indicated in Fig. 16.7a, d. Similarly, GPCC-TRMM and GPCC-CRU associations for JJA ($r = 0.92$ and $r = 0.77$, respectively) and SON ($r = 0.89$ and

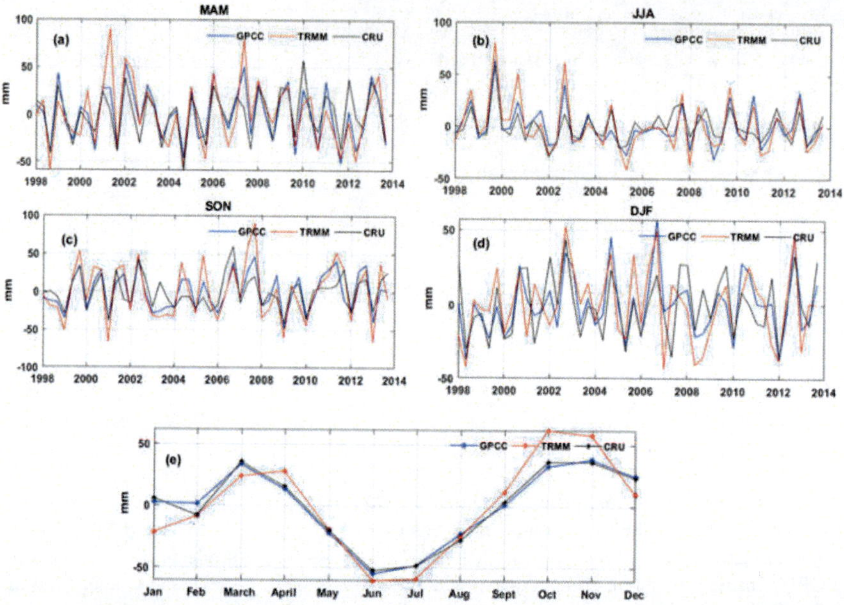

Fig. 16.7 Temporal patterns of seasonal rainfall cycles and mean monthly rainfall over the Congo basin between 1998 and 2013. **a–d** Mean seasonal cycles of precipitation of GPCC, CRU, and TRMM for March–May (MAM), June–August (JJA), September–November (SON), and December–February (DJF) **e** Mean monthly rainfall for the three products. Please note that the mean of the period has been removed from all products

$r = 0.73$, respectively) periods are relatively strong. Overall, GPCC rainfall is more associated with TRMM than with CRU rainfall over different temporal seasonal periods. Nonetheless, as indicated earlier, TRMM showed the highest error variance from the ETC scheme, which is also evident in the significant difference in mean (Fig. 16.6e).

Moreover, the seasonal monthly cycles of the three precipitation data also show some level of consistency with rainfall peaks observed twice a year ($r = 0.93$ and 0.99 for GPCC vs. TRMM and GPCC vs. CRU, respectively-Fig. 16.8e). However, TRMM shows higher precipitation estimates between September and October (Fig. 16.8e). Whereas the distribution of gauge stations used in the CRU and GPCC might not be dense compared to other tropical regions, Washington et al. (2013) mentioned that both model and satellite rainfall distribution over the Congo are inconsistent. The widely used GPCC observational reference, which is consistent with the CRU gauge data (though with fewer gauge stations), was used for subsequent analyses. In Fig. 16.8), we show the spatial distribution of historical GPCC precipitation over the Congo basin. These patterns are consistent with other reports (see, e.g., Alsdorf et al. 2016) and suggest that the preponderance of rainfall is observed between December and February in the southern region of the basin, though the rest of the basin remains generally wet

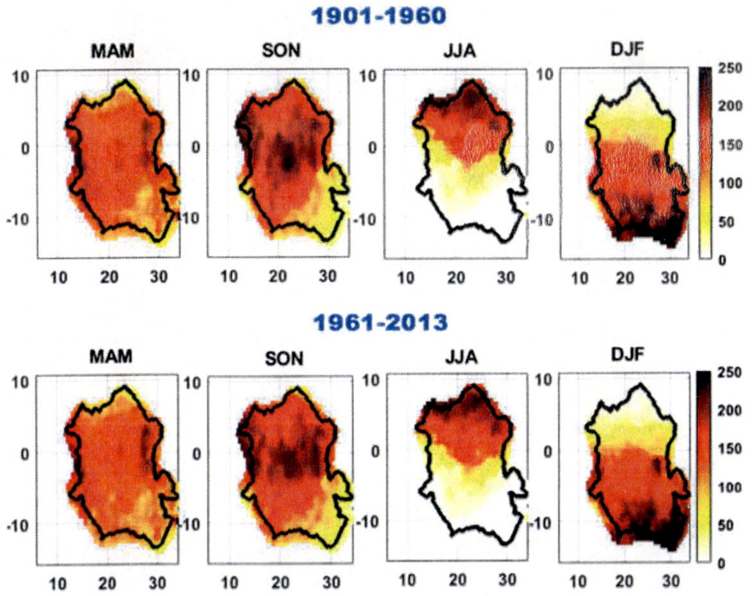

Fig. 16.8 Spatial distribution of seasonal GPCC-based precipitation over the Congo basin during 1901−1960 and 1961−2013. The seasonal windows (March–May (MAM); June–August (JJA); September–November (SON); and December–February (DJF)) are the same as those defined in Fig. 16.7. *Source* Ndehedehe et al. (2019a)

all through the year except in the July–August period (i.e., only for the southern part) (Fig. 16.8).

2. *(b) Drought indicators*: Diversity in landscape, influence of topography, uncertainties in some data, and the combined impacts of human-induced climate change and indices of oceanic variability, amongst others, have resulted in the quest for suitable and improved drought indicators (see, e.g., Ferreira et al. 2018; Van Loon et al. 2017a; Agutu et al. 2017; Ndehedehe et al. 2016; AghaKouchak 2015; Spinoni et al. 2014; Vicente-Serrano et al. 2010a). For example, Vicente-Serrano et al. (2010b) argued that Palmer drought severity index lacks the multiscalar character required not only to understand the effect of droughts on different hydro-ecological systems but for the characterization of other forms of drought, thereby introducing the standardized precipitation evapotranspiration index (SPEI). Our comparison of the SPEI with SPI for different periods is largely consistent (Fig. 16.9a–e) but the need for further assessment of the role of evapotranspiration (ET) as a significant driver of hydro-climatic extremes for some locations in the basin is echoed. While SPI in much of the basin show a strong association with SPEI, some areas indicate little or no relationship (Fig. 16.9a–d). It is unclear if the poor association between SPI and SPEI in some of those locations during different time periods (Fig. 16.9a–d) could be due to the perceived influence of land surface conditions or other drivers. This

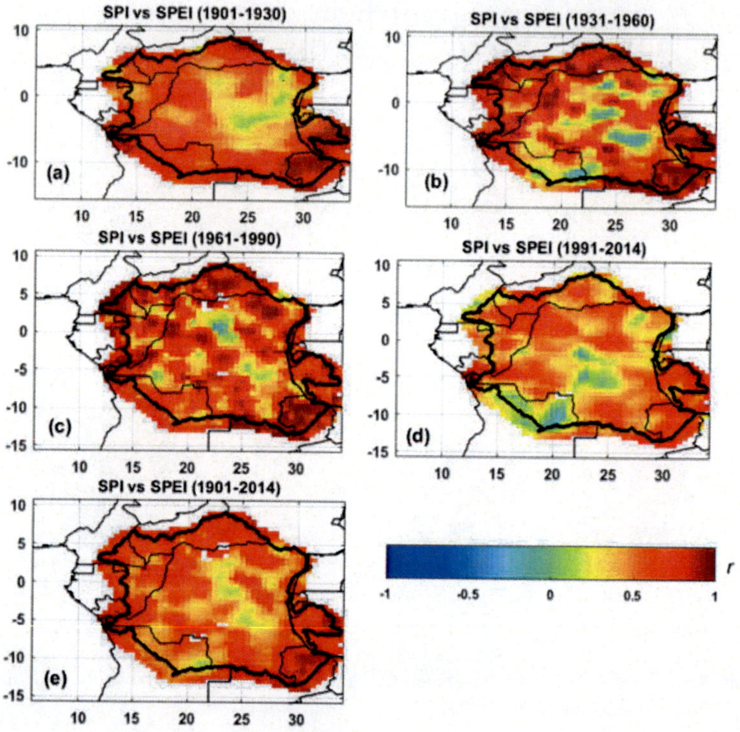

Fig. 16.9 Relationship between standardized precipitation index (SPI) and standardized precipitation evapotranspiration index (SPEI) during different climatological periods over the Congo basin. Both SPI and SPEI are based on 12-month aggregation. *Source* Ndehedehe et al. (2019a)

observed relationship is open for further discussion and can be explored in more detail in future investigations. Notably, the SPI is not only widely used but also the WMO recommended index for studies of extreme events, and is used here to help address the objectives of the study, which is more focused on the connection between extreme rainfall events (e.g., droughts) and large-scale global climate signals.

16.4.3.2 SPI Spatio-Temporal Patterns

Prior to implementing the canonical scheme, the pre-orthogonalization (regularization) results of SPI using PCA yielded several significant orthogonal modes of variability. The temporal and spatial evolutions of SPI signals (1901–2014) over the Congo basin are captured in three dominant modes of variability that represent a total variability of 29.3% (Fig. 16.10). As summarized in Table 16.1, the observed variability in the evolutions of leading SPI modes over the Congo basin for the four climatological periods is different. They are particularly found to be strongest dur-

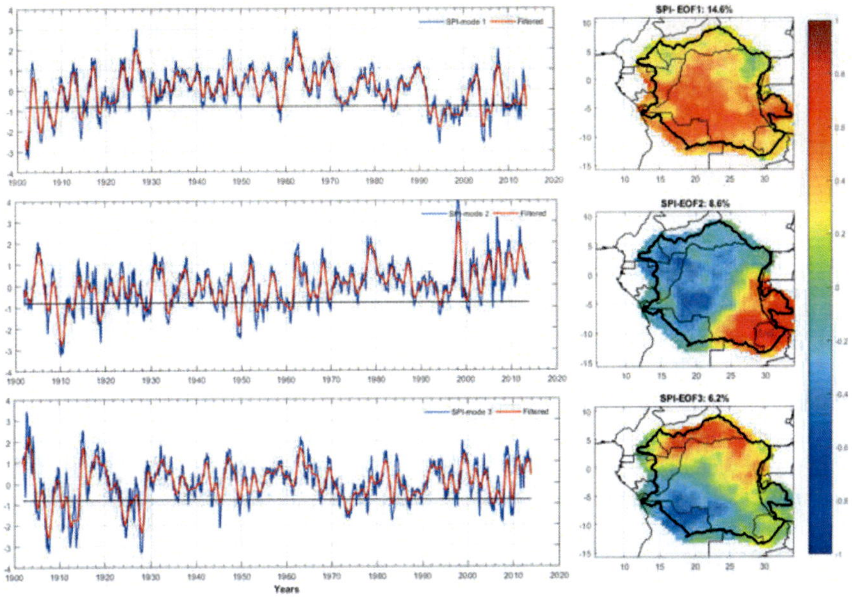

Fig. 16.10 Leading modes of spatial and temporal SPI patterns over the Congo basin. The EOFs are loadings showing spatial patterns of SPI (right) while the corresponding PCs are temporal variations (left). The SPI time series (red) are filtered to cushion the effect of residual short-term seasonal signals and also for better representation. The SPI is computed using GPCC-based precipitation for the entire period (1901–2013). *Source* Ndehedehe et al. (2019a)

Table 16.1 Observed variability (%) in the evolutions of leading SPI modes over the Congo basin in different climatological regimes during the century (1901–2014). *Source* Ndehedehe et al. (2019a)

PCA mode	1901–1930	1931–1960	1961–1990	1991–2014	1901–2014
PC/EOF 1	21.0	11.8	15.7	22.4	14.6
PC/EOF 2	13.0	9.7	9.1	12.8	8.6
PC/EOF 3	10.5	7.5	7.8	7.9	6.1
Total	**44.5**	**29.0**	**32.6**	**43.1**	**29.3**

ing periods with severe and frequent drought events (i.e., during the 1901–1930 and 1991–2014 periods as shown in Table 16.1). During these two periods (1901–1930 and 1991–2014) when drought events are significantly notable, especially for the dominant SPI mode (PC1, Fig. 16.10), the three leading SPI modes accounted for a total variability of 45.5% and 43.1%, respectively (Table 16.1) as opposed to other climatological regimes. Specifically, the temporal patterns of SPI corresponding to EOF loadings in the south-western region of the basin (PC1, Fig. 16.10) suggest that drought events were not only severe but also prevalent and frequent mostly during the 1901–1920 and 1990–2010 periods.

It should be noted that the period with the longest drought duration and other well-known drought events of the last decade (between 2003 and 2007) are observed during the 1991–2001 period (PC1, Fig. 16.10) when SPI variability is the strongest (Table 16.1). Extreme wet periods of 1925/1926 and 1960–1964 are not only the wettest (PC1, Fig. 16.10) in the century but also consistent with evolutions of SRI (Sect. 16.4.3.3). SPI spatial patterns corresponding with the second and third PCs (Fig. 16.10) show large-scale contrast with dipole patterns (Fig. 16.10). These patterns are somewhat distinct but moderately strong in the south-eastern (EOF2, Fig. 16.10) and northern (EOF3, Fig. 16.10) sections of the basin as opposed to other catchments. Although drought frequency is relatively higher in the south-eastern region of the basin (PC2/EOF2, Fig. 16.10) as opposed to the south-western part (PC1/EOF1, Fig. 16.10), SPI time series in the last decade has shown a consistent increase with a considerable wet period observed between 1998 and 1999 in the south-eastern flank of the basin (PC2, Fig. 16.10). Overall, the regionalization of SPI signals over the basin confirms that the most severe and extreme drought conditions occurred at the start of the century although droughts of the last two decades in the south-western section and Central Africa (CAR) are also obvious (PC1 and PC3, Fig. 16.10). The preponderance of wet episodes from the leading SPI modes over the Congo basin (PC1-PC3, Fig. 16.10), nonetheless, is very typical of humid tropical ecosystems largely characterized by strong amplitudes of annual and seasonal rainfall.

16.4.3.3 Drought Intensities

The temporal distribution of various drought intensities, for example, severe and extreme drought conditions over the Congo basin is conspicuously moderate or probably less (Fig. 16.11). Despite the enduring nature of droughts between 1991 and 2001 (PC1, Fig. 16.10), they were not as widespread as the extreme droughts in 1902 and 1914, which affected at least 50% of the entire basin (Fig. 16.11). Similar drought intensity but with less coverage (40%) repeated in 1994 and 1998 and is also observed between 2004 and 2006 (Fig. 16.11), one of the recent and widely acknowledged hydrological drought period in the basin. The evolutions of different drought intensities in the Congo basin certainly do not have large coverage as opposed to some neighbouring countries of the basin in East Africa where the percentage of areas affected by similar drought intensities exceeded 75% for some prominent drought years (Agutu et al. 2017). The predominance of moderate drought events over the period in the Congo basin has remained static, largely affecting about 10% of the basin as opposed to the coverage extent of extreme drought, which fluctuates between 50% at beginning of the century and 40% between 1994 and early 2006 (Fig. 16.11). A critical consideration of the last 30 years (between 1984 and 2014) shows that reoccurring severe and extreme drought episodes have affected on average at least 30% of the Congo basin (Fig. 16.11). The south-eastern end of the basin, which appears to be the main source of nourishment due to large amplitudes of rainfall (Fig. 16.9), indicates an increasing and considerable strong positive SPI values (wet conditions)

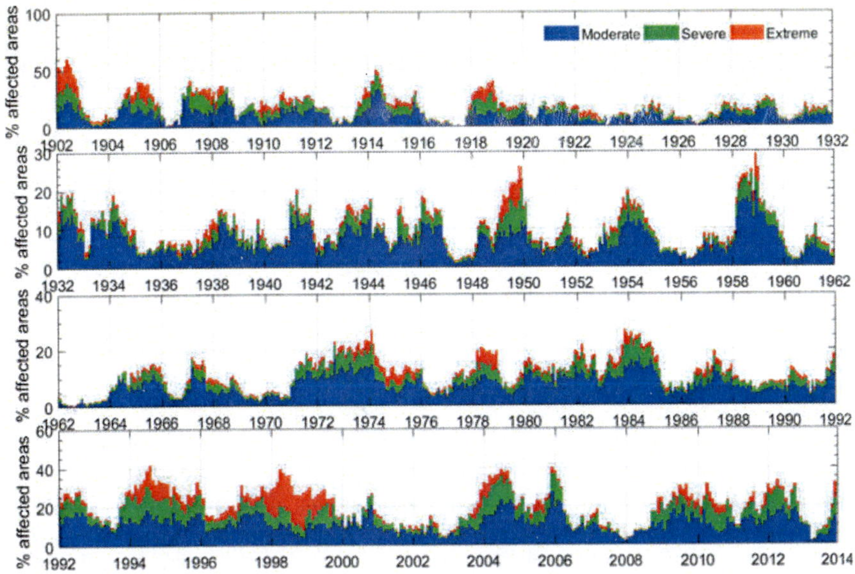

Fig. 16.11 Analysis of areas under drought over the Congo basin between 1901 and 2013. The affected areas (%) are characterized based on the classification indicated in Table 11.1, Chap. 11. *Source* Ndehedehe et al. (2019a)

since 1994 (PC2/EOF2, Fig. 16.10). However, a holistic assessment of Fig. 16.11 could imply the hydrological conditions of the basin between 1984 and 2015 that have been relatively unfavourable and would be consistent with some recent reports on the basin's terrestrial hydrology (e.g., Ndehedehe et al. 2018a). Comparatively, and based on the evolving patterns of drought intensities after 1918 (Fig. 16.11), the Congo basin may be getting drier. For instance, between 1920 and 1990 (about 70 years), extreme droughts only affected about 20% of the basin on average on five different occasions (1949, 1958–1959, 1972–1973, 1978, and 1983–1984, Fig. 16.11). But during the 1991−2014 period (i.e., 23 years), about 30% of the basin on average has been affected by extreme drought on six separate occasions (1994–1995, 1998, 2004, 2005–2006, 2010, and 2012, Fig. 16.11). This increasing intensity in extreme drought episodes can well be deduced from the dominant SPI mode shown in PC1 of Fig. 16.10. Consistent with this study, the same period (1991−2014) was characterized by widespread decline in surface vegetation greenness that coincided with a decrease in terrestrial stored water, rainfall, and above ground woody and leaf biomass water content (Zhou et al. 2014). As a side observation, we note that prior to the decomposition of the SPI signals, the leading rainfall mode over the Congo basin varied less during the 1991−2014 period (55.4%) unlike the 1931−1960 (56.3%) and 1961−1990 (57.3%) periods (see inset in Fig. 16.4). Add to that, multi-annual rainfall signals declined steadily from 16.5% at the start of the century (1901−1930) to 13.6% in the 1991−2014 period. It can be therefore argued that meteorological

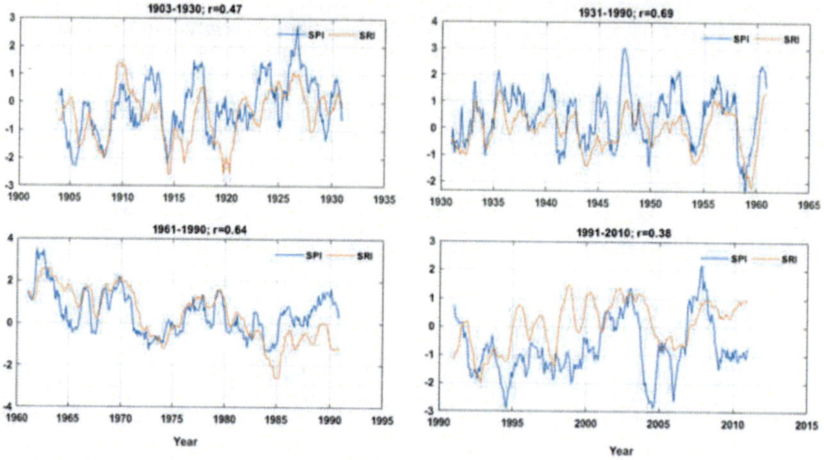

Fig. 16.12 Relationship between dominant SPI modes and SRI (1903–2010) during different climatological periods (i.e., for a 30-year interval except for the 1991–2010 period) based on Pearson correlation (r). The SRI is formulated similar to SPI, which uses an empirical probability method (Chap. 11). *Source* Ndehedehe et al. (2019a)

signals in the basin between 1991 and 2014, be it short or long term, have shifted significantly.

16.4.3.4 Time Series of SRI

The impacts of drought on the flow regime of the Congo river are also evaluated by comparing the leading SPI temporal patterns (PC1, Fig. 16.10) with time series of SRI. There is an interesting distinction between the relationship of SRI with the former during the period. On the one hand, SRI is significantly correlated with SPI ($r = 0.69$ and 0.64, $p < 0.05$) in climatological periods (1931–1990 and 1961–1990, respectively) with high positive amplitudes (wet episodes) of SPI (Fig. 16.12). On the other hand, though significant, SRI is weakly associated with SPI ($r = 0.47$ and 0.38, $p < 0.05$) during periods when the preponderance of SPI values indicates drought conditions (1903–1930 and 1991–2010) (Fig. 16.12). The relationship between SPI and SRI is poorest during climatological period (1991–2010) when drought was severe and persistent ($r = 0.38$). During this period, specifically between 1995 and 1999 (Fig. 16.12), SRI indicated positive values (except 1998) on contrary to SPI, which showed drought condition. However, on the contrary, to observe relatively higher correlations between SPI and SRI during wet periods (Fig. 16.12), for the entire period (1903–2010), SPI temporal pattern is modestly associated with SRI ($r = 0.51$).

The considerable variability in the temporal evolutions of SRI over the Congo basin is largely defined by changes in precipitation patterns and other land surface conditions such as temperature extremes, evapotranspiration rates and perhaps land-use change. A synoptic assessment of SRI and SPI largely suggests that overall, rainfall was the main driver of hydrological conditions in the basin between 1903 and 1990 (Fig. 16.12). Alsdorf et al. (2016) noted the strong variability in Congo river's discharge between 1960 and 1995, and also reported a 21% increase in the Congo river discharge during the 1960–1970 period. As shown in Fig. 16.12, SPI was considerably associated with SRI during this particular decade (1960–1970), suggesting that increased rainfall led to a rise in the Congo river discharge. The consistent fluctuations observed in both time series of SPI and SRI, especially between 1930 and 1985 changed after 1994 (Fig. 16.12). SRI remained consistently positive showing wet episodes from 1995 until 2000 (except 1998), while SPI was largely negative during the same period (Fig. 16.12). Whereas one month lag exists between rainfall and the Congo river discharge for maximum association (Ndehedehe et al. 2018a), this is not the case as there is an absolute shift in hydrological response of the Congo river after 1994.

16.4.4 Coupled Variability of Spatio-Temporal SPI Patterns with SST Anomalies

The observed coupled modes of CCA based on SPI and SST are significant (except the third CCA modes of the Indian and Pacific oceans) based on the Wilks's lambda criterion, F statistics, and chi-square statistics (Table 16.2). Figures 16.13, 16.14, and 16.15 show that the coupled modes of the observed variability in droughts over the Congo basin are linked to tropical SST fluctuations. Strong canonical positive loadings (warming) in the north Atlantic Ocean are potential predictors of variability in SPI variations in the basin (CCA-1, Fig. 16.13). The leading CCA mode indicated in Fig. 16.13, accounting for 24.7% of the total variability in the Atlantic Ocean is dominated by Atlantic Meridional Mode (AMM) and Atlantic Multi-decadal Mode (AMO) signals. The temporal patterns corresponding to the spatial maps of SST in this mode are highly correlated ($r = 0.87$) and indicate (i.e., predictors) significant relationship with AMM ($r = 0.72$) and AMO ($r = 0.65$) (Table 16.3), suggesting that these two oscillations are prominent climate forcing and oceanic indices of variability related to extreme weather fluctuations in the basin.

Note that all three CCA modes from the Atlantic Ocean show a significant relationship with SPI unlike Indian Ocean and Pacific Ocean, where only two statistically significant modes from each of the oceans were found (Table 16.2). The observed cooling of the north and south Atlantic Ocean (predictor) also shows strong interactions with SPI loadings, which show a spatial structure similar to seasonal rainfall in the basin (CCA-2, Fig. 16.13). Whereas the temporal series of the second CCA mode (CCA-2, Fig. 16.13) is also well correlated ($r = 0.76$), the third CCA mode explains some variability in ENSO ($r = 0.40$), Pacific Decadal Oscillation (PDO) ($r = 0.40$), and other teleconnections (Table 16.3).

Table 16.2 Statistical significance test for the canonical correlation model. The test is based on Wilk's lambda (λ) criterion, F-statistics, and chi-squared (χ^2). Observed canonical correlation (Canon. correlation) from the CCA model for the canonical variates (U and V) is indicated. The significance (Sig.) of F, χ^2, and the probability values (ρ) of the model are used in accepting or rejecting the null hypothesis (i.e., $r_c = 0$) that there is no association between each pair of canonical variates (i.e., the predictor-SPI and the predictand-SST). *Source* Ndehedehe et al. (2019a)

Variable	Wilk's (λ)	F	χ^2	Canon. correlation	Sig. of F	Sig. of χ^2	ρ (model)
Atlantic Ocean SST verses SPI							
CCA-1	0.036	4.52	1184.99	0.87	0.000	0.000	0.000
CCA-2	0.145	2.89	690.39	0.76	0.000	0.000	0.000
CCA-3	0.339	1.88	386.32	0.72	0.000	0.000	0.000
Indian Ocean SST verses SPI							
CCA-1	0.226	2.58	531.63	0.76	0.000	0.000	0.000
CCA-2	0.5437	1.26	218.25	0.60	0.018	0.017	0.017
CCA-3	0.8515	0.43	57.88	0.30	1.000	1.000	1.000
Pacific Ocean SST verses SPI							
CCA-1	0.1328	4.90	722.78	0.84	0.000	0.000	0.000
CCA-2	0.4673	2.21	272.65	0.69	0.000	0.000	0.000
CCA-3	0.9089	0.38	34.33	0.27	1.000	1.000	1.000

Table 16.3 Association of canonical variates (temporal series of SST-U) with climate oscillations based on Pearson's correlation. Observed correlation are statistically significant at $\alpha = 0.05$ for only values in bold fonts. *Source* Ndehedehe et al. (2019a)

Teleconnection	ENSO	PDO	AMO	AMM	IOD	MJO	QBO
Atlantic Ocean SST							
CCA-1	−0.14	**−0.31**	**0.65**	**0.72**	0.15	**−0.23**	−0.02
CCA-2	−0.17	−0.06	−0.19	0.15	**−0.24**	0.05	0.05
CCA-3	**0.40**	**0.40**	−0.20	**0.29**	−0.01	**0.29**	−0.07
Indian Ocean SST							
CCA-1	**0.40**	0.18	−0.03	−0.07	**0.44**	**0.38**	0.10
CCA-2	0.13	0.05	−0.09	**−0.26**	**0.48**	0.13	0.01
CCA-3	0.1	0.05	0.22	−0.11	0.04	0.07	0.12
Pacific Ocean SST							
CCA-1	**0.89**	**0.53**	**−0.30**	**−0.26**	**0.30**	**0.69**	0.08
CCA-2	**−0.25**	**−0.23**	0.13	0.07	−0.13	−0.15	−0.10
CCA-3	0.14	−0.04	0.18	−0.21	0.02	0.15	0.04

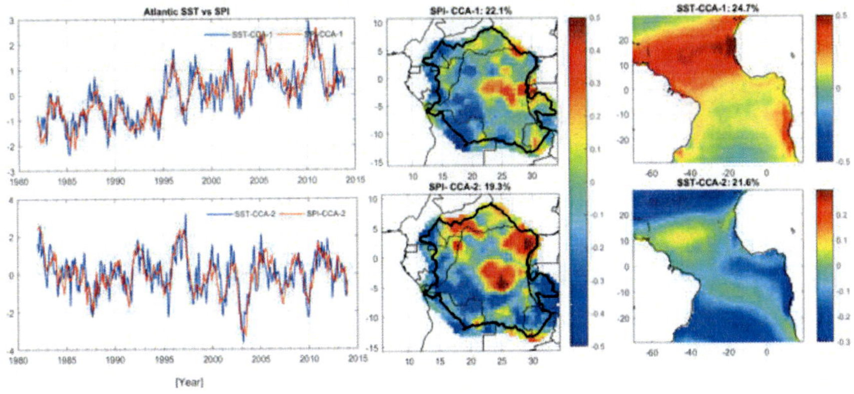

Fig. 16.13 Leading canonical modes of SST over the Atlantic Ocean and SPI over the Congo basin during the 1982 – 2014 period. The canonical component time series of SST (predictors) and SPI (predictands) are in normalized units (left). The SPI spatial patterns (middle) are in standardized unit while SST loadings (anomalies) have been adjusted to their original unit (right). Their variances (%) explained in percentages are also indicated. *Source* Ndehedehe et al. (2019a)

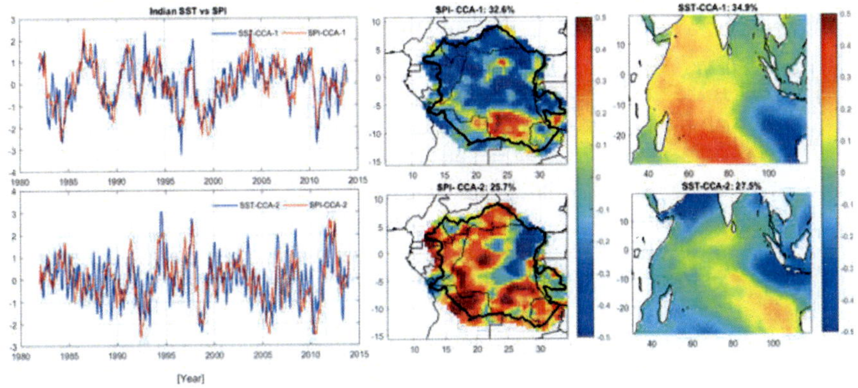

Fig. 16.14 Leading canonical modes of SST over the Indian Ocean and SPI over the Congo basin during the 1982 – 2014 period similar to Fig. 16.13. The variance (%) explained by each canonical mode in percentages is also indicated. *Source* Ndehedehe et al. (2019a)

Observed dipole pattern in the first CCA mode over the Indian Ocean is considerably associated ($r = 0.76$) with SPI signal in the Congo basin (Table 16.2 and CCA-1, Fig. 16.14). This dipole pattern of the Indian Ocean impacts considerable sections of the north and central Congo basin. Even the second CCA mode also show some weak dipole patterns that are fairly associated ($r = 0.60$) with SPI (Table 16.2 and CCA-2, Fig. 16.14). The Indian Ocean only shows moderate association with SPI over the basin and does not act alone as it only plays a complementary role. In Table 16.3, the

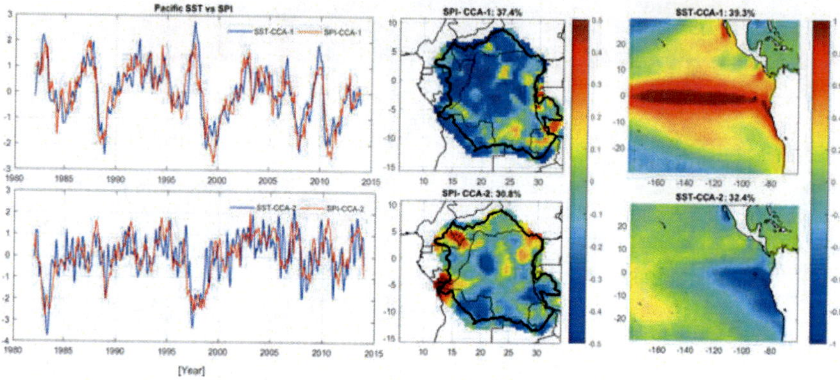

Fig. 16.15 Leading canonical modes of SST over the Pacific Ocean and SPI over the Congo basin during the 1982 – 2014 period similar to Figs. 16.13 and 16.14. The variance (%) explained by each canonical mode in percentages is also indicated. *Source* Ndehedehe et al. (2019a)

canonical variates (predictor time series of the Indian Ocean) are modestly correlated with Indian Ocean Dipole-IOD, ENSO, and Madden Julian Oscillation (MJO). Contrary to the Atlantic and Pacific oceans where the dominant signals are AMM/AMO (CCA-1, Fig. 16.13) and ENSO (CCA-1, Fig. 16.15), respectively, the modest relationship between IOD index and the predictor time series (SST) from the canonical model confirm the supporting role of the Indian Ocean in driving the strength of ENSO and teleconnections of the Atlantic Ocean. The complementary influence of the latter and the Indian Ocean on rainfall variability and the distribution of land water storage in the African sub-region has been reported (see, e.g., Ndehedehe et al. 2018a; Bader and Latif 2011; Farnsworth et al. 2011; Nicholson and Selato 2000; Giannini et al. 2003).

The coupled mode of the observed variability of SPI patterns with Pacific SST (predictor time series) shows that comparatively, the equatorial Pacific Ocean exerts the most influence on the Congo basin's extreme weather conditions. Apart from showing relatively the highest variability (39.3%) compared to the other two oceans, this predictor time series corresponding to the first canonical loadings (CCA-1, Fig. 16.15) show strong inter-annual variability and association with SPI ($r = 0.84$), i.e., the predictand time series. The predictor time series (SST) of this leading canonical mode is largely dominated by the ENSO signal, indicating a correlation of $r = 0.89$ with the ENSO index. Interestingly, the correlation results summarized in Table 16.3 also identify other low-frequency oscillations, e.g., PDO and MJO, amongst others as having a significant relationship with the predictor time series (SST). However, the structure of the equatorial Pacific SST pattern in the canonical loading whose corresponding temporal evolution (CCA-1, Fig. 16.15) is ENSO related confirms that the latter is arguably the prominent and principal force of the spatio-temporal variability of droughts over the Congo basin. Even the SST anomalies of the second canonical mode (CCA-2, Fig. 16.15) accounting for 32.4% of the total variability show a strong correlation with SPI temporal patterns ($r = 0.69$, Table 16.2) and also explain some variability (though weak) in both ENSO and PDO teleconnections (Table 16.3). Considering the several indices of oceanic variability and their association with the predictor time series of the Pacific Ocean (Table 16.3), the coupled effect of the ENSO phenomenon is emphasized on the one hand. The presence of other multiple physical mechanisms contributing to and regulating climate variability in the Pacific, on the other hand, is also insinuated.

16.4.5 Large-Scale Climatic Drivers

The considerable relationship between dominant SST canonical modes identified from the three oceans and several global teleconnection indices (Table 16.3 and Fig. 16.16a–f) confirms that the Congo basin is one of the climatic hotspots in Africa influenced by a sequence of oceanic phenomenon and litany of large-scale climatic

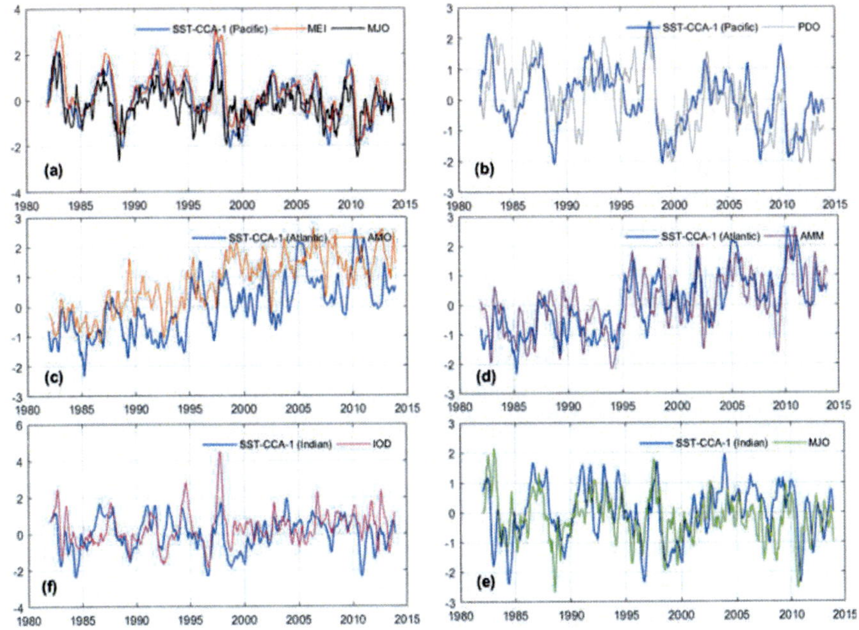

Fig. 16.16 Relationship of observed climate oscillations with leading modes of oceanic variability identified from the canonical scheme that employs SPI evolutions over the Congo basin and SST anomalies from the three oceans. Similar associations of canonical variates (temporal series of SST) with climate oscillations for the remaining canonical modes are summarized in Table 16.3. *Source* Ndehedehe et al. (2019a)

Table 16.4 Assessing the standard ANN model performance and the relationship between the predicted and observed SPI time series using Pearson's correlation (r) and coefficient of determination (R^2). The prediction of SPI is based on different climate oscillations identified in the canonical modes. Both R and R^2 are significant at $p < 0.05$. *Source* Ndehedehe et al. (2019a)

Criteria	QBO	MJO	IOD	AMO	AMM	ENSO	PDO
R	0.92	0.82	0.63	0.88	0.84	0.88	0.65
R^2	0.84	0.68	0.40	0.78	0.72	0.77	0.42

drivers. As captured in Table 16.3, the fairly strong association ($p < 0.05$) of the predictand time series obtained from the main portion of the CCA procedure (the leading CCA modes from the Atlantic, Indian, and Pacific Oceans) with ENSO/MJO (Fig. 16.16a), PDO (Fig. 16.16b), AMO (Fig. 16.16c), AMM (Fig. 16.16d), and perhaps IOD (Fig. 16.16f) are not only indicative of global climate influence on the Congo basin's extreme weather events but suggest the logicality of assimilating these large-scale climatic drivers in an experimental setup that seeks to understand and model the impact of global climate not just on drought evolutions but freshwater dynamics (e.g., surface runoff). Hence, understanding the drivers of freshwater, drought variability, and forecasting its episodes in the Congo basin requires an inclusive approach incorporating these climate teleconnections. After a robust training and validation that employed a closed-loop non-linear autoregressive network with external input, these large-scale climate modes were then used as regression covariates (e.g., Deo and Şahin 2015) for the forecast of future monthly SPI values. The statistical evaluation of the numerical results based on a wide range of prediction skill metrics showed that Quasi Biennial Oscillation (QBO), ENSO, AMO, and AMM are the leading climate modes that explained a considerable proportion of variability in SPI patterns in the Congo basin (Table 16.4 and Fig. 16.17a–d). Even MJO (Fig. 16.17b) showed higher predictive potential than IOD and PDO (Fig. 16.17f–g). Interestingly, several repeated training based on other activation functions of the non-linear autoregressive network scheme still indicated that the predicted SPI time series was considerably associated with QBO, AMO, AMM, and ENSO (Fig. 16.17a, d–f). The ANN autoregressive predictive framework shows that apart from the $3-7$ years spectral peaks in ENSO signal which is considerably associated with predicted SPI, the $2-3$ years spectral energy in QBO is more related to predicted SPI ($R^2 = 0.84$) as opposed to other teleconnections (Table 16.4; Fig. 16.17a–g). Apart from ENSO and AMO, the MJO also shows strong predictive potential ($R^2 = 0.68$) and aligns with other reports that demonstrate the significant role of MJO on convection and precipitation anomalies in the region (e.g., Berhane et al. 2015).

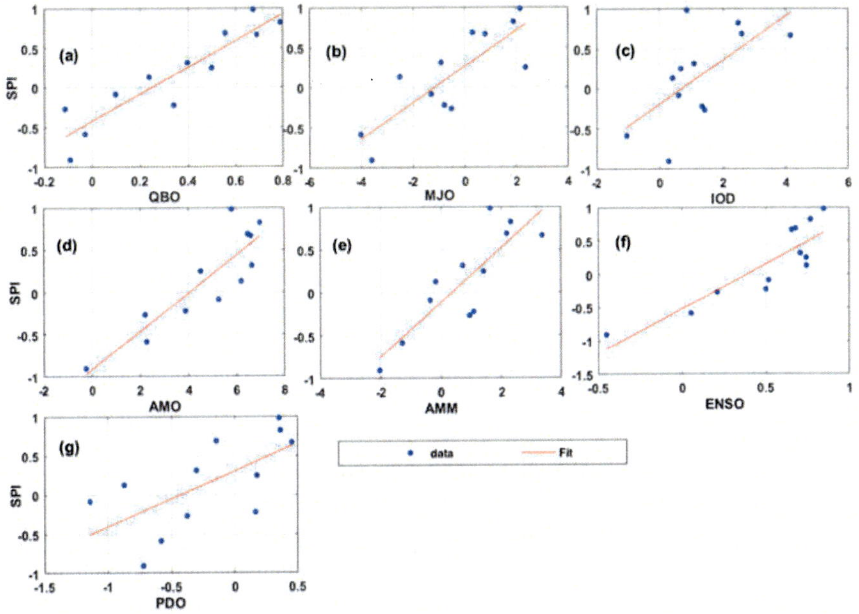

Fig. 16.17 Observed relationship between predicted SPI time series and observed global climate oscillations over the Congo basin (i.e., the prediction was for 12 time steps). The SPI and climate modes used for simulation and training were 372 time steps, i.e., from 1981 to 2012. This association is based on the 12 time steps of exogenous variable not used in the simulation. The model performance and other prediction skill metrics used in evaluating the predicted SPI are detailed in Table 16.4. *Source* Ndehedehe et al. (2019a)

16.4.6 Impacts of Droughts on Terrestrial Hydrology

In retrospect, there are two main historical periods in the Congo basin when several drought intensities occurred. During these periods (1901–1920 and 1990–2010), not only were severe and extreme droughts multi-decadal or multi-year (Fig. 16.11), the variability of dominant temporal drought patterns was relatively higher as opposed to the period between 1920 and 1980 (Fig. 16.10), which appeared to be somewhat devoid of extreme drought episodes that covered up to 15% of the basin (except for 1949, 1958, and 1973). A comparative assessment of SPI and SRI within these climatological and historical windows shows that there was a strong deviation in their temporal evolutions largely accounting for the observed modest (1903–1930) and poor (1991–2010) associations compared to other periods (Fig. 16.12). Although antecedent conditions exist, the SPI and SRI time series of the last two decades could imply that rainfall is not the only driver of hydrological conditions in the Congo basin. Strong evapotranspiration losses caused by substantial declines in soil moisture and droughts (e.g., Ndehedehe et al. 2018b; Jung 2010) can alter hydrological regimes in the Congo basin. However, the physiographic properties of tributaries that feed the

Congo river do have complex drainage systems that could create a non-stationary relationship between river discharge and rainfall.

In drought propagation studies (e.g., Loon and Laaha 2014), meteorological drought conditions impact the terrestrial hydrological cycle through a trickle-down effect that culminates in soil moisture and hydrological droughts. Processes such as seasonality effects, catchment, and climate characteristics largely affect drought propagation (e.g., Van Loon et al. 2014). Rainfall deficits in tropical climates would normally result in reduced alimentation and a temporary decrease in stream flows, storage reservoirs, and freshwater stocks. However, this was not the case in the Congo basin (1995 – 2010) as most drought episodes indicated by SPI are inconsistent with SRI, which as shown in Fig. 16.12 is largely characterized by relatively wet conditions even during periods when droughts persevered. It is unclear in the Congo basin if there are known physical and ecological processes that play key roles in drought propagation. But the basin's catchment stores (e.g., swamps, lakes, reservoirs, soil column, groundwater, etc.), which create a prolonged reservoir memory in the hydrological system could be a determinant in the delayed propagation of drought signals or even its absence in the SRI data. The SRI of the last two decades in particular alludes to the huge tropical wetland and surface water resources of the Congo basin such that despite periods of extreme and severe droughts, the Congo river nonetheless indicated wet conditions. As the only river basin that seconds the Amazon river in terms of average yearly discharge (i.e., about 40,200 $m^3\ s^{-1}$), the amount of water filling and draining the basin (111 km^3) is relatively large though still less than 50% of Amazon's storage (see Lee et al. 2011; Alsdorf et al. 2010). This kind of storage capacity increases the catchment's response time to below-normal precipitation, creating a non-linear relationship and essentially resulting in an asymmetric response of river discharge to a drought signal (e.g., Loon 2013). Overall, the observed change in the hydrological response of the Congo river to a strong meteorological shift since 1994 is not only hydrologically insightful but interesting and eminently informative for a number of reasons.

This change in the hydrological response of the Congo river suggests the threshold of non-linear interactions and complex hydrological processes in the basin. Against the backdrop of long-term declines (1982–2014) in land water storage (including soil moisture), and significant negative trends in GRACE-terrestrial water storage-TWS changes (2003 – 2013) observed over the basin (e.g., Ndehedehe et al. 2018a; Ahmed et al. 2014), the SRI time series do not completely reflect these conditions, especially between 1995 and 2010. It is therefore vital to rely on a multivariate drought index (see, e.g., Ndehedehe et al. 2016; AghaKouchak 2015; Hao and AghaKouchak 2014), or in other words, multiple drought indices derived from other water budget indicators to effectively characterize hydrological drought and model key hydrological metrics. Furthermore, it provides a mechanistic pathway to productively engage and revisit ecosystem response to extreme drought in the Congo basin. The observed drought in the Congo basin between 1991 and 2010 (PC1, Fig. 16.10 and 16.12) is consistent with the results of Spinoni et al. (2014) who observed on a global scale, prolonged and severe droughts during the same period (1991 – 2010) over the Congo basin, Amazon Forest, and north-eastern China, amongst other regions. Further, the droughts in the Congo basin during the same period triggered the pervasive decline in forest greenness in the huge vegetation blocks of Central African Republic (see Zhou et al. 2014). The Amazon basin with similar hydrology to that of the Congo was not left out. It experienced amongst other things, large basin-wide losses of biomass due to the 2005 drought and a basin-wide impact of the 2010 drought on tree growth rates across Amazonia (see Feldpausch et al. 2016). For the Congo basin, there may be some concerns on ecosystem carbon cycle as droughts are known to influence the equilibrium in terrestrial carbon by changing the rates of carbon uptake by photosynthesis and release by total ecosystem respiration, in addition to the coupling between the latter and the former (see van der Molen et al. 2011). Stressing on the impacts of long-term droughts on tropical rainforest systems, Zhou et al. (2014) argued that the persistent drying of the basin could lead to compositional and structural changes in the Congolese forest. As home to the world's second-largest rainforest block (e.g., Oslisly et al. 2013), it is critical to advance our knowledge on long-term effects of droughts on ecophysiology and ecosystem dynamics. This can be achieved in future studies through well-crafted hypothesis that seeks to understand both the physiological (e.g., preventing water loss through stomatal closure) and structural (leaf area reduction caused by early leaf senescence) response of vegetation to drought (e.g., van der Molen et al. 2011) in the Congo basin.

16.4.7 *Influence of Global Climate Modes on Drought Evolutions*

The Congo basin is one hydrologically pristine location in the globe where little is known about the influence of these indices of oceanic forcings and variability on drought development. Results from the canonical scheme suggest that drought development and its characteristics in the Congo basin are largely perturbed by the surrounding oceans. A number of independent climate oscillations (ENSO, AMO, AMM, IOD, PDO, and MJO), which are predominantly characterized by SST or explain a considerable proportion of variability in the SST anomalies were found to be significantly associated with SPI patterns over the Congo basin (Tables 16.2 and 16.3; Figs. 16.16a-e). The inter-annual or decadal variability of drought is extremely complex in the basin and suggests that several factors modulate its evolutions. These factors are in the similitude of a general warming or cooling of the SST of the north Atlantic Ocean, Indian Ocean, and the central equatorial Pacific Ocean. This conclusion is not far from other diagnostic reports on the role of SST fluctuations on drought and rainfall variability in the basin. For example, the key factors that govern the complex changes in precipitation patterns in West Central Africa as identified by Balas et al. (2007) include SST anomalies along the Benguela Coast, and warming or cooling of the three oceans (Atlantic, Indian, and Pacific Oceans). Recently, Dyer et al. (2017) suggests that understanding the link between the large-scale circulation patterns over the Indian Ocean and the local circulation over the Congo basin is key to future projections of the basin's precipitation.

Some studies have linked ENSO and IOD with drought events in much of western, eastern, and southern Africa (e.g., Gizaw and Gan 2017; Ndehedehe et al. 2016; Andam-Akorful et al. 2017; Ujeneza and Abiodun 2015). Given this influence and the impacts of the coupled variability of ENSO and IOD on Africa's precipitation, Williams and Hanan (2011) earlier advocated for the analyses of the simultaneous states of ENSO and IOD when examining influences on hydro-climatic and ecological conditions across Africa. This call was predicated on the observed influence of these teleconnections on photosynthesis across much of Africa linked to anomalies in rainfall and vegetation greenness. However, human interventions such as land-use change could also drive water deficits that impact on land surface conditions in the Congo basin. The study by Nogherotto et al. (2013) demonstrated this, suggesting that decreased evaporation over deforested areas caused heat low, which resulted in reduced precipitation.

ENSO is characterized by SST and surface pressure anomalies across the Pacific Ocean (e.g., Williams and Hanan 2011) and is the largest climate teleconnection pattern known to have substantial influence on extreme events and hydro-meteorological processes in several parts of the world. The canonical scheme implemented in the Congo basin confirms that amongst other global climate signals, the central Pacific ENSO event is a major index of oceanic variability that influences the occurrence and variability in drought episodes in the region. However, the less reported QBO pattern showed relatively stronger predictive potential than ENSO and other higher frequency oscillations. Interestingly, multi-decadal signals from the Atlantic Ocean (AMO and AMM) also indicated relatively strong association with variations in SPI. This further echoes the influence of the north Atlantic Ocean over the climate of the basin. Overall, this study reveal that large-scale climatic signals associated with extreme hydro-meteorological conditions in the Congo basin are predominantly characterized by quasi-biennial, multi-decadal, and inter-annual oscillations. Incorporating these large-scale climatic drivers in a multivariate non-parametric forecast scheme could improve the prediction of freshwater availability in the basin. Such improvement are useful in supporting drought risk mitigation studies as well as in the development of global and regional climate models.

16.5 Conclusion

Tracking and predicting ocean–atmosphere interactions and processes associated with drought episodes can improve our knowledge on drought evolutions and variability in profound ways. First, it can ensure the continued development of infrastructural and economic instruments to manage and share risks. Finally, it could also inform government and institutional efforts in raising the level of drought preparedness and promote region-specific drought risk-management policy solutions. These solutions may include mitigation strategies and complementary interventions, e.g., insurance programmes that could help cushion the effects of drought risk. Hence, drought studies are warranted to support specific policy issues relevant to food security governance, especially in non-industrialized regions where economic output and national income are considerably sensitive to hydro-climatic variability The causes of droughts in several regions of the world have been linked to multiple indices of oceanic variability, such as the ENSO (El-Niño Southern Oscillation), and other inter-decadal and multi-decadal global climate oscillations. Therefore, to manage drought risk caused by these large-scale oceanic–atmospheric patterns, the knowledge of space–time variability of extreme weather events and their drivers is crucial.

References

Achard F, Eva HD, Stibig H-J, Mayaux P, Gallego J, Richards T, Malingreau J-P (2002) Determination of deforestation rates of the world's humid tropical forests. Science 297(5583):999–1002. https://doi.org/10.1126/science.1070656

AghaKouchak A (2015) A multivariate approach for persistence-based drought prediction: application to the 2010–2011 East Africa drought. J Hydrol 526:127–135. https://doi.org/10.1016/j.jhydrol.2014.09.063

Agutu N, Awange J, Ndehedehe C, Kirimi F, Kuhn M (2019) GRACE-derived groundwater changes over Greater Horn of Africa: temporal variability and the potential for irrigated agriculture. Sci Total Environ 693:133467. https://doi.org/10.1016/j.scitotenv.2019.07.273

Agutu N, Awange J, Ndehedehe C, Mwaniki M (2020) Consistency of agricultural drought characterization over Upper Greater Horn of Africa (1982–2013): topographical, gauge density, and model forcing influence. Sci Total Environ 709:135149. https://doi.org/10.1016/j.scitotenv.2019.135149

Agutu N, Awange J, Zerihun A, Ndehedehe C, Kuhn M, Fukuda Y (2017) Assessing multi-satellite remote sensing, reanalysis, and land surface models' products in characterizing agricultural drought in East Africa. Remote Sens Environ 194:287–302. https://doi.org/10.1016/j.rse.2017.03.041

Ahmed M, Sultan M, Wahr J, Yan E (2014) The use of GRACE data to monitor natural and anthropogenic induced variations in water availability across Africa. Earth-Sci Rev 136:289–300. https://doi.org/10.1016/j.earscirev.2014.05.009

Al-Kilani MR, R. M. A.-B. J. e. a. (2021) Evaluation of remotely sensed precipitation estimates from the NASA POWER project for drought detection over Jordan. Earth Syst Environ 5:561–573. https://doi.org/10.1007/s41748-021-00245-2

Alsdorf D, Beighley E, Laraque A, Lee H, Tshimanga R, O'Loughlin F, Mahé G, Dinga B, Moukandi G, Spencer RGM (2016) Opportunities for hydrologic research in the Congo basin. Rev Geophys 54(2):378–409. https://doi.org/10.1002/2016RG000517

Alsdorf D, Han S-C, Bates P, Melack J (2010) Seasonal water storage on the amazon floodplain measured from satellites. Remote Sens Environ 114(11):2448–2456. https://doi.org/10.1016/j.rse.2010.05.020

Andam-Akorful S, Ferreira V, Ndehedehe CE, Quaye-Ballard J (2017) An investigation into the freshwater variability in West Africa during 1979 – 2010. Int J Climatol 37(S1):333–349. https://doi.org/10.1002/joc.5006

Aswathanarayana U (2001) Water resources management and the environment. CRC Press, Boca Raton

Awange J, Khandu Schumacher M, Forootan E, Heck B (2016) Exploring hydro-meteorological drought patterns over the Greater Horn of Africa (1979–2014) using remote sensing and reanalysis products. Adv Water Res 94:45–59. https://doi.org/10.1016/j.advwatres.2016.04.005

Bader J, Latif M (2011) The 1983 drought in the West Sahel: a case study. Clim Dyn 36(3–4):463–472. https://doi.org/10.1007/s00382-009-0700-y

Balas N, Nicholson SE, Klotter D (2007) The relationship of rainfall variability in West Central Africa to sea-surface temperature fluctuations. Int J Climatol 27(10):1335–1349. https://doi.org/10.1002/joc.1456

Becker A, Finger P, Meyer-Christoffer A, Rudolf B, Schamm K, Schneider U, Ziese M (2013) A description of the global land-surface precipitation data products of the Global Precipitation Climatology Centre with sample applications including centennial (trend) analysis from 1901 to present. Earth Syst Sci Data 5(1):71–99. https://doi.org/10.5194/essd-5-71-2013

Berhane F, Zaitchik B, Badr HS (2015) The Madden-Julian oscillation's influence on spring rainy season precipitation over Equatorial West Africa. J Clim 28(22):8653–8672. https://doi.org/10.1175/JCLI-D-14-00510.1

Bowman DM, Williamson GJ, Abatzoglou JT, Kolden CA, Cochrane MA, Smith AM (2017) Human exposure and sensitivity to globally extreme wildfire events. Nat Ecol Evolut 1(3):1–6. https://doi.org/10.1038/s41559-016-0058

Bretherton CS, Smith C, Wallace JM (1992) An intercomparison of methods for finding coupled patterns in climate data. J Clim 5(6):541–560. https://doi.org/10.1175/1520-0442(1992)005<0541:AIOMFF>2.0.CO;2

Charney J, Quirk WJ, Chow S, Kornfield J (1977) A comparative study of the effects of albedo change on drought in semi-arid regions. J Atmos Sci 34(9):1366–1385. https://doi.org/10.1175/1520-0469(1977)034<1366:ACSOTE>2.0.CO;2

Chen JL, Wilson CR, Tapley BD, Longuevergne L, Yang ZL, Scanlon BR (2010) Recent La Plata basin drought conditions observed by satellite gravimetry. J Geophys Res: Atmos 115(D22):D22108. https://doi.org/10.1029/2010JD014689

Cooley H, Donnelly K, Phurisamban R, Subramanian M (2015) Impacts of California's ongoing drought: agriculture. Oakland, CA, USA, Pacific Institute

Cook BI, Mankin JS, Anchukaitis KJ (2018) Climate change and drought: from past to future. Curr Clim Change Reports 4(2):164–179. https://doi.org/10.1007/s40641-018-0093-2

Cosgrove WJ, Rijsberman FR (2014) World water vision: making water everybody's business. Routledge, p 136. https://doi.org/10.4324/9781315071763

Cr S, Wr A, Am M, Jb F, F, B., G, K., M, L., K, O., JD, S., A, W., DN, H. (2017) Global patterns of drought recovery. Nature 548(7666):202–205. https://doi.org/10.1038/nature23021

Dai A (2011) Drought under global warming: a review. Wiley Interdiscip Rev: Clim Change 2(1):45–65. https://doi.org/10.1002/wcc.81

Deering DW (1978) Rangeland reflectance characteristics measured by aircraft and spacecraft sensors. Texas A & M University

Deo RC, Şahin M (2015) Application of the extreme learning machine algorithm for the prediction of monthly effective drought index in Eastern Australia. Atmos Res 153:512–525. https://doi.org/10.1016/j.atmosres.2014.10.016

Diffenbaugh NS, Swain DL, Touma D (2015) Anthropogenic warming has increased drought risk in California. In: Proceedings of the National Academy of Sciences

Dinku T, Ceccato P et al (2007) Validation of satellite rainfall products over East Africa's complex topography. Int J Remote Sens 28(7):1503–1526. https://doi.org/10.1080/01431160600954688

Diaz HF, Hoerling MP, Eischeid JK (2001) Enso variability, teleconnections and climate change. Int J Climatol 21(15):1845–1862. https://doi.org/10.1002/joc.631

Dyer ELE, Jones DBA, Nusbaumer J, Li H, Collins O, Vettoretti G, Noone D (2017) Congo basin precipitation: assessing seasonality, regional interactions, and sources of moisture. J Geophys Res: Atmos 122(13):6882–6898. https://doi.org/10.1002/2016JD026240

Enfield D, Mestas-Nuñez A (1999) Multiscale variabilities in global sea surface temperatures and their relationships with tropospheric climate patterns. J Clim 12(9):2719–2733. https://doi.org/10.1175/1520-0442(1999)012<2719:MVIGSS>2.0.CO;2

Epule ET, Peng C, Lepage L, Chen Z (2014) The causes, effects and challenges of Sahelian droughts: a critical review. Reg Environ Change 14(1):145–156. https://doi.org/10.1007/s10113-013-0473-z

Farnsworth A, White E, Williams CJ, Black E, Kniveton DR (2011)D Understanding the large scale driving mechanisms of rainfall variability over central Africa, in African climate and climate change: Physical, social and political perspectives. African Climate and Climate Change: Physical, Social and Political Perspectives, pp 101–122. https://doi.org/10.1007/978-90-481-3842-5_5

Feldpausch TR, Phillips OL et al ((2016) Amazon forest response to repeated droughts. Global Biogeochem Cycles 30(7):964–982. https://doi.org/10.1002/2015GB005133.

Ferreira V, Montecino H, Ndehedehe C, Heck B, Gong Z, Westerhaus M, de Freitas S (2018) Space-based observations of crustal deflections for drought characterization in brazil. Sci Total Environ 644:256–273. https://doi.org/10.1016/j.scitotenv.2018.06.277

Fuller DO, Ottke C (2002) Land cover, rainfall and land-surface albedo in West Africa. Clim Change 54(1):181–204. https://doi.org/10.1023/A:1015730900622

Gerdener H, Engels O, Kusche J (2020) A framework for deriving drought indicators from the Gravity Recovery and Climate Experiment (GRACE). Hydrol Earth Syst Sci 24(1):227–248. https://doi.org/10.5194/hess-24-227-2020

Getirana A (2016) Extreme water deficit in brazil detected from space. J Hydrometeorol 17(2):591–599. https://doi.org/10.1175/JHM-D-15-0096.1

Giannini A, Saravanan R, Chang P (2003) Oceanic forcing of Sahel rainfall on interannual to decadal time scales. Science 302(5647):1027–1030. https://doi.org/10.1126/science.1089357

Gidley SL (2009) Using high resolution satellite imagery to map aquatic macrophytes on multiple lakes in Northern Indiana. Unpublished MSC thesis, Indiana University. Retrived from https://core.ac.uk/download/pdf/46956355.pdf on 15th April, 2019

Gibson RN, Barnes M, Atkinson RJA (2002) Impact of changes in flow of freshwater on estuarine and open coastal habitats and the associated organisms. Oceanogr Mar Biol Ann Rev 40:233

Gizaw MS, Gan TY (2017) Impact of climate change and El Niño episodes on droughts in sub-Saharan Africa. Clim Dyn 49(1):665–682. https://doi.org/10.1007/s00382-016-3366-2

Glahn HR (1968) Canonical correlation and its relationship to discriminant analysis and multiple regression. J Atmos Sci 25(1):23–31. https://doi.org/10.1175/1520-0469(1968)025<0023:CCAIRT>2.0.CO;2

Graham NE, Michaelsen J, Barnett TP (1987) An investigation of the el niño-southern oscillation cycle with statistical models: 1. predictor field characteristics. J Geophys Res: Oceans 92(C13):14251–14270. https://doi.org/10.1029/JC092iC13p14251

Griffin D, Anchukaitis KJ (2014) How unusual is the 2012–2014 California drought? Geophys Res Lett 41(24):9017–9023. https://doi.org/10.1002/2014GL062433

Guha-Sapir D, Hargitt D, Hoyois P (2004) Thirty years of natural disasters 1974–2003: The numbers. Presses univ, de Louvain

Haile GG, Tang Q, Sun S, Huang Z, Zhang X, Liu X (2019) Droughts in East Africa: causes, impacts and resilience. Earth Sci Rev 193:146–161. https://doi.org/10.1016/j.earscirev.2019.04.015

Haley MR (2017) K-fold cross validation performance comparisons of six naive portfolio selection rules: how naive can you be and still have successful out-of-sample portfolio performance? Ann Finance 13(3):341–353. https://doi.org/10.1007/s10436-017-0301-4

Hanasaki N, Yoshikawa S, Pokhrel Y, Kanae S (2018) A global hydrological simulation to specify the sources of water used by humans. Hydrol Earth Syst Sci 22(1):789–817. https://doi.org/10.5194/hess-22-789-2018

Hao Z, AghaKouchak A (2014) A nonparametric multivariate multi-index drought monitoring framework. J Hydrometeorol 15(1):89–101. https://doi.org/10.1175/JHM-D-12-0160.1

He J, Yang X, Li Z, Zhang X, Tang Q (2016) Spatiotemporal variations of meteorological droughts in china during 1961–2014: an investigation based on multi-threshold identification. Int J Disaster Risk Sci 7(1):63–76

Hotelling H (1936) Relations between two sets of variates. Biometrika 28(3–4):321–377. https://doi.org/10.2307/2333955

Houborg R, Rodell M, Li B, Reichle R, Zaitchik BF (2012) Drought indicators based on model-assimilated Gravity Recovery and Climate Experiment (GRACE) terrestrial water storage observations. Water Resour Res 48(7):W07525. https://doi.org/10.1029/2011WR011291

Hua W, Zhou L, Chen H, Nicholson SE, Raghavendra A, Jiang Y (2016) Possible causes of the Central Equatorial African long-term drought. Environ Res Lett 11(12):124002. https://doi.org/10.1088/1748-9326/11/12/124002

Hunt ED, Hubbard KG, Wilhite DA, Arkebauer TJ, Dutcher AL (2009) The development and evaluation of a soil moisture index. Int J Climatol 29(5):747–759. https://doi.org/10.1002/joc.1749

IPCC (2014) Climate change 2014 synthesis report summary chapter for policymakers. Accessed from (https://www.ipcc.ch/site/assets/uploads/2018/05/SYR_AR5_FINAL_full_wcover.pdf) on 12thAugust 2021

Jolliffe IT (2002) Principal component analysis, 2nd edn. Springer Series in Statistics. Springer, New York

Jung M e. a. (2010) Recent decline in the global land evapotranspiration trend due to limited moisture supply. Nature 467(7318):951–954. https://doi.org/10.1038/nature09396

Khashei M, Bijari M (2010) An artificial neural network (p, d, q) model for timeseries forecasting. Expert Syst Appl 37(1):479–489. https://doi.org/10.1016/j.eswa.2009.05.044

Kiem AS, Johnson F, Westra S, van Dijk A, Evans JP, O'Donnell A, Rouillard A, Barr C, Tyler J, Thyer M, Jakob D, Woldemeskel F, Sivakumar B, Mehrotra R (2016) Natural hazards in Australia: droughts. Clim Change 139(1):37–54. https://doi.org/10.1007/s10584-016-1798-7

Kitzberger T, Falk DA, Westerling AL Swetnam TW (2017) Direct and indirect climate controls predict heterogeneous early-mid 21st century wildfire burned area across western and boreal North America. PloS one, 12(12).https://doi.org/10.1371/journal.pone.0188486

Knutti R, Sedláček J (2013) Robustness and uncertainties in the new cmip5 climate model projections. Nat Clim change 3(4):369–373. https://doi.org/10.1038/nclimate1716

Kubiak-Wójcicka K, Bąk B (2018) Monitoring of meteorological and hydrological droughts in the Vistula basin (Poland). Environ Monit Assess 190(11):1–16. https://doi.org/10.1007/s10661-018-7058-8

Lamprey HF (1988) Report on the desert encroachment reconnaissance in Northern Sudan. Desertif Control Bull 17:1–7

Latif M, Barnett T (1996) Decadal climate variability over the North Pacific and North America: dynamics and predictability. J Clim 9(10):2407–2423. https://doi.org/10.1175/1520-0442(1996)009<2407:DCVOTN>2.0.CO;2

Lee H, Beighley RE, Alsdorf D, Jung HC, Shum C, Duan J et al (2011) Characterization of terrestrial water dynamics in the Congo Basin using GRACE and satellite radar altimetry. Remote Sens Environ 115(12):3530–3538. https://doi.org/10.1016/j.rse.2011.08.015

Li B, Rodell M, Kumar S, Beaudoing H, Getirana A, Zaitchik B, e. a. (2019) Global GRACE data assimilation for groundwater and drought monitoring: advances and challenges. Water Resour Res 55(9):7564–7586. https://doi.org/10.1029/2018WR024618

Li AK, Paek H, Yu J-Y (2016) The changing influences of the AMO and PDO on the decadal variation of the Santa Ana winds. Environ Res Lett 11(6):064019. https://doi.org/10.1088/1748-9326/11/6/064019

Lin T, Horne BG, Tino P, Giles CL (1996) Learning long-term dependencies in NARX recurrent neural networks. IEEE Trans Neural Netw 7(6):1329–1338. https://doi.org/10.1109/72.548162

Long D, Scanlon BR, Longuevergne L, Sun AY, Fernando DN, Save H (2013) GRACE satellite monitoring of large depletion in water storage in response to the 2011 drought in Texas. Geophys Res Lett 40(13):3395–3401. https://doi.org/10.1002/Grl.50655

Loon AFV, Laaha G (2014) Hydrological drought severity explained by climate and catchment characteristics. J Hydrol. https://doi.org/10.1016/j.jhydrol.2014.10.059

Loon AV (2013) On the propagation of drought. how climate and catchment characteristics influence hydrological drought development and recovery. *PhD thesis, Wageningen University, Wageningen, NL* 198 p. Retrieved from http://library.wur.nl/WebQuery/wurpubs/438510 on 1st February 2018

Maity R, Suman M, Verma NK (2016) Drought prediction using a wavelet based approach to model the temporal consequences of different types of droughts. J Hydrol 539:417–428. https://doi.org/10.1016/j.jhydrol.2016.05.042

Majumdar S, Smith R, Butler Jr JJ, Lakshmi V (2020) Groundwater withdrawal prediction using integrated multitemporal remote sensing data sets and machine learning. Water Resour Res 56(11):e2020WR028059. https://doi.org/10.1029/2020WR028059.

Marengo JA, Alves LM, Alvala RC, Cunha AP, Brito S, Moraes OL (2018) Climatic characteristics of the 2010–2016 drought in the semiarid Northeast Brazil region. Anais da Academia Brasileira de Ciencias 90:1973–1985. https://doi.org/10.1590/0001-3765201720170206

Maxwell D, Fitzpatrick M (2012) The 2011 somalia famine: Context, causes, and complications. Glob Food Sec 1(1):5–12. https://doi.org/10.1016/j.gfs.2012.07.002

McKee TB, Doeskin NJ, Kieist J (1993) The relationship of drought frequency and duration to time scales. *Conference on Applied Climatology, American Meteorological Society,Boston, Massachusetts*, pp 179–184. Retrieved from www.ccc.atmos.colostate.edu/relationshipofdroughtfrequency.pdf. Accessed 27 June, 2014

McPhaden MJ (2004) Evolution of the 2002/03 el-nino. Bull Am Meteor Soc 85(5):677–696. https://doi.org/10.1175/BAMS-85-5-677

McPhaden MJ, Timmermann A, Widlansky MJ, Balmaseda MA, Stockdale TN (2015) The curious case of the el nino that never happened: a perspective from 40 years of progress in climate research and forecasting. Bull Am Meteor Soc 96(10):1647–1665. https://doi.org/10.1175/BAMS-D-14-00089.1

Mcphaden MJ, Zebiak SE, Glantz MH (2006) Enso as an integrating concept in earth science. Science 314:1740–1745

Mishra AK, Singh VP (2010) A review of drought concepts. J Hydrol 391:202–216. https://doi.org/10.1016/j.jhydrol.2010.07.012

Molina-Terrén DM, Cardil A (2016) Temperature determining larger wildland fires in ne Spain. Theoret Appl Climatol 125(1):295–302. https://doi.org/10.1007/s00704-015-1511-1

Ndehedehe CE, Agutu N, Ferreira VG, Getirana A (2020) Evolutionary drought patterns over the Sahel and their teleconnections with low frequency climate oscillations. Atmos Res 233:104700. https://doi.org/10.1016/j.atmosres.2019.104700

Ndehedehe CE, Agutu NO, Okwuashi OH, Ferreira VG (2016) Spatio-temporal variability of droughts and terrestrial water storage over Lake Chad Basin using independent component analysis. J Hydrol 540:106–128. https://doi.org/10.1016/j.jhydrol.2016.05.068

Ndehedehe CE, Anyah RO, Alsdorf D, Agutu NO, Ferreira VG (2019) Modelling the impacts of global multi-scale climatic drivers on hydro-climatic extremes (1901–2014) over the Congo basin. Sci Total Environ 651:1569–1587. https://doi.org/10.1016/j.scitotenv.2018.09.203

Ndehedehe CE, Awange J, Kuhn M, Agutu N, Fukuda Y (2017) Climate teleconnections influence on West Africa's terrestrial water storage. Hydrol Process 31(18):3206–3224. https://doi.org/10.1002/hyp.11237

Ndehedehe CE, Awange JL, Agutu NO, Okwuashi O (2018) Changes in hydro-meteorological conditions over tropical West Africa (1980 – 2015) and links to global climate. Global Planet Change 162:321–341. https://doi.org/10.1016/j.gloplacha.2018.01.020

Ndehedehe CE, Burford MA, Stewart-Koster B, Bunn SE (2020) Satellite-derived changes in floodplain productivity and freshwater habitats in northern Australia (1991–2019). Ecol Ind 114:106320. https://doi.org/10.1016/j.ecolind.2020.106320

Ndehedehe CE, Ferreira VG (2020) Identifying the footprints of global climate modes in time-variable gravity hydrological signals. Clim Change 159:481–502. https://doi.org/10.1007/s10584-019-02588-2

Ndehedehe CE, Ferreira VG, Agutu NO (2019) Hydrological controls on surface vegetation dynamics over West and Central Africa. Ecol Ind 103:494–508. https://doi.org/10.1016/j.ecolind.2019.04.032

Ndehedehe CE, Ferreira VG, Agutu NO, Onojeghuo AO, Okwuashi O, Kassahun HT, Dewan A (2021) What if the rains do not come? J Hydrol 595:126040. https://doi.org/10.1016/j.jhydrol.2021.126040

Ndehedehe CE, Ferreira VG, Onojeghuo AO, Agutu NO, Emengini E, Getirana A (2020) Influence of global climate on freshwater changes in Africa's largest endorheic basin using multi-scaled indicators. Sci Total Environ 737:139643. https://doi.org/10.1016/j.scitotenv.2020.139643

Ndehedehe CE, Haile GG, Agutu NO, Ferreira VG, Getirana A, Okwuashi O (2020) Hydrological hotspots of climatic influence in Brazil: A two-step regularization approach. Atmos Res 246:105116. https://doi.org/10.1016/j.atmosres.2020.105116

Ndehedehe CE, Okwuashi O, Ferreira VG, Agutu NO (2018) Exploring evapotranspiration dynamics over Sub-Sahara Africa (2000 – 2014). Environ Monit Assess 190(7):400. https://doi.org/10.1007/s10661-018-6780-6

Ni S, Chen J, Wilson CR, Li J, Hu X, Fu R (2017) Global terrestrial water storage changes and connections to ENSO events. Surv Geophys. https://doi.org/10.1007/s10712-017-9421-7

Nicholson S, Selato J (2000) The influence of La-Nina on African rainfall. Int J Climatol 20(14):1761–1776. https://doi.org/10.1002/1097-0088(20001130)20:14<1761::AID-JOC580>3.0.CO;2-W

Nicholson SE, Tucker CJ, Ba MB (1998) Desertification, drought, and surface vegetation: an example from the West African Sahel. Bull Am Meteorol Soc 79(5):815–829. https://doi.org/10.1175/1520-0477(1998)079<0815:DDASVA>2.0.CO;2

NJ M, SR H (2002) The pacific decadal oscillation. J Oceanogr 58(1):35–44. https://doi.org/10.1023/A:1015820616384

Nogherotto R, Coppola E, Giorgi F, Mariotti L (2013) Impact of Congo basin deforestation on the African monsoon. Atmos Sci Lett 14(1):45–51. https://doi.org/10.1002/asl2.416

Oelkers EH, Hering JG, Zhu C (2011) Water: is there a global crisis? Elements 7(3):157–162. https://doi.org/10.2113/gselements.7.3.157

O'Loughlin F, Trigg MA, Schumann GJ-P, Bates PD (2013) Hydraulic characterization of the middle reach of the Congo river. Water Resour Res 49(8):5059–5070. https://doi.org/10.1002/wrcr.20398

Oslisly R, White L, Bentaleb I, Favier C, Fontugne M, Gillet J, Sebag D (2013) Climatic and cultural changes in the West Congo Basin forests over the past 5000 years. Philos Trans Roy Soc B. Biol Sci 368:10. https://doi.org/10.1098/rstb.2012.0304

Peters E, Lanen HAJ, Bradford RB, Cruces de Abia J, Martinez Cortina L (2001) Droughts derived from groundwater heads and groundwater discharge. pp 35–39

Preisendorfer R(1988) Principal component analysis in meteorology and oceanography. Developments in atmospheric science, vol 17. Elsevier, Amsterdam

JM Q, P A (2012) Changes in the rainfall regime along the extratropical west coast of South America (Chile): 30–43° s. Atmosfera 25(1):1–22

Reynolds RW, Rayne NA, Smith TM, Stokes DC, Wang W (2002) An improved in situ and satellite SST analysis for climate. J Clim 15(3):1609–1625. https://doi.org/10.1175/1520-0442(2002)0153C1609:AIISAS3E2.0.CO;2

Riebsame WE, Changnon SA, Karl TR (2019) Drought and natural resources management in the united states: impacts and implications of the 1987–89 drought. Routledge. 192 p. https://doi.org/10.4324/9780429045332

Rippey BR (2015) The U.S. drought of 2012. Weather and climate extremes 10(Part A):57 – 64. https://doi.org/10.1016/j.wace.2015.10.004

Ruiz LGB, Cuéllar MP, Calvo-Flores MD, Jiménez MDCP (2016) An application of non-linear autoregressive neural networks to predict energy consumption in public buildings. Energies 9(9). https://doi.org/10.3390/en9090684

Savitzky A, Golay MJE (1964) Soothing and differentiation of data by simplified least squares procedures. Anal Chem 36(8):1627–1639

Schneider U, Becker A, Finger P, Meyer-Christoffer A, Ziese M, Rudolf B (2014) GPCC's new land surface precipitation climatology based on quality-controlled in situ data and its role in quantifying the global water cycle. Theoret Appl Climatol 115(1–2):15–40. https://doi.org/10.1007/s00704-013-0860-x

Seager R, Ting M, Li C, Naik N, Cook B, Nakamura J, Liu H (2013) Projections of declining surface-water availability for the Southwestern United States. Nat Clim Chang 3(5):482–486. https://doi.org/10.1038/nclimate1787

Sheffield J, Wood EF, Chaney N, Guan K, Sadri S, Yuan X, Olang L, Amani A, Ali A, Demuth S, Ogallo L (2014) A drought monitoring and forecasting system for Sub-Sahara African water resources and food security. Bull Am Meteor Soc 95(6):861–882. https://doi.org/10.1175/BAMS-D-12-00124.1

Sheffield J, Andreadis KM, Wood EF, Lettenmaier DP (2009) Global and continental drought in the second half of the twentieth century: severity-area-duration analysis and temporal variability of large-scale events. J Clim 22(8):1962–1981. https://doi.org/10.1175/2008JCLI2722.1

Sheffield J, Wood EF (2007) Characteristics of global and regional drought, 1950–2000: analysis of soil moisture data from off-line simulation of the terrestrial hydrologic cycle. J Geophys Res: Atmos 112(D17). https://doi.org/10.1029/2006JD008288

Sheffield J, Wood EF, Roderick ML (2012) Little change in global drought over the past 60 years. Nature 491(7424):435–438. https://doi.org/10.1038/nature11575

Singh A, Reager JT, Behrangi A (2021) Estimation of hydrological drought recovery based on precipitation and Gravity Recovery and Climate Experiment (GRACE) water storage deficit. Hydrol Earth Syst Sci 25(2):511–526. https://doi.org/10.5194/hess-25-511-2021

Spinoni J, Naumann G, Carrao H, Barbosa P, Vogt J (2014) World drought frequency, duration, and severity for 1951–2010. Int J Climatol 34(8):2792–2804. https://doi.org/10.1002/joc.3875

Su B, Huang J, Fischer T e. a. (2018) Drought losses in china might double between the 1.5c and 2.0c warming. Proc Natl Acad Sci 115(42):10600–10605. https://doi.org/10.1073/pnas.1802129115

Swain DL, Singh D, Horton DE, Mankin JS, Ballard TC, Diffenbaugh NS (2017) Remote linkages to anomalous winter atmospheric ridging over the northeastern pacific. J Geophys Res: Atmos 122(22):12,194–12,209. https://doi.org/10.1002/2017JD026575

Thomas AC, Reager JT, Famiglietti JS, Rodell M (2014) A GRACE-based water storage deficit approach for hydrological drought characterization. Geophys Res Lett 41(5):1537–1545. https://doi.org/10.1002/2014GL059323

Thomas BF, Famiglietti JS, Landerer FW, Wiese DN, Molotch NP, Argus DF (2017) GRACE groundwater drought index: evaluation of California Central Valley groundwater drought. Remote Sens Environ 198(Supplement C):384 – 392. https://doi.org/10.1016/j.rse.2017.06.026

Tockner K, Lorang MS, Stanford JA (2010) River flood plains are model ecosystems to test general hydrogeomorphic and ecological concepts. River Res Appl 26(1):76–86. https://doi.org/10.1002/rra.1328

Trenberth KE (1997) The definition of El Niño. Bull Am Meteorol Soc 78(12):2771–2777. https://doi.org/10.1175/1520-0477(1997)078<2771:TDOENO>2.0.CO;2

Trenberth KE (2011) Changes in precipitation with climate change. Climate Res 47(1–2):123–138. https://doi.org/10.3354/cr00953

Trenberth KE, Caron JM, Stepaniak DP, Worley S (2002) Evolution of El niño-Southern oscillation and global atmospheric surface temperatures. J Geophys Res: Atmos 107(D8):AAC5–1–AAC 5–17. https://doi.org/10.1029/2000JD000298

Trenberth KE, Dai A, Van Der Schrier G, Jones PD, Barichivich J, Briffa KR, Sheffield J (2014) Global warming and changes in drought. Nat Clim Chang 4(1):17–22. https://doi.org/10.1038/nclimate2067

Tucker CJ, Pinzon JE, Brown ME, Slayback DA, Pak EW, Mahoney R, Vermote EF, Saleous NE (2005) An extended AVHRR 8-km NDVI dataset compatible with MODIS and SPOT vegetation NDVI data. Int J Remote Sens 26(20):4485–4498. https://doi.org/10.1080/01431160500168686

Ujeneza EL, Abiodun BJ (2015) Drought regimes in Southern Africa and how well GCMs simulate them. Clim Dyn 44:1595–1609. https://doi.org/10.1007/s00382-014-2325-z

van der Molen M, Dolman A, Ciais P, Eglin T, Gobron N, Law B, Meir P et al (2011). Drought and ecosystem carbon cycling. Agric Forest Meteorol 151(7):765 – 773. https://doi.org/10.1016/j.agrformet.2011.01.018

Van Loon AF (2015) Hydrological drought explained. WIREs. Water 2:359–392. https://doi.org/10.1002/wat2.1085

Van Loon AF, Kumar R, Mishra V (2017) Testing the use of standardised indices and GRACE satellite data to estimate the European 2015 groundwater drought in near-real time. Hydrol Earth Syst Sci 21(4):1947–1971. https://doi.org/10.5194/hess-21-1947-2017

Van Loon AF, Stahl K, Di Baldassarre G, Clark J, Rangecroft S, Wanders N, Gleeson T, Van Dijk AIJM, Tallaksen LM, Hannaford J, e. a. (2016) Drought in a human-modified world: reframing drought definitions, understanding, and analysis approaches. Hydrol Earth Syst Sci 20(9):3631–3650. https://doi.org/10.5194/hess-20-3631-2016

Van Loon AF, Tijdeman E, Wanders N, Van Lanen HA, Teuling AJ, Uijlenhoet R (2014) How climate seasonality modifies drought duration and deficit. J Geophys Res: Atmos 119(8):4640–4656. https://doi.org/10.1002/2013JD020383

Verdin J, Funk C, Senay G, Choularton R (2005) Climate science and famine early warning. Philos Trans R Soc B: Biol Sci 360(1463):2155–2168. https://doi.org/10.1098/rstb.2005.1754

Vicente-Serrano SM, Beguería S, Lorenzo-Lacruz J, e. a. (2012) Performance of drought indices for ecological, agricultural, and hydrological applications. Earth Interact 16(10):1–27. https://doi.org/10.1175/2012EI000434.1

Vicente-Serrano SM, Beguería S, López-Moreno JI (2010) A multiscalar drought index sensitive to global warming: the standardized precipitation evapotranspiration index. J Clim 23(7):1696–1718. https://doi.org/10.1175/2009JCLI2909.1

Vicente-Serrano SM, Beguería S, López-Moreno JI, Angulo M, El Kenawy A (2010b). A new global 0.5° gridded dataset (1901–2006) of a multiscalar drought index: Comparison with current drought index datasets based on the palmer drought severity index. J Hydrometeorol 11(4):1033–1043. https://doi.org/10.1175/2010JHM1224.1

Vorosmarty CJ, McIntyre PB, Gessner MO, Dudgeon D, Prusevich A, Green P, Glidden S, Bunn SE, Sullivan CA, Liermann CR, Davies PM (2010) Global threats to human water security and river biodiversity. Nature 467:555–555–561. https://doi.org/10.1038/nature09440

Washington R, James R, Pearce H, Pokam WM, Moufouma-Okia W (2013). Congo Basin rainfall climatology: can we believe the climate models? Philos Trans R Soc Lond B: Biol Sci 368(1625). https://doi.org/10.1098/rstb.2012.0296

Webster KE, Kratz TK, Bowser CJ, Magnuson JJ, Rose WJ (1996) The influence of landscape position on lake chemical responses to drought in Northern Wisconsin. Limnol Oceanogr 41(5):977–984. https://doi.org/10.4319/lo.1996.41.5.0977

White WB, Gershunov A, Annis JL, McKeon G, Syktus J (2004) Forecasting Australian drought using Southern Hemisphere modes of sea-surface temperature variability. Int J Climatol 24(15):1911–1927. https://doi.org/10.1002/joc.1091

Williams CA, Hanan NP (2011) ENSO and IOD teleconnections for African ecosystems: evidence of destructive interference between climate oscillations. Biogeosciences 8(1):27–40. https://doi.org/10.5194/bg-8-27-2011

WMO and GWP (2017). Integrated drought management programme (IDMP) working paper 1. *Geneva: WMO and Stockholm: GWP*

Wolf G, Brayshaw DJ, Klingaman NP, Czaja A (2018) Quasi-stationary waves and their impact on European weather and extreme events. Q J R Meteorol Soc 144(717):2431–2448. https://doi.org/10.1002/qj.3310

Wolski P, Murray-Hudson M, Thito K, Cassidy L (2017) Keeping it simple: Monitoring flood extent in large data-poor wetlands using MODIS SWIR data. Int J Appl Earth Obs Geoinf 57:224–234. https://doi.org/10.1016/j.jag.2017.01.005

Wolter K, Timlin MS (2011) El niño/southern oscillation behaviour since 1871 as diagnosed in an extended multivariate ENSO index (MEI.ext). Int J Climatol 31(7):1074–1087. https://doi.org/10.1002/joc.2336

Yin J, Zhan X, Hain CR, Liu J, Anderson MC (2018) A method for objectively integrating soil moisture satellite observations and model simulations toward a blended drought index. Water Resour Res 54(9):6772–6791. https://doi.org/10.1029/2017WR021959

Yirdaw SZ, Snelgrove K, Agboma C (2008) GRACE satellite observations of terrestrial moisture changes for drought characterization in the Canadian Prairie. J Hydrol 356:84–92

Zeng N, Neelin JD, Lau K-M, Tucker CJ (1999) Enhancement of interdecadal climate variability in the Sahel by vegetation interaction. Science 286(5444,):1537–1540. https://doi.org/10.1126/science.286.5444.1537

Zeng N (2003) Drought in the Sahel. Science 302(5647):999–1000

Zhang G, Patuwo BE, Hu MY (1998) Forecasting with artificial neural networks: the state of the art. Int J Forecast 14(1):35–62. https://doi.org/10.1016/S0169-2070(97)00044-7

Zhang X, Obringer R, Wei C, Chen N, Niyogi D (2017) Droughts in India from 1981 to 2013 and implications to wheat production. Sci Rep 7(44552):1–11. https://doi.org/10.1038/srep44552

Zhao M, Geruo A, Velicogna I, Kimball J (2017) Satellite observations of regional drought severity in the continental United States using GRACE-based terrestrial water storage changes. J Clim 30(16):6297–6308. https://doi.org/10.1175/JCLI-D-16-0458.1

Zhou L, Tian Y, Myneni RB, Ciais P, e. a. (2014) Widespread decline of congo rainforest greenness in the past decade. Nature 509(7498):86–90. https://doi.org/10.1038/nature13265

Chapter 17
Influence of Global Climate on Terrestrial Water Storage

Abstract Identifying the coupled relationship between terrestrial stored water and the surrounding oceans is critical because these oceans are important sources of large-scale mesoscale convective systems, and other climate variability drivers (e.g., El-Nino Southern Oscillation-ENSO). Perturbations of the nearby oceans influence meteorological processes and induce extreme climatic conditions (e.g., droughts). The impacts of the latter on hydrological systems (e.g., lakes, reservoirs, floodplains) and water resources are undeniable. In this chapter, the role of these global climate teleconnection (climate variability indices) patterns on terrestrial hydrology is discussed. Methodological approaches that incorporate advanced multivariate statistics and parameter estimation techniques to quantify their contributions on terrestrial water storage (soil moisture, groundwater, surface, water) are detailed. Important hotspots and test beds (South America and Africa) where novel methodological frameworks have been implemented using Gravity Recovery and Climate Experiment satellite data are examined.

17.1 Understanding Climatic Drivers of Land Water Storage

17.1.1 The Role of Teleconnections on Water Cycle

There is evidence that rapid changes in the global hydrological cycle are driven by the influence of climate variability indices and teleconnection patterns (Phillips et al. 2012; Hurkmans et al. 2009). The influence of large-scale teleconnection patterns like the El-Nino Southern Oscillation (ENSO) events and low-frequency oscillations (e.g., Pacific Decal Oscillation-PDO) on rainfall and hydrological conditions is well known (Ni et al. 2017; Ndehedehe et al. 2017; Sheffield and Wood 2008). The eastern south Pacific has an extreme high-pressure zone as well as an equally strong low-pressure zone prevalent over the western south Pacific. Rasmusson (2019)

C. Ndehedehe, *Satellite Remote Sensing of Terrestrial Hydrology*,
https://doi.org/10.1007/978-3-030-99577-5_17

defined the southern oscillation as the pressure interchange between the west and east, and highlighted its significance in interannual climate fluctuations. The western tropical pacific-Indonesian region plays a key role in convectional rainfall. The enormous warm water pool produced from the region has strong relationships with heavy rainfall observed in the region and is seen as a major source of atmospheric warming, creating a strong driver for large-scale sub-regional convection circulation (Rasmusson 2019).

The trade winds are usually driven to the western pacific from the eastern pacific through the atmospheric pressure dissimilarities. Water moves with the direction of these trade winds towards the west along the equator, thereby preserving an accumulation of warm surface water, which directly results in an advanced sea level altitude, noticed over the western pacific. When the variance in the normalized sea-level atmospheric pressure between Society Island & Darwin, Tahiti, and Australia (often referred to as the Southern Oscillation Index) drops beyond zero, the atmospheric pressure of the eastern pacific decreases, which in turn leads to a weakening of the trade winds. Over the coast of Southern Ecuador and Peru lies the warm water pool, which is extended eastwards to that direction. This phenomenon is characterized as a symbol of the development of the El Nino event. According to Cane (1983), the observation and recording of the El Nino events have been in progress since 1726. This provides an adequate time series for the assessments of their impacts and is estimated to occur once in 4 years approximately. However, the time lag between sequential events fluctuates between 2 and 10 years. Given that no two events are similar, the patterns are consistent while their dissimilarity exists mainly in intensity, timing, extent, and duration. Four classifications of the El Nino Southern Oscillation include, extremely weak, weak, average, and strong. However, each of them consists of significant differences when compared to the other. Between 1982 and 1983, the strongest El Nino pattern on record was observed. This was also the strongest El Nino pattern in the century and had a distinctive pattern in terms of the warming series and the onset period (Cane 1983).

The El-Nino Southern Oscillation is a notable atmospheric circulation pattern that contributes to warm events especially in the tropical pacific ocean, with each warm episode having a two-decadal life cycle. As highlighted by past studies (Rasmusson and Carpenter 1983; Cane 1983), ENSO-related episodes have substantially impacted on the weather and atmospheric patterns of the United States. However, the type of El Nino contributes to the nature of impacts on weather systems. In 1941 and 1983, severe ENSO effects, which resulted in heavy rainfalls, especially in the West and Southwestern regions and in the Colorado River basin were observed. Most ENSO episodes result in dry conditions in these regions. For example, the event reported in 1986/1987. In assessing the integrated potential effects of the southern oscillations on total water resource systems, it is crucial to understand stream flow patterns. The stream flow readings produce unimpaired measurements, especially when applied in mountainous regions. The mid-latitude modulation of the tropical ENSO phenomena occurs by disrupting the large-scale atmospheric exchange, which ultimately results in an altered surface climate pattern especially in temperature, wind and rainfall patterns. Also, since the catchment space does not account for the total time reliance features experienced within basins, this directly influences the ocean, which consists of a substantial energy source, which is likely responsible for the hydrologic conservation observed in the region.

To understand the impact of the ENSO climatic modulation on stream flow patterns, it is pertinent to learn the dynamic non-linear association of rainfall patterns and stream flow. Kahya and Dracup (1993) pointed out four basic intermediate processes of this association which are evaporation, transpiration, infiltration, and sub/surface storage. The dynamic role of effective rainfall in modulating stream flow fluctuations has been highlighted by past studies (Kuhnel et al. 1990). Globally, the observed fluctuations in annual operative rainfall series is greater than fluctuation in the annual rainfall series (McMahon et al. 1987). It is therefore expected that rainfall pattern will vary from stream flow patterns observed in the mid-latitudes.

17.1.2 Teleconnection Patterns in Terrestrial Hydrology

The impact of climate variability indices such as the ENSO on changes in freshwater is mostly noticed in the evolution of extreme climatic conditions such as flood and drought events (Ndehedehe et al. 2017). The western African region, for example, has been a major recipient of drying trends and scarcity in soil moisture content, rainfall rate, and net-precipitation. This sub-regional variability pattern has therefore been connected to the varied techniques in oceanic inter-annual fluctuations, human-induced emission of greenhouse gases, as well as the warming of the tropical oceans (Andam-Akorful et al. 2017; Nicholson 2013; Oguntunde and Abiodun 2013; Giannini et al. 2008; Sheffield and Wood 2008). The competing multiple physical mechanisms as outlined by Druyan (2011) can potentially influence climatic trends, which in turn will translate to substantial changes in terrestrial water storage. Notably, such impacts on terrestrial stored water can be either directly or indirectly. Terrestrial water storage is an integration of total surface waters, soil moisture, groundwater, and canopy storage in any particular region. The impacts of these mechanisms or global teleconnection patterns on any of these hydrological stores (surface water, groundwater) or state variables (soil moisture) might be different. These impacts have been studied and evidence show that terrestrial stored water in all its forms respond to changes in the global climate through the influence of global climate patterns around the world (e.g., Phillips et al. 2012; Hurkmans et al. 2009; Malhi and Wright 2004).

The variation of aquifer levels reported in Japan was associated with changes in ENSO patterns amongst other climatic drivers including, temperature, rainfall patterns, and barometric pressure (see, Dong et al. 2015). Other regions like Africa have experienced rise in aquifer storage rates of 30 mm/yr due to the multi-annual and decadal fluctuations observed in the region (Scanlon et al. 2006). In the United States, strong impacts of the trans-Nino index during the warm phase of the pacific decadal oscillation in the Upper Klamath Lake were also reported (e.g., Kennedy et al. 2009). Rainfall and hydrological conditions in the West African region are also influenced by the low frequency modulating El-Nino activity in the region (Andam-Akorful et al., 2017). It is expected that a notable increase in measured extremes in West African rainfall pattern, especially due to substantial impacts of climate change, and effects of the tropical oceans and teleconnection patterns (Losada et al. 2010; Redelsperger and Lebel 2009; Redelsperger et al. 2006; Polo et al. 2008) will increase the susceptibility of the region to large anomalies and changes in hydrology.

The role of climatic variations through global ocean changes, mesoscale convective systems, and indices of climate fluctuations (e.g., Atlantic Multi-decadal Oscillation, ENSO, etc.) on water fluxes and precipitation rates have been widely studied over the African continent (e.g., Ndehedehe et al. 2016c; Diatta and Fink 2014; Nicholson 2013; Paeth et al. 2012; Bader and Latif 2011; Joly and Voldoire 2010; Losada et al. 2010; Ali and Lebel 2009; Giannini et al. 2008). The impacts of climatic change on hydrological conditions in Africa are thus inevitable. The ability of these impacts to hamper ecosystem services, agricultural activities as well as diagnostic studies seeking to identify and asses climate teleconnection patterns in terrestrial hydrology are also documented. Given the substantial influence of climate teleconnections on meteorological interactions, the knowledge of how these teleconnections impact on terrestrial water storage is highly crucial. This is highly essential as the availability of freshwater storage as well as its characteristic (e.g., recharge and infiltration) contributes to the hydrological functioning of any sub-region.

Fluctuations in the ENSO and low-frequency climate oscillations like the PDO, and Atlantic Multi-decadal Oscillation are responsible for the globally observed variations in soil moisture content, rainfall rates and patterns, drought events, as well as the distribution of freshwater on different spatial and scales (e.g., Andam-Akorful et al. 2017; Dai et al. 2009; Hurkmans et al. 2009; Sheffield and Wood 2008; Scanlon et al. 2006). Generally, many sub-regions in the world have experienced a notable ENSO-induced influence amongst other varied global climate teleconnection patterns on climatic variables like rainfall and runoff (e.g., Ndehedehe et al. 2016c, b; Diatta and Fink 2014; Nicholson 2013; Molion and Lucio 2013). These climate teleconnections cause substantial anomalies in the hydrological conditions of any region, including accelerating the region's water cycle. The intrinsic response of rainfall to these teleconnections is different compared to surface water hydrology (freshwater resources). This is particularly so because freshwater is composed of a soil moisture, snow, groundwater, surface water and biomass, all of which have varied and unique responses to climatic variability indices (e.g., Ndehedehe et al. 2016a; Henry et al. 2011; Grippa et al. 2011; Hinderer et al. 2009).

17.1.2.1 Climate-induced Teleconnection on West Africa's Terrestrial Water Storage

West Africa is made up of a group of 16 countries located in two geographical zones with lots of freshwater resources; the Gulf of Guinea and the Sahel (Fig. 17.1). It has an approximate area of 7.5 million km^2 with an estimated population of 330 million people (e.g., Amani et al. 2007; Ndehedehe et al. 2016a; USAID 2013). West

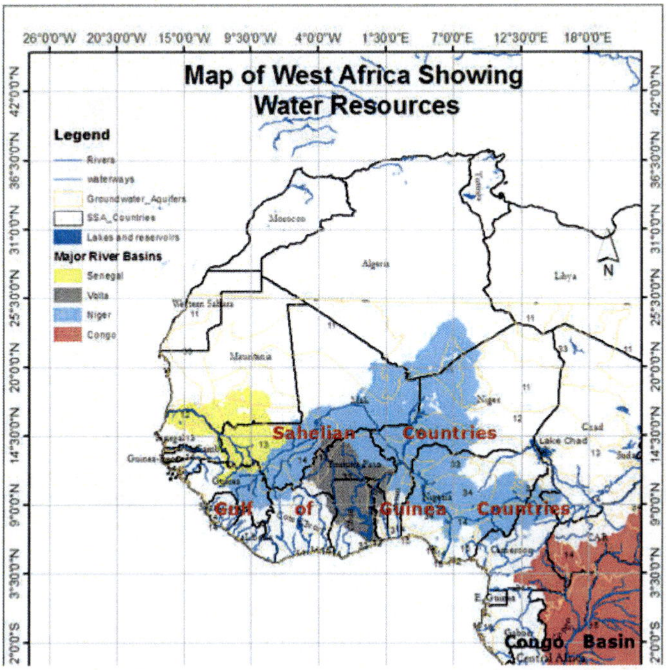

Fig. 17.1 West Africa and parts of Equatorial Africa (i.e., the Congo basin). Major river basins (e.g., Niger, Volta, Congo, and Senegal), rivers, lakes, and groundwater aquifers are also indicated. The Congo (pink) and Niger (blue) river basins are considerably large and apparently the most significant and prominent basins in the region owing to the two major rivers (Niger and Congo) that provide numerous ecosystem services. Aquifer types are described in terms of numbers; for example, the numbers ranging from 11 to 15 are found in major groundwater basins while the numbers 33 and 34 on the map, are those found in local and shallow aquifers. *Source* Ndehedehe et al. (2017)

Africa is one of the regions where extreme climatic conditions such as floods and droughts are driven by climate teleconnection patterns such as the El-Niño Southern Oscillation-ENSO. These patterns also represent considerable impact on land water storage as indicated in past studies (Ndehedehe et al. 2017; Andam-Akorful et al. 2017). The relationship of terrestrial water storage (TWS) with ENSO, Indian Ocean Dipole-IOD, and AMO) on West Africa's TWS was studied by Ndehedehe et al. (2017). This study was based on the assumption that these climate variability indices directly or remotely contribute to strong changes in hydro-meteorological conditions, including extreme wetness and dryness as implied in previous studies (e.g., Diatta and Fink 2014; Paeth et al. 2012; Nicholson et al. 2000). The contributions of these climate teleconnections (i.e., ENSO, AMO, and IOD) to TWS changes were assessed using GRACE (2002–2014) and long-term reanalysis data (MERRA, 1980–2014). The reanalysis data were used to compensate for the short term data availability of GRACE as some of these indices, e.g., the ENSO, has large energy between 2 and

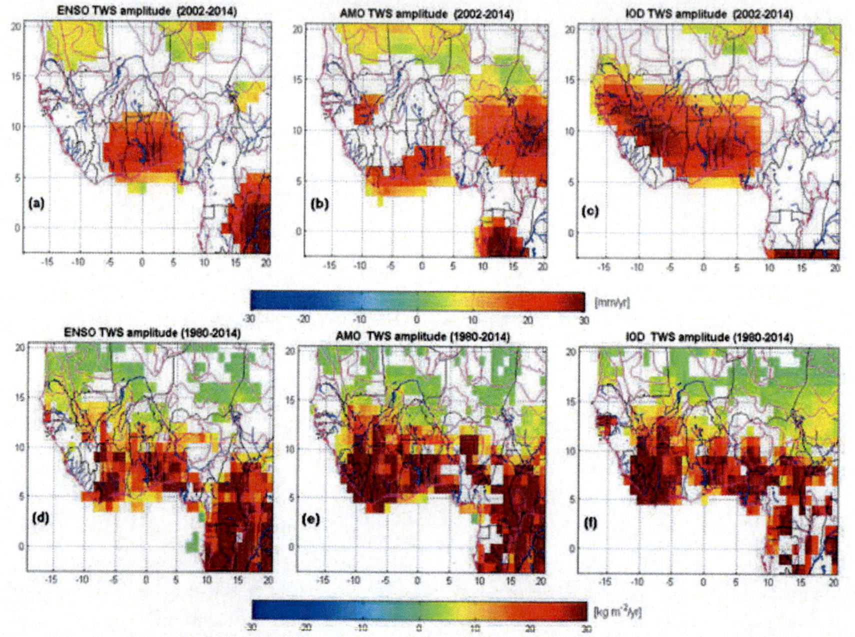

Fig. 17.2 Spatial patterns of climate-induced TWS derived from GRACE-TWS (**a–c**) and MERRA-TWS (**d–f**). These estimates (amplitudes of TWS) are derived using the MLRA technique. The regression coefficients indicated for both GRACE and MERRA data are those with statistically significant ($\alpha = 0.05$) association with climate indices (ENSO, AMO, and IOD). Groundwater aquifers (magenta), rivers and lakes (blue) have also been shown on the spatial distribution of climate-induced TWS

5 years, meaning only few oscillations will be captured in the limited GRACE data (Ndehedehe et al. 2017; Paeth et al. 2012). The other argument was around how the impacts of climate variability indices to other hydrological stores or water storage components are different. Rainfall is the key driver of TWS, and the impact of climate teleconnections on rainfall may be substantially different from how it directly impact on TWS. This is because the latter is vertical integration of hydrological stores (e.g., soil moisture, groundwater). These components of land water storage could have different responses to climate variability.

As shown in Fig. 17.2a and d, the contributions of ENSO to TWS based on GRACE and water storage from reanalysis data (MERRA) are moderately strong over the Volta and Congo basins. The influence of AMO on TWS appears to be stronger in the Guinea Coast areas and the equatorial regions where rainfall is relatively high and bimodal (Ndehedehe et al. 2016a). The decline in water levels over the Lake Volta was linked to the impact of ENSO phenomenon (Owusu et al. 2008). ENSO shows much stronger contributions to TWS in the region and the strong presence of AMO induced TWS in the Guinea coast area and much of the regions below latitude suggest the influence of both indices on TWS (Fig. 17.2). The AMO has a wider footprint and broader influence on the region's freshwater and suggests the critical role of North Atlantic temperature in the region. The relationships between TWS and teleconnections (ENSO, AMO and IOD) in much of the Sahel region are also statistically significant ($\alpha = 0.05$). Nonetheless, the contribution of IOD to TWS in the Sahel is considerably weak and may be related to the rainfall structure of the region (Fig. 17.2). Some locations in the Sahel show strong AMO contributions to GRACE-TWS compared to the observed weak AMO-related MERRA-TWS (Fig. 17.2). There are likelihood the GRACE-derived TWS–teleconnection (especially AMO) relationship could be the result of false association in some areas probably due to the limited time span of GRACE observations. However, as the records of GRACE observation increase with time, the physical mechanisms and other climate variability indices that influence TWS can be investigated much more systematically with clarity, simplicity, as well as with little uncertainties.

Presumptive evidence from the forgoing is that ENSO (i.e., depending on the phase) has more impact on TWS in West Africa. Ndehedehe et al. (2017) also showed that AMO has strong influence on the hydrology of the region. Positive phase of AMO in the Sahel coincides with above-normal rainfall or wet conditions and the negative phase with drought conditions (Ndehedehe et al. 2017). This is a further evidence that corroborates the strong presence of AMO-driven TWS in the region as shown in Fig. 17.2e. On the other hand, the impacts of IOD on TWS in the region seem to be unclear. Although the existence of IOD-induced TWS in the region appears to be complimentary, TWS over much of West Africa and countries of the Congo basin could be more influenced by the ENSO and AMO events.

17.2 Global Climate Teleconnection Patterns and Influence on TWS and Groundwater Drought

17.2.1 Climate Indices

Climate teleconnection patterns are referred to as climate modes and in some instances can be interchangeably called indices of climate variability. These climate modes are signals of regional climate variability, which are usually detected with empirical orthogonal function (EOF) analyses of one or many climate fluctuations in a specific oceanic hotspots. One of the most prominent contributors to sub-regional and global climate change is the El-Nino Southern Oscillation (ENSO), which induces several extreme climatic events ranging from heat waves, droughts, cyclones to severe flood and drought episodes in different regions (Santoso et al. 2017). Some of these identified climatic indices have been detailed and are highlighted below.

- Multivariate ENSO Index (MEI)—This is a multi-variable index proficient in the assessment of the ENSO average conditions (Wolter and Timlin 2011). Its distinction as one of the most critical element of sea surface temperatures, integrated EOFs assessment of sea level pressure, meridional and zonal modules of surface wind, as well as long-wave radiation, especially over the tropical pacific basins has been well corroborated.

- Pacific Decadal Oscillation (PDO)—Usually found in a decadal sea surface temperature oscillation pattern, the PDO is composed of a basin-wide dipole segregating the North Pacific along an emblematic northeast-southwest configuration (NJ and SR. 2002; Mantua et al. 1997). It is the primary EOF of sea surface temperature inconsistencies observed in the northern pacific region, and is most responsive when assessed with the ERSSTV5 dataset (Huang et al. 2017). The PDO is also known as the 'long-lived ENSO' given its similarity to the southern oscillation pattern in portraying long-term climatic indices.
- the North Pacific Gyre Oscillation (NPGO)—This is usually regarded as the additional mode of sea surface height fluctuations noticed especially in the northeast pacific (Di Lorenzo et al. 2008). Its correlation with PDO and ENSO is especially significant given its influence on global climate variability, as seen in its impacts on sea surface temperatures and global sea level trends (Di Lorenzo et al. 2010).
- Atlantic Multidecadal Oscillation (AMO)—the primary distinction of the AMO is in its long-term duration of at least two decades, and its possession of a cool and warm phase. Enfield et al. (2001) suggest that the AMO is contained in a multi-decadal sea surface temperature oscillation observed over the entire basin of the Atlantic.
- North Atlantic Ocean (NAO)—This index depicts the surface sea level pressure difference observed between the Subpolar lows and Subtropical Azores high (Lamb and Peppler 1987). The influence of this index on the location and the intensity of the North Atlantic storm tracks and jet streams has been identified (Woollings and Blackburn 2012). Hurrell et al. (2003) argued that the NAO efficiently moderates moisture and heat transfer, leading to a noticeable change in land/sea surface temperatures and rainfall patterns extending from North America to central Europe.
- Arctic Oscillation (AO)—It is the climatic pattern that consists of winds disseminating in an anti-clockwise direction over the Artic and is considerably correlated with NAO. Higgins et al. (2000) showed that the cold air masses are susceptible to high latitude restrictions by strong circular winds especially during positive AO phases, while the cold air masses move vibrantly southward, thereby intensifying the storminess at mid-latitudes at negative AO phases.
- Southern Annular Mode (SAM)—SAM is the primary driver of climatic fluctuations over the Southern hemisphere. Also referred to as the Antarctic oscillation (AAO), SAM has widely demonstrated great impact on storm, winds, rainfall patterns as well as temperature variations over the subtropics of the Antarctica (Gillett et al. 2006).
- Indian Ocean Dipole (IOD)—this is prevalent in interannual oscillations of the sea surface temperatures noticed over the equatorial Indian Ocean, developing an emblematic west-east dipole (Saji et al. 1999). It is also observed to be strongly related to the ENSO. Studies that have identified the footprints of IOD climate teleconnection patterns using satellite gravity solutions, like the Gravity Recovery and Climate Experiment (GRACE) mission have been reported (Ndehedehe and Ferreira 2020). This will be discussed in more detail in subsequent sections.

- Atlantic Meridional Mode-(AMM). The AMM index is a tropical Atlantic 'meridional' mode of interannual—decadal atmosphere—ocean variability (Chiang and Vimont 2004). This meridional mode is the dominant mode of variability in the Atlantic. As with the PDO, the AMM is characterized by an anomalous sea surface temperature gradient and is forced by trade wind variations in the subtropical oceans along the north (Chiang and Vimont 2004).

17.2.2 *Effects on Terrestrial Water Storage and Groundwater Drought*

Changes in the terrestrial hydrological cycle are largely a function of the large-scale seasonal fluctuations in distinct or integrated states of the atmosphere and oceans, these are the teleconnections discussed in previous sections (Martens et al. 2018; Hallett et al. 2004). For instance, the dominant variabilities in rainfall patterns coupled with its effects on terrestrial water storage (e.g., groundwater, surface water, soil moisture, etc.) have been largely associated with the El-Nino Southern Oscillations (ENSO). The latter is a well-known index of oceanic variability referred to as a primary mode of fluctuation in global atmospheric distribution (Emerton et al. 2017; Sun et al. 2016). Also, with the application of temperature and radiation fluctuations, it is possible to modify ecological and atmospheric demands, even with a substantial change to global evapotranspiration patterns and total water resources (Martens et al. 2018; Mirallles et al. 1998). The reallocation and buffering of freshwater resources among the global and regional ecosystems such as soil, plants etc., as well as terrestrial water bodies like lakes and rivers, extend beyond water budget analysis, especially given the possibility of its regulation through various climatic modes (Emerton et al. 2017; McGregor 2017; Long et al. 2014).

17.2.2.1 Water Storage Changes

Surface water being a crucial component of terrestrial water storage is affected by the influence of climate modes. Flows in the Upper Blue Nile Basin were found to be substantially associated with the ENSO indices. As detailed by Abtew et al. (2009) who assessed the link between El-Niño Southern Oscillation (ENSO) indices and the hydrology of the Blue Nile River Basin, high flows tend to occur during the La Niña phase of ENSO while the El Niño phase is associated with dry periods. Another related study focused on the northern watersheds of the South Florida also identified the relationship between the water management system of South Florida and ENSO events (Abtew and Trimble 2010). The importance of climate teleconnection assessment and their impacts on water storage changes are underpinned by its capability in aiding meaningful water management decision via providing information on a lead-time to mitigate drought or flood impacts.

On several spatial scales (globally and regionally), the relationship of terrestrial water storage and moisture/energy interactions with the surrounding the oceans, atmosphere, and regional climate has been studied (Ndehedehe and Ferreira 2020; Ni et al. 2017; Phillips et al. 2012; de Linage et al. 2014). Water storage components provide support to ecosystem and mankind, and the interannual variability in land water storage impact on regional and global terrestrial carbon sink (Humphrey et al. 2018; Fang and et al. 2017). However, it is important to disengage the corresponding influences of climatic modes to improve the forecasts of freshwater changes even under climate change. The strong relationship of ENSO with interannual terrestrial water storage (TWS) variabilities between latitude 150S and 150N has been highlighted in several studies (Humphrey et al. 2016; Phillips et al. 2012). Moreover, there have been reports confirming the significance of the integrated influences of numerous teleconnection patterns on terrestrial water storage on both regional and global scales (Anyah et al. 2018; de Linage et al. 2014). Much of these effects, however, are observed in the Northern Hemisphere, which is well known for multiple climatic interactions of varied climatic modes, especially in certain hotspots (Huang et al. 2017).

Even though it can still be argued that a noticeable gap exists on the direct techniques of measuring the influence of climate teleconnection patterns on total water storage patterns, as well as the actual contribution of each teleconnection pattern to water storage components, studies are still on going to provide an adequate evaluation of teleconnection influence on global land water storage. Atmospheric-oceanic (climate) oscillations and climate modes, which include, the Northern Atlantic Oscillation (NAO), and the El Nino-Southern Oscillation (ENSO) have strong relationship with the multi-annual fluctuation patterns observed in hydrometeorological datasets. Studies have also shown the relevance of these teleconnection signals in air temperature, rainfall patterns, river flow, and evapotranspiration (Faust et al. 2016; Kingston et al. 2006; Luković et al. 2015; Tabari et al. 2014). As will be shown later in this chapter, a case study over South America shows how the influence of several indices of oceanic variability on continental water storage changes can be identified and quantified.

17.2.2.2 Groundwater Drought

The monitoring and assessment of groundwater availability are generally limited to smaller catchments and in some regions, single annual measurements per well (Uddameri et al. 2019). This poor monitoring of groundwater is reflected in our limited knowledge of how various human actions (e.g., groundwater pumping) and extreme climate (e.g., droughts) impact on freshwater availability. On the one hand, frequent occurrence of droughts, long-term groundwater extraction, and low rates of natural recharge will lower groundwater levels within aquifers. On the other hand, climate teleconnection-driven droughts will affect recharge and surface-groundwater interconnectivity. Ultimately, this will result in a groundwater drought, i.e., a long-term decline in groundwater levels caused by reduced rainfall and recharge, low well yields and long-term extraction. Groundwater resources tend to be more resilient when the availability of surface water resources begin to plummet. The impacts of intensively over-drafted aquifers and climate change do not only increase the risk of groundwater drought, in water deficit areas or semi-arid domains, they coalesce with surface water scarcity to increase threats to water security and freshwater biodiversity (Vörösmarty et al. 2010).

The need to assess the underlying processes of these impacts on a large scale is exacerbated by changing demands in water abstraction and the high risk of groundwater drought due to changes in global climate. It is expected that groundwater resources will be easily affected by multi-annual teleconnection patterns (Townley 1995), especially due to their well-known susceptibility to long-term fluctuations in evapotranspiration and rainfall (Folland et al. 2015; Van Loon 2015; Bloomfield and Marchant 2013). This, however, has resulted in the extensive study undertaken on the detection of long-term seasonal cycles in groundwater amounts. This relationships with climatic oscillations have been reported in North America, Europe and the other parts of the world (Folland et al. 2015; Kuss and Gurdak 2014; Tremblay et al. 2011; Lee et al. 2011; Holman et al. 2009). Apart from the long-term forecasting of groundwater levels using climate teleconnections and oceanic phenomena, long-term forecast of groundwater drought with these climate teleconnections provides opportunities that underpin drought prediction and its management (Rust et al. 2018).

The growing knowledge and interest in multi-annual seasonal changes of groundwater levels are targeted efforts at improving long-term prediction of hydrogeological events (Rust et al. 2018; Kingston et al. 2006). Given that groundwater drought usually result from multi-annual moisture deficits, this prediction will be obviously useful in its assessment (Van Loon 2015; Loon 2013; Peters et al. 2006). This provides an opportunity to adequately assess seasonalities in groundwater levels, to improve understanding of their effects on teleconnections, strength of multi-annual cycles, as well as their individual contribution to groundwater drought. In groundwater teleconnection studies, it is common practice to assess the relationship between seasonal characteristics in groundwater and their interactions with climate indices. This approach helps to estimate the state and strength of the various climatic oscillations. Correlation analysis and other complex seasonality detection and assessment techniques are also popular in these quantitative procedures (Neves et al. 2019; Szolgayova et al. 2014; Knippertz et al. 2003). However, more recent techniques such as the Fourier transforms, wavelet transformations, singular spectrum analysis have been applied in this assessment (Kuss and Gurdak 2014; Fritier et al. 2012; Tremblay et al. 2011; Nakken 1999). Among these techniques, the wavelet transform is the most popular, especially due to its skillfulness in detecting multi-annual seasonal performances in noisy hydrogeological datasets. Also, its ability to identify the effect of Atlantic Multi-decadal oscillations, North Atlantic Oscillations, and El-Nino Southern Oscillation on North American groundwater levels as well as the Scandinavian and East Atlantic pattern of the European groundwater level fluctuations have been documented (Neves et al. 2019; Velasco et al. 2017; Kuss and Gurdak 2014).

We also observe the frequent use of data processing methods that alter fluctuations at the higher end of the frequency spectrum aimed at improving multi-annual seasonal assessments. For instance, the averaging or translation of the quarterly and annually observed time series dataset to accumulated mean deviations by Weber and Stewart (2004) highlighted its significance. The level of significance of the multi-annual climate fluctuations to total groundwater storage is unclear due to the adjustment of the time-series data. This is an important challenge in understanding the relationships between climate oscillations and its corresponding influence on groundwater level response rate (Rust et al. 2018). This is an existing gap in understanding the strength of multi-annual teleconnection cycles in groundwater levels on hydrogeological drought. The quantification of multi-annual fluctuations in groundwater levels provides an opportunity to sustain various water-trading platforms, efficient infrastructure security, and profound drought assessment methodologies, especially as the potential for improved long-term forecasting keeps advancing (Rey et al. 2019).

17.3 Surface and Sub-surface Storage

Multi-annual variations in hydro-meteorological datasets have significant correlation with atmospheric-oceanic oscillations. There are several of such oscillations but the prominent ones include the El-Nino Southern Oscillation (ENSO), Pacific Decadal Oscillations (PDO), Indian Ocean Dipole (IOD), Atlantic Oscillations (NAO), among other global climate indices. The influence of climate teleconnection patterns on hydrological fluxes such as air temperature, precipitation, and evapotranspiration anomalies is well documented in the literature (e.g., Luković et al. 2015). Surface and sub-surface storage systems are susceptible to this multi-scale teleconnection influence. This is particularly so due to their sensitivity to long-term changes in various hydrological parameters such as precipitation and evapotranspiration. In one way or another (directly or indirectly), all components of the global terrestrial water storage are influenced by climate teleconnection modes.

Variations in soil moisture are strongly associated with some natural and human factors ranging from pipe failures, gardening, climate change etc. However, the most significant contribution comes from climate change as it causes moisture variations up to a certain level of soil over a given period. This level depends on the soil type of the area under investigation, as the levels in arid regions (like North Africa) tend to be far lower than tropical regions (like West Africa). Some parts of Australia (Melbourne) have a specified range of 1.8–2.3 m within a 50-year period, and this has influenced the volume changes in the region, which has triggered ground vibrations and variations (Karunarathne et al. 2018). It is important to note that the process of assessing climate-induced soil moisture variations entails the thermal and hydraulic properties of the soil, whereas the process of monitoring ground movement requires understanding the non-linearity behaviour of soil as well as its moisture dynamics.

The assessment of aquifer vulnerability to climate change can be undertaken by correlating changes in groundwater levels and precipitation rates over multiple time scales (seasonally, annually, etc). Significant agreements to decadal climatic oscillations such as the PDO and the ENSO have been observed in several aquifers (e.g., Hanson and Dettinger 2005). Given the semi-confined nature of deep aquifers, an evaluation of their response to climatic factors must assess the mitigation of a time-varying recharge indicator as it penetrates the unsaturated zone, the upper layer of the aquifer's water table, and the intervening clay layers. Apparently, this could lead to an elongated recharge travel time. Factors affecting groundwater response to climatic conditions include land use/land cover (Scanlon et al. 2005), local geology (Chen et al. 2004), as well as other biogeophysical parameters, which tend to influence recharge rates and infiltration capacity. Anthropogenic influences further complicate the procedures for identifying climate-induced groundwater relations (Earman and Dettinger 2011), thus making numerical modelling an indispensable tool for evaluating irrigation impacts on important components of the global hydrological system. Climatic factors also affect unconfined and shallow groundwater reservoirs at both annual and sub-annual temporal scales, while deep confined reservoirs at longer interannual temporal scales, especially as a result of physical constraints on recharge indicator travel time. Groundwater extraction, on the other hand, causes an immediate local impact on aquifer mass changes even in deep wells. Multi-year climatic cycles have substantial influences on groundwater levels, with the most substantial correlation occurring during the 12 month lag interval between groundwater and climate indicators. This is as a result of aquifer response time and recharge travel rate (Whittemore et al. 2016).

Surface water resources that are largely deposited in lakes, rivers, and wetlands constitute a substantial percentage of the terrestrial freshwater and estuarine ecosystem. They are vital to the public health, life quality, socio-economic status and environmental sustainability of any region of interest. Aquatic ecosystems provide water and food for human consumption. They also aid water quality by replenishing, storing, and filtering our farmland nutrients through periodic flooding. With this capability, they can maintain food and habitats for wildlife, as well as provide recreational opportunities for humans. The quality of surface water is greatly affected by global warming as well as variations in resources utilized by humans to mitigate climate change impacts. This can easily be seen in the growing rate at which climate change coincides with large-scale landscape alteration and fragmentation. Climate-induced changes on the aquatic ecosystem result in increased flood and drought episodes, loss of biodiversity, degradation of drinking water quality, loss of food supplies, increased transmission of infectious diseases, as well as expensive cost of treatment. While these climate-induced variations in surface water quality, distribution and allocation may be insignificant in relation to land-use effects in several regions, notable changes in water quality are easily noticed due to the superposition of climate change on the impacts of human resource use.

17.3.1 Case Studies

17.3.1.1 Lake Chad-West Africa

Assessing the influence of large-scale hydro-climatic variations on changes in surface water hydrology is a vital component for sustainable management and allocation of freshwater resources. The Lake Chad basin houses the Lake Chad, which is a

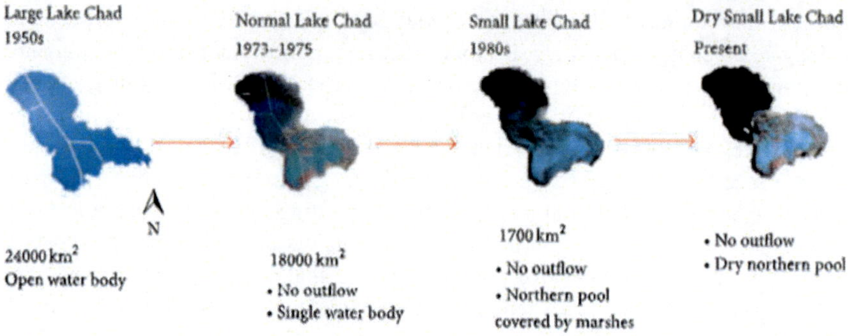

Fig. 17.3 Historical surface water transitions in the Lake Chad caused by climate teleconnection induced extreme drought events. *Source* Okonkwo et al. (2014)

confined lake at the middle of the basin. The lake has shown high susceptibility to hydro-climatic episodes over the past decade (Leblanc et al. 2011). The hydro-climatic conditions of the Lake Chad basin are dependent on several bio-geophysical factors including river discharge, runoff, precipitation rates, anthropogenic influence, and some large scale global indices and teleconnection patterns. These factors also contributed to the notable reduction in size of the lake. The Lake Chad went from a large open water body (24,000 km^2) in the 1950s to a rather small segmented lake (18,000 km^2) in the early/mid 1970s (Fig. 17.3). When the extreme and prolonged drought events of the early 1980s occurred, the lake lost another substantial portion of its surface area, shrinking to 1,700km^2 and with most areas covered by marshes (Fig. 17.3). The droughts of the 1980s, which affected the Lake Chad, were caused by climate variability indices and has been widely reported (Ndehedehe et al. 2020; Bader and Latif 2011; Giannini et al. 2003; Fontaine and Bigot 1993).

This shows that the impact of global climate change, teleconnection patterns, and anthropogenic influences goes beyond altering the quality and quantity of freshwater in large river basins but could also lead to a shrinkage and size reduction of any surface water. A significant cause of the lake's behaviour is the presence of ridge often referred to as the 'great barrier' that runs between the northern and southern parts of the lake, thus splitting the lake into two smaller lakes. Results from the study carried out by Okonkwo et al. (2014) reported small changes of less than 4 cm during the July-August-September precipitation period over the northern Lake Chad basin in 1984, which corresponds to one of the severe climate-induced droughts that plagued the Sahel region during that period. The alteration in ENSO events between the cool La Nina and the warm El Nino phases contributed to a continual reduction of the lake's water level from 1979 to 1985 in the northern pool. The results also demonstrated good interrelationship towards ENSO's role in modulating precipitation pattern of the northern part of the Lake Chad basin. The strong climate influence of the ENSO events over the Sahel region (i.e., the northern part of the lake chad basin) was also demonstrated by a clear difference in coherency between the northern and southern Lake Chad basin. This coincides with the study undertaken by

Moron and Ward (1998) who demonstrated the ENSO influence on low-frequency rainfall fluctuations over the Sahel. Precipitation over the Sahel region is largely connected to the different phases of the El-Nino and La Nina episodes (Joly et al. 2007).

17.3.1.2 Lake Victoria-East Africa

Lake Victoria has been characterized as the second-largest freshwater repository on earth. Over 30 million people depend on it for drinking, agriculture and domestic use. The climatic condition of the Lake Victoria basin can be referred to as a modified equatorial type of climate having substantial precipitation all year round, especially over the lake surface. The seasonal precipitation cycle over the basin is a bimodal cycle, which is similar to the precipitation patterns of the East African regions. This cycle is majorly regulated by the impact of the north-south migration of Inter Tropical Convergence Zone (ITCZ). The interannual variability of the Lake Victoria basin consists of seasonal events of inconsistent dry and wet conditions. Flood events of 1961/1962 and the 1997/1998 periods have been linked to the impacts of sea surface temperature gradient over the mid-troposphere westerly flow. Studies have shown that the 1997/1998 flood events strongly correlate with the strongest El-Nino and IOD phases of the last century (e.g., Omondi et al. 2012). This shows the degree of influence of climate-induced events in regulating the surface water hydrology of the river basin.

The effect of climate change on the precipitation and temperature levels of the Lake Victoria basin affects the basin's stored water. Over 80% of the Lake Victoria's water source emanates from periodic precipitation, but a substantial amount of this precipitation is lost to the atmosphere through evaporation from the surface of the lake due to high latent heat transfer (Swenson and Wahr 2009). This change in climatic conditions also affects the temperature of the lake, which is estimated to increase by 4 degrees Celsius by the end of this century, though it may not show any significant impact on precipitation regime. This impact will induce a rise in evaporation over the basin and thus result in a substantial downward trend over the lake's net basin supply as well as an upsurge in the temperature of its freshwater. A combined influence of both climatic anomalies and anthropogenic influence on the Lake Victoria's terrestrial water storage will affect the economy of the region, as well as the physical environment (e.g., land surface conditions, ecological assets). This can be as a result of decline in fish trades, and reduction in water quality that can prompt inhabitants to relocate to more suitable regions. Results from recent studies show that the effect of climate change in the Lake Victoria region is becoming worrisome, as notable drought patterns are emerging and lake levels are receding (Swenson and Wahr 2009; Awange et al. 2008). However, the advent of satellite remote sensing techniques provides tools that can be leveraged upon for adequate and intensified monitoring of the dynamics of the region's surface water resources to provide up to date information on the hydrological conditions of the region with respect to climate change and global warming.

17.4 Quantifying the Contributions of Climate Modes to Terrestrial Stored Water in South America

The warming of the climate system and other multiple forms of human stressors (e.g., deforestation, land-use changes, etc.) affect natural systems, leading to accelerations in the global hydrological cycle. (e.g., Ndehedehe 2017; Wild et al. 2008; Huntington 2006). Significant variations in the different components of the hydrological cycle are triggered by anthropogenic emissions of greenhouse gases (GHG). These gases alter the radiative balance of the atmosphere leading to considerable increase in temperature (e.g., Gleick 1989). In other words, GHG emissions are the key drivers of global warming. This in turn has the potential of providing major constraint on freshwater availability. For example, when the climate warms, the global water cycle accelerates. This rapid change in the water cycle causes a rise not only in the variability of extreme events (e.g., droughts, floods, and bushfires) but also in their duration and intensity. This increase has implications that could affect the sustainability of water resources and alter the variability of hydrological regimes (e.g., Ndehedehe 2017; Gleick 1989). Identifying the coupled relationship between terrestrial stored water and the surrounding oceans is, therefore, significant to assessing multi-scale climatic influence on continental hydrology (e.g., Ndehedehe et al. 2019). These oceans are important producers and sources of large-scale mesoscale convective systems, and other climate variability drivers (e.g., El-Niño Southern Oscillation-ENSO). Additionally, they influence meteorological processes and induce extreme hydro-climatic conditions, and their impacts on hydrological systems (e.g., lakes, reservoirs) and water resources are undeniable. This chapter presents the work of Ndehedehe and Ferreira (2020), who assessed the impacts of eleven key climate modes describing oceanic variability in the surrounding oceans on the spatial and temporal distributions of terrestrial water storage (TWS) as measured by Gravity Recovery and Climate Experiment (GRACE) (2002–2017) over South America.

The main aim of this study as undertaken by Ndehedehe and Ferreira (2020) was to assess the impacts of key climatic drivers on the spatial and temporal distributions of GRACE-derived TWS (2002–2017) over South America. To this end, a novel two-step regularization approach that combines multiple linear regression (MLR) and independent component analysis (e.g., Ndehedehe 2017; Ndehedehe et al. 2016b; Cardoso and Souloumiac 1993) was employed to assess the impacts of these climate variability indices (or climatic drivers) and quantify their contributions on continental water storage. These climatic drivers are based on eight prominent SST (sea surface temperature) products of the nearby oceans (i.e., Nino 1+2, Nino 3.0, Nino 3.4, Nino 4, North Tropical Atlantic-NTA, East Tropical Atlantic-ETA, South Tropical Atlantic-STA, and Caribbean) and three global climate teleconnections (e.g., Southern Oscillation Index-SOI and Pacific Decadal Oscillation-PDO; and ENSO). Arguably, strong ocean-land atmosphere interaction and the surrounding oceans produce these systems, which regulate precipitation patterns. The latter in turn impacts on continental TWS, including water stored in lakes and reservoirs. Notably, there are regions of the world where human water abstraction and land-use change drive hydrology. Apart from these regions, in general, this is the case for most tropical systems where changes in precipitation regulate dynamics in land water storage (e.g., Ndehedehe 2019; Rodell et al. 2018; Gal et al. 2017; Famiglietti et al. 2015).

17.4.1 South America

The South American continent is home to extremely wet river basins like the Amazon river basin and very dry regions after the similitude of the African Sahel. In these drier regions of South America, geodetic perturbations and the impacts of other physical processes such as earthquakes and land subsidence have been studied (Ferreira et al. 2018, 2019). Some of the major hydro-goelogical river basins and physiographic domains within South American are indicated in Fig. 17.4a, b. Apparently, water availability is high in tropical river basins of South America (e.g., Amazon, Orinoco, Magdalena) and shows considerable distribution of annual precipitation as opposed to other regions (Fig. 17.4a). Although about 12% of the world's available freshwater is located in the tropical South American basins, which also accounts for approximately 20% of global freshwater discharge (e.g., Ferreira et al. 2019), population in the arid regions (e.g., north-east Brazil) however, are most often plagued with water scarcity because of limited rainfall (Fig. 17.4a) and frequent extreme hydrological drought conditions (e.g., Ferreira et al. 2018). Rainfall in South America is largely the aftermath of atmospheric circulation patterns. The intertropical convergence zone regulates the seasonal behaviour of rainfall in the tropical wet-dry climatic belts of the continent. As with other tropical regions, sea surface temperature plays a critical role in the variability of rainfall in this region (Lu and Dong 2005). This variability in rainfall in turn contributes to the spatio-temporal distribution of land water storage and water fluxes (e.g., Ndehedehe 2019; Rodell et al. 2018). In other words, regions with substantial changes in annual rainfall are expected to experience stronger changes in the amplitudes and magnitude of TWS. For example, mean annual precipitation in the Amazon basin is about 4500 mm (Fig. 17.4a) and shows a rise in TWS during the same period owing to precipitation induced exchange of fluxes within the floodplain corridors (Fig. 17.4b). But this is different for the water-deficit regions of north-east Brazil where TWS shows significant decline (Fig. 17.4a, b).

17.4.2 Data

The GRACE (Gravity Recovery and Climate Experiment) mascon solutions used in this study are based on the monthly gravity field variations and are based on a 120 km wide mascon block totalling 40,692 equal-area geodesic grid tiles (Save et al. 2016; Wiese et al. 2016; Watkins et al. 2015) and was accessed from the Center for Space Research (CSR) at The University of Texas through its data portal (http://www2.csr.utexas.edu/grace/RL05_mascons.html). See Chap. 4 for more details on

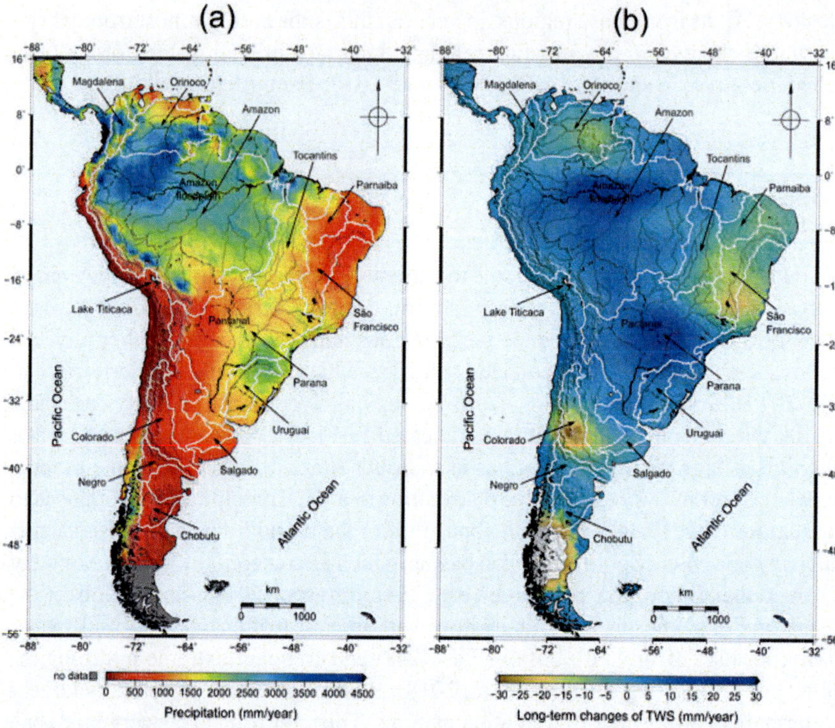

Fig. 17.4 South America showing major river basins. **a** The spatial distribution of mean annual rainfall and **b** spatial patterns of trends in observed GRACE-derived terrestrial water storage (2002–2017). The annual rainfall was estimated from the Tropical Rainfall Measuring Mission satellite (2003–2016). *Source* Ndehedehe and Ferreira (2020)

GRACE mascons as well as other GRACE solutions. The sea surface temperature (SST) products and climate teleconnection patterns used are Ninos 1+2 (0-10S, 90W-80W), 3.0 (5N-5S, 150W-90W), 3.4 (5N-5S, 170W-120W), and 4.0 (5N-5S, 160E-150W). Other SST products are those of the Atlantic ocean (North Tropical Atlantic-NTA, East Tropical Atlantic-ETA, South Tropical Atlantic-STA) and the Caribbean. These products are normalized monthly means from 2002 to 2017 (corresponding to the GRACE period) and are freely available online (https://www.esrl.noaa.gov/psd/forecasts/sstlim/atlantic_indices.html). The ENSO (El-Niño Southern Oscillation) that describes the presence of abnormally warm (El-Niño) and cold (La-Niña) sea surface temperature anomalies in the eastern Pacific are available in the National Oceanic and Atmospheric Administration's (NOAA) data portal (http://www.esrl.noaa.gov/psd/enso/mei/). Other climate teleconnections used include, the PDO (Pacific Decadal Oscillation) and SOI (Southern Oscillation Index) (https://www.ncdc.noaa.gov/teleconnections/pdo/). It is an irregular oscillation characterized by low-frequency variability and could be useful for prospective future climate outlook (e.g., Ndehedehe et al. 2019). The SOI is a measure of observed large-scale variations in air pressure occurring between the western and eastern tropical Pacific during El-Niño and La-Niña episodes. In order words, it is the atmospheric component of ENSO and basically serves as an indicator for the development and intensity of El-Niño and La-Niña events in the Pacific Ocean. Long periods of negative (positive) SOI values coincide with abnormally warm (cold) ocean waters across the eastern tropical Pacific typical of El-Niño (La-Niña) episodes. The Tropical Rainfall Measuring Mission (TRMM 3B43) (Kummerow et al. 2000) based precipitation covering the same period (2002–2017) as GRACE-TWS was used here to identify regions in South America that are vulnerable to the impacts of teleconnections. These data can be accessed through the National Aerospace and Space Administration (NASA) Goddard Space Flight Center (GSFC) website (http://disc.gsfc.nasa.gov/datacollection/TRMM3B43-V7.shtml).

17.4.3 Assessing Climatic Influence on Terrestrial Water Storage

17.4.3.1 Deseasonalization of Terrestrial Water Storage

The contributions of climate modes on continental water storage over South America were diagnosed by using time series of GRACE-TWS. These were deseasonalized (i.e., removal of trends and harmonic components), leaving the residual variability hereafter known as dTWS. Based on the multiple linear regression (MLR) technique, the trends and harmonic components of TWS (i.e., mean annual and semi-annual amplitudes) were compartmentalized from the GRACE time series $\mathbf{Y}_{TWS}$ through parameterizations as (e.g., Ndehedehe et al. 2017),

$$\mathbf{Y}(l,k,t) = \beta_0 + \beta_1 t + \beta_2 sin(2\pi t) + \beta_3 cos(2\pi t) + \beta_4 sin(4\pi t) + \beta_5 cos(4\pi t) + \beta_6 E(t + \varphi_E) + \varepsilon(t), \tag{17.1}$$

where (l, k) are the grid locations, t is the time in years, β_0 is the constant offset, β_1 is the linear trend, β_2 and β_3 account for the annual signal while β_4 and β_5 represent the semi-annual signal. The variable β_6 is the amplitude of TWS changes that contain the contributions of climate indices that describes large scale ocean-atmosphere phenomenon (e.g., ENSO, PDO). The variable E in Eq. 17.1 is the normalized time series (i.e., after removing long term mean) of each climate mode, φ_E is the phase lag between the time series of TWS and each climate index, while $\varepsilon(t)$ is the random error term. The amplitudes of TWS (i.e., mean annual and semi-annual) over the region are formulated as

$$Annual\ Amplitude = [(\beta_2)^2 + (\beta_3)^2]^{1/2},\ Semi\ Annual\ Amplitude = [(\beta_4)^2 + (\beta_5)^2]^{1/2}. \tag{17.2}$$

Removing these harmonic components (Eq. 17.2) and the linear rates in Eq. 17.1 leaves the residual component. This component arguably contains the climate teleconnection-driven signals as well as other slow dynamic climate processes (e.g., over the Pacific, Atlantic, and Caribbean) and other regional forcings. Some of the inherent residual variations, however, could also include low-frequency noise. The local least-square polynomial approximation-based filter of Savitzky and Golay (1964) was used to smoothen out such effects, normalizing its impacts across both variables (TWS and climate modes). The deseasonalized $(\mathbf{X}_{dTWS})$ signal is characterized as

$$\mathbf{X}_{dTWS} = \mathbf{Y} - [\beta_1 t + \beta_2 sin(2\pi t) + \beta_3 cos(2\pi t) + \beta_4 sin(4\pi t) + \beta_5 cos(4\pi t)]. \tag{17.3}$$

17.4.3.2 Multivariate Decomposition of Terrestrial Water Storage in Relation to Climate Modes

The time series of the data matrix $\mathbf{X}_{dTWS}$ were statistically decomposed into spatial and temporal patterns that are independent (e.g., Ndehedehe et al. 2016b, 2017; Ziehe 2005). Prior to this, X_{dTWS} was regularized using the principal component analysis (PCA, Ndehedehe et al. 2016a; Jolliffe 2002). This process is like prefiltering the data to ensure that only relevant hydrological signals are retained for the statistical decomposition. So, the significant orthogonal modes of variability from the PCA process were selected through a scree plot analysis (Martinez and Martinez 2005). To ensure that only statistically significant orthogonal modes of X_{dTWS} variability from the PCA were further analysed, the Bartlett's test statistics (e.g., Snedecor and Cochran 1989) was also used to examine the number of modes necessary to explain the non-random variations in X_{dTWS} at 95% confidence level before the implementation of the cumulant scheme, the independent component analysis (ICA). Fundamentally, as indicated in Eq. 17.4, the scheme decomposes a typical data matrix $\mathbf{Z}_P(t)$, into a mixing matrix B and a number of statistically independent source signals $s_j(t)$ as,

$$\mathbf{Z}_P(t) = \sum B_{ij} s_j(t), \qquad (i = 1, ..., n, \;\; j = 1, ..., m), \tag{17.4}$$

where t is the time index, i and j are the rows and columns, respectively. Computational details and steps for its implementation and algorithm development have been documented in the earlier works of Cardoso (1991), Cardoso and Souloumiac (1993), and Common (1994); Cardoso (1999). Some of these details are also available in Chap. 13.

Notably, the ICA decomposition approach enhances the detection of this kind of less dominant (i.e., obscured) signals, ensuring their localization in space and time (e.g., Ndehedehe et al. 2017). This makes it possible to relate their temporal evolutions concurrently with those of climate modes. After the cumulant decomposition of $\mathbf{X}_{dTWS}$, the evolving independent modes (i.e., dTWS-1, dTWS-2..., dTWS-n, representing localized hydrological signal in the region) of variability were compared with the normalized time series of each climate modes (Fig. 17.5). To recover the amplitudes of estimated climate induced TWS for each region, the evolving temporal patterns are jointly interpreted. This means the temporal series and their corresponding spatial patterns will have to be interpreted together. As an example, Fig. 17.5 shows the temporal (columns 1 and 4) and spatial (columns 2 and 3) patterns of deseasonalized TWS (hydrological signals expected to be associated with climate modes) over South America. The time series of deseasonalized TWS, that is, dTWS-1 (column 1) corresponds to the first spatial map in column 2. As can bee seen in Fig. 17.5, the GRACE signal is localized in the northern section of Amazon basin in South America. When these are jointly interpreted, the peak water storage in 2010 is approximately 150 mm in that location.

17.4.3.3 Quantifying the Contributions of Climate Modes to Changes in Terrestrial Water Storage

The contribution of each of the climate modes (i.e., ENSO, PDO, Nino 3.4, Nino 4, NTA, ETA, STA, and Caribbean) to TWS was quantified by fitting a least square to each grid location of $\mathbf{X}_{dTWS}$ in Eq. 17.3 following previous approaches (Ndehedehe et al. 2017; Phillips et al. 2012)

$$\mathbf{X}_{Climate\ mode}(x, y) = a(x, y) + b(x, y) * climate\ mode + c(x, y) * imag(H(climate\ mode)), \tag{17.5}$$

where coefficients b and c are used to estimate the climate teleconnection-induced changes in TWS $\mathbf{X}_{climate\ mode}(x, y)$ at grid locations (x, y) The imaginary part of the Hilbert transform (H) of the climate index represents the lag between TWS anomalies and climate modes. The pioneering work of Phillips et al. (2012) highlighted the importance of Hilbert transformation an important tool in Fourier analysis that move the sinusoidal Fourier components of the signal (e.g., ENSO, PDO, etc.) by 90 degrees. After the estimation of coefficients b and c in Eq. 17.5 above, the amplitude of climate-induced terrestrial stored water, dTWS was computed as

$$A = [b^2 + c^2]^{1/2}, \tag{17.6}$$

where A, is the estimated magnitude of climate mode on continental TWS. This is the predicted contribution of climate modes to observed variations in TWS.

17.4.4 Impacts of Global Climate Modes on GRACE-hydrological Signals

The novel approach in this study integrates parameter estimation method with the ICA technique sequentially. Firstly, the GRACE-hydrological signals were deseasonalized using Eqs. 17.1–17.3. This process is essential as the harmonic components and trends dominate TWS variability and would make tracking the footprints of cli-

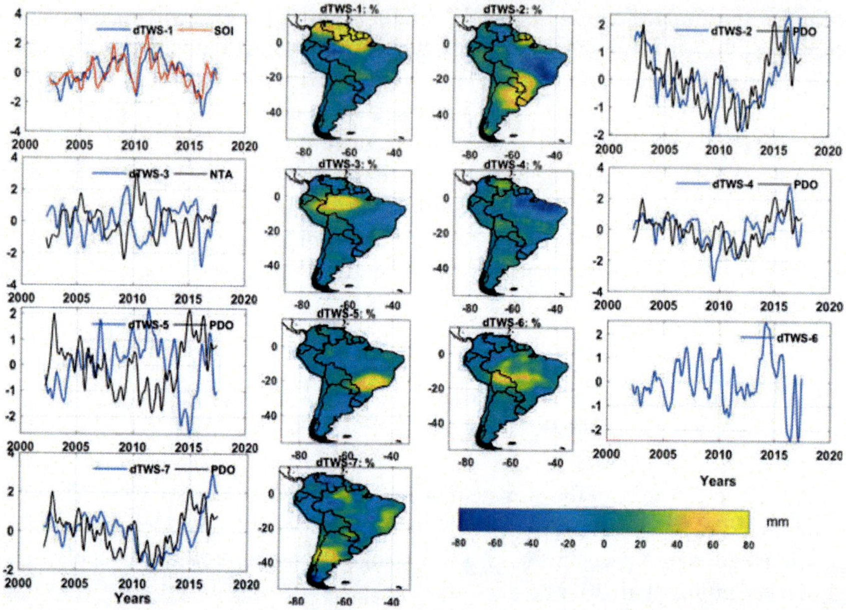

Fig. 17.5 Cumulant decomposition of TWS over South America after removing all harmonic components (mean annual and semi-annual amplitudes) and linear rates using the MLR. The spatial patterns (columns 2 and 3) have been scaled using the standard deviations of the computed independent components (columns 1 and 4). The relationship between the normalized time series of climate modes (ENSO, NTA and PDO) and the independent temporal series are also shown. *Source* Ndehedehe and Ferreira (2020)

mate modes rather difficult (Ndehedehe et al. 2017; Phillips et al. 2012). Secondly, the contributions and impacts of teleconnections and SST products on continental land water storage were assessed through a fourth order cumulant decomposition of the output from the MLR scheme (i.e., significant modes of deseasonalized TWS, i.e., dTWS) into spatio-temporal independent patterns (Fig. 17.5). The influence of these climate modes is notably strong in tropical South America. This part of South America encompasses the Amazon and northern sub-regions of the continent (dTWS-1, Fig. 17.5).

Based on Pearson's correlation (r), ENSO, PDO, SOI, SST anomalies of the Caribbean and Pacific oceans (Nino 4 and Nino 3.4) show the highest association ($r = -0.68, -0.72, -0.69, -0.67$, and -0.71 for Nino 3.4, Nino 4, PDO, Caribbean, and ENSO, respectively) with the temporal evolutions of TWS over Amazon and northern catchments (Fig. 17.6a–f and Table 17.1). Moreover, the localized spatial patterns of TWS variations over South America, in general, are largely associated with the PDO signal (dTWS-1, dTWS-3 to dTWS-5, and dTWS-7, Fig. 17.5 and Table 17.1) except for the central Amazon basin where NTA also shows a modest association (dTWS-2, Fig. 17.5) with a correlation (r) of -0.51. SST products from relatively closer oceans, Ninos 1+2 and 3.0 show significant relationship with the region tropical South

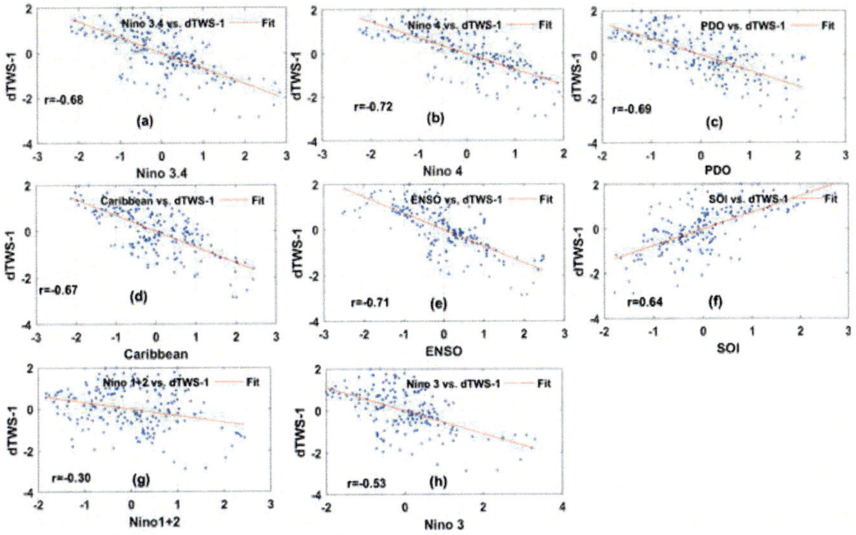

Fig. 17.6 Influence of PDO and SST products (Ninos 1+2, 3.0, 3.4, 4.0, and Caribbean) on localized TWS in the north east and upper Amazon basin regions (dTWS-1, Fig. 17.5) based on linear regression. **a** dTWS-1 versus Nino 3.4, **b** dTWS-1 versus Nino 4, **c** dTWS-1 versus PDO, **d** dTWS-1 versus Caribbean, **e** dTWS-1 versus ENSO, **f** dTWS-1 versus SOI, **g** dTWS-1 versus Nino 1+2, and **h** dTWS-1 versus Nino 3. *Source* Ndehedehe and Ferreira (2020)

Table 17.1 Correlation coefficients showing the relationship of climate teleconnections (PDO, ENSO, SOI) and SST products (Nino 3.4, Nino 4, Nino 3.0, Nino 1+2, North Tropical Atlantic-NTA, East Tropical Atlantic-ETA, South Tropical Atlantic-STA, and Caribbean) with localized TWS (i.e., dTWS) over South America. This localized TWS is the independent component (IC) associated with spatial patterns. Correlation coefficients in bold are significant at $\alpha = 0.05$

IC (dTWS)	Nino3.4	Nino4	PDO	SOI	NTA	ETA	STA	Caribbean	ENSO	Nino 1+2	Nino 3.0
1	**−0.68**	**−0.72**	**-0.69**	**0.64**	−0.15	−0.13	−0.18	**−0.67**	**−0.71**	**−0.30**	**−0.53**
2	**0.34**	**0.35**	**0.54**	**−0.28**	0.02	**0.26**	**0.33**	**0.44**	**0.36**	**0.25**	**0.28**
3	0.12	0.09	0.06	−0.14	**-0.51**	−0.23	−0.09	**−0.24**	0.16	**0.26**	0.20
4	0.21	0.25	**0.41**	**−0.25**	0.06	0.02	−0.06	**0.41**	**0.32**	−0.00	0.09
5	**-0.30**	**-0.33**	**-0.58**	**0.30**	**0.30**	0.05	0.07	−0.13	**−0.31**	−0.22	**-0.26**
6	0.13	0.12	0.02	−0.18	−0.19	−0.07	−0.05	−0.19	0.15	0.15	0.17
7	**0.25**	**0.27**	**0.46**	**−0.33**	−0.01	0.13	0.09	**0.36**	**0.32**	**0.28**	0.23

America ($r = -0.30$ and -0.53, respectively). The Caribbean SST is moderately associated with the time series of TWS in regions dTWS-2 and dTWS-4 of Fig. 17.5 and indicates $r = 0.44$ and $r = 0.41$, respectively. The relationships of these climate modes with the temporal patterns of TWS have been summarized in Table 17.1 and highlight the significant footprint of global climate in observed TWS over tropical South America.

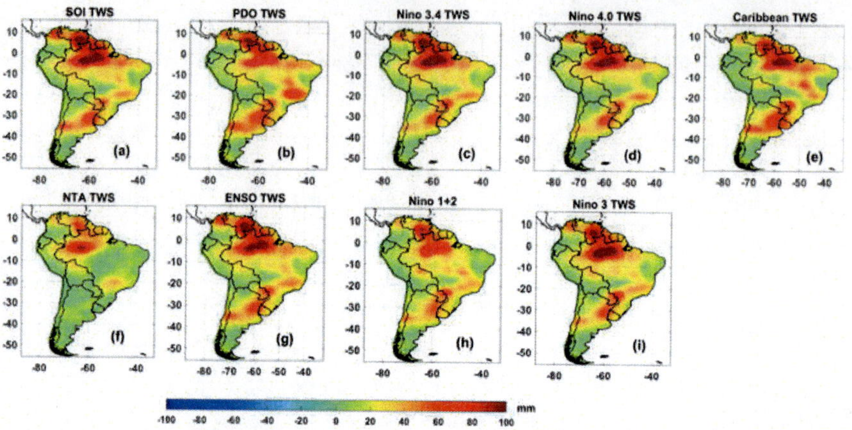

Fig. 17.7 Quantifying the contributions of climate modes and SST products on TWS over South America **a** SOI, and **b** PDO, **c**, **d** Ninos 3.4 and 4, **e** NTA, **f** Caribbean, **g** ENSO, and **h**, **i** Nino 1+2 and Nino 3, respectively. These are climate-induced amplitudes of TWS obtained using a least square fit on each grid location of dTWS (i.e., Eqns 17.5 and 17.6). *Source* Ndehedehe and Ferreira (2020)

In terms of numerical results, there is a methodological agreement in the two statistical frameworks used to assess the influence of these climate modes on land water storage dynamics over South America. From the spatial and temporal patterns obtained from statistical decomposition (dTWS-1, Fig. 17.5), relatively higher amplitude of TWS is observed in the northern-Amazon section (i.e., Orinoco and Amazon basins, Fig. 17.4). As estimated from the MLR technique, the strongest amplitude of TWS induced by these climate modes was observed in this same region (Fig. 17.7a–i). This is consistent with the statistical decomposition using the independent component analysis (Chap. 13). The influence of several climate modes (ENSO, SOI, Nino 4, and Nino 3.4) results in considerable magnitudes of more than 100 mm to the amplitudes of TWS (Fig. 17.7a, c, d, g) in the Amazon/northern regions. The contribution of these climate modes to TWS in these locations coincide with their strong temporal relationships with TWS in the same region (Fig. 17.6a–f). As Nino 4 and Nino 3.4 are signals from the ENSO region (i.e., Central/East Central Pacific), TWS variations in the Amazon and northern catchments of the continent are more driven by the ENSO phenomenon. Furthermore, the Caribbean SST is associated with relatively higher amplitudes of TWS in the south-east section of the continent (Fig. 17.7e). Bearing in mind the low association of ETA and STA with TWS (Table 17.1) in all hydrological regions as localized in Fig. 17.5, climate-induced amplitudes of TWS were only estimated for ENSO, PDO, SOI, NTA, Ninos 1+2, 3.0, 3.4, 4.0, and Caribbean SST (Fig. 17.7a–i).

With the 15-year GRACE data, this study confirms that climate variability indices like the El Niño are more likely to have direct influence on water storage in the Amazon region. The SOI shows positive correlations with dTWS-1 (Fig. 17.5) as opposed

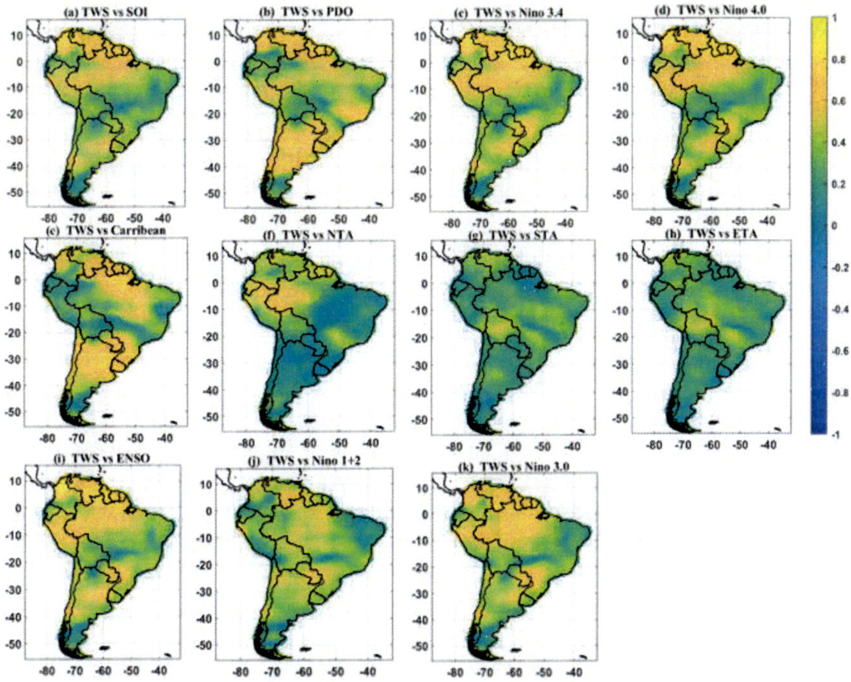

Fig. 17.8 Association between TWS (2002–2017) over South America and climate teleconnection patterns (**a**–**k**) studied. This assessment is based on the popular Pearson's correlation analysis. *Source* Ndehedehe and Ferreira (2020)

to other Ninos (1+2, 3, 3.4, and 4) and ENSO (Fig. 17.6a, b, e, g, h). However, as a reminder, they are all ENSO-related teleconnections and result in higher (lower) amplitudes during the La Niñas (El Niños) episodes as highlighted earlier. The reason is because sustained negative (positive) values of SOI below a certain threshold indicate El Niño (La Niña) events. Further, the grid-based or spatial relationships between these climate modes and TWS indicate that land water storage in much of tropical South America are impacted by the Caribbean SST, SOI, PDO, eastern-Pacific and central-Pacific-related ENSOs (Fig. 17.8a–k). Whereas STA and ETA show very minor and limited influence on TWS (Fig. 17.8g, h), PDO, ENSO, and Caribbean SST have extensive significant influence over the continent, showing relatively higher correlations ranging from about 0.7 to 1 (Fig. 17.8b, e, i,). However, the spatial correlations between Ninos 1+2 and 3 with TWS are also high but location-specific and in addition suggest the influence of SST from relatively closer oceans on TWS (Fig. 17.8j, k). The influence of these climate teleconnections to TWS in some regions is statistically insignificant while several sub-regions in the continent show significant contributions. This of course is despite the overall agreements between TWS and climate modes in South America.

17.4.5 Multi-decadal Variations in TWS

The temporal patterns of TWS are characterized by inter-annual variations (Fig. 17.5). But multi-decadal hydrological signals can also be detected in the temporal patterns of TWS in some of the regions (dTWS-1, dTWS-2, dTWS-4, dTWS-5, and dTWS-7, Fig. 17.5). The increasing (2002–2009) and decreasing (2010–2015) phases of TWS in dTWS-1 coincide with those of SOI (dTWS-1, Fig. 17.5 and Table 17.1). From the numerical results, it is obvious the PDO is in opposite phase in regions dTWS-1 and dTWS-5 while maintaining similar phase relationship with regions dTWS-2, dTWS-4, and dTWS-7 (Fig. 17.5). Generally, PDO shows modestly strong and relationship with TWS in these regions (Table 17.1). Notably, as was previously reported, the PDO appeared to have been strongly correlated with deep soil moisture in the lower Colorado River basin (Hurkmans et al. 2009). Additionally, the influence of PDO on hydro-climatic fluctuations in North America has been reported in previous studies (e.g., Hurkmans et al. 2009; MacDonald and Case 2005).

The results in this study also highlight the influence of PDO on TWS over South America. Such influence suggests that the low-frequency mode of the PDO plays a crucial role in the Northern and Southern American climate. As with ENSO, the positive PDO phase is associated with droughts. However, the negative phase coincides with extreme wet conditions (i.e., depending on the region in Fig. 17.5). Studies suggest the PDO is more associated with ENSO at inter-annual scales (Linage et al. 2013; MacDonald and Case 2005). Its links with TWS in much of South America (Fig. 17.8b) suggest a reasonable multi-decadal influence in the variability of TWS. This reaffirms the explicit role of ENSO and PDO as important multi-scale process of water availability not just in the southwestern United States (Hurkmans et al. 2009), but as significant drivers of GRACE-hydrological signal over South America.

17.4.6 Precipitation over South America: Understanding Amplitudes and Impacts of Climate Teleconnections

Teleconnection patterns (ENSO, PDO, Ninos 3.4 and 4, and the Carribean SST anomalies) around the world were also identified in the TRMM-based precipitation over South America (Fig. 17.9a, b). As with other regions, rainfall in several river basins in South America is influenced by these indices of oceanic variability (i.e., ENSO, Ninos, PDO, etc.). In South America, regions affected by these include, Orinoco, Amazon, Uruguay, Parana, amongst others (Figs. 17.9a, cf. 17.4a). From the independent patterns of rainfall highlighted in Fig. 17.9a, b, the ENSO-related teleconnections showed the strongest associations (r =0.71, 0.67, 0.66, 0.67 for Nino 3.4, Nino 4.0, ENSO, and Nino 3.0, respectively) with rainfall in these regions. However, Nino 1+2 ($r = 0.48$), PDO ($r = 0.40$), and SOI ($r = -0.58$) were also moderately associated with rainfall in this region (Fig. 17.9a). Interestingly, teleconnection patterns or climate modes affect changes in TWS over these sub-regions

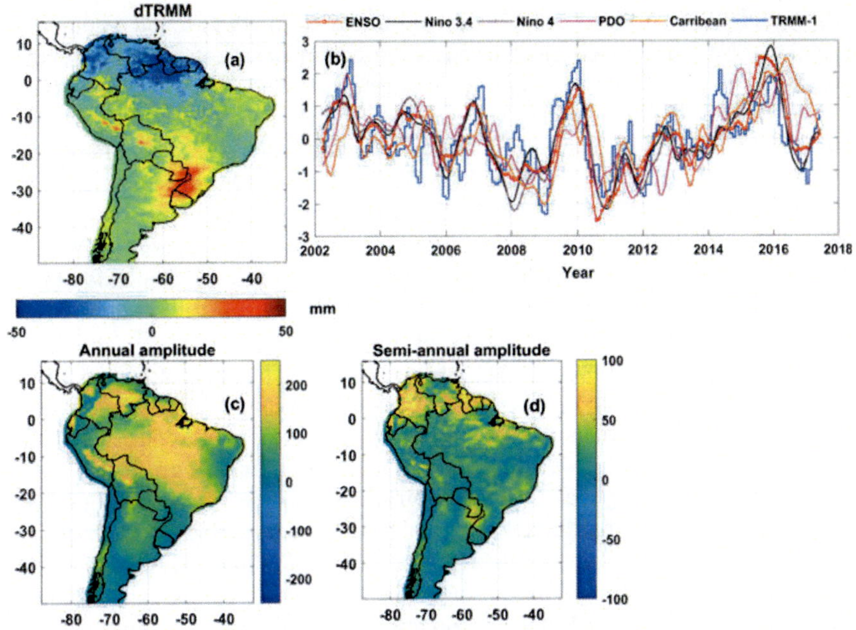

Fig. 17.9 TRMM-based precipitation patterns and role of multi-scale global climate teleconnections. **a** ICA-derived spatial patterns of rainfall, **b** relationship between independent component and climate teleconnections, **c** mean annual amplitude, and **d** mean semi-annual amplitude. *Source* Ndehedehe and Ferreira (2020)

(Fig. 17.7). On the one hand, areas with large fluctuations in rainfall appear to show strong amplitudes in climate-induced TWS. On the other hand, water-deficit and arid climates that receive relatively lower annual rainfall show little influence of climate-induced TWS (Figs. 17.9c, d, 17.7, and cf. 17.4a, b). However, TWS variations along the dry tropics of South America, particularly areas where rainfall appears to be low and ranges from about 20 mm to 100 (Fig. 17.9c) indicate significant associations with ENSO, PDO, and Carribean SST (Fig. 17.8b, e, and i).

17.4.7 Identifying Multi-scale Global Climate Footprints in GRACE-hydrological Signals

The direct and remote impacts of tropical oceans and quasi-periodic signals on rainfall extremes around the world are well known (Kumar et al. 2013; Feng et al. 2011; Lu and Dong 2005; White et al. 2004). This litany of evidence has recently been corroborated and further stresses the impacts of the surrounding oceans through strong changes on SST on hydro-meteorological processes (e.g., Ndehedehe et al. 2019). Some pioneering studies (e.g., Ndehedehe et al. 2017; Linage et al. 2013;

Phillips et al. 2012; Boening et al. 2012; Hurkmans et al. 2009) found significant impacts of ENSO and other teleconnections on continental freshwater dynamics. Changes in soil moisture across the globe have been associated with ENSO variability (e.g., Sheffield and Wood 2008). New results re-emphasizing and validating the influence of ENSO on variations in hydrological state variables are emerging (e.g., Ndehedehe et al. 2018; Anyah et al. 2018; Ni et al. 2018). In this study, ENSO, PDO, and the Caribbean SST have been identified as important climate modes with influence on TWS in the Amazon basin/northern section of South America (dTWS-1, Figs. 17.5 and 17.6). The statistical relationships of climate modes like Ninos 1+2, 3.0, 3.4, and 4.0 with TWS emphasize the influence of the tropical Pacific and around coastal South America on the continent (Figs. 17.7 and 17.8).

NTA shows relatively strong impacts only in the central Amazon basin (Fig. 17.7f). Apart from NTA, the influence of ENSO, PDO, Ninos 3.4 and 4.0 and the Caribbean SST on the spatial distributions of TWS spreads to the southern region of the continent of South America (Fig. 17.7a–e and Table 17.1). These findings align closely with the earlier work of Linage et al. (2013) in tropical South America, which of course excludes much of the region studied here (i.e., below Latitude 18S). Apart from the Pacific (Ninos 1+2, 3.0, 3.4 and 4) and Atlantic (NTA) SSTs influence on the north-eastern and Amazon sections of South America, the innovative approach of combining cumulant statistics and parameter estimation methods has further revealed the substantial contributions and influence of the Caribbean SST on the spatio-temporal distributions of TWS in the region (Figs. 17.6d and 17.7e). Given the PDO–TWS relationship, there is also a considerable evidence to suggest multi-decadal variability in land water storage over South America. Hence, incorporating these climate variability indices in a predictive scheme could improve the modelling of freshwater availability on a continental scale.

Rainfall in tropical South America is also influenced by these suite of global teleconnections, especially the ENSO-related climate oscillations (Fig. 17.9a, b). Indeed, the various mechanisms governing meteorological processes (e.g., monsoon circulations and other large scale atmospheric patterns) globally are expected to have different degrees of direct or remote contributions to TWS at local and continental scales. This is true for tropical South America where climate teleconnection footprints were identified in TRMM-based precipitation (Fig. 17.9a, b). This is consistent with areas where TWS variations are driven by global climate teleconnection patterns (Fig. 17.8a–k). Arguably, the influence of teleconnections on rainfall will be different compared to TWS. This is because the latter act as a buffer to rainfall (e.g., Ndehedehe et al. 2016a), resulting in the intensification of the water cycle.

17.5 Conclusion

Global climate change will fundamentally continue to trigger large changes in water scarcity through increased variability. Predicting freshwater systems at regional or continental scales will therefore require an assessment of the interplay between large multi-scale climatic systems and hydrological processes. To achieve this, an advance statistical decomposition method was employed to assess the complex interaction of key oceanic hotspots (Pacific, Atlantic and Caribbean) with continental freshwater dynamics over South America. Such assessment has become necessary for South America, which is one of the world's biggest water tower and accounts for nearly one fifth of global continental freshwater discharge. As detailed in this chapter, taking advantage of a 15-year GRACE data to demonstrate the efficiency of a statistical framework is critical for some reasons. Firstly, this will improve contemporary understanding of hydrological processes and the important factors that driver changes in land water storage on a regional or continental scales. Finally, in the light of limited ground observations for hydrological applications, it further articulates the potential of time-variable gravity observations as an efficient tool to assess the impacts of multi-scale climate oscillations on land water storage. As a proof of concept, the methodological framework in this study was also useful to identifying the regions in South America whose precipitation is also driven by global climate teleconnections.

References

Abtew W, Melesse AM, Dessalegne T (2009) El niño southern oscillation link to the Blue Nile River basin hydrology. Hydrol Process 23(26):3653–3660. https://doi.org/10.1002/hyp.7367

Abtew W, Trimble P (2010) El niño-southern oscillation link to south Florida hydrology and water management applications. Water Res Manag 24:4255–4271. https://doi.org/10.1007/s11269-010-9656-2

Ali A, Lebel T (2009) The Sahelian standardized rainfall index revisited. Int J Climatol 29:1705–1714. https://doi.org/10.1002/joc.1832

Amani A, Thomas J, Abou MN (2007) Climate change adaptation and water resources management in West Africa. Synthesis report WRITESHOP. UNESCO. https://www.nlcap./net/032135.070920.NCAP. Accessed 15th Apr 2014

Andam-Akorful S, Ferreira V, Ndehedehe CE, Quaye-Ballard J (2017) An investigation into the freshwater variability in West Africa during 1979–2010. Int J Climatol 37(S1):333–349. https://doi.org/10.1002/joc.5006

Anyah R, Forootan E, Awange J, Khaki M (2018) Understanding linkages between global climate indices and terrestrial water storage changes over Africa using GRACE products. Sci Total Environ 635:1405–1416. https://doi.org/10.1016/j.scitotenv.2018.04.159

Awange JL, Ogalo L, Bae KH, Were P, Omondi P, Omute P, Omullo M (2008) Falling Lake Victoria water levels: is climate a contributing factor? Clim Change 89(3):281–297. https://doi.org/10.1007/s10584-008-9409-x

Bader J, Latif M (2011) The 1983 drought in the West Sahel: a case study. Clim Dyn 36(3–4):463–472. https://doi.org/10.1007/s00382-009-0700-y

Bloomfield JP, Marchant BP (2013) Analysis of groundwater drought building on the standardised precipitation index approach. Hydrol Earth Syst Sci 17(12):4769–4787. https://doi.org/10.5194/hess-17-4769-2013

Boening C, Willis JK, Landerer FW, Nerem RS, Fasullo J (2012) The 2011 La Niña: so strong, the oceans fell. Geophys Res Lett 39(19):L19602. https://doi.org/10.1029/2012GL053055

Cane MA (1983) Oceanographic events during el niño. Science 222(4629):1189–1195. https://doi.org/10.1126/science.222.4629.1189

Cardoso J-F (1991) Super-symmetric decomposition of the fourth-order cumulant tensor, blind identification of more sources than sensors. http://perso.telecom-paristech.fr/~cardoso/Papers.PDF/icassp91.pdf. Accessed 15 Jan 2016

Cardoso JF (1999) High-Order contrasts for independent component analysis. Neural Comput 11:157–192

Cardoso JF, Souloumiac A (1993) Blind beamforming for non-gaussian signals. IEE Proc 140(6):362–370

Chen Z, Grasby SE, Osadetz KG (2004) Relation between climate variability and groundwater levels in the upper carbonate aquifer, Southern Manitoba. Canada. J Hydrol 290(1):43–62. https://doi.org/10.1016/j.jhydrol.2003.11.029

Chiang JCH, Vimont DJ (2004) Analogous pacific and Atlantic meridional modes of tropical atmosphere-ocean variability. J Clim 17(21):4143–4158. https://doi.org/10.1175/JCLI4953.1

Common P (1994) Independent component analysis, a new concept? Signal Process 36:287–314

Dai A, Qian T, Trenberth KE, Milliman JD (2009) Changes in continental freshwater discharge from 1948 to 2004. J Clim 22(10):2773–2792. https://doi.org/10.1175/2008JCLI2592.1

de Linage C, Famiglietti JS, Randerson JT (2014) Statistical prediction of terrestrial water storage changes in the Amazon Basin using tropical Pacific and North Atlantic sea surface temperature anomalies. Hydrol Earth Syst Sci 18(6):2089–2102. https://doi.org/10.5194/hess-18-2089-2014

Diatta S, Fink AH (2014) Statistical relationship between remote climate indices and West African monsoon variability. Int J Climatol 34(12):3348–3367. https://doi.org/10.1002/joc.3912

Di Lorenzo E, Schneider N, Cobb KM et al (2008) North pacific gyre oscillation links ocean climate and ecosystem change. Geophys Res Lett 35(8). https://doi.org/10.1029/2007GL032838

Di Lorenzo E, Cobb KM, Furtado JC, Schneider N, Anderson BT et al (2010) Central pacific El Nino and decadal climate change in the North Pacific ocean. Nat Geosci 3(11):762–765. https://doi.org/10.1038/ngeo984

Dong L, Shimada J, Kagabu M, Fu C (2015) Teleconnection and climatic oscillation in aquifer water level inKumamoto plain. Japan. Hydrol Process 29(7):1687–1703. https://doi.org/10.1002/hyp.10291

Druyan LM (2011) Studies of 21st-century precipitation trends over West Africa. Int J Climatol 31(10):1415–1424. https://doi.org/10.1002/joc.2180

Earman S, Dettinger M (2011) Potential impacts of climate change on groundwater resources–a global review. J Water Clim Change 2(4):213–229. https://doi.org/10.2166/wcc.2011.034

Emerton R, Cloke HL, Stephens EM, Zsoter E, Woolnough SJ, Pappenberger F (2017) Complex picture for likelihood of enso-driven flood hazard. Nat Commun 8(1):1–9. https://doi.org/10.1038/ncomms14796

Enfield DB, Mestas-Nuñez AM, Trimble PJ (2001) The Atlantic multidecadal oscillation and its relation to rainfall and river flows in the continental US. Geophys Res Lett 28(10):2077–2080. https://doi.org/10.1029/2000GL012745

Famiglietti JS, Cazenave A, Eicker A, Reager JT, Rodell M, Velicogna I (2015) Satellites provide the big picture. Science 349(6249):684–685. https://doi.org/10.1126/science.aac9238

Fang Y, Michalak AM, AM, et al (2017) Global land carbon sink response to temperature and precipitation varies with ENSO phase. Environ Res Lett 12(6):064007. https://doi.org/10.1088/1748-9326/aa6e8e

Faust JC, Fabian K, Milzer G, Giraudeau J, Knies J (2016) Norwegian fjord sediments reveal NAO related winter temperature and precipitation changes of the past 2800 years. Earth Planet Sci Lett 435:84–93. https://doi.org/10.1016/j.epsl.2015.12.003

Feng L, Hu C, Chen X, Li R, Tian L, Murch B (2011) MODIS observations of the bottom topography and its inter-annual variability of Poyang Lake. Remote Sens Environ 115(10):2729–2741. https://doi.org/10.1016/j.rse.2011.06.013

Ferreira V, Montecino H, Ndehedehe C, Heck B, Gong Z, Westerhaus M, de Freitas S (2018) Space-based observations of crustal deflections for drought characterization in brazil. Sci Total Environ 644:256–273. https://doi.org/10.1016/j.scitotenv.2018.06.277

Ferreira VG, Ndehedehe CE, Montecino HC, Yong B, Yuan P, Abdalla A, Mohammed AS (2019) Prospects for imaging terrestrial water storage in South America using daily GPS observations. Remote Sens 11(6). https://doi.org/10.3390/rs11060679

Folland CK, Hannaford J, Bloomfield JP, Kendon M, Svensson C, Marchant BP, Prior J, Wallace E (2015) Multi-annual droughts in the English lowlands: a review of their characteristics and climate drivers in the winter half-year. Hydrol Earth Syst Sci 19(5):2353–2375. https://doi.org/10.5194/hess-19-2353-2015

Fontaine B, Bigot S (1993) West African rainfall deficits and sea surface temperatures. Int J Climatol 13(3):271–285. https://doi.org/10.1002/joc.3370130304

Fritier N, Massei N, Laignel B et al (2012) Links between NAO fluctuations and inter-annual variability of winter-months precipitation in the Seine River watershed (north-western France). Comptes Rendus Geosci 344(8):396–405. https://doi.org/10.1016/j.crte.2012.07.004

Gal L, Grippa M, Hiernaux P, Pons L, Kergoat L (2017) The paradoxical evolution of runoff in the pastoral Sahel: analysis of the hydrological changes over the Agoufou watershed (Mali) using the KINEROS-2 model. Hydrol Earth Syst Sci 21(9):4591–4613. https://doi.org/10.5194/hess-21-4591-2017

Giannini A, Biasutti M, Held IM, Sobel AH (2008) A global perspective on African climate. Clim Change 90(7):359–383. https://doi.org/10.1007/s10584-008-9396-y

Giannini A, Saravanan R, Chang P (2003) Oceanic forcing of Sahel rainfall on interannual to decadal time scales. Science 302(5647):1027–1030. https://doi.org/10.1126/science.1089357

Gillett NP, Kell TD, Jones PD (2006) Regional climate impacts of the Southern Annular Mode. Geophys Res Lett 33(23). https://doi.org/10.1029/2006GL027721

Gleick PH (1989) Climate change, hydrology, and water resources. Rev Geophys 27(3):329–344. https://doi.org/10.1029/RG027i003p00329

Grippa M, Kergoat L, Frappart F, Araud Q, Boone A et al (2011) Land water storage variability over West Africa estimated by Gravity Recovery and Climate Experiment (GRACE) and land surface models. Water Res Res 47(5):W05549. https://doi.org/10.1029/2009wr008856

Hallett TB, Coulson T, Pilkington JG, Clutton-Brock TH, Pemberton JM, Grenfell BT (2004) Why large-scale climate indices seem to predict ecological processes better than local weather. Nature 430(6995):71–75. https://doi.org/10.1038/nature02708

Hanson RT, Dettinger MD (2005) Ground water/surface water responses to global climate simulations, Santa Clara-Calleguas basin, Ventura, California1. JAWRA J Am Water Res Assoc

Henry C, Allen DM, Huang J (2011) Groundwater storage variability and annual recharge using well-hydrograph and GRACE satellite data. Hydrogeol J 19:741–755. https://doi.org/10.1007/s10040-011-0724-3

Higgins RW, Leetmaa A, Xue Y, Barnston A (2000) Dominant factors influencing the seasonal predictability of US. precipitation and surface air temperature. J Clim 13(22):3994–4017. https://doi.org/10.1175/1520-0442(2000)013<3994:DFITSP>2.0.CO;2

Hinderer J, de Linage C, Boy J-P et al (2009) The GHYRAF (Gravity and Hydrology in Africa) experiment: description and first results. J Geodyn 48:172–181. https://doi.org/10.1016/j.jog.2009.09.014

Holman I, Rivas-Casado M, Howden N, Bloomfield J, Williams A (2009) Linking north Atlantic ocean-atmosphere teleconnection patterns and hydrogeological responses in temperate groundwater systems. Hydrol Process 23:3123–3126. https://doi.org/10.1002/hyp.7466

Huang W, Chen R, Yang Z, Wang B, Ma W (2017) Exploring the combined effects of the arctic oscillation and ENSO on the wintertime climate over East Asia using self-organizing maps. J Geophys Res: Atmos 122(17):9107–9129. https://doi.org/10.1002/2017JD026812

Humphrey V, Zscheischler J, Ciais P, Gudmundsson L, Sitch S, Seneviratne SI (2018) Sensitivity of atmospheric CO2 growth rate to observed changes in terrestrial water storage. Nature 560(7720):628–631. https://doi.org/10.1038/s41586-018-0424-4

Humphrey V, Gudmundsson L, Seneviratne SI (2016) Assessing global water storage variability from GRACE: trends, seasonal cycle, subseasonal anomalies and extremes. Surv Geophys 37(2):357–395. https://doi.org/10.1007/s10712-016-9367-1

Huntington TG (2006) Evidence for intensification of the global water cycle: review and synthesis. J Hydrol 319(1–4):83–95. https://doi.org/10.1016/j.jhydrol.2005.07.003

Hurrell JW, Kushnir Y, Ottersen G, Visbeck M (2003) An overview of the north Atlantic oscillation. Geophys Monogr-Am Geophysl Union 134:1–36. https://doi.org/10.1029/134GM01

Hurkmans R, Troch PA, Uijlenhoet R, Torfs P, Durcik M (2009) Effects of climate variability on water storage in the Colorado River Basin. J Hydrometeorol 10:1257–1270. https://doi.org/10.1175/2009JHM1133.1

Jolliffe IT (2002) Principal component analysis, 2nd edn. Springer Series in Statistics, Springer, New York

Joly M, Voldoire A, Douville H, Terray P, Royer JF (2007) African monsoon teleconnections with tropical SSTS: validation and evolution in a set of IPCC4 simulations. Clim Dyn 29(1):1–20. https://doi.org/10.1007/s00382-006-0215-8

Joly M, Voldoire A (2010) Role of the Gulf of Guinea in the inter-annual variability of the West African monsoon: what do we learn from CMIP3 coupled simulations? Int J Climatol 30(12):1843–1856. https://doi.org/10.1002/joc.2026

Kahya E, Dracup JA (1993) US streamflow patterns in relation to the El Niño/Southern oscillation. Water Res Res 29(8):2491–2503. https://doi.org/10.1029/93WR00744

Karunarathne AMAN, Fardipour M, Gad EF, Rajeev P et al (2018) Modelling of climate induced moisture variations and subsequent ground movements in expansive soils. Geotech Geol Eng 36(4):2455–2477

Kennedy AM, Garen DC, Koch RW (2009) The association between climate teleconnection indices and Upper Klamath seasonal streamflow: trans-niño index. Hydrol Process 23(7):973–984. https://doi.org/10.1002/hyp.7200

Kingston DG, McGregor GR, Hannah DM, Lawler DM (2006) River flow teleconnections across the northern north Atlantic region. Geophys Res Lett 33(14). https://doi.org/10.1029/2006GL026574

Knippertz P, Ulbrich U, Marques F, Corte-Real J (2003) Decadal changes in the link between el niño and springtime north Atlantic oscillation and European-north African rainfall. Int J Climatol 23(11):1293–1311. https://doi.org/10.1002/joc.944

Kuhnel I, McMahon TA, Finlayson BL, Haines A, Whetton PH, Gibson TT (1990) Climatic influences on streamflow variability: a comparison between southeastern Australia and southeastern United States of America. Water Res Res 26(10):2483–2496. https://doi.org/10.1029/WR026i010p02483

Kumar KN, Rajeevan M, Pai D, Srivastava A, Preethi B (2013) On the observed variability of monsoon droughts over India. Weather Clim Extrem 1:42–50. https://doi.org/10.1016/j.wace.2013.07.006

Kummerow C, Simpson J, Thiele O, Barnes W et al (2000) The status of the Tropical Rainfall Measuring Mission (TRMM) after two years in orbit. J Appl Meteorol 39(12):1965–1982. https://doi.org/10.1175/1520-0450(2001)040<1965:TSOTTR>2.0.CO;2

Kuss AJM, Gurdak JJ (2014) Groundwater level response in US principal aquifers to ENSO, NAO, PDO, and AMO. J Hydrol 519:1939–1952. https://doi.org/10.1016/j.jhydrol.2014.09.069

Lamb PJ, Peppler RA (1987) North Atlantic oscillation: concept and an application. Bull Am Meteorol Soc 68(10):1218–1225. https://doi.org/10.1175/1520-0477(1987)068<1218:NAOCAA>2.0.CO;2

Leblanc M, Lemoalle J, Bader J-C, Tweed S, Mofor L (2011) Thermal remote sensing of water under flooded vegetation: New observations of inundation patterns for the 'small' Lake Chad. J Hydrol 404(1):87–98. https://doi.org/10.1016/j.jhydrol.2011.04.023

Lee H, Beighley RE, Alsdorf D, Jung HC, Shum C, Duan J, Guo J, Yamazaki D, Andreadis K (2011) Characterization of terrestrial water dynamics in the Congo Basin using GRACE and satellite radar altimetry. Remote Sens Environ 115(12):3530–3538. https://doi.org/10.1016/j.rse.2011.08.015

Linage C, Kim H, Famiglietti JS, Yu J-Y (2013) Impact of pacific and atlantic sea surface temperatures on interannual and decadal variations of GRACE land water storage in tropical South America. J Geophys Res: Atmos 118(19):10,811–10,829. https://doi.org/10.1002/jgrd.50820.

Long D, Longuevergne L, Scanlon BR (2014) Uncertainty in evapotranspiration from land surface modeling, remote sensing, and GRACE satellites. Water Res Res 50(2):1131–1151. https://doi.org/10.1002/2013WR014581

Loon AV (2013) On the propagation of drought. how climate and catchment characteristics influence hydrological drought development and recovery. PhD thesis, Wageningen University, Wageningen, NL, 198 p. http://library.wur.nl/WebQuery/wurpubs/438510. Accessed 1st Feb 2018

Losada T, Rodríguez-Fonseca B, Janicot S, Gervois S, Chauvin F, Ruti P (2010) A multi-model approach to the Atlantic Equatorial mode: impact on the West African monsoon. Clim Dyn 35(1):29–43. https://doi.org/10.1007/s00382-009-0625-5

Luković J, Blagojevć D, Kilibarda M, Bajat B (2015) Spatial pattern of north Atlantic oscillation impact on rainfall in Serbia. Spat Stat 14:39–52. https://doi.org/10.1016/j.spasta.2015.04.007

Lu R, Dong B (2005) Impact of atlantic sea surface temperature anomalies on the summer climate in the western North Pacific during 1997–1998. J Geophys Res: Atmos 110(D16). https://doi.org/10.1029/2004JD005676.

MacDonald GM, Case RA (2005) Variations in the Pacific Decadal Oscillation over the past millennium. Geophys Res Lett 32(8):L08703. https://doi.org/10.1029/2005GL022478

Malhi Y, Wright J (2004) Spatial patterns and recent trends in the climate of tropical rainforest regions. Philos Trans R Soc Lond 359:311–329. https://doi.org/10.1098/rstb.2003.1433

Mantua NJ, Hare SR, Zhang Y, Wallace JM, Francis RC (1997) A pacific interdecadal climate oscillation with impacts on salmon production. Bull Am Meteorol Soc 78(6):1069–1080. https://doi.org/10.1175/1520-0477(1997)078<1069:APICOW>2.0.CO;2

Martens B, Waegeman W, Dorigo WA, Verhoest NE, Miralles DG (2018) Terrestrial evaporation response to modes of climate variability. NPJ Clim Atmos Sci 1(1):1–7. https://doi.org/10.1038/s41612-018-0053-5

Martinez WL, Martinez AR (2005) Exploratory data analysis with MATLAB. Chapman and Hall/CRC Press LLC, UK. 1-58488-366-9

McGregor G (2017) Hydroclimatology, modes of climatic variability and stream flow, lake and groundwater level variability: a progress report. Prog Phys Geograp: Earth Environ 41(4):496–512. https://doi.org/10.1177/0309133317726537

McMahon TA, Finlayson BL, Haines A, Srikanthan R (1987) Runoff variability: a global perspective. IASH-AISH 168(1987):3–11. https://iahs.info/uploads/dms/iahs_168_0003.pdf

Molion LCB, Lucio PS (2013) A note on Pacific Decadal Oscillation, El Nino Southern Oscillation, Atlantic Multidecadal Oscillation and the Intertropical Front in Sahel, Africa. Atmos Clim Sci 3:269–274. https://doi.org/10.4236/acs.2013.33028

Moron V, Ward MN (1998) Enso teleconnections with climate variability in the European and African sectors. Weather 53(9):287–295. https://doi.org/10.1002/j.1477-8696.1998.tb06403.x

Miralles DG, Van Den Berg MJ, Gash JH et al (1998) El Niño-La Niña cycle and recent trends in continental evaporation. Nat Clim Change 4(2):122–126. https://doi.org/10.1038/nclimate2068

Nakken M (1999) Wavelet analysis of rainfall-runoff variability isolating climatic from anthropogenic patterns. Environ Modell Softw 14(4):283–295. https://doi.org/10.1016/S1364-8152(98)00080-2

Ndehedehe C, Awange J, Agutu N, Kuhn M, Heck B (2016) Understanding changes in terrestrial water storage over West Africa between 2002 and 2014. Adv Water Res 88:211–230. https://doi.org/10.1016/j.advwatres.2015.12.009

Ndehedehe CE (2017) Remote sensing of West Africa's water resources using multi-satellites and models. PhD thesis, Curtin University, Bentley, Perth, Western Australia. http://hdl.handle.net/20.500.11937/59637. Accessed 12th Jan 2018

Ndehedehe CE (2019) The water resources of tropical West Africa: propblems, progress and prospect. Acta Geophysica 67(2):621–649. https://doi.org/10.1007/s11600-019-00260-y

Ndehedehe CE, Agutu NO, Okwuashi OH, Ferreira VG (2016) Spatio-temporal variability of droughts and terrestrial water storage over Lake Chad Basin using independent component analysis. J Hydrol 540:106–128. https://doi.org/10.1016/j.jhydrol.2016.05.068

Ndehedehe CE, Anyah RO, Alsdorf D, Agutu NO, Ferreira VG (2019) Modelling the impacts of global multi-scale climatic drivers on hydro-climatic extremes (1901–2014) over the Congo basin. Sci Total Environ 651:1569–1587. https://doi.org/10.1016/j.scitotenv.2018.09.203

Ndehedehe CE, Awange J, Kuhn M, Agutu N, Fukuda Y (2017) Climate teleconnections influence on West Africa's terrestrial water storage. Hydrol Process 31(18):3206–3224. https://doi.org/10.1002/hyp.11237

Ndehedehe CE, Awange JL, Agutu NO, Okwuashi O (2018) Changes in hydro-meteorological conditions over tropical West Africa (1980 – 2015) and links to global climate. Glob Planet Change 162:321–341. https://doi.org/10.1016/j.gloplacha.2018.01.020

Ndehedehe CE, Awange JL, Corner R, Kuhn M, Okwuashi O (2016) On the potentials of multiple climate variables in assessing the spatio-temporal characteristics of hydrological droughts over the Volta Basin. Scie Total Environ 557–558:819–837. https://doi.org/10.1016/j.scitotenv.2016.03.004

Ndehedehe CE, Ferreira VG (2020) Identifying the footprints of global climate modes in time-variable gravity hydrological signals. Clim Change 159:481–502. https://doi.org/10.1007/s10584-019-02588-2

Ndehedehe CE, Ferreira VG, Onojeghuo AO, Agutu NO, Emengini E, Getirana A (2020) Influence of global climate on freshwater changes in Africa's largest endorheic basin using multi-scaled indicators. Sci Total Environ 737:139643. https://doi.org/10.1016/j.scitotenv.2020.139643

Neves MC, Jerez S, Trigo RM (2019) The response of piezometric levels in Portugal to NAO, EA, and SCAND climate patterns. J Hydrol 568:1105–1117. https://doi.org/10.1016/j.jhydrol.2018.11.054

Ni S, Chen J, Wilson CR, Li J, Hu X, Fu R (2017) Global terrestrial water storage changes and connections to ENSO events. Surv Geophys. https://doi.org/10.1007/s10712-017-9421-7

Ni S, Chen J, Wilson CR, Li J, Hu X, Fu R (2018) Global terrestrial water storage changes and connections to ENSO events. Surv Geophys 39(1):1–22. https://doi.org/10.1007/s10712-017-9421-7

Nicholson S (2013) The West African Sahel: a review of recent studies on the rainfall regime and its interannual variability. ISRN Meteorol 2013(453521):1–32. https://doi.org/10.1155/2013/453521

Nicholson SE, Some B, Kone B (2000) An analysis of recent rainfall conditions in West Africa, including the rainy seasons of the 1997 El Niño and the 1998 La Niña years. J Clim 13(14):2628–2640. https://doi.org/10.1175/1520-0442(2000)013<2628:AAORRC>2.0.CO;2

Mantua NJ, Hare SR (2002) The pacific decadal oscillation. J Oceanogr 58(1):35–44. https://doi.org/10.1023/A:1015820616384

Oguntunde PG, Abiodun BJ (2013) The impact of climate change on the Niger River Basin hydroclimatology. Clim Dyn 40(1–2):81–94. https://doi.org/10.1007/s00382-012-1498-6

Okonkwo C, Demoz B, Gebremariam S (2014) Characteristics of Lake Chad level variability and links to ENSO, precipitate, and river discharge. Sci World J 13(4):13. https://doi.org/10.1155/2014/145893

Omondi P, Awange J, Ogallo L, Okoola R, Forootan E (2012) Decadal rainfall variability modes in observed rainfall records over east Africa and their relations to historical sea surface temperature changes. J Hydrol 464–465:140–156. https://doi.org/10.1016/j.jhydrol.2012.07.003

Owusu K, Waylen P, Qiu Y (2008) Changing rainfall inputs in the Volta basin: implications for water sharing in Ghana. GeoJournal 71(4):201–210. https://doi.org/10.1007/s10708-008-9156-6

Paeth H, Fink A, Pohle S, Keis F, Machel H, Samimi C (2012) Meteorological characteristics and potential causes of the 2007 flood in sub-Saharan Africa. Int J Climatol 31:1908–1926. https://doi.org/10.1002/Joc.2199

Peters E, Bier G, van Lanen H, Torfs P (2006) Propagation and spatial distribution of drought in a groundwater catchment. J Hydrol 321(1):257–275. https://doi.org/10.1016/j.jhydrol.2005.08.004

Phillips T, Nerem RS, Fox-Kemper B, Famiglietti JS, Rajagopalan B (2012) The influence of ENSO on global terrestrial water storage using GRACE. Geophys Res Lett 39(L16705):2012. https://doi.org/10.1029/2012GL052495

Polo I, Rodríguez-Fonseca B, Losada T, García-Serrano J (2008) Tropical Atlantic variability modes (1979–2002). Part I: time-evolving SST modes related to West African rainfall. J Clim 21(24):6457–6475. https://doi.org/10.1175/2008JCLI2607.1

Rasmusson EM (2019) Global prospects for the prediction of drought: a meteorological perspective. In planning for drought, pp 31–43. https://doi.org/10.4324/9780429301735

Rasmusson EM, Carpenter TH (1983) The relationship between eastern equatorial pacific sea surface temperatures and rainfall over India and Sri Lanka. Monthly Weather Rev 111(3):517–528. https://doi.org/10.1175/1520-0493(1983)111<0517:TRBEEP>2.0.CO;2

Redelsperger J-L, Lebel T (2009) Surface processes and water cycle in West Africa, studied from the AMMA-CATCH observing system. J Hydrol 375(1–2):1–2. https://doi.org/10.1016/j.jhydrol.2009.08.017

Redelsperger J-L, Thorncraft CD, Diedhiou A, Lebel T, Parker DJ, Polcher J (2006) African monsoon multidisciplinary analysis. An international research project and field campaign. Bull Am Meteorol Soc 87(12):1739–1746.https://doi.org/10.1175/BAMS-87-12-1739

Rey D, Pérez-Blanco CD, Escriva-Bou A, Girard C, Veldkamp TIE (2019) Role of economic instruments in water allocation reform: lessons from Europe. Int J Water Res Dev 35(2):206–239. https://doi.org/10.1080/07900627.2017.1422702

Rodell M, Famiglietti JS, Wiese DN, Reager JT, Beaudoing HK, Landerer FW, Lo M-H (2018) Emerging trends in global freshwater availability. Nature 557:651–659. https://doi.org/10.1038/s41586-018-0123-1

Rust W, Holman I, Corstanje R, Bloomfield J, Cuthbert M (2018) A conceptual model for climatic teleconnection signal control on groundwater variability in Europe. Earth-Sci Rev 177:164–174. https://doi.org/10.1016/j.earscirev.2017.09.017

Saji NH, Goswami BN, Vinayachandran PN, Yamagata T (1999) A dipole mode in the tropical Indian Ocean. Nature 401:360–363

Santoso A, Mcphaden MJ, Cai W (2017) The defining characteristics of ENSO extremes and the strong 2015/2016 El Niño. Rev Geophys 55(4):1079–1129. https://doi.org/10.1002/2017RG000560

Save H, Bettadpur S, Tapley BD (2016) High-resolution CSR GRACE RL05 Mascons. J Geophys Res: Solid Earth 121(10):7547–7569. https://doi.org/10.1002/2016JB013007

Szolgayova E, Parajka J, Blöschl G, Bucher C (2014) Long term variability of the Danube river flow and its relation to precipitation and air temperature. J Hydrol 519:871–880. https://doi.org/10.1016/j.jhydrol.2014.07.047

Savitzky A, Golay MJE (1964) Soothing and differentiation of data by simplified least squares procedures. Anal Chem 36(8):1627–1639

Scanlon BR, Keese KE, Flint AL, Flint LE, Gaye CB, Edmunds WM, Simmers I (2006) Global synthesis of groundwater recharge in semiarid and arid regions. Hydrol Process 20(15):3335–3370. https://doi.org/10.1002/hyp.6335

Scanlon BR, Reedy RC, Stonestrom DA, Prudic DE, Dennehy KF (2005) Impact of land use and land cover change on groundwater recharge and quality in the southwestern US. Glob Change Biol 11(10):1577–1593. https://doi.org/10.1111/j.1365-2486.2005.01026.x

Sheffield J, Wood EF (2008) Global trends and variability in soil moisture and drought characteristics, 1950–2000, from observation-driven simulations of the terrestrial hydrologic cycle. J Clim 21(3):432–458. https://doi.org/10.1175/2007JCLI1822.1

Snedecor GW, Cochran WG (1989) Statistical methods, 8th edn. Iowa State University Press

Sun Q, Miao C, AghaKouchak A, Duan Q (2016) Century-scale causal relationships between global dry/wet conditions and the state of the pacific and Atlantic oceans. Geophys Res Lett 43(12):6528–6537. https://doi.org/10.1002/2016GL069628

Swenson S, Wahr J (2009) Monitoring the water balance of Lake Victoria. East Africa, from Space 370(1–4):163–176. https://doi.org/10.1016/j.jhydrol.2009.03.008

Tabari H, Hosseinzadeh Talaee P, Shifteh Some'e B, Willems P (2014) Possible influences of north Atlantic oscillation on winter reference evapotranspiration in Iran. Glob Plan Change 117:28–39. https://doi.org/10.1016/j.gloplacha.2014.03.006

Townley LR (1995) The response of aquifers to periodic forcing. Adv Water Res 18(3):125–146. https://doi.org/10.1016/0309-1708(95)00008-7

Tremblay L, Larocque M, Anctil F, Rivard C (2011) Teleconnections and interannual variability in Canadian groundwater levels. J Hydrol 410(3):178–188. https://doi.org/10.1016/j.jhydrol.2011.09.013

Uddameri V, Singaraju S, Hernandez E (2019) Is Standardized Precipitation Index (SPI) a useful indicator to forecast groundwater droughts?-Insights from a Karst aquifer. JAWRA J Am Water Res Assoc 55(1):70–88. https://doi.org/10.1111/1752-1688.12698

USAID (2013) 2013 World population data sheet. United States Agency International Development. https://www.prb.org. Accessed 2nd Aug 2014

Van Loon AF (2015) Hydrological drought explained. WIREs. Water 2:359–392. https://doi.org/10.1002/wat2.1085

Velasco EM, Gurdak JJ, Dickinson JE, Ferré T, Corona CR (2017) Interannual to multidecadal climate forcings on groundwater resources of the US west coast. J Hydrol: Reg Stud 11:250–265. https://doi.org/10.1016/j.ejrh.2015.11.018

Vörösmarty CJ, McIntyre PB, Gessner MO, Dudgeon D, Prusevich A, Green P, Glidden S, Bunn SE, Sullivan CA, Liermann CR, Davies PM (2010) Global threats to human water security and river biodiversity. Nature 467(7315):555–561. https://doi.org/10.1038/nature09440

Watkins MM, Wiese DN, Yuan D, Boening C, Landerer FW (2015) Improved methods for observing earth's time variable mass distribution with GRACE using spherical cap Mascons. J Geophys Res: Solid Earth 120(4):2648–2671. https://doi.org/10.1002/2014JB011547

Weber K, Stewart M (2004) Critical analysis of the cumulative rainfall departure concept. Ground Water 42(6):935–938. http://info.ngwa.org/gwol/pdf/042979908.pdf

White WB, Gershunov A, Annis JL, McKeon G, Syktus J (2004) Forecasting Australian drought using Southern Hemisphere modes of sea-surface temperature variability. Int J Climatol 24(15):1911–1927. https://doi.org/10.1002/joc.1091

Whittemore DO Jr, JJB, Wilson BB, (2016) Assessing the major drivers of water-level declines: new insights into the future of heavily stressed aquifers. Hydrol Sci J 61(1):134–145. https://doi.org/10.1080/02626667.2014.959958

Wolter K, Timlin MS (2011) El niño/southern oscillation behaviour since 1871 as diagnosed in an extended multivariate ENSO index (MEI.ext). Int J Climatol 31(7):1074–1087. https://doi.org/10.1002/joc.2336

Woollings T, Blackburn M (2012) The north Atlantic jet stream under climate change and its relation to the NAO and EA patterns. J Clim 25(3):886–902. https://doi.org/10.1175/JCLI-D-11-00087.1

Wiese DN, Landerer FW, Watkins MM (2016) Quantifying and reducing leakage errors in the JPL RL05M GRACE mascon solution. Water Res Res 52(9):7490–7502. https://doi.org/10.1002/2016WR019344

Wild M, Grieser J, Schär C (2008) Combined surface solar brightening and increasing greenhouse effect support recent intensification of the global land-based hydrological cycle. Geophys Res Lett 35(L17706):L17706–1–L17706–5. https://doi.org/10.1029/2008GL034842

Ziehe A (2005) Blind source separation based on joint diagonalization of matrices with applications in biomedical signal processing. PhD thesis, Universitat Potsdam. http://en.youscribe.com/catalogue/reports-and-theses/knowledge/blind-source-separation-based-on-joint-diagonalization-of-matrices-1424347. Accessed 15 May 2015

Part XII
Eco-hydrological Processes

Chapter 18
Floodplain Productivity in Wet-Dry Tropics

Abstract The ecological, cultural and economic values and critical environmental resources provided by wetlands and freshwater habitats have been globally recognised, making them high priorities for conservation and management. The large-scale assessment of wetlands and floodplains is therefore required to improve our contemporary understanding of wetland functions and their response to the degrading impacts of anthropogenic actions and strong fluctuations in climate. In this chapter the capabilities of multi-resolution satellite systems and techniques in the assessment of floodplain productivity and wetlands are discussed. As part of efforts that underpin conservation planning and wetland management, the spatial estimations of aquatic habitats are usually required. The productivity of such habitats can be assessed by predicting the spatial distribution of aquatic plant biomass and characteristics (e.g., frequency, variability, magnitude) of floodplain inundation. A monitoring framework that was developed by Ndehedehe et al. (2020b) over a large river catchment in northern Queensland of Australia to support this kind of assessment is discussed. An important aspect of this assessment, which is detailed in this chapter is on the innovative use of satellite remote sensing biophysical indicators (vegetation and inundation) and flood water depth using a classification tree model.

18.1 Climate Variability and Floodplain Wetlands

Changes in total inundation caused by changes in the characteristics of rainfall seasonality and climate variability influence water availability on tropical floodplains (Ndehedehe et al. 2021; Hamilton et al. 2002). These floodplains are characterized by strong flows and severe floods in considerably wet seasons and then dry up or shrink during periods of limited rainfall and low flows. In the Australian wet–dry tropical areas like the Gulf region consisting of the Flinders and Gilbert catchments, these reduced flows usually happen during the dry seasons when little or no rainfall is received (Ndehedehe et al. 2020b). The extent to which these low flows or extended periods of limited rainfall during the dry season impact on floodplains will determine

C. Ndehedehe, *Satellite Remote Sensing of Terrestrial Hydrology*,
https://doi.org/10.1007/978-3-030-99577-5_18

the amount of refugia and water holes that persist for aquatic life (Hamilton et al. 2002) and the productivity of freshwater habitats. Douglas et al. (2005) defined tropical floodplains as extremely dynamic ecosystems for which periodic flood events nourish water supplies, which causes an accretion of nutrient-rich deposits that have a strong positive influence on primary production. The significant growth observed in algae and macrophytes is largely a function of the inundation frequency coupled with nutrient accessibility (Pettit et al. 2011; Finlayson 2005; Junk and Piedade 1997). Case-specific reports have highlighted the basic composition of aquatic life on tropical floodplains over the South American continent as one of the most significant as well as productive ecosystems (Silva et al. 2009; Junk and Piedade 1997). Floodplain aquatic life in South America and Australia has shown that epiphytic algae (Rasmussen 2010) is one of the major source material producers for the development of secondary floodplain biomass (Jardine et al. 2012; Douglas et al. 2005).

However, the contributions of terrestrial carbon sources and aquatic life are important in the biomass of floodplain secondary effects. A critical component of floodplain ecosystem functioning is the optimal production of floodplain algae, especially as a result of the prospects created by inundated floodplains for secondary producers like fishes to migrate to the floodplains from estuary (Jardine et al. 2012; Winemiller 2004). But the productivity of these tropical floodplains can be restricted due to several factors. Setterfield et al. (2013), for instance, highlighted numerous stressors influencing tropical floodplains around the world. This includes modified fire schemes, invasive animals, weeds, and influence of seawater intrusion on freshwater floodplains, especially due to rise in sea level caused by climate change, as well as mining (Tockner et al. 2010; Vorosmarty et al. 2010). An adequate understanding of floodplain algal production is crucial particularly due to its role in tropical aquatic food webs, as well as conservation planning. Inundation processes in various tropical floodplains range from full floodplain flooding, majorly as a function of the rainy period, and then the deep contraction period leading to isolated water bodies observed during dry seasons. This directly signifies the possibility of algae undergoing substantial periodic and inter-annual fluctuations. The assessment of spatial and temporal changes of both algal floodplain and macrophyte productivity is crucial to supporting conservative planning and management scheme. This provides an avenue to distinctly locate relevant regions ('hotspots') of floodplain productivity, which is essential for long-term sustainability of the system.

The tropical wet–dry river systems of northern Australia consists of intense seasonal streamflow (Kennard et al. 2010). Over 90% of the total annual stream flows of many rivers in this region occur during November–April, which forms a major part of the wet season (CSIRO 2009). The direct association of river systems with wetlands, particularly on floodplains impacts substantially on the longitudinal and lateral aspects of the system, as well as aids the brooding, survival, and nourishment of aquatic species, such as fish (Boulton 2007; Bunn et al. 2006). During the wet season, nourishments and productivity in freshwater resources are conveyed to the coastal waters and also provide support for economically relevant aquatic life (Buckworth et al. 2014). Other contributions of flow include its role in aquifer recharge (Boulton 2007), which in turn influences floodplain vegetation and riparian populace, and supports basic dry season base flow to downstream waterholes. Sometimes this base flow occurs after the flow from floodwater reoccurrence to river channels have stopped (Butler et al. 2007).

On the other hand, the dry season is characterized by very limited flows, which are most times completely non-existent thereby influencing aquatic life in the catchment to exist as distinct waterholes. The waterholes are usually known to undergo a drying process during the dry seasons. The size of waterholes that are connected to aquifer reserves does not undergo substantial changes. The waterholes are also known to provide important habitats for aquatic biota as well as ecological processes, especially those that persist all year round (Pearson et al. 2013; Pahl-Wostl et al. 2013; DERM 2011; Sheldon et al. 2010). The capacity of water resource developments in instigating major environmental disruptions is evident in the variation observed at basin scales, especially regarding its hydrology and limnology. Negative alterations in peak and wet season flows have the tendency of influencing downstream habitats, such as coastal territories and estuarine, including river geomorphology, groundwater recharge, as well as riverine flushing. The reduction in the volumes of waterholes is mostly experienced as a result of reduction in groundwater replenishment and/or water withdrawal rate noticed during the dry periods. Also, other water resource development activities such as agricultural practices lead to intensified sediment, agricultural, and nutrient pile to aquatic life.

Regions in northern Queensland such as the Walsh and Burdekin areas have experienced impact from irrigation. This led to a decline in water quality, contributing directly to poor faunal health, as well as negative ecosystem effects (Pearson et al. 2013; Butler et al. 2007; Butler and Burrows 2007). Studies undertaken by Leigh (2013) and Butler et al. (2009) suggested that excessive withdrawal of water related to agricultural services can lead to changes in the seasonality of river ecosystems. The availability of good water quality becomes more limiting for aquatic life, as well as ecosystem and ecological services in such waterholes. These processes are, however, highly influenced by reduced dissolved oxygen, turbidity as well as escalated temperature levels (Butler et al. 2009, 2007). Impacts that worsen water quality concerns, particularly regarding temperature and dissolved oxygen are mostly from irrigation developments. These have the tendency of resulting in damages through the dissipation of riparian vegetation or improved weed reduction and nutrient loading to surface water sources as runoff from irrigated plantations (Butler and Burrows 2007).

The Flinders catchment in northern Queensland is characterized by a semi-arid tropical climate. The average and median yearly precipitation pattern generated spatially over the catchment is 492 and 454 mm, respectively (Cuan and Yang 2013). Figure 18.1 on the other hand shows a historical yearly series for the Flinders catchment representing significant anomalies over the years (1890–2012). While 1974 experienced the maximum amount of precipitation (1310 mm) during the study period, it accounted for a triple increase in the median yearly precipitation value. The coastal north–south variation in the amount of mean annual precipitation ranged from 800 to 350 mm, respectively, in the two regions. The Flinders catchment has a

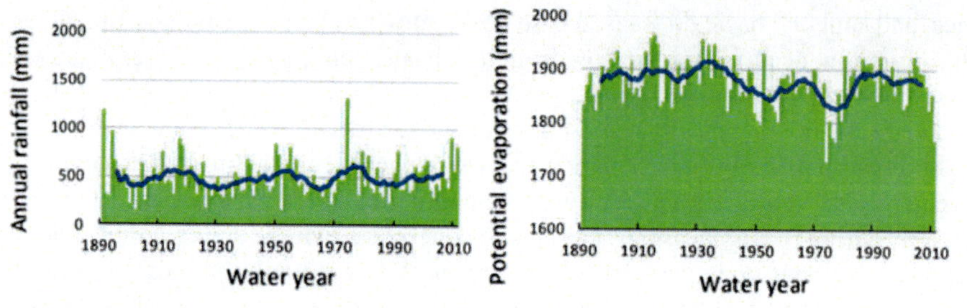

Fig. 18.1 Averaged yearly precipitation and regional potential evaporation over the Finders catchment. *Source* Cuan and Yang (2013)

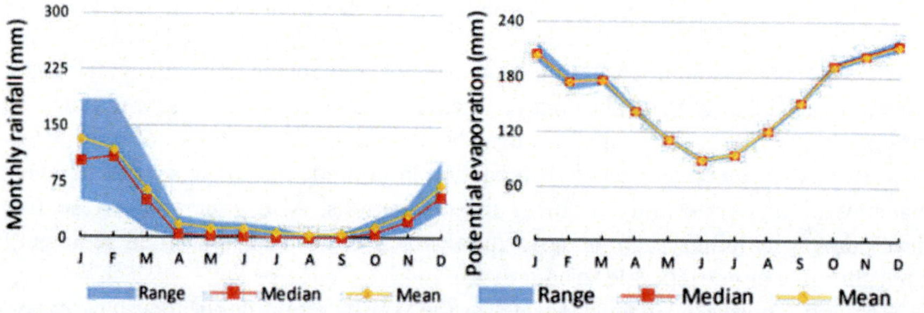

Fig. 18.2 Monthly precipitation and regional potential evaporation rate over the Finders catchment (20–80th percentile monthly precipitation range). *Source* Cuan and Yang (2013)

unique climate characteristic, especially pertaining to having over 88% of precipitation occurring from November to April (Fig. 18.2).

The months of January and February have the highest median monthly precipitation (approximately 100 mm) over the Flinders catchment (Waltham et al. 2013). Conversely, July and August are the months that have the lowest precipitation rate as observed within the study period (Fig. 18.2). While the Flinders catchment has an average yearly potential evaporation of 1862 mm, the mean dry and wet seasonal potential regional evaporation rates are 762 and 1115 mm, respectively. A large mean annual precipitation deficit of approximately 600 mm is observed in a major part of the Finders catchment and the impacts of rainfall variability on floodplain productivity is discussed in subsequent sections.

18.2 Remote Sensing of Wetlands

Wetlands are one of the most rapidly declining, endangered, as well as fragile ecosystems (Weise et al. 2020; Tickner et al. 2020). Despite the important ecosystem services they provide, such as water purification, water source, climate variation modification, carbon sequestration, and mitigation of disaster risk, they are vulnerable to the impacts of climate change (Tickner et al. 2020; Ecosystem 2005). Since the nineteenth century, it has been observed that over 70% of the globally recognized wetlands have diminished at various spatial and temporal rates across several regions (WWF 2020). Nevertheless, areas such as Canada, which accounts for an approximate 24% of the global wetlands (occupies over 14% of its landmass) have reported difficulties in the assessment and continuous monitoring of this crucial ecosystem resource (Government of Canada). However, the vulnerability of wetlands to climate variability is now increasingly being monitored and assessed leveraging remotely sensed observation. This has been demonstrated in the Grassland Natural Region of southern Alberta characterized by marshes, swamps, and shallow open water wetlands within the vicinity of agricultural fields (Onojeghuo et al. 2021).

18.2.1 Wetland Mapping with Multi-temporal and Multi-resolution Satellites

With improved sensor capabilities in new satellite missions, studies have shown the relevance of satellite remote sensing in wetland monitoring and assessment (Ozesmi and Bauer 2002; Adam et al. 2010). The spatial resolution of small-sized prairie pothole wetlands existing in Grassland Natural Regions are better mapped with the use of high-resolution imageries including commercial images like Rapid Eye (5 m), Planet Scope (3–5 m), SPOT6 (1.5 m) as well as aerial orthophotos (Wu et al. 2019). In the North American region, several studies have attempted to monitor wetland dynamics and inundations using remote sensing techniques and procedures from Earth observation satellites like Sentinel-1 Synthetic aperture radar (SAR) (Wang et al. 2020; Mahdianpari et al. 2019, 2020), Landsat (Amani et al. 2019; DeVries et al. 2017), as well as Sentinel-2 (S2). The use of SAR data, such as Radarsat-1 (Olthof and Rainville 2020) and Radarsat-2 (Jahncke et al. 2018), have also been captured in these studies. The enormous value in having a vast repository of multi-spatial and temporal Earth observation data to accurately assess regional wetlands is a direct function of the challenges related to individual wetlands. The challenge in the classification and delineation of existing habitat type is notable, hence the justification for the use of satellite remote sensing technique in its assessment. With the use of Landsat-8 imagery for a study period of 24 months (2016–2018), Amani et al. (2019) investigated the spatial extent of wetlands at a 30 m spatial resolution using the conventional Google Earth Engine (GEE) infrastructure to ease access and preprocessing of big data. Moreover, Mahdianpari et al. (2020) improved on a previous study by leveraging high-resolution Sentinel data to monitor wetlands over the Canada eco-zones using the GEE platform. These past studies argued on the suitability and efficiency of multispectral and SAR data as well as topographic data in providing accurate information on wetland and flood extent mapping. This is particularly helpful if complemented with optical, radar, and elevation data in both forested and non-forested wetlands (Jahncke et al. 2018; Niculescu et al. 2016, 2020).

With the advent of new satellite remote sensing techniques, the task of adequate and efficient monitoring and mapping of land cover with the use of both radar and optical data has been simplified (Slagter et al. 2020).

18.2.1.1 Challenges and Prospects

Several factors and wetland characteristics make high accuracy very difficult to achieve (Gallant 2015). This is because wetlands do not have a homogenous vegetation or land cover type, and the only consistent feature over the entire wetlands is the presence of water and aquatic components, existing either in the soil, below vegetation canopy, or on land surface (Slagter et al. 2020). This constraint has been identified as a major challenge while attempting to delineate wetlands based on backscattering or spectral details even from the newly developed satellite remote sensing techniques (Gallant 2015). Also, information derived from this procedure has a high tendency of varying with time with respect to flood level dynamics. Arguably, the capabilities of most satellite sensors to address these wetland complexities are usually inadequate.

The presence of flood under forests canopy impedes the efficiency of most satellites in precise monitoring of areas with large spatial coverage, especially when the regions of interest have experienced periods of frequent cloud cover or when there is a need to identify small or narrow wetlands (Ozesmi and Bauer 2002; Rosenqvist et al. 2007). Most satellite sensors or missions like MODIS often produce coarse images, which can be difficult for accurate detection of wetlands. It is also pertinent to note that some satellites have the shortest possible revisit time to accurately detect and quantify flood as some floods could recede much faster. Other challenges for large-scale wetland characterization and assessments are lack of comprehensive wetland classification structures (Mahdavi et al. 2018), and lack of uniformity, and/or poorly defined global wetlands (Rosenqvist et al. 2007). Consequently, past studies have focused on discrete aspects of wetlands during their assessments. For example, Rosenqvist et al. (2007) emphasized on characterizing a specific wetland type, while Semeniuk and Semeniuk (1997) focused on classifying a specific wetland characteristic, etc. To comprehensively determine a more generic and applicable wetland classification, two major features are necessary; this includes (i) surface water dynamics and (ii) wetland vegetation types (Keddy et al. 2009).

18.2.1.2 Optical Versus Radar Observations

The European Space Agency's ongoing developments of its Copernicus program provide an opportunity for a more precise technique for wetland mapping from space. This can be achieved by utilizing the Sentinel-1 and -2 datasets which are not only open access but includes very high-resolution sensors for adequate spatio-temporal wetland assessment. The Sentinel-1 data assimilates data in single and dual-polarimetric modules (10 m resolution) using a C-band SAR system of approximately 5.5 cm wavelength strength (Torres et al. 2012). With the use of radar remote sensing, it becomes more effective and efficient to adequately monitor wetlands given its numerous advantages, which include, enhanced moisture sensitivity, unconstrained functioning despite weather and daylight modes, as well as its potential in observing sub-canopy water bodies (Henderson and Lewis 2008).

The Sentinel-2 is furnished with a multispectral sensor, which provides varied resolutions for optical data, like 10, 20, and 60 m in the visible, near-infrared as well as short-wave infrared partitions of the electromagnetic spectrum (Drusch et al. 2012). Inasmuch as the radar observation mode is preferred to the optical observation for sub-canopy water body detection due to its wetland mapping limitation noticed in bad weather conditions and night times, optical dataset has proved to be very efficient auxiliary for the radar dataset (Bourgeau-Chavez et al. 2009; Ozesmi and Bauer 2002). Given the ability of these two Sentinel missions to efficiently acquire images with relatively wide swaths consisting of two polar orbiting satellites, it increases their revisit time to 5 and 6 days at the equator for the Sentinel-1 and -2 satellites, respectively. Some studies have already implemented the Sentinel-1 and -2 sensors for adequate wetland mapping and characterization in relation to its acclaimed functionalities. The potential of Sentinel-2 (Bhatnagar et al. 2018; Araya-López et al. 2018; Kaplan and Avdan 2017) and the Sentinel-1 + Sentinel-2 (Whyte et al. 2018; Kaplan and Avdan 2017; Chatziantoniou et al. 2017) were assessed by these studies for single-date wetland characterization so as to confirm its high-resolution property as well as its efficiency and applicability in adequate wetland assessments. The multi-temporal use of the Sentinel-1 (Xing et al. 2018; Tsyganskaya et al. 2018; Tian et al. 2017; Kaplan and Avdan 2017; Marek et al. 2016; Muro et al. 2016) and Sentinel-2 (Ludwig et al. 2019) has also been extensively reported for this purpose. The parameters of temporally dense Sentinel-1 dataset in an accurate assessment of surface water dynamics, characterization, as well as flood frequencies in distinct wetland types were highlighted. Besides, the use of Sentinel-1 for flood detection and assessment, especially in vegetated regions is increasingly becoming popular. However, the process can also be efficiently undertaken using information from optical datasets. The integration of Sentinels-1 and -2 datasets in multi-temporal assessment have also been implemented, despite its overarching limitation to generic wetland delineation characteristics (Mahdianpari et al. 2019).

18.2.1.3 Data Fusion-Multispectral and LiDAR

With the recent development in Earth observation sciences and sensor technologies, the utilization of multi-sensor-derived datasets for land use/cover and habitat monitoring over diverse landscapes is increasing. This led to the integration of spectral data from very high-resolution (VHR) multispectral and hyperspectral sensors with Light Detection and Ranging (LiDAR) information for adequate land cover and vegetation monitoring (Mücher et al. 2015). This integration of optical imagery and LiDAR data has led to a more accurate land cover assessment and habitat mapping even at very high resolutions, thus meeting the demands of end-users, including nature conservation analysts. This integration is warranted given the major limitation of the sole-utilization of optical datasets for this analysis (e.g., reliable information for vegetation height cannot be determined directly). However, with this integration, it is easier to monitor vegetation composition, which compensates by enhancing classification precision when applied for assessing single or diverse habitat ranks (Zhang 2014; Onojeghuo and Blackburn 2011). From past studies, we understand the need to spatially integrate hyperspectral and multispectral imageries with LiDAR data aimed at enhancing classification precision (Bork and Su 2007; Mücher et al. 2015; Rapinel et al. 2015; Zhang 2014; Jones et al. 2010; Ke et al. 2010). Bork and Su (2007), Rapinel et al. (2015) and Ke et al. (2010) highlighted the advantages of assimilating VHR multispectral imagery, like QuickBird with LiDAR data for adequate monitoring of varied vegetation types such as forests and wetlands. The results from these studies showed that this integration of optical and LiDAR datasets provided the best classification results. The study by Onojeghuo et al. (2021) developed a framework that utilizes freely available synthetic aperture radar and optical Sentinel-2 (optical) data fused with LiDAR-derived topographic wetness index to improve the accurate delineation of the wetlands. This framework is transferable and can enable the mapping of wetland habitats across complex landscapes.

18.3 Measuring Floods and Primary Productivity in Floodplain Wetlands

Mapping flood extent in climate-driven tropical catchments like the Flinders is important to keep track of the impacts of climate variability on surface water hydrology, especially in wetlands and floodplains that tend to be vulnerable to such influence. Maps derived from SAR data also provide a distinct and primary source of information for effective disaster management. The monitoring of inundation patterns in a timely and cost-efficient manner aids meaningful decision that underpins such management (Voigt et al. 2007). Additionally, these special-purpose maps can also provide important details on distributed calibration, as well as serve as validation data for hydraulic models of various flow processes (Mason et al. 2015; García-Pintado et al. 2015; Schumann et al. 2009). It can also serve as an alternative for creating hazard maps within the scope of flood mitigation activities, spatial planning, as well as insurance risk management (de Moel et al. 2009).

However, floodplains are identified as one of the most biodiverse ecosystems in the world, with a niche for sustaining high rates of basic production (Keddy et al. 2009). Their importance regardless, has not in any way prevented the massive and ongoing degradation and modification of most of the world's important wetlands, especially as a result of anthropogenic influences (Reis et al. 2017; Davidson 2014). The pressure on floodplains is currently on the increase, especially as a result of economic and population growth, which has the ability to alter several terrestrial components as evidenced in the tropics where considerable developments are on going (Roggeri 2013; Keddy et al. 2009). This danger posed on wetlands will continue unless there is a rapid intervention facilitated by a recognition of the ecological significance and fundamental ecosystem services they deliver (Reis et al. 2017). A crucial subsidy for organisms of higher tropical levels is facilitated by the high rate of aquatic production on floodplains (Saigo et al. 2015; Jardine et al. 2012; Bunn et al. 2003; Lewis et al. 2001).

Past studies undertaken using chemical traces integrated with stable isotopes and fatty acids revealed that algae are a primary basal food resource, especially for wetlands and river food webs across a broad range of flora and fauna. For instance, a large portion of the erect biomass (range 42–97%) for aquatic life caught in the lower part of an Australian tropical area was determined from floodplain algae (Jardine et al. 2012). Analogously, as pointed out by Mortillaro et al. (2015), the huge body mass of aquatic life in the Amazon floodplain is due to the available high-quality food sources, which include microalgae. This underpins our understanding of the basic role of algae in the aquatic food web, as well as its response to an alteration to environmental parameters and how it can be used to estimate larger ecosystems responses of varied species of aquatic organisms. Studies have identified important environmental characteristics that influence algal productivity, as well as the numerous parameters of the aquatic habitat that affects the framework and functioning of algal regions (Jia et al. 2020). The presence of sunlight has been underlined as the overarching component influencing aquatic primary biomass as well as its value in wetlands (Jia et al. 2020; Pettit et al. 2017). Factors influencing the presence of this sunlight have been highlighted in previous studies as evaporated organic carbon, turbidity, as well as shading from macrophyte concealment (Pettit et al. 2017; Yeh et al. 2011; Grimshaw et al. 1997). The presence of nutrients also contributes significantly to bottom-up control of algal productivity, especially while having phosphorus and nitrogen as primary nutrients (e.g., Elser et al. 2007). Algal productivity has also been linked to other environmental parameters that enhance its productivity, and they include the presence of aquatic plants due to the action of their stems and leaves as algae substrate, especially for algae to append and propagate in the euphotic zone (Pettit et al. 2017).

18.3.1 Normalized Difference Chlorophyll Index

Chlorophyll is a pigment responsible for the transformation of sunlight energy for photosynthesis present in various photosynthesizing life forms (Zimba and Gitelson 2006). Its assessment and estimations are functions of various models and algorithms obtained from remote sensing of phytoplankton's abundance and dissemination in coastal waters. The impact of chlorophyll and carotenoids which are referred to as accessory pigments have diminutive impacts which can be captured by multispectral remote sensing (Trees et al. 2000). The absorption of solar energy for the chlorophyll pigment occur in the red and blue wavelengths (Tedetti et al. 2010). With this high rate absorption rate in this spectrum, it therefore leads to a relatively high reflectance rate in the green wavelength (Gitelson et al. 2003). Chlorophyll also radiate in the near-infrared spectrum (700nm), especially while photosynthesizing (Gower et al. 2004; Gitelson et al. 2003). However, during global monitoring with satellites, most of the observational datasets from satellite sensors as well as the corresponding algorithms emphasis on the blue and green spectra, which are appropriate for specific regions, such as open oceans (Gower et al. 2004). As highlighted by Kamerosky et al. (2015), remote sensing of chlorophyll in the coastal surface water is more complex when compared to open ocean remote sensing due to some factors including,

- absorption by coloured evaporated organic components (Miller et al. 2002),
- rapid current from both organic and non-organic totally appended solids (Novo et al. 1991),
- deflection of radiation from surface benthic vegetation, for example, immersed aquatic vegetation (Bricaud et al. 1995),
- benthic substrates that generate substantial blast in upwelling energy (Gilerson and et al. 2010; Zimba and Gitelson 2006).

Miller et al. (2002) pointed out the high absorption rate experienced by the coloured evaporated organic components in the blue portion of the visible spectrum, whereas the totally appended solids are known to vehemently reflect in the red spectrum as well as stimulate backscattering, especially in the blue spectrum (Gitelson et al. 2007; Tzortziou et al. 2006). It is also possible for the incident ray to hit the bottom. This will lead to a reflection of the signal, which can be added to the compound upwelling spectral profile from the surface water, particularly if water clarity is considerably high with a corresponding shallow depth (Bargain et al. 2012). The immersed aquatic vegetation similar to other vascular, photosynthetic plants reflects the near-infrared waves and absorbs photosynthetically active radiation, which subsequently leads to a further change in the total water reflectance (Williams et al. 2010; Hestir et al. 2008).

18.3.2 On the Variability of Algal Productivity in Floodplains Using Satellite Observations

Principal producers like epiphyton and phytoplankton are adequate food sources for aquatic food web on floodplain rivers and also reinforce higher-order consumers (Jardine et al. 2012; Saigo et al. 2015; Bunn et al. 2015). Given the importance of algae in such ecosystem, especially as a basic food source for higher trophic levels existing in river systems, several studies have attempted to develop algorithms to estimate their production rate (Molinari et al. 2021a; Jia et al. 2020; Vis et al. 2007). Generally, more specific details are highlighted during the estimation of algal production rates in contrast with algal biomass explaining ecosystem functioning. This happens because observed biomass is the result of various procedures as depicted by Garcia et al. (2015). It is also noteworthy to mention that observed biomass depicts the residue of losses due to ingestion by invertebrates while the production rate is a function of the foodweb-based productivity as well as the uptake of nutrients from the system (Polis 1999).

This alludes to the importance of direct observations of algal production rates over inundated floodplains, as it reflects the quantity of carbon being generated per unit time. Additionally, this helps in supporting secondary producers of aquatic consumers as mentioned by Thompson and Townsend (2005). The essence of these field measurements is to provide critical information, which can be easily assimilated into various statistical algorithms and combined with remote sensing data to estimate algal productivity even on relatively smaller or local scales (Molinari et al. 2021a). This is extremely useful for large floodplain systems existing in remote regions where field observations can be a bit challenging, given the prevailing limiting access during inundation periods, but for which remotely sensed observation can be easily obtained (Ndehedehe et al. 2020a, b). This therefore provides an opportunity to use field observations in complementing remote sensing information to enable a more accurate prediction of primary production rates even at various spatial scales (Palmer et al. 2015).

The use of satellite remote sensing in enhancing spatial ecological modelling and assessing algal primary productivity dynamics is gradually emerging (e.g., Kuhn et al. 2020). These methods have been successfully implemented to directly quantify the abundance of phytoplankton as well as its productivity over large freshwater basins, inland open water regions, and floodplains (Rotta et al. 2021; Kuhn et al. 2020; Ward et al. 2016). Although studies on assessing phytoplankton using remote sensing techniques are common, few analogous studies focused on epiphyton productivity (Vis et al. 2007). The observed complication resulting from remote sensing observation of epiphyton dynamics is due to the presence of aquatic macro-vegetation, which is basically coupled with the epiphyton itself. While the aquatic macro-vegetation is a source of habitation and brooding for the epiphytic algae (Pettit et al. 2011, 2016), it also poses a distinct obstruction for easy remote sensing observation by blocking signals from space. This led to the technique of estimating epiphyton biomass and productivity by integrating remote sensing observation with modelled field data (Ward et al. 2016; Vis et al. 2007). More available techniques, algorithms, and procedures to adequately and accurately monitor algal productivity and hotspot identification for floodplain conservation and restoration are expected to unfold in the near future. This is warranted, given the importance of these primary producers to floodplain river ecology (Molinari et al. 2021a).

18.3.2.1 Flood Events: Impacts and Assessments in Floodplain Ecosystems

The rise in severe and extreme flood events, caused by climate variability are triggering more deleterious impacts on infrastructure, land use, human population, losses, and socio-economic problems (e.g., migration) (Ndehedehe et al. 2018; Najibi and Devineni 2018; Jongman et al. 2012). Under an increased acceleration of the global water cycle, more people are being affected by floods in this day and age compared to

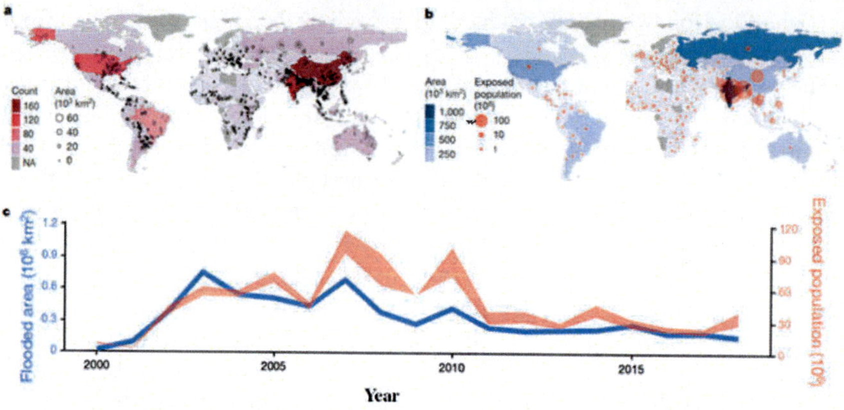

Fig. 18.3 Summarized index of the global flood database **a** country-specific cases of flood episodes in the global flood database (colour scale), including with the circular depiction of each flood event as well as their centroid positions **b** total exposed area and population per country from 2000 to 2015 **c** yearly global population estimation. *Source* (Tellman et al. 2021)

any other environmental hazard that impacts human progress and sustainable development (Fig. 18.3). With the use of satellite observation, Tellman et al. (2021) studied population exposure for more than 900 large flood events (2000–2018), showing that up to 290 million people were directly affected by floods. They observed an increase of 58–86 million people affected by flood events between 2000 and 2015. These impacts can translate to financial losses either through agricultural goods or infrastructural damages. For instance, an estimated 651 billion dollars was incurred in flood damages, which occurred from 2000 to 2019 (CRED 2020). Moreover, Thailand lost about 59 billion dollars due to the extreme 2011 flood events (Hall et al. 2014), and several flood events around the world (India, US, Bangladesh, Nepal) have left behind more economically damaging devastating impacts like the epic flood caused by Hurricane Harvey in the United States in August 2017. With more than 25 million gallons of rainfall received, this extreme flood displaced more than 17,000 people and caused power outages, among other physical damages in the region (Ndehedehe 2017).

Climate change projections for 2030 show an increasing proportion of global population will be exposed to floods (Tellman et al. 2021). This therefore heightens the relevance of investments in flood adaptation, which is basically aimed at reducing asset losses and mortality (Ward et al. 2017; Jongman et al. 2012). However, only 13% of funds meant for disaster management are allocated to mitigation, preparedness, and adaptation (Kellett and Caravani 2013). Several studies attempting to understand flood vulnerability and exposure have depended on modelled flood hazards (Winsemius et al. 2016; Tanoue et al. 2016; Jongman et al. 2012). The study undertaken by Jongman et al. (2012) predicted that population growth in a 100-year floodplain outperformed total population by 2.6% in over 20 countries from 1970 to 2010. A wide range of adverse outcomes as a result of flood events is influenced by organism vulnerability, which translates to disease, physiological/psychological trauma, death, migration, poverty, and loss of property (Rufat et al. 2015). Studies conducted using global flood database (Fig. 18.3) show trends depicting degenerations in loss of life and diminished property damage when managing hazard extents (Formetta and Feyen 2019; Tanoue et al. 2016; Jongman et al. 2015). The only region with a consistently increasing flood mortality rate since 1990 is sub-Saharan Africa (Tanoue et al. 2016), especially given its rate of frequent urban flooding, which is not expected to end anytime soon(Winsemius et al. 2016; Jongman et al. 2015). The evaluation of flood risks and their potential damage is now at the front of most hydrological studies as it is a key component that underpins flood risk management. However, the errors and uncertainties from hydrological and hazard models constitute one of the numerous limitations of model-based flood exposure and vulnerability assessments. This, however, can be due to the difficulty in incorporating anthropogenic influence to data calibration, as well as poorly captured topographic details (Ward et al. 2015).

The influence of human modification on rivers and land use, as well as re-channeling flood water is also responsible for the unreliability of these models (Shastry et al. 2020; Grill et al. 2019). The effects of these discrepancies in modelling assumptions are always in the pronounced dissimilarities between population and estimated area under exposure in each individual model (Ward et al. 2015). However, contrary to model estimates, satellite-derived observations have the potential to directly monitor inundations (DeVries et al. 2020), with an absolute depiction of climate variations, infrastructure, and land use which are hardly reflected and accounted for in modelled flood episodes.

Studies implementing remote sensing techniques in mapping and assessing submerged aquatic vegetation in rivers are few (Marcus and Fonstad 2008). But some advancements are currently ongoing in the area due to its relevance. One of the pioneering studies, which assessed the potential of applying airborne hyperspectral imaged data to monitor submerged aquatic vegetation communities in freshwater and existing rivers is that undertaken by Lee et al. (2011). This study provided the broad classification of four distinct vegetation types in relatively smaller rivers over the United States, that consisted of submerged green and brown filamentous algae regions. Similarly in the United Kingdom, Hill et al. (2009) also pioneered the assessment of estimating submerged vegetation biomass from airborne hyperspectral sensors. This was executed over the River Frome chalk stream. Even though a relatively accurate prediction was achieved from these studies, the major limitation was in the spatial resolution and data quality used in the analysis (Visser and Hill 2011). This shows the need for further advancements in this area.

Another influential technique for flood assessment is the microwave remote sensing technique. The use of such remote sensing method is ideal due to its powerful microwave capability in detecting surface water and relative imperviousness to solar illumination, cloud cover limitations, as well as atmospheric properties (Chapman et al. 2015). Also, microwave signals have more capability in distinguishing water features under vegetation relative to optical remote sensing observations. However, studies have shown that the degree of vegetation contamination and signal loss from the microwave is directly proportional to its channel frequency while maintaining a relatively larger vegetation translucency and surface water responsiveness even at diminutive microwave frequencies (Chapman et al. 2015). With the utilization of active microwave remote sensing procedures, it is possible to adequately monitor flood events under all weather conditions even at high resolutions (Bartsch et al. 2012; Bourgeau-Chavez et al. 2009). However, with inconsistent assessment from the synthetic aperture radar (SAR) based observations or inadequate spatial coverage from global navigation satellite (GNSS)-based techniques (Eroglu et al. 2019), these methods are still limited.

18.3.3 Flows in the Flinders Catchment: Impacts on Marine Fisheries

The wet–dry tropics of Australia experience substantial rainfall and runoff, which is a source of nourishment and productivity to the downstream floodplains (Ndehedehe et al. 2021). However, the availability of surface water immediately after the summer rainfall leading to runoff has created interest in water resource developments in the catchments along this Gulf region in northern Australia to support irrigated agriculture. Studies have shown the significance of water developments in the wet–dry tropics adjacent to the Great Barrier Reef (Howley et al. 2018). This progress and efforts resulted in the plans for substantial water development over three rivers in the Carpentaria Gulf region (the Gilbert, Finders, and Mitchell rivers). But these tropical rivers are very crucial to the productivity of the downstream floodplains. Robins et al. (2005) showed the influence of these downstream estuaries on commercial fisheries and threatened migratory shorebirds as well as its potential in serving as a habitat for other animal and plant species that are ill characterized. Water withdrawal and dam construction can go a long way in mitigating freshwater flows to the estuaries. The interest in assessing effect of human actions on primary production, especially in near-shore environments and estuaries is still developing. Flow events alter water resistance times, enhance turbidity, and diminish estuarine salinity content, which can influence phytoplankton and benthic algal production negatively. This occurs on a short term. Conversely, in the longer term, the reduction in annual flows and turbidity which is followed by an increase in salinity content causes nutrient inputs during the wet seasons to stimulate principal production (Auricht et al. 2017; Mallin et al. 1993).

The Flinders catchment is characterized by a series of distinct and disengaged water holes during the dry periods (Faggotter et al. 2013). The existing estuaries in the region also consists of irregular river channels integrated with small tidal creeks, fringed with a contracted mangrove line, behind which are vast salt flats (Ndehedehe et al. 2020b). The inundations of these salt flats mostly occur during the wet periods or at the utmost astronomical tides (Burford and Faggotter 2021). The freshwater flows in the region constantly influence many ecosystem services provided by the corresponding wetlands. However, it is important to understand that ecosystem freshwater needs can be in contrast to freshwater withdrawal needs, especially for human utilization (Stewart-Koster and Bunn 2016; Grey 2007). The key to sustainable freshwater utilization lies in proper accounting and balancing of the overarching demands on the finite water resources (Stewart-Koster and Bunn 2016). Studies showing the effects of over- or under-allocation of freshwater resources for upstream human needs, which contributed to substantial stress on the downstream ecosystem have been conducted notably in several large rivers including, the Murray–Darling River and the Colorado River (Kingsford et al. 2011; Kendy et al. 2017). Locations, where finite water resources influence a wide range of ecosystem services, are well captured over the Gulf of Carpentaria as well as other associated river catchments existing in the northern Australian region. The request for upstream development of water infrastructures for mining and irrigated agriculture is under consideration. Government legislations as well as regional jurisdictions are responsible for the sustainable utilization and administration of freshwater resources, which is required for directly balancing upstream and downstream water use and influence on the ecosystem as well as the services it provides (Mcgregor et al. 2018).

The coastline of the Gulf of Carpentaria (approximately 2000 km long) receives freshwater input from over 25 major catchments (approximately 647,000 km^2) (Leahy 2021). The main source of this freshwater input is rainfall and associated river flow regimes and exists in a dissimilar wet–dry cycle (Warfe et al. 2011) in relation to the monsoonal wet season existing in austral summer. The variations in rainfall between catchments are noticeable; however, its consistency between November and April with yearly mean totals of 650–1200 mm was captured by the Australian Bureau of Meteorology in 2012. This inundated floodplain therefore are a function of the flooding events and the resulting river flows, which are also responsible for linking water ways disconnected during the dry seasons, as well as invigorate productivity (Burford et al. 2020; Pettit et al. 2017). A basic driver of high-value inshore finfish and offshore penaeid shrimp fisheries over the Carpentaria Gulf is seasonal flooding (Halliday et al. 2012; Vance et al. 1985). In northern Australia, studies have shown that the criterion for proposed basic developments lies in the yield and accommodation of wet season river flows (Petheram et al. 2018). Reports have identified potential consequences of upstream development on downstream ecological services and aquatic life that depend on the same river flow (Broadley et al. 2020; Stewart-Koster and Bunn 2016; Gibson et al. 2002). One of the important fish species in the Gulf of Carpentaria is the Barramundi. This region is equally essential for recreation, social, as well as cultural activities to local communities, and sub-regional communities (Streipert et al. 2019).

The Barramundi possesses a temporally and spatially inconsistent proportion of components to aid accessing of freshwater habitats during its life time, and is popularly regarded as facultative catadromous specie (Milton et al. 2008; Milton 2018; Crook et al. 2016; Halliday et al. 2012). It has been observed that river flows make a significant contribution to the increasing Barramundi population. One of the important functions of the existing seasonal freshwater flows is to provide a link between up and downstream species. This contributes to the observed changes in food availability, which has a direct effect on habitat survival, growth, recruitment, and strength

(Robins et al. 2005). The possibility of Barramundi growth rate increases as a result of the varying river flow rate has also been highlighted in Robins et al. (2005). It is likely that water resource development could have some influence on these mechanisms. Other studies have also attempted to relate the growth rate of fishes to the corresponding regional or sub-continental atmospheric indices (Martino et al. 2019) as well as their catch rates. Another important feature of the atmospheric index is its potential to serve as a proxy for local environmental conditions over different periods. This serves as a better ecological process predictor than direct field observations of local conditions (Hallett et al. 2004). One of the techniques for the direct assessment of fish growth rate is through the mark-recapture studies (Robins et al. 2005) or the extrapolated observations of otolith increment widths (Roberts et al. 2019). The association of the otolith observations with somatic growth has been captured by several studies (Campana 2001) and has also been applied over a wide array of species to depict favorable specifications for daily, monthly and annual growth rates (Tonkin et al. 2017; Morrongiello et al. 2012; Sponaugle 2009). Even though the otolith method may not have the same accuracy as a direct observation technique, the method however, is cost-effective, and applies historic collections in its estimations, which are aimed at recreating growth rates over years for suitable samples.

18.4 Changes in Floodplain Productivity of the Flinders Catchment

Climate change drives changes in the water cycle, increasing the frequency of droughts and exacerbating dry spells that negatively affect floodplain rivers and the ecosystem services they provide. Studies that seek to understand how climate variability is contributing to increasing acceleration of the hydrological cycle and how these floodplain river ecosystems respond to myriads of other stressors including, human actions like dam construction, land cover change, water transfer, etc., are emerging. The use of remote sensing methods and integrative modelling approaches that incorporates several Earth observation data in some of these studies reinforces the concept of remote hydrology and its application in eco-hydrological assessments of freshwater catchments (Ndehedehe et al. 2021; Molinari et al. 2021a, b; Ndehedehe et al. 2020b). In this regard, this section presents the work of Ndehedehe et al. (2020b) who predicted the spatial distribution of hotspots of primary producers (aquatic plant biomass) and assessed the characteristics of large-scale inundation patterns in a large floodplain wetland (Flinders catchment) in the wet–dry tropics of Australia. These characteristics underpin knowledge on the impacts of climate change and/or anthropogenic influence on the distribution of primary production in aquatic systems within wet–dry tropical systems. This study demonstrates how satellite remote sensing indicators and ground observations of rainfall and river discharge can be combined with other data in a multivariate framework to improve understanding of the spatial and temporal variability of floodplain inundation and aquatic biomass accumulation.

18.4.1 Changes in Total Inundation and Hotspots of Primary Producers

Several ecological and socio-economic functions supported by wetlands and freshwater habitats such as regulating nutrient cycle, maintaining fishery production, and the role of aquatic plants in enhancing water clarity, among other functions have been widely reported (e.g., Chen et al. 2014b; Zhao et al. 2012; Tockner et al. 2010; Midwood and Chow-Fraser 2010; Keddy et al. 2009; Gidley 2009; Cazzanelli et al. 2008; Ozesmi and Bauer 2002). Assessing wetlands and floodplains is, therefore, required to improve contemporary understanding of wetland functions and how they response to impacts associated with anthropogenic actions and climate change. This is particularly the case for floodplains that are currently unaltered but under pressure from development that would impact their ecological function (Kingsford 2000). However, physical access to several wetlands and freshwater habitats can be restricted due to difficult topography and landscape. Inaccessibility during times of inundation and extreme floods as well as technological limitations associated with field surveys are other key factors limiting the large-scale assessments of floodplain wetlands. This inaccessibility makes in situ sampling challenging and difficult. Apart from this, aquatic primary producers such as macrophytes often thrive in freshwater habitats and floodplains that are extremely difficult to reach and perhaps unsafe (e.g., Zhao et al. 2012; Davranche et al. 2010), and thus making traditional field surveys not only expensive but nearly impossible and risky. The ongoing development of satellite technologies nonetheless is helping to overcome these challenges.

Freshwater ecosystems in the Flinders catchment are characterized by numerous isolated and discrete water holes of various sizes and extents. These aquatic ecosystems have been identified as highly productive "hotspots" of phytoplankton production, where macrophytes, periphyton, and other primary producers generate large amounts of biomass that is then eaten by aquatic herbivores (Ndehedehe 2020). This high productivity is preceded by extreme wet periods and high water flows during summer, which result in an abundance of aquatic plants (e.g., macrophytes, algae) and higher-order organisms (e.g., fish, invertebrates, and waterfowl). Rivers and freshwater bodies, including water holes, can benefit from moderately high flood events when stored nutrients are mobilized during storms and then processed by primary producers when water levels decline (Ndehedehe 2020). These nutrients later constitute a major source of support to aquatic ecosystems, stimulating primary production. This suggests that hotspots of aquatic plant biomass in freshwater habitats could develop in the presence of steady flood waters caused by moderately higher annual rainfall on the floodplain or from flows from the upstream section of the catchment. But the vulnerability of these freshwater systems and aquatic habitats to climate influence and potential land-use change reinforce the need for their assessments.

The main aim of this study is to assess changes in large-scale seasonal total inundation and predict the spatial distribution of hot spots of primary producers (aquatic plant biomass) in a large floodplain wetland in the wet–dry tropics of Australia. The specific objectives are to (i) assess large-scale spatio-temporal changes in total floodplain inundation and (ii) predict the spatial distribution of hot spots of primary producers by integrating remote sensing indicators with digital elevation model in a classification tree model. The relationships of the predicted hot spots in (ii) above with total downstream floodplain inundation, rainfall, and discharge are explored to understand the connection between ecology and wetland hydrology. Notably, remote sensing-based inundation metrics have been used in quantifying floods and mapping floodplain inundation. But the need to assess the stability of these indicators in quantifying open water features and total floodplain inundation on complex landscapes such as the Flinders catchment is essential to provide more knowledge on the role of climate variability on changes in wetland hydrology. While this has also been undertaken in this study, an empirical framework that integrates a multivariate analysis of hydrological indicators (e.g., river flow) is developed for the first time and used for the quantitative assessment of local wetland hydrology. Another key application of this framework in this study is to assess the influence of inter-annual variability of rainfall on total floodplain inundation and spatial distribution of aquatic plant biomass accumulation. Such assessment is crucial and provides insight that underpins the prioritization of future surface water developments that are currently under consideration,[1] conservation planning, and wetland management in this region.

18.4.2 The Flinders Catchment: Hydrology and Landscape

The study region is one of the southern gulf catchments in semi-arid tropical climate of northern Australia, the Flinders catchment (Fig. 18.4), and has an approximate area of 109,000 km^2 (e.g., Waltham et al. 2013). The unique geological features, topography, landscape characteristics, and hydrology of the funnel-shaped Flinders catchment (Fig. 18.4a) are interesting and include valuable resources, e.g., minerals, coal, groundwater aquifers, and soil, among others. Apart from its rich geological properties (comprising four types of rocks and sediments), the occurrence of Shale oil and gas deposits containing vanadium, molybdenum, and nickel; basalt; gas, coal, and petroleum have been reported (Petheram et al. 2013). These assets and several other potentials for large-scale agricultural developments in the catchment have created the quest for interesting new research directions that addresses critical hypothesis with focus on water resources and improving the mechanistic understanding of rivers as key indicators of floodplain productivity. As one of the areas within the Gulf region, the Flinders catchment is home to the longest river in Queensland, the Flinders River, whose largest tributary is the Cloncurry River. The catchment is also characterized by a network of creeks, isolated and intermittent freshwater bodies with the presence of aquatic vegetation (Fig. 18.4b), and rivers, whose surface waters flow downstream immediately after the wet season rainfall. The aquatic plant biomass in these freshwater habitats and water holes (Fig. 18.4b) serves as primary

[1] https://www.industry.gov.au/sites/default/files/June%202018/document/pdf/nawp-fullreport.pdf?acsf_files_redirect.

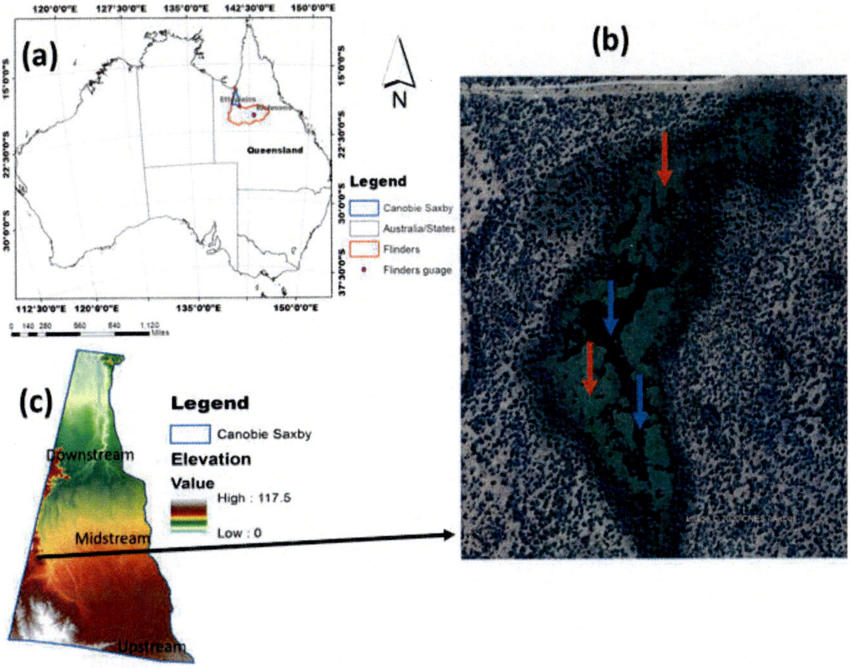

Fig. 18.4 Study region showing the **a** Flinders catchment in northern Australia and the Canobie–Saxby Roundup constriction area located between the up-stream and downstream of the catchment, **b** topographic distribution (elevation in meters) based on corrected digital elevation model, and **c** hot spots of aquatic primary producer in a typical water hole in the mid-stream region of the Flinders catchment. The red arrows illustrate a floating aquatic vegetation cover, while the blue arrows indicate a portion of open water surface in the water hole. The image in (**c**) is taken from the Google Earth scene and was captured on 22 June 2018. *Source* Ndehedehe et al. (2020b)

producers that generate food and energy sources for higher-order consumers. While its river systems are known for their unpredictable flow, which occurs during summer, strong annual rainfall totals trigger widespread flooding of the mid-stream and downstream regions of the Flinders catchment (e.g., Faggotter et al. 2013; Kennard et al. 2010). Due to the lack of elevation (Fig. 18.4c) and the impacts of river systems (Saxby, Cloncurry, and Flinders), the major flood area is around the Canobie–Saxby Roundup constriction located downstream Flinders. Considering the extensive nature of the upstream regions (about 90,000 km^2), surface water flow through this constriction, which is somewhat narrow creates higher flood water depth during major flood events.

18.4.3 Satellite and in Situ Observations

Surface water extent was estimated using Landsat 5 (Level 1T) thematic mapper (TM) and 8 OLI imageries retrieved from the United States Geological Survey data portal (https://earthexplorer.usgs.gov/). The Landsat images were used to generate multi-band spectral indicators for the prediction of aquatic plant biomass distribution (hot spots of primary production). Preprocessing chains, including masking for cloud, atmospheric correction, and calibration to standard surface reflectance values are as previously detailed in Chaps. 3 and 6. Monthly Australian gridded rainfall data and river discharge data of the Flinders River were accessed from the SILOS climate database (https://silo.longpaddock.qld.gov.au/gridded-data) and the Queensland Water Monitoring Information Portal (https://water-monitoring.information.qld.gov.au/), respectively. Discharge observations were limited to two locations on the Flinders River (Ettaplains and Richmond stations, Fig. 18.4), between the Canobie-Saxby Roundup constriction and the upstream catchment where Flinders River floods across into one of the tributaries (Alice Creek).

18.4.4 Inundation and Vegetation Metrics

The methodological framework used in this study is summarized in Fig. 18.5. This empirical framework shows a novel approach that underpins the quantitative assessment of wetland hydrology and the impacts of global climate on changes in floodplain productivity (total inundation and aquatic plant biomass accumulation in freshwater habitats).

18.4.4.1 Quantifying Spatial Extents of Hot Spots of Aquatic Primary Producers

In this study, the distribution of hot spots of aquatic primary producers (flooded aquatic vegetation and inundated areas covered with different aquatic species such as benthic algae) and extended mixed flood areas in the catchment was assessed by combining three spectral indices; the MDNWI, Optimized Soil Adjusted Vegetation Index (OSAVI, Rondeaux et al. 1996), the normalized difference water index (NDWI) proposed by Gao (1996) and flood water depth in a classification tree model (Fig. 18.5). The spectral properties of the red (R) and near-infrared (NIR) bands of Landsat, which show relatively low reflectance and high reflectance, respectively, were used to estimate OSAVI in each of the Landsat image. The value of 0.16 is the canopy background adjustment factor that provides greater soil variation for regions with limited vegetation cover and considerable sensitivity to vegetation cover (Rondeaux et al. 1996). Although a preliminary assessment of the distribution of flooded vegetation and hot spots was based on the Normalized Difference Vegetation Index

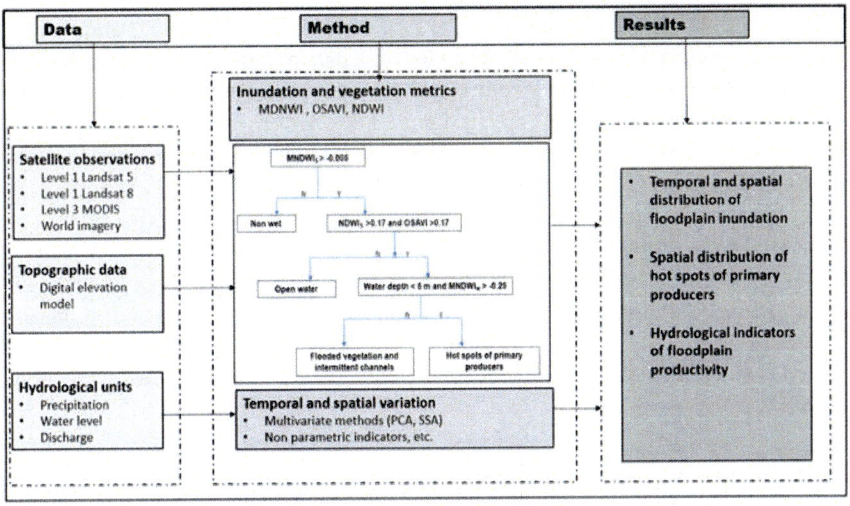

Fig. 18.5 Schematic representation of the methodological framework used in this study. Note that before deriving remote sensing indicators used in floodplain mapping (including AWEIsh and AWEInsh) and aquatic biomass distribution (hot spots of primary producers), all satellite observations were preprocessed (e.g., cloud removal using the Fmask algorithm and atmospheric correction). *Source* Ndehedehe et al. (2020b)

(NDVI) and MNDWI, the former was subsequently replaced with OSAVI in the CT model in the prediction of hot spots. Our decision to replace the NDVI with OSAVI is based on the popular notion regarding the sensitivity of the NDVI to soil background and atmospheric effects. Even recently (Fern et al. 2018), OSAVI has been deemed a more suitable vegetation metric for quantifying green biomass and vegetative cover in semi-arid regions as opposed to NDVI.

The threshold definition for pixels associated with vegetation and wet vegetation in the classification tree model were OSAVI and $NDWI_s$ (summer NDWI) values greater than 0.17 (Fig. 18.5). The thresholds defined for all the vegetation and inundation indices used for predicting hot spots of aquatic plant biomass (Fig. 18.5) are based on the expert knowledge of these target variables obtained through an understanding of their complexity and spectral properties in the Flinders landscape. The thresholds for OSAVI and $NDWI_s$ are targeted at wet vegetation within inundated precincts. They can be slightly increased (perhaps to 0.20 or more) depending on the magnitude of inundation, i.e., when the latter becomes more extensive. Indeed, threshold identification for land cover detection could be complicated and rather subjective, especially for water bodies whose spectral property has been compromised with the presence of chlorophyll and aquatic vegetation. Consequently, threshold definitions for multi-band satellite-derived inundation metrics are known to be highly variable (e.g., Worden and de Beurs 2020; Ndehedehe et al. 2020a; Pekel et al. 2016), depending on regions, characteristics, and the capability of sensors onboard satellite systems.

Floodwater depth was estimated for the Flinders catchment by combining floodplain inundation maps (Sect. 18.4.4) with elevation data obtained from a 30-m digital elevation model (DEM). Water depth less than 5 m was chosen because the water holes and the freshwater habitats are predominantly shallow. The inclusion of water depth and winter inundation map (greater than −0.25) helps in identifying areas with relatively permanent freshwater bodies while separating intermittent channels and flooded vegetation. The innovative use of these season-based water extraction indicators ($MNDWI_w$ and $NDWI_s$) with MNDWI, OSAVI, and the $NDWI_{Gao}$, which is used for remote sensing of vegetation liquid water is aimed at improving the accuracy of estimated distribution of floodplain inundation states and aquatic biomass distribution. Although the NDWI is rather complementary to NDVI, one key merit of the former is that it is less sensitive to atmospheric scattering effects as opposed to the NDVI.

18.4.5 Statistical Analyses

The spatio-temporal variability of rainfall over the Flinders catchment was undertaken through the statistical decomposition of precipitation grids using the principal component analysis (PCA, Jolliffe 2002), prior to identifying significant principal components (PCs) or orthogonal modes of rainfall with a scree plot analysis (see, Martinez and Martinez 2005). As was discussed in Chap. 13, there is no fixed criterion to determine the number of significant PCs that are best retained in a given situation. But the choice of truncation can vary from available selection rules (e.g., rules based on hypothesis testing; size of cumulative variance; size of the last retained eigenvalue, etc.) to subjective choices (e.g., *scree plot*, *log-eigenvalue* diagram, and retaining PCs that represent a 'sufficient fraction' of the variance in the original data) depending on the data and the purpose of the statistical rotation (e.g., Wilks 2011; Martinez and Martinez 2005; Jolliffe 2002). For the Flinders catchment, this was not a difficult choice as the scree plot analysis was sufficient. In the statistical rotation of rainfall using the PCA method, a deseasonalized matrix of precipitation grids (i.e., after removing the mean) masked over the Flinders catchment would be given as $\mathbf{P}_{rainfall} = [x(s_k, t)]$, where s_k is spatial locations; k = 1, 2, ..., nx, which are the number of spatial locations for $\mathbf{P}_{rainfall}$, and t is the monthly time step from 2000 to 2018 (i.e., temporal evolutions). As an illustration, the 8 km monthly precipitation grids over the Flinders for the period between 2000 and 2018, $\mathbf{P}_{m\times\kappa}$ will have the size, $m = 228$ and $\kappa = 3781$ where m and κ are the monthly time steps (January 2000 to December 2018) and observations (precipitation grids masked over the region), respectively, over the Flinders catchment. The spatial and temporal patterns of rainfall using this method can therefore be parameterized,

$$\mathbf{P}_{rainfall}(t) = \sum_{k=1}^{n} a_{(k)}\mathbf{s}_k, \qquad (18.1)$$

where $a_{(k)}(t)$ is the temporal variations also called expansion coefficients, normalized using their standard deviation to be unitless and $\mathbf{s}_k$ are the corresponding spatial maps (empirical orthogonal functions-EOF loadings). While these loadings are usually adjusted (scaled) to their original units (mm) using their principal components (PCs), they provide the weights of the original variables in the PCs. By reducing the number of orthogonal modes, the leading modes of rainfall that defines the characteristics of discharge and floodplain productivity in the catchment were identified. The singular spectrum analysis (SSA, Ghil et al. 2002) was employed to further explore the relationship of river discharge oscillations of downstream Flinders catchment with those of inter-annual variations in rainfall. See further details about this method in Chap. 13.

The multivariate standardized indicator (MSI, e.g., Farahmand and AghaKouchak 2015; Ndehedehe et al. 2016) was used to assess flood characteristics such as duration and is interpreted similar to the range of variability approach (e.g., Genz and Luz 2012; McKee et al. 1993). This interpretation uses the value of one standard deviation to set the thresholds in the characterization of hydrological conditions (e.g., values >1.5 are very wet and < -1.5 are very dry). Moreover, the cumulative departures of both discharge and water level were also compared to assess the role of discharge as indicators of downstream inundation and water-level fluctuation.

18.4.6 *Delineation of Inundated Areas and Floodplain Inundation Mapping*

18.4.6.1 Automated Surface Water Extraction Metrics

The 2009 summer flood (Fig. 18.6) in the Flinders catchment is one of the most extreme and major flood events in three decades. During this flood, considerable anomalies in water levels and discharge were recorded (Fig. 18.6a, b). Observed water levels at Ettaplains were relatively high (approximately 8 m), lasting for about 25 and 10 days in the first and second peaks, respectively (Fig. 18.6a). But the maximum (i.e., highest peak) flood duration at Ettaplains for the first peak based on the range of variability approach was approximately 7, 8 days (Fig. 18.6c). Due to the non availability of cloud-free images for the peak flood (Fig. 18.6c), which occurred in the last week of January 2009, the available Landsat image (February 27, 2009) corresponding to the second highest peak in water level and discharge of 2009 (Figs. 18.6a–c) was used in this study. Indeed, the strong association between their cumulative departures (i.e., water level and river discharge) at Ettaplains ($r = 0.99$) and Richmond stations ($r = 0.99$) confirm the strong interactions between mid-stream and downstream inundation dynamics (Fig. 18.6d, e).

The spatial distribution of flood extents for three consecutive periods in 2009 (27 February, 15 March, and 31 March) are indicated using the three automated surface water extraction metrics mentioned earlier (Fig. 18.7a–i). In the delineation and

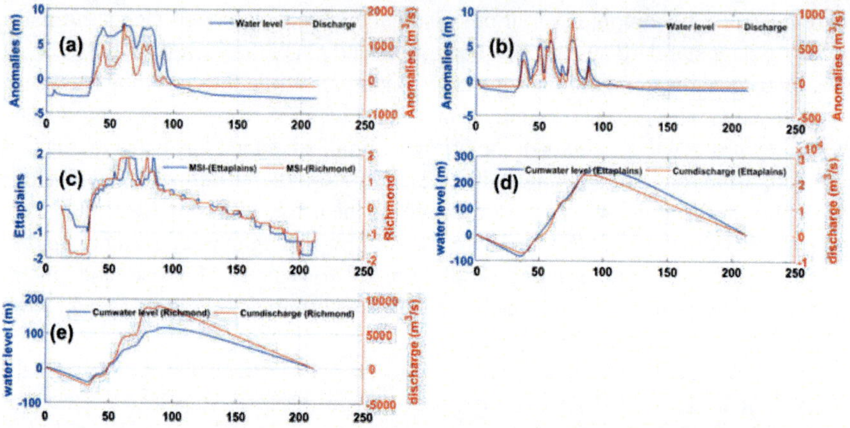

Fig. 18.6 Daily river discharge and stream water levels of 2008/2009 (i.e., 212 daily time steps (x-axis for all plots) starting from 1 December 2008 to 30 June 2009), the year with high density, mega flood in much of the wet–dry tropics of northern Australia. **a, b** The anomalies in mean river discharge and stream water levels, **c** the computed multivariate standardized indicator for the two hydrological stations, while **d–e** indicate the cumulative departures of observed water levels and discharge. The gauge records are taken from Ettaplains and Richmond stations downstream in the Flinders catchment. *Source* Ndehedehe et al. (2020b)

mapping of inundation extent, the maximum aerial extents of floods/inundation in February 2009 were 2522, 2063, 824.12 km^2 for MNDWI, AWEI$_{nsh}$ and AWEI$_{nsh}$, respectively (Fig. 18.7a–c). The MNDWI had the highest inundation extents in the two periods in March followed by the AWEI$_{sh}$ (Fig. 18.7d, e). Generally, the AWEI$_{nsh}$ appear to underestimate water features on the floodplain (Fig. 18.7c, f, and i). Using error matrices (e.g., Congalton and Green 2009), the assessment and validation of outputs from these remote sensing indicators based on Landsat image of September 2003 showed that MNDWI and AWEI$_{sh}$ were more suitable for the mapping of floodplain inundation in the Flinders catchment. Overall accuracies and kappa coefficients were 98.8% and 0.97 for MNDWI and 98.4% and 0.96 for AWEI$_{sh}$, respectively. But overall accuracy and kappa coefficient for AWEI$_{nsh}$ were 95.7% and 0.90, respectively. Comparatively, the kappa statistics for all inundation metrics show that MNDWI and AWEI$_{sh}$ are more suitable remote sensing indicators for the extraction of open water features in the Flinders catchment compared to the AWEI$_{nsh}$.

18.4.6.2 2009 Versus 2019 Mega Floods in the Flinders Catchment

Despite the magnitude and flood intensity, within 15 days of observed maximum flood of 2009 February 27 (2522 km^2), the inundation extent receded rapidly to 265.7 and 102.7 km^2 in March 15 and 31st of the same year, respectively. In the inundation mapping of the 2009 flood in the Mitchell catchment, a 15-day flood recession based on MODIS data was also reported by Ward et al. (2013). While the extreme 2009

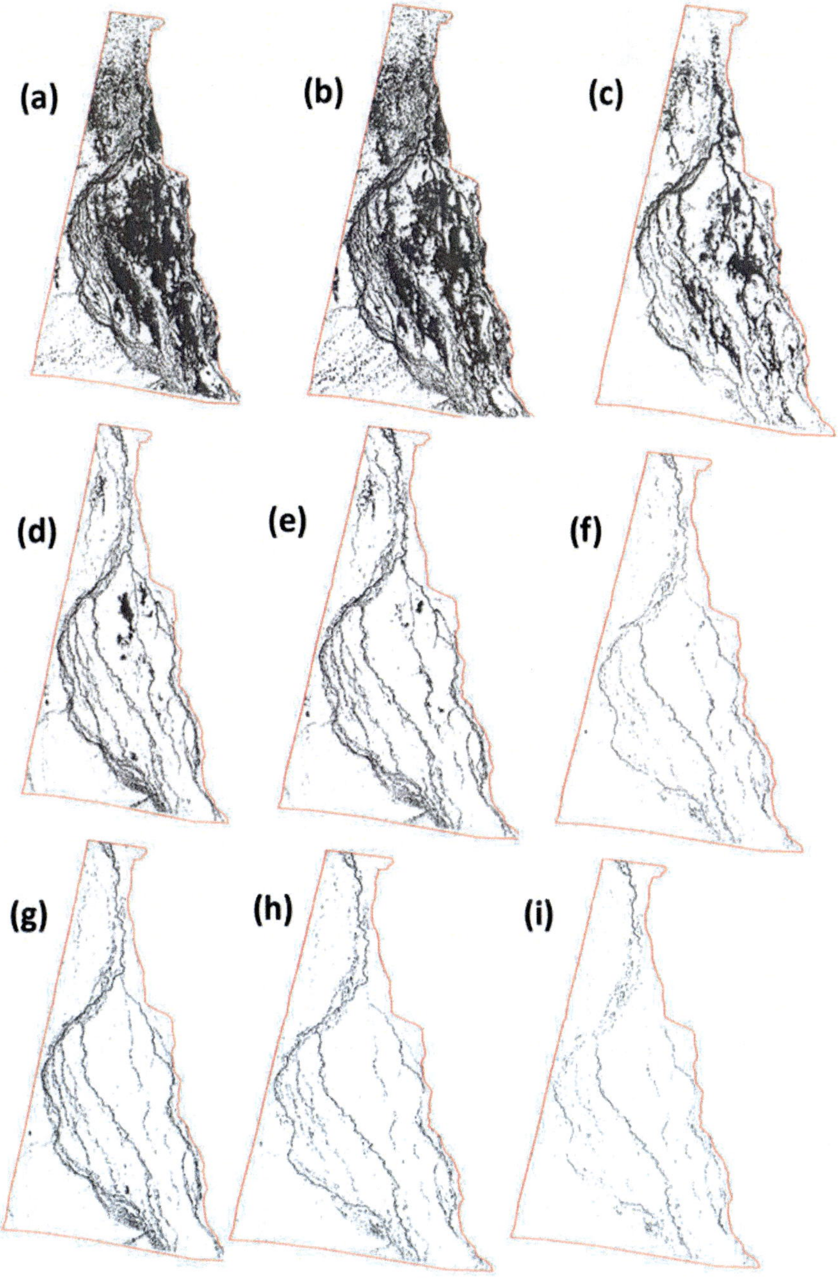

Fig. 18.7 Spatial patterns of the 2009 floods in the Flinders catchment estimated from Landsat 5 based on three automated water extraction metrics (MNDWI (a, d, g), $AWEI_{sh}$ (b, e, h), and $AWEI_{nsh}$ (c, f, i)) for 27 February 2009 (**a**–**c**), 15 March 2009 (**d**–**f**), and 31 March 2009 (**g**–**i**). *Source* Ndehedehe et al. (2020b)

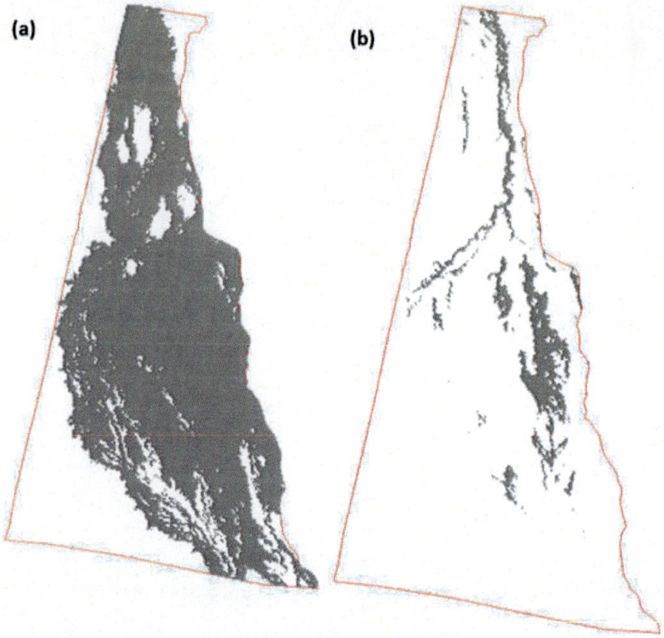

Fig. 18.8 Spatial distribution of the summer 2019 floods in the Flinders catchment estimated from 500 m spatial resolution MODIS surface reflectance product using the MNDWI for **a** 10 February 2019 and **b** 18 February 2019. *Source* Ndehedehe et al. (2020b)

flood was an event that impacted the entire coastal and wet–dry tropical regions of Queensland, the recent 1 in 50 year flood, which occurred in the first week of February 2019 appear to be a location-specific event for the Flinders catchment and environ. From our MODIS-derived inundation mapping, the recent summer 2019 flood (Fig. 18.8a) on the Flinders floodplain show stronger magnitudes compared to the summer 2009 flood (Fig. 18.7a) and is consistent with observed hydrological records (e.g., river flow) and water levels, which was more than 10 m in depth in the downstream catchment. The floodplain inundation extent as on 10 February 2019 was 7094.9 km^2 and greatly exceeded the 2009 flood (Fig. 18.8b). However, this flood receded to about 808.6 km^2 on the 18 February 2019 (Fig. 18.8b). So, within 10 days of the 2019 summer flood on the Flinders catchment, post flood declined by approximately 87%, similar to that of the 2009 summer flood. The magnitude and intensity of the 2019 summer flood tend to be significantly higher than the 2009 February flood as gauge records (stations at Richmond and Ettaplains) confirm significant difference in observed water levels during these periods. However, it should be noted that by 23 February 2019, estimated Landsat-derived inundation extent was 140.3 km^2 (Fig. 18.9) and confirms the similitude in flood recession patterns (i.e., within 15 days) in the catchment immediately after extreme 'big wet' periods.

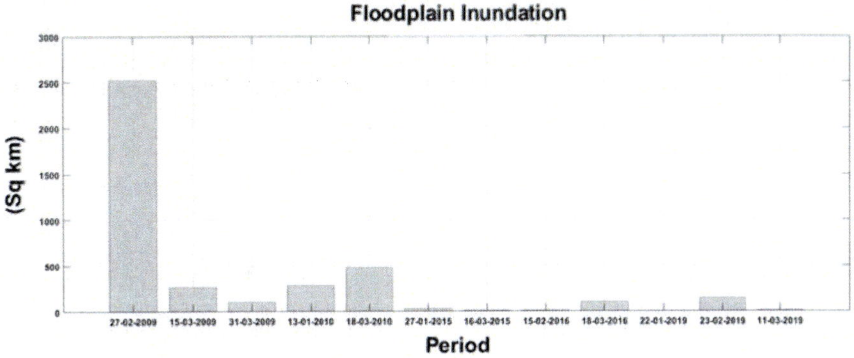

Fig. 18.9 Seasonal floodplain inundation extents based on the MNDWI for 2009–2010, 2015–2016, and 2019. *Source* Ndehedehe et al. (2020b)

Regarding the 2009 summer flood, it is noted that generally, the degree of recession from extreme wet in January/February to mildly wet in mid-March and dry in March ending is significantly variable (Fig. 18.7a–i and Supporting Information 1). Despite the massive and major flood in February 2009, by March 15 2009, flooded inundation states had receded to 265.7 km^2 (Figs. 18.7 and 18.9). But during the wet season of 2010 when the flood was merely minor, floodplain inundation extent estimated by 18 March 2010 was 478.1 km^2 (Fig. 18.9). In other words, there was more water on the floodplain towards the end of wet season in the year with a slightly above-normal flooded inundation states than an extremely flooded year. Similar scenario also played out in 2016 (Fig. 18.7) when inundation extent was higher in March (102.4 km^2) than in February (12.3 km^2) when peak rainfall usually occurs. As will be highlighted in subsequent sections, the duration or span of flooded inundation states impact on the spatial distribution of hot spots primary producers. Moreover, the spatial extent of floodplain inundation encountered substantial temporal variation. The temporal variations of wet season floodplain inundation extent in the Flinders show that 2009 was the wettest period on record during the last decade (Fig. 18.9). The spatio-temporal variability of rainfall and discharge (Sect. 18.4.8) also confirms this strong amplitude in flood distribution. Although the maximum flood extent in 2009 is considerably higher than 2010, floodplain inundation in March 2010 (478 km^2) was higher as opposed to March 2009 (265.7 km^2). It is during this period (March 2010) we found that several freshwater swamps and isolated billabongs in the upstream high elevation areas were fully inundated and flooded unlike March 2009 (Figs. 18.7 and 18.10b). But floodplain inundation extent in 2015–2016 and the period after the megaflood of 2019 (23 February) are significantly less and considerably low compared to 2009–2010 period (Fig. 18.9) consistent with hydrological records.

18.4.6.3 Wet Versus Dry Episodes

The wet season floodplain inundation of 2009 was also compared to other moderately wet (2010) and relatively dry (2015 and 2016) periods. The floodplain becomes inundated at the start of the summer rainfall, peaks in January/February, and gradually recedes towards the end of March (Figs. 18.10a–h and Supporting Information). While these dynamics are very typical of rainfall in the region, the changing patterns of wet season (January and February) floodplain inundation in the Flinders drive the spatial dynamics and connectivity of freshwater habitats (Fig. 18.11a–e). In February 2009, for example, all riverine systems and intermittent water bodies around the Canobie–Saxby Roundup constriction in the Flinders were completely inundated with over bank spilling in the main river channels (Fig. 18.11a). But by mid-March of 2009 when the flood extent had diminished with about 89% of the flooded area receding within 15 days, some isolated water bodies were barely inundated (Fig. 18.11b). As the dry season drew near (31 March 2009), approximately

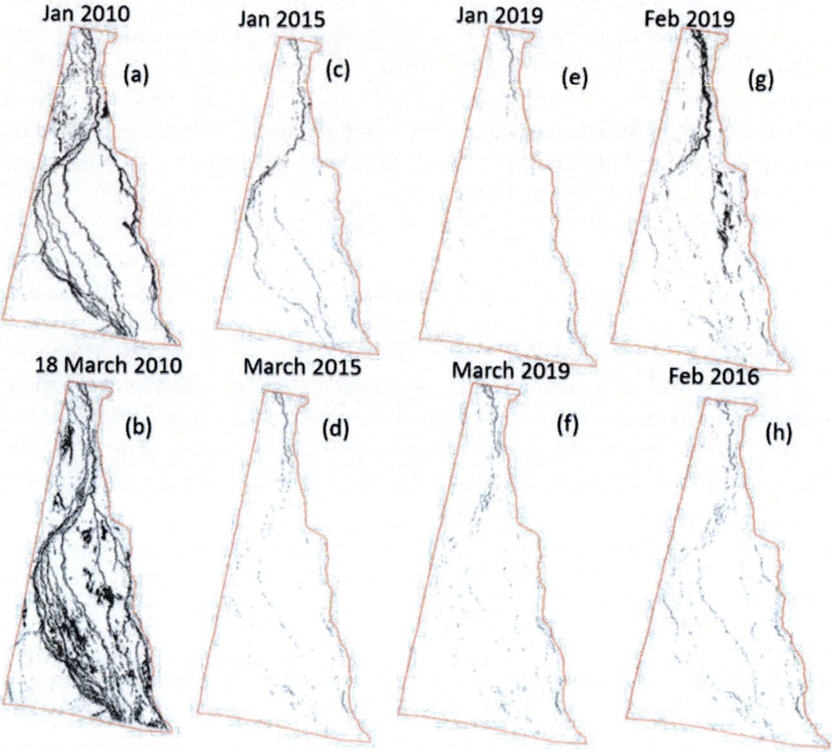

Fig. 18.10 Illustration of the changing patterns of seasonal (January and March) inundation over the Flinders for 2010, 2015, and 2019. These inundation patterns are indicated for **a** 13 January 2010 **b** 18 March 2010, **c** 27 January 2015, **d** 16 March 2015, **e** 22 January 2019, **f** 11 March 2019, **g** 23 February 2019, and **h** 16 March 2016. *Source* Ndehedehe et al. (2020b)

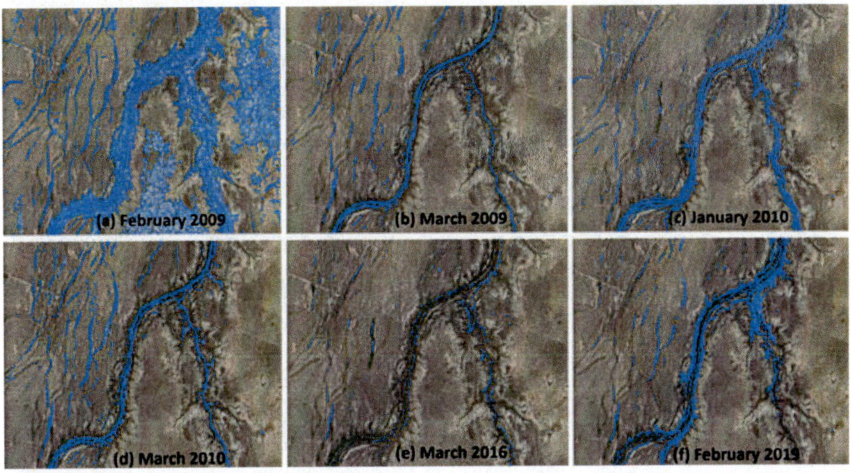

Fig. 18.11 Spatial patterns of floodplain inundation and impact on the connectivity of freshwater habitats in downstream Flinders. **a–f** Show the characteristics of inundation in riverine systems and intermittent water bodies for some wet season periods indicated in Fig. 18.9, i.e., 2009, 2010, 2016, and 2019. The region in the background high-resolution satellite imagery is that of Digital Globe and was taken on 22 May 2014. *Source* Ndehedehe et al. (2020b)

96% of post-flood surface inundation extents declined, leaving behind perennial freshwater bodies and persistent aquatic habitats. Comparatively, the distribution of inundation on the floodplain in March 2009 and 2010 are somewhat similar despite the 2009 summer being extremely wet compared to that of 2010 (Fig. 18.11b–d). In other words, the floodplain in 2010 was still considerably inundated towards the end of the dry season (March) despite a lack of extreme or severe wet conditions occurring in February or January. As with changes in inter-annual rainfall, the floodplain also reflects the impacts of changes in climatic conditions in the observed spatial patterns of inundation in March of 2009, 2010, and 2016 (Fig. 18.11b, d, and e) and February 2009 and 2019 (Fig. 18.11a and f). Similarly, inter-annual changes in the spatial patterns of floodplain inundation for January and March of 2010, 2015, and 2019 were also observed (Supporting Information 1). Generally, from the estimated inundation extents, 2015–2016 and 2019 (i.e., post-flood) periods were dry as opposed to the observed wet periods in 2009/2010 (Figs. 18.9, 18.10 and 18.11). As will be shown in the next section, the extent and magnitude of floodplain inundation have implications on the distribution of hot spots of primary producers.

18.4.7 Spatial Distribution of Aquatic Primary Production

Five variables (MNDWI, OSAVI, NDWI, water depth, including the winter floodplain inundation map of each year) were included in the classification tree model

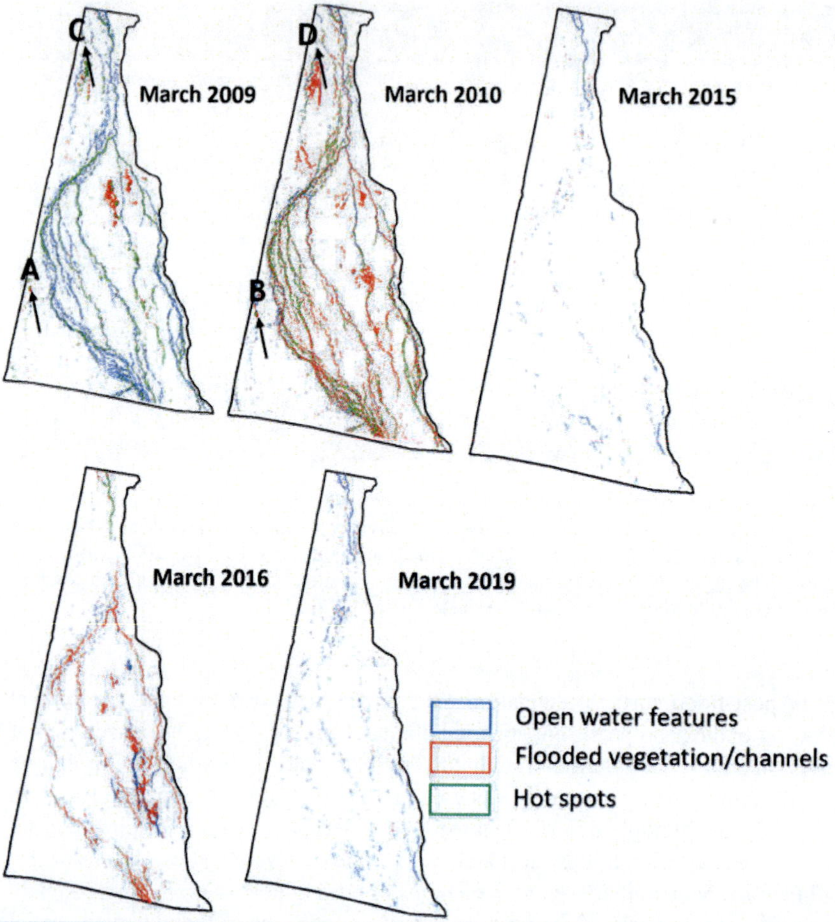

Fig. 18.12 Predicted spatial distribution of open water features, hot spots, and flooded vegetation/channels and wetlands for years with major flood events (2009 and 2010) and relatively dry periods (2015, 2016, and 2019 (i.e., period after the megaflood)) based on the classification tree model. The arrows **a–d** are illustrations, which have been zoomed-in (Figs. 18.13) to highlight the physical dynamics of aquatic habitats and primary production centres in downstream (C and D) and upstream (A and B) Flinders. The acquisition dates of all imageries in each year are nearly similar ranging from 11 to 18 of March of each year. *Source* Ndehedehe et al. (2020b)

(CTM) to predict the spatial distribution of hot spots of primary producers (Fig. 18.12). The winter inundation map and floodwater depth (estimated using elevation and flood inundation map) were introduced to isolate permanent open water features and flooded vegetation/wetlands/channels from freshwater bodies and intermittent water bodies with the presence of aquatic plant biomass. Outputs from the CTM indicate the Flinders floodplain is more productive in years characterized by heavy inundation as opposed to relatively dry years (Fig. 18.12). For example, there was

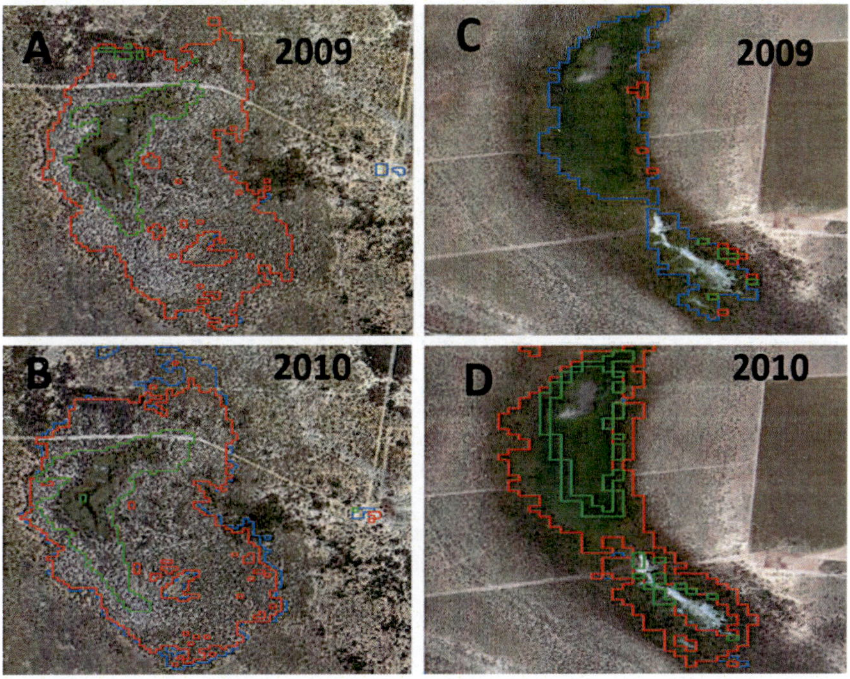

Fig. 18.13 Illustration of the spatial distribution of predicted hot spots, flooded vegetation, and surface water features in isolated water holes and intermittent floodplain water bodies for locations corresponding to **a**, **b**, **c**, and **d** in Fig. 18.12 in the upstream Flinders catchment in 2009 and 2010. The legend is the same as Fig. 18.12 above where blue, red, and green represent open water features, flooded vegetation/channel, and hot spots of primary producers, respectively. The background high-resolution satellite imagery is that of Digital Globe acquired on 14th June, 2016. *Source* Ndehedehe et al. (2020b)

a wider spread of hot spots in 2009 (24.0 km^2) and 2010 (26.5 km^2) unlike 2015 (0.2 km^2), 2016 (2.3 km^2), and 2019 (0.02 km^2). The zoomed-in spatial distribution of predicted hot spots with water features and flooded vegetation/channels (arrows A–D) are indicated in Fig. 18.13 to provide finer spatial details on the relatively wider extent of hot spot in 2010 compared to 2009. In total, distribution of open water features plus flooded vegetation/channels in 2016 March (100.1 km^2) was higher than in 2015 (8.8 km^2) and 2019 (13.5 km^2). The nourishment of existing isolated water holes and freshwater habitats will impact positively on the production of aquatic plants. As illustrated in Fig. 18.11, isolated water holes and intermittent floodplain water bodies in the Flinders are not only nourished but also more productive and appear to be connected during periods of large inundation. However, from a 30-m Landsat outlook, some water holes and freshwater bodies during dry spells and limited river flows (the gauge records of wet seasons with higher flows where consistent observed floodplain inundation) contract and diminish considerably. Because of the 2009 extensive flood, most parts of the downstream Flinders catchment were characterized by open water as opposed to 2010 when the preponderance of vegetation and seasonally inundated aquatic habitats became flooded (Fig. 18.12). This flood resulted in a relatively higher distribution of hot spots (26.5 km^2). Generally, higher freshwater floodplain productivity is very much typical of extreme wet years characterized by higher rainfall amount (Figs. 18.10, 18.11 and 18.12). While such inundation also resuscitate intermittent channels in downstream catchment (Fig. 18.11), as shown in the high-resolution background imagery acquired in June 2016 for downstream locations (i.e., A and B; Fig. 18.13 and Supporting Information), there was a strong decline in the surface extent of aquatic habitats/hot spots in relatively dry summers or years with below-average rainfall (e.g., 2015 and 2019). The spatial extents of hot spots in downstream location (A and B) in 2009 and 2010 showed hot spot extents of approximately 0.38 km^2 and 0.52 km^2, respectively (Fig. 18.13). By 16th June 2016, the spatial distribution of aquatic plant biomass in the same downstream location (i.e., A and B with a centre point coordinate of 140° 28' 33"E and 19° 04' 18.6"S) in Fig. 18.13 as observed on a 0.5 m high-spatial resolution image (World Imagery archives) was approximately 0.15 km^2. As noted in this study, the loss in surface extent of aquatic plant biomass (hot spots) is usually associated with the loss of freshwater habitats in the Flinders catchments caused primarily by limited rainfall and surface flow.

18.4.8 Hydrological Indicators of Floodplain Productivity

The relationship of predicted extents of hot spots of primary producers (Sect. 18.4.7) was compared with estimated total floodplain inundation, total monthly rainfall amounts, discharge, and NDVI (Fig. 18.14a–d). The observed relationships of hot spots with total inundation($r = 0.94$) and rainfall ($r = 0.81$) (Fig. 18.14a–b) are linear and found to be considerably strong. The association between hot spots and discharge is also linear but found to be moderately strong ($r = 0.68$). The relationships above reaffirm rainfall and discharge as key drivers of downstream total floodplain inundation and productivity. As surface vegetation and phenology responds to changes in hydrological conditions in terrestrial ecosystems (e.g., Ndehedehe et al. 2019; Chen et al. 2014a), river flows, biomass accumulation in freshwater habitats, and total floodplain inundation will be restricted in semi-arid and wet–dry tropics during periods of insufficient or limited rainfall. The marked variability in the distribution of predicted hot spots and total inundation, which aligns closely with those of rainfall and discharge (Fig. 18.14a–d) strengthens this argument. The observed correlation between predicted extents of hot spots and hydrological units and even NDVI (Fig. 18.14a–d) therefore suggest that apart from other biophysical factors, which could promote growth of aquatic primary producers in perennial and intermittent freshwater bodies, the productivity and nourishment of the Flinders downstream catchment is considerably tied to the impact of climate variability (inter-annual variations in rainfall) on local hydrology.

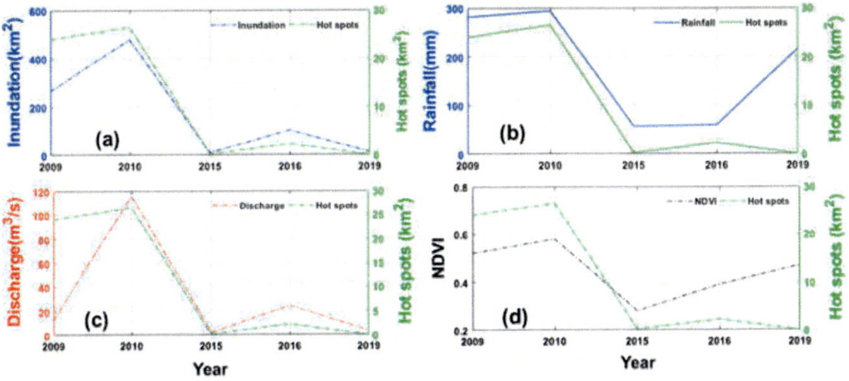

Fig. 18.14 Drivers of floodplain productivity based on the predicted spatial distribution of hot spots in Fig. ??. The relationship of predicted extents of hot spots of primary producers with inundation, total monthly rainfall amounts, discharge, and NDVI are indicated in **(a)**–**(d)**, respectively. *Source* Ndehedehe et al. (2020b)

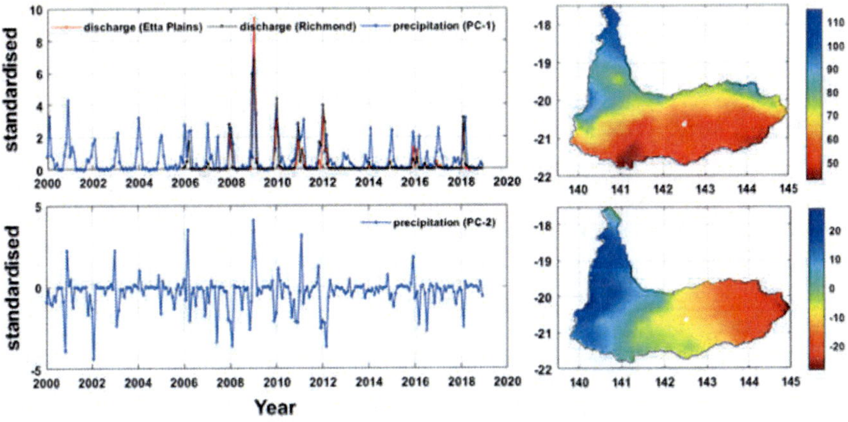

Fig. 18.15 Rainfall variability over the Flinders catchment (2000–2014). The temporal variations (right) are the principal components and have been normalized using their standard deviation to be unitless while the corresponding spatial patterns (left) are the EOF loadings (mm). Time series of river discharge at two downstream locations are also indicated. *Source* Ndehedehe et al. (2020b)

To further assess the roles of key hydrological indicators (rainfall and river flow) as drivers of freshwater floodplain, both rainfall and discharge in the Flinders were statistical decomposed using third-order statistical methods. The two orthogonal modes of rainfall in the Flinders catchment drive the variability in floodplain inundation. The spatial structures and temporal patterns of the leading modes of rainfall highlight the changes in annual and short-term seasonal signals and together accounts for approximately 93% of the total variability (Fig. 18.15). The temporal patterns (PC-1, Fig. 18.15) associated with spatial patterns indicate 2009 was the wettest year and in addition explains the considerable inter-annual variability of floodplain inun-

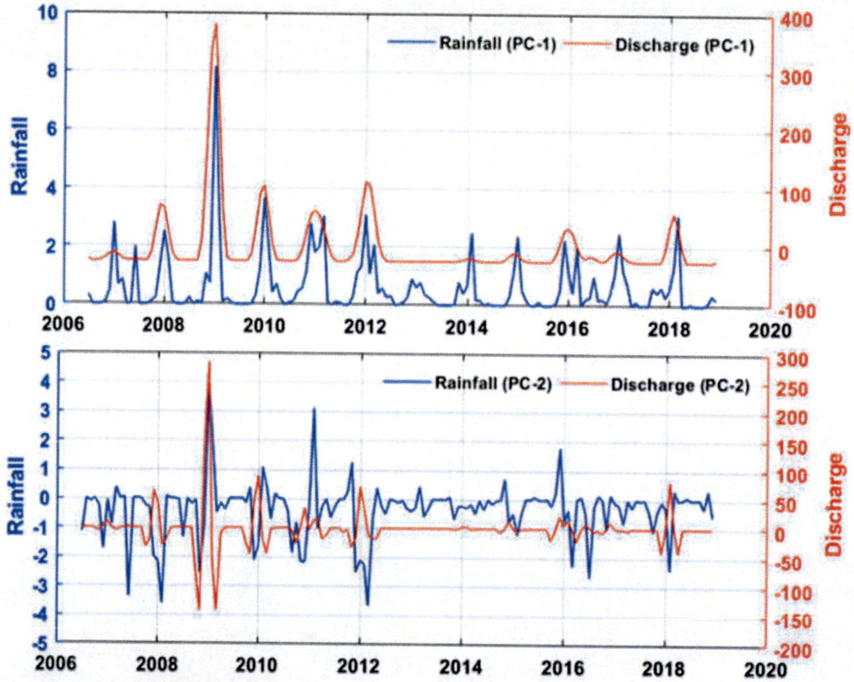

Fig. 18.16 Leading modes of river discharge time series (2006–2018) in the Flinders catchment. The annual variations and multi-annual variations of river discharge are compared with those of precipitation. The time series of river discharge is the reconstructed expansion coefficients. *Source* Ndehedehe et al. (2020b)

dation in this downstream catchment during the wet season (Fig. 18.7). Comparing the amplitudes of rainfall and discharge after the 2009 period, 2010 was relatively the wettest and coincides with the wider distribution of hot spots (26.5 km^2) and flooded vegetation/channels (135.5 km^2) unlike the 2015 – 2016 and 2019 periods (Fig. 18.12). Moreover, it should be noted that the distribution of EOF loadings (spatial patterns) are strongest in the downstream Flinders (EOF-1, Fig. 18.15) where the CTM has been deployed in the mapping of hot spots of primary producers. From the common period (2006–2018), the inter-annual variations of rainfall are linked to changes in river discharge at Ettaplains ($r = 0.72$), which is the closest gauge station in the mid-stream catchment. Even at Richmond gauge station, which is upstream Flinders catchment, the association of river discharge with dominant rainfall (PC-1, Fig. 18.15) is moderately strong ($r = 0.63$). The considerable correlation of NDVI and discharge (Fig. 18.17) with the leading mode of rainfall (PC-1, Fig. 18.15) over the Flinders is another pointer to the influence of rainfall variability on water availability and the likely constraints of water on terrestrial ecosystems. This is supported by the nature of inundation in riverine systems and intermittent water bodies as illustrated for some wet season periods in 2010, 2015, and 2019 (Fig. 18.18). Ultimately,

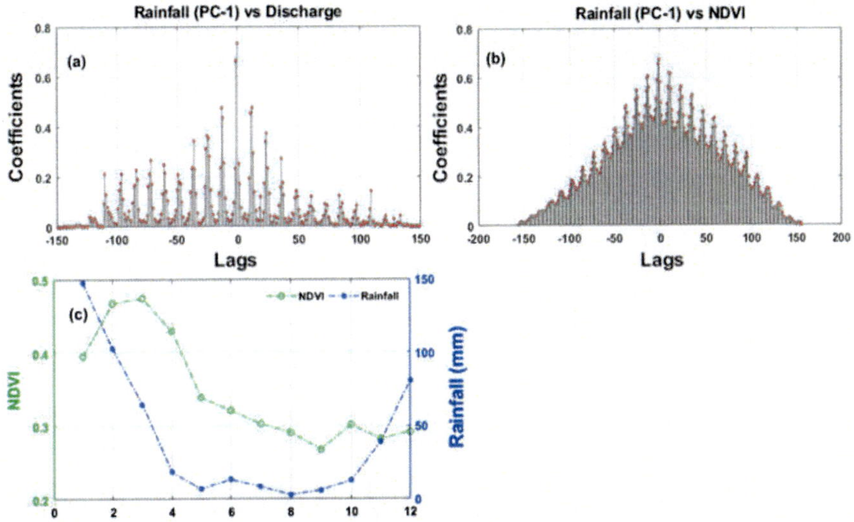

Fig. 18.17 Relationship of dominant temporal patterns of rainfall with **a** discharge and **b** NDVI. The seasonal annual cycle of NDVI is compared with that of rainfall during a common a period (2001–2014). *Source* Ndehedehe et al. (2020b)

the aforementioned correlation values imply a strong linkage between discharge and rainfall and that the nourishment of the floodplain and the connectivity of freshwater habitats are expected to be sustained by these hydrological signals.

In the singular spectral analysis of discharge, it is again confirmed that the dominant mode of river discharge explaining about 52% of the total variability and is well correlated ($r = 0.70$) with that of rainfall (Fig. 18.16). The temporal patterns of these time series are consistent indicating 2009 as the year with the strongest amplitude in rainfall and discharge while the period between 2013 and 2016 had limited or no flow because of little rainfall. Be it floodplain inundation or hot spot distribution, the two wettest periods with high productivity in the catchment based on the Landsat images analysed are 2009 and 2010 and coincides with the leading temporal series of discharge and rainfall (Figs. 18.15 and 18.16). Although identifying the key driver of floodplain inundation was not the mainstay of this study, observed amplitudes of rainfall and river discharge (PC-1, Figs. 18.15 and 18.16) appear to be consistent with the results of inundation mapping (Fig. 18.7), spatial distribution of hot spots, and flooded vegetation/channels (Figs. 18.9, 18.10 and 18.11) presented in this study. This could imply that both local rainfall (strong spatial loadings observed in dominant rainfall modes over the downstream Flinders) and river discharge play a significant role in regulating floodplain productivity of the Flinders catchment.

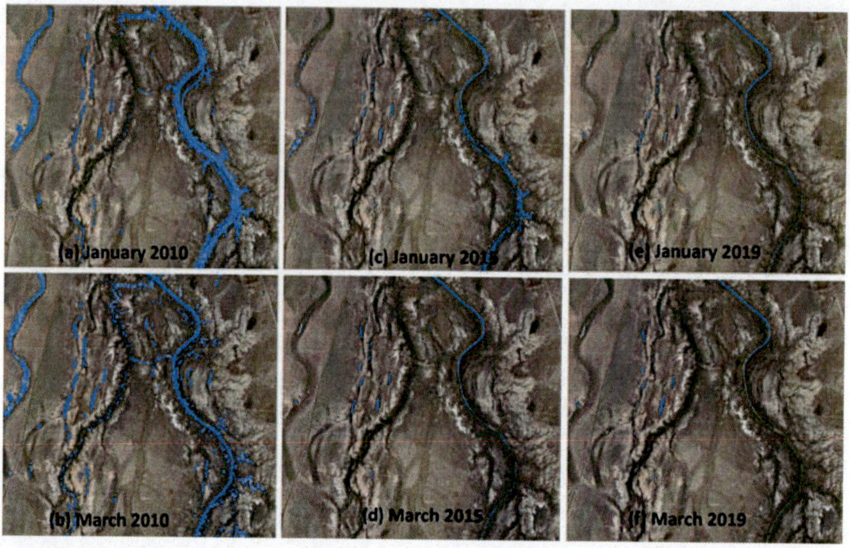

Fig. 18.18 Spatial patterns of floodplain inundation and impact on the connectivity of freshwater habitats in downstream Flinders. **a–f** Show the characteristics of inundation in riverine systems and intermittent water bodies for some wet season periods (2010, 2015 and 2019). The background high-resolution satellite imagery is that of Digital Globe captured in May 2015. *Source* Ndehedehe et al. (2020b)

18.4.9 The Role of Extreme Events on Floodplain Productivity

In this study, we show that the wetland hydrology and hydrological variability of tropical floodplain ecosystems such as the Flinders is strongly climate-driven. Observed strong spatio-temporal distribution in seasonal floodplain inundation was consistent with hydro-climatic (discharge and rainfall) fluctuations in the Flinders catchment. More specifically, the summers of 2009 and 2010 were such periods during the last decade when the Flinders catchment was extensively flooded (from the available imageries analysed) due to the impacts of climate variability. In one of the largest flood events of the last 10 years (i.e., 2009), post-flood surface extent receded by about 89 and 96% within 15 and 31 days, respectively, leaving behind perennial freshwater bodies and persistent aquatic habitats. But the 1 in 50 year flood, which occurred recently during the first week of February 2019 as estimated from MODIS data (10th February) showed a somewhat similar recession pattern as the 2009 summer flood. That is, approximately 87% of the extreme 2019 summer flood receded within 10 days. Despite being significantly higher in intensity, depth, and magnitude than the February 2009 flood, the February 2019 flood on the Flinders catchment appeared, however, to have receded a bit faster than that of 2009. The rapid decline of post-flood in 2019 is evident in the estimated floodplain inundation extent and spatial distribution of hot spots in the last week of February and early March of 2019. These were significantly lower than those of 2009. In such a case, one would argue that considering the nature of flood recession in the 2019 summer despite its higher flood intensity and magnitude, the floodplain rivers were the key channels that conveyed the tremendous amount of rainfall in the high elevation regions of the upstream Flinders catchment to the downstream Canobie-Saxby Roundup areas.

However, towards the end of the summer wet season of 2009, not only were the isolated water bodies extensively nourished, the predicted inundation translated to extensive hot spots of primary producers in the catchment. This was not the case in 2019 and also very much like the 2015–2016 period when the floodplain was less productive because of dry spells and limited rainfall in the region. We found that the loss in surface extent of aquatic plant biomass accumulation (hot spots) coincided with diminished extents of freshwater habitats in the Flinders catchments. While there is significant inter-annual variability in the distribution of floodplain inundation on a seasonal time scale, declines in spatial extents of freshwater habitats and hot spots are caused primarily by limited rainfall and surface flow. The evidence of the composite influence of local rainfall (strong spatial rainfall patterns are observed downstream) and flow dynamics on floodplain productivity in Flinders underscores the role of climate in driving the downstream floodplain inundation and hot spots of primary producers.

18.4.10 Rivers as Critical Indicators of Hot Spots of Aquatic Primary Producers

Due to the low-lying elevation of the Canobie–Saxby Roundup constriction in the Flinders (i.e., lowest elevation area with the most rainfall along the mid-stream corridor-Fig. 18.4), surface flow from the massive upstream area creates a relatively higher flood depth and highly inundated floodplain in the downstream area during and after major rainfall events such as the 2009 and 2019 episodes. A key question therefore is, how important are rivers in the Flinders catchment to its wetland hydrology and productivity? Given the considerable correlation between the dominant modes of rainfall and discharge ($r = 0.70$), which is the annual variability regulating yearly flow and floodplain inundation, discharge is obviously a critical component of the Flinders water cycle. Moreover, considering that approximately 95% of surface runoff occurs during summer when rainfall is at its peak (Waltham et al. 2013), the river systems in the Flinders could be considered not only as indicators of floodplain inundation during such times but as vehicles for the transport of nutrient.

Imagine that connectivity between rivers and intermittent water holes in the catchment is required to sustain aquatic biodiversity, fish populations, and other higher consumers (e.g., invertebrates and waterfowl) and that high levels of aquatic primary producers are generating food and energy sources for higher-order consumers during peak heavy inundation. And if food nutrients are being transported during such times, then, these floodplain rivers are expected to be essential indicators of aquatic production in Flinders. The strong flows in 2009 and 2010, which were consistent with the extensive distribution of hot spots and flooded vegetation in the catchment underscores the possible roles of rivers as critical indicators for the growth of these hot spots. Generally, we found that the spatial distribution of predicted hot spots of aquatic plant biomass coincide with peak amplitudes of dominant modes of rainfall and river discharge. Similarly, in 2015–2016 and 2019 (i.e., after the extreme summer flood) periods when annual river flow was restricted, the spatial distribution of hot spots of primary producers diminished significantly. When these connections are superimposed on the fact that more open surface water features (including those existing as discrete and intermittent water bodies) are found on the floodplain in years with stronger annual amplitudes in river discharge, the evidence distilled from this study points to rivers as critical indicators of aquatic primary producers in the Flinders catchment, in addition to rainfall.

There is a history between the existing relationship of aquatic plant biomass and higher order consumers with surrounding wetland hydrology. For example, the synthesis of fish and invertebrates response to extreme events (floods and droughts) in case-specific studies in Europe, metrics such as abundance and richness, amongst others showed declines after the occurrence of extreme events (Piniewski et al. 2017). In the Australian arid-zone, an increase in channel flows and infrequent episodes of large floods culminated in a rise in fish production while limited numbers of fish were recorded in periods with no channel flow (Balcombe and Arthington 2009). These patterns and ecological responses and processes across different spatial domains could be different because of environmental heterogeneity that may be linked to human influence, climate, topography, hydrology and perhaps catchment characteristics. Within the context of flow variability and its impacts on the spatial distribution of refugia for aquatic biota and aquatic production, these patterns (e.g., the response of primary production to flow variability) were noted by Bunn et al. (2006). In addition to highlighting the influence of changing hydro-climatology on aquatic production on floodplains, they also noted the impacts of human water resource development on ecosystem health and productivity. Generally, in much of the Australian wet–dry tropics, aquatic biodiversity and primary productivity (e.g., algal productivity), which is essential for supporting aquatic food webs in tropical rivers and floodplains tend to be sensitive to wetland hydrology (see, e.g., Karim et al. 2018; Bunn et al. 2015; Ward et al. 2016, 2013).

18.5 Conclusion

Despite the importance of wetlands, they are very fragile and continuously under threat and is evidenced in their rapid decline. Moreover, given the role of climate and human activities as key indicators of processes that may affect water availability and the productivity of floodplain ecosystems, understanding such implications is important for environmental water resource allocation and the ecological communities supported by the wetlands and freshwater systems that generate multiple strings of economic values and services. These values have generated increasing interest in water resource development among all levels of government, however, such development can itself impair them. The large-scale assessment of the Flinders floodplain using a framework that integrates remote sensing biophysical indicators (vegetation and inundation) with hydrological data (rainfall and river discharge) was therefore undertaken. Flood prediction can be extremely difficult, especially with enough time lag for landholders to change the locations of their stocks. However, given that livestock losses and damage to fences, roads, and structures can have a catastrophic effect on the floodplains of the Gulf rivers, this exercise is therefore of utmost importance.

References

Adam E, Mutanga O, Rugege D (2010) Multispectral and hyperspectral remote sensing for identification and mapping of wetland vegetation: a review. Wetlands Ecol Manag 18(3):281–296. https://doi.org/10.1007/s11273-009-9169-z

Amani M, Mahdavi S, Afshar M, Brisco B et al (2019) Canadian wetland inventory using Google Earth Engine: the first map and preliminary results. Remote Sens 11(7). https://doi.org/10.3390/rs11070842

Araya-López RA, Lopatin J, Fassnacht FE, Hernández HJ (2018) Monitoring andean high altitude wetlands in central chile with seasonal optical data: a comparison between Worldview-2 and Sentinel-2 imagery. ISPRS J Photogram Remote Sens 145:213–224. https://doi.org/10.1016/j.isprsjprs.2018.04.001

Auricht HC, Clarke KD, Lewis MM, Mosley LM (2017) Have droughts and increased water extraction from the Murray river (Australia) reduced coastal ocean productivity? Marine Freshw Res 69(3):343–356. https://doi.org/10.1071/MF17226

Balcombe SR, Arthington AH (2009) Temporal changes in fish abundance in response to hydrological variability in a dryland floodplain river. Marine Freshw Res 60:146–159. https://doi.org/10.1071/MF08118

Bargain A, Robin M, Le Men E, Huete A, Barillé L (2012) Spectral response of the seagrass zostera noltii with different sediment backgrounds. Aquatic Botany 98(1):45–56. https://doi.org/10.1016/j.aquabot.2011.12.009

Bartsch A, Trofaier AM, Hayman G, Sabel D, Schlaffer S, Clark DB, Blyth E (2012) Detection of open water dynamics with ENVISAT ASAR in support of land surface modelling at high latitudes. Biogeosciences 9(2):703–714. https://doi.org/10.5194/bg-9-703-2012

Bhatnagar S, Ghosh B, Regan S, Naughton O, Johnston P, Gill L (2018) Monitoring environmental supporting conditions of a raised bog using remote sensing techniques. Proc Int Assoc Hydrol Sci 380:9–15. https://doi.org/10.5194/piahs-380-9-2018

Bork EW, Su JG (2007) Integrating LIDAR data and multispectral imagery for enhanced classification of rangeland vegetation: a meta analysis. Remote Sens Environ 111(1):11–24. https://doi.org/10.1016/j.rse.2007.03.011

Boulton AJ (2007) Hyporheic rehabilitation in rivers: restoring vertical connectivity. Freshw Biol 52(4):632–650. https://doi.org/10.1111/j.1365-2427.2006.01710.x

Bourgeau-Chavez LL, Riordan K, Powell RB, Miller N, Nowels M (2009) Improving wetland characterization with multi-sensor, multi-temporal sar and optical/infrared data fusion. In: Advances in geoscience and remote sensing. https://doi.org/10.5772/8327

Bricaud A, Babin M, Morel A, Claustre H (1995) Variability in the chlorophyll-specific absorption coefficients of natural phytoplankton: analysis and parameterization. J Geophys Res: Oceans 100(C7):13321–13332. https://doi.org/10.1029/95JC00463

Broadley A, Stewart-Koster B, Kenyon RA, Burford MA, Brown CJ (2020) Impact of water development on river flows and the catch of a commercial marine fishery. Ecosphere 11(7):e03194. https://doi.org/10.1002/ecs2.3194

Buckworth RC, Venables WN, Lawrence E, Kompas T, Pascoe S, Chu L, Rothlisberg PC (2014) Incorporation of predictive models of banana prawn catch for MEY-based harvest strategy development for the Northern Prawn fishery. Final Report to the Fisheries Research and Development Corporation, Project 2011/239

Bunn S, Ward D, Crook D, Jardine T, Pettit N, Douglas M, Kyne P (2015) Tropical floodplain food webs—connectivity and hotspots. Darwin: Charles Darwin University

Bunn SE, Davies PM, Winning M (2003) Sources of organic carbon supporting the food web of an arid zone floodplain river. Freshw Biol 48(4):619–635. https://doi.org/10.1046/j.1365-2427.2003.01031.x

Bunn SE, Thoms MC, Hamilton SK, Capon SJ (2006) Flow variability in dryland rivers: boom, bust and the bits in between. River Res Appl 22(2):179–186. https://doi.org/10.1002/rra.904

Burford MA, Faggotter SJ (2021) Comparing the importance of freshwater flows driving primary production in three tropical estuaries. Marine Pollut Bull 169:112565. https://doi.org/10.1016/j.marpolbul.2021.112565

Burford, MA, Faggotter SJ, Kenyon R (2020) Contribution of three rivers to floodplain and coastal productivity in the gulf of carpentaria

Butler B, Burrows D, Pearson R (2007) Providing regional NRM with improved aquatic health risk assessment and monitoring tools: the nationally significant indication: dissolved oxygen

Butler B, Burrows D (2007) Dissolved oxygen guidelines for freshwater habitats of northern Australia. Department of Environment and Heritage Canberra. https://researchonline.jcu.edu.au/29265/1/29265_Butler_Burrows_2007.pdf

Butler B, Loong D, Davis A (2009) Water for bowen freshwater ecology. Australian Centre for Tropical Freshwater Research, James Cook University, Australia

Campana SE, Thorrold SR (2001) Otoliths, increments, and elements: keys to a comprehensive understanding of fish populations? Canadian J Fish Aquat Sci 58(1):30–3. https://doi.org/10.1139/f00-177

Cazzanelli M, Warming TP, Christoffersen KS (2008) Emergent and floating-leaved macrophytes as refuge for zooplankton in a eutrophic temperate lake without submerged vegetation. Hydrobiologia 605(1):113–122. https://doi.org/10.1007/s10750-008-9324-1

Chapman B, McDonald K, Shimada M, Rosenqvist A, Schroeder R, Hess L (2015) Mapping regional inundation with spaceborne l-band SAR. Remote Sens 7(5):5440–5470. https://doi.org/10.3390/rs70505440

Chatziantoniou A, Psomiadis E, Petropoulos GP (2017) Co-orbital Sentinel 1 and 2 for LULC mapping with emphasis on wetlands in a mediterranean setting based on machine learning. Remote Sens 9(12). https://doi.org/10.3390/rs9121259

Chen T, de Jeu R, Liu Y, van der Werf G, Dolman A (2014) Using satellite based soil moisture to quantify the water driven variability in NDVI: a case study over mainland Australia. Remote Sens Environ 140:330–338. https://doi.org/10.1016/j.rse.2013.08.022

Chen Y, Wang B, Pollino CA, Cuddy SM, Merrin LE, Huang C (2014) Estimate of flood inundation and retention on wetlands using remote sensing and GIS. Ecohydrology 7(5):1412–1420. https://doi.org/10.1002/eco.1467

Congalton R, Green K (2009) Assessing the accuracy of remotely sensed data: principles and practices. Taylor & Francis. 2nd

CRED (2020) Human cost of disasters.an overview of the last 20 years: 2000–2019. Scientific reports. https://reliefweb.int/report/world/human-cost-disasters-overview-last-20-years2000-2019. Accessed 1st Aug 2021

Crook DA, Buckle DJ, Allsop Q, Baldwin W, Saunders TM, Kyne PM, Douglas MM (2016) Use of otolith chemistry and acoustic telemetry to elucidate migratory contingents in barramundi latescalcarifer. Marine Freshw Res 68(8):1554–1566. https://doi.org/10.1071/MF16177

CSIRO (2009) Northern Australia land and water science review 2009 chapter summaries. Canberra

Cuan P, Yang A (2013) Climate data and their characterisation for hydrological and agricultural scenario modelling across the flinders and gilbert catchments. CSIRO

Davidson NC (2014) How much wetland has the world lost? long-term and recent trends in global wetland area. Marine Freshw Res 65(10):934–941. https://doi.org/10.1071/MF14173

Davranche A, Lefebvre G, Poulin B (2010) Wetland monitoring using classification trees and SPOT-5seasonal time series. Remote Sens Environ 114(3):552–562. https://doi.org/10.1016/j.rse.2009.10.009

DERM (2011) Waterhole persistence and productivity in two rivers in northern Queensland. Department of environment and resources management. Queensland Brisbane

de Moel H, van Alphen J, Aerts JCJH (2009) Flood maps in Europe-methods, availability and use. Nat Hazards Earth Syst Sci 9(2):289. https://doi.org/10.5194/nhess-9-289-2009

DeVries B, Huang C, Armston J, Huang W, Jones JW, Lang MW (2020) Rapid and robust monitoring of flood events using sentinel-1 and landsat data on the google earth engine. Remote Sens Environ 240:111664. https://doi.org/10.1016/j.rse.2020.111664

DeVries B, Huang C, Lang MW, Jones JW, Huang W, Creed IF, Carroll ML (2017) Automated quantification of surface water inundation in wetlands using optical satellite imagery. Remote Sens 9(8). https://doi.org/10.3390/rs9080807

Douglas MM, Bunn SE, Davies PM (2005) River and wetland food webs in Australia"s wet-dry tropics: general principles and implications for management. Marine Freshw Res 56(3):329–342. https://doi.org/10.1071/MF04084

Drusch M, Del Bello U, Carlier S, Colin O, V F et al (2012) Sentinel-2: Esa's optical high-resolution mission for GMES operational services. Remote Sens Environ 120:25–36. https://doi.org/10.1016/j.rse.2011.11.026

Ecosystem AM (2005) Ecosystems and human well-being: wetlands and water. World Resources Institute, Washington, DC

Elser JJ, Bracken ME, Cleland EE, Gruner DS et al (2007) Global analysis of nitrogen and phosphorus limitation of primary producers in freshwater, marine and terrestrial ecosystems. Ecol Lett 10(12):1135–1142. https://doi.org/10.1111/j.1461-0248.2007.01113.x

Eroglu O, Kurum M, Boyd D, Gurbuz, AC (2019) High spatio-temporal resolution cygnss soil moisture estimates using artificial neural networks. Remote Sens 11(19). https://doi.org/10.3390/rs11192272

Faggotter SJ, Webster IT, Burford MA (2013) Factors controlling primary productivity in a wet-dry tropical river. Marine Freshw Res 64:585–598. https://doi.org/10.1071/MF12299

Farahmand A, AghaKouchak A (2015) A generalized framework for deriving nonparametric standardized drought indicators. Adv Water Res 76:140–145. https://doi.org/10.1016/j.advwatres.2014.11.012

Fern RR, Foxley EA, Bruno A, Morrison ML (2018) Suitability of NDVI and OSAVI as estimators of green biomass and coverage in a semi-arid rangeland. Ecol Indic 94:16–21. https://doi.org/10.1016/j.ecolind.2018.06.029

Finlayson CM (2005) Plant ecology of Australia's tropical floodplain wetlands: a review. Ann Botany 96(4):541–555. https://doi.org/10.1093/aob/mci209

Formetta G, Feyen L (2019) Empirical evidence of declining global vulnerability to climate-related hazards. Glob Environ Change 57:101920. https://doi.org/10.1016/j.gloenvcha.2019.05.004

Gallant AL (2015) The challenges of remote monitoring of wetlands. Remote Sens 7(8):10938–10950. https://doi.org/10.3390/rs70810938

Gao B-C (1996) NDWI–a normalized difference water index for remote sensing of vegetation liquid water from space. Remote Sens Environ 58(3):257–266. https://doi.org/10.1016/S0034-4257(96)00067-3

Garcia EA, Townsend SA, Douglas MM (2015) Context dependency of top-down and bottom-up effects in a northern Australian tropical river. Freshw Sci 34(2):679–690. https://doi.org/10.1086/681106

García-Pintado J, Mason DC, Dance SL, Cloke HL, Neal JC, Freer J, Bates PD (2015) Satellite-supported flood forecasting in river networks: a real case study. J Hydrol 523:706–724. https://doi.org/10.1016/j.jhydrol.2015.01.084

Genz F, Luz L (2012) Distinguishing the effects of climate on discharge in a tropical river highly impacted by large dams. Hydrol Sci J 57(5):1020–1034. https://doi.org/10.1080/02626667.2012.690880

Ghil M, Allen MR, Dettinger MD, Ide K, Kondrashov D, Mann ME, Robertson AW, Saunders A, Tian Y, Varadi F, Yiou P (2002) Advanced spectral methods for climatic time series. Rev Geophys 40(1):3-1–3-41. https://doi.org/10.1029/2000RG000092

Gidley SL (2009) Using high resolution satellite imagery to map aquatic macrophytes on multiple lakes in Northern Indiana. Unpublished Msc thesis, Indiana University. https://core.ac.uk/download/pdf/46956355.pdf. Accessed 15th Apr 2019

Gibson RN, Barnes M, Atkinson RJA (2002) Impact of changes in flow of freshwater on estuarine and open coastal habitats and the associated organisms. Ocean Marine Biol Ann Rev 40:233

Gilerson AA, Gitelson AA et al (2010) Algorithms for remote estimation of chlorophyll-a in coastal and inland waters using red and near infrared bands. Opt Express 18(23):24109–24125. https://doi.org/10.1364/OE.18.024109

Gitelson AA, Gritz Y, Merzlyak MN (2003) Relationships between leaf chlorophyll content and spectral reflectance and algorithms for non-destructive chlorophyll assessment in higher plant leaves. J Plant Physiol 160(3):271–282. https://doi.org/10.1078/0176-1617-00887

Gitelson AA, Schalles JF, Hladik CM (2007) Remote chlorophyll-a retrieval in turbid, productive estuaries: Chesapeake Bay case study. Remote Sens Environ 109(4):464–472. https://doi.org/10.1016/j.rse.2007.01.016

Gower J, King S, Borstad G, Brown L (2004) Detection of intense plankton blooms using the meris optical imager on envisat. Gayana (Concepción) 68(2):246–251. https://doi.org/10.4067/S0717-65382004000200044

Grey D, Sadoff CW (2007) Sink or swim? Water security for growth and development. Water Policy 9(6):545–571. https://doi.org/10.2166/wp.2007.021

Grill G, Lehner B, Thieme M et al (2019) Mapping the world"s free-flowing rivers. Nature 569(7755):215–221

Grimshaw HJ, Wetzel RG, Brandenburg M et al (1997) Shading of periphyton communities by wetland emergent macrophytes: decoupling of algal photosynthesis from microbial nutrient retention. Archiv für Hydrobiologie 139(1):17–27. https://doi.org/10.1127/archiv-hydrobiol/139/1997/17

Halliday IA, Saunders T, Sellin MJ, Allsop Q, Robins JB, McLennan MF, Kurnoth P (2012) Flow impacts on estuarine finfish fisheries of the gulf of carpentaria. Technical Report FRDC Project No 2007/002

Hallett TB, Coulson T, Pilkington JG, Clutton-Brock TH, Pemberton JM, Grenfell BT (2004) Why large-scale climate indices seem to predict ecological processes better than local weather. Nature 430(6995):71–75. https://doi.org/10.1038/nature02708

Hall JW, Grey D, Garrick D, Fung F, Brown C, Dadson SJ, Sadoff CW (2014) Coping with the curse of freshwater variability. Science 346(6208):429–430. https://doi.org/10.1126/science.1257890

Hamilton SK, Sippel SJ, Melack JM (2002) Comparison of inundation patterns among major South American floodplains. J Geophys Res: Atmos 107(D20):LBA 5-1–LBA 5-14. https://doi.org/10.1029/2000JD000306

Henderson FM, Lewis AJ (2008) Radar detection of wetland ecosystems: a review. Int J Remote Sens 29(20):5809–5835. https://doi.org/10.1080/01431160801958405

Hestir EL, Khanna S, Andrew ME, Santos MJ et al (2008) Identification of invasive vegetation using hyperspectral remote sensing in the California delta ecosystem. Remote Sens Environ 112(11):4034–4047. https://doi.org/10.1016/j.rse.2008.01.022

Howley C, Devlin M, Burford M (2018) Assessment of water quality from the Normanby river catchment to coastal flood plumes on the northern great barrier reef Australia. Marine Freshw Res 69(6):859–873. https://doi.org/10.1071/MF17009

Hill R, O'Hare MT, Stillman R, Gozlan R (2009) Hyperspectral remote sensing of river macrophyte vegetation: towards an assessment of wildfowl and fish habitat quality. In: Proceedings of the RSPSoc annual conference: new dimensions in earth observation

Jahncke R, Leblon B, Bush P, LaRocque A (2018) Mapping wetlands in nova scotia with multi-beam radarsat-2 polarimetric SAR, optical satellite imagery, and lidar data. Int J Appl Earth Obs Geoinformation 68:139–156. https://doi.org/10.1016/j.jag.2018.01.012

Jardine T, Pusey B, Hamilton S, Pettit N, Davies P, Douglas et al (2012) Fish mediate high food web connectivity in the lower reaches of a tropical floodplain river. Oecologia 168(3):829–838. https://doi.org/10.1007/s00442-011-2148-0

Jia J, Gao Y, Zhou F, Shi K, Johnes PJ, Dungait JA, Ma M, Lu Y (2020) Identifying the main drivers of change of phytoplankton community structure and gross primary productivity in a river-lake system. J Hydrol 583:124633. https://doi.org/10.1016/j.jhydrol.2020.124633

Jolliffe IT (2002) Principal component analysis, 2nd edn. Springer Series in Statistics. Springer, New York

Jones TG, Coops NC, Sharma T (2010) Assessing the utility of airborne hyperspectral and LiDAR data for species distribution mapping in the coastal Pacific Northwest Canada. Remote Sens Environ 114(12):2841–2852. https://doi.org/10.1016/j.rse.2010.07.002

Jongman B, Ward PJ, Aerts JC (2012) Global exposure to river and coastal flooding: long term trends and changes. Glob Environ Change 22(4):823–835. https://doi.org/10.1016/j.gloenvcha.2012.07.004

Jongman B, Winsemius HC, Aerts JCJH, Coughlan de Perez E, van Aalst MK, Kron W, Ward PJ (2015) Declining vulnerability to river floods and the global benefits of adaptation. Proc Nat Acad Sci 112(18):E2271–E2280. https://doi.org/10.1073/pnas.1414439112

Junk WJ, Piedade MTF (1997) Plant life in the floodplain with special reference to herbaceous plants. Central Amazon Floodplain: Ecol Pulsing Syst 147–185. https://doi.org/10.1007/978-3-662-03416-3_8

Kamerosky A, Cho HJ, Morris L (2015) Monitoring of the 2011 super algal bloom in Indian river lagoon, FL, USA, using MERIS. Remote Sens 7(2):1441–1460. https://doi.org/10.3390/rs70201441

Kaplan G, Avdan U (2017) Mapping and monitoring wetlands using sentinel-2 satellite imagery. In: ISPRS annals of photogrammetry, remote sensing & spatial information sciences, vol 4. https://pdfs.semanticscholar.org/a101/515a9d639c896364cec0b589172af3649717.pdf. Accessed 1st Aug 2021

Karim F, Peña-Arancibia J, Ticehurst C, Marvanek S, Gallant J, Hughes J, Dutta D, Vaze J, Petheram C, Seo L, Kitson S (2018) Floodplain inundation mapping and modelling for the Fitzroy, Darwin and Mitchell Catchments. a technical report to the Australian government from the CSIRO northern Australia water resource assessment, part of the national water infrastructure development fund: Water resource assessments. CSIRO, Australia

Keddy PA, Fraser LH, Solomeshch AI, Junk WJ, Campbell DR, Arroyo MTK, Alho CJR (2009) Wet and wonderful: the world's largest wetlands are conservation priorities. BioScience 59(1):39–51. https://doi.org/10.1525/bio.2009.59.1.8

Ke Y, Quackenbush LJ, Im J (2010) Synergistic use of quickbird multispectral imagery and LIDAR data for object-based forest species classification. Remote Sens Environ 114(6):1141–1154. https://doi.org/10.1016/j.rse.2010.01.002

Kellett J, Caravani A (2013) Wetland ecology: principles and conservation. Cambridge University Press

Kendy E, Flessa KW, Schlatter KJ, Carlos AD et al (2017) Leveraging environmental flows to reform water management policy: Lessons learned from the 2014 Colorado River Delta pulse flow. Ecol Eng 106:683–694. https://doi.org/10.1016/j.ecoleng.2017.02.012

Kennard MJ, Pusey Bradley J, Olden JD, Mackay SJ, Stein JL, Marsh N (2010) Classification of natural flow regimes in Australia to support environmental flow management. Freshw Biol 55(1):171–193. https://doi.org/10.1111/j.1365-2427.2009.02307.x

Kingsford RT, Walker KF, Lester RE, Young WJ, Fairweather PG, Sammut J, Geddes MC (2011) A ramsar wetland in crisis-the coorong, lower lakes and murray mouth, Australia. Marine Freshw Res 62(3):255–256. https://doi.org/10.1071/MF09315

Kingsford R (2000) Ecological impacts of dams, water diversions and river management on floodplain wetlands in Australia. Austral Ecol 25(2):109–127. https://doi.org/10.1046/j.1442-9993.2000.01036.x

Kuhn C, Bogard M, Johnston SE, John A et al (2020) Satellite and airborne remote sensing of gross primary productivity in boreal Alaskan lakes. Environ Res Lett 15(10):105001. https://doi.org/10.1088/1748-9326/aba46f

Leahy SM, Robins JB (2021) River flows affect the growth of a tropical finfish in the wet-dry rivers of northern Australia, with implications for water resource development. Hydrobiologia. https://doi.org/10.1007/s10750-021-04641-7

Lee BS, McGwire KC, Fritsen CH (2011) Identification and quantification of aquatic vegetation with hyperspectral remote sensing in western Nevada rivers, USA. Int J Remote Sens 32(24):9093–9117. https://doi.org/10.1080/01431161.2010.549850

Leigh C (2013) Dry-season changes in macroinvertebrate assemblages of highly seasonal rivers: responses to low flow, no flow and antecedent hydrology. Hydrobiologia 703(1):95–112. https://doi.org/10.1007/s10750-021-04641-7

Lewis WM, Hamilton SK, Rodríguez MA, Saunders JF, Lasi MA (2001) Foodweb analysis of the orinoco floodplain based on production estimates and stable isotope data. J North Am Benthol Soc 20(2):241–254. https://doi.org/10.2307/1468319

Ludwig C, Walli A, Schleicher C, Weichselbaum J, Riffler M (2019) A highly automated algorithm for wetland detection using multi-temporal optical satellite data. Remote Sens Environ 224:333–351. https://doi.org/10.1016/j.rse.2019.01.017

Mahdavi S, Salehi B, Granger J, Amani M, Brisco B, Huang W (2018) Remote sensing for wetland classification: a comprehensive review. GIScience Remote Sens 55(5):623–658. https://doi.org/10.1080/15481603.2017.1419602

Mahdianpari M, Salehi B, Mohammadimanesh F, Homayouni S, Gill E (2019) The first wetland inventory map of newfoundland at a spatial resolution of 10 m using sentinel-1 and sentinel-2 data on the google earth engine cloud computing platform. Remote Sens 11(1). https://doi.org/10.3390/rs11010043

Mahdianpari M, Brisco B, Granger JE, Mohammadimanesh F, Salehi B, Banks S, Homayouni S, Bourgeau-Chavez L, Weng Q (2020) The second generation Canadian wetland inventory map at 10 m resolution using google earth engine. Canadian J Remote Sens 1–16. https://doi.org/10.1080/07038992.2020.1802584

Mallin MA, Paerl HW, Rudek J, Bates PW (1993) Regulation of estuarine primary production by watershed rainfall and river flow. Marine Ecol-Prog Ser 93:199. http://www.int-res.com/articles/meps/93/m093p199.pdf. Accessed 1st Aug 2021

Marcus WA, Fonstad MA (2008) Optical remote mapping of rivers at sub-meter resolutions and watershed extents. Earth Surf Process Landf 33(1):4–24. https://doi.org/10.1002/esp.1637

Marek M, Magdalena M, Magdalena F (2016) Exploitation of sentinel-1a data in one year survey of water transfer on wetlands. http://marshall.uwm.edu.pl/publications/1.pdf. Accessed 1st Aug 2021

Martino JC, Fowler AJ, Doubleday ZA, Grammer GL, Gillanders BM (2019) Using otolith chronologies to understand long-term trends and extrinsic drivers of growth in fisheries. Ecosphere 10(1):e02553. https://doi.org/10.1002/ecs2.2553

Martinez WL, Martinez AR (2005) Exploratory data analysis with MATLAB. Chapman and Hall/CRC Press LLC, UK. 1-58488-366-9

Mason D, Garcia-Pintado J, Cloke H, Dance S (2015) The potential of flood forecasting using a variable-resolution global digital terrain model and flood extents from synthetic aperture radar images. Front Earth Sci 3:43. https://doi.org/10.3389/feart.2015.00043

Mcgregor GB, Marshall JC, Lobegeiger JS, Holloway D, Menke N, Coysh J (2018) A risk-based ecohydrological approach to assessing environmental flow regimes. Environ Manag 61(3):358–374. http://marshall.uwm.edu.pl/publications/1.pdf. Accessed 1st Aug 2021

McKee TB, Doeskin NJ, Kieist J (1993) The relationship of drought frequency and duration to time scales. In: Conference on applied climatology, American meteorological society,Boston, Massachusetts, pp 179–184. https://www.ccc.atmos.colostate.edu/relationshipofdroughtfrequency.pdf. Accessed 27 June 2014

Midwood JD, Chow-Fraser P (2010) Mapping floating and emergent aquatic vegetation in coastal wetlands of eastern Georgian bay, lake Huron, Canada. Wetlands 30(6):1141–1152. https://doi.org/10.1007/s13157-010-0105-z

Miller RL, Belz M, Castillo CD, Trzaska R (2002) Determining CDOM absorption spectra in diverse coastal environments using a multiple pathlength, liquid core waveguide system. Cont Shelf Res 22(9):1301–1310. https://doi.org/10.1016/S0278-4343(02)00009-2

Milton D, Halliday I, Sellin M, Marsh R, Staunton-Smith J, Woodhead J (2008) The effect of habitat and environmental history on otolith chemistry of barramundi lates calcarifer in estuarine populations of a regulated tropical river. Estuar Coast Shelf Sci 78(2):301–315. https://doi.org/10.1016/j.ecss.2007.12.009

Milton DA, Chenery SR (2018) Movement patterns of barramundi latescalcarifer, inferred from 87sr/86sr and sr/ca ratios in otoliths, indicate non-participation in spawning. Marine Ecol Prog Ser 301:279–291. https://doi.org/10.3354/meps301279

Molinari B, Stewart-Koster B, Adame MF, Campbell MD, McGregor G, Schulz C, Malthus TJ, Bunn S (2021) Relationships between algal primary productivity and environmental variables in tropical floodplain wetlands. Inland Waters 1–11. https://doi.org/10.1080/20442041.2020.1843932

Molinari B, Stewart-Koster B, Malthus TJ, Bunn SE (2021b) Assessing spatial variation in algal productivity in a tropical river floodplain using satellite remote sensing. Remote Sens 13(9). https://doi.org/10.3390/rs13091710

Morrongiello JR, Thresher RE, Smith DC (2012) Aquatic biochronologies and climate change. Nat Clim Change 2(12):849–857. https://doi.org/10.1038/nclimate1616

Mortillaro JM, Pouilly M, Wach M, Freitas CEC, Abril G, Meziane T (2015) Trophic opportunism of central Amazon floodplain fish. Freshw Biol 60(8):1659–1670. https://doi.org/10.1111/fwb.12598

Muro J, Canty M, Conradsen K et al (2016) Short-term change detection in wetlands using sentinel-1 time series. Remote Sens 8(10). https://doi.org/10.3390/rs8100795

Mücher C, Roupioz L, Kramer H, Bogers M et al (2015) Synergy of airborne LiDAR and Worldview-2 satellite imagery for land cover and habitat mapping: a BIO SOS-EODHaM case study for the Netherlands. Int J Appl Earth Obs Geoinformation 37:48–55. https://doi.org/10.1016/j.jag.2014.09.001

Najibi N, Devineni N (2018) Recent trends in the frequency and duration of global floods. Earth Syst Dyn 9(2):757–783. https://doi.org/10.5194/esd-9-757-2018

Ndehedehe C (2020) Floodplain productivity of the Gilbert and Flinders catchments: component 3 final report. Griffith University, Brisbane.https://research-repository.griffith.edu.au/handle/10072/394768. Accessed 8th July 2020

Ndehedehe CE (2017) Remote sensing of West Africa's water resources using multi-satellites and models. PhD thesis, Curtin University, Bentley, Perth, Western Australia. http://hdl.handle.net/20.500.11937/59637. Accessed 12th Jan 2018

Ndehedehe CE, Agutu NO, Okwuashi OH, Ferreira VG (2016) Spatio-temporal variability of droughts and terrestrial water storage over Lake Chad Basin using independent component analysis. J Hydrol 540:106–128. https://doi.org/10.1016/j.jhydrol.2016.05.068

Ndehedehe CE, Awange JL, Agutu NO, Okwuashi O (2018) Changes in hydro-meteorological conditions over tropical West Africa (1980–2015) and links to global climate. Glob Planet Change 162:321–341. https://doi.org/10.1016/j.gloplacha.2018.01.020

Ndehedehe CE, Burford MA, Stewart-Koster B, Bunn SE (2020) Satellite-derived changes in floodplain productivity and freshwater habitats in northern Australia (1991–2019). Ecol Indic 114:106320. https://doi.org/10.1016/j.ecolind.2020.106320

Ndehedehe CE, Ferreira VG, Agutu NO (2019) Hydrological controls on surface vegetation dynamics over West and Central Africa. Ecol Indic 103:494–508. https://doi.org/10.1016/j.ecolind.2019.04.032

Ndehedehe CE, Onojeghuo AO, Stewart-Koster B, Bunn SE, Ferreira VG (2021) Upstream flows drive the productivity of floodplain ecosystems in tropical queensland. Ecol Indic. https://doi.org/10.1016/j.ecolind.2021.107546

Ndehedehe CE, Stewart-Koster B, Burford MA, Bunn SE (2020) Predicting hot spots of aquatic plant biomass in a large floodplain river catchment in the Australian wet-dry tropics. Ecol Indic 117:106616. https://doi.org/10.1016/j.ecolind.2020.106616

Niculescu S, Boissonnat J-B, Lardeux C, Roberts D, Hanganu J et al (2020) Synergy of high-resolution radar and optical images satellite for identification and mapping of wetland macrophytes on the danube delta. Remote Sens 12(14). https://doi.org/10.3390/rs12142188

Niculescu S, Lardeux C, Grigoras I, Hanganu J, David L (2016) Synergy between lidar, radarsat-2, and spot-5 images for the detection and mapping of wetland vegetation in the danube delta. IEEE J Sel Top Appl Earth Obs Remote Sens 9(8):3651–3666. https://doi.org/10.1109/JSTARS.2016.2545242

Novo EMLM, Steffen CA, Braga Zuccari Fernandes C (1991) Results of a laboratory experiment relating spectral reflectance to total suspended solids. Remote Sens Environ 36(1):67–72. https://doi.org/10.1016/0034-4257(91)90031-Z

Olthof I, Rainville T (2020) Evaluating simulated RADARSAT constellation mission (RCM) compact polarimetry for open-water and flooded-vegetation wetland mapping. Remote Sens 12(9). https://doi.org/10.3390/rs12091476

Onojeghuo AO, Onojeghuo AR, Cotton M, Potter J, Jones B (2021) Wetland mapping with multi-temporal Sentinel-1 & -2 imagery (2017–2020) and LiDAR data in the grassland natural region of Alberta. GIScience Remote Sens 1–23. https://doi.org/10.1080/15481603.2021.1952541

Ozesmi SL, Bauer ME (2002) Satellite remote sensing of wetlands. Wetl Ecol Manag 10(5):381–402. https://doi.org/10.1023/A:1020908432489

Onojeghuo AO, Blackburn GA (2011) Forest transition in an ecologically important region: patterns and causes for landscape dynamics in the Niger Delta. Ecol Indic 11(5):1437–1446. https://doi.org/10.1016/j.ecolind.2011.03.017

Pahl-Wostl C, Arthington A, Bogardi J, Bunn SE et al (2013) Environmental flows and water governance: managing sustainable water uses. Current Opin Environ Sustain 5(3):341–351. https://doi.org/10.1016/j.cosust.2013.06.009

Palmer SC, Kutser T, Hunter PD (2015) Remote sensing of inland waters: challenges, progress and future directions. Remote Sens Environ 157:1–8. https://doi.org/10.1016/j.rse.2014.09.021

Pearson RG, Godfrey PC, Arthington AH, Wallace J, Karim F, Ellison M (2013) Biophysical status of remnant freshwater floodplain lagoons in the great barrier reef catchment: a challenge for assessment and monitoring. Marine Freshw Res 64(3):208–222. https://doi.org/10.1071/MF12251

Pekel J-F, Cottam A, Gorelick N, Belward AS (2016) High-resolution mapping of global surface water and its long-term changes. Nature 540:418–422, 102159. https://doi.org/10.1038/nature20584

Petheram C, Watson I, Bruce C, Chilcott C (2018) Water resource assessment for the mitchell catchment. a report to the australian government from the CSIRO northern Australia water resource assessment, part of the national water infrastructure development fund: water resource assessments. CSIRO. https://www.csiro.au/en/Research/Major-initiatives/Northern-Australia/Current-work/NAWRA/Mitchell-report. Accessed 2019

Petheram C, Watson I, Stone P (2013) Agricultural resource assessment for the flinders catchment. a report to the Australian government from the CSIRO flinders and gilbert agricultural resource assessment, part of the north Queensland irrigated agriculture strategy. In: CSIRO water for a healthy country and sustainable agriculture flagships, Australia. https://www.csiro.au/en/Research/LWF/Areas/Water/Assessing-water-resources/Flinders-Gilbert/Flinders-report. Accessed 30th May 2020

Pettit N, Ward D, Adame M, Valdez D, Bunn S (2016) Influence of aquatic plant architecture on epiphyte biomass on a tropical river floodplain. Aquat Botany 129:35–43. https://doi.org/10.1016/j.aquabot.2015.12.001

Pettit NE, Bayliss P, Davies PM, Hamilton SK, Warfe DM et al (2011). Seasonal contrasts in carbon resources and ecological processes on a tropical floodplain. Freshw Biol

Pettit NE, Naiman RJ, Warfe DM, Jardine TD, Douglas MM, Bunn SE, Davies PM (2017) Productivity and connectivity in tropical riverscapes of northern Australia: ecological insights for management. Ecosystems 20(3):492–514. https://doi.org/10.1007/s10021-016-0037-4

Piniewski M, Prudhomme C, Acreman MC, Tylec L, Oglęcki P, Okruszko T (2017) Responses of fish and invertebrates to floods and droughts in Europe. Ecohydrology 10(1):e1793. https://doi.org/10.1002/eco.1793

Polis G (1999) Why are parts of the world green? Multiple factors control productivity and the distribution of biomass. Oikos 86:3

Rapinel S, Hubert-Moy L, Clément B (2015) Combined use of lidar data and multispectral earth observation imagery for wetland habitat mapping. Int J Appl Earth Obs Geoinformation 37:56–64. https://doi.org/10.1016/j.jag.2014.09.002

Rasmussen JB (2010) Estimating terrestrial contribution to stream invertebrates and periphyton using a gradient-based mixing model for δ13C. J Anim Ecol 79(2):393–402. https://doi.org/10.1111/j.1365-2656.2009.01648.x

Reis V, Hermoso V, Hamilton SK et al (2017) A global assessment of Inland Wetland conservation status. BioScience 67(6):523–533. https://doi.org/10.1093/biosci/bix045

Roberts BH, Morrongiello JR, King AJ, Morgan DL, Saunders TM, Woodhead J, Crook DA (2019) Migration to freshwater increases growth rates in a facultatively catadromous tropical fish. Oecologia 191(2):253–260. https://doi.org/10.1007/s00442-019-04460-7

Robins JB, Halliday IA, Staunton-Smith J, Mayer DG, Sellin MJ (2005) Freshwater-flow requirements of estuarine fisheries in tropical Australia: a review of the state of knowledge and application of a suggested approach. Marine and Freshw Res 56(3):343–360. https://doi.org/10.1071/MF04087

Roggeri H (2013) Tropical freshwater wetlands: a guide to current knowledge and sustainable management. Springer Sci Bus Media

Rondeaux G, Steven M, Baret F (1996) Optimization of soil-adjusted vegetation indices. Remote Sens Environ 55(2):95–107. https://doi.org/10.1016/0034-4257(95)00186-7

Rosenqvist A, Finlayson C, Lowry J, Taylor D (2007) The potential of long-wavelength satellite-borne radar to support implementation of the ramsar wetlands convention. Aquat Conserv: Marine Freshw Ecosyst 17(3):229–244. https://doi.org/10.1002/aqc.835

Rotta L, Alcântara E, Park E, Bernardo N, Watanabe F (2021) A single semi-analytical algorithm to retrieve chlorophyll-a concentration in oligo-to-hypereutrophic waters of a tropical reservoir cascade. Ecol Indic 120:106913. https://doi.org/10.1016/j.ecolind.2020.106913

Rufat S, Tate E, Burton CG, Maroof AS (2015) Social vulnerability to floods: Review of case studies and implications for measurement. Int J Disaster Risk Reduct 14:470–486. https://doi.org/10.1016/j.ijdrr.2015.09.013

Saigo M, Zilli FL, Marchese MR, Demonte D (2015) Trophic level, food chain length and omnivory in the paraná river: a food web model approach in a floodplain river system. Ecol Res 30(5):843–852. https://doi.org/10.1007/s11284-015-1283-1

Schumann G, Di Baldassarre G, Bates PD (2009) The utility of spaceborne radar to render flood inundation maps based on multialgorithm ensembles. IEEE Trans Geosci Remote Sens 47(8):2801–2807. https://doi.org/10.1109/TGRS.2009.2017937

Semeniuk V, Semeniuk CA (1997) A geomorphic approach to global classification for natural inland wetlands and rationalization of the system used by the ramsar convention-a discussion. Wetl Ecol Manag 5(2):145–158. https://doi.org/10.1023/A:1008207726826

Setterfield SA, Douglas MM, Petty AM, Bayliss P, Ferdinands KB, Winderlich S (2013) Invasive plants in the floodplains of Australia's kakadu national park. In: Plant invasions in protected areas: patterns, problems and challenges, pp 167–189. https://doi.org/10.1007/978-94-007-7750-7_9

Shastry A, Durand M, Neal J et al (2020) Small-scale anthropogenic changes impact floodplain hydraulics: simulating the effects of fish canals on the logone floodplain. J Hydrol 588:125035. https://doi.org/10.1016/j.jhydrol.2020.125035

Sheldon F, Bunn SE, Hughes JM, Arthington AH, Balcombe SR, Fellows CS (2010) Dryland river waterholes: ecological roles and threats to aquatic refugia in arid landscapes. Marine Freshw Res 61:88–895. https://doi.org/10.1071/MF09239

Slagter B, Tsendbazar N-E, Vollrath A, Reiche J (2020) Mapping wetland characteristics using temporally dense Sentinel-1 and Sentinel-2 data: a case study in the St Lucia wetlands, South Africa. Int J Appl Earth Obs Geoinform 86:102009. https://doi.org/10.1016/j.jag.2019.102009

Silva TS, Costa MP, Melack JM (2009) Annual net primary production of macrophytes in the eastern amazon floodplain. Wetlands 29(2):747–758. https://doi.org/10.1672/08-107.1

Sponaugle S (2009) Daily otolith increments in the early stages of tropical fish. Information for assessment, management and ecology, tropical fish otoliths. https://doi.org/10.1007/978-1-4020-5775-5_4

Stewart-Koster B, Bunn SE (2016) The ecology of water security. Edward Elgar Publishing, Handbook on Water Security

Streipert S, Filar J, Robins J, Whybird O (2019) Stock assessment of the barramundi (lates calcarifer) fishery in Queensland, Australia. http://era.daf.qld.gov.au/id/eprint/7003/. Accessed 22nd 2021

Tanoue M, Hirabayashi Y, Ikeuchi H (2016) Global-scale river flood vulnerability in the last 50 years. Sci Rep 6(1):1–9. https://doi.org/10.1038/srep36021

Tedetti M, Charrière B, Bricaud A, Para J, Raimbault P, Sempere R (2010) Distribution of normalized water-leaving radiances at UV and visible wave bands in relation with chlorophyll a and colored detrital matter content in the southeast Pacific. J Geophys Res 115(C010):1–12. https://doi.org/10.1029/2009JC005289

Tellman B, Sullivan JA, Kuhn C et al (2021) Satellite imaging reveals increased proportion of population exposed to floods. Nature 596:80–86. https://doi.org/10.1038/s41586-021-03695-w

Thompson RM, Townsend CR (2005) Energy availability, spatial heterogeneity and ecosystem size predict food-web structure in streams. Oikos 108(1):137–148. https://doi.org/10.1111/j.0030-1299.2005.11600.x

Tian S, Tregoning P, Renzullo LJ et al (2017) Improved water balance component estimates through joint assimilation of GRACE water storage and SMOS soil moisture retrievals. Water Res Res 53(3):1820–1840. https://doi.org/10.1002/2016WR019641

Tickner D, Opperman JJ, Abell R, Acreman M et al (2020) Bending the curve of global freshwater biodiversity loss: an emergency recovery plan. BioScience 70(4):330–342. https://doi.org/10.1093/biosci/biaa002

Tonkin Z, Kitchingman A, Lyon J et al (2017) Flow magnitude and variability influence growth of two freshwater fish species in a large regulated floodplain river. Hydrobiologia 797(1):289–301. https://doi.org/10.1007/s10750-017-3192-5

Tockner K, Lorang MS, Stanford JA (2010) River flood plains are model ecosystems to test general hydrogeomorphic and ecological concepts. River Res Appl 26(1):76–86. https://doi.org/10.1002/rra.1328

Torres R, Snoeij P, Geudtner D, Bibby D et al (2012) Gmes sentinel-1 mission. Remote Sens Environ 120:9–24. https://doi.org/10.1016/j.rse.2011.05.028

Trees CC, Clark DK, Bidigare RR, Ondrusek ME, Mueller JL (2000) Accessory pigments versus chlorophyll a concentrations within the euphotic zone: a ubiquitous relationship. Limnol Oceanogr 45(5):1130–1143. https://doi.org/10.4319/lo.2000.45.5.1130

Tsyganskaya V, Martinis S, Marzahn P, Ludwig R (2018) Detection of temporary flooded vegetation using sentinel-1 time series data. Remote Sens 10(8). https://doi.org/10.3390/rs10081286

Tzortziou M, Herman JR, Gallegos CL, Neale PJ, Subramaniam A, Harding LW, Ahmad Z (2006) Bio-optics of the chesapeake bay from measurements and radiative transfer closure. Estuarine Coastal Shelf Sci 68(1):348–362. https://doi.org/10.1016/j.ecss.2006.02.016

Vance DJ, Staples DJ, Kerr JD (1985) Factors affecting year-to-year variation in the catch of banana prawns (Penaeus merguiensis) in the Gulf of Carpentaria Australia. ICES J Marine Sci 42(1):83–97. https://doi.org/10.1093/icesjms/42.1.83

Vis C, Hudon C, Carignan R, Gagnon P (2007) Spatial analysis of production by macrophytes, phytoplankton and epiphyton in a large river system under different water-level conditions. Ecosystems 10(2):293–310. https://doi.org/10.1007/s10021-007-9021-3

Visser F, Hill R (2011) Application of hyperspectral image data for species detection and biomass estimation of submerged macrophytes in UK chalk streams. In: Proceedings of the EARSeL 7th SIG-Imaging spectroscopy workshop, 11–13 Apr 2011

Voigt S, Kemper T, Riedlinger T, Kiefl R, Scholte K, Mehl H (2007) Satellite image analysis for disaster and crisis-management support. IEEE Trans Geosci Remote Sens 45(6):1520–1528. https://doi.org/10.1109/TGRS.2007.895830

Vorosmarty CJ, McIntyre PB, Gessner MO, Dudgeon D, Prusevich A et al (2010) Global threats to human water security and river biodiversity. Nature 467:555–561. https://doi.org/10.1038/nature09440

Waltham N, Burrows D, Butler B, Wallace J, Thomas C, James C, Brodie J (2013) Waterhole ecology in the Flinders and Gilbert catchments. A technical report to the Australian Government from the CSIRO Flinders and Gilbert Agricultural Resource Assessment, part of the North Queensland Irrigated Agriculture Strategy. CSIRO Water for a Healthy Country and Sustainable Agriculture flagships, Australia

Wang X, Xiao X, Zou Z, Hou L, Qin Y et al (2020) Mapping coastal wetlands of china using time series landsat images in 2018 and google earth engine. ISPRS J Photogramm Remote Sens 163:312–326. https://doi.org/10.1016/j.isprsjprs.2020.03.014

Ward DP, Hamilton SK, Jardine TD, Pettit NE, Tews EK, Olley JM, Bunn SE (2013) Assessing the seasonal dynamics of inundation, turbidity, and aquatic vegetation in the Australian wet-dry tropics using optical remote sensing. Ecohydrology 6(2):312–323. https://doi.org/10.1002/eco.1270

Ward DP, Pettit NE, Adame M, Douglas MM, Setterfield SA, Bunn SE (2016) Seasonal spatial dynamics of floodplain macrophyte and periphyton abundance in the Alligator Rivers region (Kakadu) of northern Australia. Ecohydrology 9(8):1675–1686. https://doi.org/10.1002/eco.1757

Ward P, Jongman B, Aerts J et al (2017) A global framework for future costs and benefits of river-flood protection in urban areas. Nat Clim Change 7(9):642–646. https://doi.org/10.1038/nclimate3350

Ward P, Jongman B, Salamon P et al (2015) Usefulness and limitations of global flood risk models. Nat Clim Change 5(8):712–715. https://doi.org/10.1038/nclimate2742

Warfe DM, Pettit NE, Davies PM, Pusey BJ, Hamilton SK et al (2011) The "wet-dry" in the wet-dry tropics drives river ecosystem structure and processes in northern Australia. Freshw Biol 56(11):2169–2195. https://doi.org/10.1111/j.1365-2427.2011.02660.x

Weise K, Höfer R, Franke J, Guelmam A et al (2020) Wetland extent tools for SDG 6.6.1 reporting from the satellite-based wetland observation service (SWOS). Remote Sens Environ 247:111892. https://doi.org/10.1016/j.rse.2020.111892

Whyte A, Ferentinos KP, Petropoulos GP (2018) A new synergistic approach for monitoring wetlands using sentinels −1 and 2 data with object-based machine learning algorithms. Environ Modell Softw 104:40–54. https://doi.org/10.1016/j.envsoft.2018.01.023

Wilks D (2011) Statistical methods in the atmospheric sciences, 3rd edn. Academic Press, USA

Williams MR, Filoso S, Longstaff BJ, Dennison WC (2010) Long-term trends of water quality and biotic metrics in chesapeake bay: 1986–2008. Estuaries Coasts 33(6):1279–1299. https://doi.org/10.1007/s12237-010-9333-y

Winemiller KO (2004) Floodplain river food webs: generalizations and implications for fisheries management. In: Proceedings of the second international symposium on the management of large rivers for fisheries, vol 2, pp 285–309. http://citeseerx.ist.psu.edu/viewdoc/download?doi=10.1.1.729.280&rep=rep1&type=pdfpage=293. Accessed 1st Aug 2021

Winsemius H, Aerts J, van Beek L et al (2016) Global drivers of future river flood risk. Nat Clim Change 6(4):381–385. https://doi.org/10.1038/nclimate2893

Worden J, de Beurs KM (2020) Surface water detection in the Caucasus. Int J Appl Earth Obs Geoinform 91:102159. https://doi.org/10.1016/j.jag.2020.102159

Wu Q, Lane CR, Li X, Zhao K et al (2019) Integrating LiDAR data and multi-temporal aerial imagery to map wetland inundation dynamics using google earth engine. Remote Sens Environ 228:1–13. https://doi.org/10.1016/j.rse.2019.04.015

WWF (2020) Living planet report 2020—bending the curve of biodiversity loss. Nat Clim Change 1–159

Xing L, Tang X, Wang H, Fan W, Wang G (2018) Monitoring monthly surface water dynamics of DongtingLake using Sentinel-1 data at 10 m. PeerJ

Yeh TY, Ke TY, Lin YL (2011) Algal growth control within natural water purification systems: macrophyte light shading effects. Water Air Soil Pollut 214(1):575–586. https://doi.org/10.1007/s11270-010-0447-4

Zhang C (2014) Combining hyperspectral and lidar data for vegetation mapping in the Florida everglades. Photogramm Eng Remote Sens 80(8):733–743. https://doi.org/10.14358/PERS.80.8.733

Zimba PV, Gitelson A (2006) Remote estimation of chlorophyll concentration in hyper-eutrophic aquatic systems: model tuning and accuracy optimization. Aquaculture 256(1):272–286. https://doi.org/10.1016/j.aquaculture.2006.02.038

Zhao D, Jiang H, Yang T, Cai Y, Xu D, An S (2012) Remote sensing of aquatic vegetation distribution in Taihu Lake using an improved classification tree with modified thresholds. J Environ Manag 95(1):98–107. https://doi.org/10.1016/j.jenvman.2011.10.007

Chapter 19
Remote Sensing of Tropical Floodplain Rivers

Abstract The seasonal inundation of floodplains facilitates lateral connectivity allowing organisms to move into previously inaccessible habitats for feeding, foraging and reproduction. This serves to increase ecosystem productivity of floodplain and instream ecosystems that help maintain biodiversity and productive freshwater fisheries. Beyond freshwater habitats, the importance of flow and floodplain inundation to coastal ecosystems and marine fisheries is well documented. For example, freshwater flows to estuaries and coastal zones deliver nutrients and alter salinity regime, increasing densities of prey species as well as facilitating the migration of target species of fisheries from freshwater and floodplain systems into the marine environment. As such, conserving the ecological integrity of rivers and floodplain systems is crucial to maintaining these ecosystem services (Ndehedehe et al. (2021) Ecol Indic).

19.1 Tropical Floodplains: Flows and Productivity

Many rivers in tropical systems across the world have very strong connections with their floodplains and floodplain wetlands (Pettit et al. 2017). Inter and intra-annual variations in precipitation patterns generate runoff and river flows that deliver sediments and nutrients from upstream and help to inundate the floodplain driving aquatic primary production that supports high densities of aquatic organisms (Keddy et al. 2009; Thorp and Delong 1994; Junk et al. 1989). Seasonal floodplain inundation in large river systems may come from upstream runoff, localized rainfall, or a combination of the two interacting with physical characteristics of the catchment such as geology and topography (Poff 2018). Organisms that persist in floodplain ecosystems have evolved to take advantage of this ecohydrology (e.g., Venarsky et al. 2020) and humans often derive considerable benefit from the ecosystem services that these systems provide (Schindler et al. 2014).

C. Ndehedehe, *Satellite Remote Sensing of Terrestrial Hydrology*,
https://doi.org/10.1007/978-3-030-99577-5_19

19.1.1 Flow Variability in Tropical Systems

Variations in seasonal freshwater flows are important, given their impacts on the migrations, growth, and survival of aquatic life and freshwater resources (Duggan et al. 2014; Drinkwater and Frank 1994). The impact of freshwater flow variations goes beyond aquatic life, as its influence on human physiology, especially through direct changes in observed temperature and salinity, or indirect changes through food accessibility or habitat mobility is well known. It is important to note that the necessity of water for sustaining life does not hinder its role as a destructive force. This is particularly so in the case of severe and extreme floods (Fig. 19.1). Humans and animals need water as a life-sustaining substance but also seek protection from the cascading impacts of extreme flood. As part of flood mitigation measures and strategy, as well as water management, many rivers have been heavily modulated to aid water managers to control flows to meet human needs, like agriculture, industry, and development of historical floodplains, even while attempting to eliminate usual drought and flood episodes (Gleick 2003). Many regulated streams retain little of their own original variability by trapping floods as opposed to transporting them downstream (Poff 2018). Moreover, the management of rivers is a common practice in this day and age (Nilsson et al. 2005; Vörösmarty et al. 2004), especially given the drive to meet numerous human needs, which include flood control resources and dependable and steady water supplies to shield populated areas. The flow variability required to meet ecological demands is severely constrained by the above factors (Postel and Richter 2012; Naiman et al. 2002). The deterioration of important services provided by rivers such as recreation and water purification is expected due to substantial modification of the sediments and organic materials present (Poff et al. 1997; Richter et al. 2003). Substantial modification of river flow systems distraught material transport patterns, plant mortality, resource availability, as well as the regional ecosystem composition. Remedies to these anthropogenic problems are usually difficult to implement or very expensive.

19.1.2 Flooding in Aquatic Ecosystem: Pros and Cons

Flooding is regarded as a substantial natural hazard that could result in disease, disruption of public services, as well as loss of lives, properties, and infrastructures. The illustration in Fig. 19.1 shows the extent of inundation from flooding in Brisbane, Queensland in January 2011. These extreme floods affected about 23,000 homes and businesses in Brisbane during the period. As was discussed in Chapter 18, floods result in substantial impacts that affect a large proportion of human population and the environment in general. Floods can also cause a major disturbance to aquatic ecosystems and their corresponding services (Fig. 19.2). An estimated increase in global flood risk as a result of water cycle intensification and land-use change is likely to intensify the frequency and severity of these impacts. The impact of flooding is also

Fig. 19.1 Impact of the 2010/2011 extreme flood in Queensland. *Source The Guardian*, November 2019

felt on people through the response of freshwater aquatic ecosystem services, whose effect is always too diminutive to notice and, especially difficult to estimate. Flooding constitutes a significant environmental hazard (Fig. 19.1), given its impact on navigation and human livelihood. For example, floods can result in dangerous landslides, loss of livestock and crops, interruption of normal drainage systems, progressive discharge of urban toxic products and industrial residue, spillage of animal waste and raw sewage, and nutrients into waterways (e.g., Ogden et al. 2011; Khan 2011; Hong et al. 2007). The effect of flood on people is usually viewed as negative, especially as a result of its dramatic effect on infrastructures and regional ecosystem. However, floods could have a broad range of positive effects on ecosystems (Fig. 19.2). There are several benefits to flooding such as groundwater and wetland recharge, improving fish production, establishing new wildlife habitats, fabricating floodplains, and regenerating soil fertility (Poff 2002). Having already established that there are positive and negative impacts of flooding on aquatic ecosystem (Fig. 19.2), it is expected that these ecosystem services should as well display a combination of positive and negative outcomes as a result of several flooding episodes (Terrado et al. 2014). However, it is still unclear how different magnitudes of flood can affect losses or gains in ecosystem services, or their effect on individual ecosystems (Fig. 19.2).

The impacts of flood events are predominant when dry low-lying areas become temporarily overwhelmed with water outside their usual curbs (Rojas et al. 2013). Flooding accounts for about 33% of total natural disasters and has more negative effects on people than other types of disasters (Sivakumar 2011). However, it is expected that flood-related events will worsen in the future, especially as a result of environmental change with flood risk escalating substantially. The magnitude of flood

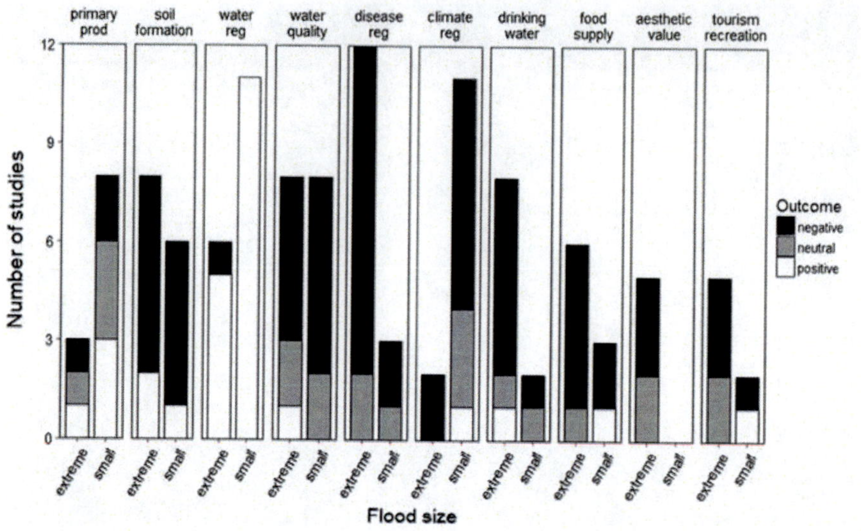

Fig. 19.2 Literature review of number of studies with positive, neutral, and negative results on ten aquatic ecosystem services monitored on small and extreme flood events. *Source* Talbot et al. (2018)

is also expected to rise as a result of the strengthened water cycling experienced from the least 1.50 C global average temperature increase (Alfieri et al. 2017). However, we must note that all floods are not equal, so their cause and consequences are also usually not unique. Periodic events of spring snowmelt or monsoon rains show the seasonality of floods. However, mechanisms like strong rainfall, storm surges, ice jams, etc. also show that they can also occur randomly. According to Douben (2006), strong or extreme rainfall accounts for an estimated 65% of total river floods. Nevertheless, regions located in the northern latitude characterized by snow cover are also susceptible to flooding caused by snowmelt and at times worsened by episodes of rainfall (Kundzewicz and et al. 2014). Further characterization of flood events based on frequency, duration, magnitude, and volume has been reported (Burn and Whitfield 2016).

The influence of floods on both humans and the aquatic ecosystem is relevant and also helps to understand its benefit to the entire ecosystem. For instance, during flood events, the volume of groundwater recharge and the extent of home and infrastructure damage can be determined using flood magnitude index. The flood magnitude index on its own is discretely one aspect of flood prediction impacts on aquatic ecosystems and ecosystem services. Ecosystem specifications before the advent of flooding events are important just as flood characteristics for assessing the ecosystem response to a flood event. Floods also contribute to the creation of unique habitats and the support of biological biodiversity and productivity, given their influences on rivers. The flood pulse concept asserts that estimable seasonal floods are valuable for riverine frameworks and have the ability to affect nutrient movement, biotic constitution, as well as sediment flow, but inestimable floods have the tendency to interrupt aquatic organisms (Junk et al. 1989). Also, various aquatic ecosystems have diminished resilience to future extreme flooding episodes, especially as a result of anthropogenic influences that include farming on floodplains, disruption of river flow, pollution, and urban expansion (Woodward et al. 2016). Some of these listed activities increase the possibility of catastrophic events to be expected from these flood events. The discrete influences of flooding on the aquatic ecosystem and their corresponding services are still complex at large, however, its importance in the balancing of ecosystem interactions in rivers has been reported (Peters et al. 2016). Most of the studies undertaken to improve understanding of flooding events take advantage of its accidental occurrence and are often less descriptive about pre-flood reference data (Poff and Zimmerman 2010). The key driver of the amount of environmental flow management and flood-related research can be attributed to the relatively sparse evidence obtained on how flooding and changes in hydrology influence the regional ecosystem (Acreman et al. 2014).

19.1.3 Flows and Connectivity

The significance of flow regimes for the maintenance of ecological interactions observed in floodplain rivers has been widely studied (Warfe et al. 2011; Hoeinghaus et al. 2007; Winemiller 2004; Junk and Wantzen 2004). Given the implications of fluvial interactions and hydrological connections on land and river capes, it is possible to identify general patterns of riverine ecosystem functioning using the layering technique (Ward et al. 2011). Furthermore, studies have highlighted the relevance of flow prediction as a determining factor of flood events, thereby improving heterogeneity and predictability, or on the other hand, as a disturbing force changing physical patterns as well as launching new recovery trajectories (Jardine and Bond 2015; Naiman et al. 2008). The conflict of flows acting to replace production or as a force of interruption is depicted in floodplain and food web features, and also through spatio-temporal connectivity. The influence of seasonal accumulation of floodplains on river productivity and resilience has been highlighted in several studies (Junk and Wantzen 2004; McClain 2002; Junk et al. 1989). Also, there is an existing transmission from seasonally accumulated floodplains of organic and nutrient matter in the form of biogeochemical substances aimed at sustaining food web system manifested as basic, secondary, and tertiary yields (Lewis et al. 2001; Tockner et al. 2000).

Food webs are expressions of riverine productivity and physical linkage among catchments, rivers, and floodplains through time, representing energy and nutrient exchange pathways among various species (Lewis et al. 2001; McCann and Rooney 2009). They transport assimilation of ecosystem interactions and biological community configuration (McCann and Rooney 2009), creating a platform to understand the biogeochemical interactions and energy transfer within aquatic frameworks (Naiman et al. 2002), as well as appearing to be driven by flow regime patterns. Undisturbed tropical riverine landscapes in the world where rivers are depicted by strong seasonality in flow regimes and high inter-annual variability in flood magnitude, duration, frequency, and timing are relatively few. For example, rivers in northern Australia (one of these few locations) are particularly helpful given their resilience to widely variable environmental disturbances. They are characterized by limited flow control, minimal human population densities, and little development in catchments (Stein et al. 2002). Cyclonic events such as the El Nino Southern Oscillation (ENSO), Indian Ocean Dipole (IOD) climatic variation, Inter-decadal Pacific Oscillation (IPO) contribute to the intensity of the inter-annual flood variability (Warfe et al. 2011; Kennard et al. 2010; Hamilton and Gehrke 2005). The fluvial attributes of the Australian tropical rivers, which include disconnection and flooding, drive patterns of linkage as well as the diverse, heterogeneous, resilient, and productive processes prevalent in the biota and biogeochemical fluxes (Ndehedehe et al. 2021; Pettit et al. 2017; Ward et al. 2016). In many river systems across the globe, floodplains are not usually deluged across their full extent each year. In other words, floodplains and waterholes could have a low connection in subsequent years (Ward et al. 2014); consumers become more dependent on locally produced resources (Jardine et al. 2012), and aquatic communities become contradictory (Warfe et al. 2011; Leigh and Sheldon 2008).

19.1.4 Importance of Flows to Marine Fisheries

The increasing threats such as water abstraction for the sole purpose of maintaining irrigated agriculture pose a significant risk to global riverine biota (Januchowski-Hartley et al. 2011). This is specifically so for extreme seasonal rivers draining catchments characterized by savannah-type ecosystems. However, such systems are predominant in densely populated regions under poor ecological conditions (McClain 2013; Woinarski et al. 2007). Regions such as the savannah ecosystems of northern Australia, which fall into this category, deviate from basic conditions. This is largely because they are sparsely populated, and considered to be in optimal ecological conditions with hydrological regimes remaining relatively competitive (Pusey et al. 2011; Woinarski et al. 2007). However, increased pressure on rivers and the biota they sustain is expected, especially as a result of increasing water demand, as pointed out by the Joint Select Committee on northern Australia in 2014. A basic example of the potential future for the extensive regions is the Daly river of the northern territory (Doughlas et al. 2011).

Contributing factors to irrigated agriculture in this catchment therefore will be a product of groundwater resources. The depletion of which may change the timing and magnitude of dry season flows and as well pressurize the river's values, consisting of its substantial subsidence and fishing activities (Jackson et al. 2014). A potential threat to the river's rich freshwater fish fauna is the altered flow regimes, as meaningful hydro-ecological connections have been reported for some species in riverine systems (Keller et al. 2019; Turschwell et al. 2019; Jardine et al. 2012; Stewart-Koster et al. 2011; Bayliss et al. 2008; Chan et al. 2012). These past studies show a strong correlation between different facets of flow regime and the copiousness of various fish species, as well as the potential future alterations in river hydrology. Little data, however, is available that on discussions about the flow-related ecology of various fish species in the river—a snag that challenges modelling flow-ecology interactions more specifically (Horne et al. 2019). Moreover, information about species distributions and habitat requirements, a distinct assemblage, or functional group status is also scarce. This generally forms an important antecedent in the anticipation of potential impacts of flow changes (Olden 2016).

The importance of flows to aquatic ecosystems cannot be over emphasized. Spatio-temporal fluctuations in freshwater flows have substantial impacts on fish and aquatic species in coastal, freshwater, or estuarine habitats, especially in regions like northern Australia (Robins et al. 2005; Vance et al. 1985). The impacts of flow variation across multiple temporal scales on the distribution and abundance of three fish species in the Daly river, the northern territory of Australia, were highlighted by Stewart-Koster et al. (2011). Their findings emphasized the importance of flow variability as a key hydrologic metric that influences the abundance of fish species among several other factors that were observed. Moreover, just like the banana prawn, several other marine species have a similar response to fluctuations in freshwater flows by modifying their positions to avoid critical conditions like salinity changes (e.g., Robins et al. 2005). Other species, on the other hand, may likely take advantage of this variation in conditions to search for prey or even reproduce (Grimes and Kingsford 1996). This performance is replicated in a species distribution in both time and space and is usually hampered by their physiological modification to environmental variations during the period of their existence (Taylor and Loneragan 2019; Vance et al. 1985). Species that live out the early parts of their lives in freshwater or estuarine habitats are specifically susceptible to variations in the magnitude and timing of freshwater flows (Vance et al. 1985). Changes in seasonal patterns of freshwater flows influence the growth, survival, conversion, and migration of fisheries of several commercially significant marine invertebrates (Duggan et al. 2014; Drinkwater and Frank 1994). Fluctuations in freshwater flows also have the capacity of influencing their physiology either indirectly through changes in habitat or food accessibility, or directly through changes in temperature and salinity. Despite this situation, flow variations are expected to influence the yield of fisheries and general aquatic life (Drinkwater and Frank 1994).

19.1.5 Water Resources Development and Floodplain Wetlands

Changes in ecosystem productivity are largely dependent on the interactions between the terrestrial and marine ecosystems. A leading cause of habitat destruction and biodiversity loss comes from the interruptions of terrestrial freshwater pathways, which is also predominant in the coastal marine ecosystems (Oczkowski et al.

2009; Vörösmarty and Sahagian 2000). However, due to the nature of environmental changes on ecological activities, the task of distinguishing between natural and human sources is always complicated (Brown 2019). These complications, however, do not hinder the assertions that accounting for these environmental variations could enhance aquatic life, as well as contribute to a better management of freshwater products (Szuwalski et al. 2015). The lack of assimilated water management over terrestrial-marine borders has contributed to the decline and collapse of freshwater resources and aquatic life, especially from dam constructions and water diversion projects (Oczkowski et al. 2009). Investigations are currently ongoing by the Australian government towards understanding the potential impacts of water resource development in northern Australia (Petheram et al. 2018). Some of these studies have examined the productivity of floodplain wetlands focusing on understanding the large-scale synergistic impacts of stressors such as climate variability and the potential impacts of water resources development on floodplains aquatic life (Ndehedehe et al. 2021, 2020c; Broadley et al. 2020).

The wet–dry tropical rivers such as those in the northern Australia have very high variable flows and lack distinctive seasonal patterns of flow as noticed in most tropical rivers (Warfe et al. 2011). These tropical floodplains like those of the Gilbert and Flinders catchment in the Gulf region are increasingly generally being affected by climate variability and are evidenced in complete floodplain flooding during the wet season (Ndehedehe et al. 2020c). When the wet season is over and the dry season begins, these floods recede and revert back to isolated but largely deep. As with these catchments, other areas in the wet–dry tropics have experience a short period of river flow rates, especially during the wet seasons, and are often accompanied by long periods of low flow or even complete stoppage during the dry seasons (Warfe et al. 2011). These dynamics can impact the seasonal and inter-annual variability in periphyton production (Ward et al. 2016). But in the advent of water infrastructure development, such impacts are likely to be exacerbated. The intense climatic fluctuations observed in the sub-region raise concerns about the impacts of water extraction in low flow years on marine fisheries. Results from some studies highlighted and emphasized the need to conserve the ecological integrity of rivers and floodplain systems given their roles in facilitating migration of fisheries into the marine environment (Broadley et al. 2020; Staunton-Smith et al. 2004).

In these floodplains (tropical northern Australia) and freshwater habitats, the movement of fish can be varied with changes in connectivity, and dam location can influence the nature of impact on fish movement (O'Mara et al. 2021). The construction of the Ord river dam in northern Australia resulted in reduced banana prawns due to the observed alteration of the river's flow regime from wet–dry to a perennial flow structure (Warfe et al. 2011). The study undertaken by O'Mara et al. (2021) showed how flow-mediated connectivity drives changes in fish community movement in an unregulated river system. Additionally, model predictions show significant reductions in annual Barramundi growth rates under three hypothetical water development scenarios (Leahy 2021). Yet, contrasting results on the factors that impact fish movements in an unregulated river have also emerged as shown by Luke et al. (2021), arguing on the frequency of fish movement in perennial, highly connected rivers and those of regulated or intermittent systems. As decreased flows due to water resource development could impact food web energetics (Venarsky et al. 2020), it is therefore important to continue assessing the scale of impacts of proposed water resource development schemes on river flows, as well as their influences on the catch and productivity of aquatic life.

19.2 Floodplain Productivity

Floodplains have been described as one of the most recognized bio diverse ecosystems on earth, maintaining high rates of basic production (Keddy et al. 2009). Despite their importance and impact on stabilizing global and regional ecosystems, much of the world's wetlands have been lost, destroyed, or heavily altered through increasing anthropogenic influences or human activities (Reis et al. 2017; Davidson 2014). Analogous to the growing concerns of population and economic trends, the pressure on floodplains is equally increasing (Keddy et al. 2009), especially in the tropics where expansions and developments are improving rapidly (Roggeri 2013). The peril and threats to wetlands have the potential to keep increasing until there is a generic acceptance of the ecological evaluations and fundamental ecosystem services they provide (Reis et al. 2017). The strong aquatic production on floodplains presents a key subsidy for organisms of higher tropical rank (Saigo et al. 2015; Bunn et al. 2003; Lewis-Beck et al. 2003). Organisms such as algae play very important roles, especially in aquatic food webs. Chemical tracer studies using stable isotopes and fatty acids have expressed the relevance of algae as a basic basal food supply for wetland food webs and rivers across a wide variety of ecospheres. For instance, a large number of standing biomass for fish used to sustain livelihood in many tropical regions is usually derived from floodplain algae (Jardine et al. 2013). Studies have shown light to be the overarching factor regulating aquatic primary biomass and productivity, especially in wetlands (Jia et al. 2020; Vadeboncoeur et al. 2014; Wetzel 2001). However, its availability is susceptible to a wide range of physiochemical factors which include dissolved organic carbon, turbidity, and biological characteristics like shading from macrophyte cover (Pettit et al. 2016; Yeh et al. 2011; Kirk 1994).

Floodplains in wet–dry semi-arid regions are exclusive ecosystem because they occur in two alternating segments under inherent specifications. They experience serious inundations, especially during the wet periods, which is not substantially different from other surrounding terrestrial ecosystems. But these inundations recede much faster due to high evaporation rates in the dry season. However, the recharge of existing shallow groundwater aquifers can be enhanced by billabongs and lakes and shallow floods driven by sporadic rainfall that improves soil moisture conditions. This recharge provides the reservoir with water, which the biota depends on till the next flood episode (Fig. 19.3). The flows and flood episodes of semi-arid rivers are characterized by strong fluctuation patterns and unpredictability (Bunn et al. 2006). Depending on the location of these terrestrial ecosystems, the overwhelming changes in flow patterns and phase may last for several months.

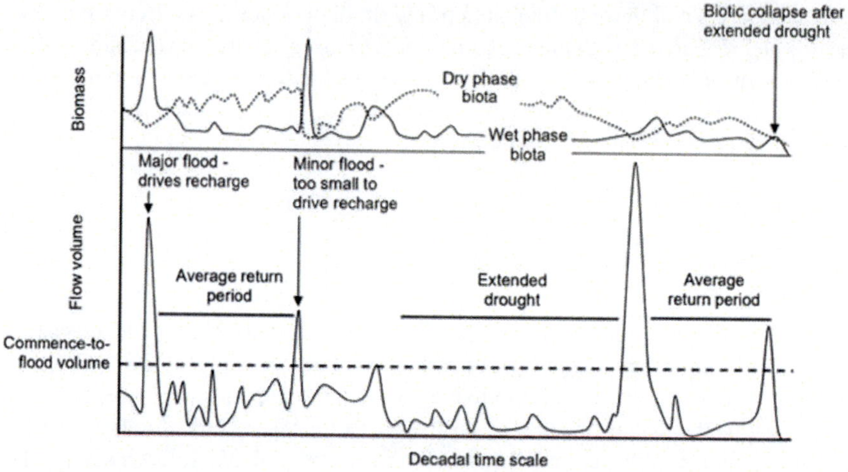

Fig. 19.3 Model depicting dry and wet phases and biomass reaction to flood and increased drought on floodplains and in wetlands in semi-arid regions. *Source* Colloff and Baldwin (2010)

Typically, the overwhelming changes in terrestrial ecosystem which is also depicted by the water loss in an aquatic ecosystem is considered a cause for serious concern for ecosystem health. However, we see a high degree of adaptation displayed by these systems in an attempt to cope with phase alterations. For instance, in the study of aquatic plants, some of these plants are well adapted and have the capacity of withstanding long drought regimes. The root depth of plants in semi-arid regions is likely to extend to the water table region, thus enhancing a complex water use mechanism during the dry season (Ndehedehe et al. 2019). But this same capability can also be seen in many higher order consumers and species. Studies have shown them possessing drought-resistant capabilities in their general life-cycle or an adaptable physiological capacity (Brock et al. 2003). For example, this same characteristic is shown among the water birds, which adapt highly dynamic techniques to monitor scarce resources. Nevertheless, studies have shown several components of the global floodplain ecosystem in some regions that are in constantly deteriorating conditions, such as the semi-arid region in south-eastern Australia (Armstrong et al. 2009). The major cause of this deterioration can be attributed partly to water diversion projects and river regulations dominant in the region (Kingsford 2000), which led to diminished flood regimes, frequency, and coverage, as well as changes in the seasonality of flooding events. Substantial swings like these have the tendency of compromising the pliability of floodplains, especially during seasons of prolonged droughts, as has been recorded in southern Australia during the past decade (Australian Bureau of Meteorology).

Under the wet- and dry-adapted conditions, species become locally extinct, which is observed directly from the reduction of their population to ranks that concede their capacity to drive ecosystem functioning and maintenance. The wetland biota fails to regenerate thereby depauperating temporary wetlands, especially in cases of successive flood events (Jenkins And Boulton 2007). Its recovery is possible only in the case of severe flood events as depicted in Fig. 19.3. This also affects the permanent and dry-phase biota, distinctively through under-storey vegetation and large-scale tree loss, as well as deficiency of soil and sediment microbial variety. In cases where the water table falls below the root zone of the under-storey and major trees, then those organisms that depend on groundwater access are likely to be replaced by organisms that can withstand episodic precipitation patterns. If lack of soil and groundwater recharge continues, the structure is expected to undergo a shift to a novel substitute in a constant state—perpetual rangeland, comprising semi-arid grasses and shrubs, including aggressive native plants where soil moisture conditions match their corresponding hydrological niche more closely.

19.2.1 Monitoring Surface Water on Floodplains

Surface water assessment is an operational requirement for understanding hydrological and ecological processes. The ongoing advances in satellite-based optical remote sensing have advanced the field of surface water monitoring in this era. Surface water encompasses all the water on the surface of the earth, like oceans, wetlands, lakes, and rivers. We might exclude the ocean in this definition given its enormous size and saltiness, even though smaller saline water bodies are regularly included. Surface water resources are critical floodplain resources, for both ecological and human benefits. They are of utmost importance in maintaining various life forms (Karpatne et al. 2016). The contribution of water to the preservation of biodiversity in riparian or wetland ecosystems is well founded in its capability to provide habitats to a surfeit of fauna and flora (Vorosmarty et al. 2010).

Water is not only indispensable to optimal ecosystem functioning, especially given its key contribution to the hydrologic cycle; it also helps provide important services to mankind, such as transportation, electricity generation, industrial purposes, agriculture, and drinking water (Vörösmarty et al. 2000). The dynamic nature of surface water bodies is well captured in their ability to change, shrink, or alter their appearance with respect to time, climate (natural conditions) and anthropogenic influences (Karpatne et al. 2016). Apart from climate variability, human behaviour indeed contributes to the changing characteristic of surface water patterns, leading to its redistribution in time and space. From water diversion projects and transfers in North Plain China and Iran to man-made lakes in the Volta basin and several other water resources development on small and large scales have resulted in water losses in one area (donor communities) and water gain in the recipient communities. Changes in water bodies influence other natural and human assets, and further impact the environment in several ways. Alterations in surface water volumes are usually followed by catastrophic effects on the environment. In intense cases, precipitous escalations of surface water can result in flooding. This, therefore, intensifies the importance of assessing its dynamics, as well as quantifying its extent.

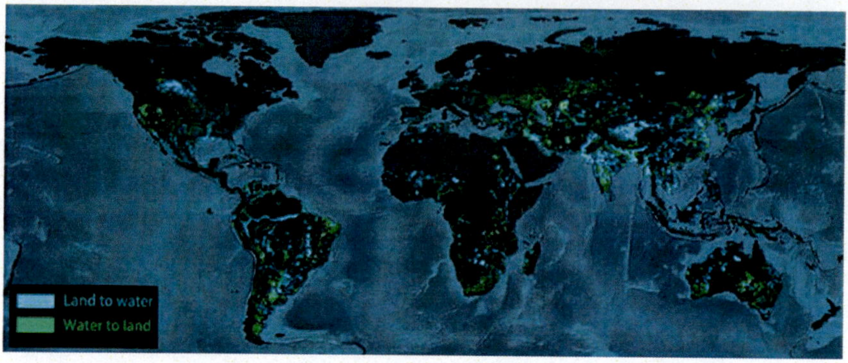

Fig. 19.4 Landsat generated images depicting heat map of global surface water and land variations. *Source* Donchyts et al. (2016)

Remote sensing techniques offer a wide variety of procedures to efficiently monitor surface water dynamics. Compared to the conventional in situ measurements, remote sensing techniques appear to be more efficient and proficient, due to their capability to monitor the Earth surface at numerous spatial (small, medium, and large) and temporal (daily, weekly, bimonthly, and monthly) scales. Remote sensing observations provide spatially explicit and temporally recurrent datasets of a wide range of physical characteristics about the Earth's surface that can be adequately controlled to map the coverage of water bodies at regional and global scales and assess their dynamics at consistent and frequent time intervals.

For instance, Fig. 19.4 shows a heat map representation of global surface water and land variations from 1985 to 2015 using Landsat images. The blue lighting depicts where land was translated to water, whereas the green lighting depicts where water was translated to land during the study period (Fig. 19.4). The spatial magnitude of variations shown in Fig. 19.4 is illustrated in the colour intensity. Information regarding surface water dynamics is needed to assist in assessment and reporting programmes associated with water management schemes. It also helps in maintaining scientific intents such as understanding the space–time inconsistency in water stored at or near the land surface (Lettenmaier and Famiglietti 2006).

However, surface water storage data are insufficient and often unavailable in many regions of the world, especially as a result of geographic isolation and/or compact data policies including the cost associated with preserving broad water monitoring programs. This case is similar to areas with scanty populations and developing countries, which tend to constrain our ability to understand the surface water balance at the global scale and therefore its influence on water planning, ecosystem sustainability, global weather prediction, and earth system simulation (Gao 2015). The suitability of satellite-based remote sensing platforms as an ideal tool for quantitative capturing and monitoring of conditions over large areas at a given point in time over long periods is well hinged on its synoptic nature (Crétaux et al. 2016; Lettenmaier et al. 2015). With the contemporary availability of open and free access satellite remote

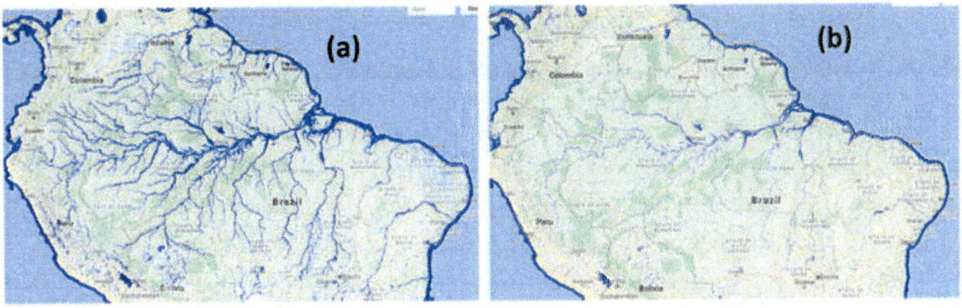

Fig. 19.5 Satellite-derived global surface water showing the spatial distribution of surface water in the Amazon basin area (1984–2015). **a** Annual distribution of surface water occurrence and **b** monthly water recurrence. The blue pixels represent 100% water availability

sensing products, users now have a wide variety of high-quality images. Also, the recent introduction of cloud computing-based platforms such as the Google Earth Engine platforms, as was discussed in Chap. 5, has created room for analysis-ready imagery at various spatial, spectral, and temporal resolutions relevant scales, particularly whilst depicting information about variations in water surface areas and elevations, and eventually storage.

As a result, the hydrology community has shown considerable interest in recent years in developing approaches to aid the improvement of global surface water assessment techniques (McCabe et al. 2017). Analysis of global water dynamic that previously would have only been achievable with relatively low spatial resolution datasets or gravimetric remote sensing such as GRACE (Humphrey et al. 2016) is now possible using moderate-resolution imagery like Landsat or the relatively high spatial resolution Sentinel-2 data. For instance, the quantification of changes in global surface water under the auspices of the European Commission Joint Research Centre's Global Surface Water project has demonstrated such potential (Pekel et al. 2016). With more than four million Landsat images (Landsat 5, 7, and 8) acquired over a 32 year period from the US geological Landsat archives, the study undertaken by Pekel et al. (2016) quantified changes in global surface water, detecting the location and persistence of surface water around the globe, including their change in time and space. The surface water data generated in their study also provided several statistics e.g., recurrence, intensity, etc. on the extent and change of open water features. As illustrated over portions of the Amazon basin, Fig. 19.5a shows the year-to-year classification of water presence based on the occurrence values detected throughout the year. Figure 19.5b shows the monthly recurrence collection, which contains 12 images. That is, the monthly presence of water based on the occurrence values detected in that month across the 32 year period (all years) (Pekel et al. 2016). The Amazon basin floodplain corridor (central part) clearly shows strong water presence all year round compared with other areas like north-east Brazil (Figs. 19.5a-b).

Additionally, the ability to monitor water levels over the open ocean and considerably large inland water bodies like the Aral Sea, the Tibetan Plateau, African Great Rift Valley lakes, and the Lake Chad at sub-regional scales is now possible leveraging satellite altimetry missions (Cai et al. 2016; Lee et al. 2011; Aladin et al. 2005; Coe and Birkett 2004). Other agencies have made substantial attempts to globally monitor surface height of large lakes and reservoirs, for example, University of Stuttgart's HydroSat, U.S. Department of Agriculture (USDA) global reservoirs and lakes monitor datasets, the database for hydrological time series of inland waters (Schwatke et al. 2015), and French space agency laboratoire d'Etudes en Geophysique et Oceanographie spatiales hydroweb database (LEGOS, Crétaux et al. 2011). Despite these efforts, global surface water storage estimation is still lacking while still having a lot unexplored (Gao et al. 2012). We expect that the new NASA mission (Surface Water and Ocean Topography, SWOT) scheduled for launch in 2021 will fill this gap in global measurement. SWOT (Chap. 1) is anticipated to provide precise estimates of water surface elevation and area on average every ten and half days, especially depending on the featured position. It is also expected to provide total estimates on surface water storage fluctuations for relatively small reservoirs and lakes with a size range of 1–2 km^2, even at a 0 height precision (Biancamaria et al. 2015).

However, until the SWOT mission becomes fully operational, the occurrence of remotely sensed long-term hydrological details for the assessment of fluctuations and changes in the terrestrial water cycle will need precise data synchronization and homogenization from existing platforms, with some techniques playing important role in achieving coherent and consistent products (McCabe et al. 2017). With the ongoing creation of high-resolution altimeters which also have nadir-pointing capabilities, long-term observations can be acquired by integrating datasets from a constellation of sensors with a range of overlapped data records. For instance, Crétaux et al. (2011) identified the potential of the Jason-2, Jason-3, France-India SARAL AltiKa mission, and the European Space Agency's Sentinel-3 constellations to provide a typical depiction of water surface elevation for almost all the lakes in the world having areas greater than 50 km^2, and encompassing 71% of the total lakes with areas greater than 10 km^2, for an approximately 40% of the global water storage of lakes on earth. The location and persistence of surface water (coastal and inland) are both influenced by natural and anthropogenic events, affecting biological variety, climate, and human welfare (WEF 2016; Gardner et al. 2015). Inventories and national descriptions, satellite imageries (Feng et al. 2016; Verpoorter et al. 2014), and statistical extrapolations of regional data have produced global datasets aimed at authenticating surface water locations (Pekel et al. 2016).

Surface water is only a component of terrestrial stored water or freshwater resources. However, it is regarded as the most available to the entire human populace, providing a wide array of ecosystem services (Postel et al. 1996). Approximately 52% of the global total surface water is predominant in the North American region, which is home to less than 5% population as recorded in 2015 (Postel et al. 1996). The continent also holds an estimated 18% of the contemporary seasonal water as highlighted in some studies. The global surface water generated by Pekel et al. (2016) provided valuable statistics on water transition across several continents and regions of the world. For instance, in North America, an increase of 17,000 km^2 in its total surface water between 1984 and 2015 was experienced. In comparison to

Asia with an estimated human population of 60%, only 9% of the total surface water appears to persist during the period while observed contemporary seasonal water is 35%. The rise in permanent surface water in Asia resulted in a gain of 71,000 km^2 and is estimated to be a 23% increment for the entire sub-region. Latin America and Africa have nearly the same proportion of the world's total surface water at an estimated 9%, irrespective of their heterogenous population, with Africa (16%) maintaining nearly double the human population of Latin America (8.6%). They further noted regarding Europe that with the inclusion of Russia they have a rather smaller population (10%) from the global population index, yet containing 22% of the total surface water and 18% of the contemporary seasonal water. A net loss of surface water was observed in Oceania despite its relatively small size of 229 km^2.

19.3 Impacts of Climate Change on Tropical Rivers

Changes in tropical rivers have been identified as one of the most significant global change concerns (Palmer et al. 2008). The vulnerability of water resources has been highlighted in many regions facing increased scarcity (Vörösmarty et al. 2000), and has the tendency of leading to serious conflicts as seen in the Syrian civil war. This conflict was the result of poor governance and over dependence on rigorous irrigation, which minimized the country's ability to cope with severe drought (Müller et al. 2016). Tropical rivers are facing a significant threat due to the synergistic impacts of multi-environmental stressors, like farming extension, pollution, and fluctuations in land use/cover (Dafforn et al. 2015; Palmer et al. 2008). Impacts of climate change and multi-year droughts are substantially evidenced in water shortages and global change scenarios.

The possibility of climate change significantly impacting the ecological chain, as well as altering the elements of hydrologic regimes in wetlands and rivers has been identified through its exacerbating influences on human water use in developed river basins (Aldous et al. 2011; Xenopoulos 2005). It is estimated that by 2050, climate change will have a significant impact on river flow characteristics, like seasonality, long-term mean discharge, and far-exceeding anthropogenic impacts like water withdrawals and dam constructions in the last few years (Fig. 19.6).

19.3.1 Improving Model Simulations to Capture Changes in Floodplain Wetlands

To improve the efficiency and faithfulness of land surface models to enable their reproduction of fluxes for general circulation models, studies have found the representation of wetland dynamics in such models considerably crucial (Getirana et al. 2021; Schrapffer et al. 2020; Penatti et al. 2015). Floodplain modelling in humid

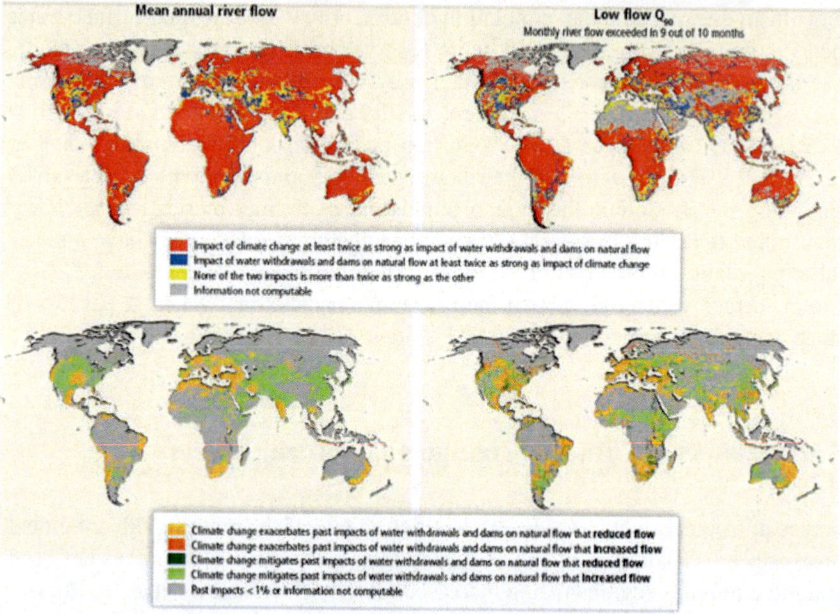

Fig. 19.6 The relative impact of climate change on water withdrawals and dams on natural course for two ecologically significant river flow characteristics (average annual river flow and monthly low flow Q90), estimated by a global water model (Döll and Zhang 2010). The estimation of the water withdrawals and reservoir impacts is done by processing the model with and without water withdrawals and dam that occurred in 2002. *Source* Döll and Bunn (2014)

regions such as the Amazon basin, coastal West Africa and Central Africa that have large floodplains characterized by considerable exchange of fluxes within their floodplain corridors can benefit from this improvement. The hydrological changes that occur in these floodplains are driven by changes in global climate through substantial atmospheric moisture contribution to local precipitation (Ndehedehe and Ferreira 2020; Sorí et al. 2017; Alsdorf et al. 2010). Despite the substantial contribution of floodplains rivers to changes in global land water storage, their representation in land surface models is poor and insufficiently captured. Representing these processes in globally recognized tropical floodplains is thus crucial to aid understanding that underpins the modelling of hydrological processes, and land surface interaction with the ocean and atmosphere. This has been the focus of new research directions and the need to integrate fluxes from tropical floodplains in land surface models has recently been advocated. For example, the importance of representing large tropical floodplains like Pantanal in model simulations to improve their capacity in reproducing fluxes and land surface conditions was articulated by Schrapffer et al. (2020). The Pantanal and other large tropical floodplains in South America have extensive surface water area and floodplain rivers that are largely driven by inter-annual changes in annual rainfall and exchange of fluxes within floodplain corridor (Ndehedehe and

Ferreira 2020; Penatti et al. 2015; Alsdorf et al. 2010). The representation of these processes in land surface models will improve the quality of global hydrological models that seek to understand the impacts of climate variability on surface water hydrology. A coupled land surface model that considers surface water availability, evapotranspiration, and hydrological stores in globally significant floodplain wetlands is thus crucial. This is particularly important in humid regions where the characteristics of rainfall, hydrological stores (e.g., soil moisture), and outgoing fluxes like evapotranspiration regulate global hydrology and changes in continental water storage.

To address these challenges, new land surface models (LSM) have been integrated with satellite data to capture the impact of surface water availability and its changes on floodplain in LSM. In particular, 2-way and 1-way coupled land surface-flood modelling experiments (i.e., 2-WAY–1-WAY techniques) have been performed and fully described by Getirana et al. (2021). Before discussing the outcome of some aspects of these experiments, we first illustrate the potential of satellite data in quantifying flood frequency of large floodplain wetlands in wet–dry tropics. Figure 19.7 shows the flood frequency map over the inner Niger delta area in West Africa based on 250 m spatial resolution MODIS observations between 2002 and 2018. The availability of other satellite missions (like those indicated with white circles in Fig. 19.7), which indicate locations where radar and laser altimeter orbits transect the Niger River (I1-I8 and H1-H4 represent ICESat and Hydrosat datasets, respectively), makes it possible to improve the understanding of the ecohydrology of such regions. The response of streamflow (red triangle in Fig. 19.7) and surface water hydrology to changes in evapotranspiration and human interference of flow (including inflow-Black triangle in Fig. 19.7), for instance, can be assessed by combining these data.

Moreover, the study Getirana et al. (2021) detailed how two-way coupling of LSM and river routing schemes impact the representation of hydrological processes in large wetlands over the inner Niger Delta region. Apparently, accounting for two-way interactions between LSM and river routing schemes improved model simulations of hydrological fluxes including floodplain inundation (Fig. 19.8). An averaged inundated area of 4,690 km^2 was estimated using the MODIS data over 2002–2018 period and showed peaks occurring between September and November (Fig. 19.8a). This peak coincides when maximum discharge is observed in the region, as shown in the inset in Fig. 19.7. Comparing the 1-WAY and 2-WAY models, estimated inundation with MODIS data shows that the 1-WAY approach overestimated water extent with approximately 61% above what was quantified from MODIS estimates (Fig. 19.8b–d). The relative bias is higher in the 1-WAY unlike the 2-WAY technique (Fig. 19.8e, f) and simulated monthly climatology of surface water extents derived from 2-WAY is considerably associated with MODIS ($r = 0.83$) compared to the 1-WAY ($r = 0.74$) approach (Fig. 19.8d). However, the coarse spatial resolution of MODIS data can lead to overestimation or underestimation in water pixels derived from MODIS.

In areas with smaller rivers, MODIS water-related pixels of smaller rivers can be excluded from the classified inundation or a misclassification can occur due to the coarse spatial resolution of MODIS (500 m). Additionally, limitations in the sensor and classification algorithm can also increase uncertainties in the classified inunda-

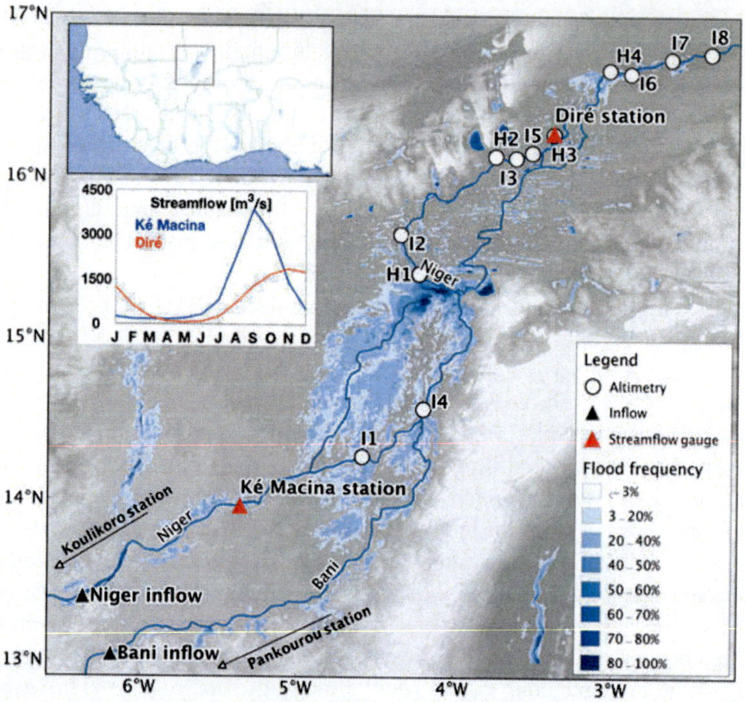

Fig. 19.7 Illustration of flood frequency maps based on MODIS satellite data. *Source* Getirana et al. (2021)

tion based on MODIS data. In wet–dry tropics, these limitations could be amplified during the dry seasons when most rivers contract and the large surface water bodies in wet seasons recede into isolated and distinct water bodies that remain undetected (Getirana et al. 2021; Ndehedehe et al. 2020a). In such cases, estimated floodplain inundation extents can either be underestimated when moderate resolution images like those of Landsat and Sentinel-2 missions are used or overestimated (caused by misclassification) when low spatial resolution data like MODIS are employed. However, misclassifications of MODIS-derived or satellite flood water extent can be caused by the spectral properties of mudflats and water quality on floodplain wetlands (Getirana et al. 2021; Ndehedehe et al. 2020a, c). The average annual wet season water extent in Fig. 19.8g shows that outputs from the 2-WAY scheme are closer to MODIS observed inundation extent. The representation of hydrological fluxes (e.g., inundation or surface water, evapotranspiration) and hydrological state variables like soil moisture was significantly better with the 2-WAY scheme (Getirana et al. 2021). However, the use of moderate or even high spatial resolution satellite data in quantifying inundation can further improve the performance of this scheme. This can be in terms of the amplitudes of actual water extent estimated, for example, during dry season when uncertainties in predicted inundation could be relatively higher.

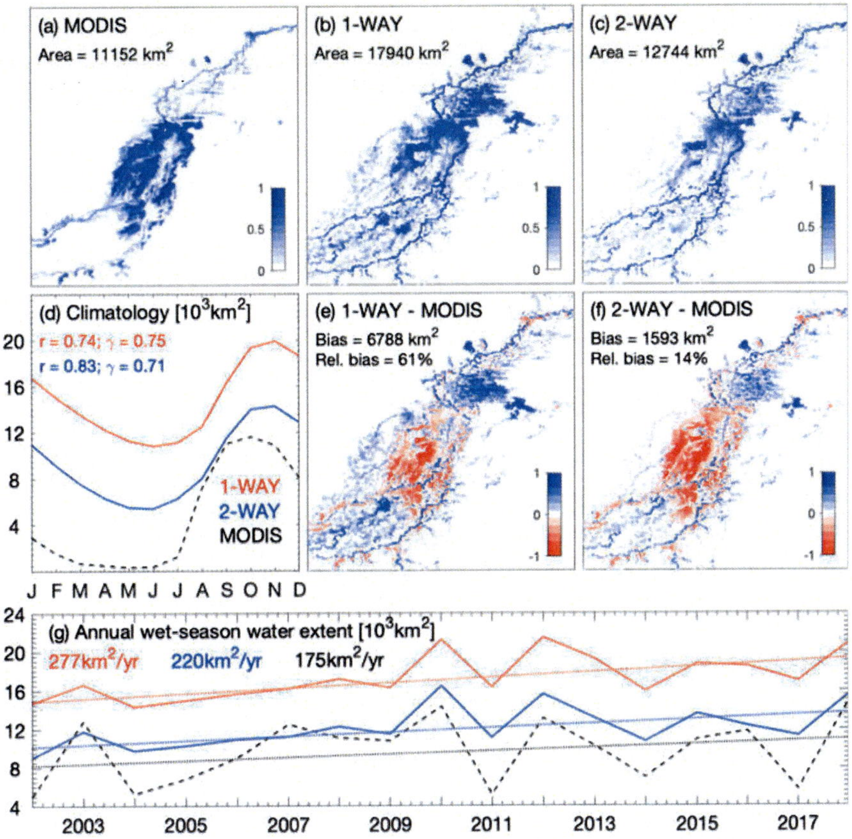

Fig. 19.8 Satellite-based inundation and analysis of flood changes during the 2002–2018 wet seasons (April–June). **a** Inundation estimates from MODIS satellite observations **b** one-way and **c** two-way coupled land surface-flood modelling experiments, **d** Mean annual climatology. Panels **e** and **f** show differences between wet season long-term averaged model simulations and satellite estimates, and **g** their annual temporal average based on wet season water extent. *Source*, Getirana et al. (2021)

19.3.2 Outcomes from Climate and Hydrological Models

With increasing impacts of climate change, about 15% of the global land area may be influenced negatively by 2050, through a 10% reduction of aquatic life in the upstream basin, as compared to only 10% of the total land area that has already experienced such decreases due to excessive water withdrawals and dam constructions (Döll and Zhang 2010). Climate change may worsen, exacerbating the deleterious impacts of dams on freshwater ecosystems. But it also has the tendency of providing benefits for dam operations, which in turn support riverine ecosystems. This predominantly the case if the amount of total runoff escalates, similar to the case in Sweden, where the annual hydrograph was similar to fluctuations in electricity requirements, that is, with a relatively reduced spring flood and improved runoff, especially during winter months (Renöfält et al. 2010). River flow fluctuations impact directly on biota levels. Substantial river flow fluctuations are often based on improved climatic conditions, which in turn influence invasive biota species (Ficke et al. 2007).

Due to the proliferated winter temperatures, a significant effect has been observed in freshwater ecosystems in basins with large snow storage, which is affected by advanced river flows in winter, primary spring peak flows, and possibly minimized summer low flows. Strongly improved winter peak flows will potentially result in a 20–40% reduction in the salmonid population as projected in the Pacific Northwest of the United States by 2050 (Döll and Bunn 2014). This, however, depends on the climate model used for the re-analysis as different models have the possibility of producing varied or relatively similar results. Minimization of low flows during summer has the tendency of improving the rivalry between irrigation water users and ecosystem services (Stewart et al. 2005). An essential scheme for ensuring that environmental flows through procuring or leasing water rights and altering reservoir release models is adopted (Palmer et al. 2009).

Models and measurements recommend that global warming influences glacier and snow-fed rivers and streams, and will potentially circulate through two distinctive stages (Jacobsen et al. 2012). In the primary stage, when river discharge is improved due to increased melting, the total variety and profusion of species may escalate. However, it is possible that fluctuations in water temperature and stream flow rate could impede narrow range endemics (Jacobsen et al. 2012). In the secondary stage, when snowfields melt on time and glaciers have retreated to the point that late-summer stream flow is diminished, negative influences are observed, with organism diversity rapidly declining once a key threshold of approximately 50% glacial cover is crossed. River discharge also affects the reaction of river temperatures to air temperature intensification. Globally averaged, air temperature intensifications ranging from 20 C to 60 C are often projected to result in an annual average river temperature upsurge of about 1.30 C–3.80 C. Emission reductions of 20 and 40% are estimated to produce an additional improvement of mean river water temperature of 0.30 C and 0.80 C. Therefore, where rivers will encounter drought more frequently in the near future, freshwater-dependent biota will experience not only drought induced by river temperature but also by altered flow conditions, as well as by associated reduced oxygen and intensified pollutant absorption.

19.3.3 Prioritization and Conservation

The need to prioritize and manage human actions and resources for biodiversity conservation within freshwater ecosystem is increasingly becoming more crucial. Given the rigorous human need for tropical rivers, this scarce commodity is becoming more threatened as a vital ecosystem service (Vorosmarty et al. 2010).

Basic interruptions that contribute to such threats include water extractions and dam construction, intrusive organisms, over-harvesting of fish, environmental pollution, and adjustments to riparian and in-stream habitats, which have significantly changed freshwater ecosystem and further proceed to impact the global ecosystem (Vorosmarty et al. 2010). Additionally, studies have shown that climate change inter-relates with land-use change to influence runoff, surface water temperature, and river flow regimes, among others (Porter et al. 2013; Anteau 2012; Nelson and Palmer 2009; Peterson and Kwak 1999). Also, conservation-proposing efforts in freshwater environments have been focused on terrestrial and marine environments, with a few studies highlighting actions for the preservation of freshwater biodiversity even while abating costs (Januchowski-Hartley et al. 2011; Stewart-Koster et al. 2011; Ticehurst et al. 2007).

19.4 Floodplain Productivity in the Australian Wet–Dry Tropic

The pioneering work of Ndehedehe et al. (2020c) in Australia showed how flow variability in one of the northern catchments of Queensland impacts total floodplain inundation and the distribution of aquatic plant biomass distribution. As further mentioned in Ndehedehe et al. (2021), flow variability is a key determinant of the ecology of wetlands, setting the conditions for various aquatic and terrestrial biota to persist under different conditions. The catchments along the Australian wet–dry tropics receive more than 90% of their annual rainfall within three months, leading to considerable flows that inundate and support the productivity of floodplains. To maximize the usage of this rainfall, water resources development through the construction of in-stream dams or water harvesting and off-stream storage are proposed to store these monsoon rains for use over the remaining dry months of the year. There is a growing interest to increase agricultural use of this rainfall through the construction of large in-stream dams and other forms of water harvesting (CommonWealth 2015). It is true that rivers in tropical systems across the world are well known for their strong connections with floodplain wetlands. However, increased water needs and changing climate could drive water management policy that leads to major changes in flow or surface water availability. These changes could have deleterious effects on ecological functions and ecosystem services provided by these freshwater ecosystems (e.g., groundwater recharge, food sources for fish, etc.). The impacts of extreme floods or droughts, for instance, can result in major alterations of physical and chemical conditions sustaining the health and development of biotic communities, thus necessitating the need to maintain flow variability in ecological communities (e.g., Ndehedehe et al. 2021). This chapter presents the work of Ndehedehe et al. (2021), who introduced a cloud-based monitoring framework using multi-resolution satellite observations (2007–2020), historical surface inundation

(1984–2015), and gauge data to better understand these connections in a large river catchment in tropical Queensland (Mitchell River catchment).

19.4.1 The Mitchell River Catchment

The Mitchell River catchment in northern Queensland, Australia (Fig. 19.9), is one of the largest river basins in the region (71,630 km^2) with a large alluvial megafan that begins below the confluence of the Palmer River and covers approximately 31,000 km^2 (Brooks et al. 2009). The headwaters of the Mitchell River rise in the south-east of the catchment near the wet-tropics region of the Great Dividing Range, though the vast majority of the catchment experiences a wet–dry tropical climate (Petheram et al. 2018). Over 80% of the annual rainfall falls across the entire catchment between December and March creating vast inundated areas on the floodplain (Brooks et al. 2009; Ward et al. 2013). The short duration of the wet season means the peak inundation on the floodplain recedes relatively rapidly; however, there are large wetlands that hold surface water year-round and the river flow of the Mitchell River mainstem is generally perennial (Karim et al. 2018; Ward et al. 2013). The land cover across the majority of the catchment is open savannah woodland and grasslands with relatively low-intensity land use, consisting primarily of light grazing,

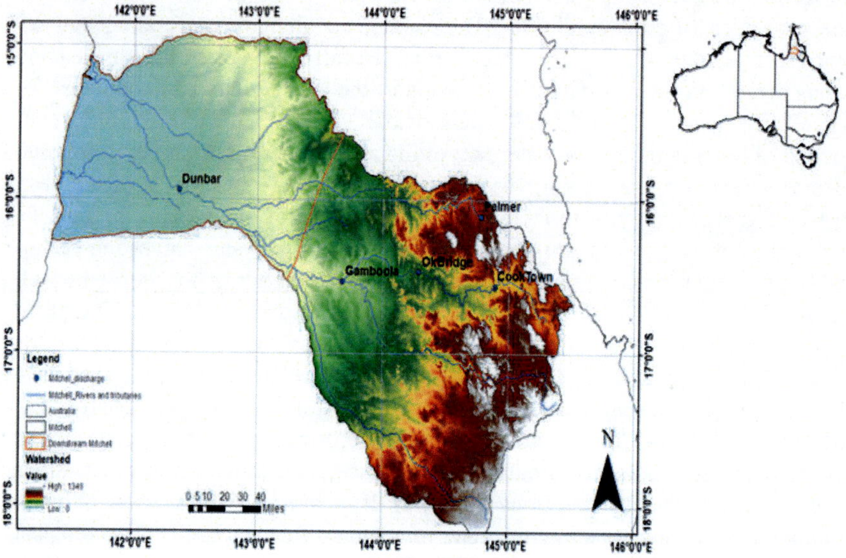

Fig. 19.9 Study region showing the Mitchell catchment in northern Queensland. The upstream and downstream of the catchment and the topographic distribution (elevation in meters) in the watershed based on corrected digital elevation model is indicated. The gauge stations for observed flow data used in this study are also indicated

traditional harvest with active fire management, and small areas of intensive agriculture and mining in parts of the headwaters (Brooks et al. 2009; Ward et al. 2013). This relatively low intensity of land use corresponds with limited water resource development, meaning most of the rivers in the catchment flow freely (Petheram et al. 2018). No major impoundments or diversions exist on the floodplain, meaning overland flows and local rainfall freely inundate wetlands across the entire floodplain and delta region (Pusey and Kennard 2009).

The goal of this study is to improve understanding of the key drivers of primary productivity of freshwater ecosystems in a large river catchment in the Australian wet–dry tropics (Mitchell River catchment), which support numerous important freshwater ecological assets, traditional harvest, and coastal fisheries. Specifically, this study combines for the first time, a planetary scale cloud computing framework with several statistical approaches including multivariate copula analysis to (i) characterize floodplain inundation and quantify seasonal changes in freshwater habitats, (ii) predict changes in aquatic plant biomass accumulation due to flow variation, in hotspots of primary production, and (iii) measure the dependence of downstream floodplain hydrology on the hydro-climatology of the upstream catchment. This assessment is required as a key logical step to (i) conservation and water planning processes, maintenance of aquatic biodiversity, and wetland management, (ii) risk assessment from large hydro-climatic fluctuations and decadal-scale droughts, and (iii) understanding implications of future land and water resource developments that are currently under consideration in key wet–dry Australian tropics (Fig. 19.10).

19.4.2 Data

This study used the Global Surface Water Dataset (GSW, Pekel et al. 2016) accessed through the GEE platform. Landsat observations for the 2007–2019 period were retrieved from GEE and used to compute total surface inundation in wet and dry seasons. These data were also used in change detection analysis for two epochs: 2007–2014 and 2014–2019. Additionally, the Sentinel-2 (S2) data were used to predict aquatic plant biomass accumulation in the downstream Mitchell catchment. The data were also accessed from the Google Earth Engine (GEE) API (Gorelick et al. 2017). Processing details are as summarized in Ndehedehe et al. (2021). Monthly Australian interpolated gridded rainfall data (http://www.bom.gov.au/climate/data/) and river discharge data of the Mitchell river were also used. Discharge observations were retrieved from five important locations in the Mitchell catchment (Dunbar/Koolatah, Cook Town, Gamboola, Okay Bridge, and Palmer). River discharge observations covering the period 2000 – 2019 were obtained from the Queensland Water Monitoring Information Portal (https://water-monitoring.information.qld.gov.au/).

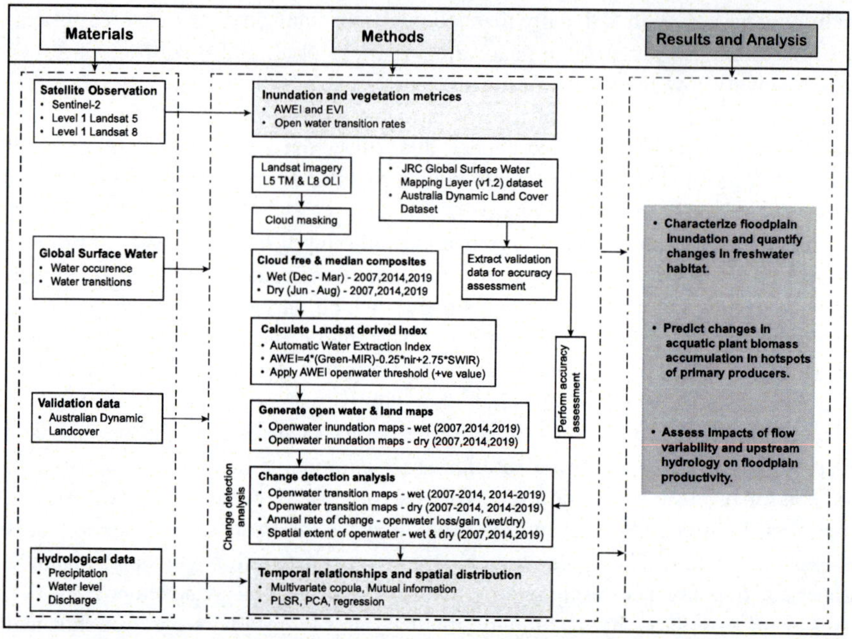

Fig. 19.10 Schematic representation of the methodological framework used in the study undertaken by Ndehedehe et al. (2021). All satellite observations, including the Global Surface Water data, were retrieved and preprocessed (e.g., cloud masking, calculating median values of pixel stacks, etc.) and analysed using the Google Earth Engine platform. *Source* Ndehedehe et al. (2021)

19.4.3 Total Inundation Mapping and Change Detection of Open Water Transition

Open water features and total floodplain inundation were quantified by applying the widely used Automated Water Extraction Index (AWEI, Feyisa et al. 2014) based on the multi-temporal Landsat data. The AWEI uses multiple bands (Green, MIR, NIR, and SWIR) and was computed as

$$AWEI_{nsh} = 4 * (Green - MIR) - (0.25 * NIR + 2.75 * SWIR). \tag{19.1}$$

Inundated areas and open water features were masked using specified threshold (ranges from 0 to 1 for water-related pixels). The average annual rates of transitions for open water loss ($T_{loss\ rate}$) and gain ($T_{gain\ rate}$) in the Mitchell downstream catchment were estimated using spatial details of open water cover area during the 2015–2020 period and are formulated as

$$T_{loss\ rate} = \frac{(\log W_{t2} - \log(W_{t1} - B)) * 100}{t_2 - t_1}\ and \tag{19.2}$$

$$T_{gain\ rate} = \frac{(\log W_{t2} - \log(W_{t1} - C)) * 100}{t_2 - t_1}, \quad (19.3)$$

where W is the open water area in hectares for t_2 (later date) and t_1 (earlier date); B is the area of open water loss, and C is the area of open water gain. Using Eqs. 19.2 and 19.3, it was possible to determine gain, loss, and persistence rates of surface inundation. The validation of these transition maps is highlighted in Ndehedehe et al. (2021). The Dynamic Landcover Dataset (DLCD) for Australia (https://www.ga.gov.au/scientific-topics/earth-obs/accessing-satellite-imagery/landcover) and the JRC (Joint Research Center) Global Surface Water Mapping Layers, v1.2 (Pekel et al. 2016) were used to perform accuracy assessment. For accuracy assessment of the total floodplain inundation mapping and change detection analysis, three measures were used. These include the overall accuracy, overall kappa statistics, and F-score results calculated using the user's and producer's accuracies for individual classes mapped in this study.

19.4.4 Prediction of Hotspots of Aquatic Plant Biomass Accumulation

The distribution of hotspots of aquatic primary production (inundated areas covered with different aquatic species such as macrophytes) was assessed by combining inundation and vegetation metrics. Specifically, Sentinel-2 data was used to compute the AWEI and Enhanced Vegetation Index. These indices were combined in a sequence of logical operations to identify hotspots of aquatic primary producers in the downstream Mitchell catchment. Locations with the presence of aquatic vegetation (e.g., macrophytes) in freshwater habitats, narrow riverine channels, and shallow water holes were identified by combining these indices. These hotspots were predicted based on defined thresholds for these indicators, leveraging on the expert knowledge of the target variables in five locations in the downstream catchment. The EVI (Enhanced Vegetation Index) uses three bands (Red, NIR, and Blue) and was computed as

$$EVI = 2.5 \times ((NIR - Red)/(NIR + 6 \times Red - 7.5 \times Blue + 1)) \quad (19.4)$$

where NIR, Red, and Blue are designated as bands 8, 4, and 2, respectively in the multiband Sentinel-2 data. Unlike NDVI, EVI is calculated with an additional wavelength of light to correct for the inaccuracies of NDVI. Important factors like differences in solar incidence angle, atmospheric conditions such as distortions in the amount of reflected light caused by particles in the air and signals from the ground cover below the vegetation tend to weaken NDVI estimates of surface vegetation. These issues are corrected by using the EVI. Essentially, the thresholds for AWEI (0.07) and EVI (0.30) were used to target freshwater hotspots and vegetation within these inundated areas. Threshold definitions for remote sensing indicators can be highly variable (e.g., Worden and de Beurs 2020; Ndehedehe et al. 2020a; Pekel et al. 2016). Such variability could depend on land surface characteristics of regions and the capability of sensors onboard satellite systems. To minimize uncertainties that could arise from this, fairly stable threshold values using visual inspection of project sites that were sampled in person were established from these metrics.

19.4.4.1 Assessing Drivers of Downstream Total Inundation

The principal component analysis (PCA, Jolliffe 2002; Preisendorfer 1988) was employed to assess the spatial and temporal patterns of rainfall over the Mitchell catchment. Mutual Information (MI) was employed as a more reliable measure of statistical dependence to explore the relationship between flow and total surface inundation, given its merits over other statistical measures that rely on the residual sum of squares (e.g., Moon et al. 1995; Sharma 2000). The coupled relationship of downstream total inundation with discharge and local rainfall was modelled using the partial least squares regression (PLSR, e.g., Chen et al. 2018; de Jong 1993). As part of the overall objective of understanding key drivers of freshwater ecosystems, a multivariate copula analysis tool that employs Markov-Chain Monte Carlo simulation within the Bayesian framework (Sadegh et al. 2017) was employed here to assess the dependence of downstream catchment on the hydrology of the upstream region.

19.4.5 The Mitchell Floodplain Hydrology

19.4.5.1 Open Water Transitions and Occurrence (1984–2015)

Over a 30-year period, the downstream catchment had more permanent water surface (4.5%) compared to the upstream region (2.7%) (Fig. 19.11a, b). The larger proportion of *new permanent* and *new seasonal* surface water classes in the upstream section (Fig. 19.11b) suggests variability in surface inundation largely due to climatic disturbance during the period (1984–2015). There is a substantially higher proportion of ephemeral seasonal surface water class in the downstream (67%) as opposed to the upstream (41.2%) (Fig. 19.11a, b). The spatial and temporal (intra-annual) variability of *water occurrence* is illustrated for randomly selected water holes and freshwater habitats in the downstream catchment (Fig. 19.12a, b). It is evident that *water occurrence* within the same water holes or freshwater habitats (locations $A - D$) varies significantly (Fig. 19.12a). The seasonal temporal patterns of *water occurrence* confirm the variability in *water occurrence* across landscapes. For example, in location A (a selected point at Racecourse Swamp), water is present for about 60% of the time from around January–February each year (Fig. 19.12b). Although some locations confirm the spatial variability in *water occurrence*, it is important to note that water persisted (i.e., 100% *water occurrence*) in much of the vicinity around Racecourse Swamp all through the year. This is clearly evidenced by the spatial pattern of *water*

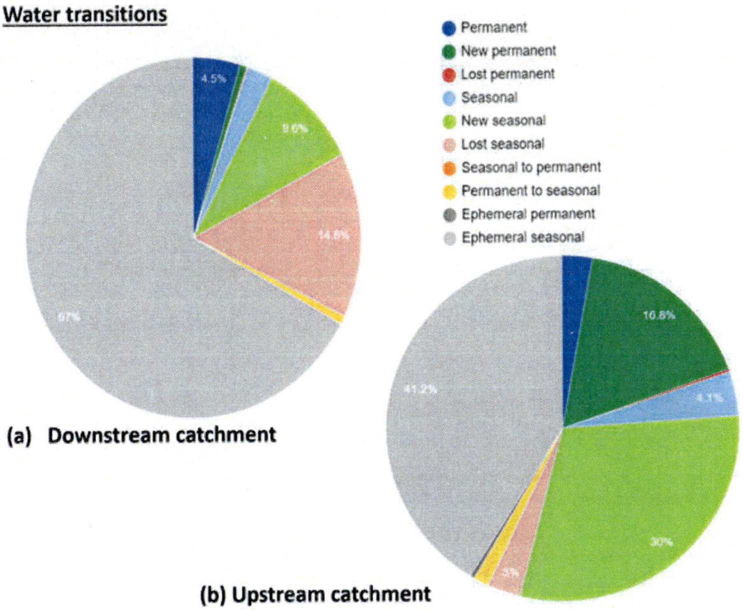

Fig. 19.11 Water transitions in the Mitchell catchment based on the global surface water data (March 1984 to October 2015). The statistics on water transitions are estimated for the **a** downstream and **b** upstream catchments

occurrence (indicates 100%) in upstream location A (Fig. 19.12b). But in location D (mid-stream), *water occurrence* is permanent (i.e., 100% of the time) (Fig. 19.12b).

19.4.5.2 Change Detection and Temporal Variations of Open Water Features (2007–2020)

Open water maps for two different periods (i.e., 2007–2014 and 2014–2019) highlight the multi-temporal change over the downstream catchment (Fig. 19.13). The summers between 2007 and 2014 were better than the 2014–2019 period (Fig. 19.13a). As opposed to the latter period, extreme surface inundation are obvious in the 2007–2014 period (Fig. 19.13b). The highest average surface inundation extent (544 km^2) observed was during the summers of 2008 followed by the summers of 2009 (259 km^2) (Fig. 19.13b). These surface inundation extents coincided with gauged records of flow in 2008 and 2009, which both showed the highest observed flows greater than 6500 m^3/s at key stations (Dunbar and Gamboola). Apart from 2014, a general observation is the fluctuations in total inundation in response to discharge. Years with a higher amount of flows were characterized by higher surface inundation. With this in mind, years with missing satellite (contaminated with clouds) like 2011, which had the highest flow observation, are expected to have average sur-

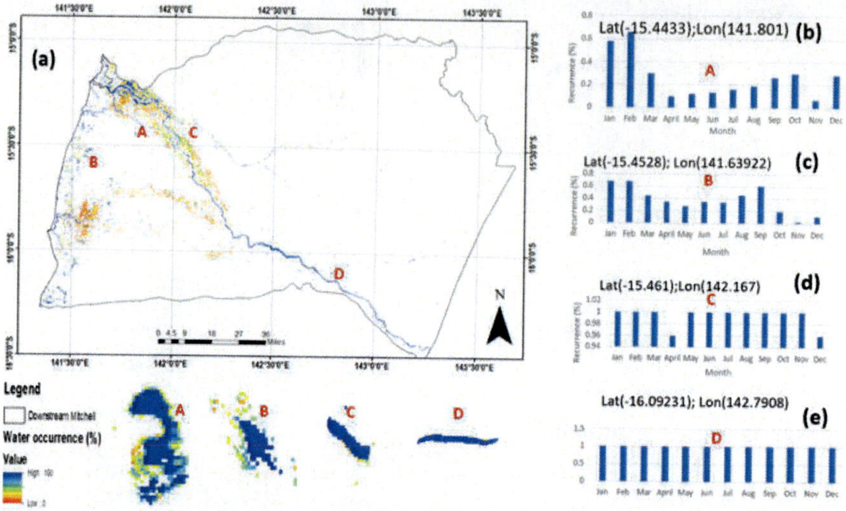

Fig. 19.12 Water occurrence (expressed in percentages with 100% representing permanent surface water) in the Mitchell catchment based on the global surface water data (1984–2015). The spatial and temporal patterns of water occurrence are illustrated using four randomly selected water holes and freshwater locations (A to D) in the downstream catchment. *Source* Ndehedehe et al. (2021)

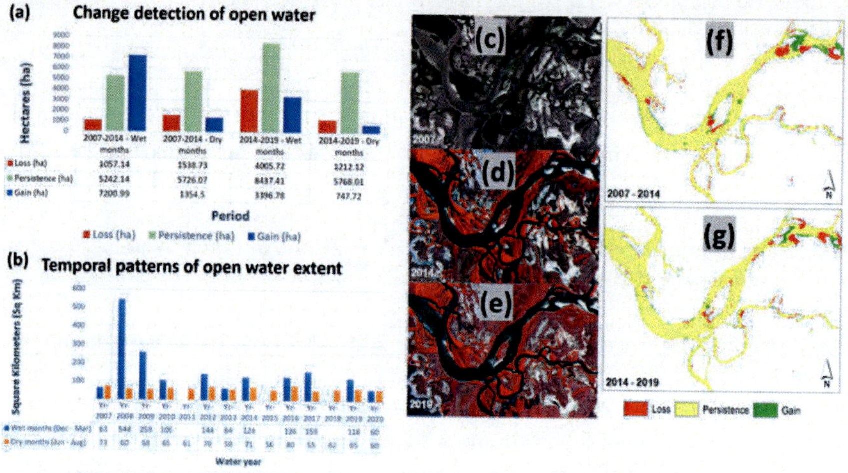

Fig. 19.13 Spatial and temporal patterns of surface water change during the 2007–2019 period. **a** Change detection analysis showing temporal characteristics of wet (December–March) and dry (June–August) seasons, **b** temporal variations of open water extent, **c**–**e** false colour composites and **f**–**g** illustration of change detection results of spatial variability in the characteristics of wet and dry seasons (loss, persistence, and gain). *Source* Ndehedehe et al. (2021)

face inundation extent exceeding that of 2008 (Fig. 19.13b). Similarly, 2015, which appears to be the driest since 2000 based on a considerably low discharge observation (297 m^3/s) caused by limited rainfall, will have the lowest total floodplain inundation. Estimated floodplain inundation extent for the dry season remained relatively stable all through the period with average surface inundation extent fluctuating between 55 and 80 km^2 (Fig. 19.13b). Furthermore, an illustration of the false colour composites and spatial patterns of change detection results in a segment of the downstream is shown in Fig. 19.13c–g. The characteristics (loss, persistence, and gain) of floodplain inundation in wet and dry seasons vary in time and space (Fig. 19.13a, f–g). These characteristics, especially the persistence rates of surface inundation, are relatively similar during dry periods. But the rates of persistence and loss are significantly different during wet seasons (Figs. 19.13a). Error matrix and accuracy assessment results for the open water classifications are generally high for all the epochs considered.

19.4.5.3 Floodplain Inundation Patterns and Connectivity of Rivers

As with changes in annual rainfall and discharge, inundation patterns are highly varied on annual and seasonal time scales. The connectivity of rivers and inundation patterns in slightly isolated water holes and other intermittent water bodies are considerably different during wet and dry seasons (Fig. 19.14a–v). In the wet season, most water holes are completely covered with inundation, while rivers remain well connected during the period (Fig. 19.14b–e, g–j). But as the dry season approaches, water holes exist independently and some sections along prominent rivers and channels become disconnected (Fig. 19.14l–r, t, and v). Results of the floodplain inundation mapping (Fig. 19.14) also align with the spatio-temporal patterns of water occurrence. Using illustration from the mid-stream section of the Mitchell catchment, floodplain inundation in some years (e.g., 2016 and 2019) is permanent and perennial (Fig. 19.14h and s, j and u). In other words, there is 100% water occurrence in those years and specific locations. Comparing 2008 and 2009 water years with those of 2016 and 2019, it appears that extreme floods during the wet season do not guarantee an all-year connection of river channels (Fig. 19.14b, c, m, n, h and s, and j and u). Another key observation is that some sections of river channels remain disconnected in both seasons (wet and dry) in years that are generally characterized by low or reduced annual rainfall. In 2013, for example, the estimated total inundation extents for the wet and dry seasons were 64 and 59 km^2, respectively (Fig. 19.13b). The spatial patterns of inundation along the midstream channel in this same year appear to be similar (Fig. 19.14). A good example is the change intensity of rainfall, which in turn influence flow variability and subsequently floodplain inundation patterns.

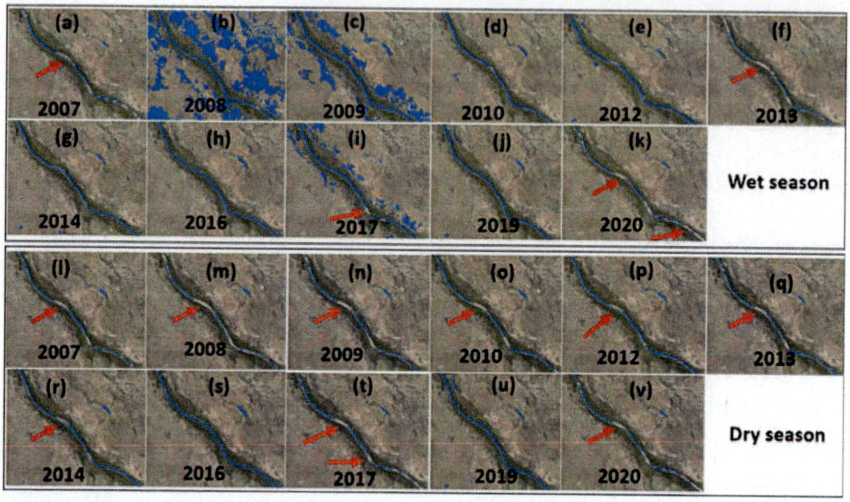

Fig. 19.14 Illustration of inundation patterns and connectivity in freshwater habitats (riverine systems and water holes) during **a–k** wet and **l–v** dry seasons (2007–2020). The wet season inundation for each year is based on the median of Landsat images for the water year (December and March) while the dry season inundation is based on the median of images between June and August of the same year. The red arrows illustrate a disconnection along the Mitchell river in the mid-stream catchment. *Source* Ndehedehe et al. (2021)

19.4.6 Aquatic Plant Biomass Accumulation in Freshwater Habitats

The distribution of aquatic plant biomass accumulation in several hotspots in the downstream catchment varies in time and space. This variability is apparently tied to climate variability, which regulates floodplain inundation patterns. Arguably, rainfall serves as the main input to any hydrological system, and as a rule of thumb, more productive years tend to be largely characterized by higher floodplain inundation as opposed to relatively dry years. This inundation, as discussed further in the next section, is driven by both local rainfall and upstream flow.

There has been a rise in the total extents of hotspots of primary production from 8.19 km^2 in 2016 to 36.9 km^2 in 2019 and 24.8 km^2 in 2020 (Fig. 19.15a–e). The zoomed-in spatial distribution of estimated hotspots in one of the locations (Racecourse Swamp-location A in Fig. 19.12) shows the spatial variability of these hotspots (Fig. 19.15f–h). As illustrated in Fig. 19.15, more hotspots were found on the floodplain in 2019 (36.9 km^2) compared to other years (2016–2018 and 2020) potentially due to higher upstream flows from all stations (Fig. 19.15a–e). There are reasonably strong linear relationships between the extent of hotspots of primary production and upstream river flows at Gamboola ($r = 0.72$), Ok Bridge ($r = 0.80$), and Cook Town ($r = 0.82$) (Fig. 19.15a, b and d). The association of hotspots with observed discharge at Palmer ($r = 0.58$) and Dunbar ($r = 0.50$) is also linear but tends to

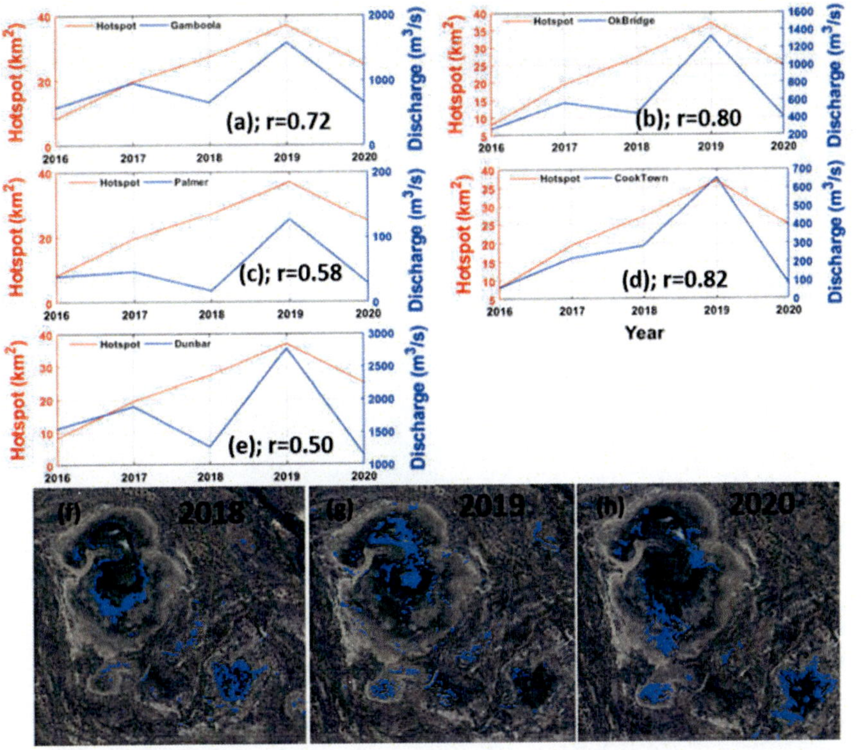

Fig. 19.15 Floodplain productivity based on predicted spatial distribution of hot spots of aquatic biomass accumulation (2016–2020) using Sentinel-2 data over the downstream Mitchell catchment. The relationship between predicted extents of hot spots and observed discharge is indicated for **a** Gamboola, **b** Ok Bridge, **c** Palmer, **d** Cook Town and, **e** Dunbar. The spatial distribution of aquatic biomass accumulation (2018–2020) for one of the sampled water holes (location A in Fig. 19.12) during field investigation is illustrated using blue polygons (**f**–**h**). *Source* Ndehedehe et al. (2021)

be moderate compared to other gauging stations. Generally, higher freshwater floodplain productivity coincides with wet years characterized by relatively higher rainfall amount and flows (Fig. 19.15a–d and 19.16a, b). In other words, the loss in surface extent of aquatic plant biomass will be the result of reduced inundation in freshwater habitats in the catchments, caused primarily by both limited surface flow and local rainfall. However, the sample size used to establish this relationship is small and should be interpreted with caution.

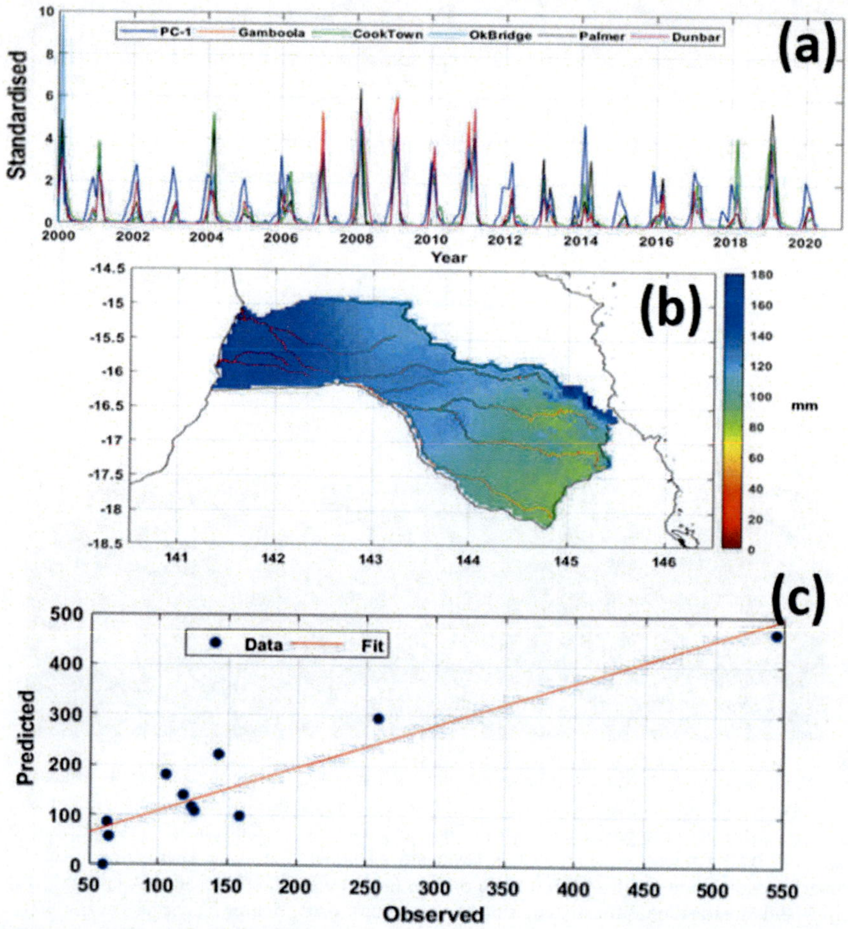

Fig. 19.16 Multivariate assessment of floodplain productivity. **a, b** Temporal and spatial variations of rainfall over the Mitchell catchment using the PCA and **c** modelling of floodplain total inundation extent using the partial least square regression. The temporal patterns of rainfall (PC-1) are standardized and its corresponding spatial pattern (b) has been adjusted to its original units (mm). Averaged monthly rainfall values over each grid in the catchment are jointly derived from PC-1 in (**a**) and spatial patterns (maps) in (**b**). River discharge observations in five gauging stations (Fig. 19.9) have been compared with PC-1, which represents the dominant temporal rainfall pattern over the catchment. Major river networks (red lines) in the catchment are indicated in (**b**). *Source* Ndehedehe et al. (2021)

19.4.7 Upstream Flows and Local Rainfall Drive the Productivity of Floodplains

19.4.7.1 Leading Rainfall Patterns vs Discharge

Two dominant patterns of rainfall in the catchment were identified via the PCA and account for 93% of rainfall variability across the catchment. The leading rainfall mode in the catchment is considerably annual and accounts for approximately 88.5% of the total variability while the second mode is a short-term seasonal signal (not shown) that accounts for about 4.8% of the total variability (Fig. 19.16a, b). The spatial loadings confirm strong spatial patterns of rainfall in the downstream segment and a section along the south-east upstream (Fig. 19.16b). A joint interpretation of averaged peak rainfall in 2014 in the downstream area from the temporal and spatial pattern is about 900 mm. But in the mid-stream region where the Gamboola gauged station is located, averaged peak rainfall in the same year is between 500 and 600 mm. The corresponding temporal evolutions of rainfall are well associated with observed discharge from five locations: Gamboola ($r = 0.73$, $p = 0.000$), Cook Town ($r = 0.66$, $p = 0.000$), Ok Bridge ($r = 0.69$, $p = 0.000$), Palmer ($r = 0.73$, $p = 0.000$), and Dunbar ($r = 0.71$, $p = 0.000$) (PC-1, Fig. 19.16a). Note that between 2007 and 2014, mean discharge values from all gauges and rainfall (PC-1, Fig. 19.16a) had higher amplitudes compared to the post 2014 period. It is during this period (2007 – 2014) that higher gain rate in surface inundation was observed and further provides a clue to the contributions of upstream rivers (Gamboola, CookTown, OkBridge, and Palmer stations) and rainfall to the downstream inundation.

19.4.7.2 Drivers of Total Floodplain Inundation

In the Mitchell, the downstream and some areas in the upstream floodplain have the highest spatial distribution of monthly summer rainfall when jointly interpreted with the corresponding temporal series (Fig. 19.16b). This relatively strong spatial pattern of rainfall in the downstream is expected to transition to variable inundation patterns. In other words, flows from upstream and local rainfall are expected to drive surface inundation in the downstream area. Observed local rainfall and river discharge data from five gauged stations resulted in six partial least square components, of which five were retained and used to develop a model to predict total floodplain inundation (Fig. 19.16c). The predicted and observed total inundation were strongly correlated ($r = 0.95$), indicating that upstream flow and local rainfall drive the productivity of the floodplains (Fig. 19.16c). When local rainfall was excluded from the model (i.e., four partial least square components were rotated), the predicted total inundation using the flow from five gauged stations was still strongly correlated ($r = 0.92$) with observed total downstream inundation. Even when fewer partial least square components were rotated with rainfall included in the model, there was still no signif-

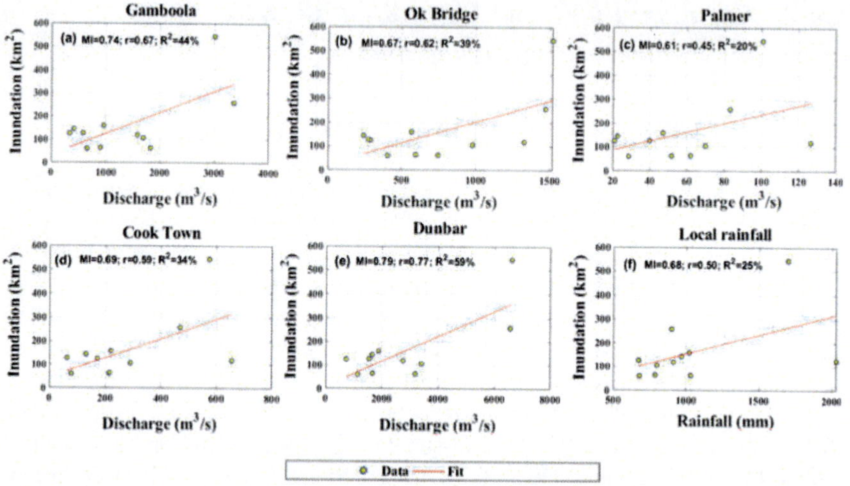

Fig. 19.17 Relationship of upstream flow (discharge) and total inundation in downstream catchment using mutual information (MI), Pearson's correlation (r), and coefficient of determination (R^2). This relationship is based on total inundation with gauged observations of total wet season discharge at **a** Gamboola, **b** Ok Bridge, **c** Palmer, **d** Cook Town, **e** Dunbar, and **f** local rainfall at Kowanyama. *Source* Ndehedehe et al. (2021)

icant difference in the relationship between observed and predicted total inundation ($r = 0.92$).

Similarly, to explore the potential impact on floodplain inundation and productivity in the event of upstream flow alteration, discharge data from Dunbar gauged station was excluded from the model. This resulted in a moderately strong ($r = 0.80$) relationship between predicted and observed total inundation, indicating about 15% lost in floodplain inundation. Given that the Gamboola discharge is the penultimate closest station to the downstream, we also tested the impact of flow alteration in this location by excluding it from the model, in addition to flows from Dunbar. This scenario provides insights as to the impact of flow alteration on the productivity of the Mitchell floodplain given that predicted and observed inundation were less associated compared to other scenarios ($r = 0.78$). While this suggests 17% lost in floodplain inundation if flows at Dunbar and Gamboola are altered, the partial least square regression coefficients indicate that flows from Ok Bridge gauge have a linear relationship with inundation and contribute substantially to floodplain inundation similar to the Dunbar station.

Interestingly, a range of metrics, including mutual information and coefficient of determination (R^2), confirm the contributions of upstream flow to downstream floodplain inundation (Fig. 19.17a–e). Observed discharge at Dunbar station explained approximately 60% of the proportion of variability in total inundation (strong mutual information and correlation of 0.79 and 0.77 ($p = 0.005$), respectively (Fig. 19.17e). Gamboola gauged station (penultimate closest station to the downstream) accounted for 44% ($r = 0.67$, $p = 0.0249$) of variability in total inundation with a moderately strong mutual information ($r = 0.74$) and correlation ($r = 0.67$, $p = 0.025$). Generally, except for Palmer-20% ($r = 0.45$, $p = 0.164$), all observed flows from the remaining gauged stations show significant (at $\alpha = 0.005$) associations with total inundation, accounting for 39% (Ok Bridge) and 34% (Cook Town) of the variability in total inundation (Figs. 19.17b-d). The mutual information between observed discharge at these stations and surface inundation ranged from 0.61 to 0.69. Moreover, local rainfall observed at the downstream region accounted for about 25% ($r = 0.50$, $p = 0.120$) of the proportion of variability in total floodplain inundation (Fig. 19.17f). Although mutual information between local rainfall and the latter was rather moderate ($r = 0.61$), its correlation was the lowest ($r = 0.50$) compared to discharge observations at Gamboola, Dunbar, Cook Town, and Ok Bridge (Figs. 19.17a–f). The scatter plot for the rainfall–inundation relationship shows an outlier (Fig. 19.17f) different from the discharge–inundation relationships (Figs. 19.17a–e). This signal appears to be that of 2014 peak summer when tropical cyclone produced a large amount of rainfall on the floodplain while upstream flows were restricted (Fig. 19.16a). The formation of Tropical Cyclone Fletcher in February 2014 deposited substantial rain on the floodplain of the Mitchell River, downstream of all flow gauges (see http://www.bom.gov.au/cyclone/history/Fletcher.shtml). Daily local rainfall at downstream location (Kowanyama) reached 361.2 mm and a total of 1432 mm was observed in February 2014, making it the wettest month at the station in 100 years since observation began. When this 2014 observation was excluded from the time series, local rainfall explained 76% ($r = 0.87$ $p = 0.001$) of the proportion of variability in the remaining total surface inundation. The dependence of downstream floodplain hydrology on upstream flow is further investigated in Sect. 19.4.8 based on a multivariate copula approach.

19.4.8 Dependence of Downstream Water Level on Upstream Flow

From the cohort of 26 bivariate models assessed using the multivariate copula toolbox (Sadegh et al. 2017), the Galambos copula was selected as the best model that describes the dependence structure between water level and discharge at the four gauged stations assessed in this study. This choice was based on the consistent performance of this model across all goodness-of-fit metrics (Akaike information criterion, BIC, Likelihood, Nash-Sutcliffe efficiency, and root-mean-square error). These criteria were also used in evaluating the dependence of water level at the downstream region (Dunbar gauged station) on upstream flows from three gauged stations (Cook Town, Palmer, and Gamboola) with no missing data during the period (2009–2020).

Table 19.1 provides a summary of the relationship between upstream discharge and downstream water levels based on different correlation metrics: Kendall rank, Spearman's rank-order, and Pearson product-moment. All correlation coefficients ($r = 0.73, 0.89$, and 0.87, with a p-value of 0.000 for Kendall rank, Spearman's

Table 19.1 Evaluating the dependence of downstream water level on upstream flows based on the best copula (Galambos) identified through several goodness-of-fit criteria. All correlation coefficients from Kendall rank, Spearman's rank-order, and Pearson product-moment categories are statistically significant at $\alpha = 0.05$. Source Ndehedehe et al. (2021)

Discharge location	Kendall rank	Spearman's rank-order	Pearson product-moment
Gamboola	0.73	0.89	0.87
Palmer	0.66	0.84	0.79
Cook Town	0.60	0.80	0.77

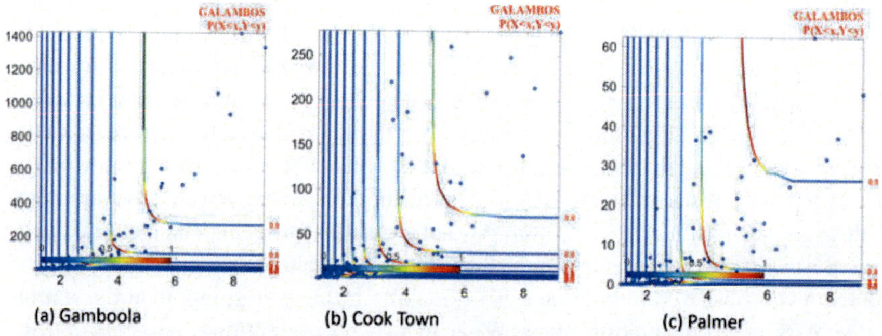

Fig. 19.18 Marginal probability distribution of downstream water level and flow from upstream locations-**a** Gamboola, **b** Cook Town, and **c** Palmer. Their joint probability isolines (flow vs water level) are indicated and the curves are colour-coded with probability densities where blue and red colours represent lower and higher densities, respectively. Analysis is based on available monthly flow (y-axis in m^3/s) and water level (x-axis in m) data (2009 – 2020). Source Ndehedehe et al. (2021)

rank order, and Pearson product-moment, respectively) confirm the significant association of downstream water levels with upstream flow from the Gamboola station. The joint probability curve confirms the dependence of water level on upstream flow (Fig. 19.18a–c). The dependence structure between water level and river discharge at Gamboola station shows a good fit with the probability of higher water levels increasing with higher flows (Fig. 19.18a). However, the dependence structure is slightly skewed for other stations (Fig. 19.18b, c) and is consistent with the nature of observed correlation between these quantities when compared to the Gamboola station. Generally, observed correlations in Table 19.1 are significant display of dependence of downstream water level on upstream flows. These relationships (Fig. 19.18a–c) are consistent with those of flow and total inundation (Fig. 19.17a–d), suggesting that flow from upstream rivers are key predictors of downstream floodplain inundation.

19.4.9 Wetland Hydrology and Aquatic Plant Biomass Accumulation

The spatial patterns of aquatic plant biomass accumulation (hotspots) are difficult to measure over large areas and longer time spans using traditional approaches. However, they are increasingly being assessed using remote sensing methods (e.g., Tortini et al. 2020; Yang et al. 2020; Slagter et al. 2020; Ndehedehe et al. 2020a). This study used several multivariate methods to improve understanding on drivers of freshwater ecosystems in the Mitchell River catchment by measuring the dependence of downstream floodplain productivity (surface inundation and biomass accumulation) on variability in upstream flow and local rainfall. Global surface water data and satellite-derived change detection analysis were combined to characterize surface water distribution in the catchment.

The persistence of surface water in freshwater habitats (palustrine, lacustrine, and riverine ecosystems) is important for the productivity of aquatic ecosystems. During summer, the downstream Mitchell floodplain wetlands are characterized by extensive floods. These floods and total inundation are lost over the course of the dry season (Fig. 19.13b). Although the rates of flood recession could differ depending on catchment characteristics (e.g., flood magnitude, topography), aquatic systems become increasingly disconnected during this period (Ward et al. 2013). Extensive flooding of landscapes during summer (wet season) and the disconnection of aquatic ecosystems during the dry season is a typical phenomenon in the wet–dry tropics of Queensland characterized by monsoonal climate (e.g., Ndehedehe et al. 2020c; Karim et al. 2018; Ward et al. 2013). Floods are essential to the productivity of ecosystems, providing a broad range of ecological functions and ecosystem services. These include contributing nutrients to aquatic ecosystems, stimulating primary production, supporting biological diversity, and providing cultural benefits such as traditional food harvest, recreation, and tourism (e.g., Ndehedehe et al. 2020a; Talbot et al. 2018; Midwood and Chow-Fraser 2010; Keddy et al. 2009). Although large high-intensity floods can impact productivity of aquatic food webs through the transport of nutrients from floodplain rivers to the nearby ocean (e.g., Talbot et al. 2018; Ndehedehe 2020), flood events are also beneficial to groundwater processes. For example, floods have impacts on groundwater recharge and the extent of surface inundation drives the quantity of flood water that infiltrates groundwater stores, with higher inundation extents culminating in extensive groundwater recharge (e.g., Talbot et al. 2018; Kazama et al. 2007).

Historically, human activities and climate change have contributed to the alteration of tropical river–floodplains and there is a substantial decline in the existing permanent open water features around the world, including water present in lakes and reservoirs (Ndehedehe et al. 2020b; Pekel et al. 2016; Tockner et al. 2010). Tracking and monitoring these variations in surface water hydrology is thus an important issue for the management of floodplain rivers. The persistence of surface inundation during the dry and wet seasons (greater than 5000 ha on average) in all periods could imply several ecological benefits for the Mitchell catchment if floodplain conditions and natural hydrological variability continue to be favourable. For instance, temporal variations of dry season inundation are consistent and maintain nearly similar peak amplitudes. These dry seasons are generally characterized by low-level flows, which are important for the sustainability of a nearby coastal commercial fishery (Broadley et al. 2020). Moreover, long-term surface water observations (1984 – 2015) show that there is more presence of permanent water (4.5% of total land area) in the downstream Mitchell compared to the upstream region (2.7% of total land area). The variability of water occurrence and its transitions in the downstream region is noted and depends on climatic factors (inter and intra-annual rainfall intensity and amplitudes) and flow variability in the upstream catchment. However, the nature of evaporation may play a significant role in duration of surface inundation and probably an all-year connection of river channels.

19.4.10 Changes in Tropical Floodplains and Freshwater Ecosystems

19.4.10.1 Inter-Annual Variations in Upstream Flows and Local Rainfall

Another key finding from this study is that upstream flow is the primary driver of changes in the downstream floodplains. Local rainfall only acts as a secondary driver of variations in productivity of floodplains. Flows at the Dunbar station accounted for approximately 60% of the variability in total surface inundation while about 25% of the proportion of variability in surface inundation was explained by local rainfall. By all measures of statistical dependence employed in this study to establish the link between floodplain total inundation and climatic variables (rainfall and discharge), upstream flows were substantially more associated with total surface inundation. Local rainfall showed no statistically significant (at 95% confidence interval) relationship with total surface inundation. Obviously, the 2014 anomalous rainfall caused by tropical cyclone in the downstream region during peak summer of 2014 skewed this relationship. Although high unprecedented amplitudes of local rainfall could result in extensive floodplain inundation, this rarely occurs.

The analyses in this study show that high magnitude flows always translate to extensive floodplain inundation. The significant association of dominant rainfall patterns over the Mitchell catchment with observed flow from upstream regions highlights the importance of rainfall in the broader region. However, the PLSR results show the existing relationship between the upstream flow and downstream floodplain inundation. Different modelling scenarios confirmed that upstream flows are dominant or primary drivers of surface inundation and floodplain productivity in tropical wet–dry catchments in Queensland. The veracity of this impact of course requires a further assessment that relies on longer time series of data with limited or no missing observations. On the other hand, failure in the downstream climate (in terms of below normal local rainfall) will have less impact on floodplain inundation if upstream flows are maintained. River flows that are driven by inter-annual variations in rainfall are transitioned into wetland inundation in the downstream area due to the high-elevation areas of the upstream. The dependence of downstream water level on upstream flows assessed using a bivariate model also confirms the significant contributions of flow from upstream rivers to downstream floodplains. The spatial and temporal distribution of rainfall in the catchment emphasizes climate variability as a key mechanism that will either mar or enhance the productivity of this floodplain. The PLSR coefficients of observed discharge suggest flows have the highest contributions to floodplain inundation, meaning rivers are important indicators of wetland conditions and can be used to assess the influence of climate variability on tropical floodplains.

The temporal patterns of total inundation in relation to those of rainfall and flow during the period studied imply climate variability will be key constraints on the productivity of freshwater habitats. In other words, climatic fluctuations will lead to variable inter-annual rainfall, which then results in flow variability. Time series of leading rainfall patterns and discharge (2000–2020) illustrates the constraints of variations in climate over the Mitchell catchment. For instance, wet years (2008–2011) with strong amplitudes in rainfall coincided with strong flows. Similarly, relatively dry years (2015–2018) were characterized by low flows. The relationships between flows and predicted hotspots of aquatic plant biomass reaffirm climatic controls (discharge) on floodplain inundation and productivity. Surface vegetation and phenology respond to changes in hydrological conditions in terrestrial ecosystems and declines in aquatic plant biomass accumulation will be in tandem with the loss of surface inundation in freshwater habitats.

Observed temporal and spatial variability in the distribution of hotspots and total inundation coincides with climate conditions, thus strengthening this argument. Furthermore, the relationship between predicted extents of hotspots and flow suggests that apart from other biophysical factors, which could promote the growth of aquatic primary producers in perennial and intermittent freshwater bodies, the productivity and inundation of tropical rivers in northern Queensland is considerably tied to the impact of climate variability (inter-annual variations in rainfall). However, this may not always be the case as human actions such as water resources development, diversions, and river management can alter flow variability, thus interfering with the observed relationship between rainfall and flow and subsequently, productivity.

19.4.10.2 Changes in Global Climate and Impacts on Flows and Surface Inundation

As with notable big wet periods (e.g., 2008, 2009, and 2011) in the Mitchell catchment, relatively strong spatio-temporal patterns of rainfall were observed in 2014 (Fig. 19.16a, b). But unlike these wet periods, the observed river flows from five gauged stations along the Mitchell river in 2014 were considerably low and significantly similar to those of dry years (Fig. 19.16a). Along the wet–dry tropics in northern Queensland where the Mitchell catchment is located, tropical cyclones and multi-scale global climate teleconnection patterns drive changes in hydrology through unprecedented flows and floods that inundate the downstream floodplain. Specifically, the 2014 signal in the dominant rainfall mode was caused by Tropical Cyclone Fletcher, resulting in one of the wettest periods since the beginning of meteorological measurements. The other big wet periods (e.g., 2008, 2009, and 2011) with stronger amplitudes of rainfall tend to be driven by tropical cyclones along the Coral Sea, which resulted in considerable amplitudes in upstream flows. These flows can also be driven by global teleconnection patterns like the La-Nina (ENSO) event as was observed in 2010 – 2011, resulting in Australia's wettest two-year period on record. Such events, be it ENSO or cyclones along the Gulf of Carpentaria and the Coral Sea, translate to extreme rainfall in the catchment. Unlike the Gulf of Carpentaria, tropical cyclones along the Coral Sea culminate in extreme floods and high magnitude flows that inundate the downstream floodplain.

In the El-Nino (ENSO) phase, which is characterized by warmer sea surface temperatures, rainfall is usually limited across the eastern and northern regions of Australia. This reduced rainfall during such periods impact on flows in catchments, which in turn restricts surface inundation along floodplains. On several spatial and temporal scales, change in surface hydrology can be influenced by interactions among climate, geology, topography and amongst other factors (Kennard et al. 2010). These interactions and potential human actions could increase flow variability leading to changes in the magnitude, frequency, duration, timing, rate of change, and predictability of discharge on different spatial and temporal scales (Naiman et al. 2008).

Flow variability impacts total floodplain inundation and the distribution of aquatic plant biomass distribution (e.g., Ndehedehe et al. 2020c; Thapa et al. 2016; Ward et al. 2014). It has been identified as a key determinant of the ecology of wetlands, setting the conditions for various aquatic and terrestrial biota to persist under different conditions (Bunn et al. 2006; Bunn and Arthington 2002; Kingsford 2000). The catchments along the Australian wet–dry tropics receive more than 90% of their annual rainfall within three months. To maximize the usage of this rainfall, water resources development through the construction of in-stream dams or water harvesting and off-stream storage are proposed to store these monsoon rains for use over the remaining dry months of the year. There is a growing interest to increase agricultural use of this rainfall through the construction of large in-stream dams and other forms of water harvesting (CommonWealth 2015). As detailed in a recent report (Kenyon et al. 2020), the development of water infrastructure *has the capacity to disproportionately modify the seasonality of flow and the magnitude of low flows*. Previous reports (e.g., CSIRO 2018) argued that water extraction from high-level river flows and major floods will have little impact on flow variability. Hydrological conditions (annual flows and dominant rainfall patterns) in the Mitchell appear to be heavily reliant on climate variability. Additionally, upstream flows have higher contributions to downstream floodplain inundation and the growth of aquatic plant biomass in water holes and freshwater habitats. The rivers of the Mitchell catchment currently flow freely with very little interruptions (CSIRO 2018), and the impacts of water extraction on flow variability will likely depend on the nature of inter-annual changes in rainfall. In other words, water extraction during drought years could have a significant impact on flows and subsequent floodplain inundation and productivity. But years with normal summer rainfall could potentially maintain the magnitude of annual flow required for the productivity of the floodplain even after water harvesting during such years. It is true that extreme events like floods have numerous ecological benefits and provide a wide range of ecosystem services (Talbot et al. 2018). The impacts of extreme floods or droughts can result in major alterations of physical and chemical conditions sustaining the health and development of biotic communities, thus necessitating the need to maintain flow variability in ecological communities (Naiman et al. 2008).

References

Acreman M, Arthington AH, Colloff MJ et al (2014) Environmental flows for natural, hybrid, and novel riverine ecosystems in a changing world. Front Ecol Environ 12(8):466–473. https://doi.org/10.1890/130134

Aladin N, Crétaux J-F, Plotnikov IS, Kouraev AV et al (2005) Modern hydro-biological state of the small aral sea. Environmetrics 16(4):375–392. https://doi.org/10.1002/env.709

Aldous A, Fitzsimons J, Richter B, Bach L (2011) Droughts, floods and freshwater ecosystems: evaluating climate change impacts and developing adaptation strategies. Marine Freshw Res 62(3):223–231. https://doi.org/10.1071/MF09285

Alfieri L, Bisselink B, Dottori F, Naumann G et al (2017) Global projections of river flood risk in a warmer world. Earth's Future 5(2):171–182. https://doi.org/10.1002/2016EF000485

Alsdorf D, Han S-C, Bates P, Melack J (2010) Seasonal water storage on the amazon floodplain measured from satellites. Remote Sens Environ 114(11):2448–2456. https://doi.org/10.1016/j.rse.2010.05.020

Anteau MJ (2012) Do interactions of land use and climate affect productivity of waterbirds and prairie-pothole wetlands? Wetlands 32(1):1–9. https://doi.org/10.1007/s13157-011-0206-3

Armstrong JL, Kingsford RT, Jenkins KM (2009) The effect of regulating the lachlan river on the booligal wetlands–the floodplain red gum swamps. Report from University of New South Wales, Sydney. https://www.researchgate.net/profile/Kate_Brandis/publication. Accessed 15th Aug 2021

Bayliss P, Bartolo R, van Dam R (2008) Quantitative ecological risk assessments for the daly river. Ecological risk assessment for Australia"s northern tropical rivers. Sub-project, vol 2, pp 271–415. http://www.environment.gov.au/system/files/resources/054092c1-5bff-4149-abd9-93b27747e55d/files/triap-sp2-chapter-4.pdf. Accessed 15th Aug 2021

Biancamaria S, Lettenmaier D, Pavelsky T (2015) The SWOT mission and its capabilities for land hydrology. Surv Geophys 37:307–337. https://doi.org/10.1007/s10712-015-9346-y

Broadley A, Stewart-Koster B, Kenyon RA, Burford MA, Brown CJ (2020) Impact of water development on river flows and the catch of a commercial marine fishery. Ecosphere 11(7):e03194. https://doi.org/10.1002/ecs2.3194

Brock MA, Nielsen DL, Shiel RJ, Green JD, Langley JD (2003) Drought and aquatic community resilience: the role of eggs and seeds in sediments of temporary wetlands. Freshw Biol 48(7):1207–1218. https://doi.org/10.1046/j.1365-2427.2003.01083.x

Brooks AP, Shellberg JG, Knight J, Spencer J (2009) Alluvial gully erosion: an example from the Mitchell fluvial megafan, Queensland, Australia. Earth Surf Process Landf 34(14):1951–1969. https://doi.org/10.1002/esp.1883

Brown CJ et al (2019) A guide to modelling priorities for managing land-based impacts on coastal ecosystems. J Appl Ecol 56(5):1106–1116. https://doi.org/10.1111/1365-2664.13331

Bunn SE, Arthington AH (2002) Basic principles and ecological consequences of altered flow regimes for aquatic biodiversity. Environ Manag 30(4):492–507. https://doi.org/10.1007/s00267-002-2737-0

Bunn SE, Davies PM, Winning M (2003) Sources of organic carbon supporting the food web of an arid zone floodplain river. Freshw Biol 48(4):619–635. https://doi.org/10.1046/j.1365-2427.2003.01031.x

Bunn SE, Thoms MC, Hamilton SK, Capon SJ (2006) Flow variability in dryland rivers: boom, bust and the bits in between. River Res Appl 22(2):179–186. https://doi.org/10.1002/rra.904

Burn DH, Whitfield PH (2016) Changes in floods and flood regimes in Canada. Canadian Water Res J/Revue Canadienne des Ressources Hydriques 41(1–2):139–150

Cai X, Feng L, Hou X, Chen X (2016) Remote sensing of the water storage dynamics of large lakes and reservoirs in the yangtze river basin from 2000 to 2014. Sci Rep 6(1):1–9. https://doi.org/10.1038/srep36405

Chan TU, Hart BT, Kennard MJ, Pusey BJ, Shenton W, Douglas MM, Valentine E, Patel S (2012) Bayesian network models for environmental flow decision making in the Daly River, Northern Territory, Australia. River Res Appl 28(3):283–301. https://doi.org/10.1002/rra.1456

Chen H, Sun Y, Gao J, Hu Y, Yin B (2018) Solving partial least squares regression via manifold optimization approaches. IEEE Trans Neural Netw Learn Syst 1–13. https://doi.org/10.1109/TNNLS.2018.2844866

Coe MT, Birkett CM (2004) Calculation of river discharge and prediction of lake height from satellite radar altimetry: example for the Lake Chad basin. Water Res Res 40(10):W10205. https://doi.org/10.1029/2003WR002543

Colloff MJ, Baldwin DS (2010) Resilience of floodplain ecosystems in a semi-arid environment. Rangel J 32:305–314. https://doi.org/10.1071/RJ10015

CommonWealth (2015) Our north, our future. White Paper on Developing Northern Australia.https://www.industry.gov.au/sites/default/files/June%202018/document/pdf/nawp-fullreport.pdf?acsf_files_redirect. Accessed 30th May 2020

Crétaux J-F, Jelinski W, Calmant S, Kouraev A, Vuglinski V, Bergé-Nguyen M, Gennero M-C, Nino F, Rio RAD, Cazenave A, Maisongrande P (2011) Sols: a lake database to monitor in the near real time water level and storage variations from remote sensing data. Adv Space Res 47(9):1497–1507. https://doi.org/10.1016/j.asr.2011.01.004

Crétaux JF, Abarca-del Río R, Berge-Nguyen M et al (2016) Lake volume monitoring from space. Surv Geophys 37(2):269–305

CSIRO (2018) Water resource assessment for the Mitchell catchment. An overview report to the Australian Government from the CSIRO Northern Australia Water Resource Assessment, part of the National Water Infrastructure Development Fund: Water Resource Assessments. CSIRO, Australia. https://www.csiro.au/en/Research/Major-initiatives/Northern-Australia/Current-work/NAWRA/Mitchell-report. Accessed 13th Nov 2020

Dafforn KA et al (2015) big data opportunities and challenges for assessing multiple stressors across scales in aquatic ecosystems. Marine Freshw Res 67(4):393–413. https://doi.org/10.1071/MF15108

Davidson NC (2014) How much wetland has the world lost? Long-term and recent trends in global wetland area. Marine Freshw Res 65(10):934–941. https://doi.org/10.1071/MF14173

de Jong S (1993) SIMPLS: an alternative approach to partial least squares regression. Chemom Intell Lab Syst 18(3):251–263. https://doi.org/10.1016/0169-7439(93)85002-x

Döll P, Bunn S (2014) Cross-chapter box on the impact of climate change on freshwater ecosystems due to altered river flow regimes. In: Climate change 2014: impacts, adaptation, and vulnerability. Part A: global and sectoral aspects. Contribution of working Group II to the fifth assessment report of the intergovernmental panel on climate change[Field CB, Barros VR, Dokken DJ, Mach KJ, Mastrandrea MD, Bilir TE, Chatterjee M, Ebi KL, Estrada YO, Genova RC, Girma B, Kissel ES, Levy AN, MacCracken S, Mastrandrea PR, White LL (eds)]. Cambridge University Press, Cambridge, UK, New York, NY, USA, pp 143–146

Döll P, Zhang J (2010) Impact of climate change on freshwater ecosystems: a global-scale analysis of ecologically relevant river flow alterations. Hydrol Earth Syst Sci 14(5):783–799. https://doi.org/10.5194/hess-14-783-2010

Donchyts G, Baart F, Winsemius H, Gorelick N, Kwadijk J, Van De Giesen N (2016) Earth's surface water change over the past 30 years. Nat Clim Change 6(9):810–813. https://doi.org/10.1038/nclimate311

Douben K-J (2006) Characteristics of river floods and flooding: a global overview, 1985–2003. Irrig Drain 55(S1):S9–S21. https://doi.org/10.1002/ird.239

Douglas M et al (2011) Northern futures: threats and opportunities for freshwater ecosystems. CDU Press

Drinkwater KF, Frank KT (1994) Effects of river regulation and diversion on marine fish and invertebrates. Aquat Cons: Marine Freshw Ecosyst 4(2):135–151

Duggan M, Connolly R, Whittle M, Curwen G, Burford M (2014) Effects of freshwater flow extremes on intertidal biota of a wet-dry tropical estuary. Marine Ecol Prog Ser 502:11–23. https://doi.org/10.3354/meps10719

Ficke A, Myrick C, Hansen L (2007) Potential impacts of global climate change on freshwater fisheries. Rev Fish Biol Fish 17(4):581–613. https://doi.org/10.1007/s11160-007-9059-5

Feng M, Sexton JO, Channan S, Townshend JR (2016) A global, high-resolution (30-m) inland water body dataset for 2000: first results of a topographic-spectral classification algorithm. Int J Digit Earth 9(2):113–133. https://doi.org/10.1080/17538947.2015.1026420

Feyisa GL, Meilby H, Fensholt R, Proud SR (2014) Automated water extraction index: a new technique for surface water mapping using landsat imagery. Remote Sens Environ 140:23–35. https://doi.org/10.1016/j.rse.2013.08.029

Gao H (2015) Satellite remote sensing of large lakes and reservoirs: from elevation and area to storage. WIREs Water 2(2):147–157. https://doi.org/10.1002/wat2.1065

Gao H, Birkett C, Lettenmaier DP (2012) Global monitoring of large reservoir storage from satellite remote sensing. Water Res Res 48(9). https://doi.org/10.1029/2012WR012063

Gardner RC et al (2015) State of the world's wetlands and their services to people: a compilation of recent analyses

Getirana A, Kumar SV, Konapala G, Ndehedehe CE (2021) Impacts of fully coupling land surface and flood models on the simulation of large wetlands' water dynamics: the case of the Inner Niger Delta. J Adv Model Earth Syst 13(5):e2021MS002463. https://doi.org/10.1029/2021MS002463

Gleick PH (2003) Global freshwater resources: soft-path solutions for the 21st century. Science 302(5650):1524–1528. https://doi.org/10.1126/science.1089967

Gorelick N, Hancher M, Dixon M, Ilyushchenko S, Thau D, Moore R (2017) Google Earth Engine: planetary-scale geospatial analysis for everyone. Remote Sens Environ 202:18–27. https://doi.org/10.1016/j.rse.2017.06.031

Grimes CB, Kingsford MJ (1996) How do riverine plumes of different sizes influence fish larvae: do they enhance recruitment? Marine Freshw Res 47(2):191–208

Hamilton SK, Gehrke PC (2005) Australia''s tropical river systems: current scientific understanding and critical knowledge gaps for sustainable management. Marine Freshw Res 56(3):243–252

Hoeinghaus DJ, Winemiller KO, Agostinho AA (2007) Landscape-scale hydrologic characteristics differentiate patterns of carbon flow in large-river food webs. Ecosystems 10(6):1019–1033. https://doi.org/10.1007/s10021-007-9075-2

Hong Y, Adler RF, Negri A, Huffman GJ (2007) Flood and landslide applications of near real-time satellite rainfall products. Nat Hazards 43(2):285–294. https://doi.org/10.1007/s11069-006-9106-x

Horne AC, Nathan R, Poff NL, Bond NR, Webb JA, Wang J, John A (2019) Modeling flow-ecology responses in the Anthropocene: challenges for sustainable riverine management. BioScience 69(10):789–799. https://doi.org/10.1093/biosci/biz087

Humphrey V, Gudmundsson L, Seneviratne SI (2016) Assessing global water storage variability from GRACE: trends, seasonal cycle, subseasonal anomalies and extremes. Surv Geophys 37(2):357–395. https://doi.org/10.1007/s10712-016-9367-1

Jackson S, Finn M, Scheepers K (2014) The use of replacement cost method to assess and manage the impacts of water resource development on Australian indigenous customary economies. J Environ Manag 135:100–109. https://doi.org/10.1016/j.jenvman.2014.01.018

Jacobsen D, Milner AM, Brown LE, Dangles O (2012) Biodiversity under threat in glacier-fed river systems. Nat Clim Change 2(5):361–364. https://doi.org/10.1038/nclimate1435

Januchowski-Hartley SR, Visconti P, Pressey RL (2011) A systematic approach for prioritizing multiple management actions for invasive species. Biol Invasions 13(5):1241–1253. https://doi.org/10.1007/s10530-011-9960-7

Jardine T, Pusey B, Hamilton S, Pettit N, Davies P, Douglas et al (2012) Fish mediate high food web connectivity in the lower reaches of a tropical floodplain river. Oecologia 168(3):829–838. https://doi.org/10.1007/s00442-011-2148-0

Jardine TD, Bond NR et al (2015) Does flood rhythm drive ecosystem responses in tropical riverscapes? Ecology 96(3):684–692. https://doi.org/10.1890/14-0991.1

Jardine TD, Hunt RJ, Faggotter SJ, Valdez D, Burford MA, Bunn SE (2013) Carbon from periphyton supports fish biomass in waterholes of a wet-dry tropical river. River Res Appl 29(5):560–573. https://doi.org/10.1002/rra.2554

Jenkins KM, Boulton AJ (2007) Detecting impacts and setting restoration targets in arid-zone rivers: aquatic micro-invertebrate responses to reduced floodplain inundation. J Appl Ecol 44(4):823–832. https://doi.org/10.1111/j.1365-2664.2007.01298.x

Jia J, Gao Y, Zhou F, Shi K, Johnes PJ, Dungait JA, Ma M, Lu Y (2020) Identifying the main drivers of change of phytoplankton community structure and gross primary productivity in a river-lake system. J Hydrol 583:124633. https://doi.org/10.1016/j.jhydrol.2020.124633

Jolliffe IT (2002) Principal component analysis, 2nd edn. Springer Series in Statistics, Springer, New York

Junk WJ, Bayley PB, Sparks RE (1989) The floodpulse concept in river-floodplain systems in DP Dodge (ed), proceedings of the international large river symposium (lars). Canadian J Fish Aquat Sci 106:110–127

Junk WJ, Bayley PB, Sparks RE (1989) The flood pulse concept in river-floodplain systems. Canadian Specl Publ Fish Aquat Sci 106(1):110–127

Junk WJ, Wantzen KM (2004) The flood pulse concept: new aspects, approaches and applications-an update. In: Second international symposium on the management of large rivers for fisheries. Food and Agriculture Organization and Mekong River Commission, FAO Regional Office for Asia and the Pacific, pp 117–149. https://ftp.cs.ru.nl/toinesmits/Recommended_readings_IWRM_2009/Water_Ecomorphological_principles/1989JunkThe%20flood%20pulse%20concept%20in.pdf. Accessed 15th Aug 2021

Karpatne A, Khandelwal A, Chen X, Mithal V, Faghmous J, Kumar V (2016) Global monitoring of inland water dynamics: state-of-the-art, challenges, and opportunities. Comput Sustain 121–147. https://doi.org/10.1007/978-3-319-31858-5_7

Khan AN (2011) Analysis of flood causes and associated socio-economic damages in the hindukush region. Nat Hazards 59(3):1239. https://doi.org/10.1007/s11069-011-9830-8

Karim F, Peña-Arancibia J, Ticehurst C, Marvanek S, Gallant J, Hughes J, Dutta D, Vaze J, Petheram C, Seo L, Kitson S (2018) Floodplain inundation mapping and modelling for the Fitzroy, Darwin and Mitchell catchments. a technical report to the australian government from the csiro northern australia water resource assessment, part of the national water infrastructure development fund: Water resource assessments. CSIRO, Australia

Kazama S, Hagiwara T, Ranjan P, Sawamoto M (2007) Evaluation of groundwater resources in wide inundation areas of the Mekong River basin. J Hydrol 340(3):233–243. https://doi.org/10.1016/j.jhydrol.2007.04.017

Keddy PA, Fraser LH, Solomeshch AI, Junk WJ, Campbell DR, Arroyo MTK, Alho CJR (2009) Wet and wonderful: the world's largest wetlands are conservation priorities. BioScience 59(1):39–51. https://doi.org/10.1525/bio.2009.59.1.8

Keller K et al (2019) Dry season habitat use of fishes in an Australian tropical river. Sci Rep 9(1):1–14. https://doi.org/10.1038/s41598-019-41287-x

Kennard MJ, Pusey Bradley J, Olden JD, Mackay SJ, Stein JL, Marsh N (2010) Classification of natural flow regimes in Australia to support environmental flow management. Freshw Biol 55(1):171–193. https://doi.org/10.1111/j.1365-2427.2009.02307.x

Kenyon R, Burford M, Jarrett A, Faggotter S, Moeseneder C, Buckworth R (2020) Addressing knowledge gaps for studies of the effect of water resource development on the future of the Northern Prawn Fishery: juvenile banana prawn abundance in the estuaries of the Mitchell, Gilbert and Flinders Rivers. CSIRO Oceans and Atmosphere, Brisbane. https://www.frdc.com.au/Archived-Reports/FRDC%20Projects/2016-047-DLD.pdf. Accessed 13th Nov 2020

Kingsford R (2000) Ecological impacts of dams, water diversions and river management on floodplain wetlands in Australia. Austral Ecol 25(2):109–127. https://doi.org/10.1046/j.1442-9993.2000.01036.x

Kirk JTO (1994) Photosynthesis in the aquatic environment. Light Photosynth Aquat Ecosyst

Kundzewicz ZW, Kanae S et al (2014) Flood risk and climate change: global and regional perspectives. Hydrol Sci J 59(1):1–28. https://doi.org/10.1080/02626667.2013.857411

Leahy SM, R. J. (2021) River flows affect the growth of a tropical finfish in the wet-dry rivers of northern Australia, with implications for water resource development. Hydrobiologia. https://doi.org/10.1007/s10750-021-04641-7

Lee H, Shum CK, Tseng KH, Jun-Yi G, Chung-Yen K (2011) Present-day lake level variation from envisat altimetry over the northeastern qinghai-tibetan plateau: links with precipitation and temperature. Terrestrial, atmospheric and oceanic sciences, vol 22(2), p 1. https://core.ac.uk/reader/194887580. Accessed 15th Aug 2021

Leigh C, Sheldon F (2008) Hydrological changes and ecological impacts associated with water resource development in large floodplain rivers in the Australian tropics. River Res Appl 24(9):1251–1270. https://doi.org/10.1002/rra.1125

Lettenmaier DP, Alsdorf D, Dozier J, Huffman GJ, Pan M, Wood EF (2015) Inroads of remote sensing into hydrologic science during the WRR ERA. Water Res Res 51(9):7309–7342. https://doi.org/10.1002/2015WR017616

Lettenmaier DP, Famiglietti JS (2006) Water from on high. Nature 444(7119):562–563. https://doi.org/10.1038/444562a

Lewis WM, Hamilton SK, Rodríguez MA, Saunders JF, Lasi MA (2001) Foodweb analysis of the orinoco floodplain based on production estimates and stable isotope data. J North Am Benthol Soc 20(2):241–254. https://doi.org/10.2307/1468319

Lewis-Beck M, Bryman A, Futing T (2003) Encyclopedia of social sciences research methods. Thousand Oaks (CA): Sage. https://www.utdallas.edu/~herve/Abdi-PLS-pretty.pdf. Accessed 4th Jan 2019

Luke C-B, Butler GL, Bond NR, Bunn SE, Kennard MJ (2021) Long-term acoustic telemetry reveals limited movement of fish in an unregulated, perennial river. Marine Freshw Res. https://doi.org/10.1071/MF21004

McCabe MF, Rodell M, Alsdorf DE, Miralles DG, Uijlenhoet R, Wagner W, Lucieer A, Houborg R, Verhoest NEC, Franz TE, Shi J, Gao H, Wood EF (2017) The future of Earth observation in hydrology. Hydrol Earth Sys Sci 21(7):3879–3914. https://doi.org/10.5194/hess-21-3879-2017

McClain ME (2002) The application of ecohydrological principles for better water resources management in South America. IAHS special, the ecohydrology of South American rivers and wetlands, p 6

McCann KS, Rooney N (2009) The more food webs change, the more they stay the same. Philos Trans R Soc B: Biol Sci 364(1524):1789–1801

McClain ME (2013) Balancing water resources development and environmental sustainability in Africa: a review of recent research findings and applications. Ambio 42(5):549–565. https://doi.org/10.1007/s13280-012-0359-1

Midwood JD, Chow-Fraser P (2010) Mapping floating and emergent aquatic vegetation in coastal wetlands of eastern Georgian bay, lake Huron, Canada. Wetlands 30(6):1141–1152. https://doi.org/10.1007/s13157-010-0105-z

Moon Y-I, Rajagopalan B, Lall U (1995) Estimation of mutual information using kernel density estimators. Phys Rev E 52:2318–2321. https://doi.org/10.1103/PhysRevE.52.2318

Müller MF, Yoon J, Gorelick SM, Avisse N, Tilmant A (2016) Impact of the Syrian refugee crisis on land use and transboundary freshwater resources. Proc Nat Acad Sci 113(52):14932–14937. https://doi.org/10.1073/pnas.1614342113

Naiman RJ, Bunn SE, Nilsson C, Petts GE, Pinay G, Thompson LC (2002) Legitimizing fluvial ecosystems as users of water: an overview. Environ Manag 30(4):455–567. https://doi.org/10.1007/s00267-002-2734-3

Naiman R, Latterell J, Pettit N, Olden J (2008) Flow variability and the biophysical vitality of river systems. Comptes Rendus Geosci 340(9–10):629–643. https://doi.org/10.1016/j.crte.2008.01.002

Nelson KC, Palmer MA et al (2009) Forecasting the combined effects of urbanization and climate change on stream ecosystems: from impacts to management options. J Appl Ecol 46(1):154–163. https://doi.org/10.1111/j.1365-2664.2008.01599.x

Ndehedehe C (2020) Floodplain productivity of the Gilbert and Flinders catchments: Component 3 final report. Griffith University, Brisbane. https://research-repository.griffith.edu.au/handle/10072/394768. Accessed 8th July 2020

Ndehedehe CE, Burford MA, Stewart-Koster B, Bunn SE (2020) Satellite-derived changes in floodplain productivity and freshwater habitats in northern Australia (1991–2019). Ecol Indic 114:106320. https://doi.org/10.1016/j.ecolind.2020.106320

Ndehedehe CE, Ferreira VG (2020) Assessing land water storage dynamics over Southern America. J Hydrol 580:124339. https://doi.org/10.1016/j.jhydrol.2019.124339

Ndehedehe CE, Ferreira VG, Agutu NO (2019) Hydrological controls on surface vegetation dynamics over West and Central Africa. Ecol Indic 103:494–508. https://doi.org/10.1016/j.ecolind.2019.04.032

Ndehedehe CE, Ferreira VG, Onojeghuo AO, Agutu NO, Emengini E, Getirana A (2020) Influence of global climate on freshwater changes in Africa's largest endorheic basin using multi-scaled indicators. Sci Total Environ 737:139643. https://doi.org/10.1016/j.scitotenv.2020.139643

Ndehedehe CE, Onojeghuo AO, Stewart-Koster B, Bunn SE, Ferreira VG (2021) Upstream flows drive the productivity of floodplain ecosystems in tropical Queensland. Ecol Indic. https://doi.org/10.1016/j.ecolind.2021.107546

Ndehedehe CE, Stewart-Koster B, Burford MA, Bunn SE (2020) Predicting hot spots of aquatic plant biomass in a large floodplain river catchment in the Australian wet-dry tropics. Ecol Indic 117:106616. https://doi.org/10.1016/j.ecolind.2020.106616

Nilsson C, Reidy CA, Dynesius M, Revenga C (2005) Fragmentation and flow regulation of the world's large river systems. Science 308(5720):405–408. https://doi.org/10.1126/science.1107887

Oczkowski AJ, Nixon SW, Granger SL, El-Sayed A-FM, McKinney RA (2009) Anthropogenic enhancement of Egypt's mediterranean fishery. Proc Nat Acad Sci 106(5):1364–1367. https://doi.org/10.1073/pnas.0812568106

Ogden FL, Raj Pradhan N, Downer CW, Zahner JA (2011) Relative importance of impervious area, drainage density, width function, and subsurface storm drainage on flood runoff from an urbanized catchment. Water Res Res 47(12). https://doi.org/10.1029/2011WR010550

Olden JD (2016) Challenges and opportunities for fish conservation in dam-impacted waters. Conserv Freshw Fishes 107–148

O'Mara K, Venarsky M, Stewart-Koster B, McGregor GB, Schulz C, Kainz M, Marshall J, Bunn SE (2021) Connectivity of fish communities in a tropical floodplain river system and predicted impacts of potential new dams. Sci Total Environ 788:147785. https://doi.org/10.1016/j.scitotenv.2021.147785

Palmer MA, Lettenmaier DP, Poff NL, Postel SL, Richter B, Warner R (2009) Climate change and river ecosystems: protection and adaptation options. Environ Manag 44(6):1053–1068. https://doi.org/10.1007/s00267-009-9329-1

Palmer MA, Reidy Liermann CA, Nilsson C, Flörke M, Alcamo J, Lake PS, Bond N (2008) Climate change and the world's river basins: anticipating management options. Front Ecol Environ 6(2):81–89. https://doi.org/10.1890/060148

Pekel J-F, Cottam A, Gorelick N, Belward AS (2016) High-resolution mapping of global surface water and its long-term changes. Nature 540:418–422, 111803. https://doi.org/10.1038/nature20584

Penatti NC, de Almeida TIR, Ferreira LG, Arantes AE, Coe MT (2015) Satellite-based hydrological dynamics of the world's largest continuous wetland. Remote Sens Environ 170:1–13. https://doi.org/10.1016/j.rse.2015.08.031

Peters D, Caissie D, Monk W, Rood S, St-Hilaire A (2016) An ecological perspective on floods in Canada. Canadian Water Res J/Revue Canadienne des Ressources Hydriques 41(1–2):288–306. https://doi.org/10.1080/07011784.2015.1070694

Peterson JT, Kwak TJ (1999) Modeling the effects of land use and climate change on riverine smallmouth bass. Ecol Appl 9(4):1391–1404. https://doi.org/10.1890/1051-0761(1999)0091391:MTEOLU2.0.CO;2

Pettit N, Ward D, Adame M, Valdez D, Bunn S (2016) Influence of aquatic plant architecture on epiphyte biomass on a tropical river floodplain. Aquat Botany 129:35–43. https://doi.org/10.1016/j.aquabot.2015.12.001

Pettit NE, Naiman RJ, Warfe DM, Jardine TD, Douglas MM, Bunn SE, Davies PM (2017) Productivity and connectivity in tropical riverscapes of northern Australia: ecological insights for management. Ecosystems 20(3):492–514. https://doi.org/10.1007/s10021-016-0037-4

Petheram C, Watson I, Bruce C, Chilcott C (2018) Water resource assessment for the Mitchell catchment. a report to the Australian government from the CSIRO northern Australia water resource assessment, part of the national water infrastructure development fund: water resource assessments. CSIRO. https://www.csiro.au/en/Research/Major-initiatives/Northern-Australia/Current-work/NAWRA/Mitchell-report. Accessed 2019

Pettit NE, Naiman RJ, Warfe DM, Jardine TD, Douglas MM, Bunn SE, Davies PM (2017) Productivity and connectivity in tropical riverscapes of northern Australia: ecological insights for management. Ecosystems 20:492–514. https://doi.org/10.1007/s10021-016-0037-4

Poff NL (2002) Ecological response to and management of increased flooding caused by climate change. Philosophical transactions of the royal society of London. Series A: mathematical, physical and engineering sciences, vol 360(1796), pp 1497–1510

Poff NL, Allan JD, Bain MB, Karr JR, Prestegaard KL, Richter BD, Sparks RE, Stromberg JC (1997) The natural flow regime. BioScience 47(11):769–784. https://doi.org/10.2307/1313099

Poff NL, Zimmerman JKH (2010) Ecological responses to altered flow regimes: a literature review to inform the science and management of environmental flows. Freshw Biol 55(1):194–205. https://doi.org/10.1111/j.1365-2427.2009.02272.x

Poff NL (2018) Beyond the natural flow regime? broadening the hydro-ecological foundation to meet environmental flows challenges in a non-stationary world. Freshw Biol 63(8):1011–1021. https://doi.org/10.1111/fwb.13038

Porter EM, Bowman WD, Clark CM, Compton JE, Pardo LH, Soong JL (2013) Interactive effects of anthropogenic nitrogen enrichment and climate change on terrestrial and aquatic biodiversity. Biogeochemistry 114(1):93–120. https://doi.org/10.1007/s10533-012-9803-3

Postel S, Richter B (2012) Rivers for life: managing water for people and nature. Island Press

Postel SL, Daily GC, Ehrlich PR (1996) Human appropriation of renewable fresh water. Science 271(5250):785–788. https://doi.org/10.1126/science.271.5250.785

Preisendorfer R (1988) Principal component analysis in meteorology and oceanography. Developments in atmospheric science, vol 17. Elsevier, Amsterdam

Pusey B, Kennard M, Burrows D, Perna C, Kyne P, Cook B, Hughes J (2011) Freshwater fish. CDU Press

Pusey, B. and Kennard, M. (2009). Aquatic ecosystems in northern AustraliaIn Northern Australia Land and Water Science Review. Final report to the northern Australia land and water taskforce. CSIRO

Reis V, Hermoso V, Hamilton SK et al (2017) A global assessment of Inland wetland conservation status. BioScience 67(6):523–533. https://doi.org/10.1093/biosci/bix045

Renöfält BM, Jansson R, Nilsson C (2010) Effects of hydropower generation and opportunities for environmental flow management in Swedish riverine ecosystems. Freshw Biol 55(1):49–67. https://doi.org/10.1111/j.1365-2427.2009.02241.x

Richter BD, Mathews R, Harrison DL, Wigington R (2003) Ecologically sustainable water management: managing river flows for ecological integrity. Ecol Appl 13(1):206–224. https://doi.org/10.1890/1051-0761(2003)013[0206:ESWMMR]2.0.CO;2

Robins JB, Halliday IA, Staunton-Smith J, Mayer DG, Sellin MJ (2005) Freshwater-flow requirements of estuarine fisheries in tropical Australia: a review of the state of knowledge and application of a suggested approach. Marine Freshw Res 56(3):343–360. https://doi.org/10.1071/MF04087

Roggeri H (2013) Tropical freshwater wetlands: a guide to current knowledge and sustainable management. Springer Science & Business Media

Rojas R, Feyen L, Watkiss P (2013) Climate change and river floods in the European Union: socio-economic consequences and the costs and benefits of adaptation. Glob Environ Change 23(6):1737–1751. https://doi.org/10.1016/j.gloenvcha.2013.08.006

Sadegh M, Ragno E, AghaKouchak A (2017) Multivariate Copula Analysis Toolbox (MvCAT): describing dependence and underlying uncertainty using a Bayesian framework. Water Res Res 53(6):5166–5183. https://doi.org/10.1002/2016WR020242

Saigo M, Zilli FL, Marchese MR, Demonte D (2015) Trophic level, food chain length and omnivory in the paraná river: a food web model approach in a floodplain river system. Ecol Res 30(5):843–852. https://doi.org/10.1007/s11284-015-1283-1

Schindler S, Sebesvari Z, Damm C et al (2014) Multifunctionality of floodplain landscapes: relating management options to ecosystem services. Landsc Ecol 29:229–244. https://doi.org/10.1007/s10980-014-9989-y

Schrapffer A, Sörensson A, Polcher J et al (2020) Benefits of representing floodplains in a land surface model: pantanal simulated with ORCHIDEE CMIP6 version. Clim Dyn 55:1303–1323. https://doi.org/10.1007/s00382-020-05324-0

Schwatke C, Dettmering D, Bosch W, Seitz F (2015) Dahiti—an innovative approach for estimating water level time series over inland waters using multi-mission satellite altimetry. Hydrol Earth Syst Sci 19(10):4345–4364. https://doi.org/10.5194/hess-19-4345-2015

Sharma A (2000) Seasonal to interannual rainfall probabilistic forecasts for improved water supply management: Part 1–a strategy for system predictor identification. J Hydrol 239(1):232–239. https://doi.org/10.1016/S0022-1694(00)00346-2

Sivakumar B (2011) Global climate change and its impacts on water resources planning and management: assessment and challenges. Stoch Environ Res Risk Assess 25(4):583–600. https://doi.org/10.1007/s00477-010-0423-y

Slagter B, Tsendbazar N-E, Vollrath A, Reiche J (2020) Mapping wetland characteristics using temporally dense Sentinel-1 and Sentinel-2 data: a case study in the St Lucia wetlands, South Africa. Int J Appl Earth Obs Geoinform 86:102009. https://doi.org/10.1016/j.jag.2019.102009

Sorí R, Nieto R, Vicente-Serrano SM, Drumond A, Gimeno L (2017) A Lagrangian perspective of the hydrological cycle in the Congo River basin. Earth Syst Dyn 8(3):653–675. https://doi.org/10.5194/esd-8-653-2017

Staunton-Smith J, Robins JB, Mayer DG, Sellin MJ, Halliday IA (2004) Does the quantity and timing of fresh water flowing into a dry tropical estuary affect year-class strength of barramundi (lates calcarifer)? Marine Freshw Res 55:787–797, 111803. https://doi.org/10.1071/MF03198

Stein J, Stein J, Nix H (2002) Spatial analysis of anthropogenic river disturbance at regional and continental scales: identifying the wild rivers of Australia. Landsc Urban Plann 60(1):1–25. https://doi.org/10.1016/S0169-2046(02)00048-8

Stewart IT, Cayan DR, Dettinger MD (2005) Changes toward earlier streamflow timing across western North America. J Clim 18(8):1136–1155. https://doi.org/10.1175/JCLI3321.1

Stewart-Koster B, Olden JD, Kennard MJ, Pusey BJ, Boone EL, Douglas M, Jackson S (2011) Fish response to the temporal hierarchy of the natural flow regime in the Daly River, northern Australia. J Fish Biol 79(6):1525–1544. https://doi.org/10.1111/j.1095-8649.2011.03072.x

Szuwalski CS, Vert-Pre KA, Punt AE, Branch TA, Hilborn R (2015) Examining common assumptions about recruitment: a meta-analysis of recruitment dynamics for worldwide marine fisheries. Fish Fish 16(4):633–648. https://doi.org/10.1111/faf.12083

Talbot C, Bennett E, Cassell K et al (2018) The impact of flooding on aquatic ecosystem services. Biogeochemistry 141:439–461. https://doi.org/10.1007/s10533-018-0449-7

Taylor MD, Loneragan NR (2019) Catchment-derived stressors, recruitment, and fisheries productivity in an exploited penaeid shrimp. Reg Stud Marine Sci 29:100628. https://doi.org/10.1016/j.rsma.2019.100628

Terrado M, Acuña V, Ennaanay D, Tallis H, Sabater S (2014) Impact of climate extremes on hydrological ecosystem services in a heavily humanized mediterranean basin. Ecol Indic 37:199–209. https://doi.org/10.1016/j.ecolind.2013.01.016

Thapa R, Thoms MC, Parsons M (2016) The response of dryland floodplain vegetation productivity to flooding and drying. J Arid Environ 129:42–55. https://doi.org/10.1016/j.jaridenv.2016.02.007

Thorp JH, Delong MD (1994) The riverine productivity model: an heuristic view of carbon sources and organic processing in large river ecosystems. Oikos 70(2):305–308. https://doi.org/10.2307/3545642

Ticehurst JL, Newham LT, Rissik D, Letcher RA, Jakeman AJ (2007) A bayesian network approach for assessing the sustainability of coastal lakes in New South Wales, Australia. Environ Modell Softw 22(8):1129–1139. https://doi.org/10.1016/j.envsoft.2006.03.003

Tockner K, Malard F, Ward JV (2000) An extension of the flood pulse concept. Hydrol Process 14(16–17):2861–2883. https://doi.org/10.1002/1099-1085(200011/12)14:16/17<3c2861::AID-HYP124>3e3.0.CO;2-F

Tockner K, Lorang MS, Stanford JA (2010) River flood plains are model ecosystems to test general hydrogeomorphic and ecological concepts. River Res Appl 26(1):76–86. https://doi.org/10.1002/rra.1328

Tortini R, Noujdina N, Yeo S, Ricko M, Birkett CM, Khandelwal A, Kumar V, Marlier ME, Lettenmaier DP (2020) Satellite-based remote sensing data set of global surface water storage change from 1992 to 2018. Earth Syst Sci Data 12(2):1141–1151. https://doi.org/10.5194/essd-12-1141-2020

Turschwell MP, Stewart-Koster B, King AJ et al (2019) Flow-mediated predator-prey dynamics influence fish populations in a tropical river. Freshw Biol 64(8):1453–1466. https://doi.org/10.1111/fwb.13318

Vadeboncoeur Y, Devlin SP, McIntyre PB, Vander Zanden MJ (2014) Is there light after depth? Distribution of periphyton chlorophyll and productivity in lake littoral zones. Freshw Sci 33(2):524–536. https://doi.org/10.1086/676315

Vance DJ, Staples DJ, Kerr JD (1985) Factors affecting year-to-year variation in the catch of banana prawns (Penaeus merguiensis) in the Gulf of Carpentaria, Australia. ICES J Marine Sci 42(1):83–97. https://doi.org/10.1093/icesjms/42.1.83

Venarsky MP, Stewart-Koster B, Hunt RJ, Jardine TD, Bunn SE (2020) Spatial and temporal variation of fish community biomass and energy flow throughout a tropical river network. Freshw Biol 65(10):1782–1792. https://doi.org/10.1111/fwb.13581

Verpoorter C, Kutser T, Seekell DA, Tranvik LJ (2014) A global inventory of lakes based on high-resolution satellite imagery. Geophys Res Lett 41(18):6396–6402. https://doi.org/10.1002/2014GL060641

Vörösmarty C, Lettenmaier D, Leveque C et al (2004) Humans transforming the global water system. Eos Trans Am Geophys Union 85(48):509–514. https://doi.org/10.1029/2004EO480001

Vörösmarty CJ, Sahagian D (2000) Anthropogenic disturbance of the terrestrial water cycle. BioScience 50(9):753–765. https://doi.org/10.1641/0006-3568(2000)050[0753:ADOTTW]2.0.CO;2

Vörösmarty CJ, Green P, Salisbury J, Lammers RB (2000) Global water resources: vulnerability from climate change and population growth. Science 289(5477):284–288. https://doi.org/10.1126/science.289.5477.284

Vorosmarty CJ, McIntyre PB, Gessner MO, Dudgeon D, Prusevich A, Green P, Glidden S, Bunn SE, Sullivan CA, Liermann CR, Davies PM (2010) Global threats to human water security and river biodiversity. Nature 467(7315):555–561. https://doi.org/10.1038/nature09440

Ward D, Petty A, Setterfield S, Douglas M, Ferdinands K, Hamilton S, Phinn S (2014) Floodplain inundation and vegetation dynamics in the alligator rivers region (kakadu) of northern Australia assessed using optical and radar remote sensing. Remote Sens Environ 147:43–55. https://doi.org/10.1016/j.rse.2014.02.009

Ward DP, Hamilton SK, Jardine TD, Pettit NE, Tews EK, Olley JM, Bunn SE (2013) Assessing the seasonal dynamics of inundation, turbidity, and aquatic vegetation in the Australian wet-dry tropics using optical remote sensing. Ecohydrology 6(2):312–323. https://doi.org/10.1002/eco.1270

Ward D, Pusey B, Brooks A, Olley J, Shellberg J, Spencer J, Tews K (2011) River landscapes and aquatic systems diversity. Aquat Biodivers Northern Australia: Patterns Threats Future 1:5–22

Ward DP, Pettit NE, Adame M, Douglas MM, Setterfield SA, Bunn SE (2016) Seasonal spatial dynamics of floodplain macrophyte and periphyton abundance in the Alligator Rivers region (Kakadu) of northern Australia. Ecohydrology 9(8):1675–1686. https://doi.org/10.1002/eco.1757

Warfe DM, Pettit NE, Davies PM et al (2011) The "wet-dry" in the wet-dry tropics drives river ecosystem structure and processes in northern Australia. Freshw Biol 56(11):2169–2195. https://doi.org/10.1111/j.1365-2427.2011.02660.x

WEF (2016) The global risks report 2016 11th edn. World economic forum. https://core.ac.uk/reader/194887580. Accessed 16th Aug 2021

Wetzel RG (2001) Limnology: lake and river ecosystems. Gulf Professional Publishing

Winemiller KO (2004) Floodplain river food webs: generalizations and implications for fisheries management. In: Proceedings of the second international symposium on the management of large rivers for fisheries, vol 2, pp 285–309. http://citeseerx.ist.psu.edu/viewdoc/download?doi=10.1.1.729.280&rep=rep1&type=pdfpage=293. Accessed 1st Aug 2021

Woinarski J, Mackey B, Nix H, Traill B (2007) The nature of northern Australia: its natural values, ecological processes and future prospects. ANU Press, p 127

Woodward G, Bonada N, Brown LE (2016) The effects of climatic fluctuations and extreme events on running water ecosystems. Philos Trans R Soc B: Biol Sci 371(1694):20150274. https://doi.org/10.1098/rstb.2015.0274

Worden J, de Beurs KM (2020) Surface water detection in the Caucasus. Int J Appl Earth Obs Geoinform 91:102159. https://doi.org/10.1016/j.jag.2020.102159

Xenopoulos MA et al (2005) Scenarios of freshwater fish extinctions from climate change and water withdrawal. Glob Change Biol 11(10):1557–1564. https://doi.org/10.1111/j.1365-2486.2005.001008.x

Yang X, Qin Q, Yésou H, Ledauphin T, Koehl M, Grussenmeyer P, Zhu Z (2020) Monthly estimation of the surface water extent in France at a 10-m resolution using Sentinel-2 data. Remote Sens Environ 244:111803. https://doi.org/10.1016/j.rse.2020.111803

Yeh TY, Ke TY, Lin YL (2011) Algal growth control within natural water purification systems: macrophyte light shading effects. Water Air Soil Pollut 214(1):575–586. https://doi.org/10.1007/s11270-010-0447-4

Part XIII
Spatial and Temporal Analysis of Extreme Events

Chapter 20
Hotspots of Climatic Influence

Abstract Extreme events like droughts and floods are important parts of human history, dating back to the Holocene. These events contributed to the collapse of ancient civilizations and continue to affect modern society. For instance, approximately 55 million people globally are affected by droughts every year. Unfortunately, the magnitude of these events is expected to increase, further subjecting human societies to perilous conditions, including financial and environmental problems. In this day and age, living with droughts is now a way of life. However, it is also a concept that must be fully understood. This understanding will underpin a blue print to enhance adaptation plans and drought resilience looking forward. This chapter explores various approaches, including data and novel techniques to measuring and assessing drought characteristics (e.g., intensity, duration, extents, etc.) to improve understanding of their impacts. Major drought hotspots are identified and used as a test-bed to illustrate the potential of advanced statistical methods and machine learning to quantify the contributions of global climate as a first step to identifying drought risk.

20.1 Hydro-Meteorological Changes and Drought Events

The influence of drought events on agricultural and economic sectors, as well as its impacts on crops and wildlife habitats has been extensively studied (Dai 2011; Sternberg 2011; Wilhite et al. 2007). Additionally, drought has considerable impacts on geothermal, hydropower, and even solar thermal generations (Forrest et al. 2018). When rainfall is restricted for a long period, it can result in low water availability in hydrological systems as well as the occurrence of heat waves, which has a high risk of triggering the spread of wildfire (e.g., AghaKouchak et al. 2018). Drought is a well-known climate phenomenon in human history. Between 1951 and 2010, increasing trends in meteorological drought intensity, duration, and frequency over the Amazon, Central America, East Asia, Western Africa, and the Mediterranean were reported by Spinoni et al. (2014). In the West African Sahel, several drought episodes in the 1970s contributed to substantial water loss in the Lake Chad.

C. Ndehedehe, *Satellite Remote Sensing of Terrestrial Hydrology*,
https://doi.org/10.1007/978-3-030-99577-5_20

Also, climate model simulations have been used to mimic ground-based observations of continental drought areas, and evidence shows substantial positive trends in droughts events (Nasrollahi et al. 2015). Apart from the Lake Chad basin, there are several climatic hotspots in the world today (e.g., the United States, India, Australia, etc.) where drought has played a major role in driving down the economy, affecting water availability and food production. The cascading impacts of droughts in these regions are unique, resulting in different water management efforts and important initiatives. These have been discussed further in this chapter and the various scientific approaches to measuring and assessing droughts and its impacts, including data and methodologies are also the main stay of this chapter. Moreover, risk management of drought that underpins an increase in resilience to climate change requires a suitable framework to support the assessment of droughts and their impact on hydro-ecological systems (Ndehedehe et al. 2021). Case-specific studies involving the use of advanced statistical methods and machine learning to quantify the contributions of global climate patterns on drought as a first step to identifying drought risk are also presented in this chapter.

20.1.1 Measuring Drought

Large hydro-climatic fluctuations and prolonged droughts affect catchments in several ways. The nature of these impacts (e.g., on water availability) which was discussed in preceding sections can vary across regions depending on several factors such as water use, topography, climate, etc. Such impacts can also affect or modify catchment properties. For instance, the impacts of drought in regions that are typically dry can lead to water use in vegetation in complex ways (Ndehedehe et al. 2019b). This complexity is evidenced in physiological as well as morphological changes in vegetation. These changes enable vegetation communities to adapt to dry or unfavourable conditions. When drought becomes prolonged it can lead to compositional changes in the structure of vegetation as was noted for the Congolese forest (Zhou et al. 2014). Given the intrinsic response of catchments and regions to drought events, choosing appropriate indicators that reflect the response of hydro-ecological, be it freshwater, agriculture to climatic and anthropogenic constraints is thus crucial (Ndehedehe et al. 2021; van Dijk et al. 2013). This will improve the understanding of the cascading impacts of drought as well as increase the opportunities for adaptation that increases drought preparedness.

Drought characterization is important, especially given the indications from future climate projections, which suggest significant increase in drought intensity, frequency, and severity across several continents like Africa, Asia, Europe, and America (Vicente-Serrano and Domínguez-Castro 2020; Spinoni et al. 2014). Studies attempting to understand the extremities of climatic influence on various hotspots have implemented both direct observations and model simulations to assess the increasing trends of average temperature and rainfall patterns. Much of these studies attributed the spike in trends to anthropogenic emissions (Easterling et al. 2016; Fischer and Knutti 2015).

However, other studies using model simulations of historical climate data have also attempted to determine these same influences even without anthropogenic forcing so as to understand if observed climatic conditions are linked to anthropogenic influences, natural variability or both (Easterling et al. 2016). Evidence on the combined influence of climate change and human intervention is converging and both drivers have contributed to the rise in anticipated potential fluctuations in the global water cycle. It is expected that fluxes leading to increased evaporation and dry periods, as well as drought intensifications will increase along with the water-holding capacity of the atmosphere, though this case will only be feasible in a warmer world. The representation of this process is largely evident in many global climate models.

20.1.1.1 Climate Models

Global climate models (GCMs) have been developed to measure the impacts of climate change like extreme drought events as well as to improve the understanding of future outlook. These models capture enhanced summer dryness and winter wetness over several regions located in the mid to high latitudes in the Northern Hemisphere, which are well-known hotspots of decline in freshwater on several spatial scales (e.g., Dai 2011; Sheffield and Wood 2008). These anomalies contribute to observed changes in terrestrial water storage (TWS) of several regions around the world through its thorough identification of areas having a similar magnitude in drought characteristics while responding to climate change. The ability to characterize and measure drought impacts helps in providing essential information for water resource management. The inherent uncertainties in GCM, however, have impeded its ability to reproduce and capture this impact on terrestrial hydrology (Hempel et al. 2013). This has therefore resulted in the use of global impact models (GIMs) forced by data from GCMs in an attempt to better understand the influences of varied climate effects on the water cycle as well as drought (Seneviratne et al. 2010). Another reason for adopting this technique was due to the lag in understanding the experience by the land atmospheric feedbacks reproduced by global models as well as the complications noticed in the full coupling of GCMs and GIMs (Taylor et al. 2012).

20.1.1.2 Drought Metrics

Drought is caused by a protracted below average rainfall and is one of the costliest natural disasters (Nasrollahi et al. 2015). This is particularly so given its catastrophic effect on ecosystem health, socio-economic development, agriculture, water resources, and its correlation with armed conflict (Mach et al. 2019). Its effect on regional water storage changes, including surface water and groundwater of any area due to its frequency and extent provides substantial information as to the nature of its impacts. This is because terrestrial water storage (TWS) represents an integrated measure of the water column, which captures all of the components of land water storage changes like soil moisture, groundwater, and water stored in lakes and reservoirs. These components together can be used to evaluate the hydrological drought condition of any region or water availability in relation to precipitation deficits (Zhao et al. 2017; Long et al. 2013). Short-term changes in climatic conditions (rainfall or evapotranspiration) can impact agriculture production and water availability in the soil (Ndehedehe et al. 2021). Several drought metrics like meteorological drought indicators prepared from 3-month aggregated normalized rainfall anomalies (McKee et al. 1995) were developed to measure drought and enable its assessment. Additionally, the difference between rainfall and evapotranspiration (water budget) has also been constructed to quantify drought or determine its onset, duration, and intensity (Vicente-Serrano et al. 2010a).

The progress in the development of several meteorological-based indices and hydrological indicators has improved our ability to measure droughts and quantify their impacts on hydro-ecological systems. Studies have reported on the use of these indicators for drought monitoring over the years. They can be meteorological-based indices or those derived from single climate variable like the standardized precipitation index (SPI, McKee et al. 1993) and standardized runoff index (SRI, Shukla and Wood 2008). Others can be based on multi-scale indicators like the Palmer drought severity index (Palmer 1965) or hydrological state variables such as the soil moisture drought index (Sheffield and Wood 2008). These critical indices are used for assessing agricultural, meteorological, and hydrological drought events (Ndehedehe et al. 2021, 2016a; Sheffield and Wood 2008; Van Loon 2015). These aforementioned indices make use of discrete storages or fluxes for their assessments. However, the introduction of terrestrial water storage drought severity index (TWS-DSI; Zhao et al. 2017) to monitor drought relative to its vertically associated water storage parameters has proved to be more effective than the conventional meteorological-based indices. The considerable correlation of the TWS-DSI to the conventional indices especially in regions with long-term water storage anomalies has been widely captured in several literatures (Zhao et al. 2017). However, it provides an integrated measure through the incarceration of the effects of slow-responding terms such as groundwater and deep soil moisture. Other studies focusing on TWS-based drought assessment have also shown the relevance, accuracy as well as the significance of integrating TWS with the conventional drought indices, especially as a result of its extensive depiction of drought impacts on vegetation growth and hydrologic systems (Du et al. 2019; Long et al. 2013).

This is possible because TWS responds directly to variations in rainfall patterns, integrates soil moisture, and regulates runoff generations, thereby providing the capacity to assess other drought types (Du et al. 2019). It is expected that by mid 2030–2060 and late 2070–2099 of the twenty-first century, the estimated TWS over the majority of the southern hemisphere, a large portion of Europe and the Mediterranean, as well as the conterminous region of the United States will decline significantly. However, northern high latitudes, northern and southern Asia, and eastern Africa are expected to experience a TWS surge during the same period (Fig. 20.1). A stronger TWS decline over the Southern Hemisphere as compared to the Northern Hemisphere through the latitudinal average is also noticed (Fig. 20.1). This is due

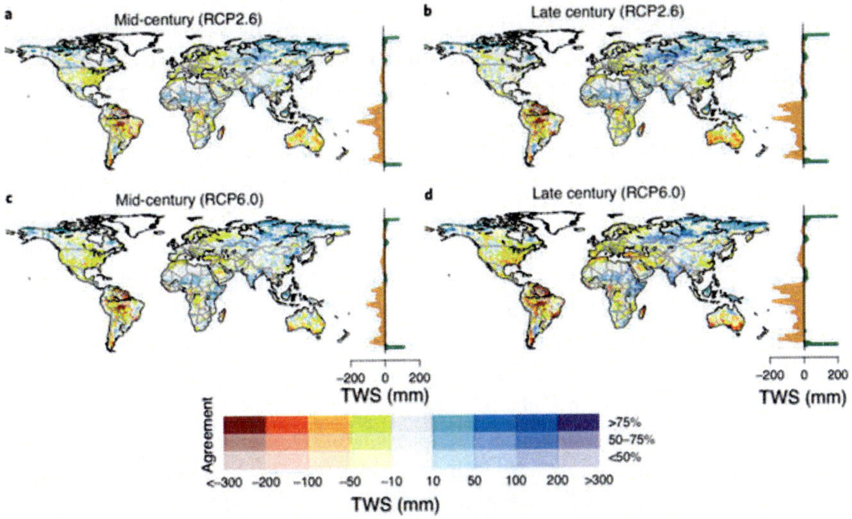

Fig. 20.1 Influence of climate change on terrestrial water storage (TWS) based on the variations in multi-model weighted mean in TWS averaged for the (**a**, **c**) mid (2030–2059) and (**b**, **d**) late (2070–2099) twenty-first century under RCP2.6 (**a**, **b**) and RCP6.0 (**c**, **d**) with respect to the mean for the historical baseline duration (1976–2005). The latitudinal mean for each figure is represented at the right-hand side of the plot. *Source* Pokhrel et al. (2021)

to the decline in the South American and Australian regions, which strongly correlates with estimated changes in rainfall patterns, while still possibly being a factor of the GCM tendency in overestimating drying trends over the Southern Hemisphere (Nasrollahi et al. 2015).

20.1.2 Remote Sensing of Drought Events

The need for a more comprehensive, efficient, and reliable drought monitoring methodology is important to aid in the planning and mitigation of global and sub-regional impacts resulting from drought events. The characterization of key drought metrics and properties such as severity, onset, duration, intensity, and frequency could be difficult and complex. For example, excessive groundwater pumping draws down the water table and under prolonged drought conditions and higher evaporation, groundwater depletion and low well yields will result in groundwater drought. This is when groundwater wells run dry or the water table is substantially lowered. Under extreme drought, impacts on surface water (its availability will be limited) will lead to increased dependence on groundwater. Moreover, discharge to streams and springs will be limited. Under such circumstances, the cascading impacts of droughts is exacerbated by anthropogenic influence such as a rise in groundwater use (Ndehedehe et al. 2021; Williams and Funk 2011). The increase in water demand during such period could also be driven by changes in human dietary conditions, leading to the expansion of irrigated agriculture and water use. In other words, there is a tendency to demand for or use more water under severe and extreme drought conditions. In this sense, comprehensive drought characterization in some regions of the world could be challenging because of the composite influence of climate and human activities. More generally, this is largely due to the reliance on the use of in situ precipitation data, which is insufficient and not truly an indicator of water availability. With the high variability in space and time in some areas, another challenge is existing gaps in precipitation records, which in turn will affect drought characterization (Naumann et al. 2014). Several studies in an attempt to compensate for this have used remotely sensed precipitation data, model outputs, and reanalysis to provide a homogenous and more consistent data with global coverage, even at varied spatial resolutions so as to suit effective drought monitoring and assessment (Damberg 2014). Past studies have also assessed these products to account for discrepancies and their limitations at sub-regional and even global scales (Agutu et al. 2017; Naumann et al. 2014; Damberg 2014).

Other methodologies have been employed in drought monitoring and assessment. Apparently, precipitation is a well-known climate variable to measure drought events. Apart from satellite observations of soil moisture or evapotranspiration, the Normalized Difference Vegetation Index (NDVI) and terrestrial water storage from the Gravity Recovery and Climate Experiment (GRACE) have been used. The latter is a new and important tool in drought assessment and prediction. Past studies that used NDVI in drought assessment showed its potential in monitoring drought and vegetation health (Kogan 1995). GRACE, on the other hand, employs the temporal measurement of changes in gravity fields to monitor TWS and has also been extensively used in drought assessment (Awange et al. 2013; Becker et al. 2010; Swenson and Wahr 2009). It is also emerging as an important tool in characterizing groundwater drought (Thomas et al. 2017). Climate models and integration of precipitation and model-based datasets, soil moisture, and/or NDVI are also popular in use for this assessment (Ndehedehe et al. 2020c; Agutu et al. 2017; AghaKouchak 2015).

Long before the launch of GRACE mission and the development of new satellite sensors, remote sensing observations have been broadly applied in assessing extreme events and climatological extremes, including drought. With the advent of the infrared observation satellite (TRIOS-1) in 1960, a new era of Earth observation and monitoring emerged. And with the development of new sensors, more accurate datasets were made available for use in understanding changes within the Earth's system. Satellite missions including advanced very-high-resolution radiometer (AVHRR), Landsat, the Geostationary Operational Environmental Satellites (GOES), as well as GRACE have together proved the effectiveness of remote sensing methods in understanding of the Earth's surface dynamics and to overcome the numerous challenges having to do with reliance on gauge-based meteorological observations for drought monitoring. These satellite observations are characterized by a wide array of advantages including

- cost-effectiveness, easy access, and availability,
- high spatial, temporal, spectral, and radiometric resolutions,
- repetitive coverage and continuous monitoring of changes in Earth's environment given their high temporal frequency (e.g., Zhang and Jia 2013).

The use of remote sensing imageries in drought monitoring and characterization has been extremely efficient, especially with the application of satellite-based drought indices, thus making them the most promising tool for drought assessment at both regional and global scales (Kogan 1997). The use of drought indices based on satellite data can be classified into two classes: (i) land surface standpoint in relation to the Earth's surface and (ii) atmospheric viewpoint with an emphasis on obtaining drought-related atmospheric parameters from satellite datasets (e.g., relative humidity and rainfall). The parameters employed in the former are implemented directly for the sole purpose of drought monitoring assessment. For example, NDVI is useful in ecosystem assessment of drought impacts. Several Earth observation data have been used over the years to monitor and quantify drought from different perspectives, and this includes vegetation, evaporation, terrestrial water storage, land surface temperature, snow cover, soil moisture, and precipitation, with each having an index for its analysis and assessment. These processes along with the modelling procedures are evident in several conventional remote sensing tools and products, which can be easily applied for the purpose of drought monitoring and assessment.

20.1.3 Hydrologic Variables for Drought Assessment Using Remote Sensing

Precipitation has a direct influence on drought. But it is difficult to create a precise spatial pattern of precipitation through point measurements, especially as a result of the relatively large spatial variability in precipitation. Since a large number of conventionally allocated hydrological models require gridded input data, it becomes pertinent to generate gridded mean areal precipitation to facilitate such assessments and further reanalysis. This, therefore, reiterates the potential of drought monitoring and hydrological modelling using satellite-based precipitation. With the launch of TIROS-1 in 1960, the observation of weather parameters even for forecasting purposes has improved. This advancement brought about the creation of the World Weather Watch program in 1963 by the World Meteorological Organization (WMO) to initiate the application of an operational geostationary network and the low Earth-orbiting meteorological satellites, which is also known as Global Observation Systems (GOS) (Kidd 2010).

To enhance the quality of observations from these dedicated satellite missions, there is a need for constant advancements and developments of new techniques, methodologies, and algorithms. Several satellite precipitation missions have generated data, which have been applied in a broad range of assessments and reanalysis. For instance, the Precipitation Estimation from Remotely Sensed Information using Artificial Neural Network (PERSIANN), Tropical Rainfall Measuring Mission (TRMM) Multi-satellite Precipitation Analysis (TMPA) (Huffman et al. 2007), Climate Predicting Center (CPC) Morphing Technique (CMORPH), Global Precipitation Climatology Project (GCPC) (Adler et al. 2003), Special Sensor Microwave Imager (SSM/I) on board the Defense Meterological Satellite Program (DMSP) platform (Ferraro 1997), CPC Merged Analysis of Precipitation (CMAP) (Xie and Arkin 1997) have been widely applied in several hydrological and climatic operations. A new era of precipitation remote sensing observations was initiated by the National Aeronautics and Space Administration (NASA) through their development of the Global Precipitation Mission (GPM, Hou et al. 2014), which has been used to develop the Integrated Multi-satellitE Retrievals for GPM (IMERG) (Huffman et al. 2015).

20.1.3.1 Precipitation

Satellite precipitation retrievals provide important high-resolution regional and global precipitation estimates. However, they could also have high uncertainty and bias (Alemohammad et al. 2015). Hydrological models or reanalysis that use or rely on these precipitation products as forcing parameters tend to inherit such uncertainties. In drought characterization, the mean differences in predicted areas under drought and different drought intensities using soil moisture products were determined by the differences in the forcing precipitation (Agutu et al. 2020). In other words, the quality of model soil moisture products will largely depend on the quality of rainfall used to force the model. As with soil moisture, much of the bias that is seen in ensemble streamflow estimation and hydrological modelling is due to input uncertainty or the uncertainty in precipitation product (Moradkhani et al. 2006). These uncertainties have an impact on the effectiveness of various drought indicators that use these products in measuring or characterizing various drought intensities (Agutu et al. 2020, 2017). The indirect measurement of satellite precipitation can be impacted by noise, which in turn lowers the accuracy of its observation in relation to ground measurements. It also experiences limitations in space and time.

The inherent bias and uncertainties being observed in satellite precipitation products have led to increased evaluation of satellite-based precipitation products on several temporal and spatial scales (Yong et al. 2016; Alemohammad et al. 2015). The goal of this evaluation and validation is to improve drought characterization and also to improve post-processing methodologies, which can be employed in bias correcting these precipitation products (Madadgar et al. 2014). The use of satellite precipitation measurements for adequate drought assessment and characterization is increasing (Zhang and Jia 2013). Out of the various drought indices which have relied on precipitation dataset is the standardized precipitation index (SPI) (McKee et al. 1993). However, for a more accurate drought characterization, the precipitation data required must be up to 30 years. This means that some conventional and recent satellite missions, such as the Tropical Rainfall Measuring Mission, cannot be used for drought indices such as SPI because of their temporal limitation (less than

30 years in data length). Nevertheless, some recent studies have employed remotely sensed precipitation for drought monitoring purposes (Zhang and Jia 2013). Several studies have evaluated the TRMM products on several spatial scales, and their potential in assessing key hydrological conditions and drought characteristics, including duration and severity, has emerged (Neto et al. 2003). Other studies have also used remotely sensed data as input for land surface models to predict soil moisture, runoff, among other fluxes, as well as assess them from a wide range of perspectives (Zulkafli et al. 2014). Notwithstanding, the availability of long-term precipitation dataset such as PERSIANN-CDR or GPCP provides a better platform for adequate drought characterization from a climatological perspective.

20.1.3.2 Surface Soil Moisture

Surface soil moisture is basically characterized as the total amount of water available in the unsaturated zone. Its usefulness spans across agricultural planning and global water–energy cycle, as it is also a key factor in water resource management (Seneviratne et al. 2010). The importance of understanding soil moisture effect on Earth surface assessment was made profound with its inclusion into the list of the top 50 most essential climate variables by the Global Climate Observing Systems (GCOS), and also maintained by the Intergovernmental Panel on Climate Change (IPCC) as well as the United Nations Framework Convention on Climate Change (UNFCC, Bojinski et al. 2014). Soil moisture is the ideal indicator for assessing water-driven changes in vegetation and measuring agricultural drought (Ndehedehe et al. 2019b; Agutu et al. 2017; Chen et al. 2014). In terms of definition, agricultural drought is defined as a lack of soil moisture availability in the root zone or a period characterized by a decline in soil moisture, resulting in crop failure. This is largely different from meteorological and hydrological droughts when precipitation deficits can result in inadequate surface water in streams, springs, lakes, etc. Given the impacts of agricultural drought on crop yield and productivity, characterizing soil moisture drought is thus important, particularly to understand how prevailing climatic conditions affect plant's biological properties and their development.

One way of measuring changes in soil moisture availability is either by satellite sensors, ground-based measurements, and model simulations. But ground-based sensor networks are unavailable in several regions, making the accurate assessment of soil moisture content difficult. Alternatively, a better spatial and temporal resolution of sub-regional soil moisture datasets across a wide range of scales can be easily obtained from remotely sensed surface soil moisture retrievals. Microwave satellites especially C, X, and L bands are known to generate important retrievals of soil moisture content given their ability to easily penetrate cloud cover, and even the top soil given the adequate amount of incident rays and the corresponding backscatter (Njoku et al. 2003). The basic principle of microwave retrievals is largely dependent on the sensitivity of the passive brightness temperature, as well as its active backscattering to the soil moisture content. Several sensors have been made available for adequate soil moisture detection since 1978. Some of which include the Soil

Table 20.1 A summary of the microwave sensors for soil moisture estimation

Sensor/missions	Spatial resolution (km)	Temporal resolution	Channel	Period
Soil Moisture Active Passive (SMAP)	3, 9, 36	2–3 days	1.4 GHz	2015–Present
Advanced Microwave Scanning Radiometer2 (AMSR2)	25	Daily	Multiple from 6.9 GHz	2012–Present
Soil Moisture Ocean Salinity Satellite (SMOS)	35	2–3 days	1.4 GHz	2009–Present
Advanced Scatterometer (ASCAT)	25	1–2 days	5.3 GHz	2006–Present
Advanced Microwave Scanning Radiometer (AMSR-E)	25	Daily	Multiple from 6.9 GHz	2002–2011
Scatterometer ERS	50	3–4 days	5.3 GHz	1992–Present
Microwave Imager TRMM	25	Daily	Multiple from 10.7 GHz	1997–2015
Special Sensor Microwave Imager (SSM/I)	25	Daily	Multiple from 19.4 GHz	1987–Present
Scanning Multichannel Microwave Radiometer (SMMR)	140	Daily	Multiple from 6.6 GHz	1978–1987

Moisture Active Passive (SMAP), the Advanced Microwave Scanning Radiometer2 (AMSR2), the Soil Moisture Ocean Salinity Satellite (SMOS), the Advanced Scatterometer (ASCAT), the Advanced Microwave Scanning Radiometer, the European Remote Sensing scatterometer, the microwave imager from the Tropical Rainfall Measuring Mission (TRMM), the Special Sensor Microwave Imager (SSM/I), and the Scanning Multichannel Microwave Radiometer (SMMR) (Wagner et al. 2013; Njoku et al. 2003; Njoku and Entekhabi 1996). In contrast to the single sensor soil moisture products, Liu et al. (2011) improved on the original concept and developed a blended soil moisture product through the integration of both active and passive soil moisture from various satellite sensors. This product is useful for effective drought monitoring and trend analysis, especially due to long-term blended soil moisture products, which have been obtained from 1978 to 2014. Table 20.1 captures some of the microwave sensors used for soil moisture predictions.

Several of these satellite soil moisture products (Table 20.1) can improve knowledge on changes in actual soil water content and volumetric soil content. However, despite constraints in operational data production of these soil moisture products (SMAP started in 2015), they can also detect water deficits in the soil, thus making them suitable in characterizing agricultural drought. Based on the SMOS satellite data, a Soil Water Deficit Index was constructed to measure agricultural drought by Martínez-Fernández et al. (2016). The SMOS soil moisture-based indicator in their study showed strong potential in agricultural drought monitoring. Even though the assimilation of satellite soil moisture like SMAP into model has been evaluated to assess potential for agricultural drought monitoring, the use of satellite remote sensing-based soil moisture products like SMAP and SMOS provides an opportunity to directly measure agricultural drought.

20.1.3.3 Evapotranspiration

Evapotranspiration (ET) is an important parameter in the hydrological cycle and is regarded as a major outgoing hydrological flux. Monitoring its changes is useful for the support of water resources management of irrigated agricultural landscapes as well as the assessment of crop stress and vegetation conditions, particularly in regions vulnerable to drought (Ndehedehe et al. 2018). Its influence on crop yield, irrigation water demand, and agriculture cannot be overemphasized. Analogous to precipitation, ET-derived ground-based observations using gauge stations are insufficiently inadequate to assess changes on a relatively large scale. This is due to their sparse nature and limited observational network. The use of remotely sensed observations has made it possible to estimate ET at very large spatial scales. Several studies have highlighted the different application of actual and potential ET products for drought assessments as well as the various different drought indices which have been developed for thorough and more accurate examination and monitoring. For instance, the Green Water Stress Index (GrWSI, Wada 2013), Green Water Scarcity Index (GWSI, Núñez and Pfister 2013), Drought Severity Index (DSI, Mu et al. 2013), Evaporative Stress Index (ESI, Anderson et al. 2013), Standardized Evapotranspiration Index (SPEI, Vicente-Serrano et al. 2010b), Evaporative Drought Index (Yao et al. 2010a), Reconnaissance Drought Index (RDI, Tsakiris and Vangelis 2005), Supply–Demand Drought Index (SDDI, Rind et al. 1990), Crop Water Stress Index (CWSI Jackson et al. 1981), Palmer Drought Severity Index (Palmer 1965) have been highlighted as drought indicators that assimilate ET into their evaluation and construction. The Moderate Resolution Imaging Spectroradiometer (MODIS), however, is one of the most used satellite instruments for extensive and practical drought assessment. Its significance is largely based on the fact that its analysis is based on ET viewpoint. Studies that have applied satellite remote sensing technique in extensive drought assessment over the United States and other regions in the world have highlighted the importance of remote sensing in drought studies (e.g., Agutu et al. 2017; Yao et al. 2010a). Mu et al. (2011) used the Penman–Monteith equation to decompose MOD16 products into global ET/PET datasets, after which they showed its use by incorporating it with the DSI to determine vegetation status (Mu et al. 2013). Yin et al. (2014) pointed out that the key parameters affecting ET are air, relative humidity, radiation, and air temperature. The rise in temperature triggers an increase in evaporation, which will drive aridity in the environment. The fall in precipitation during periods of drought impact negatively on water availability, which in turn restricts evaporation (reduction). The aftermath of declining evaporative cooling leads to a higher temperature, which drives ET (Yin et al. 2014). Other studies, on the other hand, have focused on land surface temperatures (Westermann et al. 2011) and air temperature estimates from land surface temperature measurements to monitor radiation, which is arguably another factor that influences ET rates across multiple scales (Benali et al. 2012).

20.1.3.4 Terrestrial Water Storage and Groundwater

With the advent of GRACE mission, it is now possible to quantify the amount of total water storage changes both on and below the Earth surface on a monthly and weekly time basis (Tapley et al. 2019; Reager and Famiglietti 2009). Observed changes in terrestrial water storage (TWS) are largely a function of the dynamics of surface water, soil moisture, canopy water, ice/snow as well as groundwater. The GRACE satellites are typically designed to observe monthly changes in the Earth's gravity fields to enable the tracking of variations in surface mass or continental freshwater, the oceans, cryosphere, hydrosphere, etc. As opposed to single satellite approaches with one dedicated sensor, GRACE comprises two satellites moving in tandem with each other based on a satellite-to-satellite tracking of changes in the Earth's gravitational field (Tapley et al. 2004a, b). This measurement is used to estimate global spherical harmonic model of the Earth's gravity field, which is subsequently used in quantifying mass changes on the Earth's surface. It is possible to get considerably small differences between the computed mass series determined from the GRACE measurements due to the processing choices made by the Science Data Systems (SDS) centers and their users.

To enhance the use of GRACE data in diverse applications in Earth science, SDS centres now provide quality-controlled, gridded, and basin-integrated mass change products. GRACE provides an opportunity for reliable assessment of both regional and global water availability (Velicogna et al. 2012). Its use in studying regional climatology and drought assessments, impact assessment from drought, and monitoring water availability in vegetation has been reported (Zhang et al. 2016; Ndehedehe et al. 2019b; A et al. 2015). The spatial resolution with which the GRACE mission measures monthly gravity fields ranges from an estimated 200,000 km^2 at low latitudes to approximately 90,000 km^2 near the poles (Tapley et al. 2004a). However, the GRACE mission has facilitated a comprehensive understanding of the pace of climate change impacts on continental water storage. Further, the use of GRACE has enabled the assessment of TWS changes, continental aquifers, glaciers, and ice sheets, as well as broadening our scope of large-scale mass redistribution associated with isostatic adjustments, climate change, and earthquakes.

Other satellite radar observations and laser altimetry systems measure changes in surface height of ice sheets. These technologies are, however, influenced by variations in the properties of surface layer and firm compaction, and only provide indirect measurements of the net mass change (e.g., runoff, ice discharge, evaporation, evapotranspiration, and precipitation). But with GRACE, we are able to obtain for the first time direct measurements of ice mass change. Altimetry-derived estimates are restricted to multi-annual trends of height changes due to sampling and sensitivity issues at the ice sheet surface, but the monthly temporal resolution of observations of ice sheet and glacier mass changes from GRACE satellites are exceptional. The long-term mass trends determined using GRACE for the ice sheets are minutely manipulated by sampling methods, or unknown surface processes, than other processes. However, the Arctic and Antarctica still maintain this correction problem due to the poor mass redistribution induced by the glacial isostatic adjustment (GIA). Data from GRACE mission has also been used to assess basin-scale outgoing fluxes like evapotranspiration (Rodell et al. 2004).

However, one major limitation of the GRACE data is its relatively coarse spatial resolution of over 150,000 km^2 per grid for TWS assessment. This makes it restricted as a tool for regional drought assessment, and its applications on a water management scale are also limited. Also given that the mission is relatively new having just an approximate 18 years of data, from a climatological perspective, we still see the GRACE data repository as not being sufficient for adequate drought assessment even on a global scale. This has triggered studies on the extension and reconstruction of GRACE-derived TWS datasets using algorithms such as the artificial neural network in previous years for a more feasible analysis (Zhang et al. 2016). Apart from backcasting of GRACE data to extend its time series for long-term assessment, methodologies have been developed to enhance the application of GRACE data in hydrological drought assessment despite its limited data record (Yirdaw et al. 2008). Also, the downscaling of GRACE-derived products is increasingly becoming more rampant to suit this purpose (Zaitchik et al. 2008). GRACE-derived TWS datasets are extremely useful in the investigation of global drought impacts as well as the prediction of the volume and total precipitation needed for a complete drought recovery.

20.1.3.5 Snow Cover

The impact of snow cover on the hydrologic cycle, surface radiation, and energy makes it a very crucial characteristic of the land surface (Kongoli et al. 2012). Its role as a natural water resource reservoir and a great soil moisture source is evident in its role in agriculture. Snowmelt contributes to a sizable fraction of runoff in a variety of mountainous regions. Hydrological drought can be caused by a shortage in winter snowpack, while a shortage in soil moisture content during the winter has the tendency of leading to profound agricultural drought. This shows the essence of snow monitoring and assessment for water management, not to mention of its role as a primary tool for drought estimations in several regions. Past studies have argued on the application of snow data for drought monitoring and estimation (Guan et al. 2013). Other studies have focused on the assimilation of snow observations with land surface models to enhance the precision of streamflow simulations and modelling, and by extension, hydrological drought assessment (Dong et al. 2007). Considering that there is no direct relationship between snow and drought, snow-based drought indicators are yet to be developed.

20.1.4 Climatic Hotspots of Drought Events

Drought is an intensification of the hydrological cycle that signifies a situation with below-average water availability. It usually begins with deficits in rainfall amounts. Again, this is a meteorological drought. As mentioned in the previous chapter, meteorological drought propagates through the water cycle, resulting in diminished soil moisture and limited stream flow activities, and are called agricultural drought and hydrological droughts, respectively. However, the propagation meteorological droughts can also lead to groundwater drought, a long-term decline in groundwater levels (Mishra and Singh 2010). These declines are caused by intensive water extraction, reduced rainfall and recharge, or when the water table drops below-normal levels. During drought regimes, water stress results in the retardation of crop root growth, as well as impedes agricultural productivity. For regions where livelihood is largely dependent on rain-fed agriculture, this results in a physical reduction of food availability and also increases the susceptibility of the region to drought. Over 1800 million mg of rice, maize, and wheat have been reportedly lost to global drought over the past few decades (Lesk et al. 2016). The magnitude to which different forms of drought regimes influence food security is substantially related to their spatio-temporal footprint. Also, the temporal regime of drought occurrence is essential, as the responsiveness of crop yield to water stress varies with crop growth. This calls for the need to assess drought impacts on crop yield separately at various growth stages, which is more significant for agricultural water management practices (Xu et al. 2019).

The rapid and large decline in groundwater stocks and loss of surface water across many regions of the world (Jasechko and Perrone 2021; Pekel et al. 2016) are some evidence of a water crisis. This is due to the rise in climate change, global warming, as well as population growth/intensity. Small drought magnitudes now have a significant rippling effect on drought-prone regions. This is the case even in years with normal rainfall rates, as deficits in water levels have become rampant in undeveloped, developed, arid, as well as humid regions. The propensity of relevant government agencies to act appear intensified when faced with severe drought. Unfortunately, this eagerness tends to wane when rainfall rates return to normal. However, in some cases, drought severity is often exacerbated when social safety nets and other programmes initiated by the government are degraded by societal dysfunction and/or social conflict. In 2018, for instance, an estimated 820 million people went through serious malnourishment as a result of global drought, which was also the leading cause of food crises in over 21 countries, with 14 of them suffering from intense economic recession (FAO 2019).

The urgency to improve upon some initiatives and risk-based drought management programmes in many regions has increased. This sense of urgency was, however, birthed due to the recurring and widespread nature of drought. For instance, over 14% of the United States are significantly affected by extreme drought every year. In 1934, the region recorded an estimated 65% of drought-affected areas. These values have hovered within a range of 35–40% in recent times, which explains the widespread drought occurrence over the United States in the past decade. This has led to the emergence of some drought-based national initiatives aimed at monitoring drought extent and magnitude in the United States and other drought-prone regions. Some of these initiatives include water efficiency and waging infrastructure, aquifer recharge, water reuse, desalination, and watershed sustainability. However, in other regions of the world like the sub-Saharan Africa, periodic cases of food crises are triggered by drought episodes. While these frequent drought events could be worsened by other compounding factors including communal conflict and heat waves, management efforts and programmes to cushion drought impacts are not efficient in the United States and other developed economies largely due to lack of funding for relevant agencies. In the sub-sections that follow, five climatic hotspots around the world have been identified and the impacts of droughts in these areas will be briefly discussed.

20.1.4.1 United States

Drought events are more frequent in the southern states of the United States, in particular the southwest United States. This includes Phoenix (Arizona), San Joaquin Valley (California), Colorado Springs (Colorado), Las Vegas (Nevada), etc. Of these regions, California is a well-known climatic hotspot where extreme drought events are more frequent, resulting in extensive groundwater abstraction and depletion. California's drought is typically more severe around the southern part of the state. The southern region contributes to one-fifth of the national agricultural production in the sub-region, and receives an estimated amount of 127–406 mm of rainfall every year (Ojha et al. 2019). Half of the entire Central Valley is composed of the San Joaquin valley aquifer systems, which cover an estimated 26,000 km^2 of the entire region. The region consists of the following counties: Kaweah, Kern, Tulare, Tuel, Kings, Fresno, Madera, and even Merced, and is home to an estimated 4 million people as of 2009. Increased groundwater pumping in the region's aquifer system resulting from population growth and urbanization resulted in land subsidence. In a satellite assessment of post-drought impacts of the 2012–2015 drought, subsidence rates of approximately 200–300 mm/year were reported in one of the southwest United States, which includes California (Ojha et al. 2020). This subsidence caused severe damage to infrastructures in several communities and degradation of the flow capacity and freeboard of the delta Mendota canal.

In California, extreme and recurring drought episodes are so rampant that even local communities have learned to cope with that (Woodhouse et al. 2010). However, the 2012–2015 drought has been recorded as the most severe in a millennium as a result of extremely high land surface temperatures, and reduced rainfall rates during the period. The extreme transfer of latent heat which was due to increased atmospheric levels largely correlated with human-induced climate change impacts (Diffenbaugh et al. 2015). During this period, the region was plagued with warm and dry weather, and a significant reduction in supply of surface water resources to reservoirs, lakes, and rivers was recorded. These abnormalities impacted the terrestrial and aquatic ecosystem of the California Central Valley relying on rivers, which resulted in intensified wildfire outbreaks and deficient lake levels (Williams et al. 2015). As the drought progressed to an $8.14\,km^3$ decline in surface water available for agriculture, a drought state of emergency was declared in the region; this, however, led to an intensified effort to increase groundwater pumping so as to enable survival of the inhabitants of the region. Howitt et al. (2014) reported that this drought caused an estimated USD 2.7 billion loss in the vast agricultural sector of the Californian economy, and also led to a loss of 21,000 jobs.

20.1.4.2 Australia

Australia is a hotspot for extreme climatic conditions due to the impacts of climate change in the region, as well as the influence of the nearby pacific in driving teleconnection patterns in some areas (Ndehedehe et al. 2021). Notable among these patterns are the warm El-Nino and cold La Nina phases between Australia and South America. The southeastern part of Australia is known for frequent droughts. The region consists of the well-known Murray–Darling Basin, which has experienced the most severe drought episodes over the years. In 2006, some rivers of the Murray–Darling basin were at extremely low levels, in particular the Murray river whose water is shared among three states. This resulted in the implementation of dedicated water-sharing arrangements by the Australian government to address this water stress. The subsequent inflows from 2007 to 2009 required a widespread adaptation of dry inflow contingency planning so as to guarantee that essential human water needs are met even in the coming years. Another region of interest is the northeast and west of South Australia. A large part of South Australia experienced significantly below-average rainfall patterns in recent times and above-normal temperature patterns. This led to major drought episodes given that the rainfall rate was above the lowest 5% of recorded rainfall and below the lowest 10% as documented by the country's Bureau of Meteorology (Bureau of Meteorology 2008). Over the northern Flinders, some parts recorded precipitation rates categorized over the 7% nadir of long-term measurements at a similar period. This warranted its classification as a 'significant deficiency' by the Bureau of Meteorology.

The Perth basin is another important climatic hotspot. It is regarded both as an onshore and offshore sedimentary river basin. It is composed of a 1300 km extent, which extends to the border of the Australian continent towards the southwestern region. The groundwater from this system sustains local industries, agriculture, and private domestic use in the region. Studies have also shown that the Gnangara mound nourishes substantial groundwater-dependent ecosystem that incorporates cultural, environmental, and even social values (Froend et al. 2004). As a result of the frequent drought affecting the region, a reduction in groundwater levels has been identified in recent years over the Perth basin, and this has resulted in the acidification of Jandabup, Mariginiup, as well as the well-known Gnangara lakes. The overall health of wetlands over the mound has been deteriorating (Sommer and Horwitz 2009).

Moreover, the Canning basin, which is another region of interest for drought-based studies is a huge area covering 210,000 square miles, and can be described as a region of comparative simplicity due to the very few universal sand hills and mountains. A NASA-based assessment showed that the Canning river basin in Western Australia was being used at the third-highest rate in the world. This significant depletion, which was estimated to be –9.4 mm per annum, signifies that the basin is drought-prone and this can be attributed to the climatic factors, population growth, and anthropogenic influences.

20.1.4.3 Middle East

Generally, the Middle East has a hot, arid climate, especially in the Egyptian and Peninsula regions. It is made up of several major rivers, which provide irrigation to support agriculture in limited areas like the Nile Delta in Egypt, the Euphrates, and Tigris watersheds of Kuwait, eastern Syria, and Iraq. Of all the regions in the Middle East, Iran is of paramount interest to hydrologists because of its high rank in terms of groundwater depletion rate by global groundwater assessments (Noori et al. 2021). The major source of freshwater in arid/semi-arid regions including Iran is groundwater. Groundwater provides an estimated 60% of the total water supply in Iran, whereas agricultural irrigation accounts for over 90% of groundwater withdrawal in the region. Arguably, this reliance on groundwater makes Iran the world's highest consumer of groundwater, that is, with respect to irrigation agriculture.

Using deep-well drilling techniques, it was possible to overexploit groundwater resources, while intensified surface water diversion and damming reduced groundwater recharge, and jointly contributed to the reduction of the region's water table and making groundwater harvest a bit difficult. Due to these factors, Iran is facing a state of water bankruptcy, jeopardizing its natural environs and socio-economic development. The groundwater withdrawal quotient of Iran increased from 65 million m^3 in 1965 to 132×101 million rmm^3 in 2018. This rise in groundwater withdrawal quotient is estimated to be 3.4 times the spatial extent of the famous three gorges dam in China. The problem of groundwater decline caused by excessive overexploitation of this non-renewable resource and intensifying salinity levels are present in almost all sub-basins in the region, pointing to the worsening and dire water scarcity risk across the entire region.

20.1.4.4 East Africa

Over the past five decades, losses in food production have become more rampant in all regions and across all food sectors (Cottrell et al. 2019). These losses are largely caused by extreme weather conditions with disproportionate influences on regions with minimal capacity to cope, like the capability of farmers to expand food production. For instance, the Kenya drought of 2007 was declared a national emergency. This drought led to an estimated 2.5 million people into food insecurity

(Berg and Sheffield 2018). Another drought impact in Kenya shows more than ten people died of hunger and lack of food and more than one million people were devastated and displaced by lack of water (Ndehedehe et al. 2020a). The adaptation strategies implemented by the Kenyan government, which include the response, recovery, and restoration practices helped in mitigating the impact of elevated drought risk in the region (Naumann et al. 2018). Drought intensity can be reduced through the implementation of these adaption strategies by drought-prone countries to facilitate progress towards achieving the second sustainable development goal of the UN (i.e., zero hunger).

Irrigation in sub-Saharan Africa is considered to be low. This is because the major source of water for agricultural production is rain. An estimated 5.5% of the cultivated land over the sub-Saharan region is equipped for irrigation purposes, whereas for East Africa, this only accounts for 1.5%. Out of this 1.5%, only 3.4% is set out for irrigation using groundwater resources (Agutu et al. 2019). In East Africa, groundwater use or its availability for irrigation agriculture appears to be limited. This can be attributed to (i) the presence of large areas of land with substantial potential for surface water irrigation, (ii) absence of comprehensive analysis of the region's groundwater irrigation potential, (iii) poor information on the capability of the region bound aquifers, and the yields of individual wells, (iv) poor groundwater-related knowledge for the inhabitants of the region, and (v) inadequate management plans for the sustenance of groundwater levels. However, the effect of climate change and intensifying human population on surface water sources and rain-fed agriculture of the region has diverted attention towards groundwater resources for increased food production in the region (Agutu et al. 2019; Moore and Williams 2014). The understanding that increased groundwater use could substantially improve the well-being of the inhabitants of these regions has been corroborated in the past literatures (Alley and Konikow 2015). Given the over reliance of the East African region on rain-fed subsistence agriculture, the region is increasingly becoming vulnerable to frequent devastating drought episodes as reported in the last few years (Agutu et al. 2020, 2017; Stampoulis and et al. 2016). Furthermore, drought impacts are compounded by population migration, armed conflicts, high poverty levels, and lack of social infrastructure which directly gives rise to recurring famine cycles. Given that drought is in part a naturally recurring phenomenon in the East African region, it warrants the need to plan and establish drought mitigation measures for the benefit of the local communities. Since the severity and frequency of drought is likely to become stronger with climate change, policy and decision makers can characterize drought in terms of frequency, severity, spatial extent, and duration to support national drought policies that underpin water resources management and effective allocation in the future.

20.1.4.5 India

India is a major agricultural hub underpinning food security. However, as one of the world's largest users of groundwater for agriculture (60% for irrigation), cropping intensity is projected to decrease by 68% in areas with substantial depletion of groundwater (Jain et al. 2021). The shortage of the summer monsoonal rainfall patterns is a key driver of meteorological drought in India. If this condition is prolonged, it can lead to more impactful hydrological, and agricultural drought episodes. Under an extensive human water extraction scenario as noted in India, this prolonged rainfall deficit will lead to groundwater drought. As with hydrological and agricultural droughts, groundwater drought has a high propensity of posing substantial impacts on water availability and food production in any region. In India, there are two primary monsoon seasons: (i) the Indian summer monsoon season (IMS), which is popularly referred to as the southwestern monsoon and the (ii) northeastern monsoon or the winter monsoon. The major source of rainfall in the sub-region, however, is the Indian monsoon, which lasts over the period of June to September. In India, the use of the IMS dynamics in the assessment and drought-based characterization over wet and dry years is increasing. Crop yield in India is largely related to the IMS and has shown that droughts have the capability of diminishing the country's gross domestic product by 2–5% as a result of poor agricultural yield in a season. In 1987, the IMS drought affected crop yield in approximately 59 million hectares agricultural area, which translates to 46% of the total all-India agricultural region, and therefore led to significant deficits in freshwater in much of the country (Nayak and Hassan 2021).

India's water resources are largely characterized by significant fluctuations in both spatial and temporal distribution. There is a regional mismatch in available water resources and demands. This water scarcity problem in the region led to an unsustainable exploitation of both groundwater and surface water resources (Jain 2019). As the country is still largely dependent on agriculture, fluctuations in rainfall patterns and the availability of water resources directly influence its economic and social progress, thereby increasing trends in climate change, human population growth, increasing drought and flood frequency, resource contamination, and uneven distribution of monsoon rainfall, which tends to worsen the water scarcity in many parts of the country. Due to the conflicting interest of water users, the situation is expected to worsen several unplanned anthropogenic activities to withdraw substantial amounts of water from existing sources.

20.1.5 Impacts of Drought on Surface Water Hydrology

Drought prediction and assessment is a very complex process especially due to its overarching challenges such as duration, causes, magnitude, as well as timing on various spatial and temporal scales (Kiem et al. 2016; van Dijk et al. 2013; Dai 2011). Improving knowledge of drought characteristics and climatic interactions is thus crucial. This knowledge can be leveraged to provide adequate support that aids enhancement of existing templates on risk mitigation in drought-prone regions. Several studies have shown the influence of land use change and increased water abstraction on future changes such as drought as well as the role of human-induced climatic oscillations on large-scale ocean–atmospheric interactions (Kiem et al. 2016).

20.1.5.1 The Lake Chad

Several sub-African regions (e.g., riparian countries of the Volta, Lake Chad, and Lake Victoria basins) experience severe challenges of food and water security, hydropower sector, etc. as a result of cascading drought influence (e.g., Agutu et al. 2017; Ndehedehe et al. 2016a). And given the strong reliance of these regions on precipitation as a key ecosystem function, this challenge still persists to this day (Ndehedehe et al. 2016b; Shiferaw et al. 2014; Agutu et al. 2017). Growing concerns and discussions on the utilization of satellite-based freshwater resources assessment in these regions are gaining more grounds as studies have highlighted several advantages in the application of these methods (Ndehedehe et al. 2020b; Buma et al. 2018). These advantages include the capability to improve understanding of the effects of dry years and various drought intensities on the local and sub-regional communities, especially in the arid regions of West Africa (Ndehedehe et al. 2020a; Werth et al. 2017).

The Lake Chad basin has experienced inter-state tensions, armed conflicts, and hunger caused by limited water resources resulting from prolonged drought events. This effect is pronounced and felt more in trade disruptions and price volatility due to the extended drought impacts in the West African Sahel. Drought in the Sahelian region is the result of severe perturbations of the surrounding ocean as well as human-induced climatic conditions (Epule et al. 2014). The study undertaken by Ndehedehe et al. (2020a) on the evolutionary drought patterns and the role of climate teleconnection patterns confirmed the interactions of oceanic variability with the spatial and temporal variability in extreme events like drought over the Sahel. These teleconnections triggered a prolonged drought that affected the surface water hydrology of the Sahelian countries.

Lake Chad has permanent water loss since the mid-1980s due to strong and prolonged declines in rainfall. This led to a substantial shrinkage of its surface area. This shrinkage created a direct impact on this lake, especially by restricting water inflow from the Chari/Logone river systems. Apart from drought, this shrinkage was also the direct impact of excessive water withdrawals from the lake for agricultural purposes during the water-deficit periods (Fig. 20.2, Coe and Foley 2001; Birkett 2000). The Lake Chad case is a typical example of human-modified drought in the Anthropocene where several human activities such as land-use change, deforestation, etc. and interactions between the land surface and the atmosphere, catchment storage, and climate variability have a direct impact on drought severity. The study undertaken by Van Loon et al. (2016) explains this further, highlighting urbanization as a direct factor affecting the rate of infiltration and groundwater recharge. Therefore, one can argue that the composite influence of climate and human activities, such as deforestation and land-cover change, plays a vital role in land–atmosphere interaction and coupling, and have the potential of resulting in a significant effect on hydrological processes and water resources systems.

Fig. 20.2 An overview of the Lake Chad basin depicting sections of Lake Chad

20.1.5.2 Brazil

South America is popular for its complex dynamic interactions and processes due to its geological, topographical, geophysical, and hydrological characteristics, as well as climate features (Solman 2013; Hughes et al. 2009). The continent is a well-known hotspot of geophysical processes and crustal deformations (Ferreira et al. 2019a, b). However, most hotspots in the world having the highest water scarcity and exhibiting the most profound patterns of weather and climate extremes (Buytaert and Breuer 2013) are also found in the South American continent (Ndehedehe and Ferreira 2020a; Ferreira et al. 2018; Eichler and Londoño 2013). Brazil, for instance, is one of such hotspots in South America. The region covers a surface area of 8.5×10^6 km^2 with an estimated human population of 200 million. It has the largest reserve of surface renewable water in the world. Its contribution to global inland water is estimated to be about 20% for all the oceans in the world from the Global Land Data Assimilation System (GLDAS) model (Rodell et al. 2004).

Despite the abundant surface water availability in the region, it still experiences a considerable amount of surface water variability, which is evident in the northern, southeastern and northeastern regions that produce an average precipitation rate of 6 mm/day, 3.8 mm/day, and 2.9 mm/day, respectively. This, therefore, translates to an average of 78% of total runoff in the country, with the south and northern eastern regions accounting for only 13% of the total surface amount. The highest gross domestic product from the country is obtained from the southeastern region housing the Sao Paulo and Rio de Janeiro districts, as a result of their high economic productivity. These regions have an estimated 11 million and 6 million people, respectively. Eastern Brazil with an area of 2.5×10^6 km^2, on the other hand, has experienced a terrible history with recurring drought episodes since 2012. The drought regime was succeeded by a prolonged duration of above-mean temperatures and decline in precipitation rates. This therefore led to a significant depletion of both groundwater and natural surface storages and had a substantial impact on the entire Brazilian electrical and agricultural production, irrigation as well as drinking water supply. An estimated 32% of the water in Brazil is utilized for the sole purpose of agriculture and irrigation. This is particularly common in the southern and southeastern regions of the country. The 2014 harvest in the region was massively destroyed as a result of drought and was worth millions of dollars (Getirana 2016).

This water depletion is also responsible for the negative impact on hydropower generation, which is the sole provider of an estimated 70% of the total electricity consumed in Brazil. More than 20 million residents of the Sao Paulo province together with several other residents living in eastern Brazil were also at risk, following the rate of water depletion in the region. This water crisis has led to the Sao Paulo and Sabesp state governments adopting emergency approaches to allocate water resources more efficiently within various jurisdictions and local communities (Getirana 2016). The Brazilian government has characterized drought as a natural risk, which needs urgent mitigation procedures through a crisis management technique by the establishment of short-term policy solutions. They devoted massive resources to the construction of gigantic water infrastructures and reservoirs, as well as employed the cloud seeding method in the 1950s. However, these efforts did not amount to much in terms of mitigating drought in the region (Lemos 2003). To build a substantial resilience in socio-economic conditions of any region, the approach of disaster-driven responses will be inefficient (Wilhite et al. 2014). Also, the mitigation of this crisis is known to favour poorer drought risk managers as beneficiaries of drought relief assistance (Mpelasoka et al. 2008). It is therefore necessary for drought-prone communities to implement and enact more risk-based drought management procedures and strategies to manage hazards of this nature more efficiently.

20.1.6 Drivers and Multi-scaled Impacts of Extreme Droughts

The combined impacts of climate change, increased population, urbanization, as well as other deteriorating socio-economic aspects will intensify drought hazards. Frequent exposure to drought could result in increased vulnerability of human society in the long term. This also underpins the need to manage and understand drought from a whole different perspective of complexities. It is also pertinent to consider environmental and climatic drivers along with socio-economic factors that drive a region's susceptibility to drought as well as its potential capacity to cope under such events.

Highly variable climatic conditions as observed in several regions pose a serious threat to regional eco-hydrological systems, water resources, and the productivity of floodplain ecosystems. Specifically, the impacts of droughts on agricultural yield will affect national income, resulting in financial losses. For example, the impacts of climate variability contributed to the reduction in average profits by 35% in typical Australian farms. As was further mentioned by Hughes et al. (2019), the national revenue losses from such impacts reached 1.1 billion dollars per year during the 2015–2018 period. The Millennium Drought in Australia also negatively affected the economy, resulting in relief programmes that cost about 8 billion dollars, further emphasizing the significant relationship between changes in hydrology and national income (Ndehedehe et al. 2021). Different cases of unprecedented climatic thresholds have also been highlighted in many other regions of the world. Even though these were illustrated earlier, the drying up of freshwater bodies and large endorheic lakes like Lake Urmia in Iran and Lake Chad in the West African region are clear indicators of the influence of climate variability on freshwater availability. Persistent and long drought episodes are also responsible for the dramatic decline noticed over the Lake Chad surface area (Ndehedehe 2019). The same thing happened in Lake Mead where the disappearance of the lake was associated with the impact of climate change on freshwater. A 60% decline in the lake height was noticed for few years after its last observation in 1998, when the lake was full.

When drought becomes more prolonged that could also affect groundwater availability. The impact of drought on groundwater resources is evident in the sustained groundwater loss noticed in California's Central Valley (Ojha et al. 2018). As a result of persistent human water extraction caused by prolonged drought conditions and climate change, the observed groundwater decline in several regions like Australia, India, Middle East, and the North China plain became pronounced. The expansion in energy, agriculture, and industrial water use due to population growth has a direct impact on water demands and translates to water scarcity in many sub-regions of the world. Other factors influencing water scarcity are also the contamination of water supplies and climate change. The impacts of drought are mostly felt in surface and groundwater resources and it can cause a diminished water supply, lowered water quality, poor crop yield, deteriorated range productivity, reduced power generation, interrupted riparian habitats, as well as impacts on tourism. As mentioned previously, its effect is felt in areas of social and economic activities (Riebsame et al. 2019). Studies have also identified the negative impacts of drought on water quality, because the changes that occur in climatic variations have the tendency of altering hydrological regimes, and consequently impact lake chemistry (Webster et al. 1996). Drought regimes can also interrupt the pathways created by runoffs for the purpose of transporting sediments, organic materials, and nutrients. Some climate and anthropogenic drivers of extreme drought episodes.

- Influence of land and water temperatures—the imbalance between rainfall and ET can lead to drought, especially when there is a high ET rate and low precipitation rate. This can be the result of climate change in a particular sub-region. For example, the well-known Sahara desert used to be humid thousands of years back (i.e., during the African humid period). However, climatic fluctuations have altered the hydrology of the region to what its current condition is.

- Air circulation and weather patterns—Key events like the El-Nino and La-Nina Southern Oscillations are primary drivers of climate change, and can also contribute significantly to the impacts of droughts,
- Depletion of soil moisture levels—the depletion of soil moisture brings about less evaporation of water for adequate cloud formation. This brings about an increase in surface temperature and also the need for more water, whereas less is available. This contributes to drought episodes.

Other factors such as higher water demand relative to its availability, global warming, and operational agricultural systems and practices are also responsible for several drought occurrences in many regions. In the next section, case-specific drought assessment in key hydrological hotspots (Brazil and the Lake Chad basin) are detailed.

20.2 Hydrological Hotspots in Brazil and Impacts of Global Climate Modes

Water problems in Brazil are increasingly becoming frequent due to the impacts of extreme droughts (Fig. 20.3). These problems have increased the awareness of the Brazilian population to the reality of deleterious impacts of climate (e.g., Melati et al. 2019; Ferreira et al. 2018; Erfanian et al. 2017; Getirana 2016). This awareness is the result of drought impacts on its socio-economic and hydrological systems. Large depletion rates in groundwater in the semi-arid regions of Brazil are perceived as a response to prolonged droughts that occurred during the 2010–2017 period (Melati et al. 2019; Brito et al. 2018). A rise in drought frequency and severity during the 2011–2016 period confirms the region's vulnerability to climate change (Brito et al. 2018). Several important studies that focused on drought events in Brazil have been summarized by Ferreira et al. (2018). Additionally, the stupendous impacts of variable drought conditions on freshwater in the region have been studied (e.g., Ndehedehe and Ferreira 2020a; Awange et al. 2016; Getirana 2016). Some of these studies, for example, noted the impacts of the unprecedented droughts during the 2012–2015 period on land water storage and surface displacement (e.g., Ferreira et al. 2018; Erfanian et al. 2017; Getirana 2016). Other related studies identified the influence of climate variability indices on drought events (e.g., Marengo et al. 2018; Erfanian et al. 2017). Notably, some of the drought studies were devoid of additional information on the variable nature of drought as well as its predictability (e.g., Marengo et al. 2017). As with the link between land water storage and global climate patterns (e.g., Ndehedehe and Ferreira 2020b), understanding the intrinsic response of droughts to climate indices and other large-scale oceanic processes is critical and also warranted.

The study undertaken by Ndehedehe et al. (2020c) explored this relationship over Brazil. They used a novel two-step regularization procedure that combined cumulant rotation (Ziehe 2005; Cardoso 1999; Cardoso and Souloumiac 1993) with novel

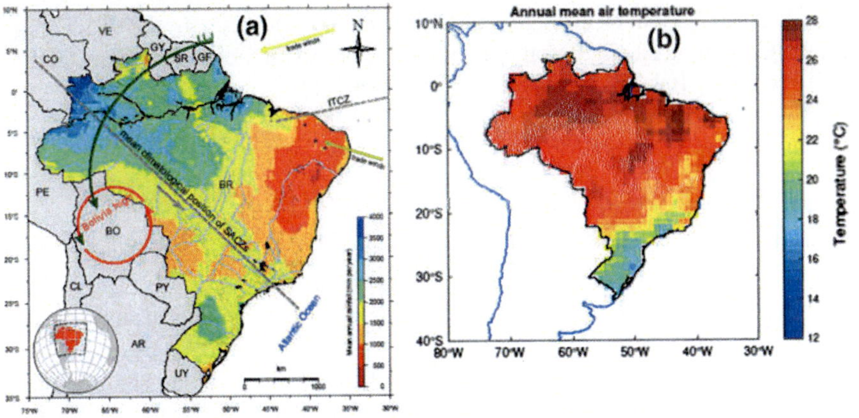

Fig. 20.3 Map showing mean annual distribution of **a** rainfall and **b** temperature over Brazil. Ndehedehe et al. (2020c)

machine learning methods, like the support vector machine regression (SVMR, Vapnik 1995). The overarching goal of their study was to improve the understanding of extreme events like floods and droughts induced by indices of oceanic variability such as El-Niño Southern Oscillation (ENSO) and Pacific Decadal Oscillation (PDO), among others (e.g., Erfanian et al. 2017). Specific aims of the study were to (i) assess the spatio-temporal patterns of droughts using meteorological-based indices and hydrological data from satellite and reanalysis, (ii) analyse drought statistics and characteristics, and (iii) identify hydrological zones where drought episodes are strongly affected by climate variability indices. The aim is to improve knowledge on the complex interactions between climate and freshwater availability. The standardized precipitation index (SPI) was computed from the Global Precipitation Climatology Centre (GPCC) (Schneider et al. 2014) precipitation and was used as a drought indicator. The Modern-Era Retrospective Analysis for Research and Applications (MERRA) was used as a hydrological indicator and several indices of climate variability were used to explore the global climate impact on extreme events like drought.

This study was undertaken in Brazil (Fig. 20.3), a country that covers the north and south hemispheres within a latitude band ranging from 5° N to 34° S. It has approximately 8.5×10^6 km^2 ranking as the world's fifth-largest country by area. About 60% of Brazil lies within the tropics. This is due to its shape and latitude range and physiography, there is considerable variation in climate across the north–south band. The climate types and characteristics are described in Fig. 20.3a–b. The Kopen of Brazil consists of rainforest (Af) in the northwest; monsoon (Am) at the central Amazon; and savanna (Aw), which covers the Brazilian Plateau over much of the eastern, southern, and central segments of the country. For instance, the hot semi-arid (BSh) covers a large part of northeast Brazil; humid subtropical and subtropical highland over the southeast; and humid subtropical and oceanic over the south.

Figure 20.3b shows that on average, annual temperature decreases gradually ranging from 28 °C in the northern region to about 18 °C or less in the southern part. More details are available in Ndehedehe and Ferreira (2020b).

20.2.1 Regularization Approach for Spatial and Temporal Patterns of Hydrological Indicators

A two-step regularization approach was employed to localize empirically derived drought indicators. To this end, principal component analysis (PCA, e.g., Jolliffe 2002) was employed to statistically decompose time series of computed drought indicator (**X**) at 6 and 12 months aggregation scales into temporal and spatial patterns. For instance, the standardized precipitation index is given as **X**= $[x(p_k, t)]$, where p_k is location in space; k = 1, 2, ..., Nx, which are the number of spatial locations for **X**, and t is the time (monthly) step from 1980 to 2014. The technique statistically decomposes the matrices containing the data (in this case the SPI or the drought indicator of interest) **X** into spatial and temporal patterns as,

$$\mathbf{X}(t) = \sum_{k=1}^{N} a_{(k)}\mathbf{p}_k, \tag{20.1}$$

where $a_{(k)}$(t) are the temporal variations, which can also be referred to as the expansion coefficients (i.e., standardized scores) and $\mathbf{p}_k$ are the spatial patterns or maps (empirical orthogonal functions (EOF) loadings) associated with these expansion coefficients. The PCA method was discussed extensively in Chap. 13. It was employed in this study primarily as a dimension reduction tool to minimize and filter the noise and random signals in the data (drought indicator). The dimension reduction capacity of this tool was aided by the use of Bartlett's test statistics (Snedecor and Cochran 1989). In particular, it was used to ensure that random variations from the PCA technique were not retained (Chap. 13). The processes regarding dimension reduction can differ. For instance, some rules based can be on hypothesis testing and others can be based on subjective choices. The latter includes *log-eigenvalue* diagram and retaining PCs that represent a 'sufficient fraction' of the variance in the original data that are available in other literatures (see, e.g., Wilks 2011; Martinez and Martinez 2005). The choice of Bartlett's test statistics for Brazil was sufficient for the pre-orthogonalization stage (step 1). In the last stage after the data has been pre-filtered, a statistical rotation was performed to localize or statistically decompose all drought indicators at 6 and 12 months aggregation scale. The drought indicators were derived from (i) the Global Precipitation Climatology Centre (GPCC, Schneider et al. 2014) rainfall data and (ii) the Modern-Era Retrospective Analysis for Research and Applications (MERRA, Rienecker et al. 2011). The rainfall data were used to compute the standardized precipitation index (SPI) while MERRA was used as a hydrological unit to measure hydrological drought.

20.2.1.1 Oceanic Hotspots Associated with Temporal Drought Patterns

To understand the influence of global climate on hydrological hotspots, the support vector machine regression (SVMR, Vapnik 1995) model was employed. A two-step procedure is adopted in the implementation of the SVMR scheme. Finding a linear hyperplane that fits the multidimensional input vectors to output values was undertaken in the first step. In the last step, the SVMR was used to predict future output values that are contained in a validation set (e.g., Okwuashi and Ndehedehe 2017; Vapnik 1995). Assuming the set of data points $\mathbf{X} = (x_i, p_i), i = 1..., n$ with x_i being the predictand data i, p_i are the actual values and n the number of data points. The linear SVMR function $f(x)$ takes the form (e.g., Vapnik 1995),

$$f(x) = wx + b. \tag{20.2}$$

The relationship between climate teleconnections and drought indicators is then assessed using Pearson correlation and coefficient of determination (R^2).

20.2.2 *Regionalization of Hydrological Hotspots Through Pre-orthogonalization*

The spatial and temporal patterns of drought indicators (independent standardized precipitation index-ISPI and independent water storage index-IWSI) indicated in Figs. 20.4, 20.5, 20.6 shows the usefulness of the statistical rotation of the pre-orthogonalized SPI and IWSI data. Previous approaches to assessing spatio-temporal drought patterns used component extraction techniques like the PCA (e.g., Agutu et al. 2017; Bazrafshan et al. 2014). But the approach implemented here is a more advanced statistical method called the independent component analysis (previously discussed in Chap. 13). This method explores the remote properties of the data matrix such as non-Gaussianity, spectral non-Flatness, and non-stationarity (e.g., Theis et al. 2005; Cardoso and Souloumiac 1993; Cardoso 1991). This is contrary to other fixed-point iteration schemes employed in the decomposition of gravity observations and geological objects identification (e.g., Yang and Cheng 2015; Frappart et al. 2011).

In much of the hydrological hotspots identified, drought frequency (the number of droughts events) appears to be higher at 6 month cumulation (Fig. 20.4) compared to 12 month cumulation (Fig. 20.5). Temporal drought patterns in northern Brazil significantly contrast with the southern section of the country (ISPI-1 and ISPI-4, Fig. 20.4). However, the frequency of drought events tends to reduce at higher aggregation scales but with longer duration (Fig. 20.5).

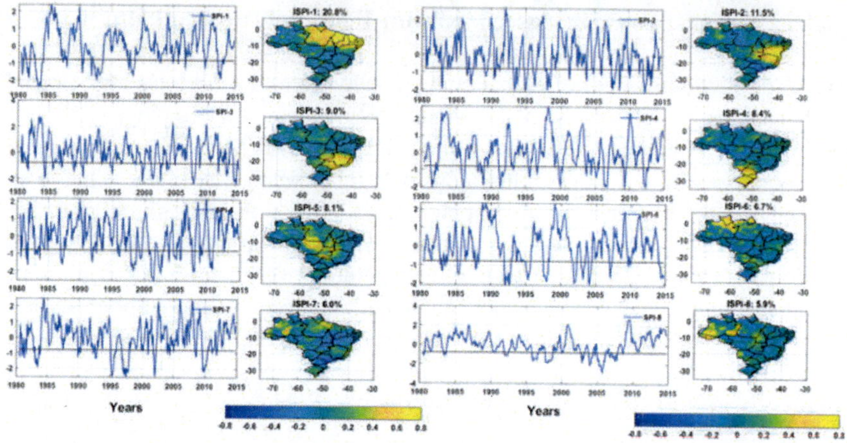

Fig. 20.4 Spatial and temporal independent standardized precipitation index (ISPI) patterns over Brazil using 6-month SPI values. SPI values are computed using precipitation for the period between 1980 and 2015. Actual values for drought classification and interpretation with respect to McKee et al. (1993) description are jointly derived from the regionalized temporal patterns (first and third columns) and their corresponding spatial patterns or maps (second and fourth columns). The black solid line indicates the adopted drought threshold for this region. *Source* Ndehedehe et al. (2020c)

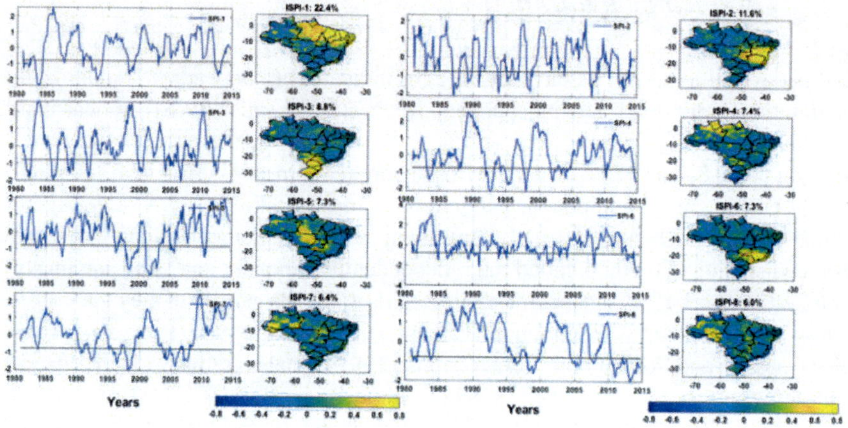

Fig. 20.5 Same as Fig. 20.4 but using 12-month SPI values. *Source* Ndehedehe et al. (2020c)

20.2.3 *Drought Properties and Statistics*

20.2.3.1 Percentage of Areas Under Drought

Droughts statistics are based on the percentage of affected areas, drought duration, frequency, and intensity. At the 6 and 12 month aggregation, evolutionary drought patterns of droughts in terms of their characteristics (moderate, severe, and extreme)

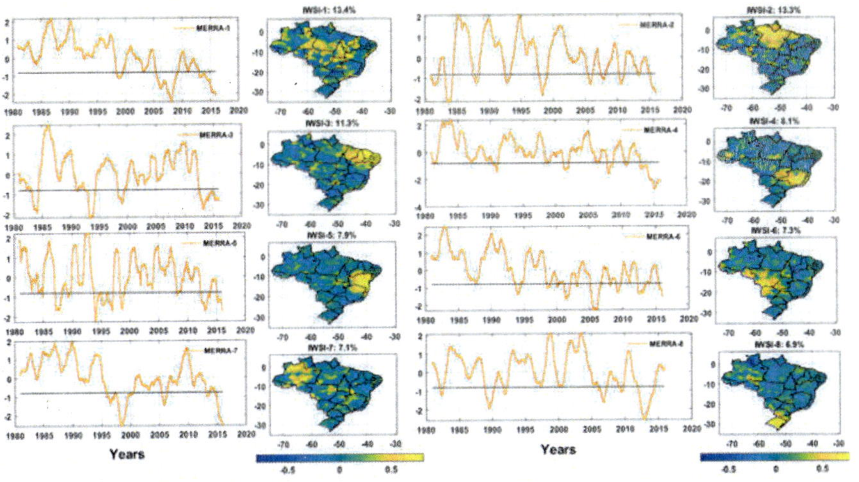

Fig. 20.6 Spatial and temporal MERRA-SPI patterns over Brazil using 12-month aggregation scale. MERRA-SPI implies that the MERRA indicator values are computed similar to SPI for the period between 1980 and 2015. The black solid line is the adopted drought threshold for this region. *Source* Ndehedehe et al. (2020c)

over Brazil have shown temporal patterns that are somewhat similar to the results in previous section. More than 50% of Brazil was affected by the 1998/1999 drought (Fig. 20.7). The drought in 1998/1999 coincides with the biggest energy crisis in the country (Getirana 2016). The MERRA data extends to 2015 and shows that the worst drought in recent times occurred during the 2012–2015 period, and fluctuated between 25% in 2012 and 70% in 2015 (Fig. 20.7). Comparatively, these three indicators (SPI-6, SPI-12, and MERRA-12) and their characterization of various drought intensities and estimation of affected areas (2007/2008, 2012/2013, and 2013/2014) tend to be consistent between SPI-6 and SPI-12 month (Fig. 20.8a–c). Nonetheless, there are some differences in predicted extents of areas affected by drought using the MERRA indicator. This dissimilarity varies during the three periods analysed (Fig. 20.8a–c). This could be linked to difference in data sources such as differences in the MERRA observed precipitation rates. Rainfall is an indirect indicator of available water and the MERRA data accounts for the sum of water stored in all reservoirs (e.g., soil moisture).

The areas affected by drought on both annual and inter-annual time scales are shown for all drought indicators (Fig. 20.9a–f). Much of the drought-affected areas occurred in the months during April to October similar to 1982 and 1998. The MERRA indicator shows this drought was an all-year (January–December) event with at least 60% areas under drought observed per month (Fig. 20.9e and f). Generally, all of these drought patterns also show the annual time series of areas under drought (Fig. 20.9b, d, and f).

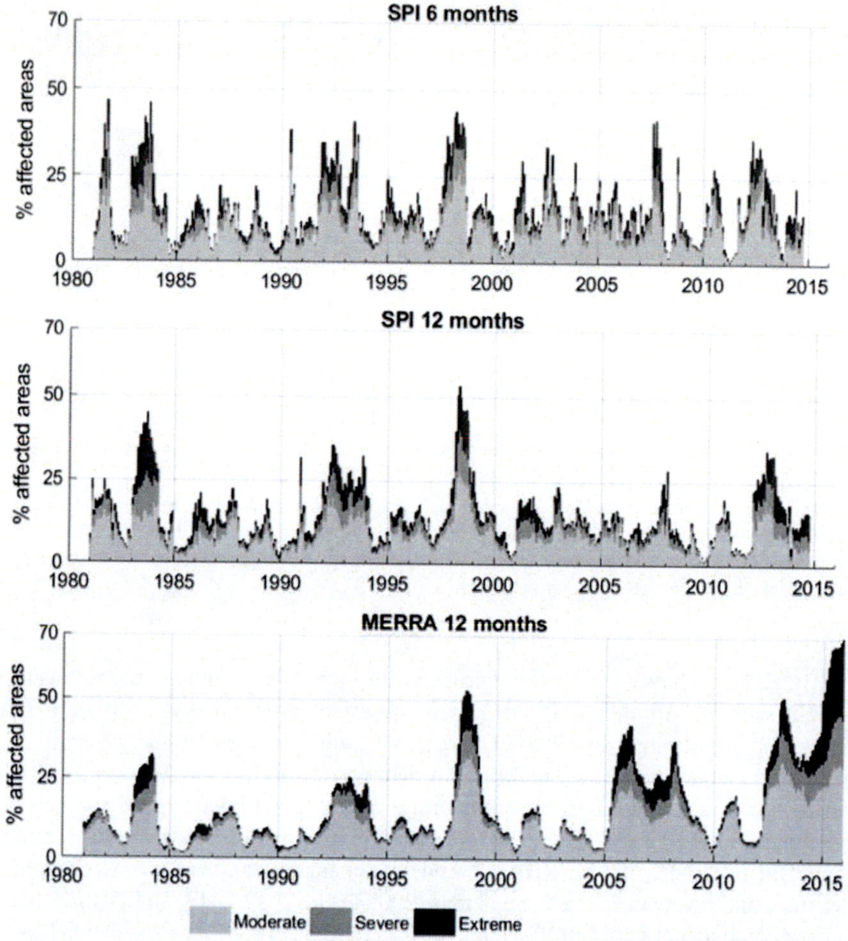

Fig. 20.7 Areas under drought over Brazil for various drought intensities (extreme, severe, and moderate) during 1980–2015 period expressed in percentage (%). The affected areas are estimated based on SPI-6, SPI-12, and MERRA-12 months aggregation. *Source* Ndehedehe et al. (2020c)

20.2.3.2 Drought Duration, Frequency, and Intensity

Drought properties such as duration, frequency, and intensity are summarized in Fig. 20.10a–i). Drought duration is more extensive and wide spread on a 6 month (SPI-6) cumulation compared to 12-month (SPI-12) cumulation (Fig. 20.10a–c). The mean duration of drought episodes appears to be prevalent around the Amazon and northwest sections of Brazil. These regions tend to have longer drought duration (on a 12-month time scale), ranging from 8–10 months compared to areas located in the south and northeast Brazil (Fig. 20.10b–c). On a relatively higher cumulation scales

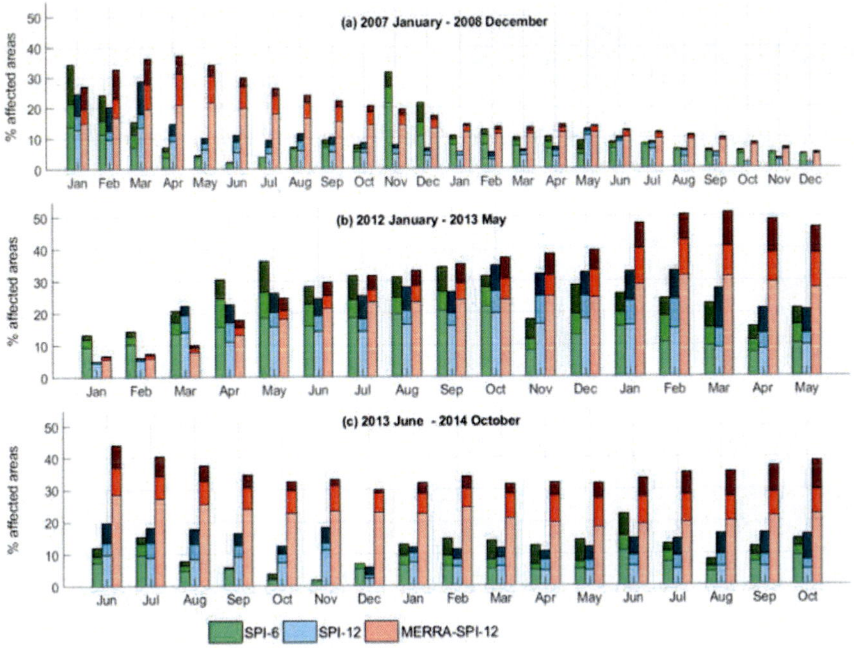

Fig. 20.8 Areas under drought over Brazil similar to Fig. 20.7 but for only specific periods (2007–2008, 2012–2013, and 2013–2014) and is expressed in percentage (%). The affected areas are characterized based on SPI-6, SPI-12, and MERRA-12 months aggregation. *Source* Ndehedehe et al. (2020c)

(SPI-12), longer drought duration may be expected unlike shorter time scales (SPI-6). Notably, there is a substantial difference in the duration of droughts quantified (Fig. 20.10b–c), especially in the south and northeast Brazil. This difference can be attributed to interactions of soil characteristics, vegetation, and topography, with droughts or their propagation process under a rising temperature.

Drought frequency tends to be unique only for SPI-6 but much similar for SPI-12 and MERRA indicators (Fig. 20.10d–f). The observed spatial patterns of drought frequency in Fig. 20.10d are not only unique but consistent with the observed temporal evolutions of drought (SPI-6) in Fig. 20.4 where drought occurrence is relatively higher when compared to SPI-12 month temporal patterns (Fig. 20.5). Additionally, drought is more frequent in the south of Brazil (i.e., SPI-6,), ranging from 16% to 15.5% than in some regions (central and northeast) where the frequency of drought falls between less than 15.5 and 14.5%. Drought seems to be generally less frequent (slightly greater than 14.75%) when higher cumulation scales (12 month) are considered. However, there are few hotspots around the Amazon basin where it is even much less frequent and shows values less than 14.55% (Fig. 20.10e). But drought intensities show the intensity of drought fairly similar over Brazil regardless of time

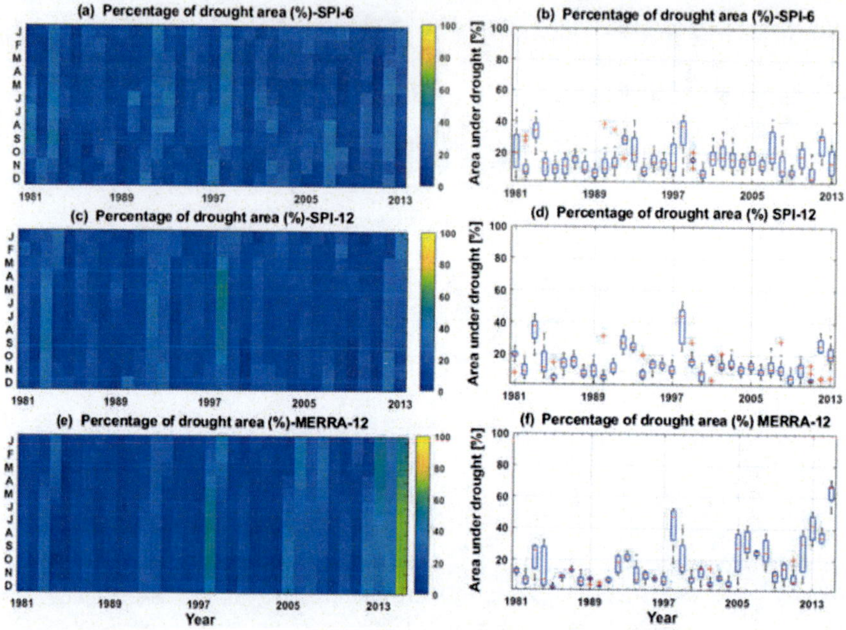

Fig. 20.9 Averaged areas under drought for annual (**a**, **c**, and **e**) and inter-annual (**b**, **d**, and **f**) time scales over Brazil for the period between 1980 and 2014. *Source* Ndehedehe et al. (2020c)

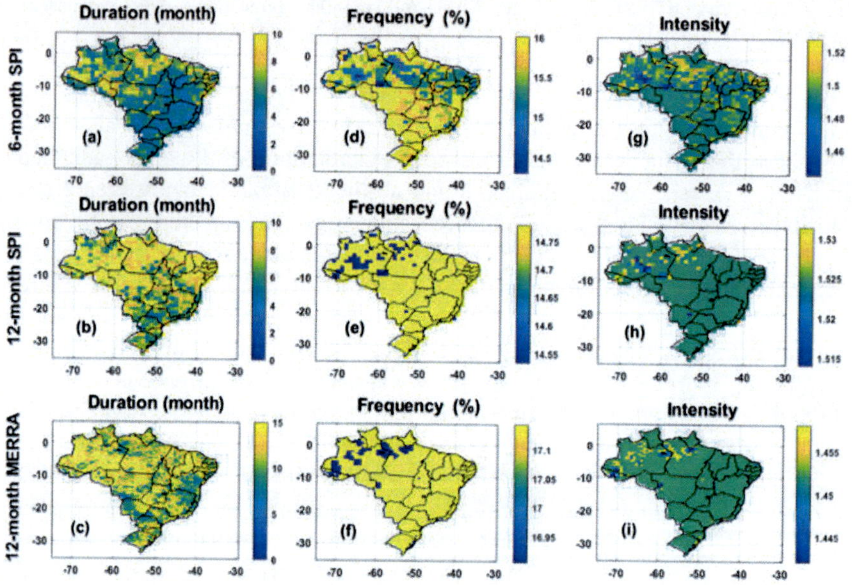

Fig. 20.10 Estimated spatial distribution of drought **a–c** duration, **d–f** frequency and **g–i** intensity between 1980 and 2014 (drought intensity is in standardized units). *Source* Ndehedehe et al. (2020c)

scales (Fig. 20.10g–i). In some regions, however, drought intensities towards northern Brazil vary significantly, particularly when evaluated using a drought indicator on shorter time scales.

20.2.4 Hydrological Hotspots Associated with Multi-scale Ocean Atmosphere Phenomena

The relationship of the temporal series of drought indicators associated with the localized spatial patterns or the hydrological hotspots identified in the previous section with climate teleconnection patterns was assessed. A relatively stronger association explains the 23.8% and 28.2%, respectively, of the variability in ENSO over northern Brazil (SPI-6, Fig. 20.11a–b). Other regions (ISPI-4, ISPI-7, Fig. 20.4) also show sig-

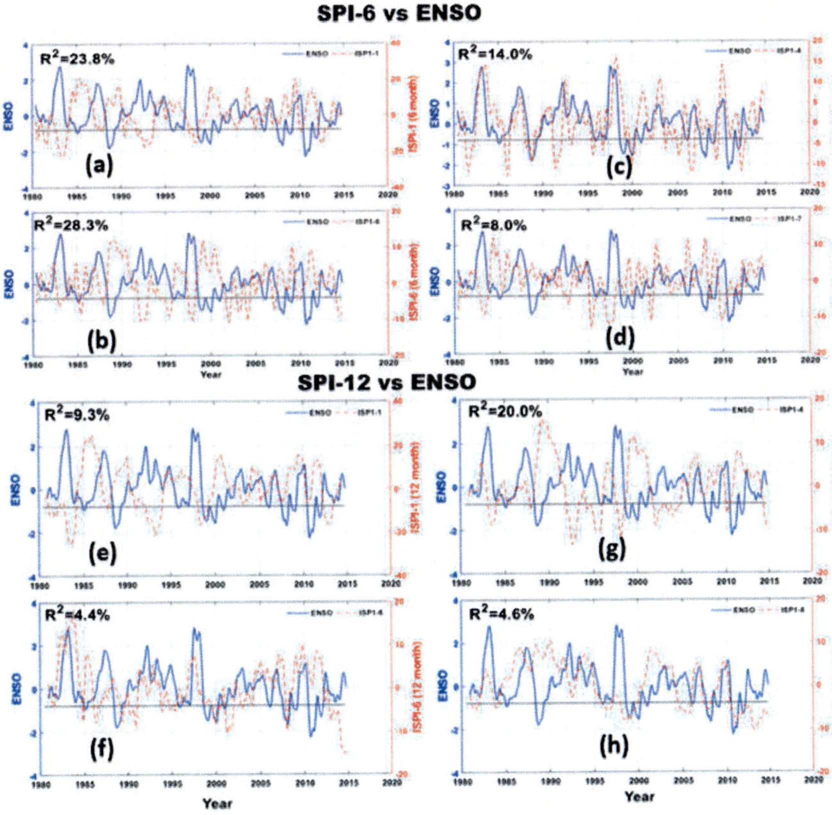

Fig. 20.11 Influence of ENSO on SPI patterns. The relationship of ENSO with **a–d** SPI-6 month and **e–h** SPI-12 month are based on linear regression. *Source* Ndehedehe et al. (2020c)

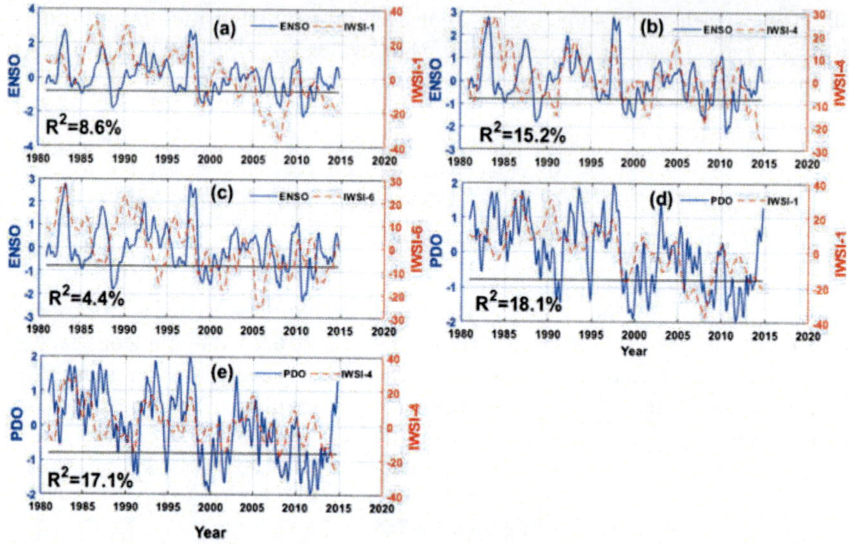

Fig. 20.12 Influence of ENSO and PDO on evolutionary patterns of MERRA-derived drought indicator. Similar to Fig. 20.11 above the **a**–**e** relationship of ENSO with the MERRA indicator is based on linear regression. *Source* Ndehedehe et al. (2020c)

nificant relationship (Fig. 20.11c–d). The influence of ENSO is also notable in SPE-12, (Fig. 20.11e–h) but relatively stronger in northern Brazil explaining 20% (R^2%) of observed variability in ENSO (Fig. 20.11g, cf. ISPI-4, Fig. 20.5). The MERRA indicator is able to extract the influence of ENSO and PDO in the region (Fig. 20.12a–e). The influence of ENSO ($R^2 = 15.1\%$) and PDO ($R^2 = 17.1\%$ and $R^2 = 18.1\%$) are more pronounced in the east and central regions of Brazil (Fig. 20.12a–b and d–e, cf. ISPI-1/ISPI-4, Fig. 20.6). This assessment confirms the existing influence of ENSO and PDO on meteorological conditions in Brazil (Fig. 20.12a–b). The relationships of these drought indicators with climate teleconnections indices as discussed here are fundamentally linear. However, the drought indicators can also interact in non-linear ways with these climate variability indices or teleconnection patterns.

For each of the hydrological hotspots that were localized, the influence of global climate was also assessed using climate modes as independent variables in the support vector machine-based regression (SVMR) scheme. The relationship between observed and predicted drought indicators is summarized in Table 20.2. There is a significant relationship between teleconnection patterns and temporal variations in drought signals (Fig. 20.13a–f), suggesting that the Brazilian sub-region is a complex web of climatic hotspot (Table 20.2). ENSO, PDO, and AMO were identified as the climate teleconnection patterns with the most influence on drought patterns in the region. However, the SVMR coefficients also revealed the influence of QBO AMM, MJO in the hotspots. Similar to the SPI-based drought metrics, the relationship between MERRA-based indicator and climate modes, confirm the influence of

Table 20.2 Relationship between observed and predicted drought indicator (SPI-12) in each hydrological region identified in Figs. 20.4, 20.5, 20.6 using all global climate teleconnection indices as predictors. Correlation values are statistically significant at $\alpha = 0.05$

Region	ISPI-1	ISPI-2	ISPI-3	ISPI-4	ISPI-5	ISPI-6	ISPI-7	ISPI-8
SPI-6	0.52	0.28	0.15	0.40	0.32	0.49	0.46	0.43
SPI-12	0.36	0.46	0.35	0.53	0.31	0.42	0.26	0.54
MERRA-12	0.72	0.71	0.52	0.31	0.24	0.44	0.62	0.57

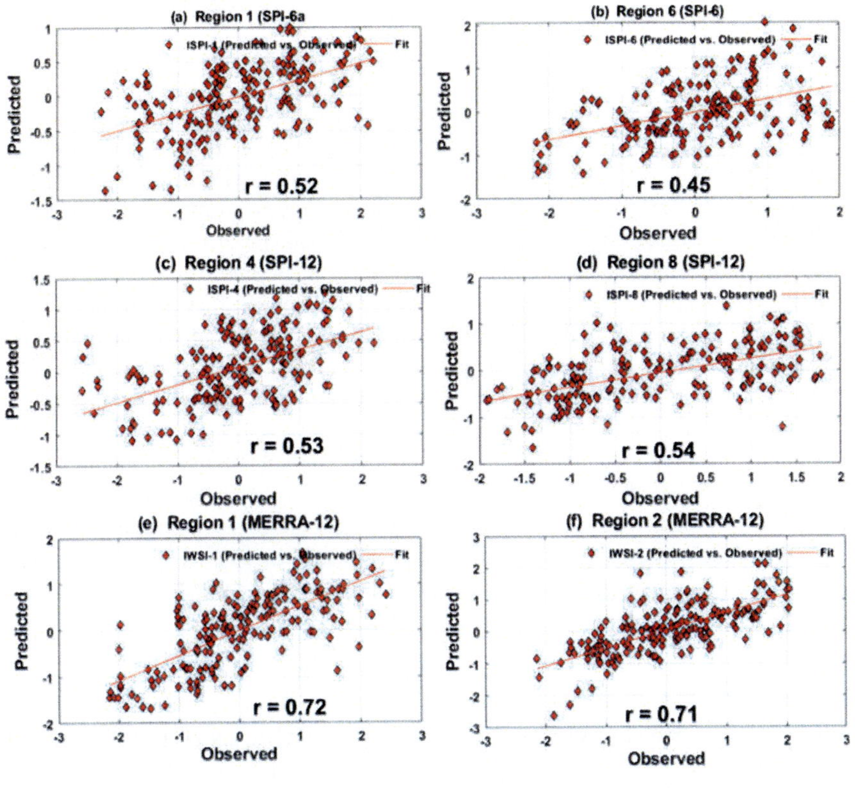

Fig. 20.13 The influence of climate oscillations on the temporal patterns of regionalized modes of SPI-6 (**a–b**), SPI-12 (**c–d**) and MERRA SPI-12 (**e–f**) indicators using SVMR. Relationship between observed and predicted drought indicator uses all seven climate teleconnection indices as predictands in the SVMR scheme based on the hold-out cross-validation partition. *Source* Ndehedehe et al. (2020c)

multiple climate modes (ENSO, AMO, MJO, and AMM) in Brazil. As illustrated in Fig. 20.13e–f, the MERRA indicator ($r = 0.72$ and $r = 0.71$) tends to be more suitable drought metric to identify the impacts of global climate on extreme events as opposed to SPIs (Fig. 20.13a–d).

20.2.5 *Interactions of Climate Teleconnection Patterns with Hydrological Hotspots*

Drought varies in time and space. The distribution of rainfall and the impacts of climate variability indices like the ENSO and climate teleconnection modes drives this variability even more. A spatio-temporal approach to drought analysis through statistical decomposition to aid the localization of these drought patterns is thus crucial. This is why the JADE rotation or independent component analysis was employed to localize and characterize spatial and temporal patterns of drought over Brazil. This study emphasized the importance and validity of such an approach over Brazil where strong spatial variability of rainfall at annual scale along climatological gradients is obvious (cf. Figure 20.3).

Drought variability from all indicators is higher in the north and northeast Brazil. While each of the sub-regions identified through the decomposition of drought indicators has unique response to climate teleconnections, time scales (i.e., SPI-6 and SPI-12) play important role in identifying the influence. This also differs with regions. As with other emerging drought indicators, the multi-scalar nature of SPI can be used to measure various forms of droughts (meteorological and agricultural). For instance, 9 and 12 months aggregation can be used to capture hydrological droughts, allowing the monitoring of impacts and understanding the characteristics and propagation. In the context of predicting drought early warning systems, this is crucial. It also underpins the assessment of important drivers of drought, be it anthropogenic or climatic, e.g., global climate teleconnection patterns.

Moreover, drought statistics could be a key step towards improving water allocation systems. When these statistics are combined with the variability of drought and its duration over Brazil in the last 35 years, they can underpin policy efforts in water management and allocation. The long history of water-related environmental problems in northeast Brazil has been largely linked to limited rainfall and prolonged drought, making it a key hotspot of climatic influence in the region.

20.3 Droughts and Freshwater Changes in the Lake Chad Basin

Lake Chad is the fourth largest lake in Africa with about 40 million people living in the Lake Chad basin. The basin is the world's largest interior drainage basin covering an approximate area of 2,500,000 km^2 and supports six riparian countries (Niger, Nigeria, Chad, Cameroon, Central African Republic, and Sudan) that depend on its water resources for economic and domestic purposes (e.g., Ndehedehe 2019). Climate change led to the contraction of Lake Chad whose surface area declined by approximately 90%. Increasing human population and demand for food resulted in food security problems. Freshwater withdrawal for agricultural was therefore a key response that contributed to the already climate-induced loss of surface water in Lake

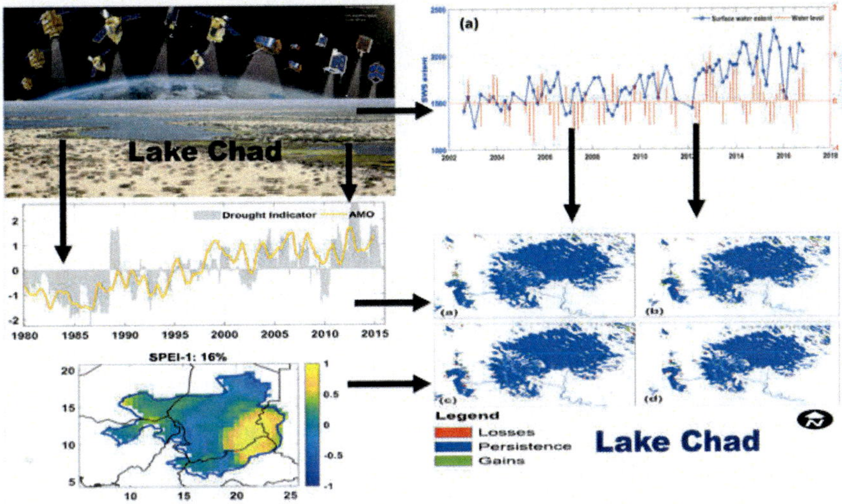

Fig. 20.14 Graphical summary illustrating the use of satellite earth observations in hydrology as highlighted in this chapter. *Source* Ndehedehe et al. (2020b)

Chad (Ndehedehe et al. 2020b). Assessing the basin's freshwater (surface water and terrestrial stored water) changes is thus crucial to improve understanding of drought impacts on its surface water hydrology.

However, this assessment is usually a challenge due to lack of in situ data in the region. This is caused by lack of funding, among other constraints, including insecurity problems (Ndehedehe 2019). Limited gauge measurements have been a major problem for hydro-climatic research in Africa. This has necessitated the need to investigate how decision makers can leverage sophisticated space-borne measurements to improve understanding of surface water hydrology. This knowledge can feed directly into water accounting processes and risk assessment from extreme droughts and their impacts. To demonstrate such potential, this chapter presents the work of Ndehedehe et al. (2020b) who combined multi-resolution satellite observations (Sentinel-2, altimetry, Landsat, GRACE, and TRMM) with data from hydrological model to assess the response of terrestrial hydrology over the Lake Chad basin (LCB) to changes in global climate (Fig. 20.14). The LCB is Africa's largest endorheic basin and this study provided additional perspective on hydrological regimes of semi-arid domains to the influence of climate variability on surface water hydrology.

The overarching objective of this study was to assess the freshwater (surface water and terrestrial stored water) changes of the LCB (Fig. 20.15) and the impacts of droughts on its surface water distribution. One of the key focus of this study is to provide information on drought properties (e.g., duration, intensity) over the LCB. To achieve this, the study undertaken by Ndehedehe et al. (2020b) combined different drought indices including multi-scaled standardized index, the standardized precipitation evapotranspiration index (SPEI Vicente-Serrano et al. 2010a, b) with

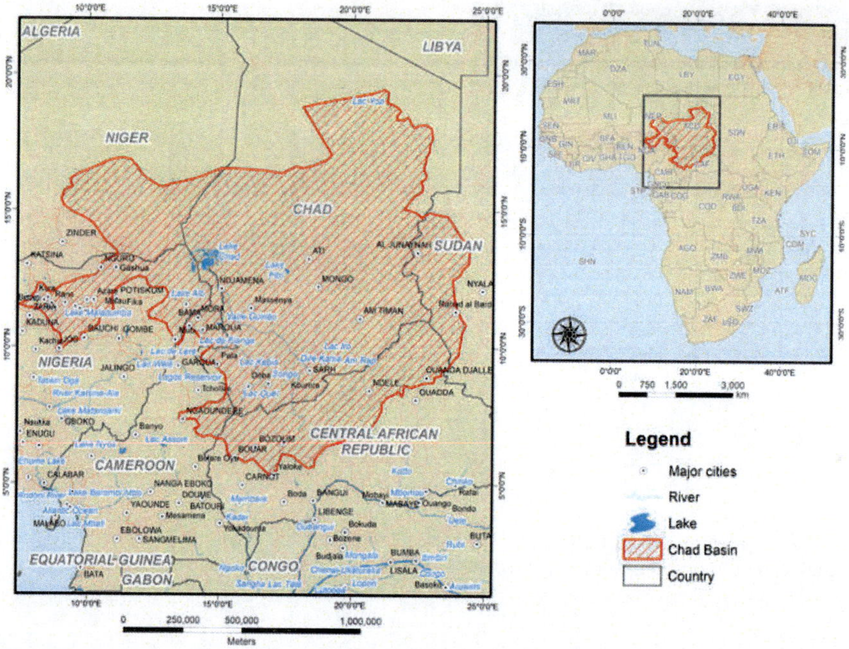

Fig. 20.15 Study area showing the Lake Chad basin and the riparian countries of the Lake Chad basin. Important river networks and open surface water (blue lines and polygons within the basin, including the current Lake Chad (Half Lake Chad) are indicated. *Source* Ndehedehe et al. (2020b)

satellite observations (see, Getirana et al. 2017a; Rodell et al. 2004). More details on the data, statistical analysis and modelling have been summarized in (Ndehedehe et al. 2020b).

20.3.1 Climate Variability in the Lake Chad Basin

The rotation of SPEI resulted in four hydrological regions in the LCB (Figs. 20.16, 20.17).

In the southern part of the basin where the lake receives about 95% of its nourishment through the Chari/Logone river, the notorious early 1980 drought episodes showed persistence on 12-month scale (Region 3, Fig. 20.17). While the 12-month cumulation shows how severe the early 1980 drought in the basin was, it has also indicated the improved marked wet conditions since 2005 unlike preceding years (Fig. 20.17). Interestingly, the classification of regions based on the varimax rotation emphasizes the spatially explicit nature of drought patterns in the basin. For instance, even though SPEI patterns show improved wet conditions in Regions 1–3 and with increasing SPEI values (Fig. 20.17), the surrounding Lake Chad indicates a drought

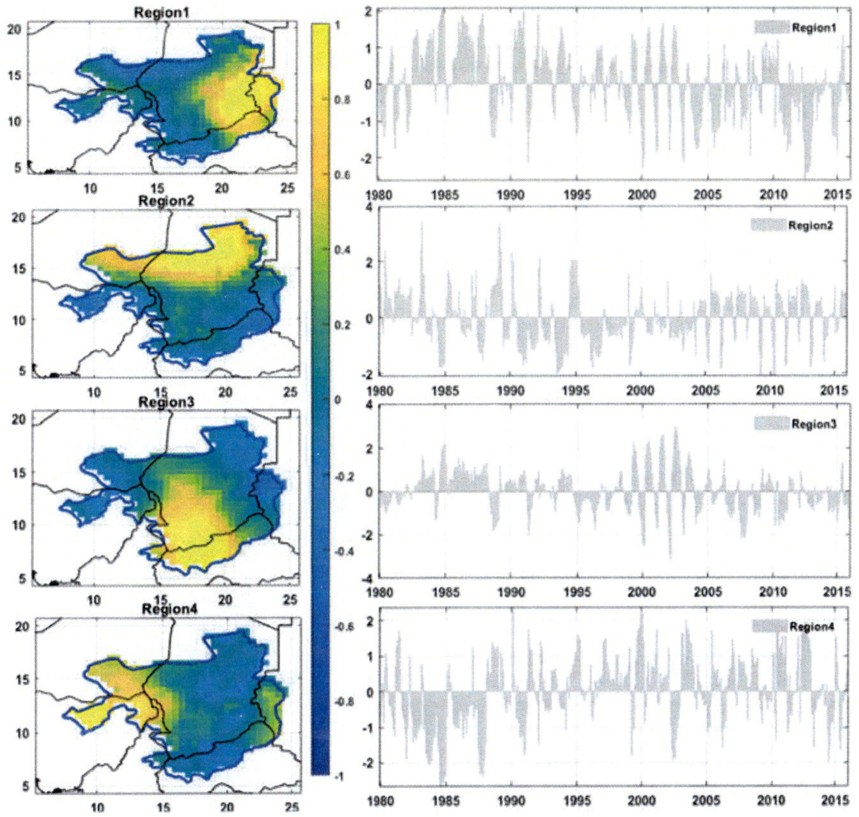

Fig. 20.16 Spatio-temporal SPEI patterns of LCB using 6-month gridded SPEI values based on the rotated principal component analysis. Localized spatial SPEI patterns (right) correspond to the temporal evolutions (left) and the actual SPEI values to be used for drought classification are jointly derived from the spatial and temporal patterns. *Source* Ndehedehe et al. (2020b)

condition (Region 4, Fig. 20.17). These SPEI patterns may have been caused by land surface conditions such as rising temperatures and large changes in annual rainfall climatology.

20.3.2 Drought Characteristics and Intensity

As can been seen from drought intensities at 6 or 12 month cumulation, extreme drought conditions of the early 1980s over the LCB are conspicuous (Fig. 20.18a–b). More than 85% of the basin was affected by these droughts, making it the worst drought in more than 50 years. Even though droughts during the 1990–1995 period were also extreme and severe, affecting between about 75 and 30% (Fig. 20.18b),

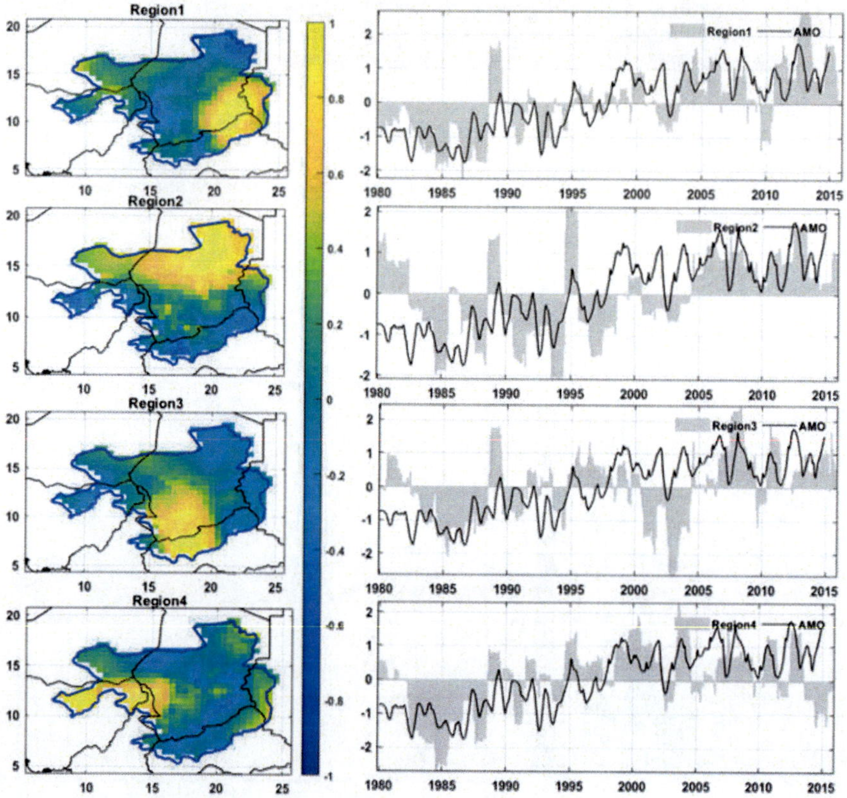

Fig. 20.17 Spatio-temporal SPEI patterns of LCB using 12-month gridded SPEI values based on the rotated principal component analysis. Localized spatial SPEI patterns (right) correspond to the temporal evolutions (left) and the actual SPEI values to be used for drought classification are jointly derived from the spatial and temporal patterns. The relationship between Atlantic Multi-decadal Oscillation in the basin and SPEI in all regions is also assessed. *Source* Ndehedehe et al. (2020b)

they were not as widespread as the extreme droughts between 1984 and 1985 period (Fig. 20.18a–b). Moderate drought events in the period before 2000 are also noted, largely affecting about 20% of the basin as opposed to the coverage extent of severe drought, which fluctuates between 20 and 60% during the period (Fig. 20.18). The time series of drought intensities (extreme, severe, and moderate) during the 1982–1999 period both from 6 to 12 month SPEI cumulation show the LCB is a climatic hotspot of unfavourable hydrological conditions.

Extreme events in the basin as captured by drought indicators correspond to periods with little or no occurrence of drought intensities, which can be clearly interpreted from Figs. 20.16 and 20.17. Interestingly, there have been considerable declines in various drought intensities in the post 2002 period (Fig. 20.18). These declines coincide with increasing patterns of AMO (Atlantic Multi-decadal Oscillation) in the 20 years on the one hand and improved rainfall conditions on the other hand. There

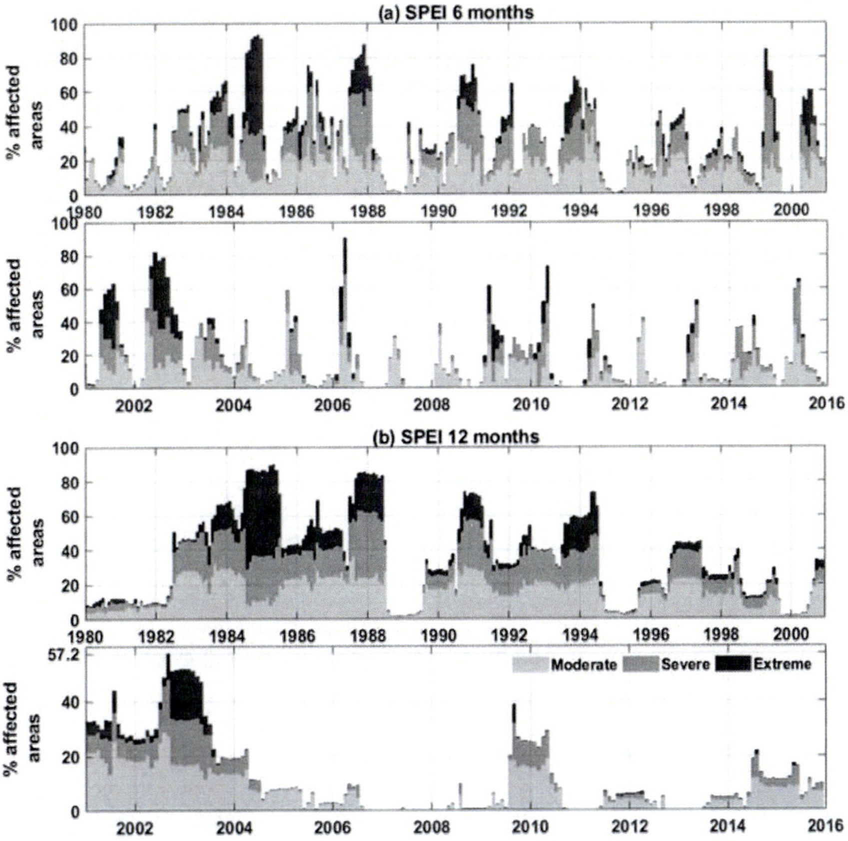

Fig. 20.18 Drought extents over the Lake Chad basin during the 1980 – 2015 period. The estimated drought extents for **a** SPEI 6 month cumulation and **b** SPEI 12 month cumulation are described in terms of their intensity (moderate, severe, and extreme). The affected areas (%) were characterized based on McKee et al. (1993) classification scheme. *Source* Ndehedehe et al. (2020b)

has been improvement in the hydrological conditions of the LCB and is consistent with results in Fig. 20.17. The declining intensity in extreme drought condition is probable indications that rainfall has improved. This also appears to coincide with the gradual rise in the existing surface water area and volume of water stored in Lake Chad.

Drought properties (duration and intensity) as well as the annual response of droughts over the LCB were computed from aerial averages. As would be expected, the mean distribution of longer drought duration appears to be widespread on a longer time scale compared to short cumulation (Fig. 20.19a–b). Even though the LCB is located in a semi-arid region, there are areas that are considerably humid regions. Drought duration in these areas ranges from 2 to 10 months based on SPEI-12 month compared to SPEI-6 month. The distinctiveness in the observed spatial patterns of

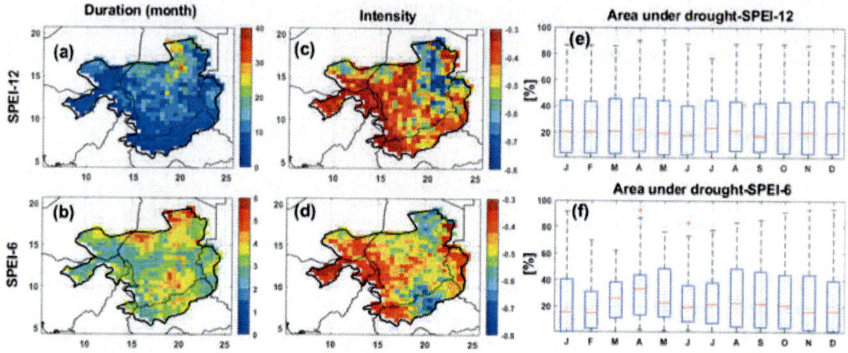

Fig. 20.19 The spatial and temporal distribution of drought properties over the LCB estimated for SPEI-6 and SPEI-12 month cumulation. These drought properties are based on **a–b** duration, **c–d** intensity, and **e–f** and annual response to drought during the 1980 – 2015 period. *Source* Ndehedehe et al. (2020b)

drought frequency in Fig. 20.19c–d suggests that these humid regions of the basin could be more susceptible to the impacts of drought. Drought intensities have a wider distribution in these regions compared to other areas and these intensities (6 or 12 month cumulation) seem to be similar in the part of the basin located in Nigeria. Furthermore, the percentage of areas affected by drought is indicated for SPEI-6 and SPEI-12 (Fig. 20.19e–f). Apart from July, drought-affected areas in other months for SPEI-12 are generally greater than 80% (Fig. 20.19e). But this seems to be different for SPEI-6. Affected drought areas using the SPEI-6 indicator shows that more than 90% but occurring mostly in April and between October and January (Fig. 20.19f).

20.3.3 Spatial and Temporal Changes in Freshwater

As with SPEI, observed terrestrial water storage (TWS) over the LCB basin was also statistically decomposed into temporal and spatial patterns using the varimax rotation. The time series of TWS in the different hydrological regions identified were then related to surface water storage (SWS) derived from the Hydrological Modeling and Analysis Platform (HyMAP) river routing scheme (Getirana et al. 2017b, 2012). The HyMAP-derived monthly surface water dynamics over the Lake Chad basin covered the same period as GRACE data. The hydrological catchments localized over the basin are associated with the model-derived surface water fluxes as shown in Fig. 20.20 and suggest that GRACE-derived TWS over the LCB is mostly driven by changes in SWS. The latter is well correlated with TWS (Fig. 20.20), showing considerably strong relationship in all the regions ($r = 0.77, 0.77$, and 0.89, for TWS-1, TWS-2, and TWS-3, respectively). Notably, these relationships (TWS with SWS) are those observed in parts of the basin that are highly critical to the nourishment of Lake Chad. For instance, TWS-1 and TWS-3 are the southern sections of the

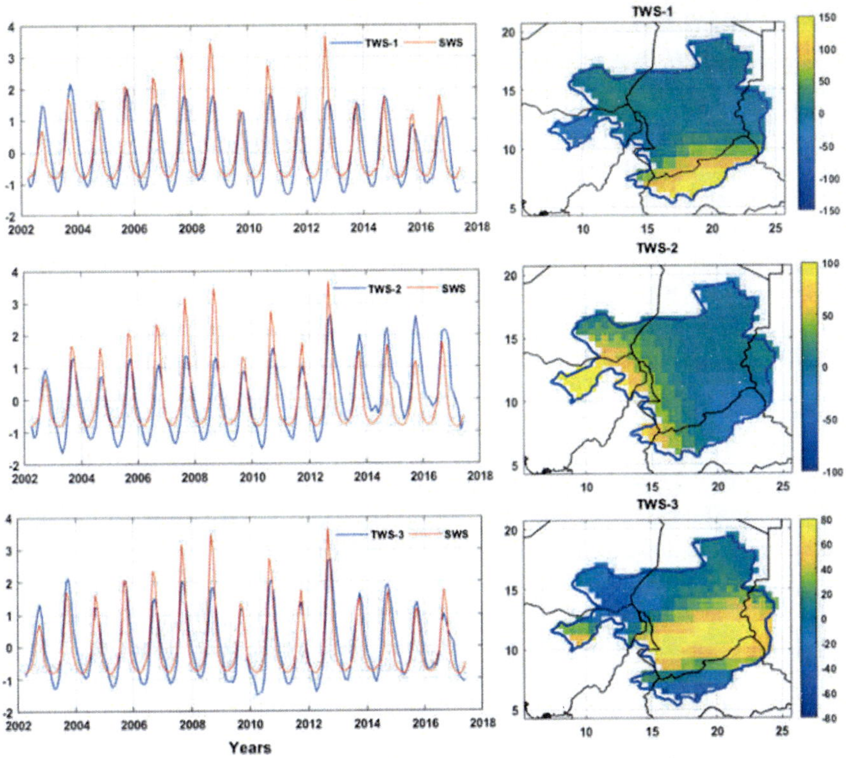

Fig. 20.20 Spatial (right) and temporal (left) patterns of TWS (2002–2017) over different hydrological zones in the LCB based on the varimax rotation of the leading modes of GRACE-derived TWS. The temporal variations of terrestrial water storage (TWS) from each hydrological zones are compared with time series of surface water storage (SWS) changes over the basin. *Source* Ndehedehe et al. (2020b)

basin, where $\sim 95\%$ of LCB's nourishment takes its origin. This region has the most rainfall and provides substantial flows into the Lake. Studies have shown that Lake Chad receives its surface water supply primarily from the Chari/Logone river system (TWS-1 and TWS-3 , Fig. 20.20). However, the Komadugu-Yobe river (TWS-2, Fig. 20.20) nourishes the northern part of the Lake, providing significantly less than 2.5% of available surface water (e.g., Ndehedehe et al. 2016a; Lemoalle et al. 2012). The rise in Landsat-derived surface water extent since 2002 is consistent with SPEI (Fig. 20.17). Strong changes in SWS which occurred in 2007–2008 and 2012 (Fig. 20.20) are also consistent with the results of drought intensities in Fig. 20.18 during the same period.

20.3.4 Satellite-Derived Open Water Features and Change Detection

Satellite observations of open water for different periods (i.e., 2015–2016, 2016–2017, 2017–2018, and 2018–2019) using the Sentinel-2 data highlight the multi-temporal change over the Lake Chad and the corresponding basin. Table 20.3 summarizes the spatial extent of open surface water for the 5-year period (2015–2019) investigated. Open surface water experienced a gradual increase in 2016 (154,925.29 ha) and peaked in 2017 (156,909.99 ha) as opposed to the 2015 threshold (153,127.66 ha). However, the surface area of open water declined from 153,523.17 ha in 2018 to 152,379.68 ha in 2019 (Table 20.3). These fluctuations in the surface water area of Lake Chad are obvious since 2002. In fact, the permanent area of the lake reported by Birkett (2000) was 1,385 km^2 (138,500 ha) and was based on satellite estimates. The rise in the Lake's surface area and inundated floodplains between 2002 and 2016 coincides with the improved rainfall conditions in the basin. The improvement in hydrological conditions coincides with reduced drought events as well as their intensities during the same period over the basin (Fig. 20.18b).

Moreover, Table 20.4 presents the change in detection results of the lake area from 2015 to 2019 as well as rates of open water loss, persistence, and gain percentages. Between 2018 and 2019, the basin had the highest level of open water loss (Fig. 20.22 and Table 20.4). The spatial distribution of this open water change detection indicates much of the lake and the Chari/Logone river are persistent (Fig. 20.22) with percentage loss and gain area ranging from 1.30 to 0.5 (Table 20.4). The percentage loss and gain rates for all years were fairly steady and are summarized in Table 20.4. This could be related to human impacts around the lake area. Regardless of the gain and loss rates, recent analyses (2017–2019) of the lake's surface area from Sentinel-2 data reveal hydrological conditions of the basin are generally not getting better relative to the pre-1984 baseline period.

Table 20.3 Spatial extent of satellite (Sentinel-2) derived land (non-water features) and open water feature classes for the Lake Chad basin (i.e., the Lake Chad and its environ) during the 2015–2019 period using the modified normalized difference water index (MNDWI). Source Ndehedehe et al. (2020b)

Year	2015	2016	2017	**2018**	**2019**
Land (ha)	374,003.56	372,205.93	370,221.23	373,608.05	374,751.54
Open water (ha)	153,127.66	154,925.29	156,909.99	153,523.17	152,379.68

Table 20.4 Results of change detection analyses of open water features in the southern pool of Lake Chad (2015–2019) based on the Sentinel-2 satellite data

Year	Losses (ha)	Persistence (ha)	Gains (ha)	Loss area (%)	Persistence area (%)	Gain area (%)	Loss rate (%)	Gain rate (%)
2015–2016	3,536.7	149,590.9	5,334.4	0.70	28.4	1.00	1.50	2.00
2016–2017	4,157.4	150,767.9	6,142.1	0.80	28.6	1.20	1.70	2.30
2017–2018	6,005.5	150,904.5	2,618.6	1.10	28.6	0.50	0.70	–0.20
2018–2019	4,726.2	148,797.0	3,582.7	1.30	28.2	0.70	1.00	0.70

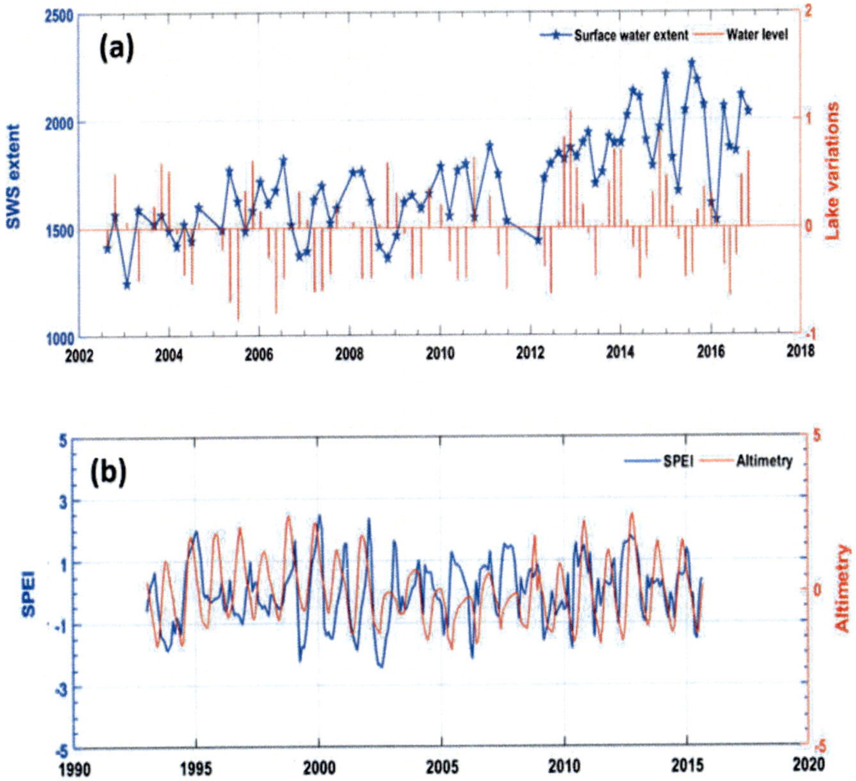

Fig. 20.21 Temporal characteristics of **a** surface water extent and standardized anomalies of altimetry-derived water levels over the lake Chad for the common period (1993–2015) and **b** SPEI over the LCB vs water level changes. *Source* Ndehedehe et al. (2020b)

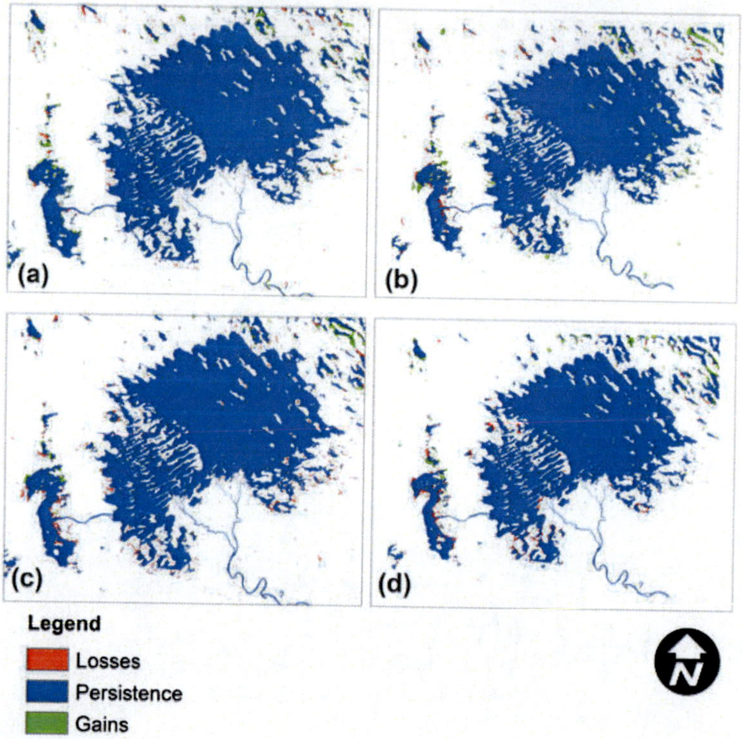

Fig. 20.22 Spatial patterns of open water change detection for **a** 2015–2016, **b** 2016–2017, **c** 2017–2018, and **d** 2018–2019. Open water features/inundation maps of the Lake Chad and environs were generated from Landsat 8 data (2015) and Sentinel-2 (2016–2019) using the Modified Normalized Difference Water Index. *Source* Ndehedehe et al. (2020b)

20.3.5 Impacts of Droughts on Freshwater Availability

The relationship between observed water levels in the lake and total inundation is insignificant (Fig. 20.21). In view of the considerable association between SWS and TWS over the LCB, findings confirm that changes in SWS in the basin is not negligible and is a key component of changes in GRACE-derived TWS. A similar conclusion was also echoed by Getirana et al. (2017a) who evaluated the contributions of SWS changes to TWS variability.

Standardized aerial averaged time series of water fluxes (rainfall, water level, TWS) obtained from satellites and models (SWS, NOAH-TWS) during the 2002–2017 period have also been compared and found to be consistent (Fig. 20.23a). First, the relationship between precipitation and SWS ($r = 0.88$), GRACE-derived TWS ($r = 0.85$), and NOAH-TWS ($r = 0.87$) was considerably strong and linear (Fig. 20.23a). Second, rainfall drives or leads TWS and SWS with a phase lag of two months but rainfall lags water-level variations by two months (i.e., at max $r = 0.82$),

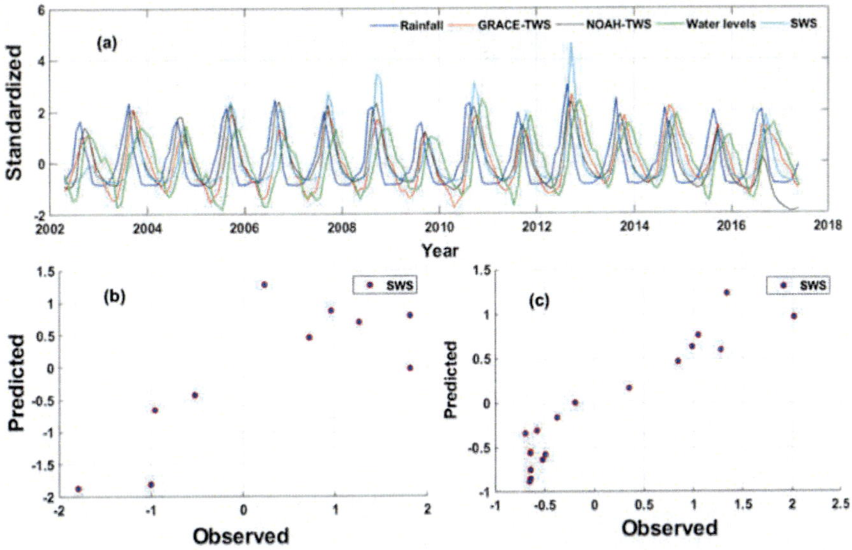

Fig. 20.23 Modelling the influence of climate on surface water dynamics using support vector machine regression. **a** The temporal variations of water fluxes (precipitation from TRMM satellite, TWS from GRACE and hydrological model (NOAH), altimetry-derived water levels and model SWS), **b** modelling of SWS using dominant modes of SPEI at 6-month aggregation, and **c** modelling SWS using precipitation, TWS from GRACE and hydrological model (NOAH), and satellite altimetry-derived water levels. The time series of water fluxes in (**a**) above were standardized using z-score. *Source* Ndehedehe et al. (2020b)

confirming that the lake's surface water hydrology is regulated by discharge from the Chari/Logone system, which emanates from watershed in the Central Africa Republic. It also receives nourishments from Komadougou-Yobe and other tributaries, but direct rainfall over the basin also directly contributes to the lake's surface water ($r = 0.79$).

Moreover, the leading SPEI modes in Fig. 20.16 were used as predictands (climate variable) to assess drivers of freshwater variability (Fig. 20.23b). By integrating the time series of water fluxes in Fig. 20.23a (excluding SWS) in the SVMR (support vector machine regression) scheme, SWS was predicted (Fig. 20.23c). The predicted and observed SWS based on drought indicator (SPEI) were considerably correlated ($r = 0.79$). The predicted and observed SWS based on water fluxes were, nonetheless, more strongly correlated ($r = 0.95$). Recall that TWS and SWS show strong linear temporal relationship (Fig. 20.20). The SPEI is derived using a climatic water balance approach, and the temporal relationship of TWS and SWS and the outputs from the SVMR scheme suggest that surface water is the most vulnerable component of TWS to climate variability. In addition to this, the results in Fig. 20.23 also confirm that freshwater in LCB will mostly be affected by changes in climatic conditions (e.g., declines in rainfall). Increased evapotranspiration caused by rising temperatures can further worsen these impacts.

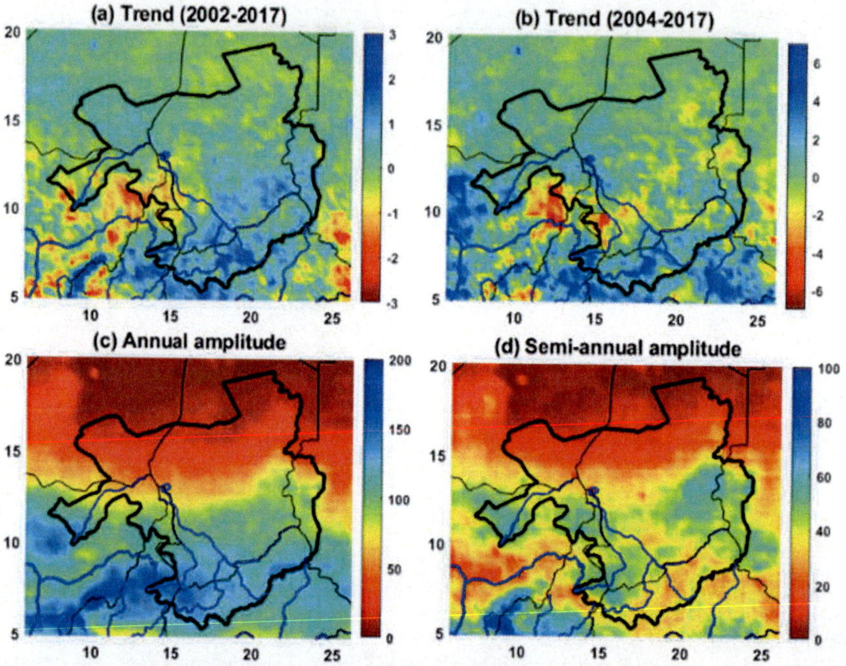

Fig. 20.24 Spatial distribution of **a**–**b** trends (mm/month) and **c**–**d** annual and semi-amplitudes of TRMM-based satellite precipitation over the Lake Chad basin and environ during the 2002–2017 period. The Chari/Logone river systems and other tributaries and other major rivers outside the basin are the blue lines. *Source* Ndehedehe et al. (2020b)

In the absence of flow alteration of the Chari/Logon river system and others, strong changes in annual rainfall in the southern catchment will contribute to the Lake's improved hydrology. The rainfall climatology over the basin confirms that the entire southern catchment is the water tower of the LCB (Fig. 20.24a–d). The spatial distribution of increasing trends in precipitation in the southern region of the basin (Fig. 20.24a–b) coincides with Landsat-derived surface water extent during the 2002–2016 period. The increasing trends in rainfall between 2004 and 2016 around the catchments (Chari/Logone and Komadougou-Yobe river systems) that nourish Lake Chad (Fig. 20.24) are noted. This rise in rainfall was observed during the period with observed significant declines in drought extents and their intensities (Fig. 20.18) (2004–2015 but excluding 2009–2010), confirming the important role of rainfall as a primary driver of surface water hydrology.

Furthermore, the spatial patterns of mean annual and semi-annual rainfall in the southern LCB (Fig. 20.24c–d) are arguably an additional evidence to the dominant role of rainfall on surface hydrology in the basin. The considerable rise in water levels and dramatic increase in surface water area underpin the significant role of climate variability in the region. Hydrologically, there is a potential for restoration in Lake Chad, and the Lake's water storage arguably will depend on the interactions between flow and rainfall, which in turn is modulated by prevailing climate and land surface conditions.

20.3.6 Extreme Events Affect Surface Water Hydrology

A well-known problem in the LCB has been the continued fluctuations in precipitation patterns. In the West African Sahel was the LCB is located, the worst drought during the twentieth century occurred during the 1982–1985 period. This was one of the worst drought of the century and perhaps the millennium, affecting more than 90% of the region between 1984 and 1985 period. However, between 1970s and 1980s, there was a slight recovery from these droughts in the Sahel (e.g., Ndehedehe 2019; Sultan and Gaetani 2016; Andam-Akorful et al. 2017; Sanogo et al. 2015; Martin and Thorncroft 2014). The SPEI results shown in this study highlight that a considerable proportion of drought conditions was caused by limited rainfall. This change in rainfall was driven by important processes of inter-annual variability (e.g., ENSO, AMO, etc.) as well as rapidly changing land surface conditions. In this regard, there is unanimous positions on the impacts of several climate teleconnections and other meteorological processes in the African Sahel (e.g., Becker et al. 2018; Ndehedehe et al. 2016b, 2019a; Anyah et al. 2018; Gizaw and Gan 2017; Dyer et al. 2017).

The influence of climate variability indices on the hydrology of LCB has been reported. Diagnostics studies undertaken in the region have shown how teleconnections patterns like the AMO, AMM (Atlantic Meridional Mode), and ENSO are associated with observed water levels and precipitation (Ndehedehe et al. 2016a; Okonkwo et al. 2014). The notorious prolonged extreme droughts of the early 1980s, are one of those events that diagnostic studies linked to abnormally-warm sea surface temperatures of the Atlantic and Indian oceans (e.g., Giannini et al. 2003). During this period, Lake Chad lost a substantial proportion of its surface water. This implies that the contraction of the Lake Chad's surface area was primarily caused by prolonged rainfall deficits. The extreme drought resulting from this long-term declines in precipitation restricted the inflow from the Chari/Logone river system, which supplies about 95% freshwater of the Lake and from other watersheds that account for the remaining in-flow. This obviously underscores the role of climate variability drivers on surface water hydrology in the basin.

20.4 Conclusions

Drought's impacts on both regional and global ecosystems can be catastrophic. Its effects on water supply, crop production, livestock, and the general ecosystem functioning have been widely captured in past studies. In recent years, an increased trend in drought frequency and severity has been observed over a wide spatial and temporal dimension. Various drivers such as global warming, influence of climatic oscillations, excessive water demand, as well as other anthropogenic influences are all contributing factors to the recurring nature of drought episodes. These various drought episodes have a long-lasting impact on the environment, and pose significant risks to both economical and societal life. This has awakened the need for extensive and proactive actions against this hazard, through the design and implementation of comprehensive drought assessment strategies as well as drought mitigation and vulnerability assessment programmes for water resource management. With new remote sensing techniques, it is now easier to monitor and assess droughts. The advent of satellite gravity observations like GRACE (Gravity Recovery and Climate Experiment) has provided timely and constant monitoring of terrestrial stored water, improving understanding of climate change impacts on water availability. This monitoring capability appears to be more efficient compared to in situ method. Drought metrics and indicators such as the standardized precipitation index, standardized soil moisture index, standardized runoff index, etc. as well as assimilation procedures are also emerging to complement satellite observations like the GRACE mission, which appears limited due to its coarse spatial resolution. This has made the assessment of drought easier, especially in high-risk zones like Lake Chad and Brazil. The Lake Chad surface has been experiencing consistent drying of its water surface area, and this has been attributed to land-use changes, increased water abstractions, as well as other forms of anthropogenic influences. Brazil, on the other hand, which is regarded as one of the regions with the most abundant surface water availability globally, has suffered from a substantial loss of surface water, which has directly affected agricultural practices, power generation, as well as water quality for the citizens and other residents who depend on them for freshwater. These investigations were possible due to the advent of the conventional remote sensing techniques, which provided good data source for scientists and hydrologists for adequate assessments that underpins policy solutions that can mitigate further depletions in these zones. Drought is inevitable as long as we are living in a world with changing climate. However, a better understanding of this environmental hazard gives more opportunity to overcome its impact on both regional and global cases.

References

Adler RF, Huffman GJ, Chang A, Ferraro R, Xie P-P, et al (2003) The version-2 Global Precipitation Climatology Project (GPCP) monthly precipitation analysis (1979-Present). J Hydrometeorol 4(6):1147–1167. https://doi.org/10.1175/1525-7541(2003)004<1147:TVGPCP>2.0.CO;2

AghaKouchak A (2015) A multivariate approach for persistence-based drought prediction: application to the 2010–2011 East Africa drought. J Hydrol 526:127–135. https://doi.org/10.1016/j.jhydrol.2014.09.063

AghaKouchak A, et al (2018) How do natural hazards cascade to cause disasters? Nature 561:458–460

Agutu N, Awange J, Zerihun A, Ndehedehe C, Kuhn M, Fukuda Y (2017) Assessing multi-satellite remote sensing, reanalysis, and land surface models' products in characterizing agricultural drought in East Africa. Remote Sens Environ 194:287–302. https://doi.org/10.1016/j.rse.2017.03.041

Agutu N, Awange J, Ndehedehe C, Kirimi F, Kuhn M (2019) GRACE-derived groundwater changes over Greater Horn of Africa: temporal variability and the potential for irrigated agriculture. Sci Total Environ 693:133467. https://doi.org/10.1016/j.scitotenv.2019.07.273

Agutu N, Awange J, Ndehedehe C, Mwaniki M (2020) Consistency of agricultural drought characterization over Upper Greater Horn of Africa (1982–2013): topographical, gauge density, and model forcing influence. Sci Total Environ 709:135149. https://doi.org/10.1016/j.scitotenv.2019.135149

Alemohammad SH, McLaughlin DB, Entekhabi D (2015) Quantifying precipitation uncertainty for land data assimilation applications. Mon Weather Rev 143(8):3276–3299. https://doi.org/10.1175/MWR-D-14-00337.1

Alley WM, Konikow LF (2015) Bringing GRACE down to earth. Groundwater 53(6):826–829. https://doi.org/10.1111/gwat.12379

Andam-Akorful S, Ferreira V, Ndehedehe CE, Quaye-Ballard J (2017) An investigation into the freshwater variability in West Africa during 1979–2010. Int J Climatol 37(S1):333–349. https://doi.org/10.1002/joc.5006

Anderson MC, Hain C, et al (2013) An intercomparison of drought indicators based on thermal remote sensing and nldas-2 simulations with u.s. drought monitor classifications. J Hydrometeorol 14(4):1035–1056. https://doi.org/10.1175/JHM-D-12-0140.1

Anyah R, Forootan E, Awange J, Khaki M (2018) Understanding linkages between global climate indices and terrestrial water storage changes over Africa using GRACE products. Sci Total Environ 635:1405–1416. https://doi.org/10.1016/j.scitotenv.2018.04.159

Awange JL, Mpelasoka F, Goncalves RM (2016) When every drop counts: analysis of droughts in Brazil for the 1901–2013 period. Sci Total Environ 566–567:1472–1488. https://doi.org/10.1016/j.scitotenv.2016.06.031

Awange J, Anyah R, Agola N, Forootan E, Omondi P (2013) Potential impacts of climate and environmental change on the stored water of Lake Victoria Basin and economic implications. Water Resour Res 49(0):8160–8173. https://doi.org/10.1002/2013WR014350

Bazrafshan J, Hejabi S, Rahimi J (2014) Drought monitoring using the multivariate standardized precipitation index (MSPI). Water Resour Mangagement 28:1045–1060. https://doi.org/10.1007/s11269-014-0533-2

Becker M, Papa F, Frappart F, Alsdorf D, Calmant S, da Silva JS, Prigent C, Seyler F (2018) Satellite-based estimates of surface water dynamics in the Congo River Basin. Int J Appl Earth Obs Geoinf 66:196–209. https://doi.org/10.1016/j.jag.2017.11.015

Becker M, LLovel W, Cazenave A, Güntner A, Crétaux J-F (2010) Recent hydrological behavior of the East African great lakes region inferred from GRACE, satellite altimetry and rainfall observations. Comptes Rendus Geosci 342(3):223–233. https://doi.org/10.1016/j.crte.2009.12.010

Benali A, Carvalho A, Nunes J, Carvalhais N, Santos A (2012) Estimating air surface temperature in Portugal using MODIS LST data. Remote Sens Environ 124:108–121. https://doi.org/10.1016/j.rse.2012.04.024

Berg A, Sheffield J (2018) Climate change and drought: the soil moisture perspective. Curr Clim Chang Rep 4(2):180–191

Birkett CM (2000) Synergistic remote sensing of Lake Chad: variability of basin inundation. Remote Sens Environ 72:218–236. https://doi.org/10.1016/S0034-4257(99)00105-4

Bojinski S, Verstraete M, Peterson TC, Richter C, Simmons A, Zemp M (2014) The concept of essential climate variables in support of climate research, applications, and policy. Bull Am Meteor Soc 95(9):1431–1443. https://doi.org/10.1175/BAMS-D-13-00047.1

Brasil-Neto RM, Santos CAG, Silva JF, et al (2003) Evaluation of the TRMM product for monitoring drought over Paraíba State, northeastern Brazil: a trend analysis. Sci Rep 11(1097). https://doi.org/10.1109/10.1038/s41598-020-80026-5

Brito SSB, Cunha APMA, Cunningham CC, Alvalá RC, Marengo JA, Carvalho MA (2018) Frequency, duration and severity of drought in the semiarid Northeast Brazil region. Int J Climatol 38(2):517–529. https://doi.org/10.1002/joc.5225

Buma WG, Lee S-I, Seo JY (2018) Recent surface water extent of Lake Chad from multispectral sensors and GRACE. Sensors 18(7). https://doi.org/10.3390/s18072082

Bureau of Meteorology B (2008) Climate glossary: Drought

Buytaert WOUTER, Breuer LUTZ (2013) Water resources in South America: sources and supply, pollutants and perspectives. In: Proceedings of the IAHS-IAPSO-IASPEI Assembly, Gothenburg, Sweden, pp 106–113. https://iahs.info/uploads/dms/15575.16-106-113-361-05-Buytaert-Gothenburg-H04CORR.pdf 26th Aug 2021

Cardoso J-F (1991) Super-symmetric decomposition of the fourth-order cumulant tensor, blind identification of more sources than sensors. http://perso.telecom-paristech.fr/~cardoso/Papers.PDF/icassp91.pdf. 15 Jan 2016

Cardoso JF (1999) High-order contrasts for independent component analysis. Neural Comput 11:157–192

Cardoso JF, Souloumiac A (1993) Blind beamforming for non-gaussian signals. IEE Proc 140(6):362–370

Chen T, de Jeu R, Liu Y, van der Werf G, Dolman A (2014) Using satellite based soil moisture to quantify the water driven variability in NDVI: a case study over mainland Australia. Remote Sens Environ 140:330–338. https://doi.org/10.1016/j.rse.2013.08.022

Coe MT, Foley JA (2001) Human and natural impacts on the water resources of the Lake Chad basin. J Geophys Res 106(D4):3349–3356. https://doi.org/10.1029/2000JD900587

Cottrell R, Nash K, Halpern B (2019) Food production shocks across land and sea. Nat Sustain 2(2):130–137

Dai A (2011) Drought under global warming: a review. Wiley Interdiscip Rev: Clim Chang 2(1):45–65. https://doi.org/10.1002/wcc.81

Damberg L, AghaKouchak A (2014) Global trends and patterns of drought from space. Theor Appl Climatol 117(3):441–448. https://doi.org/10.1007/s00704-013-1019-5

Diffenbaugh NS, Swain DL, Touma D (2015) Anthropogenic warming has increased drought risk in California. Proc Natl Acad Sci 112(13):3931–3936. https://doi.org/10.1073/pnas.1422385112

Dong J, Walker JP, Houser PR, Sun C (2007) Scanning multichannel microwave radiometer snow water equivalent assimilation. J Geophys Res: Atmos 112(D7). https://doi.org/10.1029/2006JD007209

Du J, Kimball JS, Velicogna I, Zhao M, Jones LA, Watts JD, Kim Y (2019) Multicomponent satellite assessment of drought severity in the contiguous united states from 2002 to 2017 using AMSR-E and AMSR2. Water Resour Res 55(7):5394–5412. https://doi.org/10.1029/2018WR024633

Dyer ELE, Jones DBA, Nusbaumer J, Li H, Collins O, Vettoretti G, Noone D (2017) Congo basin precipitation: assessing seasonality, regional interactions, and sources of moisture. J Geophys Res: Atmos 122(13):6882–6898. https://doi.org/10.1002/2016JD026240

Easterling DR, Kunkel KE, Wehner MF, Sun L (2016) Detection and attribution of climate extremes in the observed record. Weather Clim Extrem 11:17–27. https://doi.org/10.1016/j.wace.2016.01.001

Eichler TP, Londoño AC (2013) South American climatology and impacts of El Niño in NCEP"s CFSR data. Adv Meteorol 2013:105116. https://doi.org/10.1155/2013/492630

Epule ET, Peng C, Lepage L, Chen Z (2014) The causes, effects and challenges of Sahelian droughts: a critical review. Reg Environ Chang 14(1):145–156. https://doi.org/10.1007/s10113-013-0473-z

Erfanian A, Wang G, Fomenko L (2017) Unprecedented drought over tropical south America in 2016: significantly under-predicted by tropical SST. Sci Rep 7(5811). https://doi.org/10.1038/s41598-017-05373-2

FAO (2019) The state of food security and nutrition in the world: safeguarding against economic slowdowns and downturns

Ferraro RR (1997) Special sensor microwave imager derived global rainfall estimates for climatological applications. J Geophys Res: Atmos 102(D14):16715–16735. https://doi.org/10.1029/97JD01210

Ferreira VG, Ndehedehe CE, Montecino HC, Yong B, Yuan P, Abdalla A, Mohammed AS (2019b) Prospects for imaging terrestrial water storage in South America using daily GPS observations. Remote Sens 11(6). https://doi.org/10.3390/rs11060679

Ferreira V, Montecino H, Ndehedehe C, Heck B, Gong Z, Westerhaus M, de Freitas S (2018) Space-based observations of crustal deflections for drought characterization in brazil. Sci Total Environ 644:256–273. https://doi.org/10.1016/j.scitotenv.2018.06.277

Ferreira VG, Montecino HD, Ndehedehe CE, del Rio RA, Cuevas A, de Freitas SRC (2019) Determining seasonal displacements of Earth's crust in South America using observations from space-borne geodetic sensors and surface-loading models. Earth, Planets Space 71(1):84. https://doi.org/10.1186/s40623-019-1062-2

Fischer EM, Knutti R (2015) Anthropogenic contribution to global occurrence of heavy-precipitation and high-temperature extremes. Nat Clim Chang 5(6):560–564. https://doi.org/10.1038/nclimate2617

Forrest K, Tarroja B, Chiang F, AghaKouchak A, Samuelsen S (2018) Assessing climate change impacts on California hydropower generation and ancillary services provision. Clim Chang 151(3):395–412. https://doi.org/10.1007/s10584-018-2329-5

Frappart F, Papa F, Guntner A, Werth S, da Silva JS, Tomasella J, Seyler F, Prigent C, Rossow WB, Calmant S, Bonnet M-P (2011) Satellite-based estimates of groundwater storage variations in large drainage basins with extensive floodplains. Remote Sens Environ 115(6):1588–1594. https://doi.org/10.1016/j.rse.2011.02.003

Froend R, Loomes R, Horwitz P, Bertuch M, Storey A, Bamford M (2004) Study of ecological water requirements on the Gnangara and Jandakot mounds under section 46 of the environmental protection act. In: A report to the water and rivers commission

Getirana A (2016) Extreme water deficit in brazil detected from space. J Hydrometeorol 17(2):591–599. https://doi.org/10.1175/JHM-D-15-0096.1

Getirana ACV, Boone A, Yamazaki D, Decharme B, Papa F, Mognard N (2012) The Hydrological Modeling and Analysis Platform (HyMAP): evaluation in the Amazon basin. J Hydrometeorol 13(6):1641–1665. https://doi.org/10.1175/JHM-D-12-021.1

Getirana A, Peters-Lidard C, Rodell M, Bates PD (2017) Trade-off between cost and accuracy in large-scale surface water dynamic modeling. Water Resour Res 53(6):4942–4955. https://doi.org/10.1002/2017WR020519

Getirana A, Kumar S, Girotto M, Rodell M (2017a) Rivers and floodplains as key components of global terrestrial water storage variability. Geophys Res Lett 44(20):10,359–10,368. https://doi.org/10.1002/2017GL074684

Giannini A, Saravanan R, Chang P (2003) Oceanic forcing of Sahel rainfall on interannual to decadal time scales. Science 302(5647):1027–1030. https://doi.org/10.1126/science.1089357

Gizaw MS, Gan TY (2017) Impact of climate change and El Niño episodes on droughts in sub-Saharan Africa. Clim Dyn 49(1):665–682. https://doi.org/10.1007/s00382-016-3366-2

Guan B, Molotch NP, Waliser DE, Jepsen SM, Painter TH, Dozier J (2013) Snow water equivalent in the sierra Nevada: Blending snow sensor observations with snowmelt model simulations. Water Resour Res 49(8):5029–5046. https://doi.org/10.1002/wrcr.20387

Hempel S, Frieler K, Warszawski L, Schewe J, Piontek F (2013) A trend-preserving bias correction-the ISI-MIP approach. Earth Syst Dyn 4(2):219–236. https://doi.org/10.5194/esd-4-219-2013

Hou AY, Kakar RK, Neeck S, et al (2014) The global precipitation measurement mission. Bull Am Meteorol Soc 95(5):701–722. https://doi.org/10.1175/BAMS D 13 00164.1

Howitt R, Medellín-Azuara J, MacEwan D, Lund JR, Sumner D (2014) Economic analysis of the 2014 drought for California agriculture. University of California, Davis, CA, Center for Watershed Sciences

Huffman GJ, Bolvin DT, Braithwaite D, Hsu K, Joyce R, Xie P, Yoo SH (2015) Nasa global precipitation measurement (gpm) integrated multi-satellite retrievals for GPM (IMERG). Algorithm Theoretical Basis Document (ATBD) Version 4(26). https://gpm.nasa.gov/sites/default/files/2020-05/IMERG_ATBD_V06.3.pdf 24th Aug 2021

Huffman GJ, Adler RF, Bolvin DT, Gu G, Nelkin EJ, Bowman KP, Hong Y, Stocker EF, Wolff DB (2007) The TRMM multisatellite precipitation analysis (TMPA): quasi-global, multiyear, combined-sensor precipitation estimates at fine scales. J Hydrometeorol 8:38–55. https://doi.org/10.1175/JHM560.1

Hughes M, Hall A, Fovell RG (2009) Blocking in areas of complex topography, and its influence on rainfall distribution. J Atmos Sci 66(2):508–518. https://doi.org/10.1175/2008JAS2689.1

Hughes N, Galeano D, Hattfield-Dodds (2019) The effects of drought and climate variability on Australian farms. In: Australian bureau of agricultural and resource economics and sciences, Canberra. http://doi.org/10.25814/5de84714f6e08 16th Oct 2020

Jackson RD, Idso SB, Reginato RJ, Pinter PJ Jr (1981) Canopy temperature as a crop water stress indicator. Water Resour Res 17(4):1133–1138. https://doi.org/10.1029/WR017i004p01133

Jain SK (2019) Water resources management in India-challenges and the way forward. Curr Sci 117(4)

Jain M, Fishman R, Mondal P, Galford GL (2021) Groundwater depletion will reduce cropping intensity in India. Sci Adv 7(9):eabd2849. https://doi.org/10.1126/sciadv.abd2849

Jasechko S, Perrone D (2021) Global groundwater wells at risk of running dry. Science 372(6540):418–421. https://doi.org/10.1126/science.abc2755

Jolliffe IT (2002) Principal component analysis, 2nd edn. Springer Series in Statistics, Springer, New York

Kidd C, Levizzani V, Laviola S (2010) Quantitative precipitation estimation from earth observation satellites. Rainfall: State Sci 191:127–158. https://doi.org/10.1029/GM191

Kiem AS, Johnson F, Westra S, van Dijk A, Evans JP, O'Donnell A, Rouillard A, Barr C, Tyler J, Thyer M, Jakob D, Woldemeskel F, Sivakumar B, Mehrotra R (2016) Natural hazards in Australia: droughts. Clim Change 139(1):37–54. https://doi.org/10.1007/s10584-016-1798-7

Kogan FN (1997) Global drought watch from space. Bull Am Meteorol Soc 78(4):621–636. https://doi.org/10.1175/1520-0477(1997)078<0621:GDWFS>2.0.CO;2

Kogan F (1995) Application of vegetation index and brightness temperature for drought detection. Adv Space Res 15(11):91–100. https://doi.org/10.1016/0273-1177(95)00079-T

Kongoli C, Romanov P, Ferraro R (2012) 15 snow cover monitoring. In: Remote sensing of drought: innovative monitoring approaches, vol 359

Lemoalle J, Bader J-C, Leblanc M, Sedick A (2012) Recent changes in Lake Chad: observations, simulations and management options (1973–2011). Glob Planet Chang 80–81:247–254. https://doi.org/10.1016/j.gloplacha.2011.07.004

Lemos MCDM (2003) A tale of two policies: the politics of climate forecasting and drought relief in Ceará Brazil. Policy Sci 36(2):101–123. https://doi.org/10.1023/A:1024893532329

Lesk C, Rowhani P, Ramankutty N (2016) Influence of extreme weather disasters on global crop production. Nature 529(7584):84–87

Liu YY, Parinussa RM, Dorigo WA, et al (2011) Developing an improved soil moisture dataset by blending passive and active microwave satellite-based retrievals. Hydrol Earth Syst Sci 15(2):425–436. https://doi.org/10.5194/hess-15-425-2011

Long D, Scanlon BR, Longuevergne L, Sun AY, Fernando DN, Save H (2013) GRACE satellite monitoring of large depletion in water storage in response to the 2011 drought in Texas. Geophys Res Lett 40(13):3395–3401. https://doi.org/10.1002/Grl.50655

Mach KJ, Kraan CM, Adger WN et al (2019) Climate as a risk factor for armed conflict. Nature 571(7764):193–197. https://doi.org/10.1038/s41586-019-1300-6

Madadgar S, Moradkhani H, Garen D (2014) Towards improved post-processing of hydrologic forecast ensembles. Hydrol Process 28(1):104–122. https://doi.org/10.1002/hyp.9562

Marengo JA, Torres RR, Alves LM (2017) Drought in Northeast Brazil–past, present, and future. Theoret Appl Climatol 129(3):1189–1200. https://doi.org/10.1007/s00704-016-1840-8

Marengo JA, Alves LM, Alvala RC, Cunha AP, Brito S, Moraes OL (2018) Climatic characteristics of the 2010–2016 drought in the semiarid Northeast Brazil region. Anais da Academia Brasileira de Ciencias 90:1973–1985. https://doi.org/10.1590/0001-3765201720170206

Martin ER, Thorncroft CD (2014) The impact of AMO on the West African monsoon annual cycle. Q J R Meteorol Soc 140:31–46. https://doi.org/10.1002/qj.2107

Martinez WL, Martinez AR (2005) Exploratory data analysis with MATLAB. Chapman and Hall/CRC Press LLC, UK. 1-58488-366-9

Martínez-Fernández J, González-Zamora A, Sánchez N, Gumuzzio A, Herrero-Jiménez C (2016) Satellite soil moisture for agricultural drought monitoring: assessment of the SMOS derived soil water deficit index. Remote Sens Environ 177:277–286. https://doi.org/10.1016/j.rse.2016.02.064

McKee TB, Doeskin NJ, Kieist J (1993) The relationship of drought frequency and duration to time scales. In: Conference on applied climatology, American meteorological society, Boston, Massachusetts, pp 179–184. www.ccc.atmos.colostate.edu/relationshipofdroughtfrequency.pdf. 27 June 2014

McKee TB, Doeskin NJ, Kieist J (1995) Drought monitoring with multiple time scales. In: Conference on applied climatology, American meteorological society, Boston, Massachusetts, pp 233–236. www.southwestclimatechange.org/node/911. 13 July 2014

Melati MD, Fleischmann AS, Fan FM, Paiva RCD, Athayde GB (2019) Estimates of groundwater depletion under extreme drought in the Brazilian semi-arid region using GRACE satellite data: application for a small-scale aquifer. Hydrogeol J 27(8):2789–2802. https://doi.org/10.1007/s10040-019-02065-1

Mishra AK, Singh VP (2010) A review of drought concepts. J Hydrol 391:202–216. https://doi.org/10.1016/j.jhydrol.2010.07.012

Moore P, Williams SDP (2014) Integration of altimetry lake lavels and GRACE gravimetry over Africa: inferences for terrestrial water storage change 2003–2011. Water Resour Res 50:9696–9720. https://doi.org/10.1002/2014WR015506

Moradkhani H, Hsu K, Hong Y, Sorooshian S (2006) Investigating the impact of remotely sensed precipitation and hydrologic model uncertainties on the ensemble streamflow forecasting. Geophys Res Lett 33(12). https://doi.org/10.1029/2006GL026855

Mpelasoka F, Hennessy K, Jones R, Bates B (2008) Comparison of suitable drought indices for climate change impacts assessment over Australia towards resource management. Int J Climatol 28(10):1283–1292. https://doi.org/10.1002/joc.1649

Mu Q, Zhao M, Running SW (2011) Improvements to a MODIS global terrestrial evapotranspiration algorithm. Remote Sens Environ 115(8):1781–1800. https://doi.org/10.1016/j.rse.2011.02.019

Mu Q, Zhao M, Kimball JS, McDowell NG, Running SW (2013) A remotely sensed global terrestrial drought severity index. Bull Am Meteor Soc 94(1):83–98. https://doi.org/10.1175/BAMS-D-11-00213.1

Nasrollahi N, AghaKouchak A, Cheng L, et al (2015) How well do CMIP5 climate simulations replicate historical trends and patterns of meteorological droughts? Water Resour Res 51(4):2847–2864. https://doi.org/10.1002/2014WR016318

Naumann G, Dutra E, Barbosa P, Pappenberger F, Wetterhall F, Vogt JV (2014) Comparison of drought indicators derived from multiple data sets over Africa. Hydrol Earth Syst Sci 18(5):1625 1640. https://doi.org/10.5194/hess-18-1625-2014

Naumann G, Alfieri L, Wyser K, Mentaschi L (2018) Global changes in drought conditions under different levels of warming. Geophys Res Lett 45(7):3285–3296. https://doi.org/10.1002/2017GL076521

Nayak MA, Hassan WU (2021) A synthesis of drought prediction research over India. Water Secur 13:100092. https://doi.org/10.1016/j.wasec.2021.100092

Ndehedehe CE, Agutu N, Ferreira VG, Getirana A (2020) Evolutionary drought patterns over the Sahel and their teleconnections with low frequency climate oscillations. Atmos Res 233:104700. https://doi.org/10.1016/j.atmosres.2019.104700

Ndehedehe CE, Awange JL, Corner R, Kuhn M, Okwuashi O (2016) On the potentials of multiple climate variables in assessing the spatio-temporal characteristics of hydrological droughts over the Volta Basin. Sci Total Environ 557–558:819–837. https://doi.org/10.1016/j.scitotenv.2016.03.004

Ndehedehe CE (2019) The water resources of tropical West Africa: problems, progress and prospect. Acta Geophys 67(2):621–649. https://doi.org/10.1007/s11600-019-00260-y

Ndehedehe CE, Ferreira VG, Agutu NO, Onojeghuo AO, Okwuashi O, Kassahun HT, Dewan A (2021) What if the rains do not come? J Hydrol 595:126040. https://doi.org/10.1016/j.jhydrol.2021.126040

Ndehedehe CE, Ferreira VG (2020) Assessing land water storage dynamics over Southern America. J Hydrol 580:124339. https://doi.org/10.1016/j.jhydrol.2019.124339

Ndehedehe CE, Ferreira VG, Onojeghuo AO, Agutu NO, Emengini E, Getirana A (2020) Influence of global climate on freshwater changes in Africa's largest endorheic basin using multi-scaled indicators. Sci Total Environ 737:139643. https://doi.org/10.1016/j.scitotenv.2020.139643

Ndehedehe CE, Haile GG, Agutu NO, Ferreira VG, Getirana A, Okwuashi O (2020) Hydrological hotspots of climatic influence in Brazil: a two-step regularization approach. Atmos Res 246:105116. https://doi.org/10.1016/j.atmosres.2020.105116

Ndehedehe CE, Ferreira VG (2020) Identifying the footprints of global climate modes in time-variable gravity hydrological signals. Clim Chang 159:481–502. https://doi.org/10.1007/s10584-019-02588-2

Ndehedehe CE, Agutu NO, Okwuashi OH, Ferreira VG (2016) Spatio-temporal variability of droughts and terrestrial water storage over Lake Chad Basin using independent component analysis. J Hydrol 540:106–128. https://doi.org/10.1016/j.jhydrol.2016.05.068

Ndehedehe CE, Okwuashi O, Ferreira VG, Agutu NO (2018) Exploring evapotranspiration dynamics over Sub-Sahara Africa (2000–2014). Environ Monit Assess 190(7):400. https://doi.org/10.1007/s10661-018-6780-6

Ndehedehe CE, Anyah RO, Alsdorf D, Agutu NO, Ferreira VG (2019) Modelling the impacts of global multi-scale climatic drivers on hydro-climatic extremes (1901–2014) over the Congo basin. Sci Total Environ 651:1569–1587. https://doi.org/10.1016/j.scitotenv.2018.09.203

Ndehedehe CE, Ferreira VG, Agutu NO (2019) Hydrological controls on surface vegetation dynamics over West and Central Africa. Ecol Ind 103:494–508. https://doi.org/10.1016/j.ecolind.2019.04.032

Njoku EG, Entekhabi D (1996) Passive microwave remote sensing of soil moisture. J Hydrol 184(1):101–129. https://doi.org/10.1016/0022-1694(95)02970-2

Njoku E, Jackson T, Lakshmi V, Chan T, Nghiem S (2003) Soil moisture retrieval from AMSR-E. IEEE Trans Geosci Remote Sens 41(2):215–229. https://doi.org/10.1109/TGRS.2002.808243

Noori R, Maghrebi M, Mirchi A, et al (2021) Anthropogenic depletion of iran's aquifers. Proc Natl Acad Sci 118(25). https://doi.org/10.1073/pnas.2024221118

Núñez M, Pfister S, et al (2013) Assessing the environmental impact of water consumption by energy crops grown in Spain. J Ind Ecol 17(1):90–102. https://doi.org/10.1111/j.1530-9290.2011.00449.x

Ojha C, Shirzaei M, Werth S, Argus DF, Farr TG (2018) Sustained groundwater loss in California's central valley exacerbated by intense drought periods. Water Resour Res 54(7):4449–4460. https://doi.org/10.1029/2017WR022250

Ojha C, Werth S, Shirzaei M (2019) Groundwater loss and aquifer system compaction in San Joaquin Valley during 2012–2015 drought. J Geophys Res: Solid Earth 124(3):3127–3143. https://doi.org/10.1029/2018JB016083

Ojha C, Werth S, Shirzaei M (2020) Recovery of aquifer-systems in Southwest US following 2012–2015 drought: evidence from InSAR, GRACE and groundwater level data. J Hydrol 587. https://doi.org/10.1016/j.jhydrol.2020.124943

Okonkwo C, Demoz B, Gebremariam S (2014) Characteristics of Lake Chad level variability and links to ENSO, precipitate, and river discharge. Sci World J 13(4):13. https://doi.org/10.1155/2014/145893

Okwuashi O, Ndehedehe C (2017) Tide modelling using support vector machine regression. J Spat Sci 62(1):29–46. https://doi.org/10.1080/14498596.2016.1215272

Palmer W (1965) Meteorological drought. Weather. Bureau, U.S. Department of Commerce, Washington, Research Paper 45. D. C. (58 p)

Pekel JF, Cottam A, Gorelick N, Belward AS (2016) High-resolution mapping of global surface water and its long-term changes. Nature 540(7633):418–422. https://doi.org/10.1038/nature20584

Pokhrel Y, Felfelani F, Satoh Y, Boulange J, Burek P, GÄdeke A, Gerten D, Gosling SN, Grillakis M, Gudmundsson L, Hanasaki N (2021) Global terrestrial water storage and drought severity under climate change. Nat Clim Chang 11(3):226–233. https://doi.org/10.1038/s41558-020-00972-w

Reager JT, Famiglietti JS (2009) Global terrestrial water storage capacity and flood potential using GRACE. Geophys Res Lett 36(23). https://doi.org/10.1029/2009GL040826

Riebsame WE, Changnon SA, Karl TR (2019) Drought and natural resources management in the united states: impacts and implications of the 1987–1989 drought. Routledge, p 192. https://doi.org/10.4324/9780429045332

Rienecker MM, Suarez MJ, Gelaro R et al (2011) MERRA: NASA's modern-era retrospective analysis for research and applications. J Clim 24(14):3624–3648. https://doi.org/10.1175/JCLI-D-11-00015.1

Rind D, Goldberg R, Hansen J, Rosenzweig C, Ruedy R (1990) Potential evapotranspiration and the likelihood of future drought. J Geophys Res: Atmos 95(D7):9983–10004. https://doi.org/10.1029/JD095iD07p09983

Rodell M, Houser PR, Jambor U, Gottschalck J, Mitchell K, et al (2004) The global land data assimilation system. Bull Am Meteorol Soc 85(3):381–394. https://doi.org/10.1175/BAMS-85-3-381.R

Sanogo S, Fink AH, Omotosho JA, Ba A, Redl R, Ermert V (2015) Spatio-temporal characteristics of the recent rainfall recovery in West Africa. Int J Climatol 35(15):4589–4605. https://doi.org/10.1002/joc.4309

Schneider U, Becker A, Finger P, Meyer-Christoffer A, Ziese M, Rudolf B (2014) GPCC's new land surface precipitation climatology based on quality-controlled in situ data and its role in quantifying the global water cycle. Theoret Appl Climatol 115(1–2):15–40. https://doi.org/10.1007/s00704-013-0860-x

Seneviratne SI, Corti T, et al (2010) Investigating soil moisture-climate interactions in a changing climate: a review. Earth-Sci Rev 99(3):125–161. https://doi.org/10.1016/j.earscirev.2010.02.004

Sheffield J, Wood EF (2008) Projected changes in drought occurrence under future global warming from multi-model, multi-scenario, IPCC AR4 simulations. Clim Dyn 31(1):79–105. https://doi.org/10.1007/s00382-007-0340-z

Shiferaw B, Tesfaye K, Kassie M, Abate T, Prasanna B, Menkir A (2014) Managing vulnerability to drought and enhancing livelihood resilience in sub-Saharan Africa: technological, institutional and policy options. Weather Clim Extrem 3:67–79. https://doi.org/10.1016/j.wace.2014.04.004

Shukla S, Wood AW (2008) Use of a standardized runoff index for characterizing hydrologic drought. Geophys Res Lett 35(2):L02405. https://doi.org/10.1029/2007GL032487

Snedecor GW, Cochran WG (1989) Statistical methods. Iowa State University Press, 8th edn

Solman SA (2013) Regional climate modeling over South America: a review. Adv Meteorol https://doi.org/10.1155/2013/504357

Sommer B, Horwitz P (2009) Macroinvertebrate cycles of decline and recovery in Swan Coastal Plain (Western Australia) wetlands affected by drought-induced acidification. Hydrobiologia 624(1):191–203

Spinoni J, Naumann G, Carrao H, Barbosa P, Vogt J (2014) World drought frequency, duration, and severity for 1951–2010. Int J Climatol 34(8):2792–2804. https://doi.org/10.1002/joc.3875

Stampoulis D, K. M. A. et al (2016) Assessing hydro-ecological vulnerability using microwave radiometric measurements from windsat. Remote Sens Environ 184:58–72. https://doi.org/10.1016/j.rse.2016.06.007

Sternberg T (2011) Regional drought has a global impact. Nature 472(7342):169. https://doi.org/10.1038/472169d

Sultan B, Gaetani M (2016) Agriculture in West Africa in the twenty-first century: climate change and impacts scenarios, and potential for adaptation. Front Plant Sci 7:1262. https://doi.org/10.3389/fpls.2016.01262

Swenson S, Wahr J (2009) Monitoring the water balance of Lake Victoria, East Africa, from space 370(1–4):163–176. https://doi.org/10.1016/j.jhydrol.2009.03.008

Tapley B et al (2019) Contributions of grace to understanding climate change. Nat Clim Chang 9(5):358–369. https://doi.org/10.1038/s41558-019-0456-2

Tapley B, Bettadpur S, Watkins M, Reigber C (2004) The gravity recovery and climate experiment: mission overview and early results. Geophys Res Lett 31:1–4. https://doi.org/10.1029/2004GL019920

Tapley BD, Bettadpur S, Ries JC, Thompson PF, Watkins MM (2004) Grace measurements of mass variability in the earth system. Science 305(5683):503–505. https://doi.org/10.1126/science.1099192

Taylor CM, de Jeu RA, Guichard F, Harris PP, Dorigo WA (2012) Afternoon rain more likely over drier soils. Nature 489(7416):423–426. https://doi.org/10.1038/472169d

Theis FJ, Gruber P, Keck IR, Meyer-bäse A, Lang EW (2005) Spatiotemporal blind source separation using double-sided approximate joint diagonalization. In: Proc, EUSIPCO

Thomas BF, Famiglietti JS, Landerer FW, Wiese DN, Molotch NP, Argus DF (2017) GRACE groundwater drought index: evaluation of California central valley groundwater drought. Remote Sens Environ 198(Supplement C):384–392. https://doi.org/10.1016/j.rse.2017.06.026

Tsakiris G, Vangelis HJEW (2005) Establishing a drought index incorporating evapotranspiration. European water 9(10):3–11

van Dijk AIJM, Beck HE, Crosbie RS, de Jeu RAM, Liu YY, Podger GM, Timbal B, Viney NR (2013) The millennium drought in southeast Australia (2001–2009): natural and human causes and implications for water resources, ecosystems, economy, and society. Water Resour Res 49(2):1040–1057. https://doi.org/10.1002/wrcr.20123

Van Loon AFL et al (2016) Drought in a human-modified world: reframing drought definitions, understanding, and analysis approaches. https://doi.org/10.5194/hess-20-3631-2016

Van Loon AF (2015) Hydrological drought explained. WIREs Water 2:359–392. https://doi.org/10.1002/wat2.1085

Vapnik V (1995) The nature of statistical learning theory. Springer, New York, NY

Velicogna I, Kimball JS, Kim Y (2015) Impact of changes in GRACE derived terrestrial water storage on vegetation growth in Eurasia. Environ Res Lett 10(12):124024

Velicogna I, Tong J, Zhang T, Kimball JS (2012) Increasing subsurface water storage in discontinuous permafrost areas of the Lena River basin, Eurasia, detected from GRACE. Geophys Res Lett 39(9). https://doi.org/10.1029/2012GL051623

Vicente-Serrano SM, Beguería S, López-Moreno JI, Angulo M, El Kenawy A (2010b) A new global 0.5° gridded dataset (1901–2006) of a multiscalar drought index: comparison with current drought index datasets based on the palmer drought severity index. J Hydrometeorol 11(4):1033–1043. https://doi.org/10.1175/2010JHM1224.1

Vicente-Serrano SM, Domínguez-Castro F, et al (2020) Global characterization of hydrological and meteorological droughts under future climate change: the importance of timescales, vegetation-co2 feedbacks and changes to distribution functions. Int J Climatol 40(5):2557–2567. https://doi.org/10.1002/joc.6350

Vicente-Serrano SM, Beguería S, López-Moreno JI (2010) A multiscalar drought index sensitive to global warming: the standardized precipitation evapotranspiration index. J Clim 23(7):1696–1718. https://doi.org/10.1175/2009JCLI2909.1

Wada Y (2013) Human and climate impacts on global water resources. Doctoral dissertation, University Utrecht

Wagner W, et al (2013) The ascat soil moisture product: a review of its specifications, validation results, and emerging applications. Meteorol Z https://doi.org/10.1127/0941-2948/2013/0399

Webster KE, Kratz TK, Bowser CJ, Magnuson JJ, Rose WJ (1996) The influence of landscape position on lake chemical responses to drought in northern Wisconsin. Limnol Oceanogr 41(5):977–984. https://doi.org/10.4319/lo.1996.41.5.0977

Werth S, White D, Bliss DW (2017) GRACE detected rise of groundwater in the Sahelian Niger river basin. J Geophys Res: Solid Earth 2017JB014845. https://doi.org/10.1002/2017JB014845

Westermann S, Langer M, Boike J (2011) Spatial and temporal variations of summer surface temperatures of high-arctic tundra on Svalbard—implications for MODIS LST based permafrost monitoring. Remote Sens Environ 115(3):908–922. https://doi.org/10.1016/j.rse.2010.11.018

Wilhite DA, Svoboda MD, Hayes MJ (2007) Understanding the complex impacts of drought: a key to enhancing drought mitigation and preparedness. Water Resour Manage 21(5):763–774. https://doi.org/10.1007/s11269-006-9076-5

Wilhite DA, Sivakumar MV, Pulwarty R (2014) Managing drought risk in a changing climate: the role of national drought policy. Weather Clim Extrem 3:4–13. https://doi.org/10.1016/j.wace.2014.01.002

Wilks D (2011) Statistical methods in the atmospheric sciences, 3rd edn. Academic Press, USA

Williams AP, Funk C (2011) A westward extension of the warm pool leads to a westward extension of the walker circulation, drying eastern Africa. Clim Dyn 37(11):2417–2435. https://doi.org/10.1007/s00382-010-0984-y

Williams AP, Seager R, Abatzoglou JT, Cook BI, Smerdon JE, Cook ER (2015) Contribution of anthropogenic warming to California drought during 2012–2014. Geophys Res Lett 42(16):6819–6828

Woodhouse CA, Meko DM, MacDonald GM, Stahle DW, Cook ER (2010) A 1,200-year perspective of 21st century drought in southwestern north America. Proc Natl Acad Sci 107(50):21283–21288. https://doi.org/10.1073/pnas.0911197107

Xie P, Arkin PA (1997) Global precipitation: a 17-year monthly analysis based on gauge observations, satellite estimates, and numerical model outputs. Bull Am Meteorol Soc 78(11):2539–2558. https://doi.org/10.1175/1520-0477(1997)078<2539:GPAYMA>2.0.CO;2

Xu H, Tian Z, et al (2019) Future increases in irrigation water requirement challenge the water-food nexus in the northeast farming region of China. Agric Water Manag 213:594–604. https://doi.org/10.1016/j.agwat.2018.10.045

Yang J, Cheng Q (2015) A comparative study of independent component analysis with principal component analysis in geological objects identification. part ii: a case study of Pinghe District, Fujian, China. J Geochem Explor 149:136–146. https://doi.org/10.1016/j.gexplo.2014.11.014

Yao Y, Liang S, Qin Q, Wang K (2010) Monitoring drought over the conterminous United States using MODIS and NCEP reanalysis-2 data. J Appl Meteorol Climatol 49(8):1665–1680. https://doi.org/10.1175/2010JAMC2328.1

Yin D, Roderick ML, Leech G, Sun F, Huang Y (2014) The contribution of reduction in evaporative cooling to higher surface air temperatures during drought. Geophys Res Lett 41(22):7891–7897. https://doi.org/10.1002/2014GL062039

Yirdaw SZ, Snelgrove K, Agboma C (2008) GRACE satellite observations of terrestrial moisture changes for drought characterization in the Canadian Prairie. J Hydrol 356:84–92

Yong B, Wang J, Ren L, You Y, Xie P, Hong Y (2016) Evaluating four multisatellite precipitation estimates over the diaoyu islands during typhoon seasons. J Hydrometeorol 17(5):1623–1641. https://doi.org/10.1175/JHM-D-15-0165.1

Zaitchik BF, Rodell M, Reichle RH (2008) Assimilation of GRACE terrestrial water storage data into a land surface model: results for the mississippi river basin. J Hydrometeorol 9(3):535–548. https://doi.org/10.1175/2007JHM951.1

Zhang A, Jia G (2013) Monitoring meteorological drought in semiarid regions using multi-sensor microwave remote sensing data. Remote Sens Environ 134:12–23. https://doi.org/10.1016/j.rse.2013.02.023

Zhang D, Zhang Q, Werner AD, Liu X (2016) GRACE-based hydrological drought evaluation of the Yangtze River Basin, China. J Hydrometeorol 17(3):811–828. https://doi.org/10.1175/JHM-D-15-0084.1

Zhao M, Geruo A, Velicogna I, Kimball J (2017) Satellite observations of regional drought severity in the continental United States using GRACE-based terrestrial water storage changes. J Clim 30(16):6297–6308. https://doi.org/10.1175/JCLI-D-16-0458.1

Zhou L, Tian Y, Myneni RB, Ciais P, et al (2014) Widespread decline of Congo rainforest greenness in the past decade. Nature 509(7498):86–90. https://doi.org/10.1038/nature13265

Ziehe A (2005) Blind source separation based on joint diagonalization of matrices with applications in biomedical signal processing. Ph.D. thesis, Universitat Potsdam. http://en.youscribe.com/catalogue/reports-and-theses/knowledge/blind-source-separation-based-on-joint-diagonalization-of-matrices-1424347. Accessed 15 May 2015

Zulkafli Z, Buytaert W, Onof C, Manz B, Tarnavsky E, Lavado W, Guyot J-L (2014) A comparative performance analysis of TRMM 3B42 (TMPA) versions 6 and 7 for hydrological applications over Andean-Amazon river basins. J Hydrometeorol 15(2):581–592. https://doi.org/10.1175/JHM-D-13-094.1